Kinetics of Enzyme-Modifier Interactions

Antonio Baici

Kinetics of Enzyme-Modifier Interactions

Selected Topics in the Theory and Diagnosis of Inhibition and Activation Mechanisms

Springer

Antonio Baici
Department of Biochemistry
University of Zurich
Zurich, Switzerland

ISBN 978-3-7091-1401-8 ISBN 978-3-7091-1402-5 (eBook)
DOI 10.1007/978-3-7091-1402-5

Library of Congress Control Number: 2015942930

Springer Wien Heidelberg New York Dordrecht London

Printed on acid-free paper

Springer-Verlag GmbH Wien is part of Springer Science+Business Media (www.springer.com)

To my students, Swiss Federal Institute of Technology and University of Zurich, 1972–2012

Foreword

Many drugs in current use owe their effectiveness to the fact that they are enzyme inhibitors. In addition, going back to the studies of invertase that Leonor Michaelis and his collaborators carried out at the beginning of the twentieth century, inhibitors have played a major role in efforts to understand the mechanisms that allow enzymes to fulfil their roles as catalysts. Bringing these two aspects together, we may note that understanding how inhibitors affect individual enzymes is a necessary step in understanding what happens when an enzyme that forms part of a metabolic pathway is inhibited in a living organism. Inhibitors and activators taken together are classified as enzyme modifiers, the subject of Antonio Baici's masterly book. Activators have been less intensively studied than inhibitors over the years, but they are also important and should not be forgotten.

It surely follows, therefore, that kinetic characterization of these effects is vital both for understanding enzyme mechanisms and for drug development. Yet the treatment in most general textbooks of biochemistry (with Henry Mahler and Eugene Cordes's *Biological Chemistry*, now more than 40 years old, as an honorable exception) is nearly always superficial and sometimes even misleading. More specialized books on enzymes mostly do little more than scratch on the surface, though again there is an honorable exception, in the form of Malcolm Dixon and Edwin Webb's *Enzymes*. Some of the books specifically devoted to enzyme kinetics, such as Irwin Segel's *Enzyme Kinetics*, include considerable detail about inhibition, though not all of them do. Few of the better books are very recent, however, and J. Leyden Webb's monumental treatise *Enzyme and Metabolic Inhibitors* dates from the 1960s. Many of today's readers are unlikely to know that it even exists, especially if they do all their reading on the web, or if they think that nothing published more than 6 months ago is worth reading.

The systematic treatment of inhibition and activation by Jean Botts and Manuel Morales [*Transactions of the Faraday Society* **49**, 696–707 (1954)] is even older, and the principles they set out are now rarely taught. In the distant past, when I used their theory as the basis of teaching about enzyme modifiers, it was obvious that it was very unpopular with students, being seen as highly complex and difficult. As Antonio Baici shows, however, it can be presented in a way that allows it to be understood and applied. Moreover, it needed to be extended to take account of allosteric effects, which were essentially unknown in the 1950s.

Biochemists who believe in the comforting illusion that any inhibitor that has a structural resemblance to the substrate of the enzyme, and even one that does not, can be treated as a simple competitive inhibitor will find abundant evidence in the book that it is just that, an illusion. The author provides many examples of the kinds of behavior that people might prefer not to know about, and shows that building on Botts and Morales's pioneering work is necessary for an adequate understanding of enzyme modifiers. That in turn is necessary if one hopes to design inhibitors and activators that will be pharmacologically useful.

Marseille, France Athel Cornish-Bowden
March 2015

Preface

This monograph is neither aimed at covering all technical aspects of enzyme kinetics nor at reviewing extensively enzyme inhibition and activation. Considering that fundamental aspects of enzyme kinetics have been competently treated in several books, while countless reviews and articles dealt with specific topics, I assume that readers of *Kinetics of Enzyme-Modifier Interactions* are already familiar with the basic principles of this discipline. My last thought while writing this book was to deal with *pièces de résistance* in the classical way found in existing excellent publications by copying and pasting established theories and methods, a bare nonsense. Rather, I felt that our knowledge of enzyme-modifier interactions could benefit from a scrutiny of existing but in a way buried concepts in need to be clarified and complemented by systematization using alternative methods. Therefore, the following chapters will examine under a magnifying glass selected topics and discuss less-known, neglected, or overlooked aspects by adding a pinch of new ideas. The goal is to support investigations in vitro by suggesting logical solutions to problems of various complexity and developing theoretical aspects of mechanisms beyond those already known but likely to exist in still unexplored niches. An extension to living organisms to include the flux control in whole systems, through modification of a particular enzyme, lies outside the objectives of this book.

To discredit the rumor about enzyme kinetics as an abstruse discipline, the students in my undergraduate courses were motivated by lectures in enzyme kinetics that included material present in this book. Students in master's degree programmes grasped quickly the various mechanisms of enzyme inhibition and activation, double inhibition, and slow-onset inhibition. Therefore, not being complicated as it may appear at first sight, this book is hoped to inspire teachers, students, and investigators in academia and industry interested in exploring the still partly uncharted territory of enzyme modification.

A possible novelty to students of kinetics is the systematic approach to enzyme modification mechanisms, which will be ranked taxonomically following criteria based on individual, unique characters that make them equal to *species*. Similar to plants and animals, to which systematic names are given to distinguish them from other related or unrelated species, also enzyme-modification mechanisms deserve their own names. For users uncertain where to place their own experimental findings in the intricate labyrinth that leads from raw data to mechanisms and to the calculation of kinetic parameters, dichotomous keys will be provided as an analytical companion for the diagnosis of the basic enzyme-modification mechanisms and slow-onset enzyme-modifier interactions.

Since the customary vocabulary in enzyme inhibition and activation was insufficient in this novel perspective, the proposed nomenclature of mechanisms will follow their systematic ranking. In this respect, the most demanding group of enzyme modifiers was *mixed inhibition*, conventionally treated under this common heading, and now appearing as eight individual, well-distinguished entities flanked by five nonessential activation mechanisms and two hybrid species that have either inhibitor or activator character depending on substrate concentration. This group of fifteen mechanisms contains the strategic tools used by allosteric effectors in enzyme regulation. Knowing the details of the fine-tuning possibilities of this class of substances is not only important for interpreting physiological processes but also as support in emerging pharmacological approaches. Specific pharmacological targeting of unwanted enzyme activities deserves more attention than the estimation of a crude IC_{50} value.

In 2013 two important anniversaries have been celebrated: the 100 years of Michaelis and Menten's paper, considered the birth of modern enzyme kinetics, and the 50 years of Monod, Changeux, and Jacob's concept of allosteric regulation. A third event in 2013 passed unobserved, namely the 60th anniversary of the publication of Botts and Morales on the effects of modifiers upon the enzyme-catalyzed steady-state reaction rate. Although not highly regarded as the other mentioned illustrious publications, the work of Botts and Morales contains the basic elements for interpreting the interactions between enzymes and modifiers. The implications of their model are far-reaching and will be used in this book as the basis for a taxonomic ranking of inhibition and activation mechanisms.

Zurich, Switzerland Antonio Baici
March 2015

Acknowledgements

People who contributed with their original data to presenting worked examples will be acknowledged in the appropriate place. Below, I wish to individually acknowledge all colleagues and friends who either read and constructively commented the whole or parts of this book, or have significantly contributed to my studies in enzyme kinetics.

In 2006, the Centenary year of *The Biochemical Journal*, I was invited to write a *classics* article on papers in enzyme kinetics published in the first 100 years of the same journal. The organizers gave me only one limit: to choose just three papers, a sort of judgement of Paris that made me uneasy. After realizing that my decision was unlikely to lead to a second Trojan War, I did not hesitate in choosing *The Direct Linear Plot* co-authored by Robert Eisenthal and Athel Cornish-Bowden as one of three classics published in *The Biochemical Journal*. Besides this method that I regularly taught in my lectures, I believe to have read all papers by Athel-Cornish Bowden and his books, tacitly nominating him my teacher in enzyme kinetics. I would like to thank Athel for his engagement and enthusiasm in this fascinating discipline.

Giorgio Semenza has been the first person with whom I had the privilege of talking about enzyme kinetics when, in 1972, I started in this field at the Swiss Federal Institute of Technology in Zurich. I admire Giorgio's immense culture, dear friend, medical doctor, and excellent biochemist who published rate equations for enzyme inhibition and activation in the *Journal of Theoretical Biology*.

Stephen Bearne was very kind and helpful in sharing with me his knowledge in kinetics and thermodynamics and supporting the construction of kinetic barrier diagrams with precious suggestions.

Brigita Lenarčič has been a long-term collaborator in projects focussed on peptidases involved in the degradation and remodeling of the extracellular matrix and their interactions with naturally occurring inhibitors. I always appreciated her profound experience in this challenging topic of enzymology.

Marko Novinec, from the laboratory of Brigita Lenarčič at the University of Ljubljana, spent more than 2 years in Zurich sharing with me joys and sorrows with bench work and publishing on the kinetics of cysteine cathepsins putting in focus allosteric interactions. Marko's skill was fundamental in realizing a demanding project on the identification of allosteric sites in cathepsin K and the characterization of their interactions with a heap of amazing hyperbolic modifiers discussed in this book.

Supporting my request for help with handling multiple elementary steps between enzyme states, Feng Qi and Daniel Beard were kind enough to modify their useful algorithm KAPattern based on the method of King and Altman for the systematic generation of reaction patterns. The modified code, available online, is very helpful for deriving the rate equations of complex systems.

Patricia Schenker was unique among my students in choosing for her successful PhD work topics that were fully centered on enzyme kinetics. With her passion for mathematics, Patricia was not afraid of starting a difficult journey in the field of multiple enzyme modification or to combine molecular modeling with skillful and precise experimenting. Patricia carefully reviewed and gave constructive advice on technical details and on the structure of the entire book.

Thomas Hirt, Stefan Klauser, and Steve Rast, the information technology team at the Department of Biochemistry, University of Zurich, merit a sincere applause for supporting my demands with computer work and intervening within minutes for repairing my frequent mistakes. Thomas, who completed his Master in Biochemistry in my laboratory, actively participated in an interdisciplinary didactic project that is still used by students to simulate, by numerical integration methods, time-dependent phenomena in biochemistry, microbiology, and physics.

I thankfully acknowledge my colleagues at the Department of Biochemistry of the University of Zurich for fruitful collaborations, friendly relationships, and encouragement during my career: Hans Rudolf Bosshard, Amedeo Caflisch, Philipp Christen, Raimund Dutzler, Heinz Gehring, Sergio Gloor, Markus Grütter, Bernd Gutte, Ilian Jelezarov, Jeremias Kägi, Peter Lindner, Andreas Plückthun, Ben Schuler, Peter Sonderegger, and Milan Vašák. The scientific interaction with my colleagues and support given to their numerous students at various levels substantially contributed in enlarging my horizon in enzyme kinetics and macromolecular interactions in general.

A special sign of gratitude goes to Annamaria Camus, my mentor an the University of Trieste, who I call with affection *my scientific mother*. She hosted me in her laboratory five unforgettable years passing to me her passion for chemical sciences. After a half century, she continues to be a source of energy and sound advice.

My wife Roberta and my sons Francesco, Federico, and Luca have always been very tolerant for the time I robbed with my research and teaching activity, including the time necessary to write this book after my retirement. I hope that the time I am now dedicating with much love to my grandchildren will partly compensate my sins of omission.

Contents

Chapter 1
Basic Knowledge

The key to a knowledge of enzymes is the study of reaction velocities, not of equilibria.
J.B.S. Haldane, Enzymes, Reprint 1965, M.I.T. Press, p. 3

Abstract This introductory chapter contains a collection of concepts that serve as support for the following parts of this book. Understanding microscopic reversibility and the fundamentals of linked functions are necessary instruments for interpreting allosteric interactions between substrates and modifiers. The basic theories of enzyme kinetics are approached from the consideration that all enzyme-catalyzed reactions are reversible and that mechanisms are determined by the ensemble of kinetic barriers that preside over the quasi-equilibrium and steady-state assumptions. Emphasis is given to initial rates and their precise determination as an essential starting point in kinetic data analysis. Lying in-between, practical tools, brief refreshing of useful methods and some constructive criticism.

1.1 Introduction

As anticipated in the Preface, with the exception of particular concepts that are essential for understanding enzyme–modifier interactions, established methods and theories in general enzyme kinetics will not be replicated here. It is thus assumed that readers of this book are familiar with the basic theories of enzyme kinetics, which have been described in time-honored treatises such as *Fundamentals of Enzyme Kinetics* by Athel Cornish-Bowden, at the time of printing of this book in its fourth edition [22], *Enzyme Kinetics* by Irwin H. Segel [57], *Enzymes*, the compendium by Malcom Dixon and Edwin C. Webb [27], or *Enzyme and Metabolic Inhibitors* by J. Leyden Webb [66]. But there are more books, less acclaimed than those just cited, that contain valuable contributions to enzyme kinetics with personal ideas such as *Kinetics of Enzyme Mechanisms* by J. Tze-Fei Wong [68], *Behavior of Enzyme Systems* by John M. Reiner [56], or *Enzyme Kinetics* by Kent M. Plowman [53].

A. Baici, *Kinetics of Enzyme-Modifier Interactions*,
DOI 10.1007/978-3-7091-1402-5_1

This chapter, after presenting some tools and methods, surveys topics aimed at supporting the study of enzyme kinetic mechanisms in the presence of modifiers. Modern software for simulating kinetic processes is very powerful and attractive. However, when choosing kinetic constants and reactant concentrations to perform simulations, users should be aware that only sets of constants that obey microscopic reversibility are meaningful. Since this fundamental principle has been violated even by specialists in enzyme kinetics, microscopic reversibility will be given due emphasis.

All enzyme-catalyzed reactions are reversible. The logically minimal, monosubstrate kinetic mechanism is approached from activation-energy and rate-limiting step considerations as a paradigm for simple and complex mechanisms, discussing the conditions that apply to the treatments of Michaelis–Menten and Briggs–Haldane. Energy changes associated with enzyme-catalyzed reactions and microscopic reversibility help understanding the relationships that must be satisfied between kinetic constants when reversible processes are considered *irreversible* for practical reasons in initial rate measurements.

1.2 Tools and General Information

The subsections below contain methodology and general information, in arbitrary order, to support the following chapters.

1.2.1 Symbols, Nomenclature, and Conventions

The symbolism and terminology in enzyme kinetics approved in 1981 by the Nomenclature Committee of the International Union of Biochemistry [49][1] has not been reconsidered until the publication of this book. Cornish-Bowden reviewed the work of the organizations involved in releasing guidelines in symbolism and terminology in enzyme kinetics and discussed the salient points that should be considered in any future recommendations [21]. Harmonization in the terminology and symbolism of enzyme kinetics not only allows to recognize equations at a glance across textbooks and journal articles, but also avoids misunderstanding the identity of mechanisms when only a name is given in publications without an accompanying scheme of the mechanism and/or the pertinent rate equation.

An apparently trivial point are the symbols used to represent concentrations. The SI unit of concentration is mol m^{-3}, but for practical reasons it is expressed in units of mol dm^{-3} and denoted by the symbol M (molar concentration, molarity). The various ways the concentration of a species, say A, is represented in kinetic

[1] Also available at http://www.chem.qmul.ac.uk/iubmb/kinetics/.

equations include: A, A, a, [A], c_A. The symbol A is not acceptable because it is identical to the name of the species, while A is satisfactory as long as the difference between upright and italics symbols can be recognized without effort. The symbol a is fine as long as the difference between ab intended as the concentration of the species AB can be distinguished from the multiplication of the concentrations of A and B. c_A is traditionally used in the context of relaxation kinetics, e.g., Bernasconi [11].

[A] is unambiguous and widely used but requires three typographical signs in place of one. In view of possible confusion between lowercase letters, aimed at representing concentrations together with coefficients in the rate equations of certain mechanisms, the option [A] will be consistently used even if this makes some equations appear "larger" than they are in reality. This is particularly the case of the rate equations for double-modifier mechanisms in Chap. 6, e.g., (6.4). Uppercase subscripts further improve readability of the equations and are better legible when symbols are used in tables and figure legends.

A list of symbols is provided in Appendix 2.

1.2.2 Enzyme Nomenclature, Definitions, Acronyms, and Credits

The Names of the Enzymes mentioned in this book do not always correspond to those in the original publications because of updates in classification, nomenclature, and recognition of synonymy occurred at a later time. The EC numbers of the cited enzymes will be given as identifiers. These were taken from the Swissprot enzyme nomenclature database[2] and from BRENDA, which also contains, besides nomenclature, a large collection of enzyme functional data.[3] For peptidases and their inhibitors, recommended names are those of the MEROPS Peptidase Database, which uses a logical, hierarchical classification of the enzymes.[4]

Definitions and explanation of terms will be provided at their first occurrence and not repeated. Having renounced to write a glossary, if an unfamiliar name is found at a place where it is not defined, the *Index* should be consulted. Here, either the first entry (lowest page number) or the sought word followed by *definition* contains the information.

Acronyms and Abbreviations will be used as sparingly as possible, e.g., in tables to avoid overrunning the available space, in which case they are defined only for local use. Though, in Chap. 3 new names will be provided for the kinetic

[2]http://enzyme.expasy.org.

[3]http://www.brenda-enzymes.org.

[4]http://merops.sanger.ac.uk.

mechanisms of modifiers, which correspond in some cases to long names. In this case, acronyms are convenient for use in regular text, for cross-referencing, in tables and when redirecting from the diagnostic keys in Chap. 10 to the source, where full information is given. The acronyms of the mechanisms are logical and can be read and memorized without effort.

Modifier and Effector The term *modifier* is used in this book to denote any substance that affects the velocity of an enzyme-catalyzed reaction. A term used with more or less the same significance is *effector*. According to the recommendations (1981) of the International Union of Biochemistry (IUB)[5] [49, p. 282]:

> The terms *effector* and *modifier* are general terms that apply to substances that interact with enzymes and either increase or decrease their catalytic action. Enzyme inhibitors and enzyme activators are therefore special cases of effectors and modifiers. The term *effector* is more commonly used when the substance produces effects of physiological significance,whereas the term *modifier* is more appropriate for a substance that is artificially added to an enzyme system being studied in vitro.

In 1979, before the release of the IUB recommendations, Whitehead provided the following somewhat different definitions [67, p. 326]

> An *effector* is any substance that can affect the velocity of an enzyme reaction. The term includes substrates, inhibitors, and activators. A *modifier* is an effector which is not a substrate.

The discussion may continue with definitions found in other places. However, in the biochemical literature the terms effector and modifier have been and are used interchangeably.

Credits are first and foremost important to acknowledge the true authors of concepts and theories. Genuine citations are essential to provide direct access to the original source of information thus avoiding the tedious process of back-searching until the sought equation, buffer system, or analytical method is finally found, an unnecessary exercise that can be very costly in terms of time. A deplorable practice is self-citation of preceding work (*as described previously*) that suggests originality, and this process may be repeated a number of times until eventually discovering that the sought information came from an unrelated laboratory. Readers are also disappointed when a referenced publication does not contain at all the addressed material. Finally, when a thick book is given as reference, specification of the chapter number and possibly of the page are a welcome courtesy. All the references listed at the end of the chapters in this book have been read and accurately checked.

[5]Since 1991 International Union of Biochemistry and Molecular Biology (IUBMB).

1.2.3 Kinetic Equations, Supporting Software and Simulation

Despite not introducing in this book details of the method of King and Altman for deriving rate equations of enzyme mechanisms [42], some equations will be derived for didactic purposes showing the procedure in a hopefully intuitive way. The King–Altman method can be applied using computerized methods or by hand, reading the paths directly from a sketch of the mechanism, although the latter is more prone to mistakes. Using a method developed by Cha [18], the King–Altman procedure can be significantly simplified when steps in rapid equilibrium occur in a mechanism. Examples will be worked out in this book.

Several computerized algorithms exist, of which only two will be mentioned here. With KAPattern [54], a freely available, stand-alone program,[6] equations for steady-state and mixed steady-state/quasi-equilibrium paths can be derived, which can also contain irreversible reaction steps. Examples of the code needed by the program will be provided along with elaboration of the output for obtaining the sought rate equations.

BioKin Ltd. freely offers an efficient algorithm for online use.[7] Mechanisms can be typed within a form using an intuitive syntax and results are quickly displayed in a user-friendly manner.

In this book, for mathematical applications that included verification of algebraic expressions, calculation of limits, solution of equations, three-dimensional graphics, and others, Maple™ (Waterloo Maple Inc., Canada) was used. Simulations of progress curves for enzyme mechanisms and fitting models to experimental data by numerical integration were performed with KinTek Explorer (KinTek Corporation, U.S.A.) (more on p. 13). In some instances, computations were performed with MATLAB® (MathWorks, U.S.A.) (more on p. 13). Graphics were drawn with GraphPad Prism® (GraphPad Software Inc., U.S.A.), a software package that also includes a collection of statistical tools.

Specification of Error and Noise when explaining theories and methods in enzyme kinetics, simulated processes are of great help. To appreciate the impact of *errors* on data interpretation, such errors can be added by choosing their type and intensity, which allows comparison between error-free and error-affected data.

The topics treated in this book are concerned with two groups of experimental data: (1) *progress curves*, i.e., the time-dependent increase or decrease of measurable signals that are proportional to the concentration of one of the reactants; (2) rate measurements and the dependence of a property, usually derived from primary data obtained from progress curves, on the concentration of a selected reactant.

In the following chapters, adding *error* to simulated progress curves means adding *instrumental noise* to mimic the characteristic random spikes obtained from optical measuring devices, which are quantifiable as signal-to-noise ratio. This error

[6] http://virtualrat.org/software/kapattern.

[7] http://www.biokin.com/king-altman.

is unrelated to mistakes in sample handling. In KinTek Explorer, the function for generating fake data adds Gaussian error to simulated progress curves by entering a *sigma* value in units of the simulated signal.

Simulations of the reactant-concentration dependence of observed rates or other properties will be performed by adding error aimed at mimicking sample-handling mistakes. In this case, a multiplicative random error with specified percentage standard deviation of calculated values will be added to mimic real experiments, e.g., by using the *randn* function of MATLAB. Errors in sample handling are due to a variety of causes, the most obvious of which depend on pipetting but also include sample skipping in a series of dilutions thus ascribing wrong concentrations to otherwise correctly measured signals, or to the use of pipettes that have not been calibrated properly, e.g., passing from a finely calibrated 0–20 μL to a 0–50 μL pipette in urgent need of servicing. Errors that depend primarily on operator mistakes are manifold. They can be reduced with diligent work and planning the envisaged experiment by writing an experimental protocol. Operator mistakes due to answering incoming phone calls and SMS messages, translatable into sample skipping or grabbing the wrong pipette, cannot be taken into account in the simulation of kinetic processes but will merciless appear as outliers after plotting experimental results (see the quote at the end of Sect. 1.2.6).

1.2.4 Propagation of Error

Sums, differences, multiplications, and quotients of measured quantities with their errors must be accompanied by a flawless representation of the resulting error. Well knowing that the rules for calculating error propagation are described in a number of books, these are reproduced here for addition/substraction and multiplication/division to encourage their correct use (in place of home-made rules) in enzyme kinetics, e.g., for calculating a specificity constant as $k = k_{cat}/K_m$ from separately determined k_{cat} and K_m values with their own uncertainties. In the following expressions, A, B, C, X, Y denote measured quantities and a, b, c, x, y their uncertainties.

1.2.4.1 Rules for Addition and Subtraction

$$(A \pm a) + (B \pm b) = (A + B) \pm \sqrt{a^2 + b^2} \tag{1.1}$$

$$(A \pm a) - (B \pm b) = (A - B) \pm \sqrt{a^2 + b^2} \tag{1.2}$$

$$(A \pm a) + (B \pm b) - (C \pm c) = (A + B - C) \pm \sqrt{a^2 + b^2 + c^2} \tag{1.3}$$

1.2.4.2 Rules for Multiplication and Division

$$(A \pm a)\,(B \pm b) = (A \times B) \pm |A \times B| \sqrt{\left(\frac{a}{A}\right)^2 + \left(\frac{b}{B}\right)^2} \tag{1.4}$$

$$\begin{aligned}(A \pm a)\,(B \pm b)\,(C \pm c) &= (A \times B \times C) \\ &\pm |A \times B \times C| \sqrt{\left(\frac{a}{A}\right)^2 + \left(\frac{b}{B}\right)^2 + \left(\frac{c}{C}\right)^2}\end{aligned} \tag{1.5}$$

The inclusion of more terms in (1.5) can continue in the same way.

$$\frac{A \pm a}{B \pm b} = \frac{A}{B} \pm \left|\frac{A}{B}\right| \sqrt{\left(\frac{a}{A}\right)^2 + \left(\frac{b}{B}\right)^2} \tag{1.6}$$

$$\begin{aligned}\frac{(A \pm a)\,(B \pm b)\cdots}{(X \pm x)\,(Y \pm y)\cdots} &= \frac{A \times B \times \cdots}{X \times Y \times \cdots} \pm \left|\frac{A \times B \times \cdots}{X \times Y \times \cdots}\right| \\ &\sqrt{\left(\frac{a}{A}\right)^2 + \left(\frac{b}{B}\right)^2 + \cdots + \left(\frac{x}{X}\right)^2 + \left(\frac{y}{Y}\right)^2 + \cdots}\end{aligned} \tag{1.7}$$

1.2.5 Notation of the Intervals of Real Numbers

Particularly to manage the narrow space in some tables, short notations for the ranges taken by variables are very useful. Those used at some places in the following chapters are shown below without providing rigorous mathematical definitions.

$$(a, b) = \{x|\, a < x < b\} \quad \text{open} \tag{1.8}$$

$$(a, b] = \{x|\, a < x \le b\} \quad \text{left-open, right-closed} \tag{1.9}$$

$$[a, b) = \{x|\, a \le x < b\} \quad \text{left-closed, right-open} \tag{1.10}$$

$$[a, b] = \{x|\, a \le x \le b\} \quad \text{closed} \tag{1.11}$$

$$(a, \infty) = \{x|\, x > a\} \quad \text{left-open (right-unbounded)} \tag{1.12}$$

$$[a, \infty) = \{x|\, x \ge a\} \quad \text{left-closed (right-unbounded)} \tag{1.13}$$

See for instance Tables 6.3 and 6.5 in Chap. 6.

1.2.6 Reflections on the Least Squares Methods

An opening specification that has more than a semantic character will hopefully persuade investigators of enzyme kinetics to change their opinion: *equations are fitted to data* despite the mainstream belief in published work related to enzyme kinetics is that *data are fitted to equations*. It is not my intention to blame anybody for this lapse and will instead confess my own mistakes that occurred thee times [5, 6, 8] before reading an illuminating paper [48]. No mathematical explanation is necessary, just think about a moment with help of the following allegory. Fitting data to equations can be compared to Johnny (representing *data*) going to a department store for buying a new suit (representing the *equation*). Unfortunately, the suit of Johnny's choice is incompatible with his medium build. If you were Johnny, what would you do: put on 10 kg weight in order to feel comfortable in that suit or ask the tailor to *fit the suit* to the size of your body?

Regression methods, inseparable companions of kinetic data analysis together with their graphical representation, will be used through this book and, when applicable, commented at the appropriate place though avoiding the description of their mathematical fundament because I am not enough expert to copy–paste what others already published in the right way. The most important statistical concepts experimenters in enzyme kinetics should master have been treated by Johansen and Lumry [38] and Cornish-Bowden [22, Chap. 15]. Johansen and Lumry, discussing the statistical analysis of the Michaelis–Menten equation, put emphasis on the conditions that should be met before attempting any analysis by the least squares method in general and on appropriate weights assigned to the variables depending on the particular equation used. The conditions of applicability of least squares methods include the exclusion of unidirectional bias, i.e., systematic errors must be negligible and that errors should be normally distributed (Gaussian distribution), random, and not correlated to each other.

Cornish-Bowden surveys in a mathematical language accessible to everybody the use of weights, differences between parametric and nonparametric methods, model discrimination, residual analysis, and more. He emphasizes that *we live in the age of kits* and that using commercial software for data analysis necessitates from the part of the user prior knowledge of the underlying methods, which should not be used blindly.

For readers who do not wish to invest much time in learning lengthy statistical theories, Motulsky and Ransnas wrote a short *practical and nonmathematical review* that summarizes the essential statistical tools used in enzyme kinetics [48].

Finally, software packages that comprise graphical representation and statistical analyses are accompanied by tutorials and support on the underlying methods. For instance, GraphPad Prism (Sect. 1.2.3) provides very useful tutorials, a curve fitting guide and a statistical guide that can be consulted before attempting analyses with various methods.

A caveat on the blind use of regression methods when the data are insufficient to support the claims has been expressed by Wallace Cleland, who among his numerous skills has also been a pioneer in computerized kinetic data analysis [19, p. 138]:

> ··· Emphasis should thus be placed on improving precision and repeating experiments a number of times, rather than on more sophisticated techniques for analysis of bad data ··· And above all, it is important to use your head in making evaluations based on statistical analysis; all mathematics in the world is no substitute for a reasonable amount of common sense.

1.2.7 Numerical Integration of Progress Curves

The progress of enzyme-catalyzed reactions, typically recorded as change in the concentration of either substrates or products as a function of time, is described mathematically by differential equations. At defined values of pH, ionic strength, ionic composition, and temperature, the shape of a progress curve depends on rate constants and reactant concentrations. Additional effects, such as loss of enzyme activity for reasons unrelated to the reaction, or chemical instability of substrates and/or modifiers, may overlap the regular reaction. For diagnosing the kinetic mechanism and extracting the values of the kinetic constants from progress curves, the differential equations of the system must be integrated. However, the analytical integration of differential equations is bound to restrictive assumptions, such as the constancy of substrate and modifier concentrations during reaction. If such assumptions cannot be justified with sufficient accuracy by appropriate experimental design, results may be seriously biased. Moreover, the analytical integration of differential equations is impossible in many cases or yields exceedingly complex expressions that discourage their use.

An alternative approach consists in calculating an approximate, or numerical solution of the integral, which overcomes limitations and simplifying assumptions bypassing mathematical obstacles. Below is the short introduction I used in my lectures not having at hand a short review to pass my students. The necessary condition for calculating the integral (1.14) is to find an antiderivative of the function $f(x)$

$$\int_a^b f(x)\,. \tag{1.14}$$

In the context of enzyme kinetics this function is commonly either [S] or [P] and x is time, t. Therefore, the following descriptions will use [P] and t as variables. As illustrated in Fig. 1.1, a straightforward method for the numerical approximation of a definite integral consists in subdividing the time interval $[a, b]$ into n equal parts Δt to build the *Riemann sum*, of which there are two types, the left Riemann sum

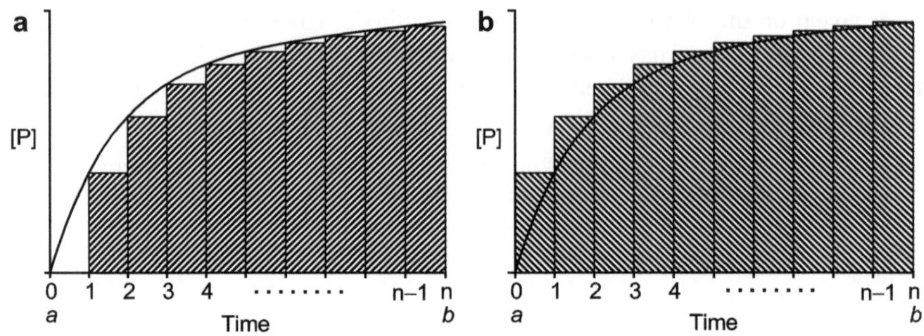

Fig. 1.1 The principle of numerical integration of a function based on the Riemann sums. Panels (**a**) and (**b**) show the left and the right Riemann sums, respectively. The progress *curve* (*continuous line* consisting of an exponential phase followed by a linear steady-state release of product) was simulated with the slow onset inhibition model described in Chap. 8

(LRS) and the right Riemann sum (RRS) . Thus, the division of the interval $[a, b]$ in n regular segments is accomplished as

$$\begin{aligned} \Delta t &= (b-a)/n \\ t_0 &= a \\ t_1 &= a + \Delta t \\ &\dots \\ t_n &= a + n\Delta t = b\ , \end{aligned}$$

and the area of the rectangles in Fig. 1.1 is calculated by multiplying their width Δt by their height $[\mathrm{P}]_i$. The LRS and the RRS can thus be calculated by adding together the areas of all rectangles:

$$\mathrm{LRS} = [\mathrm{P}]_0\Delta t + [\mathrm{P}]_1\Delta t + \cdots + [\mathrm{P}]_{n-1}\Delta t = \{[\mathrm{P}]_0 + [\mathrm{P}]_1 + \cdots + [\mathrm{P}]_{n-1}\}\,\Delta t \tag{1.15}$$

$$\mathrm{RRS} = [\mathrm{P}]_1\Delta t + [\mathrm{P}]_2\Delta t + \cdots + [\mathrm{P}]_n\Delta t = \{[\mathrm{P}]_1 + [\mathrm{P}]_2 + \cdots + [\mathrm{P}]_n\}\,\Delta t\ . \tag{1.16}$$

Another approach considers the *sum of trapezoids* instead of rectangles as illustrated in Fig. 1.2. The area of the individual trapezoids is calculated multiplying Δt by the average value of [P] between two successive time points, i.e., $1/2\,\{[\mathrm{P}]_i + [\mathrm{P}]_{i+1}\}$ and the sum of the areas of all trapezoids (ST) in Fig. 1.2 is given by

$$\mathrm{ST} = 1/2\,\{[\mathrm{P}]_0 + [\mathrm{P}]_1\}\,\Delta t + 1/2\,\{[\mathrm{P}]_1 + [\mathrm{P}]_2\}\,\Delta t + \cdots + 1/2\,\{[\mathrm{P}]_{n-1} + [\mathrm{P}]_n\}\,\Delta t\ ,$$

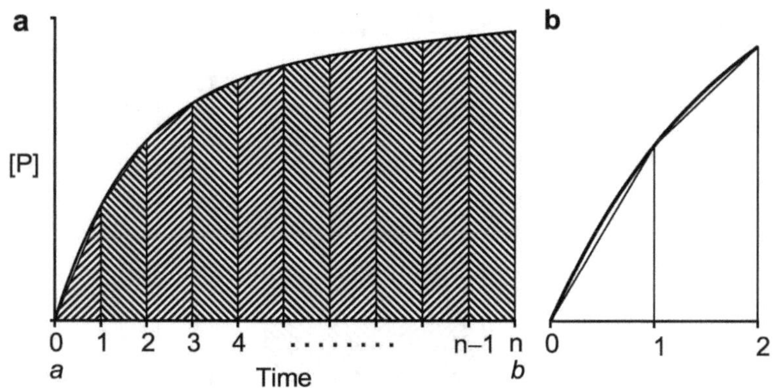

Fig. 1.2 Approximation of integrals by the trapezoid method. The area under the curve is approximated by calculating the sum of the areas of a series of trapezoids (**a**). Since in this particular example the trapezoids approximate the progress curve fairly well despite large integration steps, the intervals 0–1 and 1–2 in (**a**) are reproduced enlarged in (**b**) for closer inspection

which can be simplified to

$$\mathrm{ST} = {}^{1}\!/_{2}\{[\mathrm{P}]_0 + 2[\mathrm{P}]_1 + \cdots + 2[\mathrm{P}]_{n-1} + [\mathrm{P}]_n\}\,\Delta t\,. \tag{1.17}$$

With *Simpson's rule*, the value of a definite integral is approximated by means of polynomial interpolation. This rule can be formulated in various ways, of which the straightforward quadratic (second-order) interpolation is illustrated in Fig. 1.3. The time interval $[a, b]$ is partitioned into an n even number of intervals Δt having the same width, and to explain the method two generic intervals t_{i-2}–t_{i-1} and t_{i-1}–t_i are considered first. The calculation is performed with the aid of three fixed points with the coordinates $(t_{i-2}, [\mathrm{P}]_{i-2})$, $(t_{i-1}, [\mathrm{P}]_{i-1})$, $(t_i, [\mathrm{P}]_i)$. There is an unique parabola that passes through the three fixed points, which can be described by the following Lagrange polynomial interpolation:

$$\begin{aligned}[\mathrm{P}](t) = {} & [\mathrm{P}]_{i-2}\frac{(t-t_i)\,(t-t_{i-1})}{(t_{i-2}-t_{i-1})\,(t_{i-2}-t_i)} + [\mathrm{P}]_{i-1}\frac{(t-t_i)\,(t-t_{i-2})}{(t_{i-1}-t_i)\,(t_{i-1}-t_{i-2})} \\ & + [\mathrm{P}]_i\frac{(t-t_{i-2})\,(t-t_{i-1})}{(t_i-t_{i-2})\,(t_i-t_{i-1})}\,.\end{aligned} \tag{1.18}$$

Figure 1.3 shows three parabolas, each of which passes through the coordinates of three fixed points at times 0, 1, 2 (curve A), 2, 3, 4 (curve B), and $n-2$, $n-1$, n (curve C). The curves are shown extending beyond the fixed points for illustrative purposes. The next task is to calculate the areas under the parabolic arcs of two consecutive strips, whose positions are marked by the double-headed arrows A′, B′, and C′. Since the denominators of the three terms in the right side of (1.18) are made up by combinations of Δt, for any pair of consecutive Δt intervals this expression

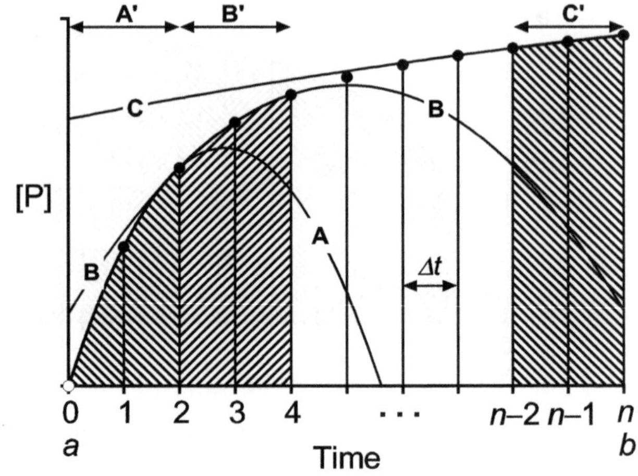

Fig. 1.3 Approximation of a definite integral by Simpson's rule. The interval $[a, b]$ is partitioned into an even number n of identical intervals Δt. A unique *parabola* can be drawn to pass through three consecutive fixed points (*circles*) and the area under the parabolic arc is calculated. The procedure is repeated for all triplets of points and the areas under the arcs summed

can be transformed as (1.19) and integrated as (1.20):

$$[\mathrm{P}](t) = \frac{1}{2(\Delta t)^2} \left\{ \begin{array}{c} [\mathrm{P}]_{i-2}\,(t-t_i)\,(t-t_{i-1}) + [\mathrm{P}]_{i-1}\,(t-t_i)\,(t-t_{i-2}) \\ +[\mathrm{P}]_i\,(t-t_{i-2})\,(t-t_{i-1}) \end{array} \right\}, \tag{1.19}$$

$$\int_{t_{i-2}}^{t_i} [\mathrm{P}](t) \approx \frac{\Delta t}{3} \{[\mathrm{P}]_{i-2} + 4[\mathrm{P}]_{i-1} + [\mathrm{P}]_i\} \,. \tag{1.20}$$

The area under the entire progress curve can now be estimated by summing expressions similar to (1.20) for all parabolic arcs through three consecutive fixed points in the a–b interval (1.21) and simplifying the resulting expression by collecting terms:

$$\begin{aligned} \int_a^b [\mathrm{P}](t) \approx \frac{\Delta t}{3} \{[\mathrm{P}]_0 + 4[\mathrm{P}]_1 + [\mathrm{P}]_2\} + \frac{\Delta t}{3} \{[\mathrm{P}]_2 + 4[\mathrm{P}]_3 + [\mathrm{P}]_4\} \\ + \frac{\Delta t}{3} \{[\mathrm{P}]_{n-2} + 4[\mathrm{P}]_{n-1} + [\mathrm{P}]_n\} \,, \end{aligned} \tag{1.21}$$

$$\int_a^b [\mathrm{P}](t) \approx \frac{\Delta t}{3} \left\{ \begin{array}{l} [\mathrm{P}]_0 + 4[\mathrm{P}]_1 + 2[\mathrm{P}]_2 + 4[\mathrm{P}]_3 + 2[\mathrm{P}]_4 \\ + \cdots + 4[\mathrm{P}]_{n-1} + [\mathrm{P}]_n \end{array} \right\} . \tag{1.22}$$

For the particular function used here as example, an exponential phase followed by a linear steady-state, and with the large integration step Δt chosen to describe the procedures, the LRS underestimates, the RRS overestimates (Fig. 1.1), while the trapezoid method (Fig. 1.2) and Simpson's rule (Fig. 1.3) yield good approximations of the area under the curve. It is intuitive that the precision of these numerical integration methods increases by decreasing the size of Δt.

Modern *solvers* for numerical integration of differential equations exploit variable integration steps, whose width is adjusted automatically. In the example shown in Figs. 1.1 and 1.2, the variable integration steps would be narrow at the beginning of the progress curve (exponential phase) and large as the curve approaches the linear steady-state. Commercially available software packages contain a variety of such solvers, which can be selected according to the particular needs of the differential equations that should be integrated.

Simulink is a graphical programming language for the simulation of dynamic processes, which depends on MATLAB.[8] It offers a graphical environment with block libraries that are used to represent symbolically time-dependent processes. Eight solvers with fixed or variable step size are available to cover a broad band of applications in any field of physics, chemistry, and engineering that requires numerical integration of differential equations. With Simulink, the user must design the whole numerical integration procedure, write differential equations based on a kinetic model (or do the same using symbolic blocks), and make connections between paths in simple or very complex mechanisms. For didactic purposes in enzyme kinetics, this compels students to *put hands on* and thus to understand the methodology in detail.

The KinTek Explorer (KinTek Corporation, U.S.A.),[9] as suggested by its name, is a software package dedicated to the simulation of kinetic processes by numerical integration [39, 41]. It does not require particular efforts from the part of the user and reaction models are input with letters connected by the "=" sign. The necessary differential equations are set up and all rate constants for forward and reverse reaction directions are displayed automatically. A graphical user interface allows to scroll the values of the parameters (for the definition of *parameter* see the note at the end of this section) to be calculated and to directly visualize the results on screen in real time. After entering experimental data, which can either be single curves, sets of curves, or sets of different experiments, simulation is performed by numerical integration. The resulting progress curves are compared with experimental data and the sum of the squared deviations is calculated. The procedure is iterated until a minimum is reached.

A key feature of KinTek is the *FitSpace Explorer*, which

> calculates the dependence of the sum square error on each pair of parameters while allowing all remaining parameters to be adjusted in seeking the best fit.

Results are displayed graphically as three-dimensional plots, which reveal all relationships between parameters and show whether the derived set of fitted parameters is unique and well constrained by the data [40].

---

[8]http://www.mathworks.com/.

[9]http://www.kintek-corp.com/KGExplorer/.

DynaFit (BioKin Ltd., Watertown, Massachusetts, U.S.A.)[10] is also dedicated to the analysis and simulation of kinetic processes by combining numerical integration and regression methods. Model equations are entered assigning any desired name to reactants. Each reaction step contains a pair of reagents separated by symbols that indicate reversible and irreversible steps and the corresponding forward and reverse kinetic constants. This symbolic input generates the differential equations that are then used for numerical integration. Models can be fitted to data with a combination of numerical integration and regression analysis.

The performance of the algorithms of MATLAB/Simulink, KinTek Explorer, and DynaFit has been compared by running simulations of several models with the three software packages, which gave perfectly overlapping results. As KinTek Explorer and DynaFit, also Simulink can be used to perform parameter optimization by combining numerical integration and nonlinear regression. However, two drawbacks of MATLAB/Simulink are exceedingly slow performance using conventional desktop computers and the lack of statistical information of the best-fit parameters.[11]

Definition of *Parameter* in Enzyme Kinetics The term *parameter* is used in various branches of science with different meanings. In enzyme kinetics, it is employed to denote *constants* such as the Michaelis constant and the catalytic constant, which are constant only under specified conditions but assume other values when pH, ionic strength, or temperature is changed. This is to remark that they differ in this respect from the *universal constants* such as the Planck constant or the speed of light in vacuum.

1.3 Microscopic Reversibility and Detailed Balance

The *principle of microscopic reversibility* has been given this name by Tolman in 1924 [61], who also critically discussed the origin of this theory in physical sciences [62] and provided a detailed treatment in his classical book on statistical mechanics [63, pp. 162–163]. The principle of microscopic reversibility states that when a system is in equilibrium, the rates of the forward and the reverse processes are identical and that the reverse reaction travels back the same mechanism as the forward path. In fact, as commented by Mahan [44], it

> ⋯ is a consequence of the invariance of the mechanical equations of motion under the operation of time reversal.

Onsager [50] added that

> the chemists are accustomed to impose a very interesting additional restriction, namely: when the equilibrium is reached each individual reaction must balance itself.

[10] http://www.biokin.com/.

[11] As per December 2014.

This is known as the *principle of detailed balance* (or detailed balancing), a concept worked up to describe several physical phenomena and historically reviewed by Fowler and Milne [33]. Detailed balance is in any case not an exclusive property of chemical reactions, having its roots in Einstein's quantum theory of radiation [29, 33].

The concepts of microscopic reversibility and detailed balance are closely related and their definitions are clear. Mahan demonstrated how detailed balance can be derived from microscopic reversibility [44]. However, the International Union of Pure and Applied Chemistry suggested to regard the two terms as synonyms because in many treatments their meanings have been interchanged [37].

Apart from subtle conceptual differences and confusion that arose between the two concepts, for the treatment of biochemical reactions it is foremost important to remember that microscopic reversibility is a fundamental principle that sets constraints between rate constants. Failure to take into account this principle may invalidate theories and lead to meaningless conclusions.

The principles of microscopic reversibility and detailed balance will now be illustrated using the *monomolecular triangle reaction* discussed by Onsager [50] shown in Scheme 1.1. The system consists of a substance S with total concentration $[S]_t$ that can exist in three distinct forms A, B, and C, each of them convertible into either of the two other forms. The rates of the conversions are governed by six positive first-order constants:

$$\frac{d[A]}{dt} = k_{-1}[B] + k_3[C] - (k_1 + k_{-3})[A] \tag{1.23}$$

$$\frac{d[B]}{dt} = k_1[A] + k_{-2}[C] - (k_{-1} + k_2)[B] \tag{1.24}$$

$$\frac{d[C]}{dt} = k_{-3}[A] + k_2[B] - (k_{-2} + k_3)[C] \; . \tag{1.25}$$

The sum of (1.23)–(1.25) is zero and the sum of any two of them corresponds to the negative of the remaining expression. Thus any one of the three equations, together with the mass conservation equation, is sufficient to characterize the system at equilibrium, which is established when the three derivatives with respect to time

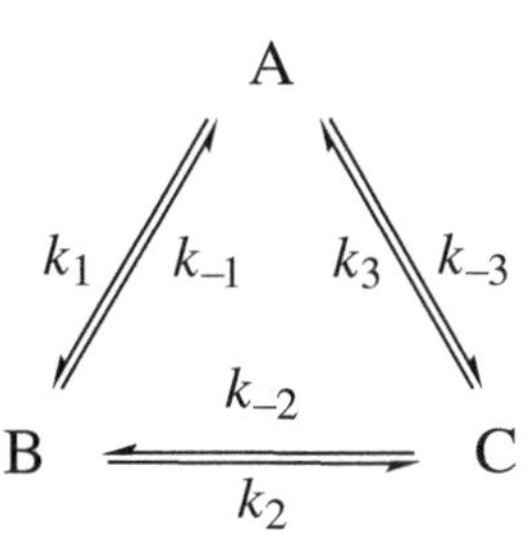

Scheme 1.1 Onsager's chemical monomolecular triangle reaction [50]. A substance exists in three different forms A, B, and C, each of which can be spontaneously converted to either of the two other forms. All reactions are first-order

are zero. We consider for instance (1.23) together with mass conservation:

$$\frac{\mathrm{d}[\mathrm{A}]_{\mathrm{eq}}}{\mathrm{d}t} = k_{-1}[\mathrm{B}]_{\mathrm{eq}} + k_3[\mathrm{C}]_{\mathrm{eq}} - (k_1 + k_{-3})\,[\mathrm{A}]_{\mathrm{eq}} = 0 \tag{1.26}$$

$$[\mathrm{A}]_{\mathrm{eq}} + [\mathrm{B}]_{\mathrm{eq}} + [\mathrm{C}]_{\mathrm{eq}} = [\mathrm{A}]_{\mathrm{z}} + [\mathrm{B}]_{\mathrm{z}} + [\mathrm{C}]_{\mathrm{z}} = [\mathrm{S}]_{\mathrm{t}}\,, \tag{1.27}$$

where the specifications of the concentrations indicate *at equilibrium* (eq) and *at time zero* (z). For the monomolecular triangle reaction, detailed balance is expressed as

$$\begin{aligned} k_1[\mathrm{A}]_{\mathrm{eq}} &= k_{-1}[\mathrm{B}]_{\mathrm{eq}} \\ k_2[\mathrm{B}]_{\mathrm{eq}} &= k_{-2}[\mathrm{C}]_{\mathrm{eq}} \\ k_3[\mathrm{C}]_{\mathrm{eq}} &= k_{-3}[\mathrm{A}]_{\mathrm{eq}}\,. \end{aligned} \tag{1.28}$$

Multiplying together the left terms and the right terms in (1.28) and equating the results it becomes apparent that detailed balance expressed by (1.28) depends on the fact that the following relationship exists between the six kinetic constants

$$k_1k_2k_3[\mathrm{A}]_{\mathrm{eq}}[\mathrm{B}]_{\mathrm{eq}}[\mathrm{C}]_{\mathrm{eq}} = k_{-1}k_{-2}k_{-3}[\mathrm{B}]_{\mathrm{eq}}[\mathrm{C}]_{\mathrm{eq}}[\mathrm{A}]_{\mathrm{eq}}\,, \tag{1.29}$$

where the concentrations cancel out leaving:

$$k_1k_2k_3 = k_{-1}k_{-2}k_{-3} \quad \text{or} \quad K_{\mathrm{eq}} = \frac{k_1k_2k_3}{k_{-1}k_{-2}k_{-3}} = 1\,. \tag{1.30}$$

Relationship (1.30), which contains only first-order constants, represents the principle of microscopic reversibility. No concentrations appear in this expression, suggesting that the principle, derived from a condition of equilibrium, is valid beyond this situation. Indeed, it retains its validity also when a system operates far from equilibrium. There are important messages in (1.30):

1. The net equilibrium constant of systems, whose mechanism is cyclic such as Onsager's monomolecular triangle and the general modifier mechanisms (Chap. 2), has the value 1.
2. Equation (1.30) and similar expressions allow calculating any one of the kinetic constants from the ensemble of the remaining constants.
3. In performing simulations of cyclic systems one must be aware that all constants can be given arbitrary (but plausible!) values, while one of the constants must be necessarily calculated from the others.

Microscopic reversibility and detailed balance have been discussed by Robert Alberty in a didactic paper using Onsager's monomolecular triangle [1]. On this track, Fig. 1.4 shows two sets of plots with the parameters and results listed in Table 1.1. As shown in the upper part of the table, the kinetic constants used for

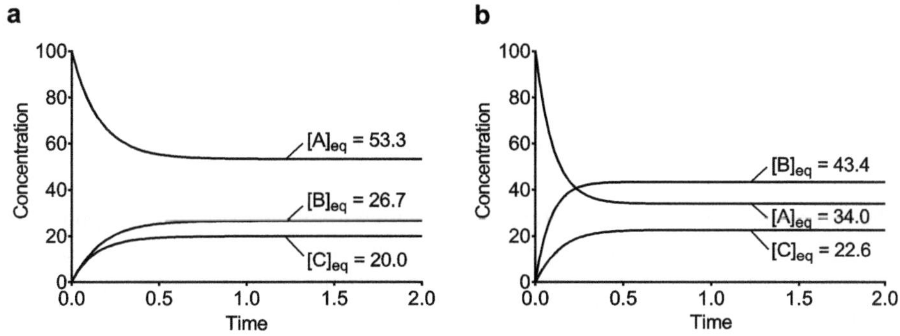

Fig. 1.4 Progress curves plotted for Onsager's monomolecular triangle reaction. Assignment of the traces to [A], [B] and [C] and their values at equilibrium as indicated. The values of the kinetic constants used for drawing the plots by numerical integration and the results of this procedure are collected in Table 1.1. The initial concentrations (at zero time in arbitrary units) were: $[A]_z = 100$, $[B]_z = [C]_z = 0$. Fulfillment (a) and violation (b) of microscopic reversibility

Table 1.1 Microscopic reversibility and detailed balance in Onsager's monomolecular triangle reaction. k_+ and k_- are the forward and reverse first-order kinetic constants of steps 1–3 used for the simulations in Fig. 1.4

	Fig. 1.4a			Fig. 1.4b		
step	k_+	k_-	K_{eq}	k_+	k_-	K_{eq}
1	1.5	3	0.5[a]	5	3	1.67[a]
2	3	4	0.75[b]	3	4	0.75[b]
3	4	1.5	2.67[c]	4	1.5	2.67[c]
1×2×3	18	18	1.0	60	18	3.33
	Fig. 1.4a			Fig. 1.4b		
1	$k_1[A]_{eq} = 80$		$k_{-1}[B]_{eq} = 80$	$k_1[A]_{eq} = 169.8$		$k_{-1}[B]_{eq} = 130.2$
2	$k_2[B]_{eq} = 80$		$k_{-2}[C]_{eq} = 80$	$k_2[B]_{eq} = 130.2$		$k_{-2}[C]_{eq} = 90.6$
3	$k_3[C]_{eq} = 80$		$k_{-3}[A]_{eq} = 80$	$k_3[C]_{eq} = 90.6$		$k_{-3}[A]_{eq} = 50.9$
1×2×3	5.12×10^5		5.12×10^5	2.00×10^6		6.01×10^5
	$K_{eq} = 1$			$K_{eq} = 3.33$		

[a]k_1/k_{-1}, [b]k_2/k_{-2}, [c]k_3/k_{-3}

drawing the plots in Fig. 1.4a fulfill the requirements of microscopic reversibility since $k_1k_2k_3 = k_{-1}k_{-2}k_{-3}$. On the contrary, the plots in (b) were drawn under deliberate violation of the principle by keeping all constants used in (a) but setting $k_1 = 5$ in place of the correct value 1.5. The set of constants gives now a calculated overall K_{eq} of 3.3 instead 1. The lower part of Table 1.1 explicitly shows that detailed balance is respected by the plots in panel (a) and violated in panel (b). The nonsensical result in (b) can be appreciated from the different values of the forward and reverse velocities at equilibrium in the single steps as well as in the overall process, which travels forward at a rate that is 3.3 times larger than in the reverse direction: a physically impossible situation. More practical aspects of microscopic reversibility will be further discussed below in this chapter and in the following chapters.

1.4 The Simplest Reversible Monosubstrate Reaction

The three-step mechanism in Scheme 1.2a, which displays explicitly the phases of substrate binding, followed by chemical transformation from ES to EP and release of product(s), represents the minimal, logical mechanism for a monosubstrate enzyme-catalyzed reaction discussed by Haldane [34, pp. 81–82].

Mechanism (b) is chemically less persuasive because its second step includes two distinct processes in one, namely the second and third step of mechanism (a). As commented by Haldane [34, p. 80] and Cornish-Bowden [22, p. 56], this mechanism is illogical and will not be further considered here. Also, Atkinson observed that mechanism (b) would not allow an extension to systems with more than one substrate and one product [4, p. 279].

Scheme 1.2c corresponds to the *original* Michaelis–Menten mechanism, which can be deduced from the description the authors gave in the summary of their famous paper [46, p. 368], where for brevity the E + S step was omitted but mentioned in the text

$$\begin{aligned}&1\,\text{Mol Sacch.} - \text{Invertin} - \text{Verbind.}\\&\quad\rightarrow 1\,\text{Mol Fructose} + 1\,\text{Mol Glucose} + 1\,\text{Mol Invertin}\,,\end{aligned} \tag{1.31}$$

where Sacch.-Invertin-Verbind. (Saccharose-Invertin-Verbindung) means the complex between sucrose and invertin, i.e., ES in mechanism (c). *Invertin* or *invertase* is the name given at that time to β-fructofuranosidase (EC 3.2.1.26).

Finally, Scheme 1.2d represents the treatment of Briggs and Haldane [14], which has been explicitly addressed to the original Michaelis–Menten mechanism (c) with the microscopic constants k_1 and k_{-1} replacing the substrate dissociation constant

a $$\mathrm{E+S} \underset{k_{-1}}{\overset{k_1}{\rightleftharpoons}} \mathrm{ES} \underset{k_{-2}}{\overset{k_2}{\rightleftharpoons}} \mathrm{EP} \underset{k_{-3}}{\overset{k_3}{\rightleftharpoons}} \mathrm{E+P}$$

b $$\mathrm{E+S} \underset{k_{-1}}{\overset{k_1}{\rightleftharpoons}} \mathrm{ES} \underset{k_{-2}}{\overset{k_2}{\rightleftharpoons}} \mathrm{E+P}$$

c $$\mathrm{E+S} \overset{K_s}{\rightleftharpoons} \mathrm{ES} \xrightarrow{k_2} \mathrm{E+P}$$

d $$\mathrm{E+S} \underset{k_{-1}}{\overset{k_1}{\rightleftharpoons}} \mathrm{ES} \xrightarrow{k_2} \mathrm{E+P}$$

Scheme 1.2 The simplest, reversible, monosubstrate mechanism (**a**). Two-step mechanism (**b**) after collapsing the second and third step of (**a**) in one. The *original* Michaelis–Menten mechanism (**c**) [46], and the Briggs–Haldane mechanism (**d**) [14]. P indicates product(s), i.e., one or more molecules

$K_S = k_{-1}/k_1$. In this paper, the authors set the basis of the steady-state hypothesis of enzyme-catalyzed reactions [7].

Michaelis and Menten, mechanism (c), and Briggs and Haldane, mechanism (d), assume the second reaction step to be irreversible. However, enzymes catalyze reactions in both directions, a point discussed by Atkinson in his analysis of the reversible mechanism (a) [4, pp. 275–282]. In a brief review of Atkinson's book written for Amazon, Ralph Osgood commented—*there is a touch of delicious heresy*. Atkinson's essay contains an unforgiving appreciation of the Briggs and Haldane theory that partly derives from a personal interpretation of the real intents of these authors. In fact, in their 1925-publication, Briggs and Haldane neither claimed that their treatment should be more general than that of Michaelis and Menten, nor that it should be valid also for the reverse reaction. The fact that textbooks and monographs may have diffused this distorted concept is another story, for which Briggs and Haldane cannot be made guilty to the point that led Atkinson to state [4, p. 282]:

> Because of its triviality, the Briggs-Haldane expression for K_m should be dropped from textbook treatments of enzyme kinetics. Until that happens, we may well ignore it ···

The importance of Atkinson's analysis resides in considering activation energy profiles to explain the relationships between the treatments of Michaelis–Menten and Briggs–Haldane and in discussing the conditions for which mechanism (a) in Scheme 1.2 describes the two models.

The steady-state rate equation will be derived for the three-step mechanism in Scheme 1.2a, which serves as a prototype for more complex systems [3]. This equation, which has been obtained algebraically by Haldane [34, p. 81], will be derived below using the method devised by Edward King and Carl Altman [42].

The system contains three enzyme *states* (E, ES, and EP) that can be conveniently represented as a triangular, cyclic linear graph as shown in Scheme 1.3a. The three vertexes, or *nodes* (E_1, E_2, and E_3), representing the enzyme states E, ES, and EP, respectively, are shown as subscheme (b) called the *master pattern*. The nodes are joined by lines, the *links* containing the information on the relationships between enzyme states. Considering that the two second-order reactions can be made pseudo-first-order by multiplication with the actual reactant concentrations, i.e., the steps governed by k_1 [S] and k_{-3} [P], this system is analogous to Onsager's chemical triangle reaction described in Sect. 1.3.

Patterns derived from the master are now drawn, each one containing one link less than the master pattern. In the present mechanism, the master has three links and therefore there are three *derived patterns* as shown in subscheme (c). Essential is that all nodes are connected in each of the derived patterns and that only links present in the master are used. All directional arrows of the original mechanism (a) must now be introduced in the derived patterns so that every node will possess a unique set of *directed patterns*. For this mechanism there are three directed patterns for each of the nodes that are displayed in subscheme (d) in three rows and three columns. The upper row contains all constants that convey to free enzyme, the

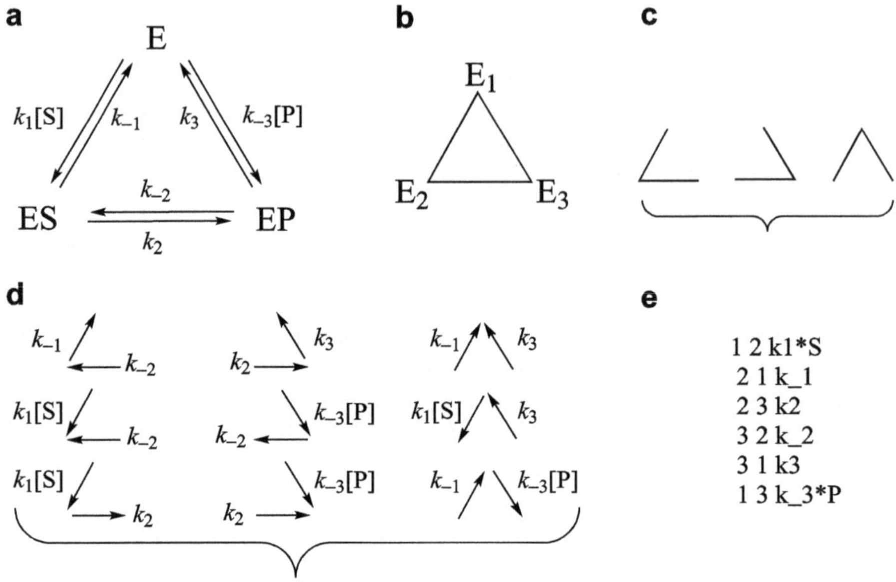

Scheme 1.3 The three-step mechanism in Scheme 1.2a and the steps of the King–Altman method for deriving the rate equation. The mechanism drawn as cyclic system (**a**), King–Altman master pattern (**b**), derived patterns (**c**), directed patterns (**d**), and code used by the KAPattern software for generating the values of the three enzyme species (**e**)

middle row contains the constants that convey to ES, and the lower row those leading to EP.

All constants appearing in the directed patterns must be first-order: k_{-1}, k_2, k_{-2} and k_3 are first-order constants, while the second-order constants k_1 and k_{-3} become pseudo-first-order by multiplication with [S] and [P], respectively. The *values* of the nodes, i.e., the sums of the multiplied pairs of constants for every species, can now either be written by inspection of the directed patterns or using a computerized method. Using KAPattern (Sect. 1.2.3), the labor necessary to generate all values for mechanism (a) consists in entering the code as .txt file shown in subscheme (e), which is self-explanatory, followed by a mouse click.

The values of the nodes for the three-step mechanism (a) are given by

$$\begin{aligned} E_1 &= k_{-1}k_3 + k_{-1}k_{-2} + k_2k_3 \\ E_2 &= k_1k_3\,[S] + k_1k_{-2}\,[S] + k_{-2}k_{-3}\,[P] \\ E_3 &= k_{-1}k_{-3}\,[P] + k_1k_2\,[S] + k_2k_{-3}\,[P]\ , \end{aligned} \tag{1.32}$$

and the fractions of the enzyme states are calculated as

$$f_E = \frac{E_1}{E_1 + E_2 + E_3}; \quad f_{ES} = \frac{E_2}{E_1 + E_2 + E_3}; \quad f_{EP} = \frac{E_3}{E_1 + E_2 + E_3}\ . \tag{1.33}$$

The net reaction rate (1.37), which comprises the contributions of the forward and of the reverse paths, is calculated from the fractions of E, ES, and EP with any of Eqs. (1.34)–(1.36):

$$v_{\text{net}} = (k_3 f_{\text{EP}} - k_{-3} f_{\text{E}}\,[\text{P}])\,[\text{E}]_{\text{t}} \tag{1.34}$$

$$v_{\text{net}} = (k_2 f_{\text{ES}} - k_{-2} f_{\text{EP}})\,[\text{E}]_{\text{t}} \tag{1.35}$$

$$v_{\text{net}} = (k_1 f_{\text{E}}\,[\text{S}] - k_{-1} f_{\text{ES}})\,[\text{E}]_{\text{t}}\;. \tag{1.36}$$

Equations (1.34)–(1.36) imply rate-limiting steps at the level of EP $\rightleftharpoons$ E + P, ES $\rightleftharpoons$ EP or E + S $\rightleftharpoons$ ES, respectively. In fact, no hypotheses have been made about the relative values of the rate constants for deriving a general expression of the rate equation, which is the same because each of the expressions (1.34)–(1.36) uses all of the six rate constants that are linked each other by the principle of microscopic reversibility. With the values (1.32) and the definitions of the fractions (1.33) we obtain the net rate equation

$$v_{\text{net}} = \frac{(k_1 k_2 k_3\,[\text{S}] - k_{-1} k_{-2} k_{-3}\,[\text{P}])\,[\text{E}]_{\text{t}}}{k_1\,(k_{-2} + k_2 + k_3)\,[\text{S}] + k_{-3}\,(k_{-1} + k_{-2} + k_2)\,[\text{P}] + k_{-1} k_3 + k_{-1} k_{-2} + k_2 k_3}, \tag{1.37}$$

which, despite its apparent complexity, can be rearranged by grouping terms that represent individual catalytic constants and Michaelis constants. To this end, the reaction rates at [P] = 0 and [S] = 0 are calculated from (1.37) to reveal the contributions of the forward and reverse reactions to the net reaction rate:

$$v_{[\text{P}]=0} = v_{\text{S}} = \frac{k_1 k_2 k_3 [\text{E}]_{\text{t}}\,[\text{S}]}{k_1\,(k_{-2} + k_2 + k_3)\,[\text{S}] + k_{-1} k_3 + k_{-1} k_{-2} + k_2 k_3} \tag{1.38}$$

$$v_{[\text{S}]=0} = v_{\text{P}} = -\frac{k_{-1} k_{-2} k_{-3} [\text{E}]_{\text{t}}\,[\text{P}]}{k_{-3}\,(k_{-1} + k_{-2} + k_2)\,[\text{P}] + k_{-1} k_3 + k_{-1} k_{-2} + k_2 k_3}\;. \tag{1.39}$$

The negative sign in (1.39) stems from the definition of the rate as the change in [P] with time, which is positive in the direction of substrate turnover and negative in the opposite direction. The negative sign in (1.39) must be considered if experiments are run in the presence of both substrate and product. The limiting rates (V), and from these the catalytic constants ($V/[\text{E}]_{\text{t}}$), are calculated as the limits of the net rate (1.37) at infinite substrate or product concentration:

$$V_{\text{S}} = \lim_{[\text{S}]\to\infty} v_{\text{net}} = \frac{k_2 k_3}{k_{-2} + k_2 + k_3} [\text{E}]_{\text{t}} \tag{1.40}$$

$$k_{\text{cat,S}} = \frac{k_2 k_3}{k_{-2} + k_2 + k_3} \tag{1.41}$$

$$V_{\mathrm{P}} = \lim_{[\mathrm{P}]\to\infty} v_{\mathrm{net}} = \frac{k_{-1}k_{-2}}{k_{-1}+k_{-2}+k_2}[\mathrm{E}]_{\mathrm{t}} \tag{1.42}$$

$$k_{\mathrm{cat,P}} = \frac{k_{-1}k_{-2}}{k_{-1}+k_{-2}+k_2}\,, \tag{1.43}$$

where in (1.42) the sign has been reverted to positive because this calculation actually refers to an experiment in the absence of substrate. The negative sign would be assigned to $k_{\mathrm{cat,P}}$ and k_{P}, which has no physical meaning.

For any mechanism, knowing the expression of V, the Michaelis constants can be evaluated by extracting its general symbolic expression from the Michaelis–Menten equation

$$v = \frac{V\,[\mathrm{S}]}{K_{\mathrm{m}}+[\mathrm{S}]}, \quad v\,(K_{\mathrm{m}}+[\mathrm{S}]) = V\,[\mathrm{S}]\,, \quad K_{\mathrm{m}} = \left(\frac{V}{v}-1\right)[\mathrm{S}]\,. \tag{1.44}$$

In case of complex rate equations, the operational definition[12] of K_{m} is exploited. Since the Michaelis constant corresponds to the substrate concentration for which the rate is one half the limiting rate, K_{m} is calculated by solving the rate equation for [S] when $v = V/2$. To avoid mistakes, calculations may be performed with the aid of mathematical software. As an example, the derivation of $k_{\mathrm{cat,S}}$ and $K_{\mathrm{m,S}}$ performed in this section is demonstrated in the Appendix using Maple. The sought expressions of the two Michaelis constants are each made up of five of the six microscopic constants of the mechanism

$$K_{\mathrm{m,S}} = \frac{k_{-1}k_{-2}+k_{-1}k_3+k_2k_3}{k_1\,(k_{-2}+k_2+k_3)} \tag{1.45}$$

$$K_{\mathrm{m,P}} = \frac{k_{-1}k_{-2}+k_{-1}k_3+k_2k_3}{k_{-3}\,(k_{-1}+k_{-2}+k_2)}\,. \tag{1.46}$$

The corresponding specificity constants are computed as the ratios $k_{\mathrm{cat}}/K_{\mathrm{m}}$:

$$k_{\mathrm{S}} = \frac{k_1k_2k_3}{k_{-1}k_{-2}+k_{-1}k_3+k_2k_3} \tag{1.47}$$

$$k_{\mathrm{P}} = \frac{k_{-1}k_{-2}k_{-3}}{k_{-1}k_{-2}+k_{-1}k_3+k_2k_3}\,. \tag{1.48}$$

With the definitions given above, the net reaction rate can be expressed compactly using specificity and Michaelis constants. This is achieved dividing the numerator

[12]The notion of *operational definition* has been introduced in 1927 by P.W. Bridgman referring to the definition of something, like a variable or an object, through the operations needed to measure it. In enzyme kinetics, operational definitions are for instance the Michaelis constant and competitive inhibition. Such definitions have acknowledged practical value because they are traceable to measurements without making assumptions on the underlying mechanisms.

and the denominator of (1.37) by $(k_{-1}k_3 + k_{-1}k_{-2} + k_2k_3)$. Without showing the cumbersome passages it can be seen that this operation converts the terms multiplying [S] and [P] in the numerator into the respective specificity constants, and the terms multiplying [S] and [P] in the denominator into reciprocal Michaelis constants, i.e.,

$$v_{\mathrm{net}} = \frac{k_{\mathrm{S}}[\mathrm{E}]_{\mathrm{t}}[\mathrm{S}] - k_{\mathrm{P}}[\mathrm{E}]_{\mathrm{t}}[\mathrm{P}]}{1 + \dfrac{[\mathrm{S}]}{K_{\mathrm{m,S}}} + \dfrac{[\mathrm{P}]}{K_{\mathrm{m,P}}}} \,. \tag{1.49}$$

Cornish-Bowden noted that writing the Michaelis–Menten equation using the specificity and the Michaelis constants (1.50) in place of catalytic and Michaelis constant (1.51)

$$v = \frac{k\,[\mathrm{E}]\,[\mathrm{S}]}{1 + \dfrac{[\mathrm{S}]}{K_{\mathrm{m}}}} \tag{1.50}$$

$$v = \frac{k_{\mathrm{cat}}[\mathrm{E}]_{\mathrm{t}}\,[\mathrm{S}]}{K_{\mathrm{m}} + [\mathrm{S}]} \,, \tag{1.51}$$

offers many advantages when discussing kinetic topics such as inhibition types and reactions with multiple substrates [22, pp. 37–38]. In fact, (1.49), which is of the type (1.50), is not only valid for mechanism (a) in Scheme 1.2 but also for mechanism (b) in the same scheme, of course with the appropriate expressions of k_{P} and K_{m}. Moreover, (1.49) applies also to a multitude of other mechanisms, including those with more than one substrate and one product and those containing modifiers (in which case the parameters assume an apparent value), because the equation has an empirical, mechanism-independent character [22, pp. 55–56].

Equation (1.49) is a rearranged form of the net flux expression from substrate(s) to product(s) used by Haldane to formulate his relationship between the thermodynamic equilibrium constant and the kinetic constants of enzyme-catalyzed reactions [34, p. 82]. When a reaction reaches equilibrium

$$\frac{[\mathrm{P}]_{\mathrm{eq}}}{[\mathrm{S}]_{\mathrm{eq}}} = K_{\mathrm{eq}} \tag{1.52}$$

and the net rate equals zero, meaning that the numerator of (1.49) is zero, i.e.,

$$k_{\mathrm{S}}[\mathrm{E}]_{\mathrm{t}}[\mathrm{S}]_{\mathrm{eq}} = k_{\mathrm{P}}[\mathrm{E}]_{\mathrm{t}}[\mathrm{P}]_{\mathrm{eq}} \,, \tag{1.53}$$

which, after introducing the explicit expressions of k_{S} (1.47) and k_{P} (1.48), becomes

$$\frac{k_{\mathrm{S}}}{k_{\mathrm{P}}} = \frac{k_1k_2k_3}{k_{-1}k_{-2}k_{-3}} = \frac{[\mathrm{P}]_{\mathrm{eq}}}{[\mathrm{S}]_{\mathrm{eq}}} = K_{\mathrm{eq}} \,. \tag{1.54}$$

Equation (1.54) is known as the *Haldane equation* or *Haldane relationship*, which links the specificity constants of forward and reverse enzyme-catalyzed reactions by the principle of microscopic reversibility (Sect 1.3). In fact, with reference to Onsager's monomolecular triangle (Scheme 1.1), the reversible three-step reaction analyzed in this section contains two second-order rate constants and the expression of microscopic reversibility (1.30) becomes now

$$\frac{k_1[\mathrm{S}]_{\mathrm{eq}}k_2k_3}{k_{-1}k_{-2}k_{-3}[\mathrm{P}]_{\mathrm{eq}}} = 1 , \tag{1.55}$$

which is the same as the Haldane relationship (1.54).

To conclude and complement this section, I recommend a paper by Donna Blackmond dealing with autocatalysis and chirality. It is an invitation to common sense when proposing theories in the style *if pigs could fly*, which overtly violate the principle of microscopic reversibility [12]. In this interesting tutorial, the author stresses that when a system such as the mechanism in Scheme 1.3a operates away from equilibrium, the observed rate constants are *a consequence of the detailed balance found at equilibrium*. Haldane's relationship (1.54) represents the fundament of this assertion.

1.5 Kinetic Barrier Diagrams

The energetics of enzyme-catalyzed reactions can be illustrated using *kinetic barrier diagrams* proposed by Burbaum et al. [16] as an alternative to Gibbs energy profiles. Kinetic barrier diagrams are helpful in managing the presence of first-order and second-order reaction segments that are part of a reaction. Second-order rate constants are converted to pseudo-first-order constants by multiplication with the appropriate reactant concentration such as the ambient concentration experienced by the enzyme in its *niche* of action. Equilibrium constants, e.g., for the reaction $\mathrm{E} + \mathrm{S} \rightleftharpoons \mathrm{ES}$, with rate constants k_1 and k_{-1} for the forward and reverse direction, respectively, have the unit M (dissociation) or M^{-1} (association). These constants can be made dimensionless by considering their *apparent*[13] value, especially in the presence of reactants in the niche of action of the enzyme. The dissociation constant K_S thus becomes $K_\mathrm{S}^{\mathrm{niche}} = k_{-1}/k_1[\mathrm{S}]_{\mathrm{niche}}$.

On suggestion and with the help of Stephen Bearne, author of two must-read didactic papers on pH effects and inhibitors on enzyme activity illustrated by Gibbs energy profiles [9, 10], kinetic barrier diagrams in this book will be drawn with the

[13]In the context of enzyme kinetics the term *apparent* applies to a parameter, such as the limiting rate and the Michaelis constant, in the form it is measured at particular reactant concentrations.

following ordinate

$$ordinate = \log\left(\frac{k_B T}{hk}\right), \tag{1.56}$$

where k_B is the Boltzmann constant ($1.381 \times 10^{-23}\,\mathrm{J\,K^{-1}}$), T is the absolute temperature (K), h the Planck constant ($6.626 \times 10^{-34}\,\mathrm{J\,s}$), and k is a first-order or a pseudo-first-order constant. At 298 K, $\log(k_B T/h) = 12.79$.

Dimensional analysis of (1.56) gives either

$$\frac{(\mathrm{J\,K^{-1}})\,(\mathrm{K})}{(\mathrm{J\,s})\,(\mathrm{s^{-1}})} \quad \text{or} \quad \frac{(\mathrm{J\,K^{-1}})\,(\mathrm{K})}{(\mathrm{J\,s})\,(\mathrm{M^{-1}\,s^{-1}})\,(\mathrm{M})}, \tag{1.57}$$

depending on whether k is a first-order or a pseudo-first-order constant, respectively, showing that the expression yields a dimensionless number. Diagrams of this type are conceptually similar to Gibbs energy profiles, in which the ordinate is $\Delta G = RT \ln(k_B T/hk)$ and thus the ordinate in (1.56) corresponds to $\Delta G/RT$. This representation has been occasionally used, e.g., by Klipp and Heinrich [43].

Kinetic barrier diagrams will be presented with a dimensionless ordinate and a horizontal line that simply defines the limits of the diagram and is not intended to represent any extent of reaction as a regular abscissa (Fig. 1.5). The labels above the horizontal line identify substrate, product, and intermediates the same way as the labels above the barriers identify transition states. The horizontal separation of the various levels is just a matter of convenient representation for immediate identification of the species associated with a given level on the ordinate. Otherwise,

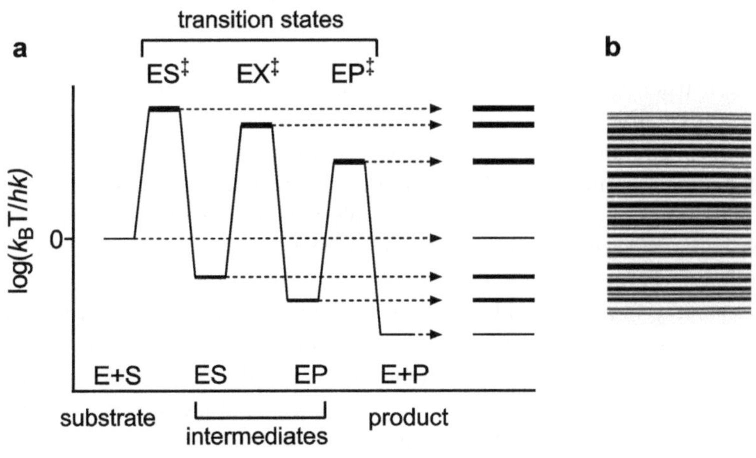

Fig. 1.5 Representation of kinetic barriers. The ordinate values belonging to each of the reactants, intermediates, and transition states can be put one above the other as shown in (**a**). Giving the levels in (**a**) different widths for ground states, intermediates, and transition states, the resulting pattern is reminiscent of a barcode (**b**). The levels of substrate and product are *ground states*

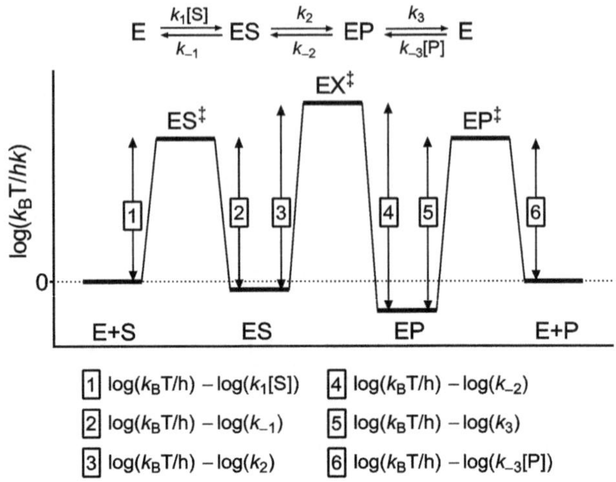

Fig. 1.6 Construction of kinetic barriers for the three-step mechanism in Scheme 1.2a. The height of the barriers labeled with *boxed numbers* is calculated as shown beneath the figure. Starting from barrier 1, to which the origin zero is assigned, the value of barrier 2 is subtracted from the value of 1, to the resulting level barrier 3 is added, and so on

a kinetic barrier diagram can be represented as a sort of *barcode* as shown by the analogy between (a) and (b) in Fig. 1.5. In panel (a), the transition states, the intermediate states and the ground states can be drawn with lines of different thickness. Together with their vertical spacing, the ensemble of these lines define the rate of an enzyme-catalyzed reaction in its actual niche because the ground states include the information of local substrate and product concentrations. After admitting that the barcode representation of kinetic barrier diagrams as in panel (a) does not allow their understanding at a glance, it represents otherwise a handy signature of the system. Moreover, it does not imply a horizontal axis interpreted as *reaction coordinate* or *reaction progress*, often mistaken for progress with time. The correct naming of the horizontal axis in the representation of *free energy surfaces* has been discussed by Cruickshank [24]. For kinetic barrier diagrams the best choice is no label at all.

The construction of a kinetic barrier diagram is illustrated in Fig. 1.6 for the three-step reversible mechanism of Scheme 1.2a. The height of every numbered flank is proportional to the logarithm of a reciprocal first-order or pseudo-first-order constant. Learning first the properties of reversible mechanisms is essential for understanding the behavior of irreversible reactions, either because the reverse reaction is thermodynamically disfavored or because a reaction is measured under initial rate conditions.

The aspect assumed by a kinetic barrier diagram depends on the rate constants, on the concentrations of reagents and on whether the reaction represents a closed or an open system. In *open systems*, continuous supply of substrates and depletion of products regulates the level of the reagents, such as in the metabolic pathways of living organisms, where a given reaction is continuously supplied with substrates (source) and another reaction consumes the products (sink). In open systems, a reaction can proceed close to equilibrium, off-equilibrium, or behave as being essentially irreversible. On the other side, *closed systems* are neither fed by a source nor they can discharge products into a sink. Enzyme-catalyzed reactions studied

$k_1 = 10\mu\text{M}^{-1}\text{s}^{-1}$ $\quad$ $K_{eq} = 1.0$

$k_{-1} = 3000\text{s}^{-1}$ $\quad$ $[\text{S}]_{\text{niche}} = 150\mu\text{M}$

$k_2 = 20\text{s}^{-1}$ $\quad$ $[\text{P}]_{\text{niche}} = 150\mu\text{M}$

$k_{-2} = 1\text{s}^{-1}$ $\quad$ $\theta = 1$

$k_3 = 150\,\text{s}^{-1}$

$$k_{-3} = \frac{k_1 k_2 k_3}{k_{-1} k_{-2}} \frac{[\text{S}]_{\text{eq}}}{[\text{P}]_{\text{eq}}} = 10\mu\text{M}^{-1}\text{s}^{-1}$$

a

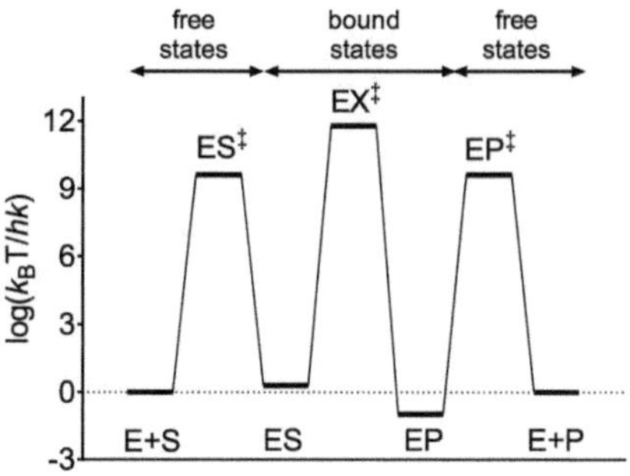

$k_1 = 20\mu\text{M}^{-1}\text{s}^{-1}$ $\quad$ $K_{eq} = 199$

$k_{-1} = 2000\text{s}^{-1}$ $\quad$ $[\text{S}]_{\text{niche}} = 100\mu\text{M}$

$k_2 = 180\text{s}^{-1}$ $\quad$ $[\text{P}]_{\text{niche}} = 30\mu\text{M}$

$k_{-2} = 20\text{s}^{-1}$ $\quad$ $\theta = 1.51 \times 10^{-3}$

$k_3 = 8000\,\text{s}^{-1}$

$$k_{-3} = \frac{k_1 k_2 k_3}{k_{-1} k_{-2}} \frac{[\text{S}]_{\text{eq}}}{[\text{P}]_{\text{eq}}} = 3.62\mu\text{M}^{-1}\text{s}^{-1}$$

b

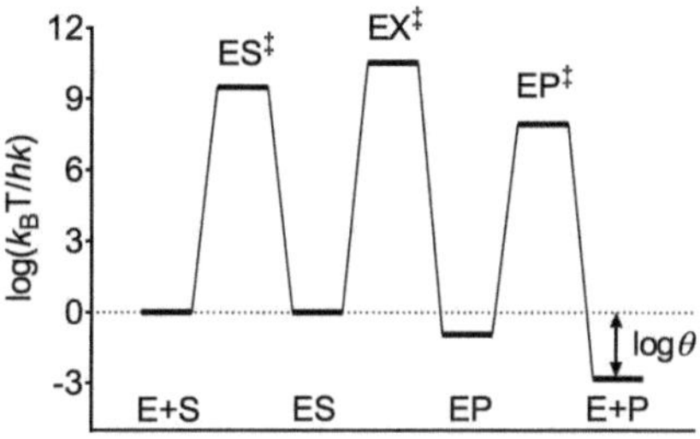

$k_1 = 8\mu\text{M}^{-1}\text{s}^{-1}$ $\quad$ $K_{eq} = 99$

$k_{-1} = 450\text{s}^{-1}$ $\quad$ $[\text{S}]_{\text{niche}} = 100\mu\text{M}$

$k_2 = 400\text{s}^{-1}$ $\quad$ $[\text{P}]_{\text{niche}} \to 0$

$k_{-2} = 2\text{s}^{-1}$ $\quad$ $\theta \to 0$

$k_3 = 2000\,\text{s}^{-1}$

$$k_{-3} = \frac{k_1 k_2 k_3}{k_{-1} k_{-2}} \frac{[\text{S}]_{\text{eq}}}{[\text{P}]_{\text{eq}}} = 71.8\mu\text{M}^{-1}\text{s}^{-1}$$

c

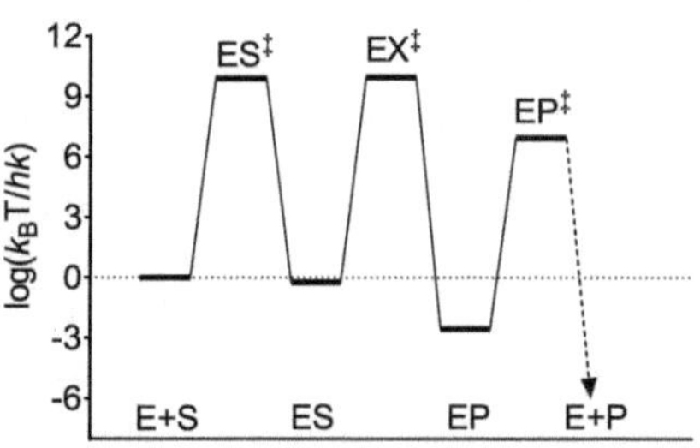

Fig. 1.7 Kinetic barriers for particular situations experienced by enzymes in their niches of action. (**a**) At or close to equilibrium, (**b**) off-equilibrium, (**c**) irreversible

in vitro fall in the category of closed systems where the niche of the considered enzyme is either a cuvette, a multiwell plate or similar accessory containing a buffer solution and all ingredients necessary to run measurements. Here, the enzyme tries to survive in an unfriendly environment, often diluted to exceedingly low concentrations, a situation that is quite different from its original physiological environment. Provided the enzyme is not inactivated, equilibrium will be reached, although some reactions behave as being irreversible and proceed to completeness, e.g., those catalyzed by hydrolytic enzymes.

These situations are illustrated in the diagrams of Fig. 1.7, which show *free and bound enzyme states* [16]. The constants and reactant concentrations used for constructing the barriers are shown on their left side. Substrate and product concentrations are those experienced by the enzyme in its niche of action. Panel (a) represents a reaction proceeding at or close to equilibrium, which is characterized by the free states E + S and E + P at the same level. Panel (b) represents the kinetic barriers of a reaction that operates off-equilibrium because a considerable proportion of the product is rapidly removed by another process. According to Burbaum et al. [16], the mark of this situation is a lower energy level for the E + P free state, which is shifted by the amount (double arrow) corresponding to the logarithm of

$$\theta = \frac{[\mathrm{P}]_{\mathrm{niche}}}{[\mathrm{S}]_{\mathrm{niche}}} \frac{1}{K_{\mathrm{eq}}} \,. \tag{1.58}$$

In this example the off-equilibrium concentration of [P] gives $\theta = 1.51 \times 10^{-3}$ and $\log\theta = -2.82$.

Panel (c) illustrates an irreversible reaction, symbolically represented with a dashed right flank of $\mathrm{EP}^{\ddagger}$. The arrow indicates $[\mathrm{P}]_{\mathrm{niche}} \rightarrow 0$ (setting $k_{-3}[\mathrm{P}]_{\mathrm{niche}} = 0$ is mathematically incorrect because the logarithm of zero is not defined). For a diagram such as that in panel (c) we can envisage three situations:

1. In a metabolic pathway the product is rapidly and almost completely withdrawn by another enzymatic reaction
2. In the test tube a coupled assay reproduces the previous scenario (see also the comments about coupled assays on p. 54)
3. An initial rate experiment is run in vitro, in which $[\mathrm{P}]_{\mathrm{niche}}$ is *assumed* to be zero, although in reality it starts to increase after the generally very short time of the pre-steady-state. This point is further discussed in Sect. 1.9.

In many textbooks and original publications, kinetic barrier diagrams and Gibbs energy profiles are very often not drawn to scale and mostly sketched by hand without ticks and labels on the ordinate. Although being impressive, the resulting mountain-valley landscapes sometimes exaggerate the differences between the height of the mountains and the depth of the valleys. This is merely a cautionary note for readers that may be disappointed in seeing such diagrams drawn to scale, *with numbers*, in the following sections and elsewhere in this book.

For the construction of kinetic barrier diagrams of the Michaelis–Menten equation, i.e., mechanism (d) in Scheme 1.2, which contains an irreversible step and only one intermediate, the method described below is practical because it allows the construction from the experimentally determined parameters k_{cat} and K_{m}. Before proceeding, the Gibbs energy profile will be illustrated first (Fig. 1.8), where the activation energies are related to the kinetic parameters by

$$\Delta G_{\mathrm{S}}^{\prime 0} = -RT \ln \frac{1}{K_{\mathrm{m}}} \tag{1.59}$$

$$\Delta G_{\mathrm{ES}}^{\prime 0\,\ddagger} = RT \left(\ln \frac{k_{\mathrm{B}}T}{h} - \ln k_{\mathrm{cat}} \right) \tag{1.60}$$

$$\Delta G_{\mathrm{E}}^{\prime 0\,\ddagger} = RT \left(\ln \frac{k_{\mathrm{B}}T}{h} - \ln \frac{k_{\mathrm{cat}}}{K_{\mathrm{m}}} \right) \,, \tag{1.61}$$

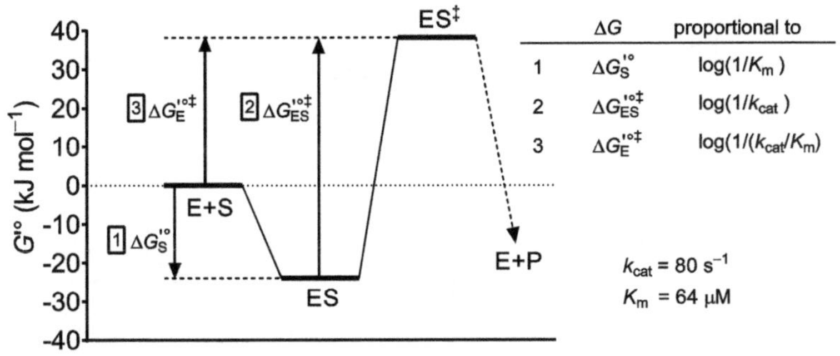

Fig. 1.8 Gibbs energy profile for a mechanism that obeys the Michaelis–Menten equation showing the relationships between activation energy and kinetic parameters. The profile was drawn to scale with (1.59)–(1.61) and the parameters shown in the figure

with the standard state $[S] = [P] = [E]_t = 1\,M$. Furthermore, the biochemical standard pH = 7 is chosen for the hydrogen ion concentration, as indicated by the prime in $\Delta G'^0$ [2], and the temperature is 25 °C = 298 K. The Gibbs energies (1.59)–(1.61) do not change under standard conditions. However, they are altered in the presence of modifiers and are thus useful for illustrating enzyme–modifier interactions [9].

The representation as kinetic barriers will now be illustrated considering the actual substrate concentration, called above $[S]_{niche}$, to convert the second-order rate constant k_1 into the pseudo-first-order constant $k_1[S]_{niche}$ that can be used to define an apparent Michaelis constant, normalized to the specified substrate concentration:

$$K_m^{niche} = \frac{k_{-1} + k_2}{k_1[S]_{niche}} = \frac{K_m}{[S]_{niche}} \,. \tag{1.62}$$

Dividing (1.59)–(1.61) on both sides by RT and taking base-10 logarithms gives:

$$\Delta_S = -\log \frac{1}{K_m^{niche}} \tag{1.63}$$

$$\Delta_{ES} = \left(\log \frac{k_B T}{h} - \log k_{cat}\right) \tag{1.64}$$

$$\Delta_E = \left(\log \frac{k_B T}{h} - \log \frac{k_{cat}}{K_m^{niche}}\right) \,. \tag{1.65}$$

Like the Gibbs energy profile in Fig. 1.8, also a kinetic barrier diagram has just one transition state but the difference is made by the position of the energy level of the intermediate ES and by the position of ES relative to ES‡, which depend on substrate concentration (Fig. 1.9). Δ_S defined by (1.63) represents a thermodynamic difference, while Δ_{ES} and Δ_E are kinetic barriers. In Fig. 1.9a the highest barrier corresponds to Δ_{ES}, proportional to the logarithm of the reciprocal specificity

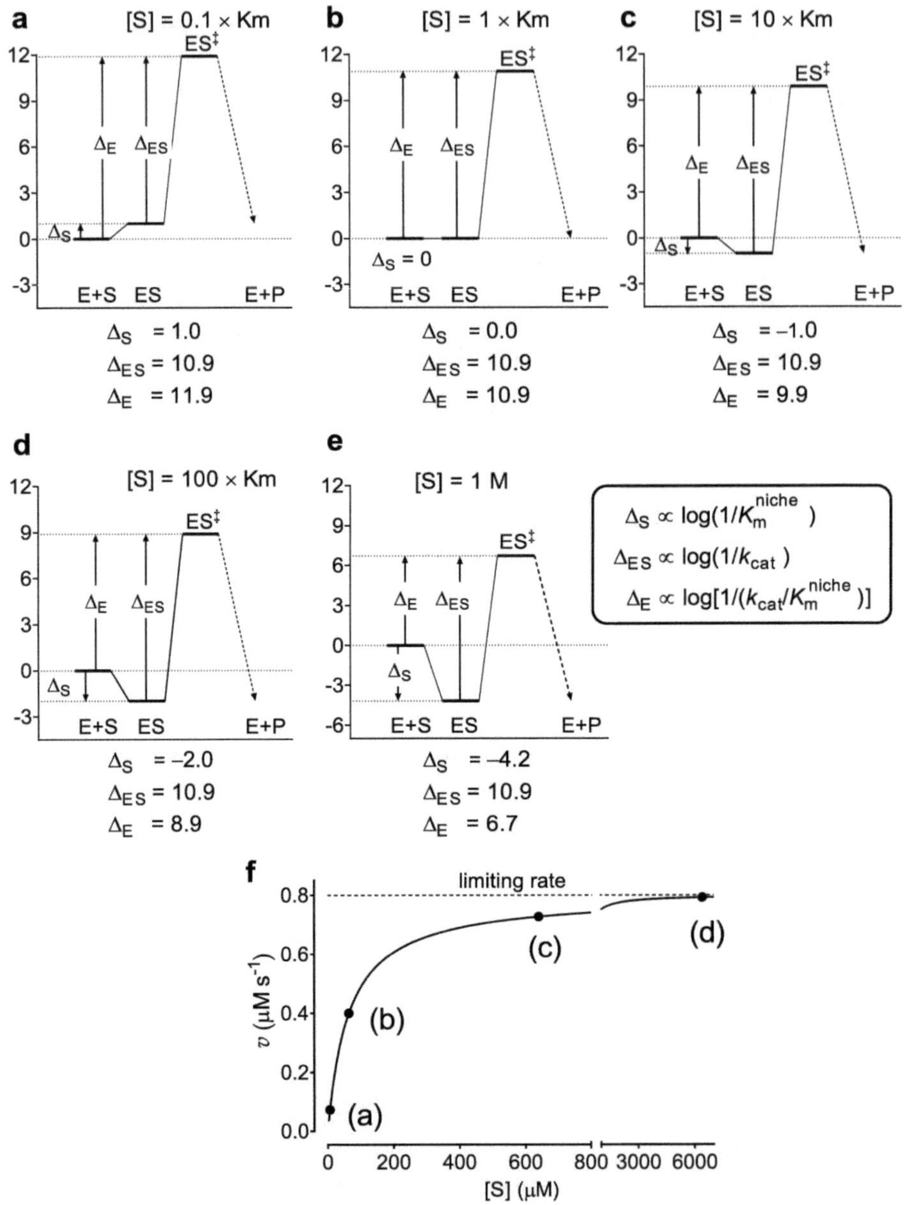

Fig. 1.9 Kinetic barrier diagrams for a mechanism that obeys the Michaelis–Menten equation. The diagrams were constructed with $k_{cat} = 80\,s^{-1}$ and $K_m = 64\,\mu M$ using (1.63)–(1.65) and various substrate concentrations ((**a**)–(**e**)). The height and sign of the barriers is indicated by *arrows* and numerically beneath the diagrams. The diagrams (**a**)–(**d**), with ordinate defined by (1.56), correspond to the points (**a**)–(**d**) in the plot of initial rate versus substrate concentration shown in (**f**), while (**e**) represents standard conditions

constant. This is the situation exploited to calculate the specificity constant from initial rate measurements under first-order conditions, i.e., $[S] \ll K_m$ labeled (a) in Fig. 1.9f:

$$\text{with } [S] \ll K_m, \quad v = \frac{k_{cat}[E]_t\,[S]}{K_m + [S]} \approx \frac{k_{cat}[E]_t\,[S]}{K_m} \quad \therefore \quad \frac{v}{[E]_t} \approx k_S\,[S]\,. \tag{1.66}$$

When $[S] = K_m$ (panel b) $\Delta_S = 0$ and $\Delta_{ES} = \Delta_E$. This indicates that the total enzyme concentration $[E]_t$ is partitioned equally between free enzyme [E] and [ES] and that the rate corresponds to one half the limiting rate (point b in panel f):

$$v = \frac{V\,[S]}{2\,[S]} = \frac{V}{2}\,. \tag{1.67}$$

The numbers beneath panels (a–e) in Fig. 1.9 indicate Δ_S, Δ_{ES} and Δ_E, showing that Δ_E corresponds to the algebraic sum of Δ_S and Δ_{ES}. Increasing the substrate concentration over K_m, as shown in panels (c–d), provokes the inversion of the sign of Δ_S, while Δ_E decreases by the same amount as Δ_S. The highest barrier is now Δ_{ES}, proportional to $\log(1/k_{cat})$. In the situation shown in panel (c), $[S] = 10\,K_m$, the enzyme is not yet saturated by substrate, with $v = 90.9\,\%$ of V, as evidenced by point (c) in panel (f).

A large excess of substrate saturates the enzyme, in which case Δ_{ES} decidedly dominates over Δ_E as shown in panel (d) with the effect on the initial rate in the corresponding point of panel (f). Under these conditions the reaction becomes zero-order being independent of [S]:

$$\text{with } [S] \gg K_m, \quad v = \frac{k_{cat}[E]_t\,[S]}{K_m + [S]} \approx \frac{k_{cat}[E]_t\,[S]}{[S]} \quad \therefore \quad \frac{v}{[E]_t} \approx k_{cat}\,. \tag{1.68}$$

Initial rate measurements under zero-order conditions described by (1.68) are very handy for evaluating the catalytic constant if the concentration of enzyme active sites is known. Likewise, the enzyme concentration can be measured if k_{cat} is known or, more practically, $[E]_t$ can be calculated in a sample after constructing a calibration curve with known enzyme concentrations. Some clinical assays are based on this principle for determining the concentration of enzymes in physiological fluids.

The diagram in Fig. 1.9e was drawn under standard conditions with [S] = 1 M, which is tantamount to a Gibbs energy profile (Fig. 1.8), just with different values on the ordinate. This representation is informative when comparing systems with different values of the kinetic parameters because the base-10 antilogarithm of Δ_S corresponds to K_m, whereas the antilogarithm of $(\Delta_{ES} - 12.79)$ corresponds to k_{cat}, and the antilogarithm of $(\Delta_E - 12.79)$ gives k_{cat}/K_m. This can be verified with (1.63)–(1.65), where $\log(k_B T/h) = 12.79$.

To render homage to all those diligent researchers who invested considerable time in determining individual rate constants in enzyme mechanisms by using methods such as kinetic isotope effects, the dependence of reaction rates on the

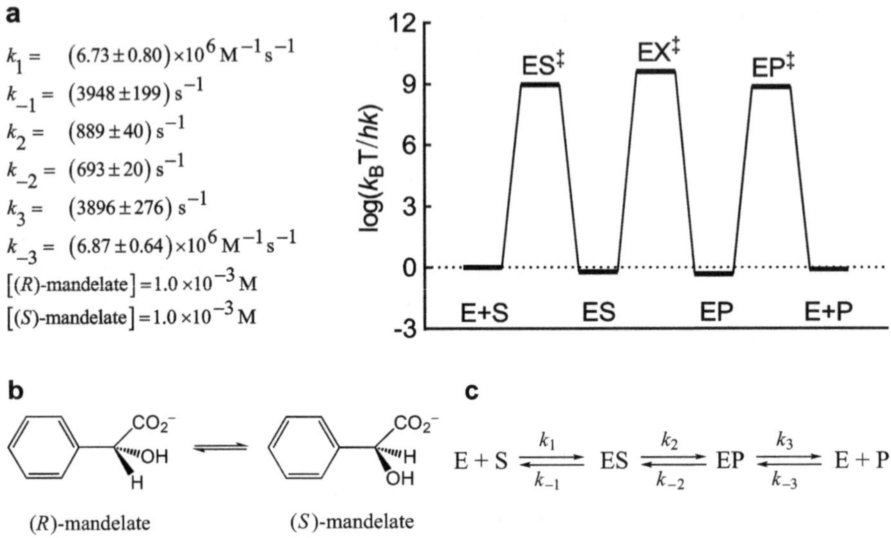

Fig. 1.10 Reaction catalyzed by mandelate racemase. (**a**) Kinetic barrier diagram drawn with the rate constants shown on the *left*. Data, at 25 °C, are from St. Maurice and Bearne [59]; reactant concentrations represent plausible physiological values. (**b**) The racemization reaction between (*R*)- and (*S*)-mandelate. (**c**) Kinetic mechanism, where S corresponds to (*R*)-mandelate and P to (*S*)-mandelate to match the listed sequence of kinetic constants. Many thanks to Steve Bearne, Dalhousie University, Halifax, Canada for providing his original data

viscosity of the medium, and the variation with temperature of k_{cat} and k_{cat}/K_m, an experimental example concludes this section: the reaction catalyzed by mandelate racemase (EC 5.1.2.2) shown in Fig. 1.10 based on individually determined rate constants.

This enzyme from the bacterium *Pseudomonas putida* catalyzes the interconversion between the enantiomers of mandelic acid (Fig. 1.10b). Kinetic constants are those of St. Maurice and Bearne [59] and reactant concentration was considered 1 mM for both enantiomers as an estimate of real values instead of using a standard state of 1 M. Considering experimental scatter, data plotted as kinetic barrier diagram produce the profile expected for a racemization reaction that operates at equilibrium (Fig. 1.10a). The experimental constants with standard deviations are shown on the left of panel (a). The relative positions of the ground states E + S and E + P are established by the equilibrium constant, K_{eq}, which depends only on the nature of the reactants and is not altered by the presence of the enzyme.

Adherence to the principle of microscopic reversibility can be verified with (1.55) using the rate constants with their standard deviations and applying the rule for error propagation (1.7). The calculated value is 1.24 ± 0.23 that, considering the complexity of the experiments with viscogenic agents employed to evaluate individual kinetic constants, is in excellent agreement with the theoretical value of 1.

1.6 Back to the Future

The theories of Michaelis–Menten, of Briggs–Haldane, and corollaries thereof will be discussed below. The concepts will not be presented as published originally but as an imaginary biochemical supplement to Robert Zemeckis's trilogy *Back to the Future*, in which the authors are engaged in rewriting their seminal papers in the 21st century. The starting point is mechanism (a) in Scheme 1.2 written in the cyclic form of Scheme 1.3a. The *original* Michaelis–Menten model (Scheme 1.2c) with related rate equation and parameters will be referred to as such to avoid confusion with the Briggs–Haldane model (Scheme 1.2d), in which the rate equation contains the accepted definition of the Michaelis constant [49]. The following subsections aim at fulfilling a triple purpose:

- to support the treatment of enzyme–modifier interactions, the core of this book, by refreshing the concepts of quasi-equilibrium and steady-state in enzyme-catalyzed turnover of substrates
- to reconcile diverging opinions on the validity of consolidated theories
- to guide readers interested in deriving rate equations under the steady-state assumption, the quasi-equilibrium assumption, or combinations of the two using the King–Altman schematic treatment [42] and facilitated variants of it.

1.6.1 The Michaelis–Menten Treatment Revisited

In the preceding section, the three-step reversible mechanism (Scheme 1.2a) was considered under a full steady-state regime, in which all individual microscopic constants were involved in the rate equation. Yet, in many enzyme-catalyzed reactions the association of substrate with enzyme to form ES and the dissociation of EP to give free enzyme and product are much faster processes than the chemical transformation events occurring at the ES $\rightleftharpoons$ EP level. Thus, in the steady-state, the reaction segments in the shaded area of Scheme 1.4a are in *quasi-equilibrium*. In the associated kinetic barrier diagram shown in Scheme 1.4b the transition state with the highest activation energy corresponds to EX‡, the chemical transformation step ES $\rightleftharpoons$ EP, which is rate-limiting.

The following definition of quasi-equilibrium is taken from Vrzheshch [65][14]

> The quasi-equilibrium assumption in enzyme kinetics is a widespread approach that implies that some intermediate enzyme forms in the mechanism of the enzyme reaction are considered to be in a rapidly established equilibrium state, and this state is practically not subjected to the disturbing effect of the reaction catalyzed by the enzyme.

An alternative name of the quasi-equilibrium assumption is *rapid-equilibrium assumption (or hypothesis)*. The quasi-equilibrium assumption does not explicitly

[14] I wish I could pronounce correctly the name of this valuable scientist.

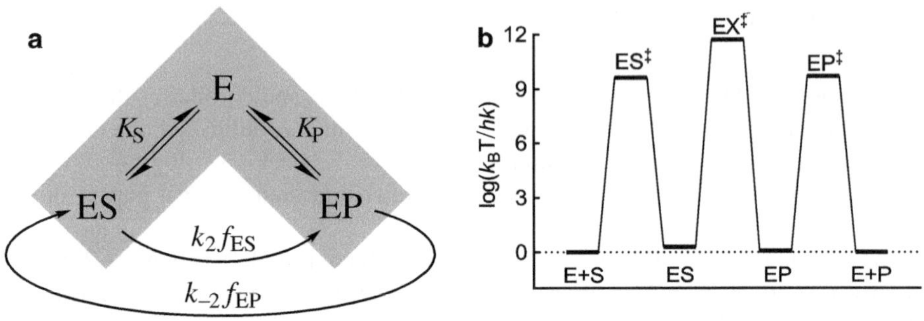

Scheme 1.4 Conditions for which the three-step reaction in Scheme 1.2a describes the reversible Michaelis–Menten mechanism. The association and dissociation steps (panel (**a**), *shaded area*) are fast processes that are in a state of quasi-equilibrium with respect to the slow chemical transformation (ES $\rightleftharpoons$ EP). Accordingly, the kinetic barrier diagram (**b**) shows that EX‡ has the highest activation energy (rate-limiting step). Kinetic barriers were drawn to scale using the constants in Fig. 1.11d and $[S]_{niche} = [P]_{niche} = 50\,\mu M$

assign numerical thresholds to rate constants in order for the hypothesis to be valid and the Latin word *quasi* means *as if* or *almost*. Therefore, accepting the quasi-equilibrium assumption depends on the precision desired in evaluating kinetic parameters.

With these premises, the reaction rate can be calculated with Cha's simplification of the King–Altman method for systems containing quasi-equilibrium segments [18]. First, the fractions of the three enzyme states are calculated from concentrations and the equilibrium constants K_S and K_P after selecting one of the enzyme states as reference. If this is E, the fraction of E in Scheme 1.4 is given by

$$f_E = \frac{[E]}{[E] + [ES] + [EP]} = \frac{1}{1 + \dfrac{[ES]}{[E]} + \dfrac{[EP]}{[E]}}. \tag{1.69}$$

The equilibrium constant of the reaction E + S $\rightleftharpoons$ ES is calculated as

$$\frac{[E]\,[S]}{[ES]} = K_S \quad \therefore \quad \frac{[ES]}{[E]} = \frac{[S]}{K_S}, \tag{1.70}$$

and similarly for EP. Thus, the sought fractions are

$$f_E = \frac{1}{1 + \dfrac{[S]}{K_S} + \dfrac{[P]}{K_P}}, \quad f_{ES} = \frac{\dfrac{[S]}{K_S}}{1 + \dfrac{[S]}{K_S} + \dfrac{[P]}{K_P}}, \quad f_{EP} = \frac{\dfrac{[P]}{K_P}}{1 + \dfrac{[S]}{K_S} + \dfrac{[P]}{K_P}}, \tag{1.71}$$

where $f_E + f_{ES} + f_{EP} = 1$, with mass conservation $[E] + [ES] + [EP] = [E]_t$. f_{ES} and f_{EP} represent the fractional concentrations needed to calculate $f_{ES}[E]_t$ and $f_{EP}[E]_t$, i.e., the concentrations of ES and EP available for catalysis in the forward and reverse direction, respectively. The rate equation can now be calculated as

$$v_{\text{net}} = (k_2 f_{ES} - k_{-2} f_{EP})\,[\text{E}]_{\text{t}} \tag{1.72}$$

$$v_{\text{net}} = \frac{\dfrac{k_2}{K_\text{S}}\,[\text{S}]\,[\text{E}]_\text{t} - \dfrac{k_{-2}}{K_\text{P}}\,[\text{P}]\,[\text{E}]_\text{t}}{1 + \dfrac{[\text{S}]}{K_\text{S}} + \dfrac{[\text{P}]}{K_\text{P}}}\,. \tag{1.73}$$

As shown below, k_2/K_S and k_{-2}/K_P represent specificity constants showing that (1.73) has the form of (1.49). The kinetic parameters for this system are

$$V_\text{S} = k_2[\text{E}]_\text{t}, \quad k_{\text{cat,S}} = k_2, \quad K_\text{S} = \frac{k_{-1}}{k_1}, \quad k_\text{S} = \frac{k_2}{K_\text{S}} \tag{1.74}$$

$$V_\text{P} = k_{-2}[\text{E}]_\text{t}, \quad k_{\text{cat,P}} = k_{-2}, \quad K_\text{P} = \frac{k_3}{k_{-3}}, \quad k_\text{P} = \frac{k_{-2}}{K_\text{P}}\,. \tag{1.75}$$

From (1.73) in the absence of product, the initial rate is given by

$$v_\text{S} = \frac{\dfrac{k_2}{K_\text{S}}[\text{E}]_\text{t}\,[\text{S}]}{1 + \dfrac{[\text{S}]}{K_\text{S}}} \quad \text{or} \quad v_\text{S} = \frac{k_2[\text{E}]_\text{t}\,[\text{S}]}{K_\text{S} + [\text{S}]}\,, \tag{1.76}$$

which is the original equation published by Michaelis and Menten [46], i.e.,

$$v = C \cdot \Phi \cdot \frac{[S]}{[S] + k}\,, \tag{1.77}$$

where $C \equiv k_2$, $\Phi \equiv [\text{E}]_\text{t}$ and $k \equiv K_\text{S}$.

The initial rate in the presence of P at [S] = 0 corresponds to

$$v_\text{P} = \frac{\dfrac{k_{-2}}{K_\text{P}}[\text{E}]_\text{t}\,[\text{P}]}{1 + \dfrac{[\text{P}]}{K_\text{P}}} \quad \text{or} \quad \frac{k_{-2}[\text{E}]_\text{t}\,[\text{P}]}{K_\text{P} + [\text{P}]}\,, \tag{1.78}$$

showing that the reverse reaction mirrors the forward reaction.

The progress curves in Fig. 1.11 illustrate how the concentrations of E, ES, EP, and P vary from reaction start until reaching equilibrium. Panel (a) shows the first ten milliseconds of reaction revealing a rapid increase of [ES], paralleled by a decrease of free enzyme [E], while [EP] is very small, resembling zero in this figure. Following this rapid change of concentrations, a steady-state is established that is characterized by constant values of [E], [ES] and [EP] in panel (a). These concentrations remain apparently constant in panel (b) that shows the progress of

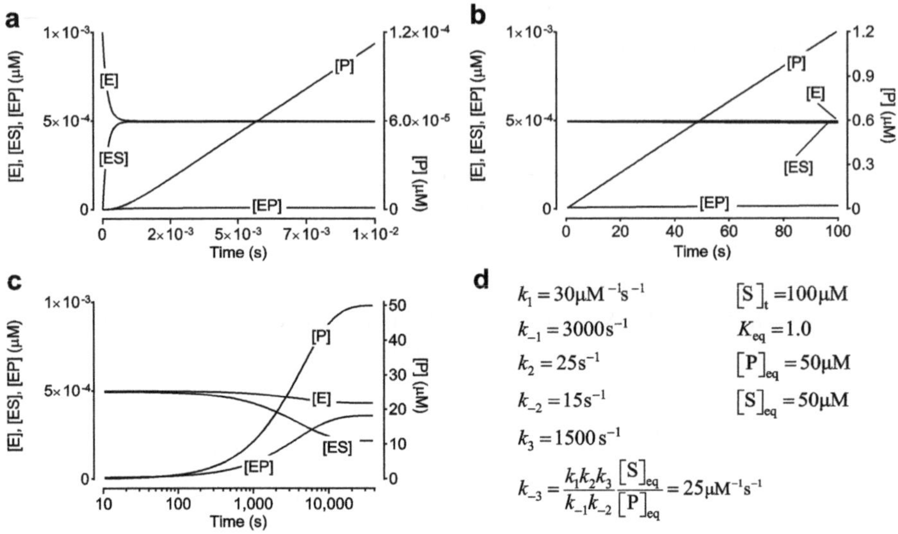

Fig. 1.11 Progress curves for the reversible mechanism in Scheme 1.4. Simulations were performed for three time intervals (**a**)–(**c**) with the constants listed in (**d**). Since the microscopic constants are constrained by microscopic reversibility, k_{-3} was calculated from the other constants and the equilibrium concentrations of substrate and product for $K_{eq} = 1$ as shown in (**d**) (Sect. 1.3). Simulations were then performed by numerical integration with $[S]_t = 100\ \mu M$

reaction till 100 s. In reality, [E] and [EP] slowly increase and [ES] decreases. During the same time [P] increases in an apparently linear fashion but closer examination reveals a slight deviation from linearity, which is barely perceptible in this figure. Should this be a real experiment, the time range in panel (b) would be ranked as fine for the acquisition of reasonable initial velocities because in this time interval only about 1 % of the total substrate concentration is used up. Furthermore, small deviations from linearity would be masked by experimental scatter (but see the comment on this issue in Sect. 1.9). This system reaches equilibrium after about 11 h (c).

While this simulated example is instructive, [E], [ES] and [EP] are generally not directly measurable and the possibility to detect the short lag phase in the formation of product (Fig. 1.11a) depends on the dead time of the flow apparatus used. An experiment such as that in panel (c) requires several hours and there is little hope that the enzyme will remain stable and that other parameters will also not change. Increasing the enzyme concentrations accelerates the reaction but still enzyme inactivation should be carefully checked (Sect. 1.8).

The kinetic barrier diagram in Scheme 1.4b is in line with the original treatment of Michaelis–Menten [46], valid under initial rate conditions. In this case, the reaction proceeds through the transition-state $EX^{\ddagger}$ to the intermediate EP, which immediately dissociates into E + P having no chances to travel back to ES for two reasons: (1) the activation energy of $EP^{\ddagger}$ is much lower than that of $EX^{\ddagger}$, (2) there is no product for the reverse reaction to take place. This fact can be appreciated in the progress curves of Fig. 1.11c, where a very small concentration of EP (in this example just 0.8 % of $[E]_t$) is maintained during the steady-state. This behavior is predicted by Eq. (1.72), which corresponds to the original Michaelis–Menten equation if $f_{EP} \approx 0$. In the diagram of Scheme 1.4b, the levels of the intermediates ES and EP will vary depending on rate constants and concentrations.

1.6.2 The Briggs–Haldane Treatment Revisited

Starting from the reversible three-step mechanism in Scheme 1.2a, this section discusses the Briggs and Haldane mechanism [14] represented as Scheme 1.2d. When only the $E + P \rightleftharpoons EP$ step in the reaction of Scheme 1.2a equilibrates rapidly with respect to the other two steps, the derivation of the rate equation with the King–Altman method [42] implemented by Cha [18] can further benefit from a rule proposed by Volkenstein and Goldstein [64]. Rule *a* in this paper establishes that parallel branches in a system of reactions can be added becoming a single branch, a procedure called *node compression*. The reaction mechanism can be sketched as shown in Scheme 1.5a, where the shaded area indicates the quasi-equilibrium segment. ES maintains a relationship with the E/EP ensemble through four first-order processes: two leading from ES to E/EP (k_{-1} and k_2) and two leading from E/EP to ES (k_1 [S] and k_{-2}). The $E \rightleftharpoons EP$ quasi-equilibrium is practically undisturbed by the events outside the shaded area and the partition between E and ES can thus be calculated as

$$f_E = \frac{1}{1 + \dfrac{[P]}{K_P}}, \quad f_{EP} = \frac{\dfrac{[P]}{K_P}}{1 + \dfrac{[P]}{K_P}}. \tag{1.79}$$

After adding the parallel branches, the mechanism can be sketched as shown in subscheme (b), which consists of the two nodes E_1 and E_2, and where the values of the first-order processes from E_1 toward E_2 have been appropriately multiplied by the fractions (1.79). Comparing Scheme 1.5b with Scheme 1.3d reveals that the nine directed patterns have been reduced to just two.

For writing the rate equation one information more is needed: the proportion of each enzyme state relative to all enzyme-containing species of the system, which can be calculated from the values of the two nodes, ES being represented by E_2

$$f_{E_1} = \frac{E_1}{E_1 + E_2} = \frac{k_{-1} + k_2}{(k_1\,[S] f_E + k_{-2} f_{EP}) + (k_{-1} + k_2)} \tag{1.80}$$

$$f_{E_2} = \frac{E_2}{E_1 + E_2} = \frac{k_1\,[S] f_E + k_{-2} f_{EP}}{(k_1\,[S] f_E + k_{-2} f_{EP}) + (k_{-1} + k_2)}. \tag{1.81}$$

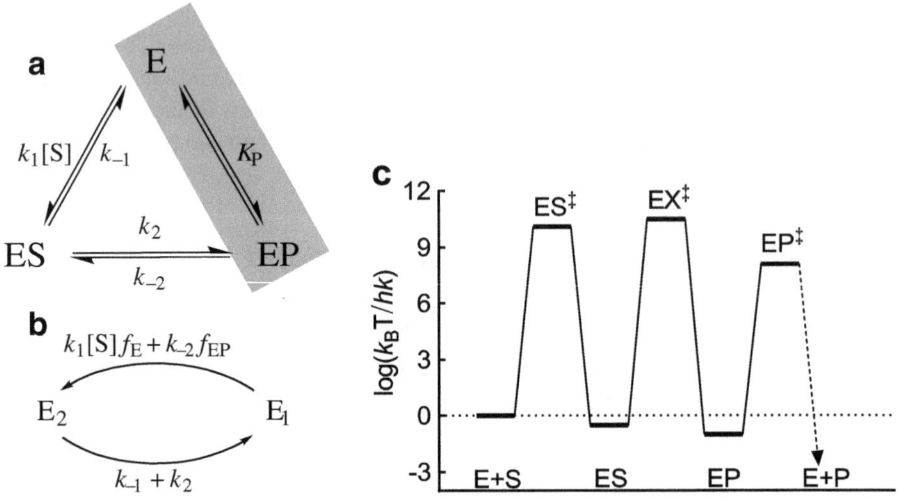

Scheme 1.5 The three-step mechanism in Scheme 1.2a when the dissociation of product is much faster than the other steps. Cyclic representation with the rapidly equilibrating segment shown in the *shadowed area* (**a**) and derived pattern with *directional arrows* labeled with their values after node compression (**b**). The kinetic barrier diagram for this system (panel (**c**)) is drawn to scale with the data in Fig. 1.12d, $[S]_{niche} = 50\ \mu M$ and $[P]_{niche} \to 0$

The net rate equation can be calculated from either one of the following equivalent expressions

$$v_{net} = (k_2 f_{ES} f_{E_2} - k_{-2} f_{EP} f_{E_1})\,[E]_t \tag{1.82}$$

$$v_{net} = (k_1\,[S] f_E f_{E_1} - k_{-1} f_{ES} f_{E_2})\,[E]_t\,, \tag{1.83}$$

and the explicit expression is finally obtained after introducing the values of the appropriate fractions

$$v_{net} = \frac{\left(k_1 k_2\,[S] - k_{-1} k_{-2} \dfrac{[P]}{K_P}\right)[E]_t}{(k_{-1} + k_2 + k_1\,[S]) + (k_{-1} + k_2 + k_{-2})\dfrac{[P]}{K_P}}\,. \tag{1.84}$$

The forward and backward rate equations deduced from (1.84) are given by

$$v_S = \frac{k_2[E]_t\,[S]}{\dfrac{k_{-1} + k_2}{k_1} + [S]}, \quad v_P = \frac{\dfrac{k_{-1} k_{-2}}{k_{-1} + k_2 + k_{-2}}[E]_t\,[P]}{\dfrac{(k_{-1} + k_2)\,K_P}{k_{-1} + k_2 + k_{-2}} + [P]}\,, \tag{1.85}$$

with parameters defined as

$$V_S = k_2[\mathrm{E}]_t, \quad k_{cat,S} = k_2, \quad K_{m,S} = \frac{k_{-1} + k_2}{k_1}, \quad k_S = \frac{k_1 k_2}{k_{-1} + k_2}, \tag{1.86}$$

$$V_P = \frac{k_{-1}k_{-2}[\mathrm{E}]_t}{k_{-1} + k_{-2} + k_2}, \quad k_{cat,P} = \frac{k_{-1}k_{-2}}{k_{-1} + k_{-2} + k_2}$$
$$K_{m,P} = \frac{(k_{-1} + k_2)\,K_P}{k_{-1} + k_{-2} + k_2}, \quad k_P = \frac{k_{-1}k_{-2}}{(k_{-1} + k_2)\,K_P}. \tag{1.87}$$

The expressions of the forward and backward velocities (1.85) are asymmetric, a property that can be deduced from the kinetic barrier diagram in Fig. 1.5c. The barrier $\mathrm{EP}^{\ddagger}$ is lower than both $\mathrm{ES}^{\ddagger}$ and $\mathrm{EX}^{\ddagger}$ and the rate-limiting step is the same as in the Michaelis–Menten mechanism, i.e., $k_{cat,S} \equiv k_2$. However, the reverse reaction is faced with different kinetic barriers with respect to the forward reaction rendering it similar to the forward reaction of the general case discussed in Sect. 1.4. These analogies are treated and discussed in Sect. 1.6.3.

The relationship between initial rate and substrate concentration in the absence of product, v_S in (1.85), using the definitions in (1.86), is characteristic of enzyme-substrate systems that follow *Michaelis–Menten kinetics* as defined in the recommendations 1981 of the Nomenclature Committee of the International Union of Biochemistry [49, Eq. 10]. Written without subscripts, the equation is

$$v = \frac{V\,[\mathrm{S}]}{K_m + [\mathrm{S}]}, \tag{1.88}$$

where V is the *limiting rate* and K_m the *Michaelis constant*. Equation (1.88) is traditionally called the *Michaelis–Menten equation*, which differs from the original expression (1.77) for the presence of K_m in place of K_S.

Simulated progress curves for enzyme species and product are shown in Fig. 1.12. In this example, the concentration of ES remains more or less constant for no longer than about 20 s and then declines progressively with time. The linearity of [P] with time in (b) is only apparent, having an impact on the calculation of the initial rate as discussed in Sect 1.9. Equilibrium is reached after about 1 h maintaining the product concentration at 99 % of the initial substrate concentration.

In accord with the predictions from the kinetic barrier diagram (Scheme 1.5c), until there is no appreciable accumulation of product, the concentration of EP is maintained at a very low level, nearly zero, which implies a net rate governed by k_2. This can be ascertained examining (1.82) and setting $f_{EP} \approx 0$.

Brocklehurst's Criterion for the Validity Of The Quasi-Equilibrium Assumption K_m is numerically close to K_S when $k_2 \ll k_{-1}$, which, as pointed out by Keith Brocklehurst [15], is a consequence of another inequality, i.e., $k_S \ll k_1$. This relationship between the specificity constant and k_1 can be deduced by recasting the

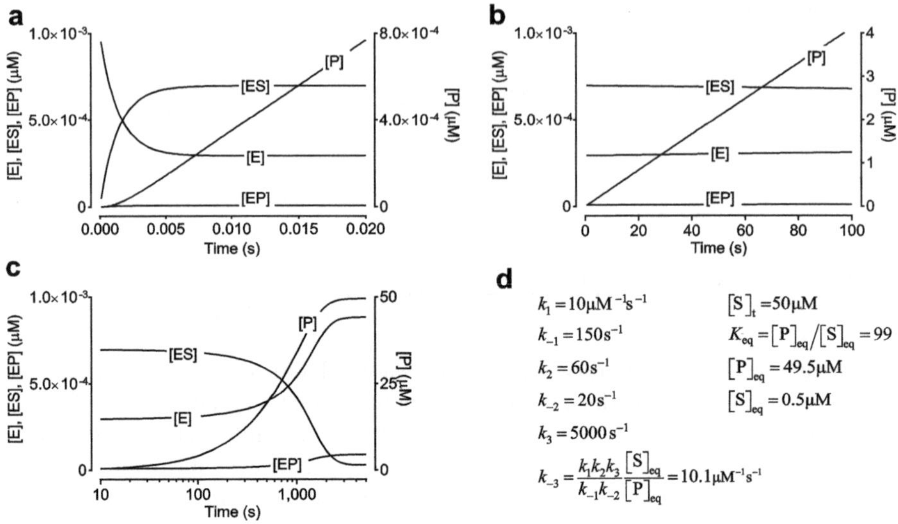

Fig. 1.12 Progress curves for the reversible mechanism in Scheme 1.5. Three time intervals (**a**)–(**c**) with *curves* in (**c**) displayed on a logarithmic time scale. The constants and other conditions used for the simulation are shown in (**d**). The equilibrium constant of 99 imposes the value of k_{-3} calculated under control of microscopic reversibility

expression of k_S in (1.86)

$$\frac{k_S}{k_1} = \frac{k_2/k_{-1}}{1 + k_2/k_{-1}} \Rightarrow \frac{k_S}{k_1} + \frac{k_S}{k_1}\frac{k_2}{k_{-1}} = \frac{k_2}{k_{-1}} \Rightarrow \frac{k_S}{k_1} = \frac{k_2}{k_{-1}} - \frac{k_S}{k_1}\frac{k_2}{k_{-1}}$$

$$\frac{k_2}{k_{-1}} = \frac{k_S/k_1}{1 - k_S/k_1}. \tag{1.89}$$

In (1.89), the condition $k_S \ll k_1$ implies $k_2 \ll k_{-1}$. This conclusion has a practical significance whenever the limiting rate, and hence k_2, cannot be measured experimentally, for instance because the substrate is poorly soluble, while the specificity constant can be measured anyway at very low substrate concentration so that (1.88) reduces to $v = (V/K_m)\,[S]$ (see first-order kinetics on p. 31). The applicability of the condition $k_2 \ll k_{-1}$ can be judged by comparing the measured specificity constant with a hypothetical value of k_1 that has a lower limit in the order of $10^6\,M^{-1}\,s^{-1}$ and falling as an average in the range 10^6–$10^8\,M^{-1}\,s^{-1}$ [32, pp. 164–165]. Brocklehurst sagely suggested $10^7\,M^{-1}\,s^{-1}$ as a comparison value. This criterion can be considered safe when no other means are available for judging the applicability of the quasi-equilibrium assumption.

1.6.3 *Quasi-Equilibrium between Enzyme and Substrate*

The system described here is the same as that discussed in Sect. 1.6.2 with reversed roles for substrate and product. What matters is to show the reverse reaction of the preceding mechanism now as a protagonist in the forward direction. To avoid confusion with the numbering of rate constants, the rate equation will be derived again but without extensive description. The mechanism is shown in Scheme 1.6a with the two nodes and associated values labeling the directional arrows (b).

$$f_{\mathrm{E}} = \frac{1}{1+\dfrac{[\mathrm{S}]}{K_{\mathrm{S}}}}, \quad f_{\mathrm{EP}} = \frac{\dfrac{[\mathrm{S}]}{K_{\mathrm{S}}}}{1+\dfrac{[\mathrm{S}]}{K_{\mathrm{S}}}} \tag{1.90}$$

$$f_{\mathrm{E}_1} = \frac{\mathrm{E}_1}{\mathrm{E}_1+\mathrm{E}_2} = \frac{k_{-2}+k_3}{(k_{-3}\,[\mathrm{P}]f_{\mathrm{E}}+k_2 f_{\mathrm{ES}})+(k_{-2}+k_3)} \tag{1.91}$$

$$f_{\mathrm{E}_2} = \frac{\mathrm{E}_2}{\mathrm{E}_1+\mathrm{E}_2} = \frac{k_{-3}\,[\mathrm{P}]f_{\mathrm{E}}+k_2 f_{\mathrm{ES}}}{(k_{-3}\,[\mathrm{P}]f_{\mathrm{E}}+k_2 f_{\mathrm{ES}})+(k_{-2}+k_3)} \tag{1.92}$$

There are two equivalent expressions for calculating the net rate

$$v_{\mathrm{net}} = (k_3 f_{\mathrm{EP}} f_{\mathrm{E}_2} - k_{-3} f_{\mathrm{E}} f_{\mathrm{E}_1}\,[\mathrm{S}])\,[\mathrm{E}]_{\mathrm{t}} \tag{1.93}$$

$$v_{\mathrm{net}} = (k_2 f_{\mathrm{ES}} f_{\mathrm{E}_1} - k_{-2} f_{\mathrm{EP}} f_{\mathrm{E}_2})\,[\mathrm{E}]_{\mathrm{t}} \tag{1.94}$$

$$v_{\mathrm{net}} = \frac{\left(k_2 k_3 \dfrac{[\mathrm{S}]}{K_{\mathrm{S}}} - k_{-2}k_{-3}\,[\mathrm{P}]\right)[\mathrm{E}]_{\mathrm{t}}}{(k_{-2}+k_3+k_{-3}\,[\mathrm{P}])+(k_{-2}+k_3+k_2)\dfrac{[\mathrm{S}]}{K_{\mathrm{S}}}}\,. \tag{1.95}$$

The initial velocities derived from (1.95) in the absence of product (v_{S}) and in the absence of substrate (v_{P}) are given by

$$v_{\mathrm{S}} = \frac{\dfrac{k_2 k_3}{k_{-2}+k_2+k_3}[\mathrm{E}]_{\mathrm{t}}\,[\mathrm{S}]}{\dfrac{(k_{-2}+k_3)\,K_{\mathrm{S}}}{k_{-2}+k_2+k_3}+[\mathrm{S}]}, \qquad v_{\mathrm{P}} = \frac{k_{-2}[\mathrm{E}]_{\mathrm{t}}\,[\mathrm{P}]}{\dfrac{k_{-2}+k_3}{k_{-3}}+[\mathrm{P}]} \tag{1.96}$$

with the following parameters

$$V_{\mathrm{S}} = \frac{k_2 k_3 [\mathrm{E}]_{\mathrm{t}}}{k_{-2}+k_2+k_3}, \quad k_{\mathrm{cat,S}} = \frac{k_2 k_3}{k_{-2}+k_2+k_3} \tag{1.97}$$

$$K_{\mathrm{m,S}} = \frac{(k_{-2}+k_3)\,K_S}{k_{-2}+k_2+k_3} = \frac{k_{-1}k_{-2}+k_{-1}k_3}{k_1\,(k_{-2}+k_2+k_3)} \tag{1.98}$$

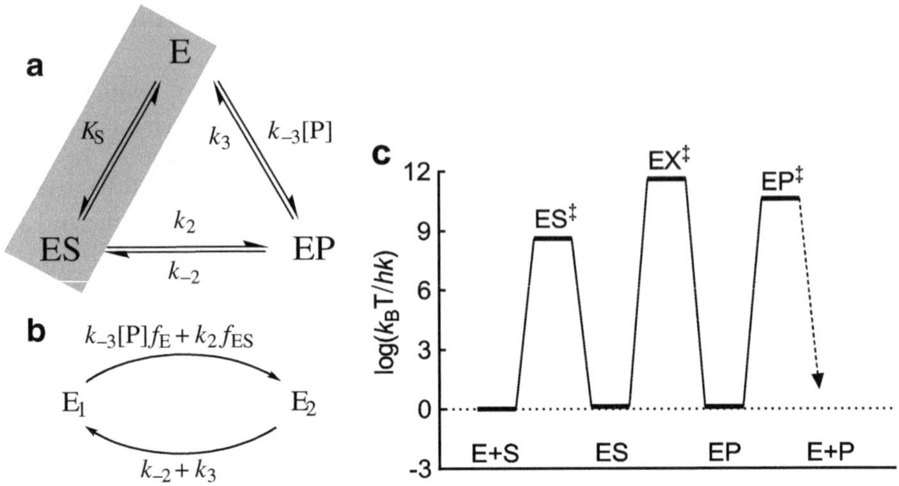

Scheme 1.6 The three-step mechanism Scheme 1.2a when the association of substrate is a fast process. Cyclic representation with the rapidly equilibrating segment shown in the *shadowed area* (**a**). Derived pattern with *directional arrows* labeled with their values after node compression (**b**). Kinetic barrier diagram (**c**) drawn to scale using the data in Fig. 1.13d, $[S]_{niche} = 300\,\mu M$ and $[P]_{niche}$ left to the destiny you prefer to see

$$k_S = \frac{k_2 k_3}{(k_{-2} + k_3)\,K_S} = \frac{k_1 k_2 k_3}{k_{-1}k_{-2} + k_{-1}k_3} \tag{1.99}$$

$$V_P = k_{-2}[E]_t, \quad k_{cat,P} = k_{-2}, \quad K_{m,P} = \frac{k_{-2} + k_3}{k_{-3}}, \quad k_P = \frac{k_{-2}k_{-3}}{k_{-2} + k_3}. \tag{1.100}$$

The reverse reaction, with kinetic parameters given in (1.100), corresponds to the Briggs–Haldane mechanism treated in Sect. 1.6.2. In the forward reaction, the kinetic barrier diagram (Scheme 1.6c) shows that the E + S $\rightleftarrows$ ES association and dissociation occur rapidly and that this step is thus in quasi-equilibrium while the other steps are in a regime of steady-state. The $EP^{\ddagger}$ transition state is higher than $ES^{\ddagger}$ and is nearly as high as $EX^{\ddagger}$ with the consequence that, contrary to the Michaelis–Menten and Briggs–Haldane mechanisms, ES is now formed both from E + S (forward) and EP (backwards). The EP intermediate must therefore accumulate to some extent and its concentration is no longer nearly zero. This can be appreciated from the simulated progress curves in Fig. 1.13. The net rate is thus given by the difference of the two terms in (1.94).

The expression of $k_{cat,S}$ (1.97) for the mechanism discussed here is identical to (1.41) derived for the generic case, while the expression of $K_{m,S}$ (1.98) differs from (1.45) for missing the term k_2k_3 in the numerator. Consequently, the same term is also absent in the denominator of (1.99) compared with (1.47). The difference is due to the quasi-equilibrium condition for substrate association/dissociation in Scheme 1.6 as opposed to Scheme 1.3, in which for none of the three steps the quasi-equilibrium assumption had been invoked.

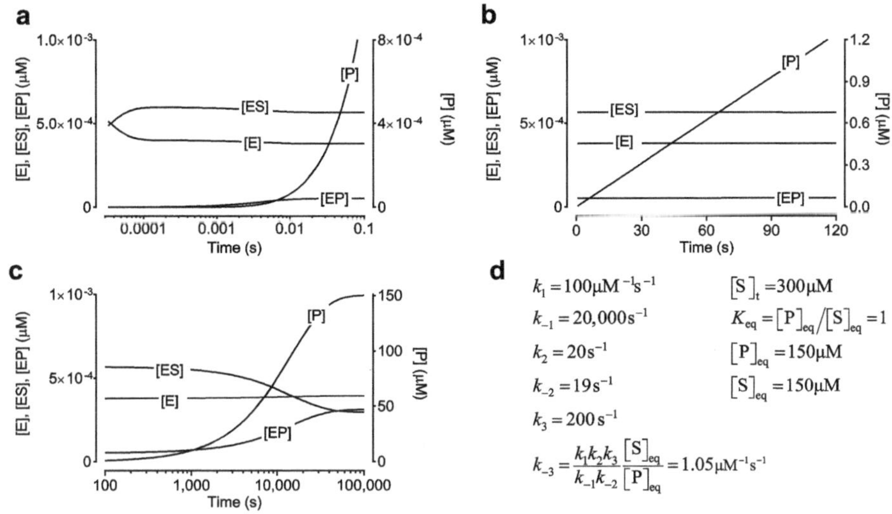

Fig. 1.13 Progress curves for the reversible mechanism in Scheme 1.6 shown in three time intervals (**a**)–(**c**). The constants and other conditions used for the simulation are shown in (**d**). Note that in the steady-state [EP] is not approximately zero

1.7 The Success of the Michaelis–Menten Equation

The core of this book deals with enzyme–modifier interactions using the reaction in Scheme 1.7b to represent the enzyme-substrate(s) system in the absence of modifiers. Since this policy has been occasionally criticized, e.g., [26], clarification is needed. In Sect. 1.6, the characteristics of the simplest monosubstrate kinetic mechanism E + S $\rightleftharpoons$ ES $\rightleftharpoons$ EP $\rightleftharpoons$ E + P were discussed demonstrating that, considering initial velocities in the absence of product, the reaction scheme accommodates the treatments of Michaelis–Menten and Briggs–Haldane. The same reaction has been used as a prototype in theoretical studies on the evolutionary optimization of enzyme catalytic efficiency. During evolution, such optimization is accomplished by natural selection by adjusting the microscopic rate constants to fulfil the needs of enzymes in their metabolic niches. A rich literature exists in this fascinating field pioneered by Jeremy Knowles, John Albery and Reinhart Heinrich. Representative publications that give access to further relevant literature can be found in [16, 35, 36] with criticism by Pettersson [52]. Extensions of this three-step mechanism have been successfully developed to include more intermediates and reactants in complex systems.

In the following chapters dealing with enzyme–modifier interactions, enzyme–substrate reactions will be formally presented with an irreversible catalytic step using the shortcut Scheme 1.7b, i.e., the rate equations will be presented for initial rate conditions. Unless particular methods are exploited to measure individual rate constants, such as isotope effects and measurements in the presence of viscogenic

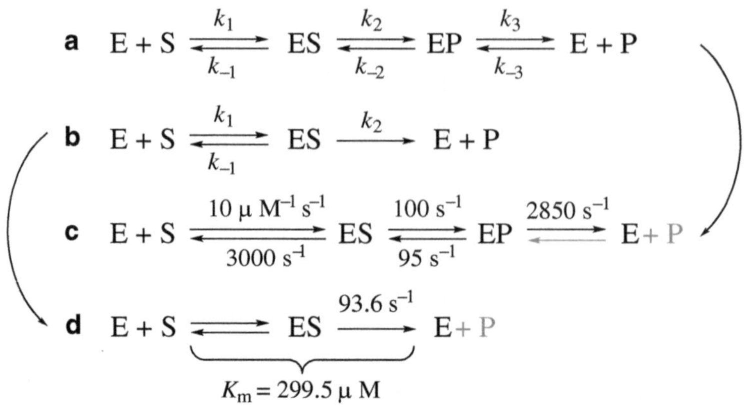

Scheme 1.7 "Translation" of kinetic information between a three-step and a two-step mechanism by considering appropriate kinetic constants. Dealing with initial rate measurements, thus assuming [P] = 0, k_{-3} [P] in (**a**) can be neglected as shown in (**c**). The simplified reaction scheme (**b**) can be used in place of (**a**) by combining the rate constants shown in (**c**) to give the parameters shown in (**d**)

agents, the steady-state approach gives access to k_{cat} and K_m, not to the individual constants whose combinations make up the expressions of these parameters. Mechanism (b) is numerically equivalent to (a) considering that $k_2 \equiv k_{cat}$ in (b) is intended to be $k_{cat} = (k_2 k_3)/(k_{-2} + k_2 + k_3)$ in (a), i.e. (1.41). With this expression, the numbers in (c) give $93.6\,s^{-1}$ for k_2 in (d). Similarly, K_m in (d) corresponds to $(k_{-1}k_{-2} + k_{-1}k_3 + k_2k_3)/[k_1\,(k_{-1} + k_{-2} + k_2)]$ in (c), which is (1.45). The initial rate, v_S, calculated for [S] = $300\,\mu M$ and $[E]_t = 1\,nM$ using (1.38) for (c) and the Michaelis–Menten equation for (d), is the same: $v_S = 0.047\,\mu M\,s^{-1}$.

In practice, after measuring the kinetic parameters K_m and V (or k_{cat}) by steady-state methods in the absence of additional information, the underlying mechanism, sketched as Scheme 1.7b could either correspond to the sequence of reactions in Scheme 1.7a or to another mechanism containing more intermediates. The individual values of the rate constants remain in any case inaccessible.

The "squeezing" performed above to obtain a slender expression can be applied by analogy to other systems. In the presence of modifiers, reactions involving multiple substrates and products cannot be treated in one block. Rather, the effect of a modifier is tested toward one substrate at a time, whose concentration is varied while the concentration(s) of the other substrate(s) is/are kept constant. This is equivalent to tacitly using reaction Scheme 1.7b for each of the varied substrates and the same approach is valid when two modifiers act concomitantly on one enzyme molecule (Chap. 6). After specifying the concentrations of the nonvaried reactants, the determined parameters are said to be *apparent*.

Anticipating Sect. 2.4.3 and (2.35), the Michaelis–Menten equation (1.88) can be written as

$$v = \frac{V^{app}\,[S]}{K_m^{app} + [S]},$$

which is a successful general expression that equally applies to a large variety of simple and complex systems.

Despite being more than 100 years old, the Michaelis–Menten treatment is amazingly young and continues to stimulate the interest of investigators. Tributes with excellent articles written by leading experts have been paid in occasion of the hundredth anniversary of the Michaelis–Menten publication [46] in a special issue of FEBS Letters (A century of Michaelis–Menten kinetics, volume 587, issue 17, September 2013) and in another special issue of the FEBS Journal to commemorate again the centenary of Michaelis and Menten together with the fiftieth anniversary of the paper by Monod, Changeux, and Jacob [46] (Enzyme Catalysis and Allostery: A Century of Advances in Molecular Understanding, Volume 281, issue 2, January 2014). The articles in these special issues are recommended readings.

1.8 Loss of Enzyme Activity During the Assay: The Selwyn-Test

Enzyme inactivation in diluted solutions is a well-known drawback, which deserves particular attention for the consequences it may have on the quality of kinetic measurements and their interpretation. The problem is not limited to recording progress curves over long times but also applies to short-term initial rate measurements, when for instance an enzyme working solution is prepared by dilution from a concentrated stock solution and the working solution is finally used in the assay with further dilution. In a series of measurements, in which the same working solution is used over a given time, the enzyme may lose activity with consequences on data quality. Whether this is the case can be checked by running control measurements at regular time intervals.

Loss of enzyme activity often occurs while measuring reactions for extended periods of time to exploit the information contained in entire progress curves. This is for instance useful for evaluating product inhibition. If not taken into account, enzyme inactivation occurring concomitantly with product inhibition may lead to erroneous conclusions. Therefore, proper analysis of data requires surveillance of enzyme stability during assays. Starting from an equation published by Michaelis and Davidsohn in their study of the hydrogen ion concentration on the activity of invertin [45], Michael Selwyn implemented a useful method for detecting enzyme inactivation during assays [58]. Selwyn observed that the velocity of enzyme-catalyzed reactions can be expressed with the general form

$$\frac{\mathrm{d}\,[\mathrm{P}]}{\mathrm{d}t} = [\mathrm{E}] \times f\,\{[\mathrm{S}]\,,[\mathrm{A}]\,,[\mathrm{I}]\,,[\mathrm{P}]\}\ , \tag{1.101}$$

in which [E] represents the concentration of enzyme active sites and [S], [A], [I], and [P] are the concentrations of substrate, any activator (if present), any inhibitor (if present), and product, respectively. If assays are performed under the same conditions of pH, ionic strength, and temperature, with constant concentrations of substrate and any added modifier, the general integral of (1.101) is given by

$$[\mathrm{E}] \times t = f\{[\mathrm{P}]\} \ , \tag{1.102}$$

which has the same meaning of Michaelis and Davidsohn's expression [45, unnumbered, p. 398]

$$\varphi \times t = f\,(a, x) \tag{1.103}$$

where φ is $[\mathrm{E}]_t$, a is $[\mathrm{S}]_t$ and x is [P].

Michaelis and Davidsohn needed a method to quantify enzyme concentration while studying the pH-dependence of invertin and demonstrated that, plotting data obtained at different enzyme concentrations as [P] against $[\mathrm{E}]\times t$, all points fell on the same curve. Selwyn added the observation that this is possible only if the enzyme does not lose activity during the time needed for assays. If however the enzyme is inactivated, [E] becomes time-dependent and the measured product concentrations fall on different curves.

Regardless of the analytical form of the function in (1.102), the product concentration depends exclusively on enzyme concentration multiplied by time. Intuitively, doubling or triplicating [E] causes the reaction rate to double/triplicate. For instance, the product generated by [E] = 5 nM in 100 s will be the same as with [E] = 10 nM in 50 s.

Selwyn discussed situations for the decrease in reaction rate with time, in which (1.102) is or is not obeyed [58]. The principal reasons raised by Selwyn are listed and complemented below. A case of practical importance is discussed under point (c), in which the Selwyn-test passes although the enzyme undergoes inactivation by denaturation or other causes.

(a) After performing measurements at different enzyme concentrations and plotting results as [P] against $[\mathrm{E}] \times t$, all observations fall on one curve, (1.102) is obeyed: *the Selwyn-test passes*

1. Depletion of substrate. Obvious in all closed systems in which the substrate consumed is not replaced
2. Inhibition by reaction product(s), provided the interaction with the enzyme is *rapid* and time-dependent terms are not present in the rate equation. The same is valid for reactions in the presence of purposely added inhibitors as long as they do not introduce time-dependent terms
3. Reaction approaches equilibrium. Valid only if a parallel nonenzymatic process for the same reaction is absent
4. An enzyme acts on two or more substrates. Should any lag time be observed, this is inversely proportional to $[\mathrm{E}]_t$
5. Two enzymes act at the same time on the same or different substrates
6. Changes in pH as reaction progresses because of poor buffer capacity of the buffering solution

(b) After performing measurements at different enzyme concentrations and plotting results as [P] against $[E] \times t$, observations fall on different curves, (1.102) is not obeyed: *the Selwyn-test does not pass*

7. Loss of enzyme activity during the assay, e.g., by denaturation or adsorption of enzyme to vessel walls
8. The measuring system is too slow to capture *in time* changing reactant concentrations. The reaction rate is represented improperly
9. Diffusion-limited delivery of substrates in systems containing two or more enzymes acting in sequence
10. Slow-onset inhibition by one of the products
11. Steady-state is reached only after an appreciable lag time. This case can be distinguished from case 4 above because the lag time does not depend on enzyme concentration
12. A nonenzymatic reaction runs in parallel and cannot be corrected by subtracting a blank

(c) The Selwyn-test applied to *slow-onset inhibition* progress curves represents a special case discussed in more detail below

The two examples in Fig. 1.14 show experiments performed during optimization of the assays conditions for further studies. The assay of human elastase-2 in panels (a) and (b) cannot be yet considered optimal since the Selwyn-test did not pass. The buffer solution (50 mM Tris/HCl, 150 mM NaCl, pH = 7.50, 25 °C) did not contain stabilizing agents. Enzyme activity in diluted elastase-2 solutions can however be improved by adding Triton X-100 to 0.01 % v/v. Conversely, HIV-1 retropepsin was remarkably stable for more than 1 h, the buffer (50 mM sodium acetate, 1 mM EDTA, 2.5 mM dithiothreitol, 1M NaCl, pH = 4.70) also contained 2.5 % glycerol (v/v), 10 % dimethyl sulfoxide (v/v) and 0.1 % (v/v) of the nondenaturing detergent nonidet P-40. This concoction is far away from the composition of the natural milieu but certainly an efficient cocktail to preserve enzyme activity during assays since the Selwyn-test passed. To finely tuning the final assay, inspection of Fig. 1.14d suggests that enzyme concentrations lower than 20 nM are better avoided in this assay.

1.8.1 The Selwyn-Test and Slow-Onset Inhibition

Point 11 of the items above established that when the steady-state is reached slowly following a lag-time, the Selwyn-test does not pass. There is however a special situation that needs clarification, namely slow-onset inhibition treated in Chap. 8, in which progress curves are typically characterized by an exponential phase preceding the steady-state. During the exponential phase, that may last from seconds to several minutes, (1.102) is not valid.

Morrison and Stone commented that if the enzyme *is not* subjected to time-dependent inactivation during the measurements, progress curves obtained at

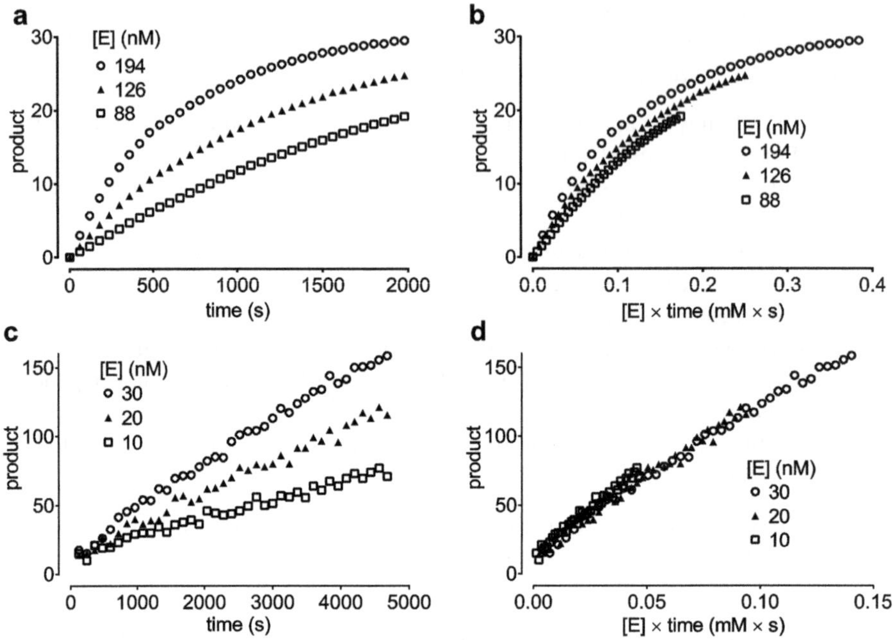

Fig. 1.14 The Selwyn-test. In (**a**) and (**b**) assays were performed with three concentrations of human elastase-2 (EC 3.4.21.37) and (**b**), (**c**) show three assays with the Q7K mutant of HIV-1 retropepsin (HIV-1 protease, EC 3.4.23.16) using appropriate substrates. (**a**) and (**c**) are primary data of [P] against time, (**b**) and (**d**) are the respective transforms as [P] against enzyme concentration multiplied with time; [P] is expressed as relative fluorescence units. HIV-1 retropepsin passes the Selwyn-test while elastase-2 does not. Data were kindly provided by Heidi Roschitzki-Voser and Andreas Flütsch, Department of Biochemistry, University of Zurich

different enzyme concentrations and plotted as [P] against [E] × t are characterized by exponential phases followed by steady-state lines that are all parallel with one another instead of falling on a common line [47]. While this assertion is true as shown below, the question arises about the outcome of the Selwyn-test when an enzyme *undergoes inactivation* in the course of a slow-onset inhibition experiment, an event not treated in the cited paper of Morrison and Stone. The simulations shown in Figs. 1.15 and 1.16 aim at illustrating this concept with an example.

First, we make the hypothesis that the enzyme is stable when assayed alone (Fig. 1.15a,b) as well as in the presence of the slow-onset inhibitor (Fig. 1.16a,b). The Selwyn-test passes for the enzyme assayed alone, as demonstrated by the points falling on the same curve (Fig. 1.15b) and by points, which in the steady-state lie on parallel lines (Fig. 1.16b). The last result can be understood from the fact that the *bursts* of the exponential phases preceding the steady-state have different amplitudes proportional to enzyme concentration.

Now we make the hypothesis that the enzyme is unstable both in the presence and absence of a slow-onset inhibitor and here a multitude of additional questions

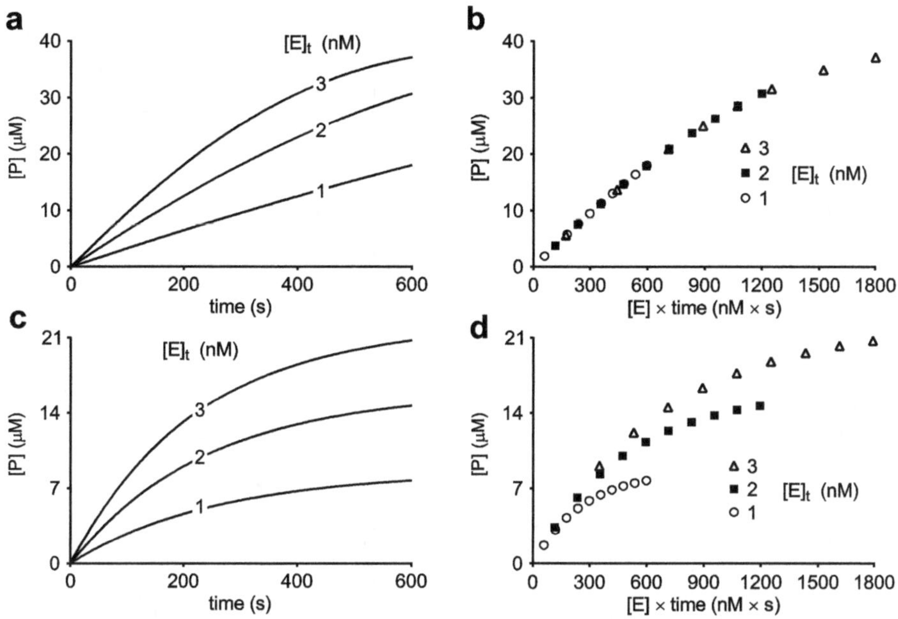

Fig. 1.15 The Selwyn-test in a simulated experiment. Progress curves for an enzyme-catalyzed reaction when the enzyme is stable ((**a**), (**b**)) or unstable during the assay ((**c**), (**d**)). Simulations performed with the following constants and concentrations (reference to Scheme 1.8): $k_1 = 100\,\mu M^{-1}\,s^{-1}$, $k_{-1} = 1950\,s^{-1}$, $k_2 = 50\,s^{-1}$, [S] $= 40\,\mu$M, $[E]_t = 1, 2, 3$ nM (all panels). Additionally $j_1 = 0.005\,s^{-1}$, $j_2 = 0.003\,s^{-1}$ in panels (**c**),(**d**), the remaining constants in Scheme 1.8 being zero

arise because enzyme instability may be due to multiple causes summarized in Scheme 1.8 for the case under examination, which is also the simplest one. The first-order rate constants j_1–j_3, slowly and irreversibly lead to the species labeled with an asterisk, which represent inactive forms of the enzyme that can be interpreted in various ways. For instance, EI* might further give E* + I. If $j_1 = j_2 = j_3 = 0$, the enzyme is stable both in the absence and presence of inhibitor and progress curves obtained at various enzyme concentrations, with the corresponding Selwyn-plots, are those in Figs. 1.15a,b and 1.16a,b, respectively. When only one, two, or all three of the *j*-constants have finite values, there is time-dependent loss of enzyme activity in addition to slow-onset inhibition. Here just an example in which all three inactivation constants are nonzero is shown.

In the absence of inhibitor, the enzyme decays to E* and ES* with the progress curves shown in Fig. 1.15c and the Selwyn-test in panel (d) clearly diagnoses loss of enzyme activity since in the plot of [P] against [E] $\times$ *t* the points fall on different curves. The same simulation run in the presence of the inhibitor produces the progress curves in Fig. 1.16c with the Selwyn-test shown in (d), which does not sound as an alarm announcing enzyme inactivation during the assay because in the [P] against [E] $\times$ *t* plot the steady-state curves seem to run parallel to one another.

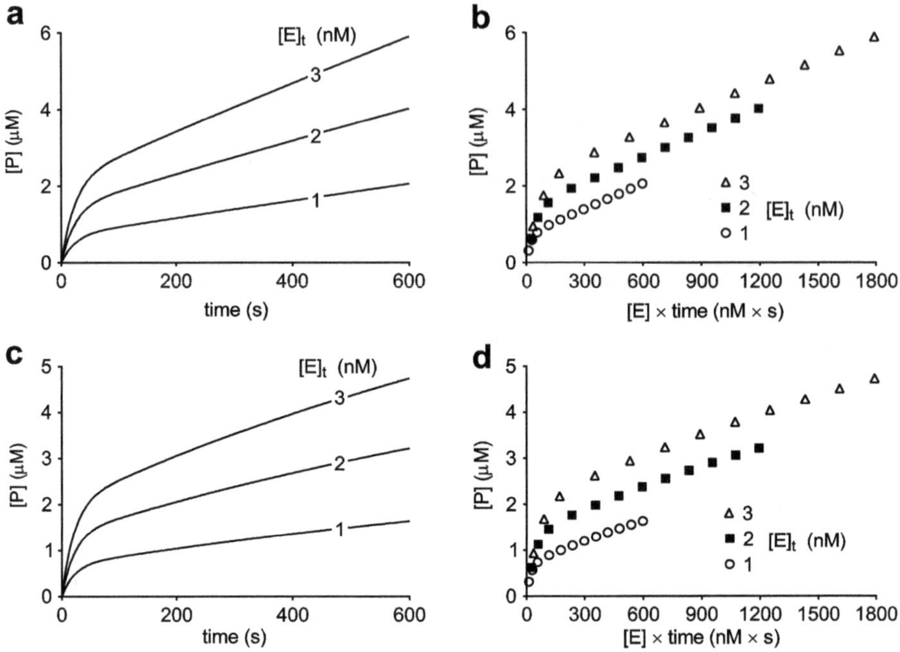

Fig. 1.16 Slow-onset inhibition for the mechanism described in Sect. 8.5 in the presence of a stable ((**a**), (**b**)) or unstable enzyme ((**c**), (**d**)). Simulations performed with the following constants and concentrations (reference to Scheme 1.8): $k_1 = 100\,\mu\text{M}^{-1}\,\text{s}^{-1}$, $k_{-1} = 1950\,\text{s}^{-1}$, $k_2 = 50\,\text{s}^{-1}$, $k_3 = 0.03\,\mu\text{M}^{-1}\,\text{s}^{-1}$, $k_{-3} = 0.003\,\text{s}^{-1}$, $[\text{I}]_t = 4\,\mu\text{M}$, $[\text{S}] = 40\,\mu\text{M}$, $[\text{E}]_t = 1, 2, 3\,\text{nM}$ (all panels). Additionally, $j_1 = 0.005\,\text{s}^{-1}$, $j_2 = 0.003\,\text{s}^{-1}$ and $j_3 = 0.0005\,\text{s}^{-1}$ in panels (**c**), (**d**)

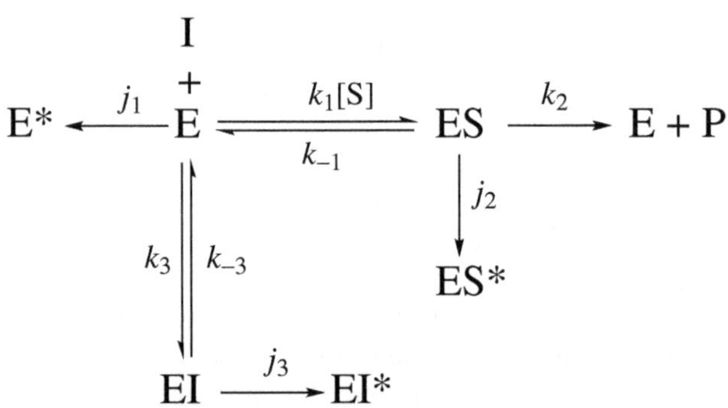

Scheme 1.8 Possible effects of enzyme instability on slow-onset inhibition. The rate constants j_1, j_2, and j_3 refer to irreversible steps of enzyme inactivation

Other simulations, in which only $j_1 \neq 0$ or both j_1 and $j_2 \neq 0$ produced seemingly parallel steady-state lines in plot (d) of Fig. 1.16 as well.

However, close examination of Fig. 1.16c, d reveals that the *steady-state* portions of the progress curves are not perfectly linear but bend progressively downwards with increasing reaction time. The same curves simulated with $j_1 = 0.005\,\mathrm{s}^{-1}$, $j_2 = 0.003\,\mathrm{s}^{-1}$ and $j_3 = 0$ consisted of perfectly straight lines. Parallel lines or curved downwards lines in Fig. 1.16d that were equally bent were obtained with a variety of values for the inactivation constants. It can be concluded that the Selwyn-test passes anyway, independently of enzyme inactivation, demonstrating that for systems containing time-dependent terms in the rate equation this test for enzyme instability does not work.

The conclusion of this analysis suggests that slow-onset inhibition experiments, with their characteristic *burst* followed by linear steady-state curves, may harbor enzyme inactivation that is hard to recognize. The only way to discover loss of enzyme activity when performing slow-onset inhibition experiments is to trust analysis of progress curves in the absence of inhibitor. If the phenomenon cannot be cured by changing reaction conditions, e.g., with the help of stabilizing agents or running experiments over short times with high enzyme concentrations, it is advisable to look prudently at results in the presence of slow-binding inhibitors.

1.9 On the Quest for the True Initial Velocity: When a Straight Line Is Insufficiently Straight

The accurate measurement of initial velocities is not a trivial issue in enzymology and considerable work has been done to find solutions to this problem. A handy approach that exploits the properties of the direct linear plot [30] has been proposed by Cornish-Bowden [20]. This method is particularly useful when a reaction can be followed only by a discontinuous method, with a few samplings from a reacting mixture at time intervals.

Boeker devised a method for uncatalyzed first-order reactions in which the change of [P] with time is plotted as $\Delta\,[\mathrm{P}]/t$ against [P] [13]. The plot consists of a straight line in about the first 50 % of reaction and gives the initial rate by extrapolating the resulting curve for $[\mathrm{P}] = 0$. With a negligible error, Boeker's method is also valid for enzyme-catalyzed reactions.

Another method based on fitting a variant of the integrated Michaelis–Menten equation to progress curves has been developed by Duggleby [28]. The equation proposed is

$$t = \frac{[\mathrm{P}]_t}{v_z} + \frac{2\,v_z\,t_{1/2} - [\mathrm{P}]_\infty}{v_z\,(1 - \ln 4)} \left[\frac{[\mathrm{P}]_t}{[\mathrm{P}]_\infty} + \ln\left(1 - \frac{[\mathrm{P}]_t}{[\mathrm{P}]_\infty}\right) \right], \qquad (1.104)$$

where the subscript t in $[\mathrm{P}]_t$ denotes *time* indicating that this is the concentration of product generated in the reaction, which may differ from the total product concentration when this is added in known amount to the reacting solution.

In irreversible reactions, $[P]_\infty$ corresponds to the initial substrate concentration provided the enzyme remains stable during the entire reaction. This equation is fitted to progress curves by nonlinear regression that yields estimates of v_z, $[P]_\infty$ and $t_{1/2}$, where the initial rate v_z is the parameter of interest. The symbol v_z is used here and in other chapters to denote a velocity measured at *zero time*. Duggleby showed that this method for evaluating initial rates from progress curves yields good results also in the presence of errors and enzyme inactivation during the reaction.

Polynomial regression or approximations by the cubic spline function, e.g., [25, 31], have not found wide application because of inherent problems in the presence of experimental error, instrumental noise and assumptions to be made in weighting measurements. The mission of such methods is to find as many possible contacts between a fitted curve and any points in its neighborhood, no matter if these are data or noise because the polynomial is not instructed to fit a precise model. Particularly critical for polynomials is to find the starting point of a reaction, and hence an estimate of the initial rate, when a substantial part of the progress curve is *missing at the beginning*. A method based on the use of a Taylor series in reaction time aimed at finding the first product concentrations in a progress curve is theoretically appealing but unfortunately inadequate in the presence of errors [55].

Depending on the particular reaction studied, measuring reaction progress using a limited number of points may constitute the unique opportunity to study the kinetics of the reaction. Progress curves may cover a substantial portion of the reaction, if not the entire course. In such cases, the methods mentioned above are valid companions to estimate initial rates from reaction traces ([P] or [S] against time) in place of ruler and pencil.

Photometric and fluorometric measurements performed in multiwell plates have their own problems. In my experience, I rarely saw acceptable initial rate measurements performed with this method by people that asked my help for rescuing their data. My answer was "hopeless" and in most cases I convinced frustrated experimenters to repeat their measurements with a classical photometer or fluorometer or, not to miss precious information at reaction start, with a stopped-flow apparatus. When this was the case, after seeing the new results their emoticon shortcut changed from :(to :). Even if the operator is the world champion in pipetting, the major problem with the estimation of initial rates using multiwell plates resides in the delay between pipetting the reagent that initiates the reaction (often the enzyme) and the start of readings. If a large number of wells is used for recording data on the same plate, dead times of various lengths will precede data collection and loss of precious, unrecoverable data. However, a method to allegedly fill these gaps and to remove other systematic errors in kinetic measurements with multiwell plates has been developed and validated [51].

1.9.1 An Alternative Method to Calculate Initial Rates

What follows refers to measurements of a very limited portion of progress curves by continuous recording, especially using rapid-mixing devices equipped to collect hundreds of data points in a short time interval. Deviations from linearity in steady-state measurements, due to substrate depletion, approach to equilibrium, product inhibition, enzyme inactivation or combinations of these effects, are well known to enzymologists. Therefore, provided one of the above-described methods for progress curves covering a substantial part of reaction is not used, initial rates can be measured for a short time at the beginning of the reaction, when the rise of [P] and the decrease of [S] are expected to be linear with time. But which are the criteria to define the beginning of a reaction? This is an ill-defined concept because typical steady-state measurements are performed in the time range of seconds to minutes, while rapid-mixing devices allow measurements starting in the range of milliseconds, extending to seconds and optionally to minutes.

Modern recording devices include software to perform data manipulation and calculations based on regression analysis. Despite being very practical, such methods should be carefully supervised by the experimenter. It can happen that an instrument is capable of very precise measurements that are however not output with the same precision. Rounding errors artificially produced by insufficiently precise numerical representation of otherwise excellent primary measurements are deleterious in parameter calculation and model discrimination. Cárdenas and Cornish-Bowden showed that this type of errors can be detected by analysis of the residuals, which have a characteristically truncated distribution [17].

Assuming that primary data aimed at measuring initial rates have been collected and stored electronically with the necessary precision, a further cautionary note regards the way initial rates are estimated. We deal here with measurements over a short period, during which the rise of [P] and decrease of [S] are, as said above, expected to be linear with time. Blind acceptance of results from automated procedures such as linear regression of recorded reaction profiles may lead to errors that depend on the time-scale data are collected. Despite accurate experiment design, even small deviations from linearity may affect the calculated slope with consequences for the accuracy of the kinetic parameters. The deviations may originate from substrate depletion, product accumulation possibly accompanied by a contribution of the reverse reaction (for reversible reactions) and product inhibition. In a series of measurements at various substrate concentrations, reaction traces at low [S]-values are more affected than those at high [S]-values. This phenomenon depends on the rate constants of the system and can only partly be controlled by optimizing the experimental protocol.

If one of the reaction products is the major cause of troubles, an elegant solution is to use a *coupled assay* with an enzyme that utilizes the unwanted product as substrate thus pushing the reaction in the forward direction. The only limitation is the availability of an enzyme for fast removal of the product. Examples of this

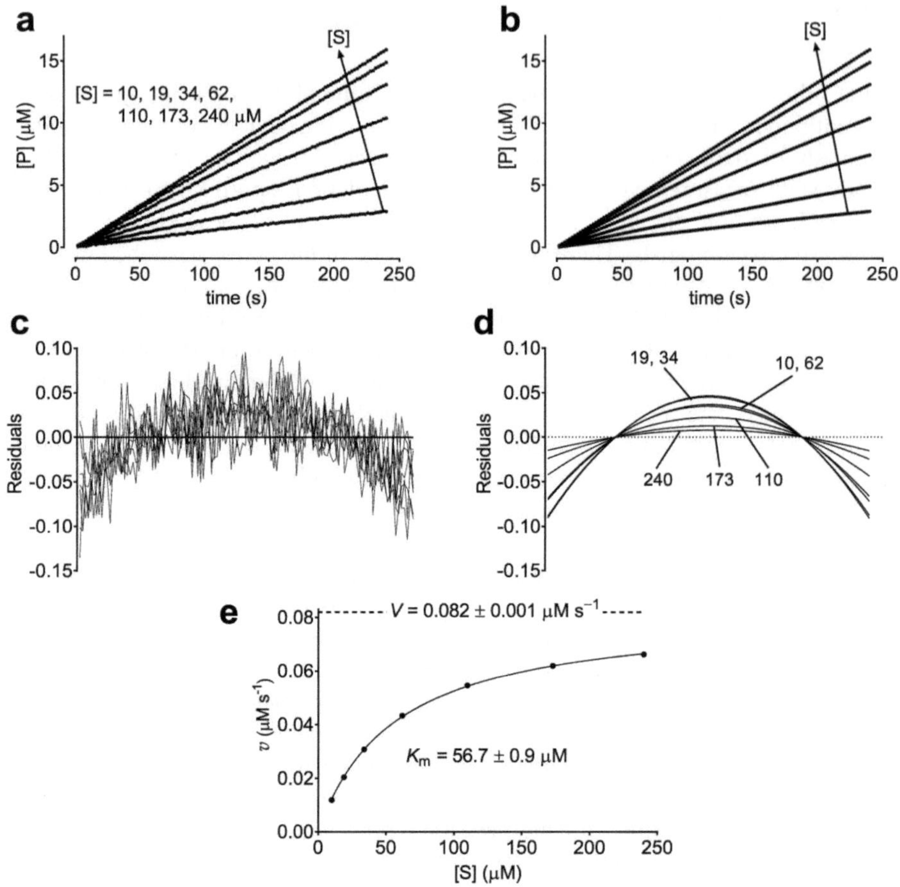

Fig. 1.17 Estimation of initial rates from a set of simulated data and calculation of kinetic parameters for the Briggs–Haldane mechanism in Scheme 1.2d. Kinetic constants and concentrations used for the simulation by numerical integration: $k_1 = 10\,\mu\text{M}^{-1}\,\text{s}^{-1}$, $k_{-1} = 400\,\text{s}^{-1}$, $k_2 = 80\,\text{s}^{-1}$, $[\text{E}]_t = 1\,\text{nM}$, [S] as shown in (**a**). Each curve simulated with 120 points and 240 s. In (**a**) Gaussian noise ($\sigma = 0.025$) was added, (**b**) without noise. The curves were analyzed by linear regression with residuals (**c**) and (**d**) corresponding to (**a**) and (**b**), respectively. V and K_m calculated by fitting the Michaelis–Menten Eq. (1.88) to initial rates (**e**). The slopes of the curves were identical for both sets of data

method have been published by Tang and Leyh [60], who used cloned, purified enzymes to remove reaction products in four representative reactions.

As an example, simulations with and without added Gaussian noise are shown in Fig. 1.17 for the Briggs–Haldane mechanism (Scheme 1.2d). The noise, in units of the measured variable, is an imitation of instrumental noise that does not include errors originating from dilutions, pipetting, and other factors derived from handling of reagents.

The curves in Fig. 1.17, which contain each 120 points simulated for 240 s, are apparently linear (panels a and b with and without noise, respectively). Unweighed linear regression of these lines afforded slopes to which the Michaelis–Menten equation was fitted to obtain V and K_m. Results from simulated data containing or not Gaussian noise were identical (panel e). The estimate of V (theoretical value 0.08 μM s^{-1}) is good while the estimate of K_m is not (calculated 56.7 μM, theoretical 48.0 μM). Despite correlation coefficients > 0.99 for all curves, this suggests the presence of an imperceptibly slight curvature as demonstrated by the systematic deviation of the residuals, (c) and (d), beneath their primary data. Due to the large number of points used in the simulation, the deviations from linearity are appreciable also in the presence of noise (c). The noise-free simulation in (d) shows deviations increasing for decreasing substrate concentration.

Evidence from a large number of initial rate measurements performed by the author with a stopped-flow apparatus, which afforded kinetic parameters that depended on the measuring time (typically 20–200 s) inspired the method described here. To explain the procedure we consider again the data in Fig. 1.17b and proceed as illustrated in Fig. 1.18 with the following four steps.

1 Divide the family of n progress curves, corresponding to n substrate concentrations, in m equally spaced time intervals (in Fig. 1.18 $n = 7$ and $m = 6$)
2 Approximating the segments to straight lines, calculate the slopes of each of them by linear regression

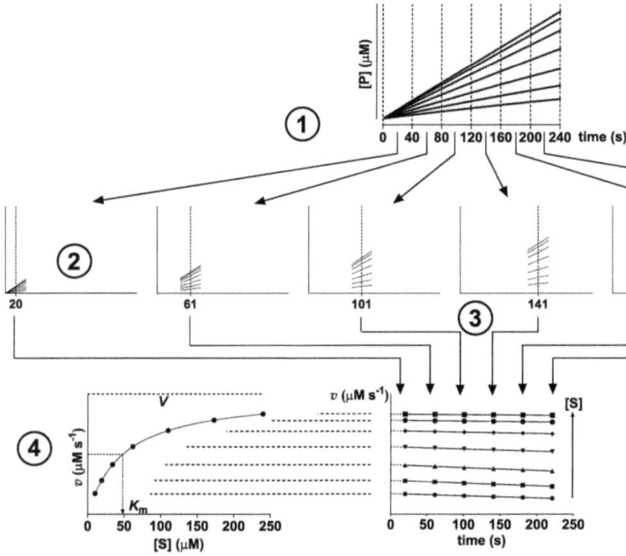

Fig. 1.18 Calculation initial rates from progress curves measured over a short time. The numbered steps refer to the itemized explanations in the text

3 Plot the slopes against the mean time of the respective intervals (in this example the mean times of the six intervals are 20, 61, 101, 141, 181 and 221 s). By linear regression of the resulting lines calculate their ordinate intercepts at time = 0
4 The ordinate-intercepts from step [3] correspond to the best estimates of the initial rates for each of the *n* substrate concentrations. Plot these optimized data against [S] and fit the Michaelis–Menten equation to them to calculate V and K_m.

The segment method is *heuristic* and is not based on an underlying mechanism. It derives from the observation that, in most practical cases investigated, the progress curves measured over short times were apparently linear but analysis of residuals revealed instead a slight curvature, which was less pronounced as the time to record progress curves was shortened. The number of segments chosen for dividing the progress curves in discrete pieces is arbitrary and 4–6 such segments have been proven sufficient.

The procedure is apparently complicated and time-consuming but closer inspection of the steps in Fig. 1.18 reveals that all calculations can be performed automatically using a spreadsheet program. After transferring raw data into a spreadsheet, the total time is divided in equal parts as shown in Fig. 1.18 and linear regression of the data in the chosen segments is performed. The resulting slopes are then arranged in a table against the mean times of the segments and again linear regression is run to calculate the rates at time zero. These optimized initial rates and the respective substrate concentrations can now be used for further processing. Applying this procedure, the simulated data in Fig. 1.17 gave the results summarized in Table 1.2.

Now a logical question is expected from the part of the reader: *is this additional work necessary*? The answer is *yes* for lovers of precision. With reference to Table 1.2, the discrepancy between the value of K_m calculated from the original, apparently linear initial part of the progress curves ($\approx$ 57 μM), and the theoretical value (48 μM) amounts to 18 %. Discrepancies can be less but also more than this value depending on the kinetic constants of the system and experimental design, particularly on the chosen substrate and enzyme concentrations. In the presence of instrumental noise, the estimated values of the kinetic parameters may vary and the standard deviation can become more or less important. In the example above (second row in Table 1.2), the segment method yields a satisfactory estimate with modest standard error. Simulations with larger noise gave good estimates of the parameters

Table 1.2 Calculation of initial rates

Data in Fig. 1.17: (a) with noise, (b) without noise	V (μM s^{-1})	K_m (μM)
Both (a) and (b), all points from 1 to 240 s	0.082 ± 0.001	56.7 ± 0.9
Segment method, data (a) with noise	0.080 ± 0.001	47.3 ± 1.4
Segment method, data (b) without noise	0.080 ± 0.000	48.1 ± 0.1
Theoretical values	0.080	48.0

albeit with larger standard errors, indicating that performing optimization of the initial rate is a safe procedure.

What has not been taken into account yet are errors derived from dilutions of the reagents, precision in determining their concentrations, pipetting, temperature oscillations in a series of measurements, and so on. These factors are deemed to be significant in determining deviations between estimated and true kinetic parameters. Thus, if we have the possibility of correcting at least one part of the total error, we should do it. A practical worked example is discussed in Sect. 1.9.2.

1.9.2 Calculation of Initial Rates: A Practical Example

The human cysteine endopeptidase cathepsin K (EC 3.4.22.38) is involved in pathophysiological processes such as bone remodeling and one of its physiological substrates is insoluble collagen present in connective tissues. Kinetic studies in vitro to evaluate the effects of modifiers on this enzyme are performed using synthetic, soluble, low-molecular mass substrates, whose hydrolysis can be followed by photometry or fluorometry. The following example illustrates the calculation of initial rates for the hydrolysis of *N*-carbobenzyloxy-L-phenylalanyl-L-arginyl-*p*-nitroanilide (Z-FR-pNA) catalyzed by recombinant human cathepsin K as part of measurements in the presence of various inhibitors. The aim was to compare the values of inhibition constants using an initial rate approach with global fitting of complete progress curves. For minimizing the problem of enzyme inactivation during long measuring times, experiments were performed at a relatively high final enzyme concentration that allowed following the reaction not only for short times but also to completeness within a few minutes. Enzyme stability during the measuring time was confirmed by the Selwyn-test.

Reactions were monitored with a Hi-Tech SF-61 stopped-flow apparatus (TgK Scientific) under conditions specified in the legend of Fig. 1.19. The advantage of rapid-mixing technology is a short dead time and the possibility of running several measurements in sequence with the same reagent solutions. Accumulating and averaging several traces reduce the impact of instrumental noise because the signal originating from the reaction either increases or decreases with time, while the noise, which occurs at high frequency, is random and can therefore be partly compensated by averaging.

The cumulative error in this example comprises instrumental noise and effects due to reagent handling. The noise could be kept within a limited range so that the deviations from linearity detected in Fig. 1.19b as an arc extending over the whole time interval of the experiment can be ascribed to substrate depletion. The discrepancy between the values of the Michaelis constant calculated from original data and with the segment method is mainly due to an overestimated limiting rate in the first case.

As a comparison with the parameter estimates in panel (d) calculated by nonlinear regression, the *direct linear plot* [30] applied to the same initial rates

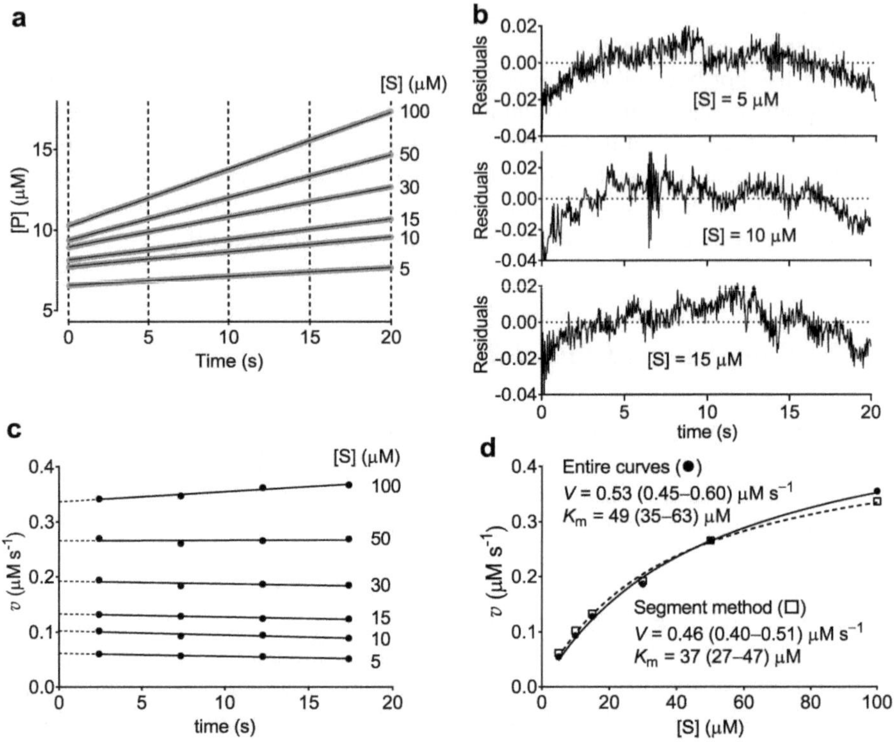

Fig. 1.19 Initial rates and kinetic parameters in a practical case. Hydrolysis of Z-FR-pNA catalyzed by human recombinant cathepsin K (original data by M. Novinec and A. Baici). Experiments were performed with a stopped-flow photometer mixing equal volumes of substrate and enzyme solution at pH = 6.00 and 25 ± 1 °C. After mixing, the concentrations of substrate were as shown in panel (**a**) and $[E]_t$ was 20 nM of titrated active sites. Each trace measured for 20 s consisted of 512 points with signals averaged from five consecutive runs. The raw data are displayed in panel (**a**), which also shows the subdivision of the traces in four time segments: data in *gray* and linear regression fit as a *black line*. Since only the slopes of the progress curves were of interest in this case, raw data were not corrected for blank. Panel (**b**) displays the residuals from linear fitting for three out of six substrate concentrations as indicated. Panel (**c**) shows the linear regression of the slopes from the segments in (**a**) following the method in Fig. 1.18. In (**d**) the initial rates calculated from original data are plotted against [S] in comparison with those optimized with the segment method. The Michaelis–Menten equation fitted to initial rates afforded the values of the parameters displayed as best fit with 95 % confidence interval

is shown in Fig. 1.20 with results summarized in Table 1.3 together with those in Fig. 1.19d. The results from the direct linear plot are more trustworthy than those from nonlinear regression because of the robustness associated with nonparametric statistics and handling of outliers. In Fig. 1.19c, the slopes of the lines are negative for [S] = 5–50 μM and are expected to tend to zero by further increasing [S]. However, the line for [S] = 100 μM is anomalous because it is slightly positive, which sounds as an outlier.

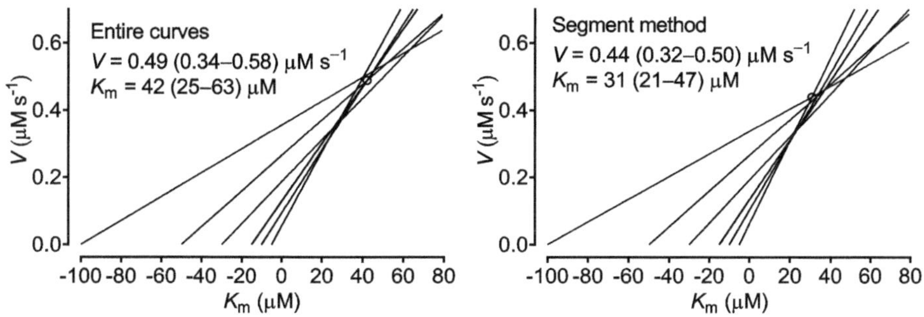

Fig. 1.20 Calculation of Michaelis–Menten parameters with the direct linear plot [30] using the data in Fig. 1.19d. The *small circles* indicate the medians. Results are displayed as median with the ≥95 % confidence interval [23]

Table 1.3 Impact of the quality of initial rates and of the calculation method on Michaelis–Menten parameters

	Method	V (μM s^{-1})	K_m (μM)
Entire curves	DLP	0.49 (0.34–0.58)	42 (25–63)
Segment method	DLP	0.44 (0.32–0.50)	31 (21–47)
Entire curves	NLR	0.53 (0.45–0.60	49 (35–63)
Segment method	NLR	0.46 (0.40–0.51)	37 (27–47)

Comparison between values calculated with the direct linear plot [30] (Fig. 1.20) and by nonlinear regression (Fig. 1.19d) *DLP* direct linear plot (median and ≥95 % confidence interval [23]), *NLR* nonlinear regression (best fit and 95 % confidence interval)

The simulated and the real example discussed above reveal how errors and the use of different approaches for data analysis may influence results. Even small underestimated or overestimated values of the limiting rate may impact K_m and other parameters if enzyme–modifier interactions are measured. Results published on the same system by different groups may thus differ somewhat despite diligent work.

Appendix 1

Derivation of Kinetic Parameters Using Mathematical Software

Redirected from Sect. 1.4. The following example (Fig. 1.21) demonstrates the derivation of $k_{cat,S}$ and $K_{m,S}$ using Maple. Mathematical software is a helpful alternative to paper and pencil in rearranging and extracting information from rate equations. A more cumbersome example is discussed in Sect. 2.5, Eq. (2.57), where the apparent Michaelis constant of the general modifier mechanism depends on modifier concentration.

$$v_{net} := -\frac{E_t\left(-k_3 k_1 k_2 S + k_{-3} P k_{-1} k_{-2}\right)}{k_{-1} k_3 + k_{-1} k_{-2} + k_2 k_3 + k_1 k_3 S + k_1 k_{-2} S + k_{-2} k_{-3} P + k_{-1} k_{-3} P + k_1 k_2 S + k_2 k_{-3} P}$$

$$v_S := eval(v_{net}, P = 0)$$

$$\frac{E_t k_3 k_1 k_2 S}{k_{-1} k_3 + k_{-1} k_{-2} + k_2 k_3 + k_1 k_3 S + k_1 k_{-2} S + k_1 k_2 S} \quad \textbf{(1)}$$

$$V_S := limit(v_S, S = \infty)$$

$$\frac{E_t k_3 k_2}{k_3 + k_{-2} + k_2} \quad \textbf{(2)}$$

$$k_{cat,\,S} := \frac{V_S}{E_t}$$

$$\frac{k_3 k_2}{k_3 + k_{-2} + k_2} \quad \textbf{(3)}$$

$$V_{S,\,half} := \frac{1}{2} \cdot limit(v_S, S = \infty)$$

$$\frac{1}{2}\,\frac{E_t k_3 k_2}{k_3 + k_{-2} + k_2} \quad \textbf{(4)}$$

$$K_{m,\,S} := solve(v_S = V_{S,\,half}, S)$$

$$\frac{k_{-1} k_3 + k_{-1} k_{-2} + k_2 k_3}{k_1\left(k_3 + k_{-2} + k_2\right)} \quad \textbf{(5)}$$

Fig. 1.21 Derivation of equations in Sect. 1.4 using Maple

Appendix 2

List of Symbols

E	Enzyme
$[E]_t$	Total (stoichiometric) concentration of enzyme catalytic centers
EI, ES	Adsorptive enzyme-inhibitor or enzyme-substrate complex
E·I	Reversible enzyme-inhibitor complex, used only when disambiguation is necessary, e.g., in reaction schemes of slow-onset inhibition
E-I	Denotes the covalent compound between enzyme and an irreversible inhibitor, used to disambiguate the notation of enzyme-containing complexes
I, [I]	Reversible or irreversible inhibitor and its concentration
k_n	Rate constant of step n (forward reaction)
k_{-n}	Rate constant of step n (reverse reaction)
K_m, K_m^{app}	Michaelis constant, apparent Michaelis constant
K_m^0	Michaelis constant in the absence of modifier
K_m^∞	Michaelis constant at saturating modifier concentration
K_I, K_X	Inhibition/modification constant of an inhibitor or a generic modifier, respectively
L, [L]	Liberator and its concentration

P, [P]	Product and its concentration
S, [S]	Substrate and its concentration
t	Time
v	Generic reaction rate
v_{I}, v_{X}	Rate in the presence of inhibitor or generic modifier, respectively
v_{IX}	Used in double-modifier interactions to denote the rate in the presence of two modifiers I and X
v_{s}	Rate at steady-state (lowercase s)
v_{z}	Rate at time zero
$v_{\mathrm{z,S}}$	Rate at time zero in reactions started by adding substrate (uppercase S)
$v_{\mathrm{z,E}}$	Rate at time zero in reactions started by adding enzyme
v_0	Reaction rate measured in the absence of modifiers (the subscript is zero)
X	Generic modifier X
[X]	Concentration of the free species X
$[\mathrm{X}]_{\mathrm{t}}$	Total concentration of species X (the subscript t is upright)
$[\mathrm{X}]_t$	Concentration of species X at a given time t (the subscript t is in italics)
$[\mathrm{X}]_{\mathrm{z}}$	Concentration of species X at time zero
V, V^{app}	Limiting rate, apparent limiting rate
V^0, V^{∞}	Limiting rate in the absence of modifier and at saturating modifier concentration, respectively
λ	Frequency constant, customary unit s^{-1}
σ	$[\mathrm{S}]/K_{\mathrm{m}}$, dimensionless

References

1. Alberty RA (2004) Principle of detailed balance in kinetics. J Chem Educ 81:1206–1209
2. Alberty RA, Cornish-Bowden A, Goldberg RN, Hammes GG, Tipton K, Westerhoff HV (2011) Recommendations for terminology and databases for biochemical thermodynamics. Biophys Chem 155:89–103. doi:10.1016/j.bpc.2011.03.007
3. Albery WJ, Knowles JR (1976) Evolution of enzyme function and the development of catalytic efficiency. Biochemistry 15:5631–5640. doi:10.1021/bi00670a032
4. Atkinson DE (1977) Cellular energy metabolism and its regulation. Academic Press, New York
5. Baici A (1987) Graphical and statistical analysis of hyperbolic, tight-binding inhibition [This paper contains a mathematical mistake: see correction in Szedlacsek S et al (1988) Biochem J 254:311–312]. Biochem J 244:793–796
6. Baici A (1990) Interaction of human leukocyte elastase with soluble and insoluble protein substrates. A practical kinetic approach. Biochim Biophys Acta 1040:355–364
7. Baici A (2006) Enzyme kinetics: the velocity of reactions. Biochemist 28:36–39
8. Baici A, Pelloso R, Hörler D (1990) The kinetic mechanism of inhibition of human leukocyte elastase by MR889, a new cyclic thiolic compound. Biochem Pharmacol 39:919–924
9. Bearne SL (2012) Illustrating enzyme inhibition using Gibbs energy profiles. J Chem Educ 89:732–737. doi:10.1021/ed200395n
10. Bearne SL (2013) Illustrating the effect of pH on enzyme activity using Gibbs energy profiles. J Chem Educ 91:84–90. doi:10.1021/ed400229g
11. Bernasconi CF (1976) Relaxation kinetics. Academic Press, New York
12. Blackmond DG (2009) "If pigs could fly" chemistry: a tutorial on the principle of microscopic reversibility. Angew Chem Int Ed 48:2648–2654

13. Boeker EA (1982) Initial rates. A new plot. Biochem J 203:117–123
14. Briggs GE, Haldane JBS (1925) A note on the kinetics of enzyme action. Biochem J 29:338–339
15. Brocklehurst K (1979) The equilibrium assumption is valid for the kinetic treatment of most time-dependent protein-modification reactions. Biochem J 181:775–778
16. Burbaum JJ, Raines TR, Albery WJ, Knowles JR (1989) Evolutionary optimization of the catalytic effectiveness of an enzyme. Biochemistry 28:9293–9305. doi:10.1021/bi00450a009
17. Cárdenas ML, Cornish-Bowden A (1993) Rounding error, an unexpected fault in the output from a recording spectrophotometer: implications for model discrimination. Biochem J 292:37–40
18. Cha S (1968) A simple method for derivation of rate equations for enzyme-catalyzed reactions under the rapid equilibrium assumption or combined assumptions of equilibrium and steady-state. J Biol Chem 243:820–825
19. Cleland WW (1979) Statistical analysis of enzyme kinetic data. Meth Enzymol 63:103–138
20. Cornish-Bowden A (1975) The use of the direct linear plot for determining initial velocities. Biochem J 149:305–312
21. Cornish-Bowden A (2006) The IUBMB recommendations of symbolism and terminology in enzyme kinetics. In: Hicks MG, Kettner C (eds) Proceedings of the 2nd international Beilstein symposium on experimental standard conditions on enzyme characterizations, Logos Verlag, Berlin, pp 35–50
22. Cornish-Bowden A (2012) Fundamentals of enzyme kinetics, 4th edn. Wiley, Weinheim
23. Cornish-Bowden A, Porter WR, Trager WF (1978) Evaluation of distribution-free confidence limits for enzyme kinetic parameters. J Theor Biol 74:163–175
24. Cruickshank FR, Hyde AJ, Pugh D (1977) Free energy surfaces and transition state theory. J Chem Educ 54:288. doi:10.1021/ed054p288
25. Dagys R, Pauliukonis A, Kazlauskas D, Mankevicius M, Simutis R (1986) Determination of initial velocities of enzymic reactions from progress curves. Biochem J 237:821–825
26. Dewolf W, Segel IH (2000) Simplified velocity equations for characterizing the partial inhibition or nonessential activation of bireactant enzymes. J Enzyme Inhib 15:311–333
27. Dixon M, Webb EC (1979) Enzymes, 3rd edn. Longman, London
28. Duggleby RG (1985) Estimation of the initial velocity of enzyme-catalysed reactions by non-linear regression analysis of progress curves. Biochem J 228:55–60
29. Einstein A (1916) Zur Quantentheorie der Strahlung. Mitt Physikal Ges Zürich 18:47–62
30. Eisenthal R, Cornish-Bowden A (1974) The direct linear plot. A new graphical procedure for estimating enzyme kinetic parameters. Biochem J 139:715–720
31. Elmore DT, Kingston AE, Shields DB (1963) The computation of velocities and kinetic constants of reactions, with particular reference to enzyme-catalysed processes. J Chem Soc No volume:2070–2078. doi:10.1039/JR9630002070
32. Fersht A (1999) Structure and mechanism in protein science. A guide to enzyme catalysis and protein folding. Freeman, New York
33. Fowler RH, Milne EA (1925) A note on the principle of detailed balancing. Proc Natl Acad Sci USA 11:400–402
34. Haldane JBS (1965) Enzymes (reprint of the 1930 edition), 2nd edn. M.I.T. Press, Cambridge, MA
35. Heinrich R, Hoffmann E, Holzhütter HG (1990) Calculation of kinetic parameters of a reversible enzymatic reaction in states of maximal activity. Biomed Biochim Acta 49:891–902
36. Heinrich R, Schuster S, Holzhütter HG (1991) Mathematical analysis of enzymic reaction systems using optimization principles. Eur J Biochem 201:1–21
37. International Union of Pure and Applied Chemistry (1996) A glossary of terms used in chemical kinetics, including reaction dynamics. Pure Appl Chem 68:149–192
38. Johansen G, Lumry R (1961) Statistical analysis of enzymic steady-state rate data. C R Trav Lab Carlsberg 32:185–214
39. Johnson KA (2009) Fitting enzyme kinetic data with Kintek Global Kinetic Explorer. Meth Enzymol 467:601–626

40. Johnson KA, Simpson ZB, Blom T (2009) FitSpace Explorer: an algorithm to evaluate multidimensional parameter space in fitting kinetic data. Anal Biochem 387:30–41
41. Johnson KA, Simpson ZB, Blom T (2009) Global Kinetic Explorer: a new computer program for dynamic simulation and fitting of kinetic data. Anal Biochem 387:20–29
42. King EL, Altman C (1956) A schematic method of deriving the rate laws for enzyme-catalyzed reactions. J Phys Chem 60:1375–1378
43. Klipp E, Heinrich R (1994) Evolutionary optimization of enzyme kinetic parameters; effect of constraints. J Theor Biol 171:309–323. doi:10.1006/jtbi.1994.1234
44. Mahan BH (1975) Microscopic reversibility and detailed balance – an analysis. J Chem Educ 52:299–302
45. Michaelis L, Davidsohn H (1911) Die Wirkung der Wasserstoffionen auf das Invertin. Biochem Z 35:386–412
46. Michaelis L, Menten ML (1913) Die Kinetik der Invertinwirkung. Biochem Z 49:333–369
47. Morrison JF, Stone SR (1985) Approaches to the study and analysis of the inhibition of enzymes by slow- and tight-binding inhibitors. Comments Mol Cell Biophys 2:347–368
48. Motulsky HJ, Ransnas LA (1987) Fitting curves to data using nonlinear regression: a practical and nonmathematical review. FASEB J 1:365–374
49. Nomenclature Committee of the International Union of Biochemistry (1982) Symbolism and terminology in enzyme kinetics. Recommendations 1981. Eur J Biochem 128:281–291
50. Onsager L (1931) Reciprocal relations in irreversible processes. I. Phys Rev 37:405–426
51. Perdicakis B, Montgomery HJ, Guillemette JG, Jervis E (2004) Validation and characterization of uninhibited enzyme kinetics performed in multiwell plates. Anal Biochem 332:122–136
52. Pettersson G (1991) Why do many Michaelian enzymes exhibit an equilibrium constant close to unity for the interconversion of enzyme-bound substrate and product? Eur J Biochem 195:663–670
53. Plowman KM (1972) Enzyme kinetics. McGraw-Hill, New York
54. Qi F, Dash R, Han Y, Beard D (2009) Generating rate equations for complex enzyme systems by a computer-assisted systematic method. BMC Bioinformatics 10:238. doi:10.1186/1471-2105-10-238
55. Rakitzis ET (1997) Kinetic analysis of chemical or enzymic reactions: an algorithm for the determination of the initial velocity of product formation by the use of a Taylor series in reaction time. J Theor Biol 188:387–389
56. Reiner JM (1969) Behavior of enzyme systems, 2nd edn. Van Nostrand-Reinhold, New York
57. Segel IH (1975) Enzyme kinetics. Behavior and analysis of rapid equilibrium and steady-state enzyme systems. Wiley, New York
58. Selwyn MJ (1965) A simple test for inactivation of an enzyme during assay. Biochim Biophys Acta 105:193–195
59. St Maurice M, Bearne SL (2002) Kinetics and thermodynamics of mandelate racemase catalysis. Biochemistry 41:4048–4058. doi:10.1021/bi016044h
60. Tang Q, Leyh TS (2010) Precise, facile initial rate measurements. J Phys Chem B 114:16131–16136. doi:10.1021/jp1055528
61. Tolman RC (1924) Duration of molecules in upper quantum states. Phys Rev 23:693–709. doi:10.1103/PhysRev.23.693
62. Tolman RC (1925) The principle of microscopic reversibility. Proc Natl Acad Sci USA 11:436–439
63. Tolman RC (1938) The principles of statistical mechanics. Oxford University Press, Oxford
64. Volkenstein MV, Goldstein BN (1966) A new method for solving the problems of the stationary kinetics of enzymological reactions. Biochim Biophys Acta 115:471–477
65. Vrzheshch PV (2008) Quasi-equilibrium assumption in enzyme kinetics. Necessary and sufficient conditions and accuracy of its application for single-substrate reactions. Biochemistry-Moscow 73:1114–1120
66. Webb JL (1963) Enzyme and metabolic inhibitors. General principles of inhibition, vol 1. Academic, New York

67. Whitehead EP (1970) The regulation of enzyme activity and allosteric transition. Progr Biophys Mol Biol 21:321–397. doi:http://dx.doi.org/10.1016/0079-6107(70)90028-3
68. Wong JTF (1975) Kinetics of enzyme mechanisms. Academic, London

Chapter 2
The General Modifier Mechanism

... the proposed scheme, though very simple, includes as special cases the types of interaction commonly considered in enzyme kinetics, and in addition leads to a more general expression, capable of describing interactions of a sort not hitherto analyzed.
Botts J, Morales M (1953) Trans Faraday Soc 49:696–707 (p. 701)

Abstract Even for a relatively simple enzyme-catalyzed reaction comprising one substrate and one modifier, the rate equation derived under the steady-state assumption contains squared concentration terms if both complexes, enzyme–substrate and enzyme–substrate-modifier, are catalytically active. This chapter discusses the properties of the general rate equation for a one-substrate, one-modifier mechanism and the conditions under which it can be simplified by dropping squared concentration terms to render it manageable in practice. The risk of making mistakes in the calculation of kinetic parameters when applying the quasi-equilibrium assumption in data analysis without knowing whether or not any simplification of the rate equation is legitimate, is critically analyzed with systems of practical interest.

2.1 Introduction

In 1953 Jean Botts and Manuel Morales provided a fine analysis of the action of modifiers on enzyme activity and derived an exact rate equation under the steady-state assumption using a model now called *the general modifier mechanism* because it accommodates the action of inhibitors and activators [4]. This paper was followed by two corollaries published independently by Botts [3] and Morales [28]. The authors have been active in a scientifically stimulating environment at the U.S. Naval Medical Research Institute in Bethesda, MD, U.S.A., where they collaborated with Terrell Hill, the prolific author of numerous papers and books in statistical mechanics, thermodynamics, and much more.

Despite having been derived for unireactant enzymes in the presence of one modifier, the general modifier mechanism can be extended to more complex systems as will be discussed in Chap. 4. Botts and Morales, introducing the interactions of

A. Baici, *Kinetics of Enzyme-Modifier Interactions*,
DOI 10.1007/978-3-7091-1402-5_2

substrate and modifier with the enzyme, specifically referred to

> ··· an enzymatic 'patch' upon the enzyme molecule, consisting of an enzymatic site which binds substrate molecules and an adjacent site which binds modifier molecules.

The *adjacent site* is reminiscent of the *allosteric site* defined 10 years later by Monod, Changeux and Jacob [26]. Indeed, the general modifier mechanism is an excellent model for the kinetic analysis of enzymatic reactions in the presence of substances that do not bind to the same site as the substrate.

The model of Botts and Morales comprises two substrate-bound intermediates. In line with this fact, the steady-state rate equation contains squared substrate and modifier concentrations. While this property points to deviations from Michaelis–Menten kinetics, it should not suggest cooperativity in the classical sense. As will be shown in this chapter, the $[S]^2$ and $[X]^2$ terms are due to the simultaneous presence of the ES and ESX complexes. Situations in which the squared terms can be dropped after making legitimate assumptions of practical interest, thus reducing the rate equation to first degree in [S] and [X], have been discussed by Frieden [17].

The above-mentioned theory of allosteric[1] interactions published in 1963 [26] and consolidated 2 years later by Monod, Wyman and Changeux [27], experienced in the following years important extensions. An innovation has been the finding that allosteric interactions are not an exclusive property of oligomeric proteins containing two or more subunits but also of monomeric proteins. A convincing proof has been the discovery of the allosteric behavior of myoglobin [16], a classical example of a nonallosteric protein in textbooks. It has also been suggested that any nonfibrous protein may theoretically be allosteric [18]. A further adaptation of the original definition of allosteric interactions regards the change of shape induced at an active site of a protein following binding of a ligand at another site [26]. This alternative view of the *change of shape* does not predict the creation of a new shape. Instead, the allosteric effector brings about a redistribution of preexisting ensembles of protein conformations by changing the ratios between them [38].

Other properties related to ligand-induced conformational changes have been treated under the heading of allostery but represent unorthodox deviations from the original concept of Monod and coworkers. As commented by Fenton [15], interactions such as those between two substrates within the same active site and covalent modifications at sites distant from the active center do not fit into the concept of allosteric interactions.

This chapter deals with the general modifier mechanism, which embraces the kinetics of allosteric interactions but also serves as a basis for nonallosteric mechanisms. Cooperativity and allostery may coexist in proteins but an allosteric interaction does not necessarily imply the existence of cooperative effects. Kinetic cooperativity, neither in the context of substrate binding nor regarding the behavior of inhibitors and activators, will be treated in this book. However, methods aimed at

---

[1]The significance of the term *allosteric* is explained in Appendix 1.

discriminating between cooperative and noncooperative systems in the presence of modifiers will be discussed in Sect. 4.5.

2.2 Linkage at Equilibrium

The Botts–Morales treatment of enzyme–modifier interactions is based on the model shown in Scheme 2.1 that considers a ternary complex between a macromolecule (E) and two ligands (S and X). The dissociation constants for the four binding steps are defined as

$$K_1 = \frac{k_{-1}}{k_1}, \quad K_4 = \frac{k_{-4}}{k_4}, \quad K_3 = \frac{k_{-3}}{k_3}, \quad K_5 = \frac{k_{-5}}{k_5}. \tag{2.1}$$

In his analysis of *the interdependence of structure and function in the heme proteins ⋯ and the nature of the heme-protein linkage*, Jeffries Wyman introduced in 1948 the concept of *linked functions* in macromolecular systems [48]. Wyman's work, consolidated in subsequent years [49–52], has been instrumental in the birth of the theory of allosteric interactions [22, 26, 27]. When two ligands are simultaneously bound to the same protein, reciprocal interactions can occur between them. With reference to Scheme 2.1, if binding of X exerts an effect on the binding of S, then binding of S influences the binding of X by the same extent (*principle of reciprocity*). The relationships between dissociation constants quantify the *allosteric linkage* between S and X. Linkage has been put in relation with the coupling free energy between ligands by Gregorio Weber [42], in which paper the word *allosteric* has however not been mentioned.

According to the principle of reciprocity, the free energy of binding is independent of the particular way ESX is formed, i.e., through the path S + E $\rightleftharpoons$ ES $\rightleftharpoons$

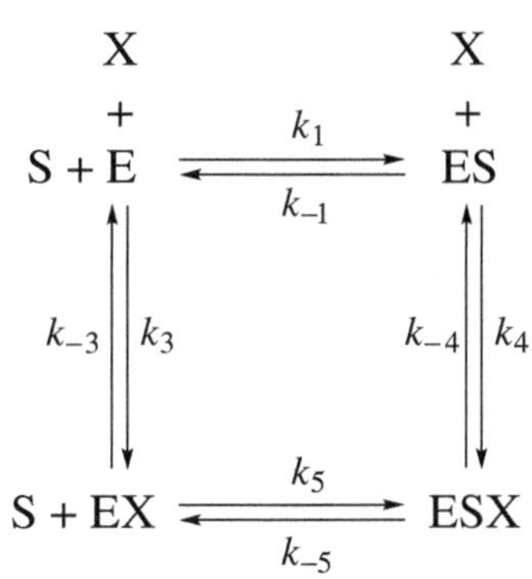

Scheme 2.1 Binding of two ligands, S and X, to a macromolecule E. For internal consistency, symbols and topographical arrangement are the same used elsewhere in this and other chapters, including the numbering of constants. An enzyme (E) is taken here as example but this may be another acceptor as well

ESX or X + E $\rightleftharpoons$ EX $\rightleftharpoons$ ESX. The *coupling free energy* (ΔG_c) between S and X is thus defined as the difference between the free energies of binding of S with free enzyme and with enzyme saturated with X, and vice versa:

$$\Delta G_c = -RT \ln \frac{K_1}{K_5} = -RT \ln \frac{K_3}{K_4} \quad \text{or} \quad \Delta G_c = -RT \ln \frac{k_{-1}k_5}{k_1 k_{-5}} = -RT \ln \frac{k_{-3}k_4}{k_3 k_{-4}} , \tag{2.2}$$

which implies

$$\frac{K_1}{K_5} = \frac{K_3}{K_4} = \frac{1}{\alpha} , \tag{2.3}$$

and

$$\Delta G_c = -RT \ln \frac{1}{\alpha} . \tag{2.4}$$

The relationships between coupling free energy and the free energies for the individual binding processes of S and X are better understood in the diagram shown in Fig. 2.1 drawn by Weber to interpret energy issues in ligand binding to polypeptides [42]. The figure contains the definitions of the displayed energy differences. ΔG_c defined by (2.4) can be negative ($\alpha < 1$), positive ($\alpha > 1$) or zero ($\alpha = 1$). With $\Delta G_c < 0$ modifier and substrate interact by reciprocally promoting each other's binding, while $\Delta G_c > 0$ denotes an interaction with reciprocal binding hindrance and $\Delta G_c = 0$ means no interaction, i.e., substrate and modifier ignore each other when bound to the enzyme. It is important *not to confuse* these concepts with positive and negative cooperativity.

Scheme 2.1 represents a *thermodynamic box*, or *thermodynamic energy cycle*, and each of the two ratios between dissociation constants in (2.3) is called *allosteric coupling constant*, here denoted $1/\alpha$ to maintain consistent symbolism in the following chapters and to agree with past and present kinetic literature. $1/\alpha$ is the same as Q in [30].

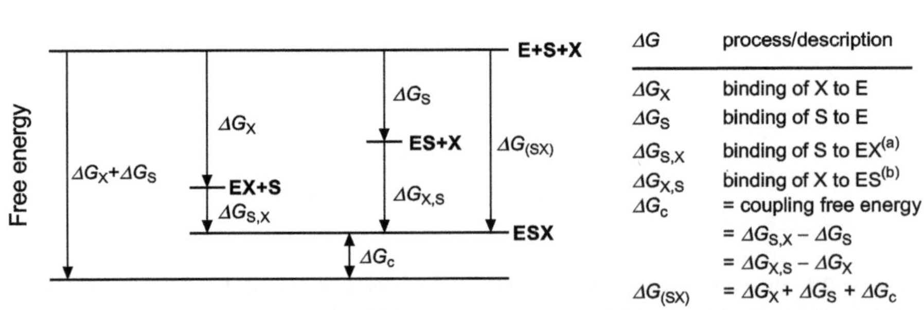

Fig. 2.1 Definition of the coupling free energy for the system in Scheme 2.1 (after Weber [42]). (a) = E saturated with X, (b) = E saturated with S

At equilibrium, detailed balance can be written for the cyclic Scheme 2.1 in the form of (1.28) as

$$\begin{aligned} k_1\,[\mathrm{E}]\,[\mathrm{S}] &= k_{-1}\,[\mathrm{ES}] \\ k_4\,[\mathrm{ES}]\,[\mathrm{X}] &= k_{-4}\,[\mathrm{ESX}] \\ k_{-5}\,[\mathrm{ESX}] &= k_5\,[\mathrm{EX}]\,[\mathrm{S}] \\ k_{-3}\,[\mathrm{EX}] &= k_3\,[\mathrm{E}]\,[\mathrm{X}]\ . \end{aligned} \tag{2.5}$$

The overall equilibrium constant can be calculated by multiplying together the left parts and equating the result with the multiplied right parts of (2.5). All concentration terms cancel out leaving

$$k_1k_4k_{-5}k_{-3} = k_3k_5k_{-4}k_{-1} \quad \therefore \quad K_{\mathrm{eq}} = \frac{k_1k_4k_{-5}k_{-3}}{k_3k_5k_{-4}k_{-1}} = 1\ . \tag{2.6}$$

The relationship (2.6) between the kinetic constants within the thermodynamic box obeys the principle of microscopic reversibility (Sect. 1.3). In Scheme 2.1 there are four second-order (k_1, k_3, k_4, k_5) and four first-order rate constants ($k_{-1}, k_{-3}, k_{-4}, k_{-5}$), where the same concentration terms that multiply the second-order constants appear in the numerator and in the denominator of (2.6).

Although the following relationships are obvious from the expressions above in this section, they are explicitly shown for easy referencing from other sections ($K_{\mathrm{S}} = k_{-1}/k_1$):

$$\frac{K_4}{K_3} = \frac{K_5}{K_1} = \frac{K_5}{K_{\mathrm{S}}} = \frac{k_3k_{-4}}{k_{-3}k_4} = \frac{k_1k_{-5}}{k_{-1}k_5} = \alpha \tag{2.7}$$

Didactic Note The concepts about linked functions and free energy of coupling briefly described here are treated in a language that does not require special mathematical knowledge in *Free energy transduction and biochemical cycle kinetics* by Terrell Hill [19], of which at least the first of the three chapters is a recommended reading. This small book, an elementary presentation *for a class or for self-study*, has been adapted by the author, in abbreviated form, from a preceding version with the purpose of explaining the principles of energy transduction in biology without cumbersome mathematics.

2.3 Linkage at Steady-State: The Treatment of Botts and Morales

The reactant binding steps of the general modifier mechanism (Scheme 2.2), which comprises a substrate and a modifier, are the same as in Scheme 2.1 with the addition of two catalytic steps. Only the initial rate is considered, i.e., [P] = 0, so that the

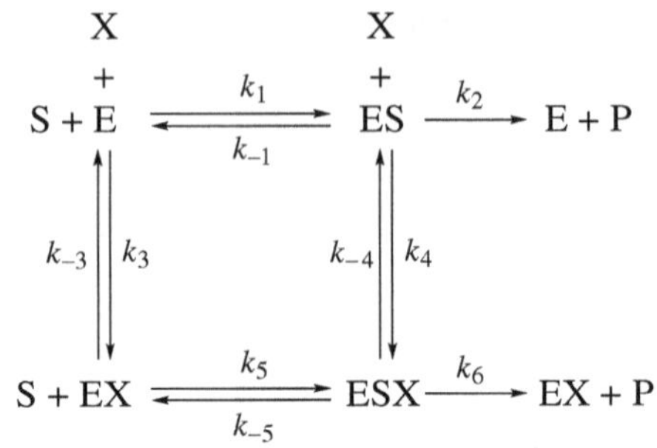

Scheme 2.2 The general modifier mechanism in the symbolism of this book. The original scheme and equation of Botts and Morales are reproduced in Appendix 2

reverse reactions can be neglected and k_{-2} and k_{-6} are not shown. The equation of Botts and Morales with its original symbols is shown in Appendix 2. In the same Appendix, the inventive method used by the authors for deriving the rate equation is explained step-by-step while in this section we discuss the properties of the general rate equation.

The *unperturbed* terms in (2.85) derived in Appendix 2 can be rearranged and the perturbation terms in (2.90), called perturbation$_\mathrm{N}$ (numerator) and perturbation$_\mathrm{D}$ (denominator), added to give the following symbolic equivalent of (2.90):

$$v = \frac{\left\{ \dfrac{k_2 + k_6 \dfrac{[\mathrm{X}]}{K_4}}{1 + \dfrac{[\mathrm{X}]}{K_4}} [\mathrm{S}] + \text{perturbation}_\mathrm{N} \right\} [\mathrm{E}]_\mathrm{t}}{K_\mathrm{m}^0 \times \dfrac{1 + \dfrac{[\mathrm{X}]}{K_3}}{1 + \dfrac{[\mathrm{X}]}{K_4}} + [\mathrm{S}] + \text{perturbation}_\mathrm{D}} . \tag{2.8}$$

Upon expansion of (2.90), besides [S] and [X] the equation contains also terms in [S] [X], $[\mathrm{S}]^2$, $[\mathrm{X}]^2$, $[\mathrm{S}]^2$ [X] and [S] $[\mathrm{X}]^2$. Botts [3] rewrote the cumbersome Eq. (2.90) in compact form as follows (just the symbol X for the modifier is used here instead of Y in the original)

$$v = \frac{(A + B\,[\mathrm{X}])\,[\mathrm{S}]^2 + \left(C + D\,[\mathrm{X}] + G[\mathrm{X}]^2\right)[\mathrm{S}]}{(H + J)\,[\mathrm{S}]^2 + \left(L + N\,[\mathrm{X}] + Q[\mathrm{X}]^2\right)[\mathrm{S}] + \left(R + U\,[\mathrm{X}] + W[\mathrm{X}]^2\right)} [\mathrm{E}]_\mathrm{t} . \tag{2.9}$$

By further grouping terms in (2.9) depending on whether in an experiment either [S] or [X] are kept constant together with constant $[\mathrm{E}]_\mathrm{t}$, simpler expressions can be obtained. If [X] and $[\mathrm{E}]_\mathrm{t}$ are constant, (2.9) can be rewritten as

$$v = \frac{a_1[\mathrm{S}]^2 + b_1\,[\mathrm{S}]}{g_1[\mathrm{S}]^2 + h_1\,[\mathrm{S}] + m_1} , \tag{2.10}$$

while if [S] and $[E]_t$ are constant, (2.9) can be written similarly as

$$v = \frac{a_2[X]^2 + b_2\,[X] + c_2}{g_2[X]^2 + h_2\,[X] + m_2}\,. \tag{2.11}$$

The coefficients in (2.10) and (2.11) consist of combinations of individual kinetic constants multiplied together and contain the concentrations of the species that are held constant. For (2.10) and (2.11), the full expressions of a_1–m_1 and a_2–m_2 are given in (2.46) and (2.47), respectively. The total concentration of enzyme catalytic centers, $[E]_t$, is embedded in the numerators. These expressions can be used for several purposes to handle various *simple models of enzyme action* [3], and can also be exploited to perform simplifications of various kinds and to calculate kinetic parameters under limiting conditions of reactant concentrations as described in Sect. 2.5.

2.4 Three Simplifications of the General Rate Equation

Both the numerator and the denominator of (2.90) and its symbolic equivalent (2.8) consist of the sum of two terms. The denominators of the long fractional terms in *Numerator* and *Denominator* are the same and both the numerators and the denominators of these fractional parts contain common factors. Therefore, the structure of (2.90) and (2.8) is strategically organized for simplifying the velocity expression under particular circumstances. The situations that allow simplification of (2.90) in mechanisms of inhibition or activation containing all constants shown in Scheme 2.2 are discussed in the following three sub-sections. The first two simplifications (Sects. 2.4.1 and 2.4.2) depend on the relationships between four constants

$$1 + \frac{k_2}{k_{-1}} \quad \text{and} \quad 1 + \frac{k_6}{k_{-5}} \tag{2.12}$$

that appear in numerators and denominators of the perturbation terms in (2.90). The third simplification (Sect. 2.4.3) is based on the assumption of quasi-equilibrium for the binding of modifier to E and ES, while the fluxes around ES and ESX are maintained in a steady-state regime. The simplifications discussed in Sects. 2.4.1 and 2.4.2 have been considered by Botts and Morales [4] and later by Morales [28].

2.4.1 Quasi-Equilibrium Conditions for All Binding Steps

If the two terms in (2.12) are very close to 1, the following inequalities are true:

$$\frac{k_2}{k_{-1}} \ll 1\,, \quad \frac{k_6}{k_{-5}} \ll 1\,, \quad \left(\frac{k_2}{k_{-1}} - \frac{k_6}{k_{-5}}\right) \ll 1 \tag{2.13}$$

and the two perturbation terms in (2.90) become small with respect to the other terms. In the limit that (k_2/k_{-1}) and (k_6/k_{-5}) are so small to practically vanish to zero, (2.90) and (2.8)) simplify to

$$v = \frac{\dfrac{k_2 + k_6\dfrac{[\mathrm{X}]}{K_4}}{1+\dfrac{[\mathrm{X}]}{K_4}}[\mathrm{E}]_t\,[\mathrm{S}]}{K_\mathrm{S} \times \dfrac{1+\dfrac{[\mathrm{X}]}{K_3}}{1+\dfrac{[\mathrm{X}]}{K_4}} + [\mathrm{S}]}\,, \tag{2.14}$$

which is the same as *unperturbed* (2.8) with K_S replacing K_m^0 because

$$K_\mathrm{m}^0 = \frac{k_{-1}+k_2}{k_1} = \frac{\dfrac{k_{-1}}{k_{-1}} + \dfrac{k_2}{k_{-1}}}{\dfrac{k_1}{k_{-1}}}\,,$$

$$\text{and with } \frac{k_2}{k_{-1}} \ll 1 \quad \therefore \quad K_\mathrm{m}^0 \approx K_\mathrm{S} = \frac{k_{-1}}{k_1}\,. \tag{2.15}$$

Under this simplification, the apparent values of k_cat, K_m and specificity constant are as follows:

$$k_\mathrm{cat}^\mathrm{app} = \frac{k_2 + k_6\dfrac{[\mathrm{X}]}{K_4}}{1+\dfrac{[\mathrm{X}]}{K_4}} \tag{2.16}$$

$$K_\mathrm{m}^\mathrm{app} = K_\mathrm{S} \times \frac{1+\dfrac{[\mathrm{X}]}{K_3}}{1+\dfrac{[\mathrm{X}]}{K_4}} \tag{2.17}$$

$$k^\mathrm{app} = \frac{1}{K_\mathrm{S}} \times \frac{k_2 + k_6\dfrac{[\mathrm{X}]}{K_4}}{1+\dfrac{[\mathrm{X}]}{K_3}}\,. \tag{2.18}$$

The validity of the inequalities (2.13) does not depend on the relative magnitudes of k_2 to k_{-1} and of k_6 to k_{-5} and, if they are true, the claim of quasi-equilibrium binding of substrate is made, which results in the relationship (2.15). If this is the case, quasi-equilibrium binding of E to X and ES is implicit as a consequence of microscopic reversibility, as evident from (2.7).

Essence of the Simplification (2.13) If (2.13) is valid, the substrate is nonsticky, the four enzyme-containing species in Scheme 2.2 are in quasi-equilibrium and the rate equation does not contain squared concentration terms. Analysis of initial velocity as a function of [S] by fitting (2.35) to data (see below) is expected to yield parameter estimates close to the true values with any deviation depending mostly on experimental error. A numerical example is shown in Sect. 2.7.1.

Definition of Sticky Substrate Cleland defined *sticky substrates* those that *dissociate more slowly from the enzyme than they react to give products* and *stickiness* the ratio k_2/k_{-1} [8, pp. 313, 331], with further comments in [9, p. 403]. Instead of thinking on how much k_2/k_{-1} is less than 1, the following equivalent definition means that a substrate is nonsticky if the value is close to 1 and sticky if it is $\gg 1$.

$$\text{stickiness} = 1 + \frac{k_2}{k_{-1}}. \tag{2.19}$$

The same definition can be extended to k_6/k_{-5}.

2.4.2 *Generalized Microscopic Reversibility*

This simplification relies on the independence of the ratios of the constants for the breakdown of ES and ESX on modifier concentration (E + S ← ES → E + P) and (EX + S ← ESX → E + P) , i.e.,

$$1 + \frac{k_2}{k_{-1}} = 1 + \frac{k_6}{k_{-5}}\,. \tag{2.20}$$

As special cases of (2.20), also the following conditions, of which the first is obvious, lead to the same result:

$$\frac{k_2}{k_{-1}} = \frac{k_6}{k_{-5}} \tag{2.21}$$

$$k_2 = k_{-1} \quad \text{and} \quad k_6 = k_{-5} \tag{2.22}$$

$$k_2 = k_6,\ k_{-1} = k_{-5} \quad \text{and} \quad k_1 \neq k_5 \tag{2.23}$$

Also for this simplification no assumptions are necessary about the relative magnitudes of k_2 to k_{-1} and of k_6 to k_{-5}. Introducing (2.20) in (2.90) and (2.8) sets the two perturbation terms to zero. This simplification differs from that discussed in Sect. 2.4.1, for which the perturbation terms can be ignored with respect to other terms only as long as they are sufficiently small. The conceptual difference between the two simplifications is that in the second case (this section) the rate equation is the same as (2.14) but it contains K_m^0 instead of K_S because (2.15) does not apply. In mechanistic terms, (2.20) claims quasi-equilibrium for $E + X \rightleftharpoons EX$ and $ES + X \rightleftharpoons ESX$, while the constraint of quasi equilibrium *is not compelling required* for $E + S \rightleftharpoons ES$ and $EX + S \rightleftharpoons ESX$, i.e., these two paths can operate in a regime of steady-state or not.

The relationship (2.21), equivalent to $k_2 k_{-5} = k_6 k_{-1}$, is implicit in (2.20) and the inequalities (2.13) are also satisfied when this is the case. Botts and Morales, who first discussed this property [4], remarked that this is a *rather coincidental case*, while Christopher Topham and Keith Brocklehurst, defined this as *a sufficient but not necessary condition for the quasi-equilibrium binding of X* [37].

Under the assumption (2.20), $k_{\mathrm{cat}}^{\mathrm{app}}$ is the same as (2.17) and the apparent values of K_{m} and specificity constant become

$$K_{\mathrm{m}}^{\mathrm{app}} = K_{\mathrm{m}}^{0} \frac{1 + \dfrac{[\mathrm{X}]}{K_3}}{1 + \dfrac{[\mathrm{X}]}{K_4}} \tag{2.24}$$

$$k^{\mathrm{app}} = \frac{1}{K_{\mathrm{m}}^{0}} \frac{k_2 + k_6 \dfrac{[\mathrm{X}]}{K_4}}{1 + \dfrac{[\mathrm{X}]}{K_3}} \, . \tag{2.25}$$

The relation (2.21), $k_2/k_{-1} = k_6/k_{-5}$, has been used in 1976 by Whitehead in his formulation of the principle of *generalized microscopic reversibility* in a paper dedicated to the simplification of complex steady-state rate equations [44]. The foundation of this concept stems from rule *a* of Volkenstein and Goldstein [40], which states that parallel branches in a system of reactions can be added becoming a single branch. This rule (node compression), already discussed in Sect. 1.6.2, can be applied to the general modifier mechanism to derive the rate equation as explained in Appendix 4. With reference to Scheme 2.2, generalized microscopic reversibility is defined as

$$k_1 k_4 (k_{-5} + k_6) k_{-3} = k_3 k_5 k_{-4} (k_{-1} + k_2) \, . \tag{2.26}$$

Considering equilibrium and Michaelis constants and using the symbolism of this book, (2.26) is identical with

$$K_{\mathrm{m}}^{\infty} K_3 = K_{\mathrm{m}}^{0} K_4 \, , \tag{2.27}$$

a relationship published in 1952 by Segal, Kachmar and Boyer in a paper that dealt with the rate equation of the general modifier mechanism [32, top of p. 189],[2] where the authors limited their treatment to the derivation of rate equations for this system

[2]In this paper, which will be reconsidered in Chaps. 3 and 5, all equations are unnumbered.

with all steps at equilibrium and a catalytically inactive ESX complex. In (2.27), $K_m^\infty = (k_{-5} + k_6)/k_5$, see (2.58).

Compared with the relationship for microscopic reversibility (2.6) (left), (2.26) contains the composite constants $(k_{-5} + k_6)$ and $(k_{-1} + k_2)$. In the general modifier mechanism, the condition of generalized microscopic reversibility is satisfied *only if* $k_2/k_{-1} = k_6/k_{-5}$ and holds independently of fast equilibration. Claims of exceptions to this law [2] were erroneously founded on the violation of microscopic reversibility, i.e. (2.6), from which (2.26) depends [46]. In systems where generalized microscopic reversibility does not apply, microscopic reversibility is valid anyway.

Generalized microscopic reversibility has been used by Whitehead in a graph-theoretical algorithm aimed at determining common factors in the numerator and denominator of rate equations [45] and by Jacques Ricard in an interesting discussion of enzyme evolution [31]. In this paper, Ricard analyzes in depth generalized microscopic reversibility and its relationships to kinetic cooperativity. Among other results, he demonstrates that, if generalized microscopic reversibility applies, no kinetic cooperativity can be manifested by the *slow-transition model*, whose reaction paths are analogous to those of the general modifier mechanism in which two enzyme conformations occupy the positions of EX and ESX in Scheme 2.2. Also, in polymeric enzymes in which interactions between subunits occur without inducing structural changes of the liganded active site, generalized microscopic reversibility must apply.

A clear-cut demonstration of the existence of generalized microscopic reversibility is provided by the interaction between thrombin and hir$^{55-65}$, the C-terminal fragment of hirudin, a natural inhibitor of this enzyme [12]. This example will be discussed in Sect. 4.2 (pp. 177–178). Beliefs that generalized microscopic reversibility is a theoretical curiosity or a nonexisting phenomenon, e.g., [35, footnote 5 on p. 396], represent statements that have not been validated. The overwhelming majority of published studies in which Michaelis–Menten behavior has been *assumed* in data analysis *did not* scrutinize in detail the mechanistic origin of this assumption by measuring individual rate constants. It is thus possible that generalized microscopic reversibility is frequently represented besides the largely accepted condition $k_2/k_{-1} \ll 1$ and $k_6/k_{-5} \ll 1$.

Essence of the Simplification (2.20) If (2.20) is valid, the substrate may or may not be sticky. Even in cases of pronounced stickiness, this is equally present in both catalytic steps of Scheme 2.2. As a consequence, the rate equation does not contain squared concentration terms and the system operates *at equilibrium*. Analysis of initial velocity as a function of [S] by fitting Eq. (2.35) to data (see below) provides parameters whose deviations from the true values depend only on experimental error as will be demonstrated in Sect. 2.7.5.

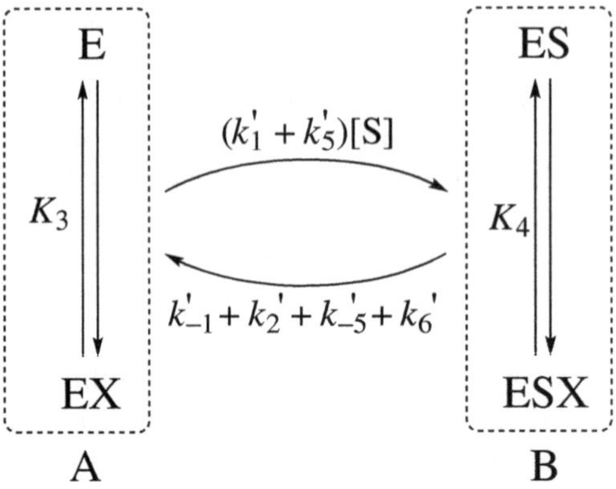

Scheme 2.3 The general modifier mechanism shown with condensed steps in quasi-equilibrium. The four nodes in Scheme 2.2 are compressed in just two nodes A and B, whose significance is shown in the *dashed boxes*. The *primed symbols* represent constants multiplied by their corresponding fractional values as defined in the text

2.4.3 Quasi-Equilibrium Assumption for the Binding of Modifier

The simplifications in the two preceding sections depend on the validity of assumptions (2.13) and (2.20), which may or may not apply. A rate equation, in which terms in $[S]^2$ are absent but retain terms in $[X]^2$ and $[S]\,[X]^2$, can be obtained by assuming quasi-equilibrium for the reactions involving binding of modifier to E and ES, while the fluxes around ES and ESX are treated under the steady-state assumption. It is not easy to arrive at the sought rate equation starting from (2.8) but this assumption allows the application of Cha's method for deriving rate equations under combined conditions of quasi-equilibrium and steady-state [6]. The simplification that arises using this method is shown in Scheme 2.3, in which the equilibria in the dashed boxes are treated as composite species (nodes) A and B. Binding and dissociation steps within A and B are much faster than the processes indicated by bent arrows occurring between nodes A and B. The resulting King–Altman patterns and the corresponding enzyme states are in this case just two, which can be directly deduced from Scheme 2.3:

$$E_A = k'_{-1} + k'_2 + k'_{-5} + k'_6 \qquad\qquad E_B = \left(k'_1 + k'_5\right)[S] \; , \qquad (2.28)$$

where the primed constants indicate the constants themselves multiplied by the fractional concentration of their associated enzyme-containing species. Within nodes A and B, the fractional concentrations of the four species can be calculated from dissociation constants:

$$f_E = \frac{1}{1+\dfrac{[X]}{K_3}}, \quad f_{EX} = \frac{\dfrac{[X]}{K_3}}{1+\dfrac{[X]}{K_3}}, \quad f_{ES} = \frac{1}{1+\dfrac{[X]}{K_4}}, \quad f_{ESX} = \frac{\dfrac{[X]}{K_4}}{1+\dfrac{[X]}{K_4}}, \qquad (2.29)$$

where $f_E + f_{EX} = 1$ and $f_{ES} + f_{ESX} = 1$.

The primed constants in (2.28) can now be converted to explicit values using fractions (2.29), and the resulting expressions simplified and rearranged:

$$\mathrm{E_A} = (k_{-1} + k_2) f_{\mathrm{ES}} + (k_{-5} + k_6) f_{\mathrm{ESX}} \qquad \mathrm{E_B} = (k_1 f_{\mathrm{E}} + k_5 f_{\mathrm{EX}})\,[\mathrm{S}]\ , \tag{2.30}$$

$$\mathrm{E_A} = \frac{k_{-1} + k_2 + (k_{-5} + k_6)\dfrac{[\mathrm{X}]}{K_4}}{1 + \dfrac{[\mathrm{X}]}{K_4}} \qquad \mathrm{E_B} = \frac{k_1 + k_5\dfrac{[\mathrm{X}]}{K_3}}{1 + \dfrac{[\mathrm{X}]}{K_3}}\,[\mathrm{S}]\ . \tag{2.31}$$

Using these enzyme states, $f_{\mathrm{A}}, f_{\mathrm{B}}$ and the rate equation can be calculated as

$$f_{\mathrm{A}} = \frac{\mathrm{E_A}}{\mathrm{E_A} + \mathrm{E_B}}, \quad f_{\mathrm{B}} = \frac{\mathrm{E_B}}{\mathrm{E_A} + \mathrm{E_B}}\ , \tag{2.32}$$

$$v = (k_2 f_{\mathrm{ES}} + k_6 f_{\mathrm{ESX}})\ f_{\mathrm{B}}[\mathrm{E}]_{\mathrm{t}}\ . \tag{2.33}$$

The terms within parentheses in (2.33) represent the fractions of ES and ESX within node B, each multiplied by the respective catalytic constants. The result of the first operation is converted from fractions to concentrations through multiplication by the fraction of enzyme concentration available in node B giving

$$v = \frac{\dfrac{k_2 + k_6\dfrac{[\mathrm{X}]}{K_4}}{1 + \dfrac{[\mathrm{X}]}{K_4}}\,[\mathrm{E}]_{\mathrm{t}}\,[\mathrm{S}]}{\dfrac{k_{-1} + k_2 + (k_{-5} + k_6)\dfrac{[\mathrm{X}]}{K_4}}{k_1 + k_5\dfrac{[\mathrm{X}]}{K_3}} \times \dfrac{1 + \dfrac{[\mathrm{X}]}{K_3}}{1 + \dfrac{[\mathrm{X}]}{K_4}} + [\mathrm{S}]}\ . \tag{2.34}$$

Expansion of (2.34), besides terms in [S] and [X], reveals the presence of terms in $[\mathrm{X}]^2$, [S] [X] and [S] $[\mathrm{X}]^2$ but no longer terms in $[\mathrm{S}]^2$. Equation (2.34) has the general form of the Michaelis–Menten equation:

$$v = \frac{k_{\mathrm{cat}}^{\mathrm{app}}\,[\mathrm{E}]_{\mathrm{t}}\,[\mathrm{S}]}{K_{\mathrm{m}}^{\mathrm{app}} + [\mathrm{S}]}\ . \tag{2.35}$$

In this generic equation, the apparent parameters $k_{\mathrm{cat}}^{\mathrm{app}}$ and $K_{\mathrm{m}}^{\mathrm{app}}$ assume different forms according to the particular kinetic mechanism in the presence of modifiers. In the case treated here:

$$k_{\mathrm{cat}}^{\mathrm{app}} = \frac{k_2 + k_6\dfrac{[\mathrm{X}]}{K_4}}{1 + \dfrac{[\mathrm{X}]}{K_4}} \tag{2.36}$$

$$K_{\mathrm{m}}^{\mathrm{app}} = \frac{k_{-1} + k_2 + (k_{-5} + k_6)\dfrac{[\mathrm{X}]}{K_4}}{k_1 + k_5\dfrac{[\mathrm{X}]}{K_3}} \times \frac{1 + \dfrac{[\mathrm{X}]}{K_3}}{1 + \dfrac{[\mathrm{X}]}{K_4}} . \tag{2.37}$$

This expression of $K_{\mathrm{m}}^{\mathrm{app}}$ can be written in compact form as

$$K_{\mathrm{m}}^{\mathrm{app}} = K_{\mathrm{m}}^{\mathrm{S}} \times \frac{1 + \dfrac{[\mathrm{X}]}{K_3}}{1 + \dfrac{[\mathrm{X}]}{K_4}} , \tag{2.38}$$

where $K_{\mathrm{m}}^{\mathrm{S}}$ is named after Topham [36]

$$K_{\mathrm{m}}^{\mathrm{S}} = \frac{k_{-1} + k_2 + (k_{-5} + k_6)\dfrac{[\mathrm{X}]}{K_4}}{k_1 + k_5\dfrac{[\mathrm{X}]}{K_3}} . \tag{2.39}$$

The Michaelis *constant* of the general modifier mechanism assuming quasi-equilibrium for modifier binding corresponds thus to $K_{\mathrm{m}}^{\mathrm{S}}$, which is not a constant because it depends on [X] and is equivalent to (2.40) previously reported by Topham [36, 37] as explained in Appendix 3:

$$K_{\mathrm{m}}^{\mathrm{S}} = \frac{\dfrac{k_{-1}}{k_1}\left\{\left(\dfrac{k_{-1} + k_2}{k_{-1}}\right)k_1 + \left(\dfrac{k_{-5} + k_6}{k_{-5}}\right)k_5\dfrac{[\mathrm{X}]}{K_3}\right\}}{k_1 + k_5\dfrac{[\mathrm{X}]}{K_3}} . \tag{2.40}$$

While (2.39) is a slender, handy form that can be used for display and simulation purposes, (2.40) is useful for making simplifications. So, if $k_2 \ll k_{-1}$ and $k_6 \ll k_{-5}$, it can be determined by inspection that (2.40) reduces to k_{-1}/k_1 but the same operation is not immediate using (2.39).

If $K_{\mathrm{m}}^{\mathrm{S}}$ is not a constant because it depends on [X], $K_{\mathrm{m}}^{\mathrm{app}}$ even contains terms in $[\mathrm{X}]^2$, a property that is better appreciated after expanding and factorizing (2.37):

$$K_{\mathrm{m}}^{\mathrm{app}} = \frac{\left(\dfrac{k_{-5} + k_6}{K_4K_3}\right)[\mathrm{X}]^2 + \left(\dfrac{k_{-1} + k_2}{K_3} + \dfrac{k_{-5} + k_6}{K_4}\right)[\mathrm{X}] + (k_{-1} + k_2)}{\dfrac{k_5}{K_3K_4}[\mathrm{X}]^2 + \left(\dfrac{k_1}{K_4} + \dfrac{k_5}{K_3}\right)[\mathrm{X}] + k_1} , \tag{2.41}$$

which can be symbolically expressed as a ratio of two quadratic polynomials

$$K_{\mathrm{m}}^{\mathrm{app}} = \frac{a[\mathrm{X}]^2 + b\,[\mathrm{X}] + c}{d[\mathrm{X}]^2 + e\,[\mathrm{X}] + f} \tag{2.42}$$

where the coefficients *a–f* represent all rate constants of the system. This expression is useful for interpreting the dependence of $K_{\mathrm{m}}^{\mathrm{app}}$ on [X] (see, e.g., Sect. 2.8.1).

The apparent specificity constant, obtained from (2.36) and (2.37), contains terms in [X] and $[\mathrm{X}]^2$ that are explicitly revealed after expanding the expressions:

$$k^{\mathrm{app}} = \frac{k_2 + k_6\dfrac{[\mathrm{X}]}{K_4}}{k_{-1} + k_2 + (k_{-5} + k_6)\dfrac{[\mathrm{X}]}{K_4}} \times \frac{k_1 + k_5\dfrac{[\mathrm{X}]}{K_3}}{1 + \dfrac{[\mathrm{X}]}{K_3}}\,. \tag{2.43}$$

Essence of the Assumption of Quasi-Equilibrium for Modifier Binding With this logical assumption, the rate equation does not contain $[\mathrm{S}]^2$ terms but retains terms in $[\mathrm{X}]^2$ and $[\mathrm{S}]\,[\mathrm{X}]^2$. The parameters obtained by analysis of initial velocity as a function of [S] by fitting (2.35) to data deviate from the true values depending on substrate stickiness and experimental error. Numerical examples are discussed in Sects. 2.7.2–2.7.4.

2.5 The Rate Equation of the General Modifier Mechanism: King–Altman Method

The complete rate equation for the general modifier mechanism, considering also reverse reactions, is derived in Appendix 4 using the King–Altman method. The resulting expression (2.105) has little practical value for data analysis but represents a useful starting point when formulating physically meaningful simplifications and can be exploited for precise simulations. Initial rates in the absence of product make it easier to handle the general modifier mechanism with a less cumbersome rate equation due to the absence of all terms containing [P] and $[\mathrm{P}]^2$ as shown in Eq. (2.44), which is obtained from (2.105) in Appendix 4 for [P] = 0. The list of the coefficients n_1–n_5 and d_1–d_8 is given in (2.45).

$$v = \frac{\left(n_1\,[\mathrm{S}] + n_2\,[\mathrm{S}]\,[\mathrm{X}] + n_3[\mathrm{S}]^2\,[\mathrm{X}] + n_4\,[\mathrm{S}]\,[\mathrm{X}]^2 + n_5[\mathrm{S}]^2\right)[\mathrm{E}]_{\mathrm{t}}}{d_1\,[\mathrm{S}] + d_2\,[\mathrm{S}]\,[\mathrm{X}] + d_3[\mathrm{S}]^2\,[\mathrm{X}] + d_4\,[\mathrm{S}]\,[\mathrm{X}]^2 + d_5[\mathrm{S}]^2 + d_6[\mathrm{X}]^2 + d_7\,[\mathrm{X}] + d_8}\,. \tag{2.44}$$

$$n_1 = k_1k_2k_{-3}k_{-4} + k_1k_2k_{-3}k_{-5} + k_1k_2k_{-3}k_6$$

$$n_2 = k_1k_{-3}k_4k_6 + k_{-1}k_3k_5k_6 + k_2k_3k_{-4}k_5 + k_2k_3k_5k_6$$

$$n_3 = k_1k_4k_5k_6$$

$$n_4 = k_3k_4k_5k_6$$

$$n_5 = k_1k_2k_{-4}k_5$$

$$
\begin{aligned}
d_1 &= k_1k_{-3}k_{-5} + k_{-1}k_{-4}k_5 + k_2k_{-4}k_5 + k_1k_{-3}k_6 + k_1k_{-3}k_{-4} \\
d_2 &= k_1k_4k_{-5} + k_{-1}k_3k_5 + k_1k_{-3}k_4 + k_1k_4k_6 + k_3k_{-4}k_5 + k_2k_3k_5 \\
d_3 &= k_1k_4k_5 \\
d_4 &= k_3k_4k_5 \\
d_5 &= k_1k_{-4}k_5 \\
d_6 &= k_3k_4k_6 + k_3k_4k_{-5} \\
d_7 &= k_{-1}k_3k_{-5} + k_{-3}k_4k_{-5} + k_2k_3k_{-4} + k_{-3}k_4k_6 + k_2k_3k_6 + k_{-1}k_3k_6 + k_{-1}k_3k_{-4} \\
&\quad + k_2k_3k_{-5} \\
d_8 &= k_{-1}k_{-3}k_{-5} + k_2k_{-3}k_{-5} + k_2k_{-3}k_{-4} + k_{-1}k_{-3}k_{-4} + k_{-1}k_{-3}k_6 + k_2k_{-3}k_6
\end{aligned}
\tag{2.45}
$$

Equation (2.44), obtained with the King–Altman method, and (2.90) derived by Botts and Morales have different forms but are exactly the same. With (2.44) and (2.45), (2.10) and (2.11) can be rewritten to reveal the identity of their coefficients:

$$
v = \frac{\overbrace{(n_5 + n_3\,[\mathrm{X}])\,[\mathrm{E}]_\mathrm{t}}^{a_1}\,[\mathrm{S}]^2 + \overbrace{\left(n_1 + n_2\,[\mathrm{X}] + n_4[\mathrm{X}]^2\right)[\mathrm{E}]_\mathrm{t}}^{b_1}\,[\mathrm{S}]}{\underbrace{(d_5 + d_3\,[\mathrm{X}])}_{g_1}\,[\mathrm{S}]^2 + \underbrace{\left(d_1 + d_2\,[\mathrm{X}] + d_4[\mathrm{X}]^2\right)}_{h_1}[\mathrm{S}] + \underbrace{\left(d_8 + d_7\,[\mathrm{X}] + d_6[\mathrm{X}]^2\right)}_{m_1}}\,,
\tag{2.46}
$$

$$
v = \frac{\overbrace{(n_4\,[\mathrm{S}]\,[\mathrm{E}]_\mathrm{t})}^{a_2}\,[\mathrm{X}]^2 + \overbrace{\left(n_2\,[\mathrm{S}] + n_3[\mathrm{S}]^2\right)[\mathrm{E}]_\mathrm{t}}^{b_2}\,[\mathrm{X}] + \overbrace{\left(n_1\,[\mathrm{S}] + n_5[\mathrm{S}]^2\right)[\mathrm{E}]_\mathrm{t}}^{c_2}}{\underbrace{(d_6 + d_4\,[\mathrm{S}])}_{g_2}\,[\mathrm{X}]^2 + \underbrace{\left(d_7 + d_2\,[\mathrm{S}] + d_3[\mathrm{S}]^2\right)}_{h_2}[\mathrm{X}] + \underbrace{\left(d_8 + d_1\,[\mathrm{S}] + d_5[\mathrm{S}]^2\right)}_{m_2}}\,.
\tag{2.47}
$$

These two expressions are particularly useful for performing exact simulations of rates, plotting results and making simplifications when either the concentration of substrate or modifier is held constant.

Even in the absence of products, an appreciation of the underlying mechanism from (2.44) is not straightforward as from equations written with grouped terms, using dissociation constants for steps considered in quasi-equilibrium and possibly showing familiar parameters such as limiting rates, Michaelis constants and modifier dissociation constants. However, the expressions of rates and Michaelis constants at particular concentrations of S and X can be obtained from (2.44) and hence from the Botts–Morales equation (2.90). For this purpose, the rate is renamed $v_{(S,X)}$ as explicit statement of its dependence on S and X with symbolism used by Botts and Morales [4, p. 700]. In the expressions of v that will follow, the first subscript refers to substrate and the second to modifier: S and X indicate finite values of concentrations, while 0 and ∞ symbolize *in the absence of* and *at saturating concentration*, respectively. In Michaelis constants, specificity constants, and limiting rates, the symbols 0 and ∞ are set as superscripts.

- Rate at finite [S] and [X], i.e., Eq. (2.44)

$$v_{(S,\,X)}$$

- Rate at finite [S] and [X] = 0 (Michaelis–Menten equation), using (2.44)

$$v_{(S,\,0)} = \lim_{[X]\to 0} v_{(S,\,X)} = \frac{k_2[E]_t\,[S]}{\dfrac{k_{-1}+k_2}{k_1} + [S]} \tag{2.48}$$

- Rate at finite [S] and saturating [X], using (2.44)

$$v_{(S,\,\infty)} = \lim_{[X]\to\infty} v_{(S,\,X)} = \frac{k_6[E]_t\,[S]}{\dfrac{k_{-5}+k_6}{k_5} + [S]} \tag{2.49}$$

- Limiting rate at [X] = 0, using (2.48)

$$V^0 = v_{(\infty,\,0)} = \lim_{[S]\to\infty} v_{(S,\,0)} = k_2[E]_t \tag{2.50}$$

- Limiting rate at finite [X], using (2.44)

$$V^{app} = v_{(\infty,\,X)} = \lim_{[S]\to\infty} v_{(S,\,X)} = \frac{\left(k_2 + k_6\dfrac{k_4}{k_{-4}}\,[X]\right)[E]_t}{1 + \dfrac{k_4}{k_{-4}}\,[X]} \tag{2.51}$$

- Limiting rate at saturating [X], using (2.51)

$$V^{\infty} = v_{(\infty,\,\infty)} = \lim_{[X]\to\infty} v_{(\infty,\,X)} = k_6[E]_t \tag{2.52}$$

- Catalytic constant at [X] = 0, using (2.50)

$$k^0_{cat} = \frac{V^0}{[E]_t} = k_2 \tag{2.53}$$

- Catalytic constant at finite [X], using (2.51), identical to (2.17)

$$k_{\mathrm{cat}}^{\mathrm{app}} = \frac{V^{\mathrm{app}}}{[\mathrm{E}]_{\mathrm{t}}} = \frac{k_2 + k_6 \dfrac{k_4}{k_{-4}}[\mathrm{X}]}{1 + \dfrac{k_4}{k_{-4}}[\mathrm{X}]} = \frac{k_2 + k_6 \dfrac{[\mathrm{X}]}{K_4}}{1 + \dfrac{[\mathrm{X}]}{K_4}} \tag{2.54}$$

- Catalytic constant at saturating [X], using (2.52)

$$k_{\mathrm{cat}}^{\infty} = \frac{V^{\infty}}{[\mathrm{E}]_{\mathrm{t}}} = k_6 \tag{2.55}$$

- Michaelis constant at [X] = 0, using (2.50) and (2.48)

$$K_{\mathrm{m}}^{0} = \left(\frac{v_{(\infty,0)}}{v_{(\mathrm{S},0)}} - 1\right)[\mathrm{S}] = \frac{k_{-1} + k_2}{k_1} \tag{2.56}$$

- Michaelis constant at finite [X]. This is a complicated expression deduced from (2.10) as the substrate concentration for which $v = 0.5 \times V^{\mathrm{app}}$. The operation yields two roots of which the following has physical meaning:

$$K_{\mathrm{m}}^{\mathrm{app,SS}} = \frac{1}{2}\,\frac{-2b_1g_1 + a_1h_1 + \sqrt{4b_1^2g_1^2 - 4a_1b_1g_1h_1 + a_1^2h_1^2 + 4a_1^2g_1m_1}}{a_1g_1} \tag{2.57}$$

The superscript SS has been added to this expression to emphasize its derivation under steady-state conditions without any simplifying assumption. The full equation showing all explicit constants of (2.57), listed in (2.45), (2.46), occupies a half page and it makes no sense showing it here. These constants, listed in (2.45) and (2.46), can be used for numerical calculations with the help of mathematical software.

- Michaelis constant at saturating [X], using (2.49) and (2.52)

$$K_{\mathrm{m}}^{\infty} = \left(\frac{v_{(\infty,\infty)}}{v_{(\mathrm{S},\infty)}} - 1\right)[\mathrm{S}] = \frac{k_{-5} + k_6}{k_5} \tag{2.58}$$

- Specificity constant at [X] = 0, using (2.53) and (2.56)

$$k^{0} = \frac{k_{\mathrm{cat}}^{0}}{K_{\mathrm{m}}^{0}} = \frac{k_1k_2}{k_{-1} + k_2} \tag{2.59}$$

- Specificity constant at finite [X], using (2.54) and (2.57)

$$k^{app} = \frac{\left(k_2 + k_6 \frac{k_4}{k_{-4}}[X]\right) \Big/ \left(1 + \frac{k_4}{k_{-4}}[X]\right)}{\frac{1}{2} \frac{-2b_1 g_1 + a_1 h_1 + \sqrt{4b_1^2 g_1^2 - 4a_1 b_1 g_1 h_1 + a_1^2 h_1^2 + 4a_1^2 g_1 m_1}}{a_1 g_1}} \quad (2.60)$$

This overly complex equation can be used for simulation purposes but cannot be exploited in practical cases unless simplifying assumptions are made (see below).
- Specificity constant at saturating [X], using (2.55) and (2.58)

$$k^{\infty} = \frac{k_{cat}^{\infty}}{K_m^{\infty}} = \frac{k_5 k_6}{k_{-5} + k_6} . \quad (2.61)$$

k_{cat}^{app} depends on [X] according to (2.54) and varies between the limits k_2 and k_6 at [X] = 0 and saturating [X], respectively. From this expression, identical under the steady-state and quasi-equilibrium assumptions, k_2, k_6 and K_4 can be calculated as will be illustrated with examples in Figs. 2.4, 2.5, 2.6, 2.7, and 2.8. However, in computing the specificity constant, the full expression (2.57) of the apparent Michaelis constant must be introduced rendering the resulting equation useless in practice.

Helpfully, Enrico Di Cera and coworkers [13] suggested a much simpler form of k^{app} than (2.60), which has proven to describe experimental observations in practical cases in which one of the approximations discussed in Sects. 2.4.1 and 2.4.2 applies thus annihilating quadratic terms. Considering that the function described by (2.60) spans between the limits k^0 and k^{∞} given by (2.59) and (2.61), respectively, the following equation has been proposed:

$$k^{app} = \frac{k^0 - k^{\infty} \frac{[X]}{K_3}}{1 + \frac{[X]}{K_3}} . \quad (2.62)$$

2.6 Verification of the Quasi-Equilibrium Assumption

How can one prove the validity of quasi-equilibrium assumptions? In the general modifier mechanism there are *two* quasi-equilibrium assumptions that can be considered: one regarding the catalytic paths and the other regarding the binding of modifier. Figure 2.2 summarizes the simplifications based on the assumptions discussed in Sect. 2.4.

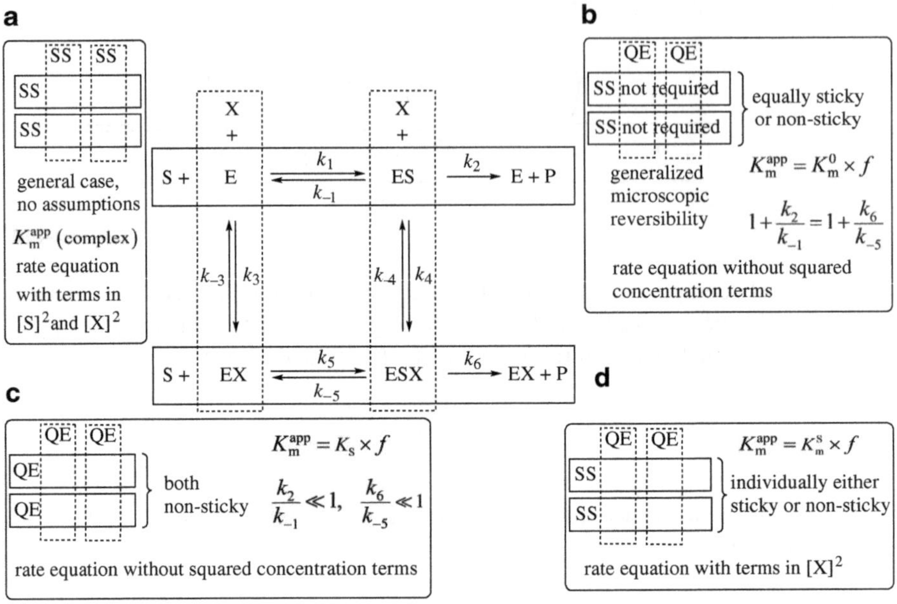

Fig. 2.2 Overview of simplifying assumptions of the general modifier mechanism. The small schemes in the boxes refer to the mechanism in the middle of the figure and show the status of the indicated paths: SS (steady-state) or QE (quasi-equilibrium). In the K_m^{app} equations, $f = (K_3 + [X]) / (K_3 + [X]/\alpha)$. (**a**) No assumptions made. (**b**) Assumption of generalized microscopic reversibility (2.20). (**c**) Assumption (2.13), all enzyme-forms in quasi-equilibrium. (**d**) Assumption of quasi-equilibrium for modifier binding (Sect. 2.4.3)

Equation (2.38) can be written in an alternative way considering the definition of the coupling constant in (2.3) to give

$$K_m^{app} = K_m^S \times \frac{K_3 + [X]}{K_3 + \dfrac{[X]}{\alpha}} \tag{2.63}$$

if the assumption in Fig. 2.2d is made. Under the assumption $k_2/k_{-1} \ll 1$ and $k_6/k_{-5} \ll 1$ summarized in Fig. 2.2c, (2.63) becomes

$$K_m^{app} = K_S \times \frac{K_3 + [X]}{K_3 + \dfrac{[X]}{\alpha}}, \tag{2.64}$$

and when the assumption $1 + k_2/k_{-1} = 1 + k_6/k_{-5}$ applies (Fig. 2.2b), Eq. (2.63) becomes

$$K_m^{app} = K_m^0 \times \frac{K_3 + [X]}{K_3 + \dfrac{[X]}{\alpha}}. \tag{2.65}$$

The algebra to arrive at (2.64) and (2.65) from the less restrictive of the three simplifying assumptions, i.e. (2.63), is demonstrated in Appendix 3.

Equations (2.64) and (2.65) may give the impression to be useful for nonlinear regression fitting. In practice, however, an unambiguous determination of K_3, α and the Michaelis constant (as either K_S or K_m^0 depending on the simplification that applies, if any), depends on the degree by which the parameters are intertwined, i.e., on how much each parameter depends from the others. This fact becomes evident after examining the covariance matrix following nonlinear regression analysis and by performing the test for mutual dependency of the parameters. These statistical tests are part of software packages for graphical and statistical analysis, e.g., GraphPad Prism. With this program, besides performing analysis of the residuals and the runs test, it is highly recommended to ascertain whether the parameters are intertwined, redundant or skewed (Hougaard's test for skewness). The simulations that will be shown in Sect. 2.7 suggest that α, K_m and K_3 are intertwined, with K_3 being the constant that can be determined with less precision using (2.65). Nevertheless, the fitting procedure provides the limiting values of K_m^{app} at $[X] = 0$ and $[X] = \infty$, i.e., K_m^0 and K_m^∞, respectively, which do not depend on whether the system operates at steady-state or at equilibrium.

Considering (2.65), the limit of K_m^{app} for $[X] \to \infty$ is given by

$$K_m^\infty = \lim_{[X]\to\infty} K_m^{app} = \alpha K_m^0 \quad \therefore \quad \frac{K_m^\infty}{K_m^0} = \alpha \,. \tag{2.66}$$

The analytical expressions of K_m^0 and K_m^∞ are given by (2.56) and (2.58), respectively, from which their ratio corresponds to

$$\frac{K_m^\infty}{K_m^0} = \frac{k_{-5} + k_6}{k_5} \times \frac{k_1}{k_{-1} + k_2} \,, \tag{2.67}$$

and, as established by (2.7), this ratio is approximately equal to α if $k_6 \ll k_{-5}$ and at the same time $k_2 \ll k_{-1}$. If this is true, the Michaelis constants at $[X] = 0$ and $[X] = \infty$ approximate numerically K_S and K_5, respectively

$$\frac{K_m^\infty}{K_m^0} \approx \frac{k_{-5}k_1}{k_5 k_{-1}} \approx \frac{K_5}{K_S} \approx \alpha \,. \tag{2.68}$$

The best service of (2.65) is thus that of providing α as a measure of the ratio K_m^∞ / K_m^0 while not seriously considering K_3.

Based on (2.68), Symcox and Reinhart proposed a test for the "verification of the rapid-equilibrium assumption in allosteric enzymes" [35]. The same concept had been published by Frieden in 1964 [17, p. 3525] in the section *rapid equilibrium treatment*. The method relies on comparing the reciprocal allosteric coupling constant α, obtained as (2.68), with the ratio K_4/K_3. If these two values are similar within the limits of experimental error, the quasi-equilibrium assumption for all binding steps of the system is valid.

Symcox and Reinhart remarked that the method can fail if $k_2/k_{-1} = k_6/k_{-5}$, i.e., if the approximation in Fig. 2.2b applies [35]. This statement is incorrect because in this case the principle of generalized microscopic reversibility (2.26) must be invoked, from which we can extract the following information

$$\frac{k_{-5}+k_6}{k_5} \times \frac{k_1}{k_{-1}+k_2} = \frac{k_3 k_{-4}}{k_{-3} k_4} = \frac{K_4}{K_3} = \alpha \,, \tag{2.69}$$

which is equivalent to

$$\frac{K_{\mathrm{m}}^{\infty}}{K_{\mathrm{m}}^{0}} = \alpha \,, \tag{2.70}$$

thus providing even an *exact correspondence* between the ratio of limiting Michaelis constants and the ratio of the dissociation constants of the modifier binding steps. Without additional information, if this test passes in practical experiments, it is impossible to decide which one of the assumptions, (b) or (c) in Fig. 2.2, applies.

To be pedantic, a passing test, i.e., $K_{\mathrm{m}}^{\infty}/K_{\mathrm{m}}^{0} = K_4/K_3$, rather than demonstrating the validity of the quasi equilibrium assumption, it demonstrates the absence of quadratic concentration terms in the rate equation. In fact, assumption (b) in Fig. 2.2 is compatible with both the steady-state and the quasi-equilibrium assumption for the horizontal steps.

A practical comment on the precision of this test is necessary because α, as an estimator of $K_{\mathrm{m}}^{\infty}/K_{\mathrm{m}}^{0}$ (2.68), is obtained by fitting (2.65) to data and we should consider that K_{m} is the least precise of the Michaelis–Menten parameters. The statistical reasons for this important issue have been discussed in detail by Cornish-Bowden [11, pp. 436–438]. We should also consider the precision of K_3 and K_4 that are obviously determined from the same set of primary data from which K_{m}^{0} and K_{m}^{∞} are obtained. The impact of errors on these and other calculations are discussed with simulated data in Figs. 2.4, 2.5, 2.6, 2.7, and 2.8.

As last remark, the full expression of $K_{\mathrm{m}}^{\mathrm{app}}$, i.e., $K_{\mathrm{m}}^{\mathrm{app,\,SS}}$ in (2.57), predicts a nonhyperbolic dependence on [X] showing a maximum or a minimum in some cases of substrate stickiness described in Sects. 2.7 and 2.8.1. In such cases, (2.65) cannot be used.

2.7 The Practical Impact of Simplifications

When analyzing the kinetics of enzyme-catalyzed reactions in the presence of substrates, any cofactors and modifiers, a customary approach consists in choosing an alleged model and fitting it to data. If the underlying mechanism is complex and the rate equation possibly contains squared or higher order terms in reactant concentrations, simplifying assumptions are necessary for managing the calculation of kinetic parameters. Since theoretical considerations are important but examples

Table 2.1 Rate constants for the simulations in Figs. 2.4, 2.5, 2.6, 2.7, and 2.8. $[E]_t = 1.0$ nM in all five cases

	Units	Set 1	Set 2	Set 3	Set 4	Set 5
k_1	$\mu M^{-1}s^{-1}$	10	10	10	10	30
k_{-1}	s^{-1}	800	100	100	400	100
k_2	s^{-1}	50	120	150	30	300
k_3	$\mu M^{-1}s^{-1}$	15.714	18.514	20.571	2.0571	10
k_{-3}	s^{-1}	600	60	60	60	60
k_4	$\mu M^{-1}s^{-1}$	22	12	12	12	11
k_{-4}	s^{-1}	2,100	70	70	70	39.6
k_5	$\mu M^{-1}s^{-1}$	10	10	10	10	15
k_{-5}	s^{-1}	2,000	180	200	80	30
k_6	s^{-1}	10	12	60	90	90
α		2.50	1.80	2.00	0.20	0.60
β		0.20	0.10	0.40	3.00	0.30

The apparently exaggerated precision in displaying some constants aims at respecting microscopic reversibility without compromises

are understood with less effort, the numerical examples described below aim at evaluating the errors that can be made in applying the assumptions discussed in Sect. 2.4 to the general modifier mechanism.

Simulations will be performed using the complete steady-state equation (2.44) to mimic curves of rate against [S] making only the logical assumption that the modifier association–dissociation reactions are fast with respect to substrate binding and turnover. The rate constants used in the simulations, which feature five representative cases, are listed in Table 2.1. This is not equivalent to simulating data using (2.34), a simplifying assumption that does not take into account minor contributions of some terms that are otherwise present in the complete steady-state Eq. (2.44) as will be discussed in Sect. 2.8.2.

To avoid repetitions in the legends of Figs. 2.4, 2.5, 2.6, 2.7, and 2.8, the following itemized list (1a–d, 2e–h) is common to all five figures to explain the methods employed and the goal of the simulations. Lettering (a–h) in the list corresponds to lettering in the figures that are all drawn with the same layout. The analyses of simulated error-free data are represented in panels (a–d), while the analyses of error-affected data are shown in panels (e–h). Theoretical limiting values at [X] = 0 and [X] = ∞ of k_{cat}^{app}, K_m^{app} and k^{app} are shown in panels (b–d) as dashed arrows with the corresponding values. Calculated parameters are shown with redundant precision with the sole scope of demonstrating in detail the effects of the quasi-equilibrium assumption for selected reaction steps.

(1) Artificial, error-free data were generated using (2.44) and 200 substrate concentrations. Equation (2.35), which has the form of the Michaelis–Menten equation in the presence of the modifier, was fitted to primary fake data to extract estimates of k_{cat}^{app} and K_m^{app} for further analysis of their dependence on

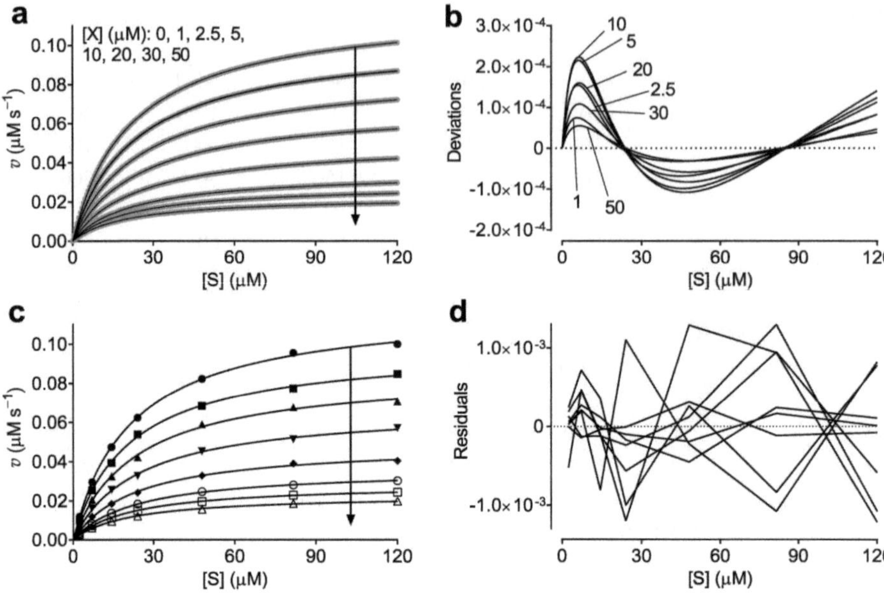

Fig. 2.3 Primary data as v against [S] simulated using the constants of set 2 in Table 2.1. (**a**) Error-free simulations with (2.44) and 200 [S]-values plotted as *gray dots* that form a continuum to which (2.35) was fitted to give best-fits represented by the *continuous black curves*. (**b**) Deviations of the fitted curves from theoretical data in (**a**). (**c**) Simulation with only seven substrate concentrations to mimic a real experiment in the presence of a multiplicative random error corresponding to a 2.5 % standard deviation of the simulated rates. The curves are best fits of (2.35) to data. (**d**) Residuals from the fit in (**c**) with more details in Fig. 2.5e. The seven modifier concentrations used for simulating data are shown in (**a**) and are the same in (**c**). The increasing direction of [X] is indicated by the *vertical arrows*

modifier concentration. To avoid figure overload, primary simulated data as rate against substrate concentration are shown only for example 2 in Fig. 2.3 assuring that the remaining examples behave in an analogous way. In Fig. 2.3a the deviations from theoretical values are not appreciable by eye suggesting perfect fit as black curves superimposed to theoretical data. Our satisfaction in front of this brilliant result is soon frustrated after reading the report of nonlinear regression analysis, which tells that deviation from the model is significant. This verdict is indisputable but the consolation comes after examining the deviations reported in Fig. 2.3b, from which we realize that these are yes systematic but very small. This is why the black lines in Fig. 2.3a remain within the gray traces representing theoretical values. The same statement is valid for all examples 1–5.

(a) The fitting procedure to primary data, as exemplified in Fig. 2.3a, aims at ascertaining the sensitivity of (2.35), the general form of the Michaelis–Menten equation, in *feeling* the presence of any squared concentration terms we know to be possibly present in the simulated data. These become manifest by calculating the differences between fitted curves and theoretical values and the purpose of using 200 substrate concentrations is to create a continuum of points to reveal deviations in the entire substrate concentration range. When performing data analysis in this way, without prior knowledge of the mechanism and thus of the expressions of k_{cat}^{app} and K_m^{app} in (2.35), the validity of any simplification is still unknown. Therefore, this procedure gives information on deviations[3] from the steady-state model (2.44) if simplifying assumptions are made. Deviations from the fitting procedure on primary data are shown in panels (a) of Figs. 2.4, 2.5, 2.6, 2.7, and 2.8. Since simulations were performed error-free, any systematic trends of the deviations indicate the presence of quadratic concentration terms in the rate equation, which are not *sensed* by (2.35). Numbers labeling the curves indicate [X] and $|D^{max}|$ represents the maximal absolute percent deviation relative to V^0: $\left(|\text{max deviation}|/V^0\right) \times 100$.

(b) Apparent catalytic constant obtained from the primary fit plotted against modifier concentration. The open squares and the solid curve through them represent theoretical values. Black circles represent values from the primary analysis to which (2.54) was fitted giving a best-fit curve practically superimposed to the theoretical one and the parameters shown explicitly.

(c) Apparent Michaelis constant: black circles are values from regression analysis of primary data and open squares represent theoretical data calculated with (2.57). The solid line is a curve plotted with the same equation. Only when fitting (2.65) to the black circles makes sense (examples 1, 4 and 5), the best-fit curve almost coincides with the theoretical curve and in such cases best-fit values of K_m^{∞}/K_m^0 and K_m^0 from this procedure are displayed explicitly.

(d) Apparent specificity constant plotted against modifier concentration. Also here, black circles and open squares represent values calculated by non-linear regression and theoretical values, respectively. The theoretical curve shown as continuous line is indistinguishable from the best-fit of (2.60) to the black circles. The parameters calculated from this procedure are shown explicitly.

(2) Simulations using (2.44), the complete rate equation, with seven substrate concentrations to mimic a real experiment (results in panels e–h of Figs. 2.4, 2.5, 2.6, 2.7, and 2.8). A multiplicative random error with 2.5 % standard deviation of calculated rates is added to mimic a real experiment using the *randn* function of MATLAB. A sample of the v against [S] plot with residuals is shown in Fig. 2.3c,d.

---

[3] Since deviations are calculated as difference between fitted curves and theoretical values, they cannot be called *residuals* but represent something like the *disturbance* (also called *statistical error* or *error*), which in statistics represents the deviation of an observed value from a theoretical (unobservable) value. They will be called here *deviations*.

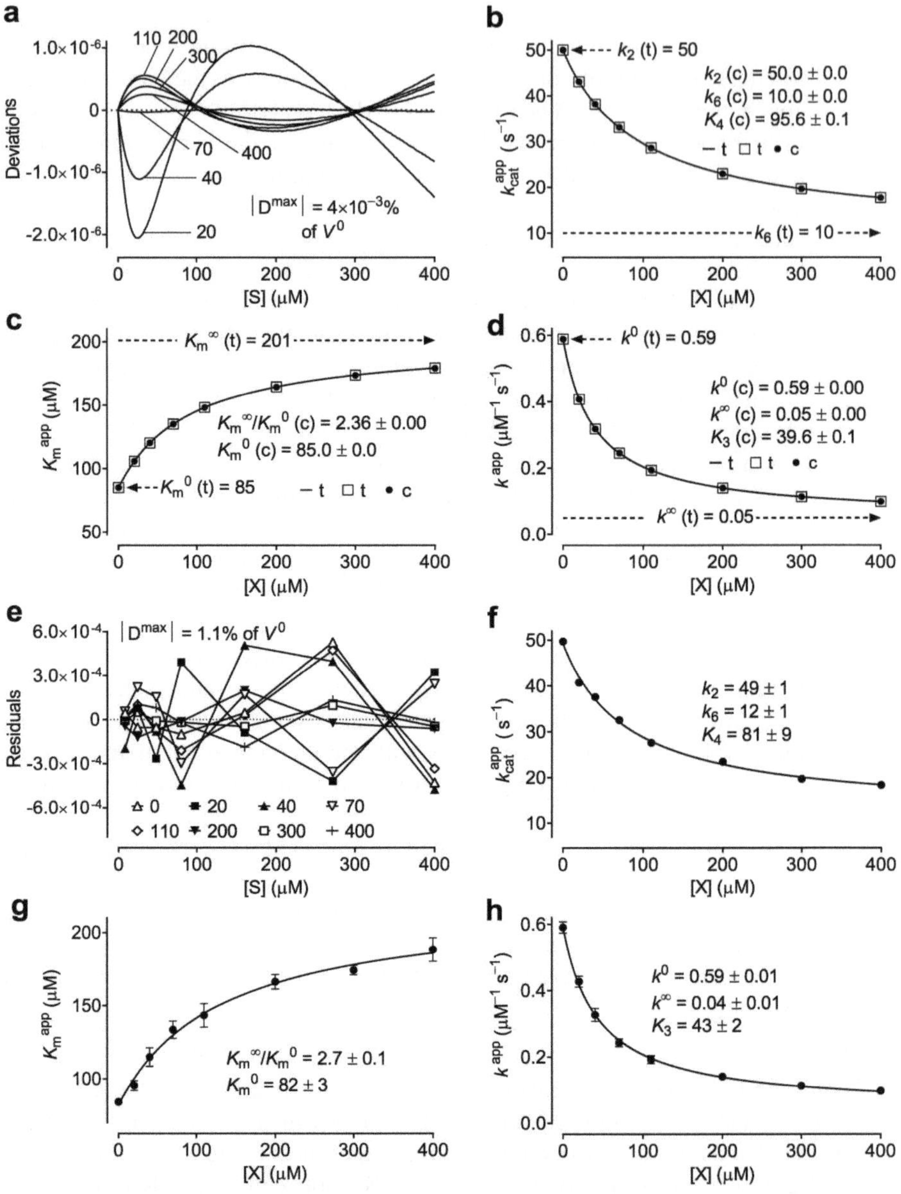

Fig. 2.4 Simulations with set 1 in Table 2.1 with model (2.44). (**a**)–(**d**) Error-free simulations (t = theoretical values as *continuous lines* and *open squares*) and parameters calculated from primary data by fitting (2.35) (*black circles*, c = calculated). (**e**)–(**h**) Simulations with the same [X]-values and constants as in (**a**)–(**d**), seven [S]-values and error added; best-fit curves together with parameters calculated by fitting the appropriate equations. Numbers in (**a**) and (**e**) denote [X] (μM). More details on pp. 87–89. Summary of results and units, omitted for reasons of space, in Table 2.2

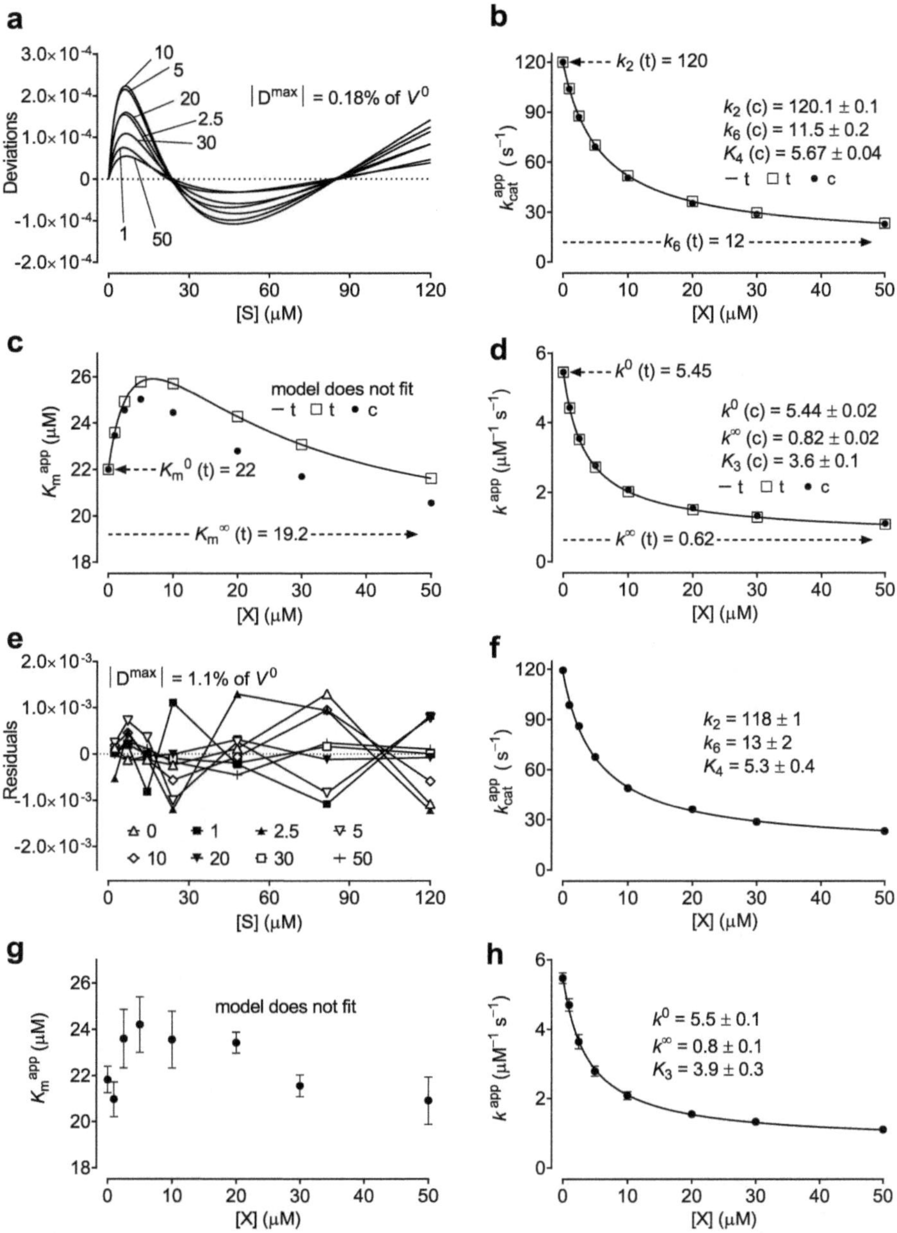

Fig. 2.5 Simulations with set 2 in Table 2.1 with model (2.44). (**a**)–(**d**) Error-free simulations (t = theoretical values as *continuous lines* and *open squares*) and parameters calculated from primary data by fitting (2.35) (*black circles*, c = calculated). (**e**)–(**h**) Simulations with the same [X]-values and constants as in (**a**)–(**d**), seven [S]-values and error added; best-fit curves together with parameters calculated by fitting the appropriate equations. Numbers in (**a**) and (**e**) denote [X] (μM). More details on pp. 87–89. Summary of results and units, omitted for reasons of space, in Table 2.3

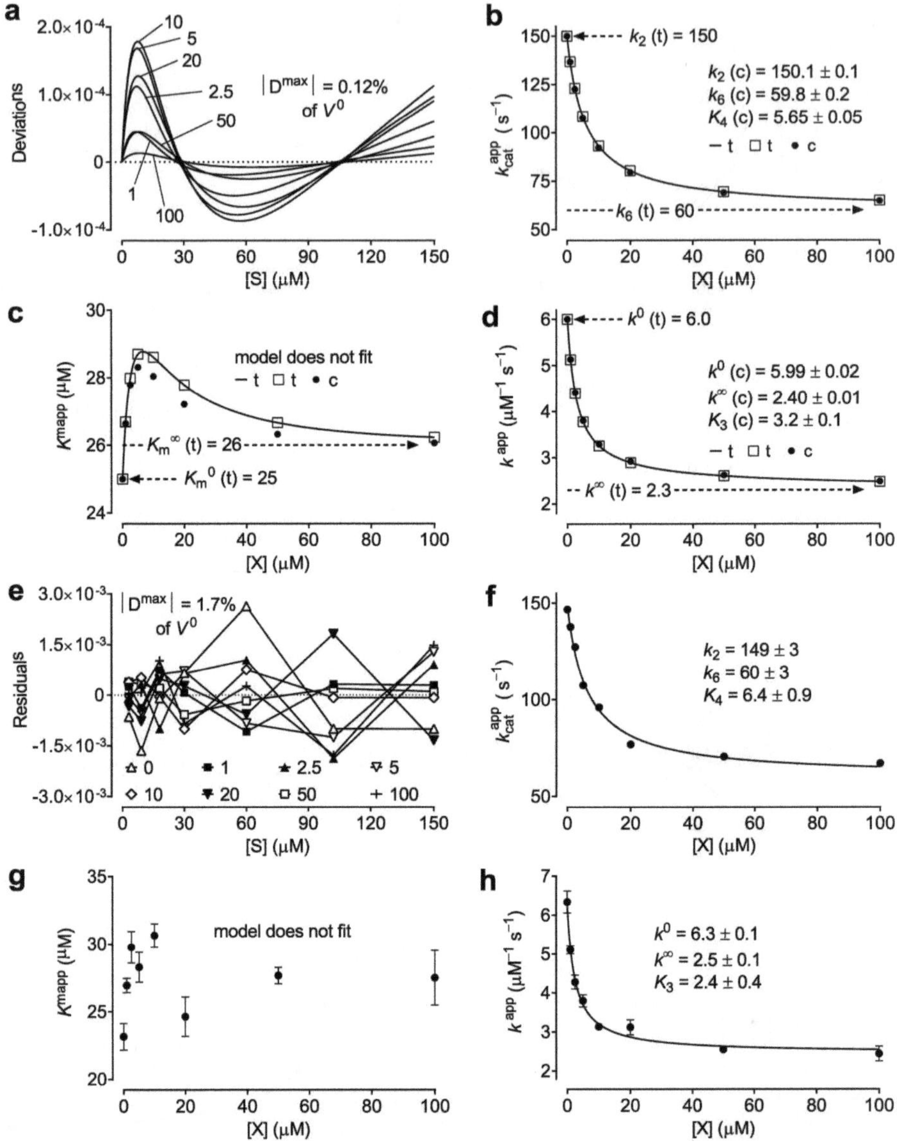

Fig. 2.6 Simulations with set 3 in Table 2.1 with model (2.44). (**a**)–(**d**) Error-free simulations (t = theoretical values as *continuous lines* and *open squares*) and parameters calculated from primary data by fitting (2.35) (*black circles*, c = calculated). (**e**)–(**h**) Simulations with the same [X]-values and constants as in (**a**)–(**d**), seven [S]-values and error added; best-fit curves together with parameters calculated by fitting the appropriate equations. Numbers in (**a**) and (**e**) denote [X] (μM). More details on pp. 87–89. Summary of results and units, omitted for reasons of space, in Table 2.4

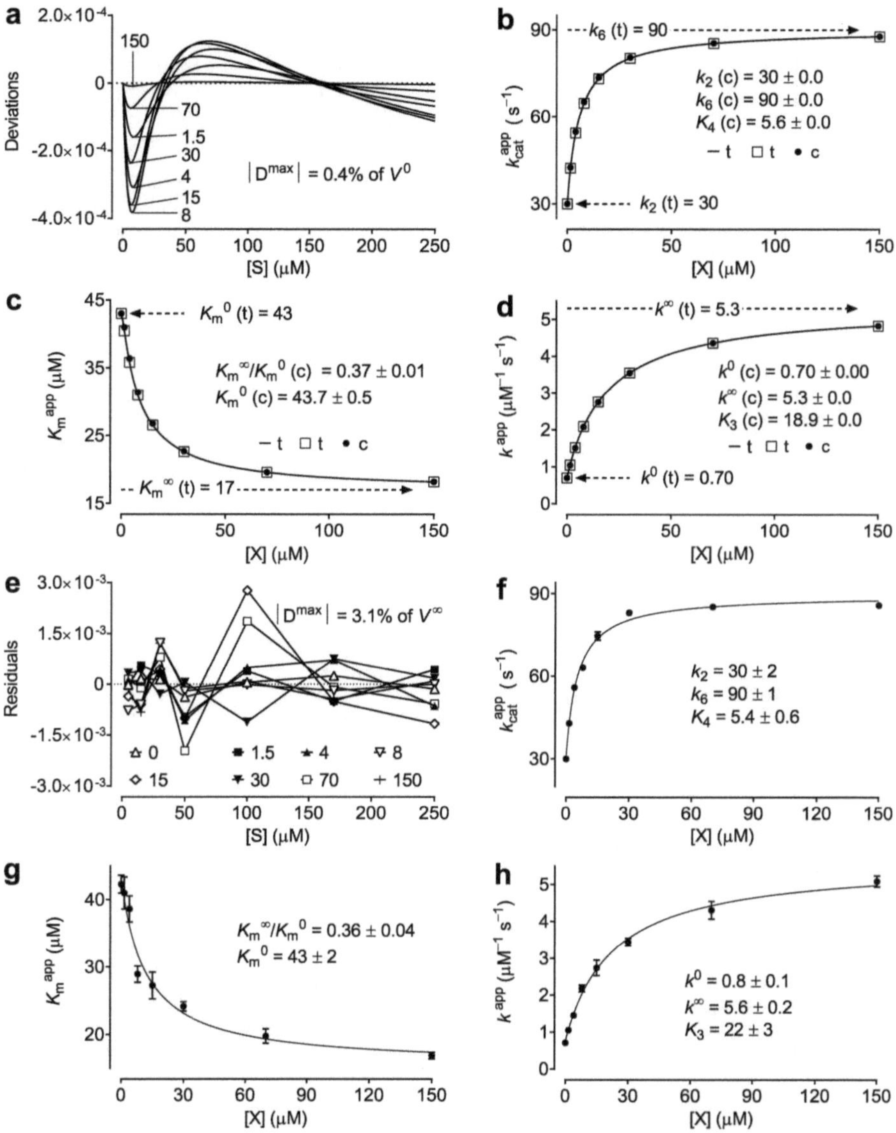

Fig. 2.7 Simulations with set 4 in Table 2.1 with model (2.44). (**a**)–(**d**) Error-free simulations (t = theoretical values as *continuous lines* and *open squares*) and parameters calculated from primary data by fitting (2.35) (*black circles*, c = calculated). (**e**)–(**h**) Simulations with the same [X]-values and constants as in (**a**)–(**d**), seven [S]-values and error added; best-fit curves together with parameters calculated by fitting the appropriate equations. Numbers in (**a**) and (**e**) denote [X] (μM). More details on pp. 87–89. Summary of results and units, omitted for reasons of space, in Table 2.5

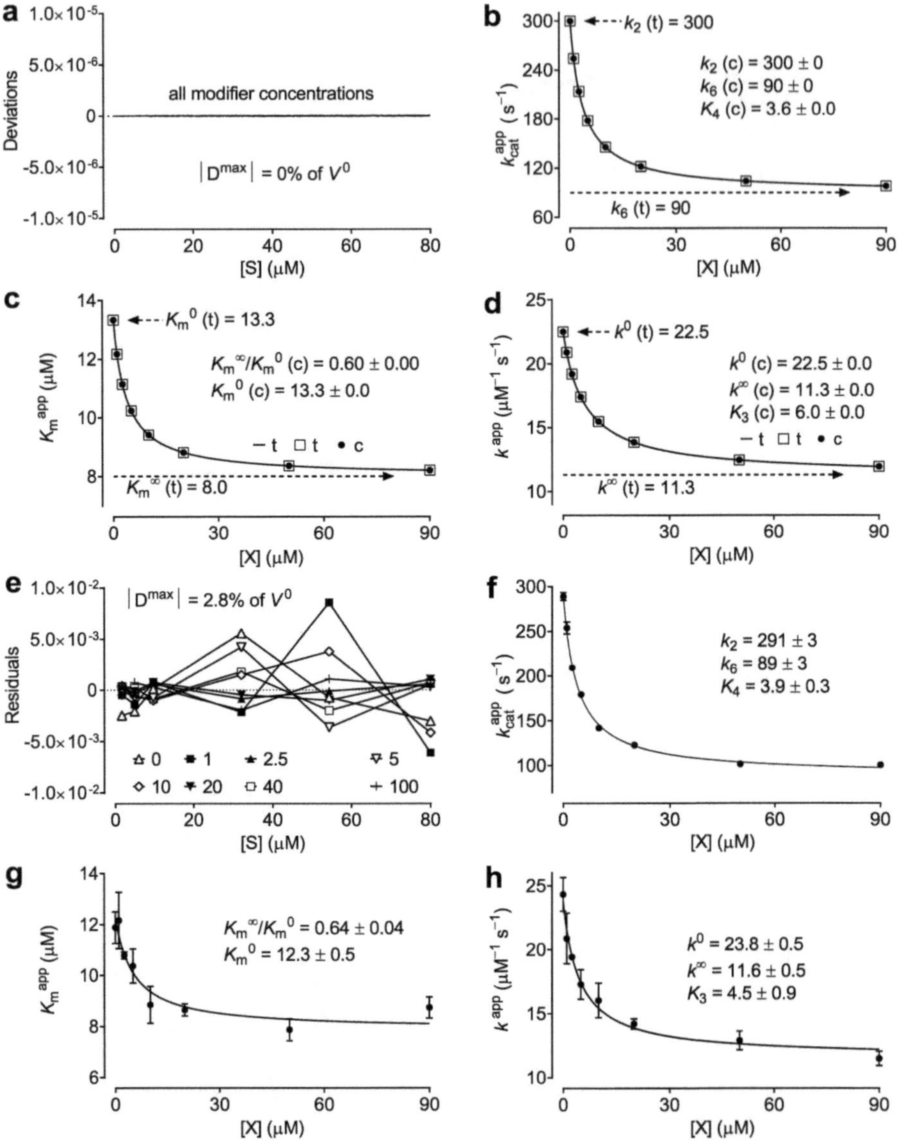

Fig. 2.8 Simulations with set 5 in Table 2.1 with model (2.44). (**a**)–(**d**) Error-free simulations (t = theoretical values as *continuous lines* and *open squares*) and parameters calculated from primary data by fitting (2.35) (*black circles*, c = calculated). (**e**)–(**h**) Simulations with the same [X]-values and constants as in (**a**)–(**d**), seven [S]-values and error added; best-fit curves together with parameters calculated by fitting the appropriate equations. Numbers in (**a**) and (**e**) denote [X] (μM). More details on pp. 87–89. Summary of results and units, omitted for reasons of space, in Table 2.6

(e) Residuals from primary data analysis for comparison with the deviations in panel (a) to assess the impact of systematic deviations from the model and experimental error.
(f) Equation (2.54) fitted to the k_{cat}^{app} dependence on [X] to obtain k_2, k_6 and K_4.
(g) Equation (2.65) fitted to the K_m^{app} dependence on [X] to calculate the ratio K_m^{∞}/K_m^0, i.e., α in (2.65) and K_m^0.
(h) Equation (2.62) fitted to the dependence of the apparent specificity constant on [X] to calculate k^0, k^{∞} and K_3.

2.7.1 *Example 1*

Results from the simulation of error-free data using the complete rate equation (2.44) and the constants in Table 2.1, set 1, are shown in Fig. 2.4a–d. This example of the rapid equilibrium assumption $k_2/k_{-1} \ll 1$, $k_6/k_{-5} \ll 1$, illustrates the behavior of a hyperbolic mixed inhibitor with predominantly specific character that will be discussed in Sect. 5.3.2. Fitting the general form of the Michaelis–Menten equation (2.35) to simulated data to calculate V^{app} and K_m^{app} shows systematic deviations as seen in panel (a). The fit is perfect for [X] = 0 (curve coincident with the dotted line) and almost so for [X] = 70 μM. When [X] is increased toward saturation the deviations become smaller. The deviations are due to the presence of $[X]^2$ terms in the rate equation, as we know from the purposely chosen rate constants, but these are not *seen* by (2.35). Yet, the deviations are very small, with the largest absolute value being only 0.004 % of V^0. These small deviations illustrate numerically the meaning of quasi-equilibrium versus equilibrium, which had produced zero deviations.

In (b)–(d), k_{cat}^{app}, K_m^{app}, and k^{app}, respectively, are plotted as a function of [X]. The agreement between calculated and theoretical values of the parameters is satisfactory as can be judged visually from the overlapping filled circles and open squares. The calculated limiting values of these parameters can be compared with their theoretical counterparts indicated by dashed arrows.

The same kinetic constants of the simulations in panels (a–d) were used to produce fake data with seven substrate concentrations and a multiplicative random error added to velocities to mimic a real experiment. The residuals of this simulation, which include both systematic deviations for the presence of quadratic concentrations in the rate equation and random error (panel e), are larger than the deviations from the model in panel (a). Thus, a modest experimental error is sufficient to overwhelm and mask systematic deviations from the model.

At the theoretical level, the proof that applying the quasi-equilibrium assumption for performing calculations is reasonable for this example can be gathered by comparing the results in the column "error-free" with those in the "theoretical values" column of Table 2.2. At the practical level, i.e., when only results as those in the column "error-affected" are available, the ratio K_m^{∞}/K_m^0 as an estimator of $K_4/K_3 = \alpha$ is inconclusive because we cannot state that the two values are similar.

Table 2.2 Results from simulated data with the rate constants of set 1 in Table 2.1

Constants	Units	Theoretical values	Calculated from error-free simulation	Calculated from error-affected simulation	Equations fitted
$1 + k_2/k_{-1}$[a]	–	1.062			
$1 + k_6/k_{-5}$[a]	–	1.005			
α	–	2.50			
k_2	s^{-1}	50.0	50.0 ± 0.0	49 ± 1	
k_6	s^{-1}	10.0	10.0 ± 0.0	12 ± 1	(2.54)
K_4	μM	95.5	95.6 ± 0.1	81 ± 9	
K_m^{∞}/K_m^0[b]	–	2.36	2.36 ± 0.00	2.7 ± 0.1	
K_m^0	μM	85.0	85.0 ± 0.0	82 ± 3	(2.65)
k^0	$\mu M^{-1} s^{-1}$	0.59	0.59 ± 0.00	0.59 ± 0.01	
k^{∞}	$\mu M^{-1} s^{-1}$	0.05	0.05 ± 0.00	0.04 ± 0.01	(2.62)
K_3	μM	38.2	39.6 ± 0.1	43 ± 2	
K_4/K_3	–	2.50	2.41 ± 0.01[c]	1.88 ± 0.23[c]	
$k^0/10^7$[d]	–	0.059[e]	0.059	0.059	
$k^{\infty}/10^7$[d]	–	0.005[f]	0.005	0.004	

[a]Cleland's criterion of substrate stickiness (p. 73)
[b]Estimator of K_4/K_3 (Sect. 2.6)
[c]From best-fit values with (2.54) and (2.62)
[d]Brocklehurst's criterion for the validity of the quasi-equilibrium assumption (p. 39)
[e]In this case the same as k^0/k_1
[f]The same as k^{∞}/k_5 used in the simulation

The failure of this test is essentially due to experimental error in determining K_3 and K_4, which is neither a surprise nor a novelty.

By repeating simulations a number of times, equivalent to performing more experiments in a real case, these values change randomly at every run but the test remains inconclusive in any case. However, Brocklehurst's criterion shown in the last two rows of Table 2.2 reveals that the specificity constants in the absence of and at saturating modifier are much smaller than hypothetical second-order rate constants for substrate binding as large as $10^7\,M^{-1}s^{-1}$. If we trust this criterion, we conclude that the substrate is nonsticky in both catalytic paths and that the kinetic parameters obtained with the error-affected simulation can be trusted too.

2.7.2 *Example 2*

This is again an example of hyperbolic mixed, predominantly specific inhibition. Simulations were performed with the constants in Table 2.1 (set 2) aimed at analyzing the effects of the quasi-equilibrium assumption when $k_6/k_{-5} \ll 1$ but not so k_2/k_{-1}. Error-free fake data are shown in Fig. 2.5a–d and error-affected data are plotted in panels (e–h). The salient difference with example 1 is the dependence

Table 2.3 Results from simulated data with the rate constants of set 2 in Table 2.1

Constants	Units	Theoretical values	Calculated from error-free simulation	Calculated from error-affected simulation	Equations fitted
$1 + k_2/k_{-1}$ [a]	–	2.20			
$1 + k_6/k_{-5}$ [a]	–	1.07			
α	–	1.80			
k_2	s^{-1}	120.0	120.1 ± 0.1	118 ± 1	
k_6	s^{-1}	12.0	11.5 ± 0.2	13 ± 2	(2.54)
K_4	µM	5.83	5.67 ± 0.04	5.3 ± 0.4	
K_m^{∞}/K_m^0 [b]	–	0.87	No fit	No fit	
K_m^0	µM	22.0	No fit	No fit	(2.65)
k^0	$\mu M^{-1}s^{-1}$	5.45	5.44 ± 0.02	5.5 ± 0.1	
k^{∞}	$\mu M^{-1}s^{-1}$	0.62	0.82 ± 0.02	0.8 ± 0.1	(2.62)
K_3	µM	3.24	3.6 ± 0.1	3.9 ± 0.3	
K_4/K_3	–	1.80	1.6 ± 0.1 [c]	1.4 ± 0.2 [c]	
$k^0/10^7$ [d]	–	0.55 [e]	0.54	0.55	
$k^{\infty}/10^7$ [d]	–	0.06 [f]	0.08	0.08	

[a] Cleland's criterion of substrate stickiness (p. 73)
[b] Estimator of K_4/K_3 (Sect. 2.6)
[c] From best-fit values with (2.54) and (2.62)
[d] Brocklehurst's criterion for the validity of the quasi-equilibrium assumption (p. 39)
[e] In this case the same as k^0/k_1
[f] The same as k^{∞}/k_5 used in the simulation

of K_m^{app} on [X], which exhibits a maximum. While this behavior is a regular one and is predicted from theory (Sect. 2.8.1), fitting (2.65) to data is impossible. Thus, the quasi-equilibrium assumption is not fully legitimate in this case.

To support this assertion, Brocklehurst's criterion in the last two rows of Table 2.3 suggests that the specificity constant in the absence of modifier may not be much smaller than $10^7\,M^{-1}s^{-1}$ but that this is reasonably the case at saturating [X] and we know that this statement is true because simulation was performed with second-order rate constants $k_1 = k_5 = 10^7\,M^{-1}s^{-1}$. The conclusion is that the moderate substrate stickiness in the unmodified reaction path determines the K_m^{app}-behavior in Fig. 2.5c,g for the presence of squared concentration terms in its expression.

The residuals in Fig. 2.5e are larger by an order of magnitude than the deviations in (a) and exhibit a random distribution that masks the smaller, systematic deviations from the model. Comparing the theoretical parameters with those calculated from the error-free simulation allows judging deviations from the model, while the same comparison with the error-affected simulation gives information on the cumulated effects of systematic deviations and experimental error. Applying the quasi-equilibrium assumption to these data has a modest underestimating effect on k_6 and K_4, while k^{∞} and K_3 are slightly overestimated. Experimental error does not alter the values of the parameters in a substantial way and the mechanism of inhibition can be unmistakably diagnosed because from the error-affected simulation we can deduce $K_4/K_3 = \alpha = 1.4 \pm 0.2$ and $k_6/k_2 = \beta = 0.11 \pm 0.02$, a combination of characters that is typical of hyperbolic mixed, predominantly specific inhibition (Sect. 5.3.2).

2.7.3 *Example 3*

This is again an example of hyperbolic mixed, predominantly specific inhibition but now the substrate is sticky in both catalytic steps. The rate constants (set 3 in Table 2.1) do not fulfill the conditions for the validity of the quasi-equilibrium assumption $k_2/k_{-1} \ll 1$ and $k_6/k_{-5} \ll 1$. However, Fig. 2.6a shows that systematic deviations from the model are small. The dependence of K_m^{app} on [X], as was the case in example 2, first increases with [X], it reaches a maximum and then declines to the asymptote K_m^{∞} at saturating [X]. Again, this behavior is due to an important contribution of $[X]^2$-containing terms in the expression of K_m^{app} indicating that (2.63) is more appropriate than (2.65) to describe these data but, unfortunately, (2.63) cannot be used for analyzing the dependence of K_m^{app} on modifier concentration because of too many constants in the expression of K_m^S.

The overall quality test for the analysis of this case is summarized in Table 2.4, where the deviations from the model by making the quasi-equilibrium assumption are insufficient to impair the quality of the analysis as can be judged after comparing results in the *error-free* column with the *theoretical* column. The same conclusion is reached after examining the simulation in the presence of error (column *error-affected*), whose results are not dramatically different from the respective theoretical values. This welcome and beneficial outcome must be sought in the stickiness of substrate that is not polarized toward only one of the catalytic steps but is instead present in both steps. As will be evident from the fifth example of this series and the discussion that will follow (Sect. 2.7.5), independently of the magnitude of substrate stickiness, the effects exerted on rate constant and K_m^{app} by the presence of $[X]^2$-containing terms become progressively smaller the closer are the ratios k_2/k_{-1} and k_6/k_{-5}.

2.7.4 *Example 4*

Figure 2.7, simulated with the constants listed in Table 2.1 (set 4), features an example of hyperbolic mixed, predominantly specific activation (Sect. 5.4.3). Regarding substrate stickiness, in this example the roles of the two catalytic paths are inverted with respect to example 2 discussed in Sect. 2.7.2 because, in the absence of modifier, the substrate is nonsticky but it becomes sticky when the modifier is present. In fact, we are now dealing with an activator that triplicates the catalytic constant.

Table 2.4 Results from simulated data with the rate constants of set 3 in Table 2.1

Constants	Units	Theoretical values	Calculated from error-free simulation	Calculated from error-affected simulation	Equations fitted
$1+k_2/k_{-1}$ [a]	–	2.5			
$1+k_6/k_{-5}$ [a]	–	1.3			
α	–	2.00			
k_2	s^{-1}	150.0	150.1 ± 0.1	149 ± 3	
k_6	s^{-1}	60.0	59.8 ± 0.2	60 ± 3	(2.54)
K_4	μM	5.83	5.65 ± 0.05	6.4 ± 0.9	
K_m^{∞}/K_m^0 [b]	–	1.04	No fit	No fit	
K_m^0	μM	25.0	No fit	No fit	(2.65)
k^0	$\mu M^{-1}s^{-1}$	6.00	5.99 ± 0.02	6.3 ± 0.1	
k^{∞}	$\mu M^{-1}s^{-1}$	2.3	2.40 ± 0.01	2.5 ± 0.1	(2.62)
K_3	μM	2.92	3.2 ± 0.1	2.4 ± 0.4	
K_4/K_3	–	2.00	1.8 ± 0.1[c]	2.7 ± 0.6[c]	
$k^0/10^7$ [d]	–	0.60[e]	0.60	0.63	
$k^{\infty}/10^7$ [d]	–	0.23[f]	0.24	0.25	

[a]Cleland's criterion of substrate stickiness (p. 73)
[b]Estimator of K_4/K_3 (Sect. 2.6)
[c]From best-fit values with (2.54) and (2.62)
[d]Brocklehurst's criterion for the validity of the quasi-equilibrium assumption (p. 39)
[e]In this case the same as k^0/k_1
[f]The same as k^{∞}/k_5 used in the simulation

At variance with the dependence of K_m^{app} on modifier concentration seen in example 2, in example 4 this dependence is hyperbolic, monotonically decreasing for increasing [X] and fitting (2.65) to data is thus possible. The reciprocal allosteric coupling constant α calculated by fitting this equation to data overestimates the theoretical value of 0.2 obtained as the K_4/K_3 ratio. This result, together with Brocklehurst's criterion $k^{\infty}/10^7$ in the last row of Table 2.5 suggests that the quasi-equilibrium assumption is not fully legitimate. Nevertheless, the summary of results in Table 2.5 shows that all parameters can be calculated with satisfactory precision with the exception of K_3 and K_4.

For this example that deals with an activator, the limiting rate at saturating modifier concentration, V^{∞}, has been considered as reference for calculating the percent maximal deviations in panels (a) and (e) in place of V^0. The systematic deviations from the model, Fig. 2.7a, which are due to the application of the quasi-equilibrium assumption, are one order of magnitude smaller than the randomly distributed residuals of the simulation in the presence of error (panel e). Yet, although masked in panel (e), the systematic deviations in (a) are not so small to be neglected and in examples 1–4 they are indeed the largest of this type. In Table 2.5, comparing the calculated parameters from the error-free simulation with their corresponding theoretical values provides the error made in their determination

Table 2.5 Results from simulated data with the rate constants of set 4 in Table 2.1

Constants	Units	Theoretical values	Calculated from error-free simulation	Calculated from error-affected simulation	Equations fitted
$1 + k_2/k_{-1}$[a]	–	1.08			
$1 + k_6/k_{-5}$[a]	–	2.13			
α	–	0.20			
k_2	s^{-1}	30.0	30.0 ± 0.0	30 ± 2	
k_6	s^{-1}	90.0	90.0 ± 0.0	90 ± 1	(2.54)
K_4	μM	5.83	5.6 ± 0.0	5.4 ± 0.6	
K_m^{∞}/K_m^0[b]	–	0.40	0.37 ± 0.01	0.36 ± 0.04	
K_m^0	μM	43.0	43.7 ± 0.5	43 ± 2	(2.65)
k^0	$\mu M^{-1}s^{-1}$	0.7	0.7 ± 0.0	0.8 ± 0.1	
k^{∞}	$\mu M^{-1}s^{-1}$	5.3	5.3 ± 0.0	5.6 ± 0.2	(2.62)
K_3	μM	29.2	18.9 ± 0.0	22 ± 3	
K_4/K_3	–	0.20	0.30 ± 0.00[c]	0.25 ± 0.04[c]	
$k^0/10^7$[d]	–	0.07[e]	0.07	0.08	
$k^{\infty}/10^7$[d]	–	0.53[f]	0.53	0.56	

[a]Cleland's criterion of substrate stickiness (p. 73)
[b]Estimator of K_4/K_3 (Sect. 2.6)
[c]From best-fit values with (2.54) and (2.62)
[d]Brocklehurst's criterion for the validity of the quasi-equilibrium assumption (p. 39) [e]In this case the same as k^0/k_1
[f]The same as k^{∞}/k_5 used in the simulation

due to deviation from the fitted model: K_4 is slightly underestimated, while K_3 is patently underestimated.

Tracking the residuals in Fig. 2.7e for the individual modifier concentrations does not give the impression that these may be related by a systematic trend, as in panel (a) but that they are rather due to random error. The results in the *error-affected* column of Table 2.5 show that the most affected parameter is again K_3. As will be discussed in Sect. 2.8.2, it remains to be established if this sort of discrepancy with the real values of the parameters may have a major practical bearing.

2.7.5 Example 5

The impact of simplifications on the correspondence of calculated parameters with their true values after applying the assumptions summarized in boxes (c) and (d) of Fig. 2.2 have been discussed in examples 1–4. The fifth and last example discussed here deals with a case in which generalized microscopic reversibility is obeyed, i.e., the assumption summarized in Fig. 2.2b. The condition $1 + k_2/k_{-1} = 1 + k_6/k_{-5}$ does not impose restrictions on the magnitude of the individual rate constants, which can take any value provided the equality holds. If this is the case, independently

Table 2.6 Results from simulated data with the rate constants of set 5 in Table 2.1

Constants	Units	Theoretical values	Calculated from error-free simulation	Calculated from error-affected simulation	Equations fitted
$1 + k_2/k_{-1}$ [a]	–	4.0			
$1 + k_6/k_{-5}$ [a]	–	4.0			
α	–	0.60			
k_2	s^{-1}	300	300 ± 0	291 ± 3	
k_6	s^{-1}	90	90 ± 0	89 ± 3	(2.54)
K_4	μM	3.6	3.6 ± 0.0	3.9 ± 0.3	
K_m^{∞}/K_m^0 [b]	–	0.6	0.6 ± 0.0	0.64 ± 0.04	
K_m^0	μM	13.3	13.3 ± 0.0	12.3 ± 0.5	(2.65)
k^0	$\mu M^{-1} s^{-1}$	22.5	22.5 ± 0.0	23.8 ± 0.5	
k^{∞}	$\mu M^{-1} s^{-1}$	11.3	11.3 ± 0.0	11.6 ± 0.5	(2.62)
K_3	μM	6.0	6.0 ± 0.0	4.5 ± 0.9	
K_4/K_3	–	0.6	0.6 ± 0.0[c]	0.9 ± 0.2[c]	
$k^0/10^7$ [d]	–	2.3 (0.75)[e]	2.3	2.4	
$k^{\infty}/10^7$ [d]	–	1.1 (0.75)[f]	1.1	1.2	

[a] Cleland's criterion of substrate stickiness (p. 73)
[b] Estimator of K_4/K_3 (Sect. 2.6)
[c] From best-fit values with (2.54) and (2.62)
[d] Brocklehurst's criterion for the validity of the quasi-equilibrium assumption (p. 39)
[e] k^0/k_1
[f] k^{∞}/k_5, the true values used in the simulation (set in parentheses)

on whether the substrate is sticky or nonsticky all complexes are no longer at quasi-equilibrium but *at equilibrium*, which is the fundamental consequence of the principle of generalized microscopic reversibility formulated by Whitehead [44]. This happens because squared concentrations completely disappear from the rate equation.

The constants used for simulating error-free and error-affected data of example 5 are listed in Table 2.1 as set 5, with graphics in Fig. 2.8 and results are summarized in Table 2.6. The first important result is shown in Fig. 2.8a that does not show deviations from the fitted model. Therefore, no wonder if the calculated parameters in Fig. 2.8b–d coincide with their theoretical values and have zero standard errors.

The comparison of error-free simulated data in Fig. 2.8a–d with error-affected data (e–h) allows to clearly distinguish the impact of experimental error, which in this case is not accompanied by overlapping deviations from the model, on the quality of the calculated parameters. If fact, the residuals in panel (e) are determined exclusively by the multiplicative error added to simulated data to mimic a real experiment. This allows to establish that K_4 is only slightly overestimated, while K_3 is more affected by experimental error.

Brocklehurst's criterion needs a comment in this particular case because the system operates at equilibrium if generalized microscopic reversibility is obeyed.

Thus, the tests ratios $k^0/10^7$ and $k^\infty/10^7$ can take values smaller or larger than one, with the system being at equilibrium in any case. Provided experimental errors are not very large, comparing $K_\mathrm{m}^\infty/K_\mathrm{m}^0$ with K_4/K_3 may be invoked as criterion for the validity of the quasi-equilibrium assumption. In fact, the two values are identical (in the absence of experimental error) only when generalized microscopic reversibility applies (Sect. 2.6).

2.8 Steady-State Versus Quasi-Equilibrium

In the general modifier mechanism, the origin of the differences between rates calculated under the steady-state or the quasi-equilibrium assumption for modifier-binding can be appreciated examining the expressions of the nodes (E1–E4) used in the derivation of the rate equation by the King–Altman method (Appendix 4). The same expressions are rewritten in (2.71) in the absence of product with rate constants grouped in a different way.

$$
\begin{aligned}
\mathrm{E1} &= (k_{-1}+k_2)\,k_{-3}k_{-4} + k_{-3}k_4\,(k_{-5}+k_6)\,[\mathrm{X}] \\
&\quad \underbrace{+k_{-3}\,(k_{-1}+k_2)\,(k_{-5}+k_6) + (k_{-1}+k_2)\,k_{-4}k_5\,[\mathrm{S}]} \\
\mathrm{E2} &= k_1k_{-3}k_{-4}\,[\mathrm{S}] + k_3k_{-4}k_5\,[\mathrm{S}]\,[\mathrm{X}] \\
&\quad \underbrace{+k_1k_{-3}\,(k_{-5}+k_6)\,[\mathrm{S}] + k_1k_{-4}k_5[\mathrm{S}]^2} \\
\mathrm{E3} &= k_1k_{-3}k_4\,[\mathrm{S}]\,[\mathrm{X}] + k_3k_4k_5\,[\mathrm{S}]\,[\mathrm{X}]^2 \\
&\quad \underbrace{+\,(k_{-1}+k_2)\,k_3k_5\,[\mathrm{S}]\,[\mathrm{X}] + k_1k_4k_5[\mathrm{S}]^2\,[\mathrm{X}]} \\
\mathrm{E4} &= (k_{-1}+k_2)\,k_3k_{-4}\,[\mathrm{X}] + k_3k_4\,(k_{-5}+k_6)\,[\mathrm{X}]^2 \\
&\quad \underbrace{+k_3\,(k_{-1}+k_2)\,(k_{-5}+k_6)\,[\mathrm{X}] + k_1k_4\,(k_{-5}+k_6)\,[\mathrm{S}]\,[\mathrm{X}]}
\end{aligned}
\qquad (2.71)
$$

Comparing (2.71) with the patterns in Scheme 2.5 and the corresponding directed patterns in Scheme 2.6 (Appendix 4), reveals that patterns (e) and (f) are absent if the quasi-equilibrium assumption for modifier-binding is made so that all terms underscored by braces in (2.71) are canceled. The full expressions E1–E4 apply thus to a steady-state regime, while the terms underscored by braces are absent if the modifier operates at quasi-equilibrium. Since product is absent, (2.102) in Appendix 4 simplifies to

$$v = k_2\,[\mathrm{ES}] + k_6\,[\mathrm{ESX}]\ , \qquad (2.72)$$

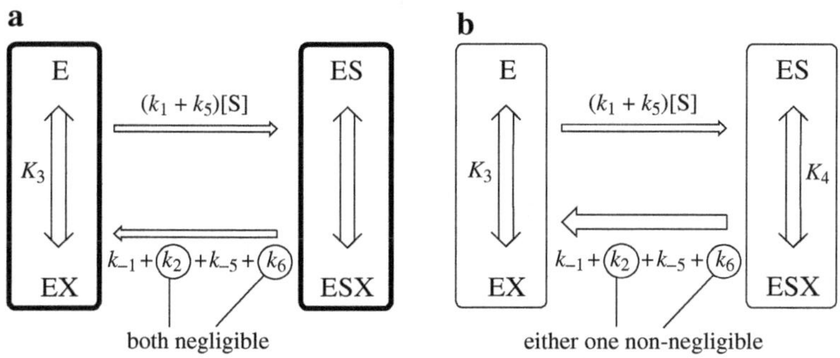

Fig. 2.9 Quasi-equilibrium assumption for modifier binding. (**a**) As long as *both* k_2 *and* k_6 are negligible with respect to $k_{-1}+k_{-5}$, the vertical paths remain isolated from and do not appreciably interfere with the horizontal paths as symbolized by the *thick boxes*. The rate equation contains terms in $[X]^2$, while terms in $[S]^2$ are very small. (**b**) If *either* k_2 *or* k_6 are not negligible, the vertical paths mix up and interfere with the horizontal paths being no longer well isolated from them (symbolized by the *thinner boxes*). The rate equation contains terms in $[X]^2$ but terms in $[S]^2$ become appreciable

and in the expressions of [ES] and [ESX], made up by the elements of (2.71), the underscored terms are absent at the same time in the numerator and in the denominator of the rate equation. This deletion does not mean that any rate constant is eliminated from the expressions, just combinations of rate constants are no longer present.

An important point that had not been anticipated in Sect. 2.4.3 (while discussing the quasi-equilibrium assumption for modifier binding) are the limits for its validity, in particular the role played by substrate stickiness. Excluding the case of generalized microscopic reversibility, in which all connections between enzyme complexes in the general modifier mechanism are at equilibrium, the underscored terms in (2.71) do not disappear *totally*: they can just become very small with respect to other terms as illustrated in Fig. 2.9. However, the quasi-equilibrium assumption under discussion settles the matter once and for all by considering all *negligible* terms equal to zero.

A comparison between the rates calculated using the steady-state assumption or the quasi-equilibrium approximation using the rate constants of example 2 in Table 2.1 is shown in Fig. 2.10. The continuous gray curves were plotted using the complete steady-state expression (2.44) and with the simpler quasi-equilibrium expression (2.35) for a modifier that is assumed to equilibrate rapidly (dashed lines). Equation (2.44) uses the full set of constants in expressions (2.71), while in (2.35) the underscored terms are completely omitted. There is no difference between the rates calculated with either equation in the absence of modifier or when this is present at a saturating concentration but sizable differences are present at finite modifier concentrations. The differences between the curves depend nonlinearly on [S] as seen in Fig. 2.10b suggesting that they originate from a small contribution of terms containing $[S]^2$ because, in this example, k_2 *is larger than* k_{-1} and thus

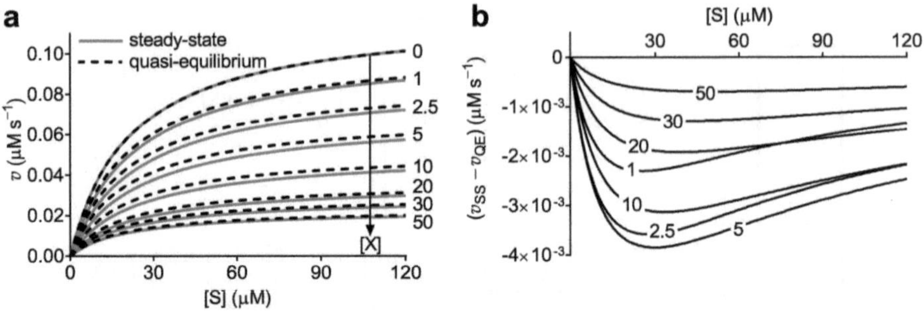

Fig. 2.10 Curves of v against [S] plotted using the constants of set 2 in Table 2.1. (**a**) Curves plotted with the complete steady-state rate equation (2.44) and 200 [S]-values shown as *gray dots* that form a continuum. Curves plotted with the quasi-equilibrium equation (2.35) (*black, dashed*). (**b**) Differences between the steady-state and quasi-equilibrium curves in panel (**a**). The numbers indicate seven modifier concentrations (μM) increasing as shown

not negligible in the sum $(k_{-1} + k_2 + k_{-5} + k_6)$. This is the situation described in Fig. 2.9b, in which the vertical paths are not sufficiently isolated from the events occurring in the horizontal paths and thus interfere with catalysis. Depending on the mechanism, the differences between curves such as those shown in Fig. 2.10 can be negative or positive.

It should be remarked that the differences between the steady-state and the quasi-equilibrium curves in Fig. 2.10 are roughly one order of magnitude larger than the deviations shown in Fig. 2.5a. The reason is that, while the plot in Fig. 2.10 for the quasi-equilibrium approximation was performed under full exclusion of the underscored terms in (2.71), fitting (2.35) to data simulated with steady-state conditions, produces curves nearing data as close as possible thanks to the flexibility in adjusting the parameters $k_{\mathrm{cat}}^{\mathrm{app}}$ and $K_{\mathrm{m}}^{\mathrm{app}}$ in (2.35). These considerations explain the rationale followed in designing examples 1–5 in Sect. 2.7.

Alleged macroscopic deviations arising from applying the quasi-equilibrium assumption to systems for which the steady-state approach would be required have been analyzed by Walter [41] using the original rate equation of Botts and Morales [4] and v against $v/[\mathrm{S}]$ plots. The results of the examples in this paper show largely overestimated deviations from linearity. Close examination of these examples reveals that calculations have been performed with sets of arbitrary rate constants that grossly violate the principle of microscopic reversibility, thus making useless predictions that passed however unobserved in accounts that cited this work [20, 47]. For instance, in [41, Fig. 2], the rate constants used to simulate the line labeled with open circles predict that, at equilibrium, the systems runs in clockwise direction three times faster than in the opposite direction, which is physically impossible.

2.8.1 *Properties of the Michaelis "Non-Constant"*

Depending on the rate constants in the general modifier mechanism, plotting $K_{\mathrm{m}}^{\mathrm{app}}$ against [X] results in variously shaped curves that can either increase or decrease for increasing [X] within the boundaries $K_{\mathrm{m}}^{0} = (k_{-1} + k_2)/k_1$ at [X] = 0 and $K_{\mathrm{m}}^{\infty} = (k_{-5} + k_6)/k_5$ for $[\mathrm{X}] \to \infty$. The increase or the decrease of $K_{\mathrm{m}}^{\mathrm{app}}$ with [X] can be monotonic as in Fig. 2.4c or nonmonotonic as in examples 2 and 3 illustrated in Figs. 2.5c and 2.6c, respectively, which show a maximum. Plots of $K_{\mathrm{m}}^{\mathrm{app}}$ against [X] can also exhibit a *minimum* in some hyperbolic activation mechanisms (see an example in Fig. 2.12 at the end of this section). The abscissa coordinate at which a maximum or a minimum occurs can be calculated by differentiating (2.57) with respect to [X], equating the result to zero and solving for [X]. Depending on the rate constants, the roots can be negative and thus maxima or minima occur in a physically meaningless quadrant, some roots can be complex numbers composed of a real and an imaginary part but results of practical interest are those in which there is one positive real root as in the example of Fig. 2.5c, in which the maximum occurs at [X] = 6.8 μM.

Any nonmonotonic behavior of $K_{\mathrm{m}}^{\mathrm{app}}$ against [X] is a sign that quasi-equilibrium assumptions do not apply properly. In this case the situation to be considered is that shown in Fig. 2.9b, where small values of either k_2 or k_6 that are not properly negligible do not allow the vertical steps to remain enough isolated in such a way that catalysis remains unaffected. As discussed above, in this situation the effect on the rate is modest because the same terms of elements of (2.71) appear in the numerator and in the denominator of the velocity expression. However, this is not the case in the expression of $K_{\mathrm{m}}^{\mathrm{app}}$ (2.57), where the effect on the numerator is not compensated by a similar effect on the denominator causing a nonmonotonic dependence of the parameter on [X].

This complex behavior can be deciphered with some patience from (2.57) with the coefficients in (2.46) , where $[\mathrm{X}]^2$-containing terms contribute differently to numerator and denominator. Figure 2.11 illustrates this concept graphically. The four dependences of $K_{\mathrm{m}}^{\mathrm{app}}$ on [X] in panels (a–d) were simulated with (2.57), i.e., all steps in steady-state represented by dashed lines, and with (2.37) assuming quasi-equilibrium behavior shown by continuous lines. The constants for the simulations, listed in Table 2.7, are those of example 2 in Sect. 2.7.2. Of these, only β was progressively increased determining an increase of k_6 while all other constants were the same in the four panels. The effect of increasing k_6 corresponds to an increase of $(1 + k_6/k_{-5})$ that becomes closer to $(1 + k_2/k_{-1})$. The dramatic difference between the two simulations in panel (a) becomes smaller by increasing k_6 until disappearing in panel (d) when substrate stickiness of the two catalytic steps is comparable (the two last rows in Table 2.7).

The question may arise why $K_{\mathrm{m}}^{\mathrm{app}}$ calculated assuming a quasi-equilibrium condition for [X]-binding (continuous lines in Fig. 2.11) does not show a maximum as the steady-state curve. The difference is due in part to the lack of some constants in the quasi-equilibrium expression of $K_{\mathrm{m}}^{\mathrm{app}}$, as discussed above, and in part to the

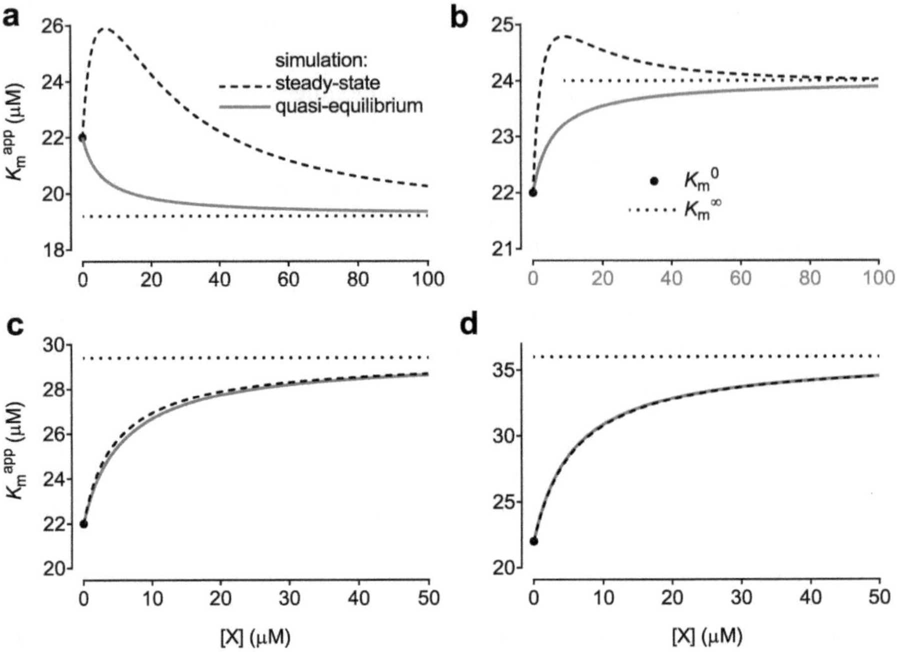

Fig. 2.11 K_m^{app} as function of modifier concentration depending on the relative substrate stickiness of the catalytic paths. Simulations were performed with the constants in Table 2.7 and the full-steady-state Eq. (2.57) (*dashed lines*), as well as (2.37) for the quasi-equilibrium condition. The curves in (**a**)–(**d**) were generated varying only the parameter β that determines the value of k_6, while all other constants were the same. How $1 + k_6/k_{-5}$ varies with respect to $1 + k_2/k_{-1}$ is shown in the last two rows of Table 2.7

relative values of the dissociation constants of the modifier binding steps, K_3 and K_4. This concept is best appreciated from the expression of K_m^{app} rearranged as ratio of quadratic polynomials (2.4.3), from which we can deduce that the function can represent a hyperbola as well as curves showing either a minimum or a maximum.

An example of the K_m^{app} against [X] plot exhibiting a minimum is shown in Fig. 2.12 that features the mechanism *hyperbolic mixed, predominantly catalytic activation* discussed in Sect. 5.4.6. In this case, the curves for the dependence of K_m^{app} on [X] of both the steady-state and the quasi-equilibrium assumption show a minimum due to the strong effect of the $[X]^2$-containing terms. For this example, using the constants in the legend of Fig. 2.12, (2.4.3) becomes

$$K_m^{app} = \frac{96.6[X]^2 + 800\,[X] + 1200}{0.92[X]^2 + 11.9\,[X] + 10}\,, \tag{2.73}$$

which shows how increasing [X] determines sizably different effects by [X] and $[X]^2$ in the numerator and in the denominator.

Table 2.7 Rate constants and parameters for the simulations in Fig. 2.11

	Units	a–d	a	b	c	d
k_1	$\mu M^{-1}s^{-1}$	10				
k_{-1}	s^{-1}	100				
k_2	s^{-1}	120				
k_3	$\mu M^{-1}s^{-1}$	18.514				
k_{-3}	s^{-1}	60				
k_4	$\mu M^{-1}s^{-1}$	12				
k_{-4}	s^{-1}	70				
k_5	$\mu M^{-1}s^{-1}$	10				
k_{-5}	s^{-1}	180				
k_6	s^{-1}		12	60	114	180
α		1.8				
β			0.10	0.50	0.95	1.50
$1 + k_2/k_{-1}$		2.20				
$1 + k_6/k_{-5}$			1.07	1.33	1.63	2.00

Letters a–d refer to the panels in the figure

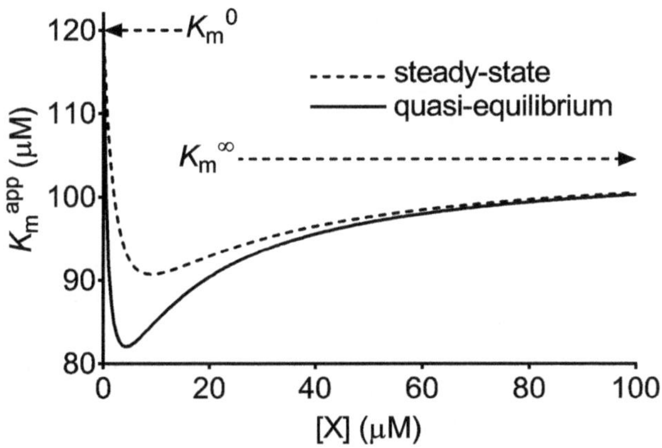

Fig. 2.12 K_m^{app} as function of modifier concentration showing a minimum. The curves were plotted with the full-steady-state (2.57) (*dashed line*) and (2.37) for the quasi-equilibrium condition (*continuous line*). Rate constants: $k_1 = 10$, $k_{-1} = 400$, $k_2 = 800$, $k_3 = 126.67$, $k_{-3} = 800$ ($K_3 = 6.3$), $k_4 = 100$, $k_{-4} = 1{,}200$ ($K_4 = 12.0$), $k_5 = 70$, $k_{-5} = 5{,}320$, $k_6 = 2{,}000$. The units of k_1, k_3, k_4 and k_5 are $\mu M^{-1}s^{-1}$, those of k_{-1}, k_{-3}, k_{-4}, k_{-5}, k_2 and k_6 are s^{-1}, and that of K_3, K_4 is μM. $\alpha = 1.9$, $\beta = 2.5$

2.8.2 *Weighing and Accepting Compromises*

Examples 1–5 discussed above dealt with the precision of kinetic parameters calculated by applying the quasi-equilibrium assumption to data analysis of the general modifier mechanism. What we learnt is summarized in the following points.

1. When applying the quasi-equilibrium assumption by fitting (2.35) to data, deviations from this model do not depend on substrate stickiness per se but on how different from one another are the values of $1 + k_2/k_{-1}$ and $1 + k_6/k_{-5}$.
2. When both terms are close to one, the quasi equilibrium assumption affords kinetic parameters whose deviations from the true values depend almost only on experimental error.
3. A certain discrepancy between calculated and true parameters is present when either one of $1 + k_2/k_{-1}$ or $1 + k_6/k_{-5}$ is larger than one, while the other is close to one. In this case, the bias of the calculated parameters depends both on deviations from the model *and* experimental error, with a preponderance of the second factor.
4. If generalized microscopic reversibility applies, i.e., $1 + k_2/k_{-1} = 1 + k_6/k_{-5}$, all enzyme complexes are at equilibrium independently of the values of the two terms and the precision of the calculated kinetic parameters depends exclusively on experimental error.
5. The parameter most prone to error is K_3.
6. Two criteria can be used to judge if, in a given experiment, the quasi-equilibrium assumption is valid: to compare K_m^{∞}/K_m^0 with K_4/K_3, and to calculate whether the value of the specificity constant is or is not much smaller than a reference value of $10^7\,M^{-1}\,s^{-1}$ for the second-order rate constant of ES formation from E and S. The first of these methods is very sensitive to experimental error and therefore of limited utility. Conversely, the second method is a handy and valid test in all cases with the exception of systems that obey generalized microscopic reversibility, in which case the test may not pass despite the system operates at equilibrium.
7. A nonhyperbolic dependence of K_m^{app} on [X], exhibiting a maximum or a minimum, indicates nonadherence to the quasi-equilibrium assumption.
8. Even if the quasi-equilibrium assumption can clearly not be accepted, this is not the major cause of discrepancies between true and calculated kinetic parameters because experimental error plays the protagonist role.

From point 8 above we can draw the conclusion that deviations from the model due to the presence of squared concentration terms in the rate equation may play a subordinated role with respect to even modest mistakes in reagent handling that include preparing solutions, imprecise pipetting, instrumental noise and others. In the impossibility of drawing such a conclusion from published data, this statement has been validated, using plausible kinetic constants, with simulations like those in examples 1–5 for the mechanisms that will be described in Chaps. 3 and 5.

The analysis of the examples in Sect. 2.7 revealed that using the quasi-equilibrium assumption for extracting kinetic information from primary data of systems obeying the general modifier mechanism is a rational procedure if due attention is paid in checking the limits of its applicability. The successful application of approximations depends on modifier binding and dissociation in a rapid fashion with respect to the catalytic steps and the agreement between calculated and true parameters depends on substrate stickiness.

The clinching argument that settles any doubt on the validity of the quasi-equilibrium assumption is performing dedicated experiments aimed at determining the individual rate constants of the mechanism. With this information available, assumptions are no longer needed. For the general modifier mechanism (Scheme 2.2), this means gathering good estimates at least of k_1, k_{-1}, k_2, k_5, k_{-5}, and k_6. For more complex mechanisms, to which the general treatment discussed in this chapter can be extended (Chap. 4), the same information can also be gathered. Yet, the common practice in most published studies, including those of the author, has been to proceed straight to evaluating kinetic parameters based on steady-state measurements exploiting a variety of graphical and mathematical methods under the quasi-equilibrium assumption without worrying about subtleties. This approach is equivalent to accepting that measured kinetic parameters can differ from their true counterparts and, should a calculated constant be double or half the true value, the result *is not the end of the world.* As illustrated by the five examples in Sect. 2.7, experimental error lays a mercy veil on our sins by hiding systematic deviations from models, which pass undetected except in extreme cases.

Notable exceptions in the literature exist, in which the individual kinetic constants of the catalytic steps have been determined exploiting the properties of systems in solutions of different viscosity. According to Einstein's theory of Brownian motion [14], only steps that depend on diffusion are affected by the viscosity of the solution. Nakatani and Dunford applied this concept to enzymology by studying the viscosity-dependence of apparent rate constants for horseradish peroxidase with *p*-nitrobenzoate as substrate [29]. The authors demonstrated that the formation of the ES complex from E and S proceeds as $\mathrm{E} + \mathrm{S} \rightleftharpoons \mathrm{E} \cdots \mathrm{S} \rightarrow \mathrm{ES}$, where the bimolecular association in the first step and the corresponding reverse reaction are diffusion-controlled processes followed by a diffusion-independent reaction to ES.

Provided the integrity of enzyme and reactants is preserved and other artifacts can be excluded, performing measurements in buffers containing or not a viscogenic agent allows calculating the individual kinetic constants. In particular, the viscosity method is a practical tool for determining the stickiness of the substrate and thus for judging the applicability of the quasi-equilibrium approach. For sticky substrates, increasing the relative viscosity of the reaction medium causes the specificity constant to decrease. Changes in relative viscosity have however no influence on the specificity constant of nonsticky substrates.

The viscosity method has been used in several studies, for instance by Brouwer and Kirsch on chymotrypsin [5], by Kurz et al. on adenosine deaminase [23], by St. Maurice and Bearne on mandelate racemase [33] and by Stone et al. on thrombin to locate the rate-limiting step of the amidolytic reaction [34]. A study relevant to the topic of this chapter is the meticulous work of Wells and Di Cera on thrombin activation by Na^+ that behaves as an allosteric modifier [43]. Di Cera and collaborators further discussed activation of thrombin by Na^+ and inhibition by hir^{55-65}, the C-terminal fragment of hirudin [12]. The authors performed steady-state measurements in solutions of variable viscosity and were able to calculate the rate constants for the steps in a steady-state regime. Under these premises, kinetic parameters determined for the allosteric modifier can be trusted with a high degree of probability. These studies from the Di Cera laboratory will be discussed in Sect. 4.2.

Another method exploits the temperature-dependence of kinetic parameters, which allows the determination of some individual rate constants [24]. This method has been applied by Ayala and Di Cera to thrombin with extension to serine peptidases in general and other enzymes obeying a similar mechanism [1]. A third method is based on isotope effects [10], which requires the synthesis of appropriately radiolabeled substrates.

Appendix 1

On the Meaning of the Word Allosteric

Redirected from Sect. 2.1.

In the original definition of Monod, Changeux, and Jacob [26, p. 307], the "functional structures of controlling proteins" are defined as follows:

> These proteins are assumed to possess two, or at least two, stereospecifically different, nonoverlapping receptor sites. One of these, the *active site*, binds the substrate and is responsible for the biological activity of the protein. The other, or *allosteric site*, is complementary to the structure of another metabolite, the *allosteric effector*, which it binds specifically and reversibly. The formation of the enzyme-allosteric effector complex does not activate a reaction involving the effector itself: it is assumed only to bring about a discrete reversible alteration of the molecular structure of the protein or *allosteric transition*, which modifies the properties of the active site, changing one or several of the kinetic parameters which characterize the biological activity of the protein.

Monod and his colleagues did not explain the etymology of the word allosteric but the first sentence of the quotation above contains *stereospecifically different, non-overlapping effector sites*. The term *Stereochemie* (English stereochemistry), from which close epithets, e.g., stereospecific, have been derived, has been coined in 1888 by Victor Meyer with a further detailed account in 1890 [25]. Meyer wanted to condense in a single word definitions such as "Chemie im Sinne der räumlichen Betrachtungsweise," "Lehre von der geometrischen Isomerie" (Chemistry from a spatial point of view, The doctrine of geometric isomery) previously used by the German chemist Johannes Adolf Wislicenus, to whom Jacobus Henricus van't Hoff dedicated the title of his book "La chimie dans l'espace" (Chemistry in space) [39].

The word *allosteric* derives from Greek and is composed of ἄλλος (allos), which means *another*, and στερεός (stereos) which has different meanings depending on the context. The basic meaning is *solid, rigid, hard, firm* and this is cited in some accounts of allostery although it makes no sense since what happens in allosteric interactions has nothing to do with rigid objects but exactly the contrary! An additional meaning is however στερεὸν εἶδος (stereon eidos) or τὸ στερεὸν, terms used by the philosopher and mathematician Plato to indicate three dimensional geometric figures. It suffices to think at stereophonic sound, stereo microscope or stereoscopy to understand that these terms describe three-dimensional perceptions of sound or images. Thus, allostery in proteins means *having a different three-dimensional organization*, referring to differences in the distribution of the atoms of amino acid side chains of the substrate binding pocket and that (those) of the effector(s). No matter their relative distance along the protein surface, important is that the substrate and effector binding sites do not overlap. In the original definition quoted above, binding of an allosteric effector at its receptor site causes an *allosteric transition* that affects the spatial organization of atoms in the substrate binding pocket.

Appendix 2

The Original Equation of Botts and Morales

Redirected from Sect. 2.3.

The method used by Botts and Morales for deriving the rate equation of the general modifier mechanism [4] is illustrated here step-by-step. The original reaction paths are shown in Scheme 2.4a in comparison with the notation of this book (b). A comparison of the different symbolism is made in the following table.

The original rate equation is reproduced as (2.74) with the original symbols. Note that Botts and Morales used association constants, including the Michaelis constant, instead of dissociation constants.

Naming of constants by Botts and Morales (**a**) and in this book (**b**)

a	k_1	k_{-1}	k_2	k_0	k_{-0}	k'_0	k'_{-0}	k'_1	k'_{-1}	k'_2
b	k_1	k_{-1}	k_2	k_3	k_{-3}	k_4	k_{-4}	k_5	k_{-5}	k_6
a	K_0	K'_0		$\bar{K}_1$	$\bar{K}'_1$		r	r'		$[E_0]$
b	$1/K_3$	$1/K_4$		$1/K_m^0$	$1/K_m^\infty$		k_2/k_{-1}	k_6/k_{-5}		$[E]_t$

$$\bar{K}_1 = \frac{k_1}{k_{-1} + k_2} \qquad \bar{K}'_1 = \frac{k'_1}{k'_{-1} + k'_2}$$

$$K_m^0 = \frac{k_{-1} + k_2}{k_1} \qquad K_m^\infty = \frac{k_{-5} + k_6}{k_5}$$

$$v = \frac{\left[\begin{array}{l} k_2\bar{K}_1[S] + k'_2K'_0\bar{K}_1[S][Y] \\ + \dfrac{k_{-0}k'_{-1}(r - r')\{k_2k'_{-0} + k'_2\langle k'_0[Y] + k_{-1}(1 + r)\rangle\}K'_0\bar{K}_1[S][Y]}{k_{-0}k'_{-1}(1 + r')\{k'_0[Y] + k_{-1}(1 + r)\} + k'_{-0}k_{-1}(1 + r)\{k'_1[S] + k_{-0}\}} \end{array}\right][E_0]}{1 + K_0[Y] + \bar{K}_1[S] + K'_0\bar{K}_1[S][Y]}$$

$$+ \frac{k'_{-1}(r - r')\{(k_{-0} - k'_{-0})k_{-1}(1 + r) + k_{-0}k'_{-0}(1 + K'_0[Y])\}K'_0\bar{K}_1[S][Y]}{k_{-0}k'_{-1}(1 + r')\{k'_0[Y] + k_{-1}(1 + r)\} + k'_{-0}k_{-1}(1 + r)\{k'_1[S] + k_{-0}\}} \tag{2.74}$$

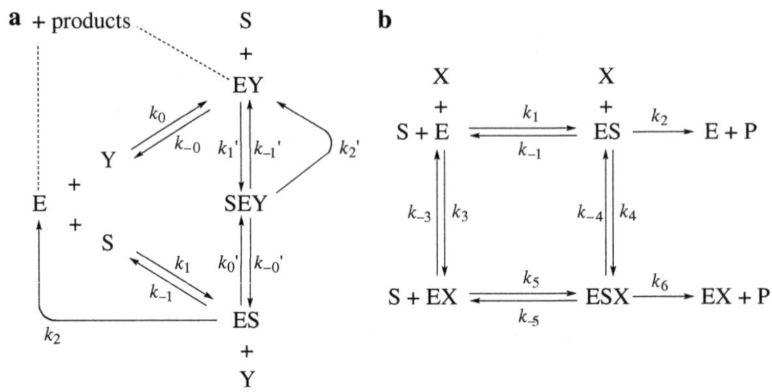

Scheme 2.4 Reaction paths for the general modifier mechanism as originally published by Botts and Morales [4], where Y is the modifier (**a**). The same mechanism in the notation used in this book (**b**)

This equation is not that complicated as it appears to be. It is mainly a matter of legibility due to the particular symbolism that comprises subscripts and primes.[4]

Necessary assumptions for deriving the rate equation are constant [S] and [X]. The authors used dimensionless ratio variables for solving the system of equations that lead to the rate equation, i.e.,

$$\begin{aligned} u_{\mathrm{EX}} &= [\mathrm{EX}]/[\mathrm{E}] \\ u_{\mathrm{ES}} &= [\mathrm{ES}]/[\mathrm{E}] \\ u_{\mathrm{ESX}} &= [\mathrm{ESX}]/[\mathrm{E}]\,. \end{aligned} \tag{2.75}$$

The concentration of free enzyme, [E], and similarly the concentrations of the three enzyme-containing complexes can thus be calculated as

$$[\mathrm{E}] = \frac{[\mathrm{E}]_{\mathrm{t}}}{1 + u_{\mathrm{EX}} + u_{\mathrm{ES}} + u_{\mathrm{ESX}}}\,. \tag{2.76}$$

The general form of the sought rate equation is

$$v = k_2\,[\mathrm{ES}] + k_6\,[\mathrm{ESX}] \tag{2.77}$$

which, using (2.76), can be rewritten as

$$v = \frac{k_2 u_{\mathrm{ES}} + k_6 u_{\mathrm{ESX}}}{1 + u_{\mathrm{EX}} + u_{\mathrm{ES}} + u_{\mathrm{ESX}}}[\mathrm{E}]_{\mathrm{t}}\,. \tag{2.78}$$

[4]Equation (14) in the paper of Botts and Morales [4] contains a typographic error, namely r in the denominator of the left part should be r'. There is also a superfluous, unbalanced curled brace in the right term of this equation.

In the steady-state, the concentration of the four enzyme-containing species does not change with time, a condition expressed as

$$\frac{d\,[\mathrm{E}]}{dt} = 0, \quad \frac{d\,[\mathrm{EX}]}{dt} = 0, \quad \frac{d\,[\mathrm{ES}]}{dt} = 0, \quad \frac{d\,[\mathrm{ESX}]}{dt} = 0\,, \tag{2.79}$$

where any three of these expressions are linearly independent, for instance the first three. They can be written explicitly using the corresponding rate constants and the ratio variables (2.75) as

$$k_{-3}\left(u_{\mathrm{EX}} - \frac{[\mathrm{X}]}{K_3}\right) + (k_{-1} + k_2)\left(u_{\mathrm{ES}} - \frac{[\mathrm{S}]}{K_{\mathrm{m}}^0}\right) = 0 \tag{2.80a}$$

$$(k_{-5} + k_6)\left(u_{\mathrm{ESX}} - \frac{[\mathrm{S}]}{K_{\mathrm{m}}^{\infty}} u_{\mathrm{EX}}\right) - k_{-3}\left(u_{\mathrm{EX}} - \frac{[\mathrm{X}]}{K_3}\right) = 0 \tag{2.80b}$$

$$k_{-4}\left(u_{\mathrm{ESX}} - \frac{[\mathrm{X}]}{K_4} u_{\mathrm{ES}}\right) - (k_{-1} + k_2)\left(u_{\mathrm{ES}} - \frac{[\mathrm{S}]}{K_{\mathrm{m}}^0}\right) = 0\,, \tag{2.80c}$$

where the K_{m} expressions are $K_{\mathrm{m}}^0 = (k_{-1} + k_2)/k_1$ and $K_{\mathrm{m}}^{\infty} = (k_{-5} + k_6)/k_5$ in the absence of modifier and at saturating modifier concentration, respectively. The tactic used in (2.80a)–(2.80c) is to collect terms so that the expressions in large parentheses are multiplied by first-order constants. As help to better understanding the strategy behind (2.80a)–(2.80c) we write (2.80a) in the customary way and then modify it with algebraic tricks:

$$\frac{\mathrm{d}\,[\mathrm{E}]}{\mathrm{d}t} = k_{-3}\,[\mathrm{EX}] - k_3\,[\mathrm{E}]\,[\mathrm{X}] + (k_{-1} + k_2)\,[\mathrm{ES}] - k_1\,[\mathrm{E}]\,[\mathrm{S}] = 0$$

$$k_{-3}\frac{[\mathrm{EX}]}{[\mathrm{E}]} - k_3\,[\mathrm{X}] + (k_{-1} + k_2)\,\frac{[\mathrm{ES}]}{[\mathrm{E}]} - k_1\,[\mathrm{S}] = 0$$

$$k_{-3}u_{\mathrm{EX}} - k_{-3}\frac{[\mathrm{X}]}{\dfrac{k_{-3}}{k_3}} + (k_{-1} + k_2)\,u_{\mathrm{ES}} - (k_{-1} + k_2)\,\frac{[\mathrm{S}]}{\dfrac{(k_{-1} + k_2)}{k_1}} = 0$$

$$k_{-3}\left(u_{\mathrm{EX}} - \frac{[\mathrm{X}]}{K_3}\right) + (k_{-1} + k_2)\left(u_{\mathrm{ES}} - \frac{[\mathrm{S}]}{K_{\mathrm{m}}^0}\right) = 0$$

To proceed, the terms within large parentheses in (2.80a)–(2.80c), are condensed as simpler symbols that are then introduced into (2.80a)–(2.80c):

$$\begin{aligned} w_{\mathrm{EX}} &= u_{\mathrm{EX}} - [\mathrm{X}]/K_3 \\ w_{\mathrm{ES}} &= u_{\mathrm{ES}} - [\mathrm{S}]/K_{\mathrm{m}}^0 \\ w_{\mathrm{ESX}} &= u_{\mathrm{ESX}} - [\mathrm{X}]\,[\mathrm{S}]/K_4 K_{\mathrm{m}}^0\,, \end{aligned} \tag{2.81}$$

$$k_{-3} w_{\mathrm{EX}} + (k_{-1} + k_2)\, w_{\mathrm{ES}} = 0 \tag{2.82a}$$

$$-\left\{ \frac{(k_{-5} + k_6)\,[\mathrm{S}]}{K_{\mathrm{m}}^{\infty}} + k_{-3} \right\} w_{\mathrm{EX}} + (k_{-5} + k_6)\, w_{\mathrm{ESX}} = \mathscr{D} \tag{2.82b}$$

$$-\left\{ k_{-4} \frac{[\mathrm{X}]}{K_4} + (k_{-1} + k_2) \right\} w_{\mathrm{ES}} + k_{-4} w_{\mathrm{ESX}} = 0 \tag{2.82c}$$

$\mathscr{D}$ in (2.82b) is a constant term that is not multiplied with either w_{EX} or w_{ESX}

$$\mathscr{D} = \left(\frac{k_2}{k_{-1}} - \frac{k_6}{k_{-5}} \right) k_{-5} \frac{[\mathrm{X}]}{K_4} \frac{[\mathrm{S}]}{K_{\mathrm{m}}^0}\ . \tag{2.83}$$

Only when $\mathscr{D} = 0$ the matrix of coefficients is zero, in which case the solutions of (2.82a)–(2.82c) are $w_{\mathrm{EX}} = w_{\mathrm{ES}} = w_{\mathrm{ESX}} = 0$. In this case, from (2.81),

$$\text{when } \mathscr{D} = 0 \qquad \begin{aligned} u_{\mathrm{EX}} &= [\mathrm{X}]/K_3 \\ u_{\mathrm{ES}} &= [\mathrm{S}]/K_{\mathrm{m}}^0 \\ u_{\mathrm{ESX}} &= [\mathrm{X}]\,[\mathrm{S}]/K_4 K_{\mathrm{m}}^0 \end{aligned} \tag{2.84}$$

and the rate equation (2.78) becomes

$$v = \frac{\left(k_2 \dfrac{[\mathrm{S}]}{K_{\mathrm{m}}^0} + k_6 \dfrac{[\mathrm{S}]}{K_{\mathrm{m}}^0} \dfrac{[\mathrm{X}]}{K_4} \right) [\mathrm{E}]_{\mathrm{t}}}{1 + \dfrac{[\mathrm{X}]}{K_3} + \dfrac{[\mathrm{S}]}{K_{\mathrm{m}}^0} + \dfrac{[\mathrm{S}]}{K_{\mathrm{m}}^0} \dfrac{[\mathrm{X}]}{K_4}}\ . \tag{2.85}$$

Equation (2.83) is written in a way that allows recognizing at a glance the conditions under which $\mathscr{D} = 0$. These include

1. The modifier is absent, i.e., $[\mathrm{X}] = 0$: the expression of the rate reduces to the Michaelis–Menten equation.
2. The enzyme binds exclusively either substrate or modifier, i.e., $K_4 = \infty$, $k_{-5} = 0$. This is the case of linear specific (competitive) inhibition (Sect. 5.2.1).
3. $k_2/k_{-1} = k_6/k_{-5}$: this is discussed in Sect. 2.4.2.

A note is necessary for readers interested in performing the derivation above. Depending on the algebraic method used, the following expression is obtained instead of (2.83):

$$\mathscr{D}' = -(k_{-5} + k_6) \left(\frac{[\mathrm{X}]\,[\mathrm{S}]}{K_4 K_{\mathrm{m}}^0} - \frac{[\mathrm{X}]\,[\mathrm{S}]}{K_3 K_{\mathrm{m}}^{\infty}} \right), \tag{2.86}$$

which seems to be a different result but $\mathscr{D}$ and $\mathscr{D}'$ *are the same*. I will not tell here why this is so and the demonstration is left to the reader as an exercise.[5]

Using (2.85) as the starting point, the general case can now be considered, namely when $\mathscr{D} \neq 0$, by adding *perturbation terms* to (2.84), i.e.,

$$\text{when } \mathscr{D} \neq 0 \quad \begin{aligned} u_{\mathrm{EX}} &= [\mathrm{X}]/K_3 + \Delta_{\mathrm{EX}}/\Delta \\ u_{\mathrm{ES}} &= [\mathrm{S}]/K_{\mathrm{m}}^0 + \Delta_{\mathrm{ES}}/\Delta \\ u_{\mathrm{ESX}} &= [\mathrm{X}]\,[\mathrm{S}]/K_4 K_{\mathrm{m}}^0 + \Delta_{\mathrm{EXS}}/\Delta\,. \end{aligned} \tag{2.87}$$

Δ, Δ_{EX}, Δ_{ES}, and Δ_{ESX} are obtained as the solutions of the following matrix equation that represents (2.82a)–(2.82c)

$$\begin{bmatrix} k_{-3} & k_{-1} + k_2 & 0 \\ -\dfrac{(k_{-5}+k_6)\,S}{K_{\mathrm{m}}^{\infty}} - k_{-3} & 0 & k_{-5} + k_6 \\ 0 & -\dfrac{k_{-4}X}{K_4} - k_{-1} - k_2 & k_{-4} \end{bmatrix} \begin{bmatrix} u_{\mathrm{EX}} \\ u_{\mathrm{ES}} \\ u_{\mathrm{ESX}} \end{bmatrix} = \begin{bmatrix} 0 \\ \mathscr{D} \\ 0 \end{bmatrix}. \tag{2.88}$$

$$\Delta = k_{-3}k_{-5}\left(1 + \frac{k_6}{k_{-5}}\right)\left\{k_4\,[\mathrm{X}] + k_{-1}\left(1 + \frac{k_2}{k_{-1}}\right)\right\} + k_{-1}k_{-4}\left(1 + \frac{k_2}{k_{-1}}\right)\{k_5\,[\mathrm{S}] + k_{-3}\} \tag{2.89a}$$

$$\Delta_{\mathrm{EX}} = -k_{-1}k_{-4}\left(1 + \frac{k_2}{k_{-1}}\right)\left(\frac{k_2}{k_{-1}} - \frac{k_6}{k_{-5}}\right)k_{-5}\frac{[\mathrm{X}]}{K_4}\frac{[\mathrm{S}]}{K_{\mathrm{m}}^0} \tag{2.89b}$$

$$\Delta_{\mathrm{ES}} = k_{-3}k_{-4}\left(\frac{k_2}{k_{-1}} - \frac{k_6}{k_{-5}}\right)k_{-5}\frac{[\mathrm{X}]}{K_4}\frac{[\mathrm{S}]}{K_{\mathrm{m}}^0} \tag{2.89c}$$

$$\Delta_{\mathrm{ESX}} = k_{-3}\left\{k_4\,[\mathrm{X}] + k_{-1}\left(1 + \frac{k_2}{k_{-1}}\right)\right\}\left(\frac{k_2}{k_{-1}} - \frac{k_6}{k_{-5}}\right)k_{-5}\frac{[\mathrm{X}]}{K_4}\frac{[\mathrm{S}]}{K_{\mathrm{m}}^0} \tag{2.89d}$$

The solutions (2.89a)–(2.89d) are now substituted into (2.87) and the resulting expressions of u_{EX}, u_{ES} and u_{ESX} used to construct the numerator and the denominator of the rate equation with (2.78):

$$\textit{Numerator} = k_2\frac{[\mathrm{S}]}{K_{\mathrm{m}}^0} + k_6\frac{[\mathrm{S}]}{K_{\mathrm{m}}^0}\frac{[\mathrm{X}]}{K_4}$$

[5]The answer is given in Appendix 3 (second item).

$$+\frac{k_{-3}k_{-5}\left(\frac{k_2}{k_{-1}}-\frac{k_6}{k_{-5}}\right)\left\{k_2k_{-4}+k_6\left[k_4\,[\mathrm{X}]+k_{-1}\left(1+\frac{k_2}{k_{-1}}\right)\right]\right\}\frac{[\mathrm{S}]}{K_\mathrm{m}^0}\frac{[\mathrm{X}]}{K_4}}{k_{-3}k_{-5}\left(1+\frac{k_6}{k_{-5}}\right)\left[k_4\,[\mathrm{X}]+k_{-1}\left(1+\frac{k_2}{k_{-1}}\right)\right]+k_{-1}k_{-4}\left(1+\frac{k_2}{k_{-1}}\right)(k_5\,[\mathrm{S}]+k_{-3})}$$

$$Denominator = 1+\frac{[\mathrm{X}]}{K_3}+\frac{[\mathrm{S}]}{K_\mathrm{m}^0}+\frac{[\mathrm{S}]}{K_\mathrm{m}^0}\frac{[\mathrm{X}]}{K_4} \tag{2.90}$$

$$+\frac{k_{-5}\left(\frac{k_2}{k_{-1}}-\frac{k_6}{k_{-5}}\right)\left\{(k_{-3}-k_{-4})\,k_{-1}\left(1+\frac{k_2}{k_{-1}}\right)+k_{-3}k_{-4}\left(1+\frac{[\mathrm{X}]}{K_4}\right)\right\}\frac{[\mathrm{S}]}{K_\mathrm{m}^0}\frac{[\mathrm{X}]}{K_4}}{k_{-3}k_{-5}\left(1+\frac{k_6}{k_{-5}}\right)\left[k_4\,[\mathrm{X}]+k_{-1}\left(1+\frac{k_2}{k_{-1}}\right)\right]+k_{-1}k_{-4}\left(1+\frac{k_2}{k_{-1}}\right)(k_5\,[\mathrm{S}]+k_{-3})}\,. \tag{2.91}$$

The rate equation is finally given by

$$v=\frac{Numerator}{Denominator}[\mathrm{E}]_\mathrm{t}\,, \tag{2.92}$$

which is identical to (2.74).

Appendix 3

Equivalent expressions

When alternative, equivalent pathways exist in a reaction mechanism for the application of King and Altman's method [21] or improved variants of the original method [6, 7, 40], the constants appearing in the final equations depend on the selected path and rearrangements. Any differences between particular terms in the final equations are due to different choices in selecting the reference node during derivation. When the King–Altman method is not used, there may be alternative approaches to derive the equations, e.g., by linear algebra, and also in such cases the final expressions may considerably differ. However, despite containing different constants, such expressions are identical. For the general modifier mechanism, the principle of microscopic reversibility helps in differentiating between equivalent expressions and possible mistakes.

Apparent Michaelis Constant in the General Modifier Mechanism

Redirected from Sect. 2.4.3.

We compare here the expression of K_m^{app} focusing on the first term on the right of (2.37) with that published by Topham [36, 37], below as (2.94), which may not immediately be recognized as being identical. However, invoking the principle of microscopic reversibility, which implies $k_1 k_4 k_{-5} k_{-3} = k_3 k_5 k_{-4} k_{-1}$, they result to be the same. In fact, the individual rate constants can be grouped in various ways, of which the following is useful for the present purpose:

$$\frac{k_{-1}k_5}{k_1 k_{-5}} = \frac{k_{-3}k_4}{k_3 k_{-4}} = \frac{K_3}{K_4}\,. \tag{2.93}$$

Thus, starting from Topham's expression of K_m^S

$$K_m^s = \frac{\frac{k_{-1}}{k_1}\left\{\left(\frac{k_{-1}+k_2}{k_{-1}}\right)k_1 + \left(\frac{k_{-5}+k_6}{k_{-5}}\right)k_5\frac{[X]}{K_3}\right\}}{k_1 + k_5\frac{[X]}{K_3}}\,, \tag{2.94}$$

this can be transformed with the following passages using (2.93)

$$\begin{aligned}
K_m^s &= \frac{\left\{k_{-1} + k_2 + \left(\frac{k_{-1}k_{-5}k_5 + k_{-1}k_5k_6}{k_1 k_{-5}}\right)\frac{[X]}{K_3}\right\}}{k_1 + k_5\frac{[X]}{K_3}} \\
&= \frac{k_{-1} + k_2 + \left(\frac{k_{-1}k_5}{k_1 k_{-5}}k_{-5} + \frac{k_{-1}k_5}{k_1 k_{-5}}k_6\right)\frac{[X]}{K_3}}{k_1 + k_5\frac{[X]}{K_3}} \\
&= \frac{k_{-1} + k_2 + k_{-5}\frac{K_3}{K_4}\frac{[X]}{K_3} + k_6\frac{K_3}{K_4}\frac{[X]}{K_3}}{k_1 + k_5\frac{[X]}{K_3}} \\
&= \frac{k_{-1} + k_2 + (k_{-5}+k_6)\frac{[X]}{K_4}}{k_1 + k_5\frac{[X]}{K_3}}\,.
\end{aligned} \tag{2.95}$$

In conclusion, (2.94) by Topham [36, 37] is the same as (2.95). To demonstrate the simplification of (2.95) to $K_{\mathrm{m}}^{0} = (k_{-1} + k_2)/k_1$ mentioned in Sect. 2.4.3, it suffices to travel back from (2.95) to (2.94) and to set

$$\frac{k_{-1} + k_2}{k_{-1}} = \frac{k_{-5} + k_6}{k_{-5}}, \quad \text{i.e.} \quad 1 + \frac{k_2}{k_{-1}} = 1 + \frac{k_6}{k_{-5}},$$

which is the condition of generalized microscopic reversibility (Sect. 2.4.2).

The simplification of (2.95) to $K_{\mathrm{S}} = k_{-1}/k_1$ described in Sect. 2.4.1 is accomplished in the same way by setting

$$\frac{k_2}{k_{-1}} \ll 1 \quad \text{and} \quad \frac{k_6}{k_{-5}} \ll 1 .$$

The Constant Term in Eq. (2.83)

Redirected from Appendix 2 (footnote on p. 115).

The proof that expressions (2.83) and (2.86) are identical can be provided by expanding the two equations, inserting the explicit expressions of K_{m}^{0}, K_{m}^{∞}, K_3 and K_4. If the ratio of the two transformed expressions is 1, the proof is provided.

$$\mathscr{D} = \frac{k_1 k_2 k_4 k_{-5}\,[\mathrm{X}]\,[\mathrm{S}]}{k_{-1} k_{-4}\,(k_{-1} + k_2)} - \frac{k_1 k_4 k_6\,[\mathrm{X}]\,[\mathrm{S}]}{k_{-4}\,(k_{-1} + k_2)} \tag{2.96}$$

$$\mathscr{D}' = -\frac{k_1 k_4 k_{-5}\,[\mathrm{X}]\,[\mathrm{S}]}{k_{-4}\,(k_{-1} + k_2)} + \frac{k_3 k_5 k_{-5}\,[\mathrm{X}]\,[\mathrm{S}]}{k_{-3}\,(k_{-5} + k_6)} - \frac{k_1 k_4 k_6\,[\mathrm{X}]\,[\mathrm{S}]}{k_{-4}\,(k_{-1} + k_2)} + \frac{k_3 k_5 k_6\,[\mathrm{X}]\,[\mathrm{S}]}{k_{-3}\,(k_{-5} + k_6)} \tag{2.97}$$

$$\begin{aligned}
\frac{\mathscr{D}}{\mathscr{D}'} &= \frac{k_6\,(k_1 k_{-1} k_{-3} k_4) - k_2\,(k_1 k_{-3} k_4 k_{-5})}{k_{-1}\left(\underbrace{k_1 k_{-3} k_4 k_{-5}} + k_1 k_{-3} k_4 k_6 - k_2 k_3 k_{-4} k_5 - \underbrace{k_{-1} k_3 k_{-4} k_5}\right)} \\
&= \frac{k_6\,(k_1 k_{-1} k_{-3} k_4) - k_2\,(k_1 k_{-3} k_4 k_{-5})}{k_1 k_{-1} k_{-3} k_4 k_6 - k_{-1} k_2 k_3 k_{-4} k_5} \\
&= \frac{k_6\,(k_1 k_{-1} k_{-3} k_4) - k_2\left(\underbrace{k_1 k_{-3} k_4 k_{-5}}\right)}{k_6\,(k_1 k_{-1} k_{-3} k_4) - k_2\left(\underbrace{k_{-1} k_3 k_{-4} k_5}\right)} = 1
\end{aligned} \tag{2.98}$$

Here again it is the principle of microscopic reversibility that fixes the apparent discrepancy because the terms underscored by braces in (2.98) are equivalent.

Appendix 4

Systematic Derivation of the Rate Equation for the General Modifier Mechanism

Redirected from Sect. 2.5.

This is shown here for didactic purposes, to demonstrate how rate equations may be obtained with different methods. The derivation of the rate equation for the general modifier mechanism comprising also the reverse reactions as shown in Scheme 2.5a, is best accomplished using the systematic method of King and Altman [21]. It is useful to observe that the system contains parallel branches between E and ES and between EX and ESX as evidenced in the alternative representation of Scheme 2.5b. In this case, the number of King–Altman patterns is reduced from 12 to 4 as shown in Scheme 2.5d–g, derived from the master pattern (c). Here, node compression according to Volkenstein and Goldstein has been applied [40]. Each of the four nodes of the mechanisms (E1, E2, E3, and E4, corresponding to E, ES, ESX, and EX, respectively) owns four *directed* patterns shown in Scheme 2.6.

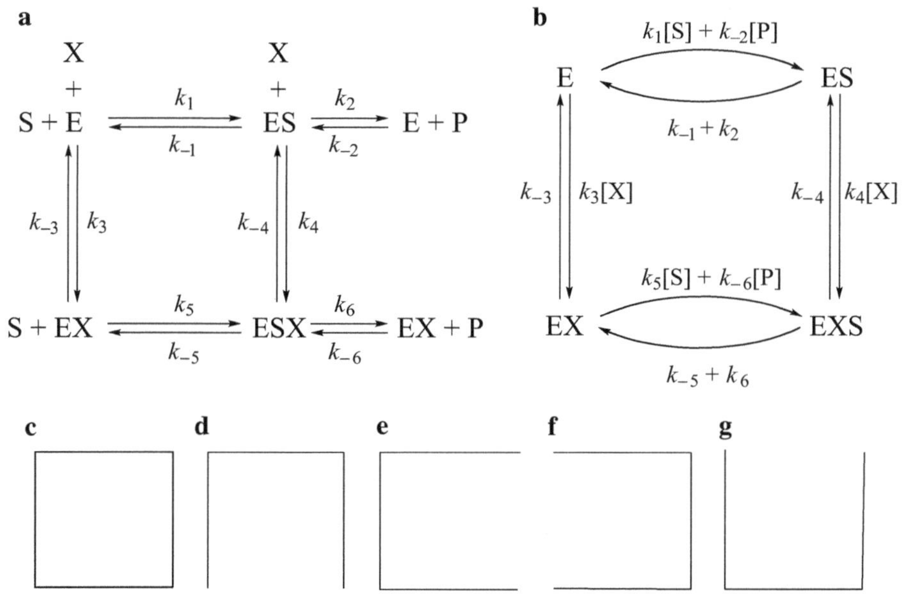

Scheme 2.5 The general modifier mechanism represented with all rate constants. In (**a**) the nature of the parallel paths in the fluxes around ES and ESX may not be perceived immediately, while this is evident in the compact representation (**b**). The master pattern for the application of the King–Altman method (**c**), and the four patterns belonging to the master pattern ((**d**)–(**g**)).

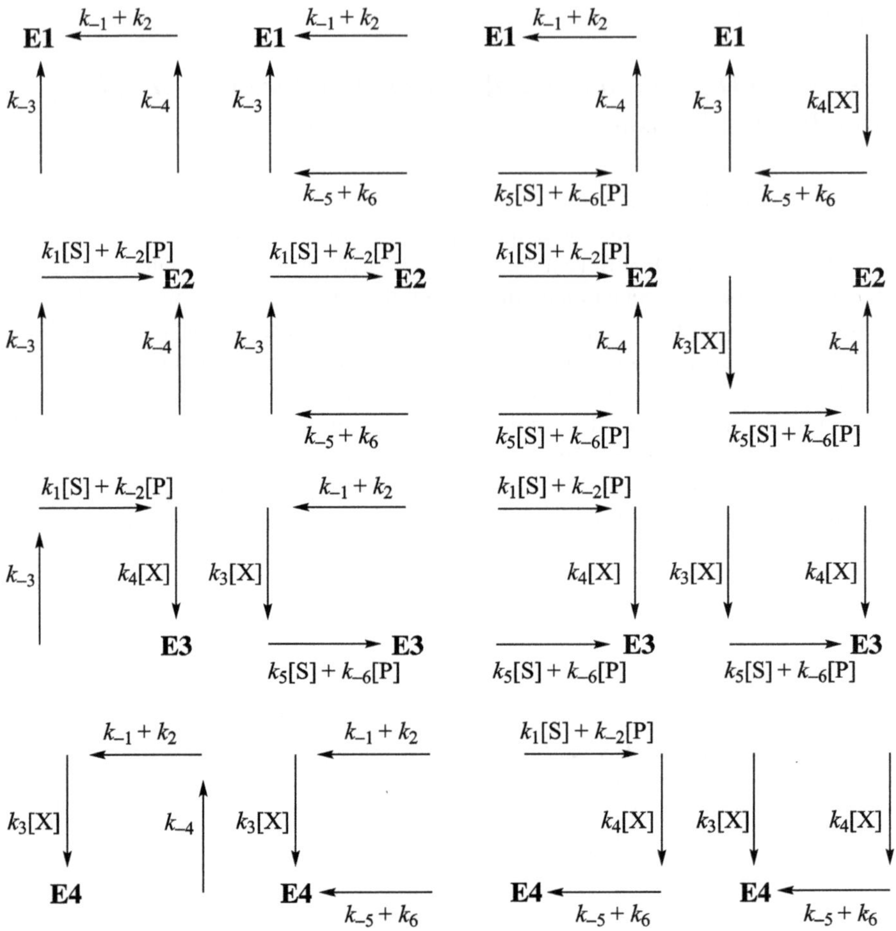

Scheme 2.6 Directed patterns for the four patterns (d–g) in Scheme 2.5

$$
\begin{aligned}
\mathrm{E1} &= (k_{-1} + k_2)\, k_{-3}k_{-4} + (k_{-1} + k_2)\, k_{-3}\, (k_{-5} + k_6) \\
&\quad + (k_{-1} + k_2)\, k_{-4}\, (k_5\,[\mathrm{S}] + k_{-6}\,[\mathrm{P}]) + k_{-3}\, (k_{-5} + k_6)\, k_4\,[\mathrm{X}] \\
\mathrm{E2} &= (k_1\,[\mathrm{S}] + k_{-2}\,[\mathrm{P}])\, k_{-3}k_{-4} + (k_1\,[\mathrm{S}] + k_{-2}\,[\mathrm{P}])\, k_{-3}\, (k_{-5} + k_6) \\
&\quad + (k_1\,[\mathrm{S}] + k_{-2}\,[\mathrm{P}])\, k_{-4}\, (k_5\,[\mathrm{S}] + k_{-6}\,[\mathrm{P}]) + k_3\,[\mathrm{X}]\, (k_5\,[\mathrm{S}] + k_{-6}\,[\mathrm{P}])\, k_{-4} \\
\mathrm{E3} &= k_{-3}\, (k_1\,[\mathrm{S}] + k_{-2}\,[\mathrm{P}])\, k_4\,[\mathrm{X}] + k_3\,[\mathrm{X}]\, (k_{-1} + k_2)\, (k_5\,[\mathrm{S}] + k_{-6}\,[\mathrm{P}]) \\
&\quad + (k_5\,[\mathrm{S}] + k_{-6}\,[\mathrm{P}])\, k_4\,[\mathrm{X}]\, (k_1\,[\mathrm{S}] + k_{-2}\,[\mathrm{P}]) + k_3k_4[\mathrm{X}]^2\, (k_5\,[\mathrm{S}] + k_{-6}\,[\mathrm{P}]) \\
\mathrm{E4} &= k_3\,[\mathrm{X}]\, (k_{-1} + k_2)\, k_{-4} + k_3\,[\mathrm{X}]\, (k_{-1} + k_2)\, (k_{-5} + k_6) \\
&\quad + (k_1\,[\mathrm{S}] + k_{-2}\,[\mathrm{P}])\, k_4\,[\mathrm{X}]\, (k_{-5} + k_6) + k_3k_4[\mathrm{X}]^2\, (k_{-5} + k_6) \qquad (2.99)
\end{aligned}
$$

In each of the four rows in Scheme 2.6, the directional arrows lead to a particular node and are labeled with the first-order or pseudo-first-order constants from Scheme 2.5b. Beneath Scheme 2.6, the values of the four nodes E1-E4 are listed (2.99). These are assembled by multiplying the three values in each of the four patterns belonging to a node and adding together the four results. To facilitate reading of the four equations in (2.99) and comparison with the corresponding patterns in Scheme 2.6, the sequence left-to-right in the scheme is respected.

The fractional concentrations of E, ES, ESX, and EX are calculated from (2.99)

$$f_{\mathrm{E}} = \frac{\mathrm{E1}}{D}, \quad f_{\mathrm{ES}} = \frac{\mathrm{E2}}{D}, \quad f_{\mathrm{ESX}} = \frac{\mathrm{E3}}{D}, \quad f_{\mathrm{EX}} = \frac{\mathrm{E4}}{D}, \tag{2.100}$$

where $D = \mathrm{E1}+\mathrm{E2}+\mathrm{E3}+\mathrm{E4}$ and the concentrations of enzyme-containing species are

$$[\mathrm{E}] = f_E[\mathrm{E}]_\mathrm{t}, \quad [\mathrm{ES}] = f_{ES}[\mathrm{E}]_\mathrm{t}, \quad [\mathrm{ESX}] = f_{ESX}[\mathrm{E}]_\mathrm{t}, \quad [\mathrm{EX}] = f_{EX}[\mathrm{E}]_\mathrm{t}\,. \tag{2.101}$$

The sum of the fractions (2.100) is 1 and the sum of the concentrations (2.101) corresponds to $[\mathrm{E}]_\mathrm{t}$. The rate of reaction must consider two forward and two reverse steps with positive and negative signs, respectively:

$$v = k_2\,[\mathrm{ES}] + k_6\,[\mathrm{ESX}] - k_{-2}\,[\mathrm{E}]\,[\mathrm{P}] - k_{-6}\,[\mathrm{EX}]\,[\mathrm{P}]\;, \tag{2.102}$$

in which (2.101) and the values (2.99) are introduced and expanded producing the numerator and the denominator of the sought rate equation:

$$\begin{aligned}
&\mathit{Numerator}\\
= \; & -(k_{-1}k_{-2}k_{-3}k_{-4} + k_{-1}k_{-2}k_{-3}k_{-5} + k_{-1}k_{-2}k_{-3}k_6)\,[\mathrm{P}]\\
& + (k_1k_2k_{-3}k_{-4} + k_1k_2k_{-3}k_{-5} + k_1k_2k_{-3}k_6)\,[\mathrm{S}]\\
& + (k_{-2}k_4k_5k_6 - k_1k_4k_{-5}k_{-6})\,[\mathrm{P}]\,[\mathrm{S}]\,[\mathrm{X}]\\
& + (k_1k_{-3}k_4k_6 + k_{-1}k_3k_5k_6 + k_2k_3k_{-4}k_5 + k_2k_3k_5k_6)\,[\mathrm{S}]\,[\mathrm{X}]\\
& - (k_{-2}k_{-3}k_4k_{-5} + k_2k_3k_{-5}k_{-6} + k_{-1}k_3k_{-4}k_{-6} + k_{-1}k_3k_{-5}k_{-6})\,[\mathrm{P}]\,[\mathrm{X}]\\
& + k_1k_4k_5k_6[\mathrm{S}]^2\,[\mathrm{X}]\\
& - k_{-2}k_4k_{-5}k_{-6}[\mathrm{P}]^2\,[\mathrm{X}] + k_3k_4k_5k_6\,[\mathrm{S}]\,[\mathrm{X}]^2\\
& - k_3k_4k_{-5}k_{-6}\,[\mathrm{P}]\,[\mathrm{X}]^2 + (k_1k_2k_{-4}k_{-6} - k_{-1}k_{-2}k_{-4}k_5)\,[\mathrm{P}]\,[\mathrm{S}]\\
& + k_1k_2k_{-4}k_5[\mathrm{S}]^2 - k_{-1}k_{-2}k_{-4}k_{-6}[\mathrm{P}]^2
\end{aligned} \tag{2.103}$$

Denominator

$$
\begin{aligned}
= &+ (k_{-2}k_{-3}k_{-4} + k_{-1}k_{-4}k_{-6} + k_{-2}k_{-3}k_6 + k_{-2}k_{-3}k_{-5} + k_2k_{-4}k_{-6})\,[\mathrm{P}] \\
&+ (k_1k_{-3}k_{-5} + k_{-1}k_{-4}k_5 + k_2k_{-4}k_5 + k_1k_{-3}k_6 + k_1k_{-3}k_{-4})\,[\mathrm{S}] \\
&+ (k_1k_4k_{-6} + k_{-2}k_4k_5)\,[\mathrm{P}]\,[\mathrm{S}]\,[\mathrm{X}] \\
&+ (k_1k_4k_{-5} + k_{-1}k_3k_5 + k_1k_{-3}k_4 + k_1k_4k_6 + k_3k_{-4}k_5 + k_2k_3k_5)\,[\mathrm{S}]\,[\mathrm{X}] \\
&+ (k_3k_{-4}k_{-6} + k_2k_3k_{-6} + k_{-1}k_3k_{-6} + k_{-2}k_{-3}k_4 + k_{-2}k_4k_6 + k_{-2}k_4k_{-5})\,[\mathrm{P}]\,[\mathrm{X}] \\
&+ k_1k_4k_5[\mathrm{S}]^2\,[\mathrm{X}] \\
&+ k_{-2}k_4k_{-6}[\mathrm{P}]^2\,[\mathrm{X}] \\
&+ k_3k_4k_5\,[\mathrm{S}]\,[\mathrm{X}]^2 + k_3k_4k_{-6}\,[\mathrm{P}]\,[\mathrm{X}]^2 \\
&+ (k_{-2}k_{-4}k_5 + k_1k_{-4}k_{-6})\,[\mathrm{P}]\,[\mathrm{S}] \\
&+ k_1k_{-4}k_5[\mathrm{S}]^2 + k_{-2}k_{-4}k_{-6}[\mathrm{P}]^2 \\
&+ (k_3k_4k_6 + k_3k_4k_{-5})\,[\mathrm{X}]^2 \\
&+ \begin{pmatrix} k_{-1}k_3k_{-5} + k_{-3}k_4k_{-5} + k_2k_3k_{-4} + k_{-3}k_4k_6 + k_2k_3k_6 + k_{-1}k_3k_6 \\ +k_{-1}k_3k_{-4} + k_2k_3k_{-5} \end{pmatrix} [\mathrm{X}] \\
&+ (k_{-1}k_{-3}k_{-5} + k_2k_{-3}k_{-5} + k_2k_{-3}k_{-4} + k_{-1}k_{-3}k_{-4} + k_{-1}k_{-3}k_6 + k_2k_{-3}k_6)
\end{aligned}
\tag{2.104}
$$

The rate equation is calculated with (2.103) and (2.104) as:

$$
v = \frac{\mathit{Numerator}}{\mathit{Denominator}}[\mathrm{E}]_{\mathrm{t}}\,, \tag{2.105}
$$

which contains first- and second-order terms in [S], [X] and [P]. While this derivation was performed here pedantically to demonstrate the method, writing all directed patterns in Scheme 2.6 is not necessary. Instead, the complete expressions of the four nodes (2.99) can be written by taking the values directly from Scheme 2.5b and arranging them according to paths (c–f) in the same scheme.

Equation (2.105), which is not subjected to other restrictions than constancy of [S] and [X], is useful for performing simulations with appropriate sets of kinetics constants and reactant concentrations.

Simplifications can be applied to (2.105) to obtain rates, limiting rates, and Michaelis constants under particular sets of reactant concentrations, the most obvious of which is the rate in the absence of product.

The derivation of rate equations for complex systems, including the general modifier mechanism, can be accomplished by computerized methods. The code for obtaining the four expressions (2.99) using the KAPattern program is the following:

```
1 2 k1*S
1 2 k_2*P
2 1 k_1
2 1 k2
1 4 k3*X
4 1 k_3
2 3 k4*X
3 2 k_4
4 3 k5*S
4 3 k_6*P
3 4 k_5
3 4 k6
```

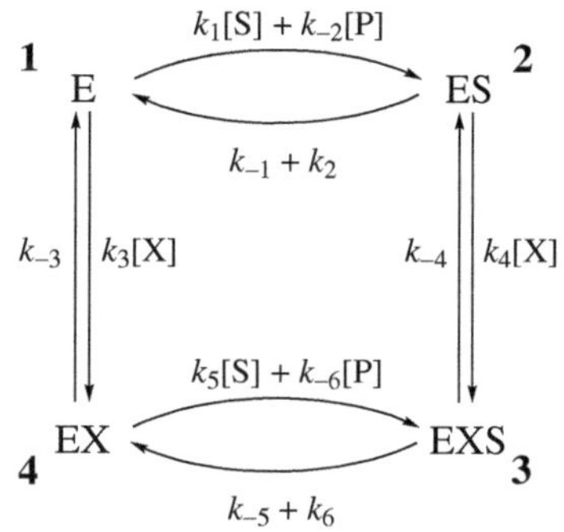

Omitting the code lines 1 2 k_2*P and 4 3 k_6*P generates the enzyme states in the absence of product.

References

1. Ayala YM, Di Cera E (2000) A simple method for the determination of individual rate constants for substrate hydrolysis by serine proteases. Protein Sci 9:1589–1593
2. Bardsley WG, Leff P, Kavanagh J, Waight RD (1980) Deviations from Michaelis-Menten kinetics. The possibility of complicated curves for simple kinetic schemes and the computer fitting of experimental data for acetylcholinesterase, acid phosphatase, adenosine deaminase, arylsulphatase, benzylamine oxidase, chymotrypsin, fumarase, galactose dehydrogenase, beta-galactosidase, lactate dehydrogenase, peroxidase and xanthine oxidase. Biochem J 187:739–765
3. Botts J (1958) Typical behaviour of some simple models of enzyme action. Trans Faraday Soc 54:593–604. doi:10.1039/tf9585400593
4. Botts J, Morales M (1953) Analytical description of the effects of modifiers and of enzyme multivalency upon the steady state catalyzed reaction rate. Trans Faraday Soc 49:696–707. doi:10.1039/TF9534900696
5. Brouwer AC, Kirsch JF (1982) Investigation of diffusion-limited rates of chymotrypsin reactions by viscosity variation. Biochemistry 21:1302–1307
6. Cha S (1968) A simple method for derivation of rate equations for enzyme-catalyzed reactions under the rapid equilibrium assumption or combined assumptions of equilibrium and steady-state. J Biol Chem 243:820–825
7. Chou KC (1990) Application of graph theory to enzyme kinetics and protein folding kinetics. Steady and non-steady-state systems. Biophys Chem 35:1–24
8. Cleland WW (1977) Determining the chemical mechanisms of enzyme-catalyzed reactions by kinetic studies. Adv Enzymol Relat Areas Mol Biol 45:273–387
9. Cleland WW (1982) Use of Isotope effects to elucidate enzyme mechanism. Crit Rev Biochem Mol Biol 13:385–428. doi:10.3109/10409238209108715
10. Cleland WW (1995) Isotope effects: determination of enzyme transition state structure. Meth Enzymol 249:341–373. doi:10.1016/0076-6879(95)49041-8
11. Cornish-Bowden A (2012) Fundamentals of enzyme kinetics, 4th edn. Wiley, Weinheim

12. Di Cera E, Dang QD, Ayala Y, Vindigni A (1995) Linkage at steady state: allosteric transitions of thrombin. Meth Enzymol 259:127–144
13. Di Cera E, Hopfner KP, Dang QD (1996) Theory of allosteric effects in serine proteases. Biophys J 70:174–181
14. Einstein A (1905) Über die von der molekularkinetischen Theorie der Wärme geforderte Bewegung von in ruhenden Flüssigkeiten suspendierten Teilchen. Annalen der Physik 322:549–560. doi:10.1002/andp.19053220806
15. Fenton AW (2008) Allostery: an illustrated definition for the 'second secret of life'. Trends Biochem Sci 33:420–425. doi:10.1016/j.tibs.2008.05.009
16. Frauenfelder H, McMahon BH, Austin RH, Chu K, Groves JT (2001) The role of structure, energy landscape, dynamics, and allostery in the enzymatic function of myoglobin. Proc Natl Acad Sci USA 98:2370–2374
17. Frieden C (1964) Treatment of enzyme kinetic data. I. The effect of modifiers on the kinetic parameters of single substrate enzymes. J Biol Chem 239:3522–3531
18. Gunasekaran K, Ma B, Nussinov R (2004) Is allostery an intrinsic property of all dynamic proteins? Proteins 57:433–443
19. Hill TL (1989) Free energy transduction and biochemical cycle kinetics. Springer, New York
20. Keleti T (1968) Effect of steric changes in the protein on the kinetics of enzymic reactions. II. Steady-state treatment of reactions with one substrate. Acta Biochim Biophys Acad Sci Hung 3:247–258
21. King EL, Altman C (1956) A schematic method of deriving the rate laws for enzyme-catalyzed reactions. J Phys Chem 60:1375–1378
22. Koshland DE, Némethy G, Filmer D (1966) Comparison of experimental binding data and theoretical models in proteins containing subunits. Biochemistry 5:365–385
23. Kurz LC, Weitkamp E, Frieden C (1987) Adenosine deaminase: viscosity studies and the mechanism of binding of substrate and of ground- and transition-state analog inhibitors. Biochemistry 26:3027–3032. doi:10.1021/bi00385a012
24. Laidler KJ, Peterman BF (1979) Temperature effects in enzyme kinetics. Meth Enzymol 63:234–257
25. Meyer V (1890) Ergebnisse und Ziele der stereochemischen Forschung. Ber Deut Chem Ges 23:567–619
26. Monod J, Changeux JP, Jacob F (1963) Allosteric proteins and cellular control systems. J Mol Biol 6:306–329
27. Monod J, Wyman J, Changeux JP (1965) On the nature of allosteric transitions: a plausible model. J Mol Biol 12:88–118
28. Morales MF (1955) If an enzyme-substrate modifier system exhibits non-competitive interaction, then, in general, its Michaelis constant is an equilibrium constant. J Amer Chem Soc 77:4169–4170. doi:10.1021/Ja01620a066
29. Nakatani H, Dunford HB (1979) Meaning of diffusion-controlled association rate constants in enzymology. J Phys Chem 83:2662–2665. doi:10.1021/j100483a023
30. Reinhart GD (2004) Quantitative analysis and interpretation of allosteric behavior. Meth Enzymol 380:187–203. doi:10.1016/S0076-6879(04)80009-0
31. Ricard J (1978) Generalized microscopic reversibility, kinetic co-operativity of enzymes and evolution. Biochem J 175:779–791
32. Segal HL, Kachmar JF, Boyer PD (1952) Kinetic analysis of enzyme reactions. I. Further considerations of enzyme inhibition and analysis of enzyme activation. Enzymologia 15:187–198
33. St Maurice M, Bearne SL (2002) Kinetics and thermodynamics of mandelate racemase catalysis. Biochemistry 41:4048–4058. doi:10.1021/bi016044h
34. Stone SR, Betz A, Hofsteenge J (1991) Mechanistic studies on thrombin catalysis. Biochemistry 30:9841–9848. doi:10.1021/bi00105a005
35. Symcox MM, Reinhart GD (1992) A steady-state kinetic method for the verification of the rapid-equilibrium assumption in allosteric enzymes. Anal Biochem 206:394–399. doi:10.1016/0003-2697(92)90384-j

36. Topham CM (1990) A generalized theoretical treatment of the kinetics of an enzyme-catalysed reaction in the presence of an unstable irreversible modifier. J Theor Biol 145:547–572
37. Topham CM, Brocklehurst K (1992) In defence of the general validity of the Cha method of deriving rate equations. The importance of explicit recognition of the thermodynamic box in enzyme kinetics. Biochem J 282:261–265
38. Tsai CJ, del Sol A, Nussinov R (2008) Allostery: Absence of a change in shape does not imply that allostery is not at play. J Mol Biol 378:1–11
39. van't Hoff JH (1875) La chimie dans l'espace. P. M. Bazendijk, Rotterdam
40. Volkenstein MV, Goldstein BN (1966) A new method for solving the problems of the stationary kinetics of enzymological reactions. Biochim Biophys Acta 115:471–477
41. Walter C (1962) The effect of the steady-state on the kinetic analysis of enzyme inhibitors that are not competitive. Biochemistry 1:652–658
42. Weber G (1972) Ligand binding and internal equilibria in proteins. Biochemistry 11:864–878
43. Wells CM, Di Cera E (1992) Thrombin is a Na+-activated enzyme. Biochemistry 31:11721–11730. doi:10.1021/bi00162a008
44. Whitehead EP (1976) Simplifications of the derivations and forms of steady-state equations for non-equilibrium random substrate-modifier and allosteric enzyme mechanisms. Biochem J 159:449–456
45. Whitehead EP (1979) The structure of steady-state enzyme kinetic equations: a graph-theoretical algorithm for obtaining conditions for reduction in degree by common-factor cancellation. J Theor Biol 80:355–381
46. Whitehead EP (1981) On an alleged exception to generalized microscopic reversibility as a condition for reduction in degree of steady-state kinetic equations. Biochem J 199:461–462
47. Wong JTF (1965) The possible role of polyvalent carriers in cellular transports. Biochim Biophys Acta 94:102–113
48. Wyman J (1948) Heme proteins. Adv Protein Chem 4:407–531. doi:10.1016/S0065-3233(08)60011-X
49. Wyman J (1964) Linked functions and reciprocal effects in hemoglobin: a second look. Adv Protein Chem 19:223–286. doi:10.1016/S0065-3233(08)60190-4
50. Wyman J (1965) The binding potential, a neglected linkage concept. J Mol Biol 11:631–644
51. Wyman J (1967) Allosteric linkage. J Amer Chem Soc 89:2202–2218
52. Wyman J (1981) The cybernetics of biological macromolecules. Biophys Chem 14:135–146. doi:10.1016/0301-4622(81)85014-4

Chapter 3
Taxonomy of Enzyme–Modifier Interactions and the Specific Velocity Plot

Mixed inhibition as defined here encompasses such a broad range of behaviour that it may sometimes be helpful to subdivide it further.
Nomenclature Committee of the International Union of Biochemistry (1982). Symbolism and terminology in enzyme kinetics. Recommendations 1981. Eur. J. Biochem. 128, 281–291 (p. 287)

Abstract The mechanisms of enzyme–modifier interactions are treated as taxons and ranked in a way similar to that used in the biological classification of animals and plants. Two basic characters are used in the systematization of mechanisms: the allosteric linkage between substrate and modifier, if present, and the factor by which the modifier affects the catalytic constant. The states of these two characters can be combined giving rise to 17 unique combinations that are assigned to individual kinetic mechanisms embracing inhibition and nonessential activation. Essential activation is ranked as a separate group. The specific velocity plot, a method based on the same basic characters used in taxonomy, is exploited for the differential diagnosis of mechanisms during data analysis.

3.1 Introduction

In 1981, the International Union of Biochemistry (IUB) released recommendations on symbolism and terminology in enzyme kinetics [12], which at the date of printing this book have not yet been updated. Omissions in these recommendations and the need of their revision because of progress made after their publication have been discussed by Cornish-Bowden [6]. This chapter focuses on the classification and diagnosis of the interactions occurring between enzymes and modifiers using an approach that is more familiar to Biologists than Biochemists: by attempting a mechanistically based approach to the *systematics* of enzyme–modifier interactions. In the study of plants, animals, and microorganisms, the terms systematics, taxonomy, and nomenclature have been used with different meanings by scientists in

A. Baici, *Kinetics of Enzyme-Modifier Interactions*,
DOI 10.1007/978-3-7091-1402-5_3

Europe and North America. Waiving a philosophical discussion, I shall consider the following relationship between the three concepts [16, p. 23]:

$$\textit{Systematics} = \textit{Taxonomy} + \textit{Nomenclature}.$$

The lack of systematics in the study of enzyme–modifier interactions is due to shortage of both taxonomy and nomenclature. Regarding nomenclature, epithets applied to enzyme inhibitors, such as *competitive*, are based on operational definitions in accordance with IUB recommendations [12], whereas mechanistic criteria based on the interpretation of observed properties would improve nomenclature issues [6], [7, p. 136]. Regardless of the appropriateness of the terms competitive and uncompetitive used to describe enzyme inhibitors, problems arise with their application to activators. Among these, the designation *competitive activation* makes little sense to describe essential activation [12], or which name should be given a modifier that behaves as inhibitor at one substrate concentration and as activator at another concentration? Unfortunately, definitions have been and continue to be applied inappropriately, with *noncompetitive inhibition* leading the ranking list of ambiguous appliance, a fact that prompted the IUB to state ··· *it is discouraged for all purposes* [12].

In textbooks, the nomenclature of enzyme–modifier interactions has been interpreted in various ways. So, Sidney Bernhard [4, p. 82] called uncompetitive inhibition the mechanism generally known as mixed inhibition that also has been and continues to be called noncompetitive inhibition.

While collecting information on the kinetics of enzyme–modifier interactions for this book, and putting concepts together with the aim at finding any correlation between mechanisms, the first hurdle that presented was nomenclature. For some mechanisms, the way was straight, for others a fork was found in the road that needed a decision in defining a logical path. Thus, nomenclature was provisionally laid aside while attempting a taxonomic approach based on mechanistic characters. Reconsideration of a large number of kinetic studies conducted on allosteric modifiers revealed the existence of mechanisms that have been considered marginally, put in an undefined pool of special cases, or were new at all.

Inhibition and activation examples that would be lumped together in the pot of *mixed* effects will be considered as distinct kinetic entities, possessing a definite taxonomic position, to which a logical name has been given subsequently. The short account of allosteric interactions given in Sect. 2.1 highlighted the importance of allosteric modifiers for the richness of kinetic strategies used by this category of inhibitors and activators, which are the target of this chapter.

In the cause of internal consistency, the terms *competitive* and *uncompetitive* in the following taxonomic assessment, which comprises inhibition and activation in the same context, will be replaced by *specific* and *catalytic*, respectively. Inspiration in doing so came from a paper by Cárdenas and Cornish-Bowden, who were faced with the same problem [5]. Still in the interest of systematics, the epithet *balanced* will be used for mechanisms in which the specific and the catalytic components are

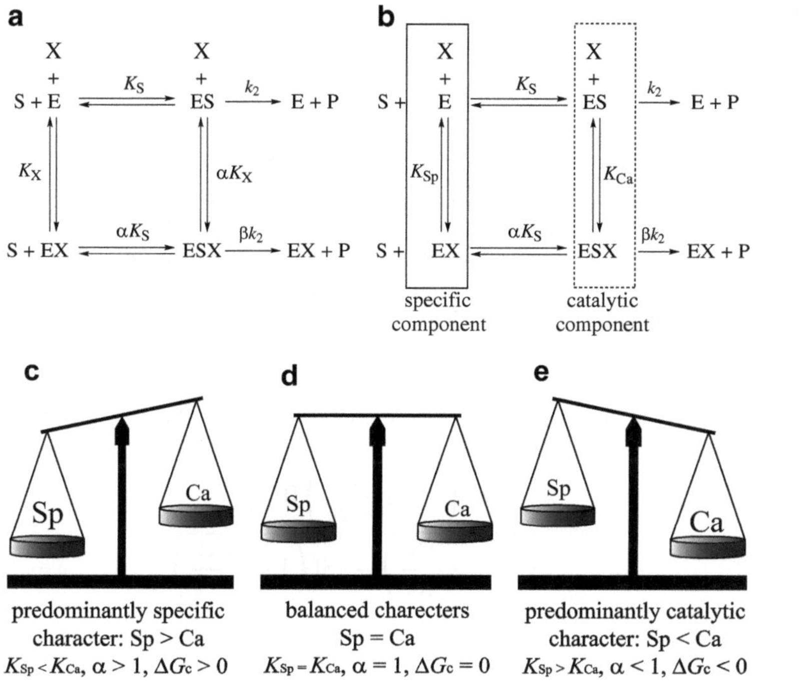

Scheme 3.1 Abridged representations of the general modifier mechanism. (**a**) Using the reciprocal of the allosteric coupling constant (α) in the formation of ESX. (**b**) K_{Sp} and K_{Ca} explicitly mark the specific and the catalytic branches of the general modifier mechanism. When E, ES, EX, and ESX exist, the two schemes represent mixed enzyme modification. Mixed modification can have *predominantly specific character* (Sp > Ca) if $K_{Sp} < K_{Ca}$, i.e., $\alpha > 1$ (**c**), *predominantly catalytic character* (Sp < Ca) if $K_{Sp} > K_{Ca}$, i.e., $\alpha < 1$ (**e**), or the specific and catalytic components can be equally represented (*balanced*, Sp = Ca), i.e., $\alpha = 1$ (**d**). The coupling energy according to (2.4) is therefore positive when the specific character dominates, negative when the catalytic character dominates, or zero if Sp = Ca

equally represented in inhibition and activation mechanisms. In such cases, the term *noncompetitive* would make no sense and cause confusion.

3.2 Alternative Symbolism

A widespread representation of the general modifier mechanism (Chap. 2) is shown in Scheme 3.1, where dissociation constants replace the corresponding couples of forward and reverse rate constants. The principle of microscopic reversibility cannot be violated and the ratios of dissociation constants for the cycle defined by the four enzyme-containing species discussed in Sect. (2.2) retain their validity whether or not the fluxes around ES and ESX are at equilibrium. The relationships (2.3) between dissociation constants become now

$$\frac{K_4}{K_3} = \frac{K_5}{K_1} = \frac{\alpha K_X}{K_X} = \frac{\alpha K_S}{K_S} = \frac{K_{Ca}}{K_{Sp}} = \alpha \,. \tag{3.1}$$

For reasons explained in the Introduction, K_{Sp} and K_{Ca} are used in place of the symbols K_{ic} and K_{iu} [12], respectively, to denote the *specific* (= competitive) and the *catalytic* (= uncompetitive) modification constants. Using the symbols K_X and αK_X makes sense only when both constants occur together in the rate equation. In Scheme 3.1 either the specific or the catalytic component may be absent, in which case the constant αK_X is better represented as K_{Ca}.

Note that in (3.1) the reciprocals of the relationships (2.3) are used. So, α is the *reciprocal of the allosteric coupling constant*. The rate equation (2.34) and the expressions of the apparent kinetic parameters therein can thus be rewritten as:

$$v = \frac{k_2\left(1+\beta\dfrac{[X]}{\alpha K_X}\right)[E]_t\,[S]}{K_m^S\left(1+\dfrac{[X]}{K_X}\right)+[S]\left(1+\dfrac{[X]}{\alpha K_X}\right)}\,, \tag{3.2}$$

$$k_{cat}^{app} = k_2 \times \frac{1+\beta\dfrac{[X]}{\alpha K_X}}{1+\dfrac{[X]}{\alpha K_X}}\,, \tag{3.3}$$

$$K_m^{app} = K_m^S \times \frac{1+\dfrac{[X]}{K_X}}{1+\dfrac{[X]}{\alpha K_X}}\,, \tag{3.4}$$

$$k^{app} = \frac{k_2}{K_m^S} \times \frac{1+\beta\dfrac{[X]}{\alpha K_X}}{1+\dfrac{[X]}{K_X}}\,. \tag{3.5}$$

The mechanism in Scheme 3.1a or its equivalent in panel (b) is appropriate for analyzing the kinetic behavior of allosteric inhibitors and activators. The general case of this reaction scheme comprises two catalytic paths: depending on the coefficient β, the ESX complex can be turned over to EX and product(s) at a lower ($\beta < 1$), at the same ($\beta = 1$), at a higher rate than ES ($\beta > 1$) or not at all if $\beta = 0$. If only ESX decomposes into EX + P while ES does not, the scheme corresponds to *essential activation* discussed in Sect. 3.4. Considering all possible combinations, definitions of the degrees of enzymatic inhibition and activation attainable in the presence of modifiers are necessary. Thus, the adjectives *linear* and *hyperbolic* will be consistently used in naming enzyme-modification mechanisms depending on the shape of the plots of $1/k_{\text{cat}}^{\text{app}}$ or $1/k^{\text{app}}$ versus [X]:

$$\frac{1}{k_{\text{cat}}^{\text{app}}} = \frac{1}{k_2} \times \frac{1 + \dfrac{[\text{X}]}{\alpha K_{\text{X}}}}{1 + \beta \dfrac{[\text{X}]}{\alpha K_{\text{X}}}}, \tag{3.6}$$

$$\frac{1}{k^{\text{app}}} = \frac{K_{\text{m}}^{\text{S}}}{k_2} \times \frac{1 + \dfrac{[\text{X}]}{K_{\text{X}}}}{1 + \beta \dfrac{[\text{X}]}{\alpha K_{\text{X}}}}. \tag{3.7}$$

Depending on the mechanism, *both plots* of the dependent variables against [X] according to (3.6) and (3.7) can be linear or hyperbolic or either one can be *independent* of [X] while the other is linear or hyperbolic. A modifier mechanism is called linear if both dependences of $1/k_{\text{cat}}^{\text{app}}$ and $1/k^{\text{app}}$ are linear or one is linear and the other *does not depend* on [X]. A modifier mechanism is said to be hyperbolic if both dependences of $1/k_{\text{cat}}^{\text{app}}$ and $1/k^{\text{app}}$ are hyperbolic or one is hyperbolic while the other *does not depend* on [X].

As a support to the systematics of enzyme–modifier interactions in the following sections, Fig. 3.1 shows graphically the many ways $k_{\text{cat}}^{\text{app}}$, $K_{\text{m}}^{\text{app}}$ and k^{app} in (3.3)–(3.5) depend on modifier concentration. Instead of using generic plots, examples with representative values of α and β were plotted assuming quasi-equilibrium for all steps. In this case (3.2)–(3.5) are the same with K_{m}^{S} replaced by K_{m}^{0}, which can be simply represented as K_{m} if there is no ambiguity about its significance:

$$v = \frac{k_2 \left(1 + \beta \dfrac{[\text{X}]}{\alpha K_{\text{X}}}\right) [\text{E}]_{\text{t}} [\text{S}]}{K_{\text{m}} \left(1 + \dfrac{[\text{X}]}{K_{\text{X}}}\right) + [\text{S}] \left(1 + \dfrac{[\text{X}]}{\alpha K_{\text{X}}}\right)}. \tag{3.8}$$

On any account, the curves in Fig. 3.1 retain the same trends also when drawn using the full steady-state equation. The only exception is $K_{\text{m}}^{\text{app}}$, which may show a maximum or a minimum if substrate stickiness considerably differs in the two catalytic steps (Sect. 2.8.1).

The reciprocal allosteric coupling constant α can take any positive value, i.e., $0 < \alpha \leq \infty$. The apparent Michaelis constant depends on α, K_{m} and K_{X}: it decreases hyperbolically with increasing [X] when $\alpha < 1$, it is independent of [X] when $\alpha = 1$, it increases hyperbolically with [X] when $\alpha > 1$ and it increases linearly with [X] only if it takes very large values (symbolically when $\alpha = \infty$). These properties are summarized in the middle panel of Fig. 3.1d.

The dependences of the kinetic parameters on [X] for $0 < \alpha < 1$ are shown in row (a) of Fig. 3.1 that shows calculations performed with $\alpha = 0.5$ and five values

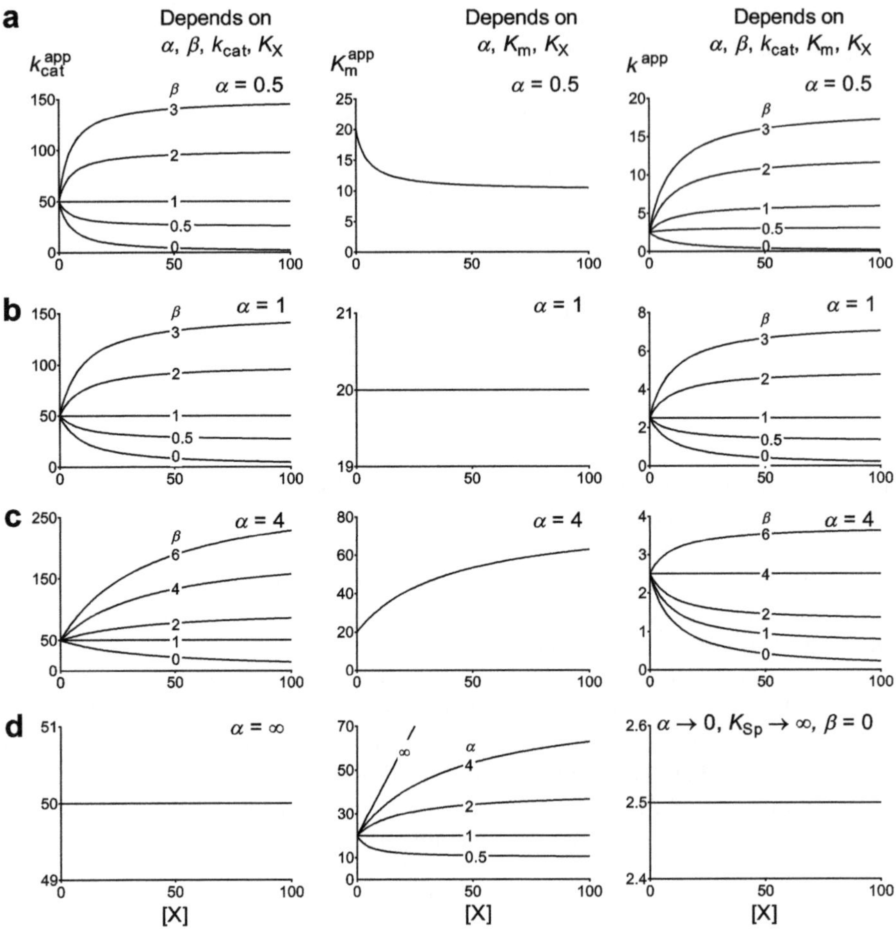

Fig. 3.1 Dependences of the apparent kinetic parameters on modifier concentration for the general modifier mechanism as a function of α and β. The plots were generated using (3.3)–(3.5) with $k_{cat} = 50\,\mathrm{s}^{-1}$, $K_m = 20\,\mu\mathrm{M}$ and $K_X = 10\,\mu\mathrm{M}$. The apparent catalytic, Michaelis and specificity constants are displayed vertically in the three arrays of graphics. Horizontally in (a)-(c): dependences on three finite values of α, while (d) illustrates limiting cases

of β, including zero. The catalytic constant increases hyperbolically with [X] when $\beta > 1$, is independent of [X] when $\beta = 1$ and decreases hyperbolically with [X] when $0 < \beta < 1$. The specificity constant has a similar behavior but becomes independent of [X] when $\beta = \alpha$.

The dependences of the kinetic parameters on [X] when α is equal to or larger than 1, shown in Fig. 3.1b,c, can be commented in the same way. The panels for k_{cat}^{app} and k^{app} in two limiting cases are shown in (d). $\alpha = \infty$ characterizes linear specific inhibition, for which k_{cat}^{app} does not depend on [X], whereas the right panel shows the independence of the specificity constant on [X], characteristic of linear catalytic inhibition.

Altogether, the information obtained by measuring how the kinetic parameters depend on [X] can not only be exploited for diagnostic and quantitative purposes but also for defining taxonomically meaningful rankings. As will be evident in the following sections, unique combinations of α and β exist for mechanisms that deserve the promotion to individual, unambiguously identified entities instead of being treated as *special cases* in a generic pool of mechanisms. It is hoped that this approach will shed light in present conflicting issues around enzyme inhibition and activation by providing unambiguous quantitative tools for identifying and naming the action of modifiers using a common language.

3.3 Taxonomy of Enzyme–Modifier Interactions

The symbolism in Scheme 3.1a, with dimensionless multiplication coefficients α and β, can be exploited for ranking enzyme–modifier interactions in a way analogous to the binomial system introduced by Carl Linnaeus, which forms the basis of the modern systematics of plants and animals. A major aim of this book is to provide a logical basis for the identification of kinetic mechanisms starting from measurements. In this endeavor, *diagnostic keys*[1] will be used as those used by Biologists for the identification of animals and plants. The goal of a botanical excursion is collecting and analyzing specimens to finally assign them to a *taxon*, which owns a name and a rank within a genus, family, and higher groupings. To do this, characters are sought that are specific of a given taxon so that a group of characters applies to just one entity. Similarly, the interactions between enzymes and modifiers possess their own characters, which can be measured, analyzed in every accessible detail, and finally assigned to a given mechanism, which corresponds to a taxon.

Compared with biological classification, which may be either *phylogenetic* (based on characters that developed by evolution from an ancestor) or *phenetic* (based on observing present-day characters), the ranking of modifier mechanisms in this book is purely phenetic, with no claims of ancestral lineages in an evolutionary sense. References made below to mechanisms as *descendants of a common ancestor mechanism* are based solely on measurable characters and logical relationships.

Individual mechanisms that can be derived from the general modifier mechanism possess two basic *characters*: the ratio between the catalytic and the specific dissociation constants, $\alpha = K_{Ca}/K_{Sp}$, corresponding to the reciprocal of the allosteric coupling constant between modifier and substrate, and the coefficient β as the factor by which the modifier affects the catalytic constant. These two characters

---

[1]For practical reasons, all keys are grouped in Chap. 10 because of linkages between modification mechanisms described in different chapters.

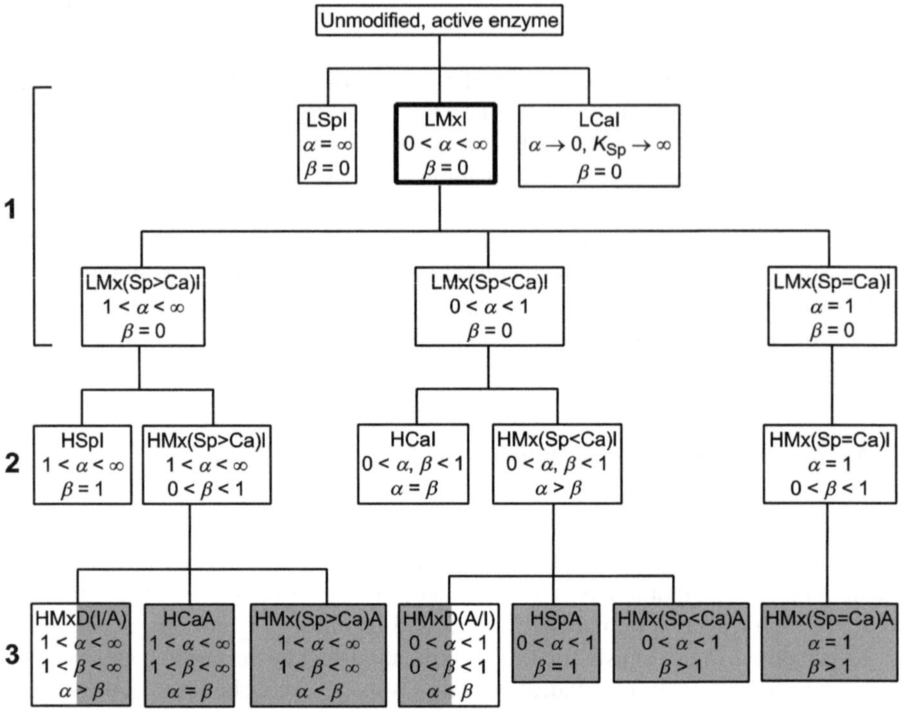

Fig. 3.2 Taxonomy of enzyme–modifier interactions. Full names are given in Figs. 3.3, 3.4, and 3.5. The *characters* that form the basis of this classification are the coefficients α and β in the general modifier mechanism (Scheme 3.1), i.e., the reciprocal of the allosteric coupling constant and the factor by which the catalytic constant is decreased or increased by the modifier, respectively. The dendrogram is organized in three levels with the root on the *top* that represents the unmodified enzyme and the branches descending to the *bottom*. The acronyms are composed of the following abbreviations: A = activation, Ca = catalytic, D = dual character (inhibition or activation), H = hyperbolic, I = inhibition, L = linear, Mx = mixed, Sp = specific, (Sp>Ca) = predominantly specific, (Sp<Ca) = predominantly catalytic, (Sp=Ca) = equally contributing specific and catalytic components (*balanced*). In level 1, linear mixed inhibition (LMxI) shown in a *box with thicker border* comprises three mechanisms typified by different values of α. The total number of modification mechanisms is thus 17. *Boxes with completely white background* indicate inhibition and entirely *gray-shaded boxes* represent activation at any substrate and modifier concentration. *Boxes with half white/half gray background* represent dual modifiers that behave as inhibitors or activators depending on the critical substrate concentration $[S]^0$ (Sect. 3.7). Modifiers that belong to HMxD(I/A) in level 3 are inhibitors when $[S] < [S]^0$ and activators when $[S] > [S]^0$. In the HMxD(A/I) mechanism, activation occurs at $[S] < [S]^0$ and inhibition at $[S] > [S]^0$

exhibit various states, whose combinations are used here for taxonomic purposes (Fig. 3.2). Full names and acronyms of these mechanisms are shown in Figs. 3.3, 3.4, and 3.5, while descriptions, more details, and published examples will be provided in Chap. 5.

In Fig. 3.2, from the *root* of the tree located on the top as unmodified enzyme, *branches* descend vertically. According to the characters chosen as taxonomic

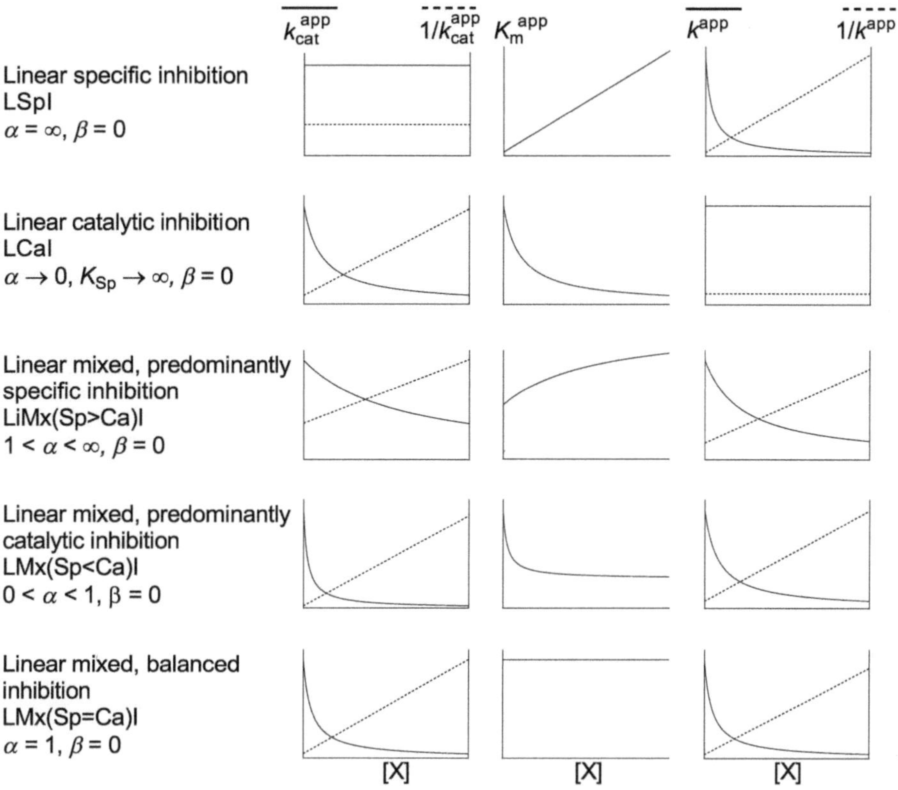

Fig. 3.3 Dependences of apparent k_{cat}, K_m and specificity constant on modifier concentration for the five linear inhibition mechanisms in taxonomic level 1. The ordinates, labeled on the top panels, refer all five plots in the vertical series

criteria, linear mixed inhibition is the *ancestor* of a large progeny: the branches follow the dominant character α, which remains unchanged within three groups containing a total of 15 entities in taxonomic levels 1–3. Horizontally in the tree at levels 2 and 3 it is the factor β that differentiates among related mechanisms.

Taxonomically, all mechanisms in Fig. 3.2 are in line with the IUBMB recommendations of symbolism and terminology in enzyme kinetics [12, pp. 287–288]. As shown in the synoptic dependences of k_{cat}^{app}, K_m^{app}, and k^{app} upon modifier concentration in Figs. 3.3, 3.4, and 3.5, the effects exerted by the modifier on the specificity constant and on the catalytic constant as a basis of the definitions of competitive (specific) and uncompetitive (catalytic) effects are fully respected. An extension was necessary for the activation and dual mechanisms in level 3, which have not been exhaustively treated in the recommendations [12]. The dependences of k_{cat}^{app}, K_m^{app} and k^{app} upon modifier concentration are *unique* and thus *diagnostic* of the individual mechanisms.

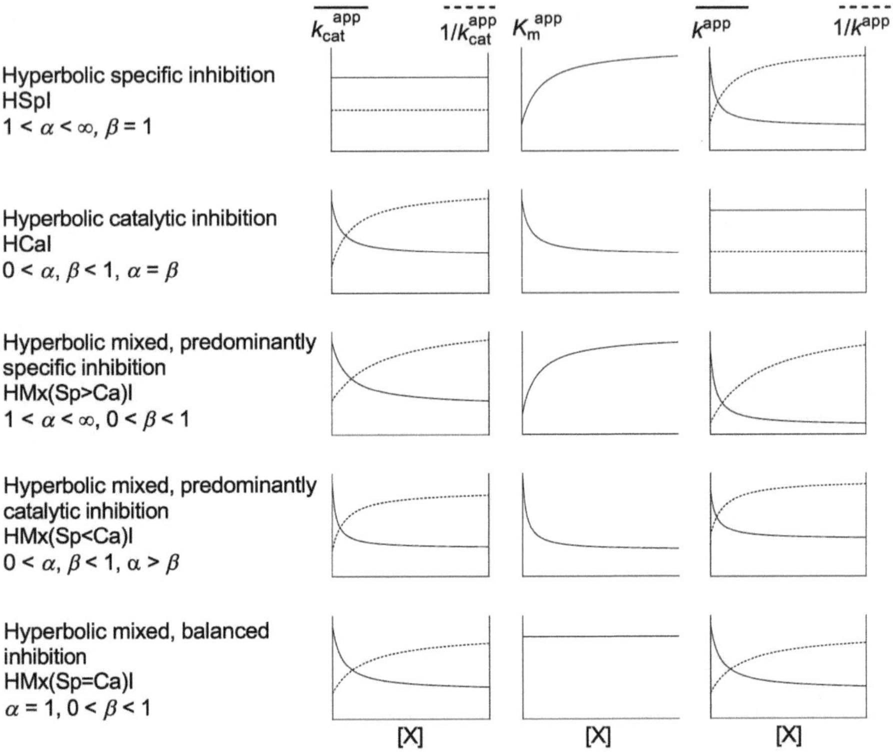

Fig. 3.4 Dependences of apparent k_{cat}, K_m, and specificity constant on modifier concentration for the five hyperbolic inhibition mechanisms in taxonomic level 2. The ordinates, labeled on the top panels, refer all five plots in the vertical series

All mechanisms listed in Fig. 3.2 will be presented briefly in the following subsections and their properties illustrated by the dependences of the kinetic parameters on modifier concentrations. More details about individual mechanisms, accompanied by practical examples, will be provided in Chap. 5.

3.3.1 Taxonomic Level 1: Linear Inhibition

In Fig. 3.2, with the unmodified enzyme as the starting point (*root*), the first taxonomic level contains five linear inhibition mechanisms. The enzyme at the root is catalytically active to distinguish the present taxonomic tree from that of essential activation (Sect. 3.4). Linear specific inhibition and linear catalytic inhibition do not belong to the general modifier mechanism because they lack allosteric linkage between substrate and modifier (Sects. 2.2 and 2.3). They remain thus isolated from other species in the taxonomic tree and have no descendants. However, from a purely

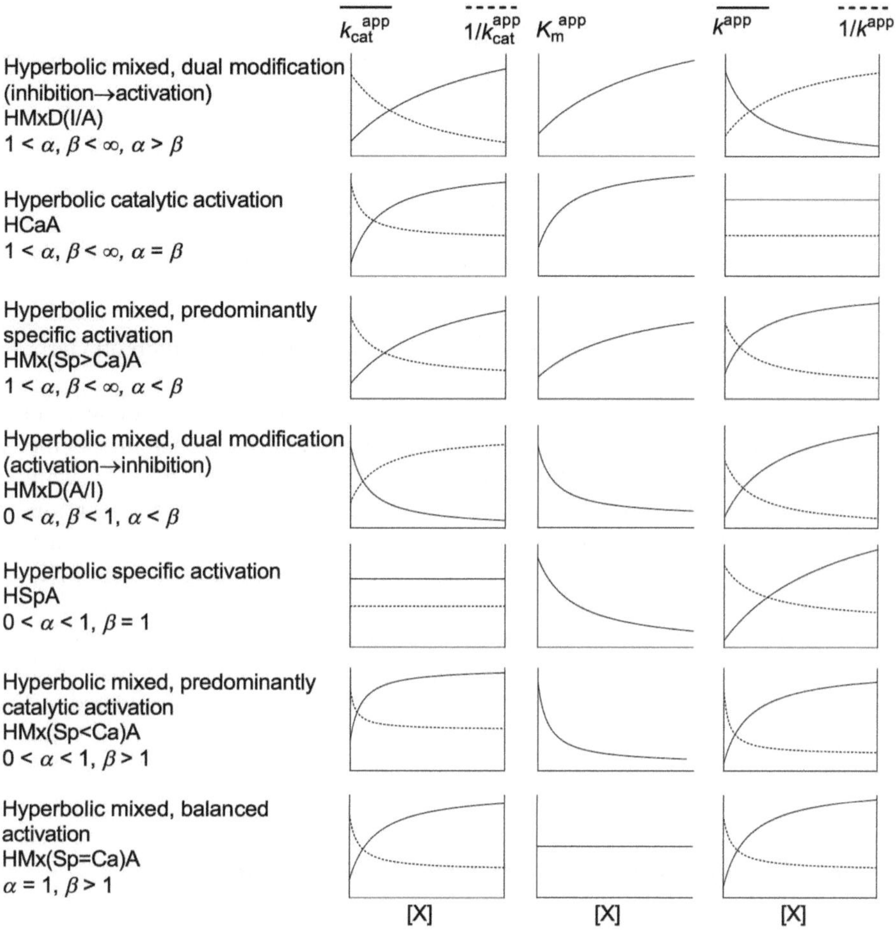

Fig. 3.5 Dependences of the apparent kinetic parameters for the mechanisms in taxonomic level 3 of Fig. 3.2. The ordinates, labeled on the top panels, refer to all plots in three vertical series

formal point of view and with constants defined as shown in the corresponding boxes, LSpI and LCaI can be deduced from the basic scheme of the general modifier mechanism (Scheme 3.2). The thick box around linear mixed inhibition indicates that it consists of three species distinguished by the dominant character α. In LMxI, the ESX complex is catalytically inactive.

The importance of linear mixed inhibition for enzyme regulation can be appreciated from its progeny, which makes up all remaining mechanisms in levels 2 and 3. It can have either predominantly specific [LMx(Sp>Ca)I] or catalytic character [LMx(Sp<Ca)I], well-defined mechanisms known from numerous examples in the literature. Linear mixed, balanced inhibition [LMx(Sp=Ca)I], although representing a definite taxonomic position, has been and is often confused with either one of

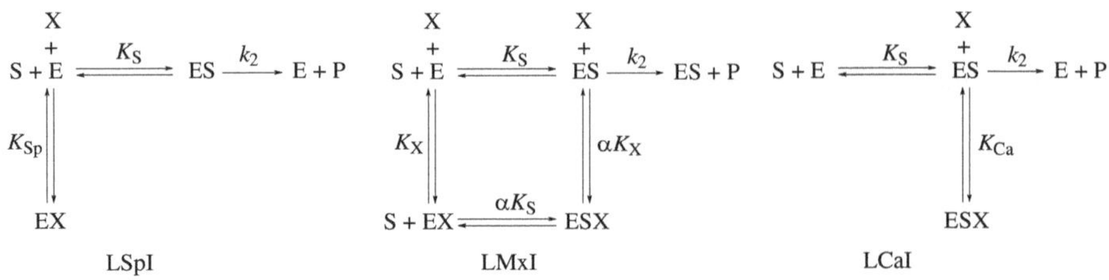

Scheme 3.2 Linear inhibition mechanisms in taxonomic level 1 of Fig. 3.2

the preceding mechanisms. *Mixed inhibition* has been used as an umbrella term to cover many of the mechanisms in levels 1–3 of Fig. 3.2.

Internal consistency within the five linear mechanisms in taxonomic level 1 is demonstrated in Fig. 3.3. In linear specific inhibition, the apparent specificity constant decreases with [X], while k_{cat}^{app} is unaffected. Conversely, in linear catalytic inhibition k_{cat}^{app} decreases with and k^{app} is independent of [X]. In linear mixed inhibition with predominantly specific character, $K_X < \alpha K_X$, more enzyme lingers during reaction in EX than in ESX, a property reflected by a larger effect of [X] on k^{app} than on k_{cat}^{app}. The contrary is valid when the catalytic character dominates ($K_X > \alpha K_X$). With equal contributions of the specific and catalytic components, linear mixed, balanced inhibition is characterized by the independence of K_m^{app} on [X]. The criterion for ascribing all five mechanisms to the linear type is validated by the linear dependences of $1/k_{cat}^{app}$ and $1/k^{app}$ upon [X] (dashed lines in the corresponding panels of Fig. 3.3).

3.3.2 Taxonomic Level 2: Hyperbolic Inhibition

The characteristic dependences of the apparent kinetic parameters on modifier concentration are shown in Fig. 3.4, where the five hyperbolic mechanisms are displayed in the same order (top to bottom) as their linear counterparts in Fig. 3.3. The dependences of the parameters on [X] for the hyperbolic mechanisms is the same as for the corresponding linear cases with the exception of the K_m^{app}-dependence, which is hyperbolic in HSpI and linear in LSpI. However, the hyperbolic mechanisms can be unequivocally identified from the dependences of reciprocal apparent catalytic constant and reciprocal specificity constant on [X], which increases hyperbolically in the five mechanisms of Fig. 3.4 (the plots $1/k_{cat}^{app}$ for hyperbolic specific inhibition and $1/k^{app}$ for hyperbolic catalytic inhibition are exceptions because there is no dependence).

Hyperbolic specific inhibition (HSpI), recognized by Segal, Kachmar and Boyer as *another plausible mechanism* [13, p. 190] (more in Sect. 5.3.1) shares with linear specific inhibition the independence of the catalytic constant of modifier concentration (Fig. 3.4). The catalytic constant is not affected by the modifier ($\beta = 1$) and the hyperbolic nature of this mechanisms, which may be confused with linear specific inhibition, is due to sequestration of the enzyme in the EX and ESX complexes thus leaving less enzyme available for the E $\rightarrow$ ES $\rightarrow$ P catalytic path. Hyperbolic catalytic inhibition (HCaI), in which α and β have the same numerical value, the specificity constant is independent of modifier concentration because both the catalytic and the Michaelis constant are affected by the same factor. Hyperbolic mixed, predominantly specific inhibition [HMx(Sp>Ca)I] and its predominantly catalytic counterpart [HMx(Sp<Ca)I] are frequently found in the biochemical literature although recognizing their identity even in recent literature is not always immediate when insufficient information is provided. The fifth mechanism (left-to-right in taxonomic level 2) is hyperbolic mixed, balanced inhibition [HMx(Sp=Ca)I], the only offspring of its linear predecessor. HMxI is the acronym we can give hyperbolic mixed inhibition as a pool of HMx(Sp>Ca)I, HMx(Sp<Ca)I and HMx(Sp=Ca)I, used for instance in multiple enzyme–modifier interactions (Chap. 6).

3.3.3 Taxonomic Level 3: Nonessential Activation

The branches of the taxonomic tree end up at level 3 with five hyperbolic activation mechanisms and two mechanisms with dual behavior as either inhibitors or activators depending on substrate concentration. The increase of the initial velocity in enzyme-catalyzed reactions discussed in this section belongs to *nonessential activation*, meaning that the enzyme is active also in the absence of modifiers [12].

The presence of hyperbolic catalytic activation (HCaA) in a predominantly specific lineage and hyperbolic specific activation (HSpA), as descendant of a predominantly catalytic inhibition mechanism, may appear odd at first sight. However, the taxonomic position of these two mechanisms is correct along their branches in the tree that follows the dominant character α and the question may be only on nomenclature. The answer comes after examining the dependences of the catalytic and of the specificity constant in Fig. 3.5: the specificity constant of HCaA and the catalytic constant of HSpA do not depend on modifier concentration, which mirror the definitions of catalytic and specific inhibition mechanisms, respectively, seen in linear (Fig. 3.3) and hyperbolic inhibition (Fig. 3.4).

That both HMx(Sp>Ca)A and HMx(Sp<Ca)A behave as activators at any modifier concentration can be deduced by comparing the effect on the catalytic constant defined by β, which is larger than 1, compared with $\beta < 1$ in the two inhibition mechanisms in level 2 from which they descend. Spotting these mechanisms in the literature is not easy when only results in tabular form are available. Also, nonessential activation is far less popular than enzyme inhibition in general.

Ranking the two dual modification mechanisms (half white/half gray boxes in Fig. 3.2) using specific and catalytic lineages would not immediately suggest their changing behavior depending on substrate concentration. These modifiers can in fact either behave as inhibitors or activators because a physically meaningful critical substrate concentration, $[S]^0$, can be calculated (Sect. 3.7). Therefore, the full names shown in Fig. 3.5 and acronyms HMxD(I/A) and HMxD(A/I) are proposed here, where I/A and A/I denote the transition from inhibition to activation and activation to inhibition when the substrate concentration is increased. When $[S] < [S]^0$ these modifiers act as inhibitors [HMxD(I/A)] or activators [HMXD(A/I)], they have no effect on enzyme activity if $[S] = [S]^0$ and invert their behavior when $[S] > [S]^0$. Further details will be provided in Sects. 5.4.1 and 5.4.4.

The seventh mechanism in level 3 [HMx(Sp=Ca)A] predicts activation at any substrate concentration because $\beta > 1$ and $\alpha = 1$. Thus, the apparent Michaelis constant does not change, while the apparent catalytic and specificity constants increase with modifier concentration.

Levels 2 and 3 in the taxonomic tree of Fig. 3.2, which contain mechanisms with hyperbolic character, represent the *realm of allosteric modifiers*. In virtue of their extended functionality and flexibility, modifiers ranked in these categories constitute powerful tools for the control of enzyme activity. The behavior of these mechanisms emphasizes that interactions take place between three partners: enzyme, modifier, and substrate. The substrate can either have a monotonous effect on the action of the modifier, increasing or decreasing its effects as the substrate concentration increases, or a dual effect, determining if the modifier should behave as inhibitor or as activator.

3.3.4 Extended Applicability of the Taxonomic Tree

The systematic ranking of enzyme–modifier interactions in Fig. 3.2 is not limited to the reversible mechanisms discussed in this chapter. Also modifiers that inactivate enzymes irreversibly (Chap. 9) are either specific, mixed, or catalytic and they can be linear or hyperbolic much as their reversible counterparts [17, 18]. The basic reaction schemes of the reversible, slow-onset modification mechanisms treated in Chap. 8 have their roots in Fig. 3.2. Actually, any slow-onset inhibition or activation is irrelevant to the underlying type of interaction between enzyme, substrate, and modifier. Likewise, a tight-binding condition often present when measuring enzyme–modifier interactions is merely an experimental issue that is independent of the kinetic mechanisms (Sect. 4.4).

Substrate inhibition, which will be discussed in Chap. 4, can in principle belong to any of the inhibition mechanisms listed in Fig. 3.2. Though, the palette of mechanisms known for substrate inhibition is not so broad.

The kinetic treatment of the general modifier mechanism, the dependences of the kinetic parameters on modifier concentration, and the specific velocity plot as a diagnostic method for discriminating among the basic mechanisms can be extended to enzyme-catalyzed reactions that involve more than one substrate. This transfer of methods is legitimate only if three criteria formulated by Frieden are fulfilled [9]:

1. The modifier interferes with only one of the substrates of the reaction. Accordingly, initial velocity measurements are performed by varying the concentration

of the affected substrate while keeping the concentrations of the other substrate(s) at a saturating level
2. The Michaelis constant of a substrate does not depend on the concentration of the other substrate(s)
3. The effect exerted by the modifier on the limiting rate does not imply a change of the rate-limiting step observed in the absence of modifier.

It is implicit that the kinetic mechanism of reactions that involve two or more substrates in the absence of modifiers is already known before attempting measurements in the presence of modifiers.

3.3.5 K-Systems and V-Systems

Monod, Wyman, and Changeux subdivided the activating and inhibitory effects exerted by modifiers on *allosteric enzymes* in two groups depending on which kinetic parameter, K_m or V, is affected by the modifier [11]. Although formulated in the frame of cooperative phenomena, the authors did not restrict this definition to cooperative systems, asserting the compatibility of their definitions with other mechanisms.

- *K-systems*: those in which the modifier exerts an effect on the apparent affinity of the substrate for the enzyme, and conversely
  - K-inhibition ($K_m \uparrow$): K_m^{app} increases for increasing [X]
  - K-activation ($K_m \downarrow$): K_m^{app} decreases for increasing [X]
- *V-systems*: those in which the modifier exerts an effect on the apparent limiting rate (and consequently on k_{cat})
  - V-inhibition ($V \downarrow$): V^{app} decreases for increasing [X]
  - V-activation ($V \uparrow$): V^{app} increases for increasing [X]

This subdivision of allosteric effects has been and is widely used and, more commonly, the two classes are called *K-type* inhibition or activation and *V-type* inhibition or activation. If consistently applied *and* accompanied by sufficiently documented experimental evidence, this nomenclature is satisfactory. However, reporting that a modifier causes K-type activation and V-type inhibition is ambiguous without further details. In fact, such a modifier can belong to hyperbolic catalytic inhibition, hyperbolic mixed, predominantly catalytic inhibition, or hyperbolic mixed dual modification (activation to inhibition) because in all three cases V and K_m decrease for increasing [X] as shown in Figs. 3.4 and 3.5. Disambiguation is therefore necessary by stating which one of the two effects increases or decreases faster than the other when the concentration of the modifier is increased. This is equivalent to showing the dependence of the specificity constant, which is independent of [X] in HCaI, decreases with increasing [X] in HMx(Sp<Ca)I and increases with [X] in HMxD(A/I). For the sake of clarity, Table 3.1 compares

Table 3.1 K-systems and V-systems [11]

K-type		V-type	Nomenclature of this chapter
Activation (K_m $\downarrow$)	and	Activation (V $\uparrow$)	HMx(Sp<Ca)A
Activation (K_m $\downarrow$)	and	Inhibition (V $\downarrow$)	HCaI, HMx(Sp<Ca)I, HMxD(A/I)
Inhibition (K_m $\uparrow$)	and	Activation (V $\uparrow$)	HMxD(I/A), HCaA, HMx(Sp>Ca)A
Inhibition (K_m $\uparrow$)	and	Inhibition (V $\downarrow$)	HMx(Sp>Ca)I
Inhibition (K_m $\uparrow$)	only	–	HSpI
–	only	Inhibition (V $\downarrow$)	HMx(Sp=Ca)I
Activation (K_m $\downarrow$)	only	–	HSpA
–	only	Activation (V $\uparrow$)	HMx(Sp=Ca)A

Comparison of the nomenclature of enzyme-modifiers proposed in this chapter with the effects described as combinations of K-type and V-type. The arrows mean decreasing ($\downarrow$) and increasing ($\uparrow$), respectively, for increasing [X]

the possible combinations of K- and V-type interactions with the nomenclature proposed in this chapter. In six cases disambiguation is necessary because reporting just the K-type and V-type effects is insufficient.

In essence, unless *all dependences* of the kinetic parameters on modifier concentration are shown as in Figs. 3.4 and 3.5, or the allosteric coupling constant and the effect of the modifier on catalysis are explicitly stated, the nature of the interaction between enzyme and modifier is insufficiently typified.

3.4 Essential Activation

Some enzymes are catalytically inert on their substrates but become catalytically active in the presence of a further substance, e.g., a metal ion or a small molecule. This behavior is known as *essential activation* [12] or *obligatory activation* [20] because the rate is zero in the absence of the activator. Essential activation differs from the nonessential counterpart discussed in Sect. 3.3.3, where all members in the group (taxonomic level 3) possess finite activity in the absence of their activators. The mechanisms that correspond to substrate turnover occurring *only* with the help of an additional substance belong to a taxonomic group distinct from that of the modifiers in the main tree of Fig. 3.2.

An early account of essential activation, treated together with nonessential activation, has been published by Segal, Kachmar, and Boyer [13], while Segel dedicated the entire Chap. 5 and part of Chap. 6 of his *Enzyme Kinetics* book to a comprehensive survey of essential and nonessential activation [14]. A collection of rate equations involving one to four activator molecules/ions, accompanied by three-dimensional graphics, has been provided by Alberty [1, Chap. 9]. In their comprehensive treatment *Enzymes*, Dixon and Webb considered essential and nonessential activation in several systems [8, pp. 381–399] providing the rate equations under equilibrium conditions but not the dependences of kinetic parameters on the concentration of the activators, indispensable tools in the differential diagnosis of mechanisms.

The effective substrate of an enzyme may not be the molecule that is subjected to transformation (S) but a complex between this molecule and an activator (X). A classical example is ATP as substrate and Mg^{2+} as activator. The reaction that takes place is then $E + SX \rightarrow \cdots$ but the enzymes that operate with this mechanism exploit manifold strategies to cover their regulatory needs. Segel [14, pp. 242–271] treated in detail the following mechanisms that sit astride nonessential and essential activation:

1. only SX binds to the enzyme
2. SX and S bind to the enzyme
3. SX and X bind to the enzyme
4. SX, S, and X bind to the enzyme
5. X is an essential activator that allows binding of S, X, and SX to the catalytic site
6. X is an essential activator that allows the binding of only SX to the catalytic site
7. X is an essential activator, the binding of X, S, and SX to E and EX occurs at random
8. X is a nonessential activator and only SX binds to the catalytic site
9. both S and SX are substrates

A systematic reconsideration of this complex topic lies outside the envisaged goals of the present chapter but we will come back to this point in Chap. 6 and 7 that deal with multiple enzyme–modifier interactions. The methods described there are suitable for representing and interpreting simultaneous interactions that can include various types of essential and nonessential activation.

This section surveys the four prototypes of essential activation that, in combination with the seven nonessential mechanisms described in Sect. 3.3.3, are indispensable for understanding the variations on the theme.

3.4.1 Linear Mixed Activation

Scheme 3.3a represents *linear mixed activation*. The name has been recommended [12] and the mechanism has been discussed by several authors, e.g., [7, 13, 14]. The steady-state rate equation of linear mixed activation can be derived in the same way described for the general modifier mechanism in Appendix 4 in Chap. 2 setting $k_2 = 0$. This equation, which is not shown here, contains terms that are quadratic both in the substrate and in the modifier concentration. The assumption of quasi-equilibrium for the binding of modifier eliminates all $[S]^2$-containing terms while retaining terms in $[X]^2$. A bonus with respect to the corresponding situation in the general modifier mechanism is that, in essential activation, there is only one catalytic path, a property that avoids the problems related to differences in substrate stickiness when two catalytic paths are present. The derivation method is shown in Sect. 2.4.3 and the final result is shown here. Depending on personal taste, rearrangements of

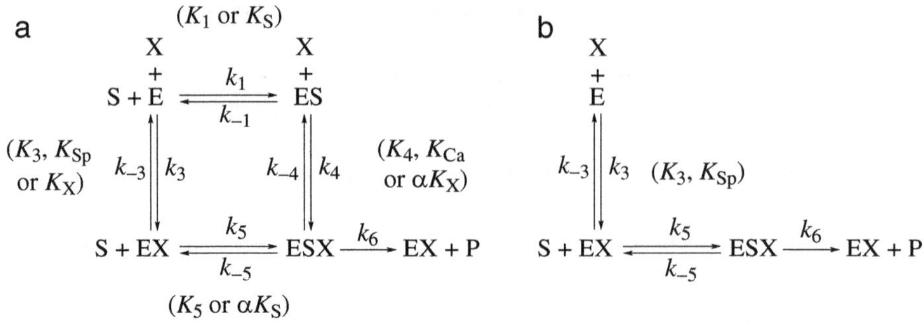

Scheme 3.3 Mechanisms of enzyme activation. (**a**) Linear mixed activation, (**b**) linear specific activation. The dissociation constants indicated in parentheses show alternative notations

the rate equation can be performed in equivalent ways, one of which is

$$v = \frac{\dfrac{k_6}{1+\dfrac{K_4}{[\mathrm{X}]}}[\mathrm{S}]\,[\mathrm{E}]_t}{\dfrac{k_{-1}\dfrac{K_4}{[\mathrm{X}]}+(k_{-5}+k_6)}{k_1\dfrac{K_3}{[\mathrm{X}]}+k_5}\times\dfrac{1+\dfrac{K_3}{[\mathrm{X}]}}{1+\dfrac{K_4}{[\mathrm{X}]}}+[\mathrm{S}]}, \tag{3.9}$$

which can be rearranged as (3.11) considering that microscopic reversibility dictates

$$k_1K_3 = \frac{k_{-1}k_5}{k_{-5}}K_4\,,\quad k_{-1}K_4 = \frac{k_1k_{-5}}{k_5}K_3\,,\quad \frac{k_1K_3}{k_5} = \frac{k_{-1}K_4}{k_{-5}}\,, \tag{3.10}$$

$$v = \frac{\dfrac{k_6}{1+\dfrac{K_4}{[\mathrm{X}]}}[\mathrm{S}]\,[\mathrm{E}]_t}{\dfrac{k_{-5}\left(1+\dfrac{k_1K_3}{k_5\,[\mathrm{X}]}\right)+k_6}{k_5\left(1+\dfrac{k_{-1}K_4}{k_{-5}\,[\mathrm{X}]}\right)}\times\dfrac{1+\dfrac{K_3}{[\mathrm{X}]}}{1+\dfrac{K_4}{[\mathrm{X}]}}+[\mathrm{S}]}$$

$$= \frac{\dfrac{k_6}{1+\dfrac{K_4}{[\mathrm{X}]}}[\mathrm{S}]\,[\mathrm{E}]_t}{\dfrac{k_{-5}\left(1+\dfrac{k_1K_3}{k_5\,[\mathrm{X}]}\right)+k_6}{k_5\left(1+\dfrac{k_1K_3}{k_5\,[\mathrm{X}]}\right)}\times\dfrac{1+\dfrac{K_3}{[\mathrm{X}]}}{1+\dfrac{K_4}{[\mathrm{X}]}}+[\mathrm{S}]}\,. \tag{3.11}$$

The apparent kinetic parameters in (3.11) are

$$k_{\text{cat}}^{\text{app}} = \frac{k_6}{1 + K_4/[\text{X}]} \tag{3.12}$$

$$K_{\text{m}}^{\text{app}} = \underbrace{\frac{k_{-5}\left(1 + \dfrac{k_1 K_3}{k_5\,[\text{X}]}\right) + k_6}{k_5\left(1 + \dfrac{k_1 K_3}{k_5\,[\text{X}]}\right)}}_{K_{\text{m}}} \times \frac{1 + \dfrac{K_3}{[\text{X}]}}{1 + \dfrac{K_4}{[\text{X}]}} \tag{3.13}$$

$$k^{\text{app}} = \frac{k_5 k_6 \left(1 + \dfrac{k_1 K_3}{k_5\,[\text{X}]}\right)}{\left\{k_{-5}\left(1 + \dfrac{k_1 K_3}{k_5\,[\text{X}]}\right) + k_6\right\}\left(1 + \dfrac{K_3}{[\text{X}]}\right)} \tag{3.14}$$

Equation (3.11) contains $[\text{X}]^2$-terms in the denominator but, if the quasi-equilibrium assumption is made also for the catalytic step, i.e., $k_6 \ll k_{-5}$, it simplifies to

$$v = \frac{\dfrac{k_6}{1 + \dfrac{K_4}{[\text{X}]}}\,[\text{S}]\,[\text{E}]_{\text{t}}}{\dfrac{k_{-5}}{k_5} \times \dfrac{1 + \dfrac{K_3}{[\text{X}]}}{1 + \dfrac{K_4}{[\text{X}]}} + [\text{S}]}\,. \tag{3.15}$$

With this assumption, $k_{\text{cat}}^{\text{app}}$ remains the same as (3.12), while the apparent Michaelis and specificity constants become

$$K_{\text{m}}^{\text{app}} = \frac{k_{-5}}{k_5} \times \frac{1 + \dfrac{K_3}{[\text{X}]}}{1 + \dfrac{K_4}{[\text{X}]}}, \quad k^{\text{app}} = \frac{k_5}{k_{-5}} \times \frac{k_6}{1 + \dfrac{K_3}{[\text{X}]}}\,. \tag{3.16}$$

The form of the kinetic parameters in (3.16) is analogous to that resulting from the assumption of quasi-equilibrium described in Sect. 2.4.1 for the general modifier mechanism with the important difference that the terms by which the kinetic parameters are multiplied and divided are of the type $1 + K/[\text{X}]$ instead of $1 + [\text{X}]/K$.

The rate equation of mixed activation can be reduced to that of a system operating at equilibrium in the same way of the general modifier mechanism by applying generalized microscopic reversibility (Sect. 2.4.2). To this end, (2.27) proposed by

Segal and coworkers [13] can be adapted to mixed activation in Scheme 3.3a as follows:

$$k_1 k_4 (k_{-5} + k_6) k_3 = k_{-1} k_3 k_5 k_{-4}$$

$$k_1 \left(\frac{k_{-5} + k_6}{k_5} \right) K_3 = k_{-1} K_4 \quad \therefore \quad k_1 K_3 = \frac{k_{-1} K_4}{\dfrac{k_{-5} + k_6}{k_5}} . \tag{3.17}$$

Substituting the expressions (3.17) into (3.9) gives

$$v = \frac{\dfrac{k_6}{1 + \dfrac{K_4}{[\mathrm{X}]}} [\mathrm{S}] [\mathrm{E}]_\mathrm{t}}{\dfrac{k_1 \left(\dfrac{k_{-5} + k_6}{k_5} \right) \dfrac{K_3}{[\mathrm{X}]} + (k_{-5} + k_6)}{k_1 \dfrac{K_3}{[\mathrm{X}]} + k_5} \times \dfrac{1 + \dfrac{K_3}{[\mathrm{X}]}}{1 + \dfrac{K_4}{[\mathrm{X}]}} + [\mathrm{S}]}$$

$$= \frac{\dfrac{k_6}{1 + \dfrac{K_4}{[\mathrm{X}]}} [\mathrm{S}] [\mathrm{E}]_\mathrm{t}}{\dfrac{k_{-5} + k_6}{k_5} \times \dfrac{1 + \dfrac{K_3}{[\mathrm{X}]}}{1 + \dfrac{K_4}{[\mathrm{X}]}} + [\mathrm{S}]} = \frac{\dfrac{k_6}{1 + \dfrac{K_4}{[\mathrm{X}]}} [\mathrm{S}] [\mathrm{E}]_\mathrm{t}}{K'_\mathrm{m} \times \dfrac{1 + \dfrac{K_3}{[\mathrm{X}]}}{1 + \dfrac{K_4}{[\mathrm{X}]}} + [\mathrm{S}]} . \tag{3.18}$$

In (3.18), $(k_{-5} + k_6)/k_5$ represents the Michaelis constant, K'_m, which indicates that the system operates *at equilibrium*. Since in mixed activation a difference between Michaelis constants in the absence and in the presence of saturating modifier cannot be defined, it would make little sense calling K^∞_m the expression $(k_{-5} + k_6)/k_5$ because there is no need to differentiate this term from K^0_m because the latter does not exist. What is measured experimentally is a value of "K_m" and, without deeper investigation, it is unknown whether or not it approximates or equals the dissociation constant of the reaction S + E $\rightleftharpoons$ ESX, i.e., k_{-5}/k_5. Therefore, the two simplifications discussed above are conceptually different but the *measured* numerical values of the kinetic parameters will be the same.

The discussion of mixed activation above is admittedly annoying but it may clarify fuzzy concepts. In Scheme 3.3a, a choice of symbols is shown for dissociation constants to satisfy personal taste. In line with the policy of this chapter, the following expressions will be further used:

$$v = \frac{\dfrac{k_6}{1 + \dfrac{\alpha K_\mathrm{X}}{[\mathrm{X}]}}\,[\mathrm{S}]\,[\mathrm{E}]_\mathrm{t}}{K'_\mathrm{m} \times \dfrac{1 + \dfrac{K_\mathrm{X}}{[\mathrm{X}]}}{1 + \dfrac{\alpha K_\mathrm{X}}{[\mathrm{X}]}} + [\mathrm{S}]} \tag{3.19}$$

$$k_\mathrm{cat}^\mathrm{app} = \frac{k_6}{1 + \dfrac{\alpha K_\mathrm{X}}{[\mathrm{X}]}} \tag{3.20}$$

$$K_\mathrm{m}^\mathrm{app} = K'_\mathrm{m} \times \frac{1 + \dfrac{K_\mathrm{X}}{[\mathrm{X}]}}{1 + \dfrac{\alpha K_\mathrm{X}}{[\mathrm{X}]}} \tag{3.21}$$

$$k^\mathrm{app} = \frac{1}{K'_\mathrm{m}} \times \frac{k_6}{1 + \dfrac{K_\mathrm{X}}{[\mathrm{X}]}} \tag{3.22}$$

Equations (3.19)–(3.22) reveal that linear mixed activation comprises three distinct entities that are the counterparts of the linear mixed inhibition mechanisms discussed in Sect. 3.3.1. These are: linear mixed, predominantly specific activation, with acronym LMx(Sp>Ca)A, defined by $\alpha > 1$, linear balanced activation, LMx(Sp=Ca)A, with $\alpha = 1$, and linear mixed, predominantly catalytic activation, LMx(Sp<Ca)A, in which $\alpha < 1$.

The differential diagnosis of the three linear mixed activation mechanisms can be made exploiting the dependences of the kinetic parameters in (3.20)–(3.22). In this case, activation, the tests of linearity are not based on dependences on [X] as in the corresponding inhibition mechanisms but on $1/[\mathrm{X}]$. These properties are collected in Fig. 3.6. The dependences of $k_\mathrm{cat}^\mathrm{app}$, k^app and their reciprocals on $1/[\mathrm{X}]$ for the activators are the same as those of the corresponding inhibitors on [X] shown in Fig. 3.3, while the Michaelis constants of the predominantly specific and predominantly catalytic mechanisms show opposite directions in their hyperbolic dependences.

3.4.2 *Linear Specific Activation*

The mechanism in Scheme 3.3b is the counterpart of linear specific inhibition named *linear specific activation* and given the acronym LSpA. To the term specific activation, recommended by authorities [12], the adjective *linear* has been added

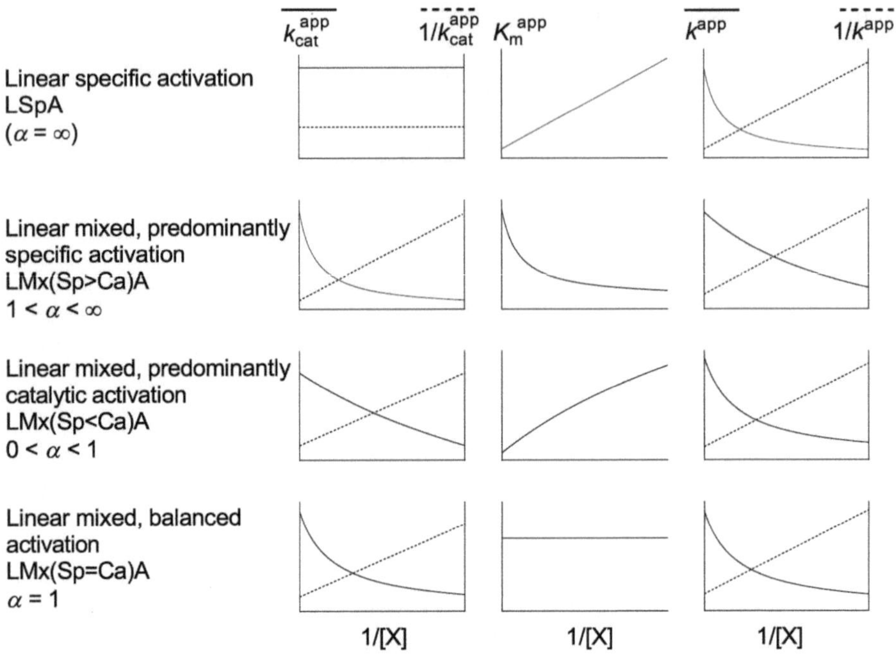

Fig. 3.6 Dependences of the kinetic parameters on the reciprocal activator concentration for linear specific and linear mixed activation

here for consistency with the nomenclature of inhibitors. This nomenclature also avoids confusion since

> ··· linear activation must be essential, and nonessential activation must be non-linear, but essential activation can in principle be non-linear. [12, p. 288]

Nonlinear essential activation can occur in some of the cases itemized on p. 143. The rate equation of this mechanism can either be deduced from (3.19) by setting $\alpha K_X = 0$ or, more elegantly, using the King–Altman method. The latter, demonstrated for didactic purposes in parallel with the derivation of linear specific inhibition in Appendix 1 in Chap. 5, is reproduced here together with the expressions of the apparent kinetic parameters ($K'_m = (k_{-5} + k_6)/k_5$):

$$v = \frac{k_6[\mathrm{E}]_t\,[\mathrm{S}]}{K'_m\left(1 + \dfrac{K_{Sp}}{[\mathrm{X}]}\right) + [\mathrm{S}]} \tag{3.23}$$

$$k_{cat}^{app} = k_6 \tag{3.24}$$

$$K_m^{app} = K'_m\left(1 + \frac{K_{Sp}}{[\mathrm{X}]}\right) \tag{3.25}$$

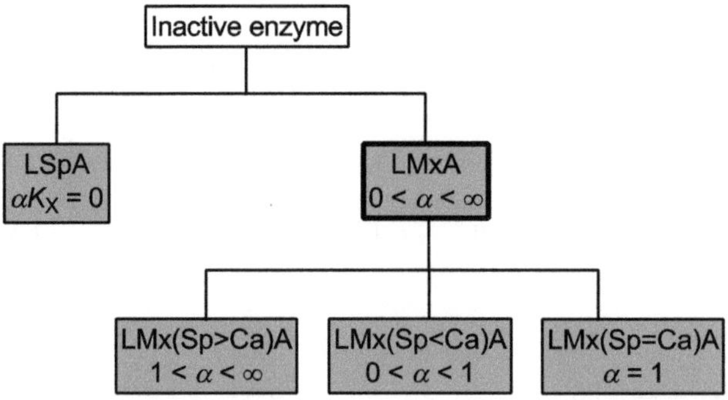

Fig. 3.7 Taxonomy of essential activation. The three is based on only one character, i.e., allosteric linkage, with no further possibilities of differentiation because linear mixed activation has only one catalytic path

$$k^{\text{app}} = \frac{k_6}{K'_{\text{m}}\left(1 + \frac{K_{\text{Sp}}}{[\text{X}]}\right)} \tag{3.26}$$

As seen in Fig. 3.6, the catalytic constant is not affected by the activator, while $K_{\text{m}}^{\text{app}}$, k^{app} and $1/k^{\text{app}}$ depend on $1/[\text{X}]$ exactly as the same parameters depend on [X] in linear specific inhibition (Fig. 3.3). These characteristics distinguish linear specific activation from the mixed mechanisms shown in the same figure.

3.4.3 *Taxonomy of Essential Activation*

The mechanisms of essential activation, as shown above, are just four. In line with the taxonomic ranking of modifiers in Fig. 3.2 the essential activators are displayed in Fig. 3.7. The *root* of the tree is an inactive enzyme, a basic characteristic that delimits essential activation from all the remaining modifier mechanisms. Distinction between the four members in Fig. 3.7 is essentially made on the basis of allosteric coupling. In linear specific activation there is no linkage between substrate and activator because the system is noncyclic. Linkage is however possible in the mixed mechanisms in the same way as in the inhibitory counterparts. We distinguish thus predominantly specific ($\alpha > 1$) and predominantly catalytic ($\alpha < 1$), mixed activation and linear balanced activation with $\alpha = 1$.

There are no offsprings of linear mixed activation because, with just one catalytic branch, the mechanisms cannot be further differentiated.

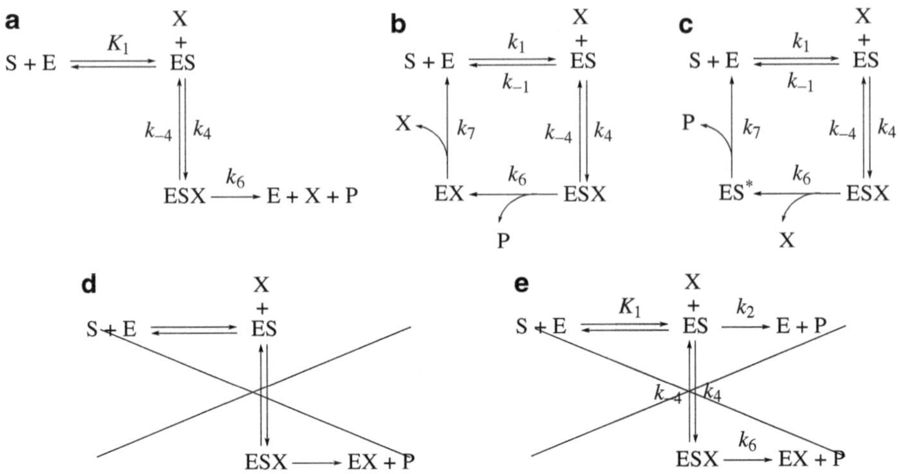

Scheme 3.4 Alternative representations of essential activation reaction paths. (**a**) The reaction scheme of Case III by Segal et al. [13], with two alternative mechanisms,(**b**) and (**c**), that behave as (**a**). (**d**) Inconsistent reaction pattern. (**e**) Mechanism Ib from Frieden [9]. The barred reactions are invalid

3.4.4 Equivalent Reaction Schemes and a Comment on "Uncompetitive Activation"

Linear specific, predominantly specific, and balanced activation do not pose interpretative questions but when the activation character is predominantly catalytic, the reaction scheme can be represented in different ways that nevertheless correspond to equivalent rate equations. This is the case of a mechanism discussed by Segal et al. summarized in Table I of their paper as case III [13]. The reaction paths are those in Scheme 3.4a and the rate equation under equilibrium conditions provided by the authors is[2]:

$$v = \frac{\dfrac{k_6}{1 + K_m/[X]}\,[S]\,[E]_t}{\dfrac{K_1}{1 + [X]/K_m} + [S]}, \tag{3.27}$$

where $K_m = (k_{-4} + k_6)/k_4$ and the following are the apparent kinetic parameters

$$k_{cat}^{app} = \frac{k_6}{1 + K_m/[X]}, \quad K_m^{app} = \frac{K_1}{1 + [X]/K_m}, \quad k^{app} = \frac{k_6\,[X]}{K_1 K_m}. \tag{3.28}$$

[2]Rewritten from a double-reciprocal equation with symbols used in this book.

The point with Scheme 3.4a is how to represent the products of catalysis and the way shown here is fine because ESX is represented as decaying to product and two species already present before catalysis has taken place. Conversely, the representation of the same mechanism as shown in Scheme 3.4d *creates* the EX complex that is not utilized further. What is depicted in panel (d) is an enzyme-inactivation reaction that for the sake of clarity is barred to indicate that it should not be used to represent essential activation.

The representation in (a) as a termolecular reaction may cause problems if the reverse reaction is considered because this requires the simultaneous collision of three molecules with the necessary energy. This is a theoretical statement because essential activation is logically tailored to proceed forwards. In any case, the two reaction paths shown in panels (b) and (c) can be envisaged as alternatives to (a). In mechanism (b), ESX generates product and EX, which is then rapidly decomposed to free enzyme and modifier. In mechanism (c), the modifier is released first giving an activated ES* complex, which then decomposes to product and free enzyme. Both mechanism (b) and (c) have the same King–Altman patterns, from which the terms in (3.29) derive, and the resulting rate equation is the same, i.e. (3.30) with the difference that in (b) the rate corresponds to $v = k_6\,[\mathrm{ESX}]\,[\mathrm{E}]_\mathrm{t}$ while in (c) it is $v = k_7\,[\mathrm{ES}^*]\,[\mathrm{E}]_\mathrm{t}$. This has no numerical consequences in calculating the rate with (3.30). Important in mechanisms (b) and (c) is that k_6 is rate-limiting and a ratio $k_7/k_6 = 100$ is fine for this to be true. Under these premises, equation (3.27) derived under the equilibrium hypothesis, compared with the steady-state Eq. (3.30), overestimates the rate by 4 % at $[\mathrm{S}] = 0.5K_\mathrm{m}$, by 2 % at $[\mathrm{S}] = 2K_\mathrm{m}$ and there is no difference at $[\mathrm{S}] = 5K_\mathrm{m}$, a result in line with what we have learnt in Chap. 2 about the significance of approximations.

$$\begin{aligned} \mathit{Numerator} &= k_1k_4k_6k_7\,[\mathrm{S}]\,[\mathrm{X}] \\ \mathit{Denominator} &= (k_1k_4k_7 + k_1k_4k_6)\,[\mathrm{S}]\,[\mathrm{X}] + k_4k_6k_7\,[\mathrm{X}] \\ &\quad + (k_1k_{-4}k_7 + k_1k_6k_7)\,[\mathrm{S}] + (k_{-1}k_{-4}k_7 + k_{-1}k_6k_7) \end{aligned} \tag{3.29}$$

$$v = \frac{\mathit{Numerator}}{\mathit{Denominator}}[\mathrm{E}]_\mathrm{t}\,. \tag{3.30}$$

Which activation type does correspond to mechanisms (a)–(c) in Scheme 3.4? This is with no doubt linear activation but the dependence of the kinetic parameters on modifier concentration in (3.28) does not correspond to a *hypothetical* linear catalytic activation mechanism, as may be suggested by Scheme 3.4a, because the activator affects the catalytic and the Michaelis constant differently and the specificity constant does therefore depend on the concentration of the activator (a purely catalytic character is diagnosed by the independence of k^{app} on [X]).

The diagnostic property is implicit in $K_\mathrm{m}^{\mathrm{app}}$, which decreases for increasing [X] and *increases* with 1/[X] (3.28), i.e., we have again the characteristics of linear mixed, predominantly catalytic activation, LMx(Sp<Ca)A, shown in Fig. 3.6. Is there a need to create a new taxon in Fig. 3.7 for this mechanism? The answer

is no because, experimentally, an ensemble of characters are evaluated, in this case the dependences of the apparent kinetic parameters on activator concentration. When the ensemble matches that of a known mechanism, such as the plots for LMx(Sp<Ca)A in Fig. 3.6, we are not able to attribute the findings to mechanism (a) in Scheme 3.3 or to mechanism (a) in Scheme 3.4. In this case the mathematical expressions of the apparent parameters are not identical but their identity cannot be distinguished experimentally because they generate the same plots. This property, known as *kinetic homeomorphism*, can be problematic in making inference on mechanisms from kinetic experiments.

The *uncompetitive activation* found in the biochemical literature needs a comment. Comparing Figs. 3.3 and 3.6 we see that a mechanism corresponding to linear catalytic inhibition (Fig. 3.3) is missing in Fig. 3.6 among the activators. From the discussion above it can be realized that there is no purely catalytic activation mechanism, whose apparent specificity constant is not affected by the activator.

All cases of uncompetitive activation mentioned in theoretical, e.g., by Laidler [10], and experimental work, e.g., by Woods et al. [21] and Shikita et al. [15] belong in fact to any one of the hyperbolic mechanisms discussed in Sect. 3.3.3. Once more we must realize the necessity of a uniform systematization in the classification and nomenclature of modifier action.

Occasionally, reaction schemes are found in the biochemical literature that do not either correspond to the rate equation that accompanies the mechanism or to the description of their properties. It is instructive to comment one of these cases if nothing else to stimulate beginners in enzyme kinetics to critically reading and evaluating published theories and experimental results. Thus, Scheme 3.4e, taken from Frieden [9] (IB in the paper) depicts a mechanism with the same problem as mechanism (d). For this system Eq. (3.31) has been provided [9, Eq. 4]. The EX complex is produced only from ESX and is not further utilized in the reaction thus behaving as a sink for enzyme that is sequestered irreversibly. This system can be analyzed with an integrated rate equation that takes into account time-dependent enzyme inactivation with methods described in Chap. 9.

$$\frac{v_{\mathrm{X}}}{[\mathrm{E}]_{\mathrm{t}}} = \frac{\dfrac{k_2\left(1+\dfrac{k_6\,[\mathrm{X}]}{k_2 K_{\mathrm{Ca}}}\right)}{1+\dfrac{[\mathrm{X}]}{K_{\mathrm{Ca}}}}}{1+\dfrac{K_{\mathrm{m}}}{[\mathrm{S}]}\left(\dfrac{1}{1+\dfrac{[\mathrm{X}]}{K_{\mathrm{Ca}}}}\right)}\,. \tag{3.31}$$

Simulations with (3.31) produce modification patterns that depend only on the ratio k_6/k_2. If $k_6 = 0$ the equation generates the patterns of linear catalytic inhibition (LCaI); $0 < k_6/k_2 < 1$ gives the dual modification mechanism HMxD(A/I); $k_2 = k_6$ corresponds to hyperbolic specific activation (HSpA) and when $k_6 > 1$ the patterns are those of hyperbolic mixed, predominantly catalytic activation, HMx(Sp<Ca)A. These results are incompatible with the reaction scheme because the simulation affords modification patterns characteristic of a system with allosteric coupling between substrate and modifier, which does not exist in Scheme 3.4d.

There is no homeomorphism in this case and the mechanism is best dropped in favor of the general modifier mechanism with E, ES, EX and ESX linked with one another. Trials to produce valid King–Altman patterns from Scheme 3.4e end up with invalid patterns that do not connect all enzyme forms [7, pp. 113–117]. This situation is similar to that of a car traveling with three wheels because one has been stolen.

3.5 Overview of Allosteric Interactions

The properties of the allosteric modifiers discussed in Chap. 2 and up to this point in the present chapter allow to draw a graphical summary of their multifaceted kinetic strategies. The two basic characters, α and β in Scheme 3.1, used for taxonomic purposes, represent the abscissa and the ordinate in Fig. 3.8, respectively. The ordinate and the abscissa are labeled with the origin and the value 1 leaving the positive direction open in order to embrace symbolically any positive number. In practice however, α and β attain finite, relatively small positive values.

In Fig. 3.8, the mechanisms are located in their action domains defined by the range of α and β in the α-β space. Six mechanisms are located within the areas delimited by rectangles and triangles, suggesting an ample palette of possible effects. Linear mixed, balanced inhibition occupies the position of the black circle, linear mixed, predominantly catalytic inhibition exerts its action in the interval (0,1) of the abscissa, while linear mixed, predominantly specific inhibition operates in the interval (1,∞) of the abscissa. Hyperbolic catalytic inhibition exerts its functions with coefficients α and β along the diagonal of the square with edges (0,1)/(0,1), and hyperbolic catalytic activation along the diagonal with edges (1,∞)/(1,∞). Hyperbolic specific activation and hyperbolic specific inhibition have their action range in the domains (0,1) and (1,∞) along the horizontal lines at ordinate 1, respectively. The vertical lines at abscissa 1 represent the domains of action of hyperbolic mixed, balanced inhibition (0,1) and hyperbolic mixed, balanced activation (1,∞), respectively.

The black diamond at coordinates (1,1) in Fig. 3.8 represents the *liberator*, a special modifier that will be discussed in Sect. 7.2

3.6 The Specific Velocity Plot

Various graphical methods, each one with advantages and disadvantages, have been used for analyzing enzyme inhibition and activation. The disadvantages include either statistical weakness or inapplicability to hyperbolic modifiers. The specific

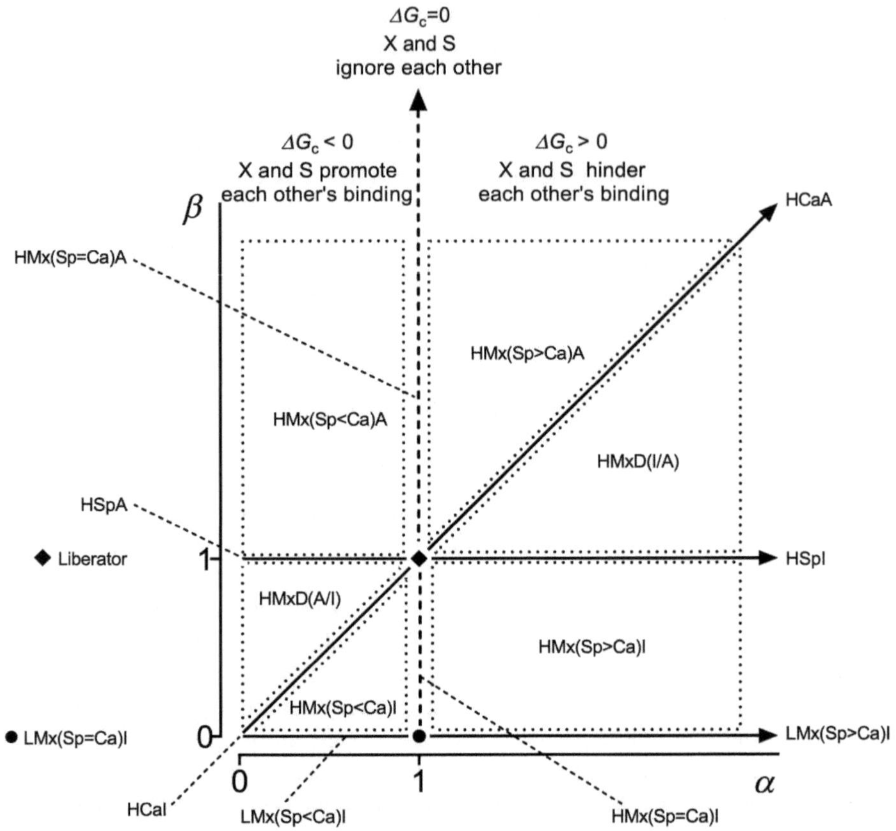

Fig. 3.8 Synoptic representation of the effects exerted by allosteric modifiers on the affinity between enzyme and substrate. The positive direction of the axes is *top* and *right* for the ordinate and abscissa, respectively, and there is no numbering apart 1 to signify that, in the frame of allosteric interactions, both variables can attain large positive values. The kinetic mechanisms that correspond to various combinations of the allosteric coupling constant (the reciprocal of α) and the modifier effect on the catalytic constant (β) are shown in their characteristic action domains in the α–β space. The action domains are either located within the areas delimited by the *dotted rectangles* and *triangles*, or along lines that demarcate their boundaries. The liberator, characterized by $\alpha = \beta = 1$ (*black diamond*), LMx(Sp=Ca)I (*black circle*), HMx(Sp=Ca)I and HMx(Sp=Ca)A are all located on the vertical line at abscissa $\alpha = 1$. For these modifiers, the change in free energy of coupling between modifier and substrate is therefore zero

velocity plot, purposely created as a *graphical* method for the analysis of enzyme–modifier interactions, offers the experimenter an immediate representation of the underlying mechanism, whether linear or hyperbolic [2]. This plot is particularly sensitive in detecting inhibitors and activators with hyperbolic mixed character, notably allosteric modifiers. Above, the emphasized adjective *graphical* underlines its main purpose: to serve as diagnostic tool for identifying mechanisms and to provide preliminary values of the sought kinetic parameters without statistical pretenses. Once the mechanism of action has been identified, refinement of the analysis can continue by fitting the appropriate rate equation to data using the preliminary parameter estimates.

As demonstrated in Sect. 2.7, large tolerance can be granted systems that are not strictly in quasi-equilibrium. Assuming a prevalence of quasi-equilibrium conditions, (3.2) can be rewritten calling the rate v_X to explicitly indicate a velocity in the presence of the modifier. In rearranging this equation with the approximation $K_m^S \approx K_m^0 \approx K_S$ for reasons described in Sects. 2.4.1 and 2.4.2, the Michaelis constant will be named with its customary notation, K_m. Indeed, steady-state kinetic experiments in the absence of modifiers allow to measure K_m, which is numerically close to K_S if quasi-equilibrium conditions prevail. Thus, rearrangement of (3.2) gives

$$\frac{v_X}{V} = \frac{[S] + \beta \dfrac{[X]}{\alpha K_X}[S]}{K_m\left(1 + \dfrac{[X]}{K_X}\right) + [S]\left(1 + \dfrac{[X]}{\alpha K_X}\right)} = \frac{\dfrac{[S]}{K_m}\left(1 + \beta \dfrac{[X]}{\alpha K_X}\right)}{1 + \dfrac{[X]}{K_X} + \dfrac{[S]}{K_m}\left(1 + \dfrac{[X]}{\alpha K_X}\right)}, \tag{3.32}$$

where $V = k_2[E]_t$. Using the same symbolism, the Michaelis–Menten equation reads

$$\frac{v_0}{V} = \frac{[S]/K_m}{1 + [S]/K_m}, \tag{3.33}$$

in which v_0 explicitly indicates the rate in the absence of modifiers. The relative velocity is obtained dividing (3.32) by (3.33)

$$\frac{v_X}{v_0} = \frac{\left(1 + \dfrac{[S]}{K_m}\right)\left(1 + \beta \dfrac{[X]}{\alpha K_X}\right)}{\dfrac{[S]}{K_m}\left(1 + \dfrac{[X]}{\alpha K_X}\right) + 1 + \dfrac{[X]}{K_X}}, \tag{3.34}$$

which, naming $\sigma = [S]/K_m$, becomes

$$\frac{v_X}{v_0} = \frac{(1 + \sigma)\left(1 + \beta \dfrac{[X]}{\alpha K_X}\right)}{\sigma\left(1 + \dfrac{[X]}{\alpha K_X}\right) + 1 + \dfrac{[X]}{K_X}}. \tag{3.35}$$

The reciprocal of (3.35), after collecting the σ-containing terms, results in a normalized equation that has been called *specific velocity equation* because $\sigma/(1 + \sigma)$ is known as *specific velocity* [2]

$$\frac{v_0}{v_X} = \frac{[X]\left(\dfrac{1}{\alpha K_X} - \dfrac{1}{K_X}\right)}{1 + \beta\dfrac{[X]}{\alpha K_X}} \frac{\sigma}{1+\sigma} + \frac{1 + \dfrac{[X]}{K_X}}{1 + \beta\dfrac{[X]}{\alpha K_X}} . \tag{3.36}$$

The ratio $\sigma/(1+\sigma)$ can be written as $[S]/(K_m + [S])$ as well, but the first form will be consistently employed in line with the use of the handy symbol σ for $[S]/K_m$ at many places in this book.

The *specific velocity plot*, in which v_0/v_X is plotted against $\sigma/(1+\sigma)$, consists of straight lines for the 17 mechanism in Fig. 3.2, independently on whether they are linear or hyperbolic, with specific, catalytic or mixed character, and whether the modifier behaves as activator or inhibitor. Obviously, the essential activation mechanisms in Sect. 3.4, for which $v_0 = 0$ in the absence of modifier, cannot be analyzed with the specific velocity plot.

When plotting data according to (3.36) it is convenient to draw two ordinates crossing the abscissa at 0 and 1, which are the limits for $\sigma = 0$ and $\sigma \to \infty$, respectively. When $[S] = K_m$, $\sigma/(1+\sigma) = 0.5$ and a typical range of substrate concentrations in kinetic experiments is between $[S] \approx 0.2\,K_m$ and $[S] \approx 5\,K_m$, i.e., $\sigma/(1+\sigma) \approx 0.2$ and $\sigma/(1+\sigma) \approx 0.8$, respectively.

The method is illustrated in Fig. 3.9a using constants characteristic of hyperbolic mixed inhibition with predominantly specific character [HMx(Sp>Ca)I]. The solid lines, which intersect at a common point, show an experimentally typical substrate concentration range, while the dashed parts of the lines are extrapolated. The specific velocity plot is thus a normalized representation of initial velocity measurements in the presence of modifiers, with dependent and independent variable as pure numbers varying in a discrete range. The properties of the specific velocity plot can be summarized as follows.

- The *primary plot*, Fig. 3.9a, consists of straight lines independently of the mechanism. When $\alpha \neq 1$, the family of lines obtained at various modifier concentrations intersect at a common point with coordinates $v_0/v_X = 1$ on the ordinate and $\sigma/(1+\sigma) = (\alpha - \beta)/(\alpha - 1)$ on the abscissa, which represents the *critical specific velocity*. At the intersection point, in which the rates in presence and absence of modifier are the same, a *critical substrate concentration* can be calculated as described in Sect. 3.7.
- For inhibitors and activators, the slopes of the straight lines in the primary plot
  - are *negative* if $\alpha > 1$ (purely specific or the specific character dominates over the catalytic)
  - consist of lines *parallel* to the abscissa (slope zero) if $\alpha = 1$ (the specific and the catalytic components are equally represented)
  - are *positive* if $\alpha < 1$ (purely catalytic or the catalytic character dominates over the specific)

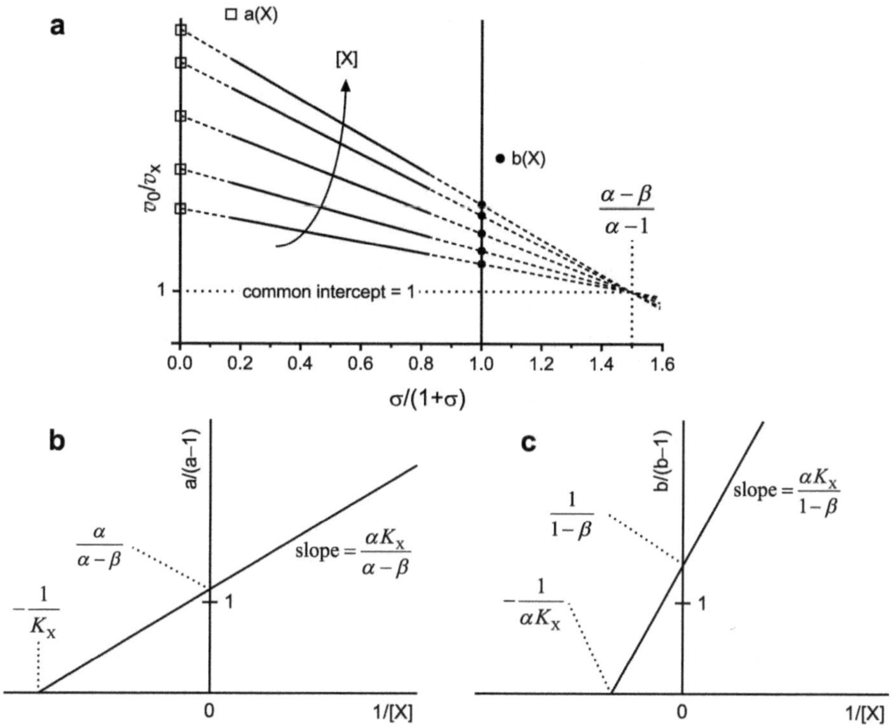

Fig. 3.9 The specific velocity plot. (**a**) Primary plot with the *arrow* showing the direction in which [X] increases; (**b**) and (**c**) secondary plots

- When extrapolated, the straight lines obtained at various modifier concentrations (solid segments in Fig. 3.9a), intersect the two ordinates crossing the abscissa at its values 0 and 1 at points that depend on [X] called a(X) and b(X) and indicated in the figure as open squares and black circles, respectively. These intercepts, whose analytical expressions are obtained from (3.36), can be rearranged as

$$\frac{\mathrm{a}}{\mathrm{a}-1} = \frac{\alpha K_X}{\alpha-\beta}\frac{1}{[X]} + \frac{\alpha}{\alpha-\beta}\,, \tag{3.37}$$

$$\frac{\mathrm{b}}{\mathrm{b}-1} = \frac{\alpha K_X}{1-\beta}\frac{1}{[X]} + \frac{1}{1-\beta}\,. \tag{3.38}$$

Secondary plots of a/(a −1) and b/(b −1) against 1/[X] yield straight lines with distinctive slopes and intercepts shown in Fig. 3.9b,c, from which α, β and K_X can be calculated.

- The ordinate intersections of the replots (Fig. 3.9a,b) assume the values
  - 1 for all linear mechanisms, as can be verified by setting $\beta = 0$ in the expressions of the ordinate intersections $1/(1-\beta)$ and $\alpha/(\alpha-\beta)$.
  - > 1 for all hyperbolic inhibition mechanisms
  - < 1 for all hyperbolic activation mechanisms
  - indeterminate for either a/(a−1) or b/(b−1) if the intersection point of the lines in the primary plot falls at abscissa values 0 or 1, respectively

In the presence of experimental error, the lines have a maximum of $n\,(n-1)/2$ intersection points. The best estimate of the ordinate and abscissa of the critical specific velocity $(\alpha-\beta)/(\alpha-1)$ is given by the median of all intersection points as illustrated in Fig. 3.10.

The specific velocity plot is *not suitable* for extracting statistically valid information because $\sigma/(1+\sigma) = v_0/V$. The errors in v_0 appear thus on both sides of (3.36), which precludes application of regression analysis that demands stochastic independence of the dependent and the independent variable (Sect. 1.2.6). Any least squares fit performed to avoid subjectivity in drawing lines through experimental points should not be accompanied by statistical claims.

Experimental error is mercilessly revealed in the specific velocity plot and there is no discount for bad data at any reactant concentration. Instead of deleting or ignoring points that do not apparently fit in the ensemble of data, bad-looking plots should be interpreted as an invitation to going back to the laboratory for repeating measurements with the due care after checking the working protocol and possibly redesigning the experiment. Recalcitrant systems that produce sets of data with different scatter in different runs represent a positive finding, namely that error occurs at random and that deviations are not systematic. Experience has shown that in such cases the trends are reproducible and the underlying mechanism can

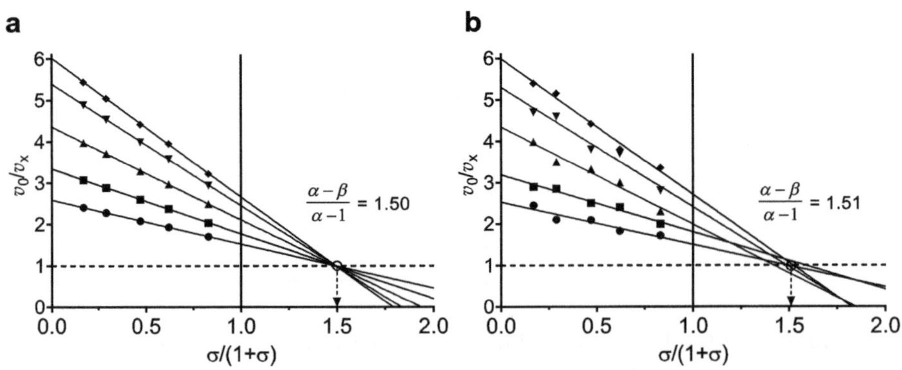

Fig. 3.10 Effect of errors on the specific velocity plot. (**a**) Error-free primary plot simulated with $\alpha = 2.4$ and $\beta = 0.3$; the lines intersect at the common point with coordinates 1.50 and 1.00 (*open circle*). (**b**) Random error was added to the same data as in (**a**). The five lines have ten intersection points whose median is indicated by the *open circle* at coordinates 1.51 and 0.98

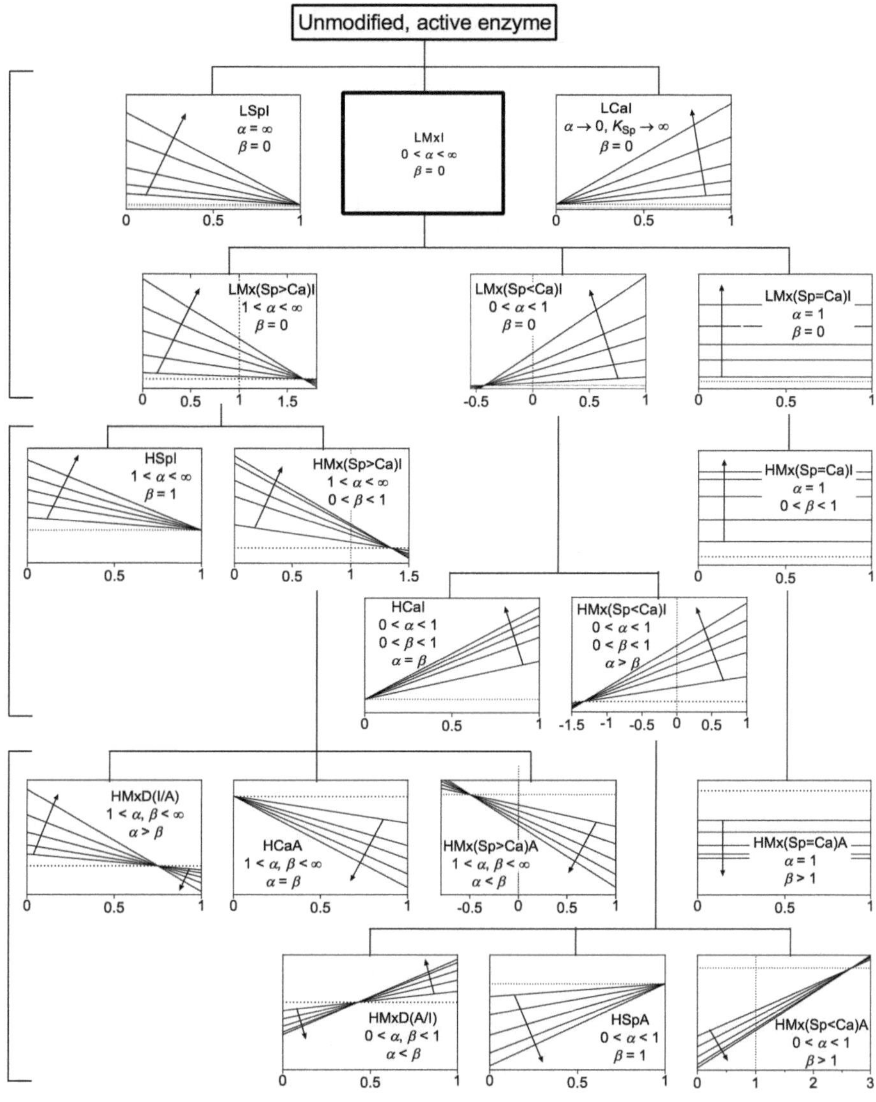

Fig. 3.11 Taxonomic tree of modification mechanisms represented by *primary specific velocity plots* organized as in Fig. 3.2. Mechanisms in levels 2 and 3 are distributed in two rows each for space reasons. Axes labeling (not shown) is as in Fig. 3.9a, with v_0/v_X on the ordinate (positive direction *bottom to top*) and $\sigma/(1+\sigma)$ on the abscissa. *Dotted horizontal lines* represent the ordinate value $v_0/v_X = 1$. *Arrows* indicate the direction in which the modifier concentration increases

be diagnosed. Refinement of data analysis by fitting the rate equation to titration curves using preliminary parameter estimates finally leads to the sought results. An example of this procedure is shown in Sect. 5.3.2, Fig. 5.11.

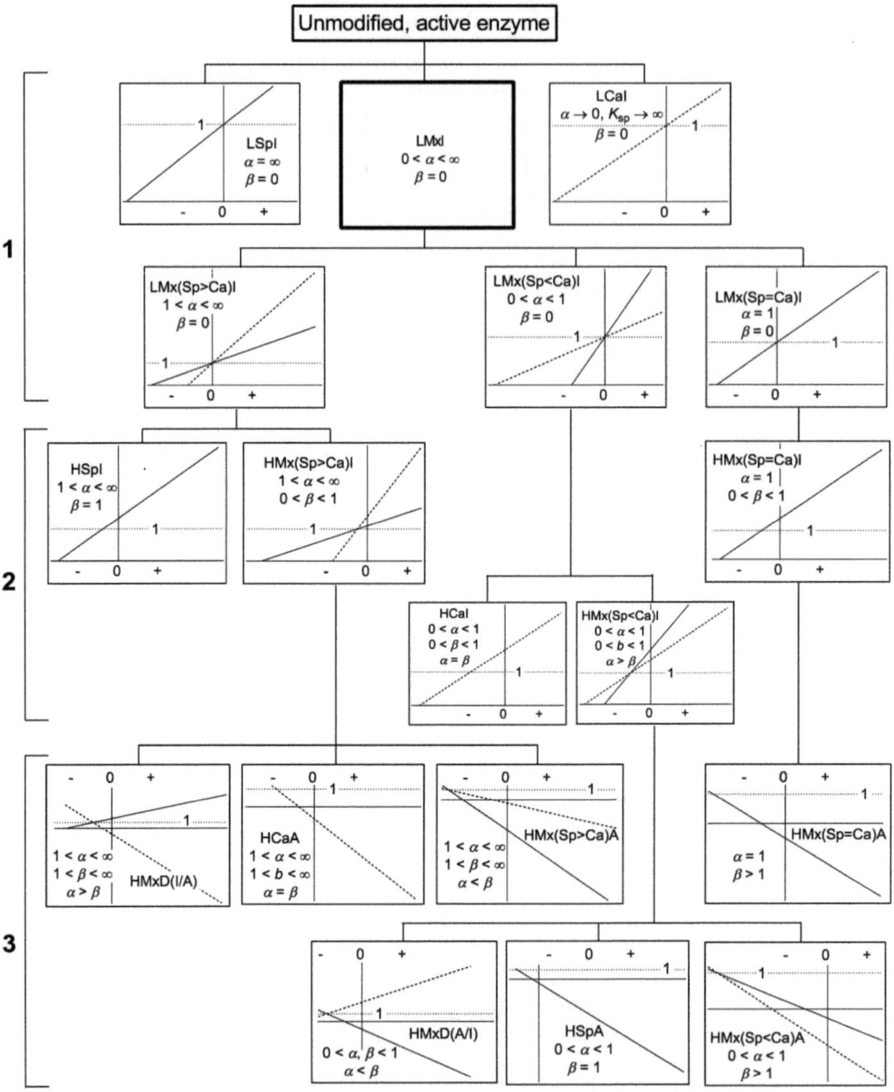

Fig. 3.12 *Secondary plots* of the corresponding primary specific velocity plots in Fig. 3.11. Axes labeling is in all cases as shown in Fig. 3.9b,c: the abscissa is 1/ [X] and the ordinate is either a ⁄ (a − 1) (*continuous lines*) or b ⁄ (b − 1) (*dashed lines*). The origin (0) and sign of the abscissa values are shown schematically with the ordinate value of 1 represented by a *horizontal dotted line*

The properties of primary and secondary specific velocity plots for the mechanisms in the taxonomic tree of Fig. 3.2 are illustrated in Figs. 3.11 and 3.12, respectively, where the same topological layout of Fig. 3.2 is maintained. It is worth to dedicate a few minutes to examine attentively Fig. 3.11, which stimulates *eidetic memory* in recognizing enzyme-modification mechanisms by inspection.

3.7 The Critical Substrate Concentration

If $v_X = v_0$, numerator and denominator on the right side of (3.34) are the same, i.e.,

$$\left(1+\frac{[S]}{K_m}\right)\left(1+\beta\frac{[X]}{\alpha K_X}\right)=\frac{[S]}{K_m}\left(1+\frac{[X]}{\alpha K_X}\right)+1+\frac{[X]}{K_X}\,. \tag{3.39}$$

Rearranging (3.39) and solving for [S] yields the *critical substrate concentration*, which corresponds to the value of [S] for which the rate is the same in the presence or absence of modifier

$$[S]^0 = K_m\frac{\alpha-\beta}{\beta-1}\,. \tag{3.40}$$

A dimensionless and more practical variant of (3.40), because it can be directly related to the specific velocity plot, is the *relative critical substrate concentration*

$$\sigma^0 = \frac{[S]^0}{K_m} = \frac{\alpha-\beta}{\beta-1}\,. \tag{3.41}$$

For the 14 enzyme–modifier mechanisms in which $\alpha \neq 1$ (Fig. 3.11), the intersection point of the straight lines in the specific velocity plot, Fig. 3.9a, can be redirected to the relative critical substrate concentration σ^0 since the abscissa value of the intercept corresponds to the *critical specific velocity*

$$\frac{\sigma^0}{1+\sigma^0} = \frac{\alpha-\beta}{\alpha-1}\,, \tag{3.42}$$

which is very helpful in the diagnosis of mechanisms and for calculating α and β.

Equations (3.40)–(3.42) represent the same concept but only (3.42) can be measured as the intersection point of the experimental lines in the specific velocity plot. The critical specific velocity is undefined for the three balanced mechanisms, it falls in the experimentally accessible substrate concentration range for the dual mechanisms HMxD(I/A) and HMxD(A/I) (as far as substrate solubility or other reasons do not hamper measurements), and otherwise it is an extrapolated constant with values in the ranges indicated in Table 3.2.

The specific velocity plot takes four basic patterns that we discuss with the help of Fig. 3.13, where the panels (a–d) are divided into four quadrants numbered 1–4 for practical referencing. The delimiters of the quadrants (dotted lines) are $(\alpha-\beta)/(\alpha-1)$ on the abscissa and 1 on the ordinate. Experimentally accessible

Table 3.2 Relative critical substrate concentration and critical specific velocity for 17 enzyme–modifier mechanisms

Mechanism	σ^{0a}	$\sigma^0/(1+\sigma^0)^b$
LSpI	$-\infty$	1
LCaI	0	0
LMx(Sp>Ca)I	< 0	> 1
LMx(Sp<Ca)I	< 0	< 0
LMx(Sp=Ca)I	−1	Undefined
HSpI	Undefined	1
HMx(Sp>Ca)I	< 0	> 1
HCaI	0	0
HMx(Sp<Ca)I	< 0	< 0
HMx(Sp=Ca)I	−1	Undefined
HMxD(I/A)	> 0	(0,1)
HCaA	0	0
HMx(Sp>Ca)A	< 0	< 0
HMxD(A/I)	> 0	(0,1)
HSpA	Undefined	1
HMx(Sp<Ca)A	< 0	> 1
HMx(Sp=Ca)A	1	Undefined

The entries *undefined* mean division by zero
[a]Relative critical substrate concentration (3.41)
[b]Critical specific velocity (3.42)

areas are shaded in gray, arrows indicate the increasing concentration of the modifier, and the mechanisms corresponding to the illustrated specific velocity patterns are listed beneath or above the gray-shaded areas. Please note that quadrants 2 and 4 in panel (a) and quadrants 1 and 3 in panel (c) represent each a specific velocity plot and the junction between them is therefore fictional, whereas panels (b) and (d) are themselves whole specific velocity plots as can be judged by the adjacent, experimentally accessible gray zones.

LSpI is present in Fig. 3.13, panel (a) just for sharing with the other members the main character $\alpha > 1$ but, being incapable of allosteric coupling, is not a member of the family and the same holds for LCaI in panel (c).

Considering the *patterns of inhibitors* in quadrant 2 of panels (a) and (b), the extension of the lines beyond their intersection point results in *activator patterns* in quadrant 4, which thus represents the logical continuance of the kinetic properties in quadrant 2. The same reasoning is valid for the *activation patterns* in quadrant 3 of panels (c) and (d), where the extended lines result in *inhibition patterns* in quadrant 1. We can thus distinguish four types of specific velocity plots:

1. Panels (c, d), quadrant 1: the lines *descend* toward their intersection point *right-to-left*. This group comprises two linear and two hyperbolic inhibition mechanisms, a well as the inhibitory domain of the dual mechanism HMxD(A/I).
2. Panels (a, b), quadrant 2: the lines *descend* toward their intersection point *left-to-right*. Members of the group are two linear and two hyperbolic

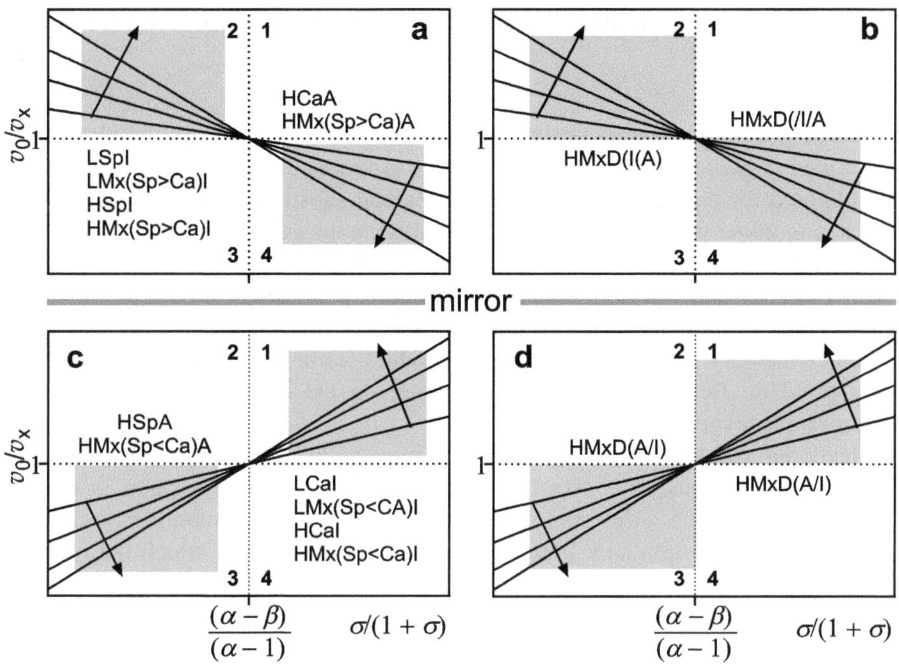

Fig. 3.13 The *primary specific velocity plot*. Significance of the relative critical substrate concentration and of the critical specific velocity. The condensed views illustrate the intersection points of the primary specific velocity plots for the 14 modifier mechanisms with $\alpha \neq 1$ in the taxonomic tree of Fig. 3.2. *Gray-shaded areas* represent experimentally accessible ranges and *arrows* indicate increasing modifier concentration. Imagining to put a mirror in the position indicated, the *upper panels* result to be the specular image of the *lower panels*

inhibition mechanisms and the inhibitory domain of the dual HMxD(I/A) mechanism.

3. Panels (c, d), quadrant 3: the lines *ascend* toward their intersection point *left-to-right*. The group comprises two hyperbolic activation mechanisms and the activating domain of HMxD(A/I).
4. Panels (a, b), quadrant 4: the lines *ascend* toward their intersection point *right-to-left*. In this group we find two hyperbolic activation mechanisms and the activating branch of mechanism HMxD(I/A).

The specific velocity plot corroborates the systematics proposed in Sect. 3.3. Imagining to put a mirror as indicated by the gray horizontal line in Fig. 3.13, it can be seen that the lower panels are the specular image of the upper panels: the upper panels correspond to the *specific* lineage starting from linear mixed, predominantly specific inhibition in the taxonomic tree, whereas the lower panels contain all members of the *catalytic* lineage with linear mixed, predominantly catalytic inhibition as ancestor. The taxonomic location of HCaA and HSpA is thus confirmed.

3.8 Enzyme Modification Illustrated with Kinetic Barrier Diagrams

The energetics of enzyme-catalyzed reactions can be represented graphically as described in Sect. 1.5, where the advantages of kinetic barrier diagrams have been highlighted for their ability to include the actual substrate concentration. On this track, a modifier will be now included considering the general modifier mechanism as a model. In the most general case described in Sect. 2.5 the modifier interferes with catalysis and the rate equation contains two product-forming steps. Since, in the most formal sense, the horizontal axis of kinetic barrier diagrams refers to an elementary step, representing two product-forming steps on a single profile is a problem. Below, a method is discussed on the assumption of generalized microscopic reversibility.

Invoking the principle of generalized microscopic reversibility (Sect. 2.4.2), equilibrium dissociation constants replace the individual rate constants in the rate equation. This is the *only condition* under which the general modifier mechanism operates at equilibrium and, importantly, the only situation in which the modifier binding steps *do not interfere* at all with the catalytical steps. Assuming quasi equilibrium for modifier binding gives a similar result in which the resulting rate equation is not exact, but *approximates* the true equation when $k_2/k_{-1} \ll 1$ and $k_6/k_{-5} \ll 1$.

Generalized microscopic reversibility can be expressed with (2.27), which allows treating the system with only three constants since

$$K_{\mathrm{m}}^{\infty} = \frac{K_{\mathrm{m}}^{0} K_4}{K_3}\,, \tag{3.43}$$

and since $K_4 = \alpha K_3$, $K_{\mathrm{m}}^{\infty} = \alpha K_{\mathrm{m}}^{0}$. Exploiting node compression with rule (a) of Volkenstein and Goldstein [19] as demonstrated in Sect. 2.4.3, the general modifier mechanism, under generalized microscopic reversibility conditions, can now be represented in the succinct form shown Scheme 3.5, where A represents E $\rightleftharpoons$ EX and B represents ES $\rightleftharpoons$ ESX. In this situations the two catalytic paths of the general modifier mechanism can be compared with two rivers as emissaries of the same lake, which eventually rejoin their waters flowing as one river.

The rate equation and the expressions of the apparent parameters, reported earlier in this chapter, are rewritten here in more convenient forms:

$$\frac{v_{\mathrm{X}}}{[\mathrm{E}]_{\mathrm{t}}} = \frac{k_{\mathrm{cat}}^{\mathrm{app}}\,[\mathrm{S}]}{K_{\mathrm{m}}^{\mathrm{app}} + [\mathrm{S}]} = \frac{k_{\mathrm{cat}}^{\mathrm{app}}}{1 + \dfrac{K_{\mathrm{m}}^{\mathrm{app}}}{[\mathrm{S}]}} = \frac{k_{\mathrm{cat}}^{\mathrm{app}}}{1 + K_{\mathrm{m}}^{\mathrm{app,niche}}} \tag{3.44}$$

$$\overbrace{\mathrm{S + A} \rightleftharpoons \mathrm{B} \xrightarrow{k_{\mathrm{cat}}^{\mathrm{app}}}}^{K_{\mathrm{m}}^{\mathrm{app}}} \mathrm{A + P}$$

Scheme 3.5 Compressed form of the general modifier mechanism. A and B represent the rapid equilibria $\mathrm{E} \rightleftharpoons \mathrm{EX}$ and $\mathrm{ES} \rightleftharpoons \mathrm{ESX}$, respectively

$$k_{\mathrm{cat}}^{\mathrm{app}} = \frac{k_{\mathrm{cat}}\left(1 + \beta\frac{[\mathrm{X}]}{\alpha K_{\mathrm{X}}}\right)}{1 + \frac{[\mathrm{X}]}{\alpha K_{\mathrm{X}}}} \tag{3.45}$$

$$K_{\mathrm{m}}^{\mathrm{app}} = K_{\mathrm{m}}^{0}\frac{1 + \frac{[\mathrm{X}]}{K_{\mathrm{X}}}}{1 + \frac{[\mathrm{X}]}{\alpha K_{\mathrm{X}}}} \tag{3.46}$$

$$K_{\mathrm{m}}^{\mathrm{app,niche}} = K_{\mathrm{m}}^{0,\mathrm{niche}}\frac{1 + \frac{[\mathrm{X}]}{K_{\mathrm{X}}}}{1 + \frac{[\mathrm{X}]}{\alpha K_{\mathrm{X}}}} \tag{3.47}$$

$$k^{\mathrm{app,niche}} = \frac{k_{\mathrm{cat}}\left(1 + \beta\frac{[\mathrm{X}]}{\alpha K_{\mathrm{X}}}\right)}{K_{\mathrm{m}}^{0,\mathrm{niche}}\left(1 + \frac{[\mathrm{X}]}{K_{\mathrm{X}}}\right)}. \tag{3.48}$$

The rate equation in the absence of modifier can be expressed similarly as

$$\frac{v_0}{[\mathrm{E}]_{\mathrm{t}}} = \frac{k_{\mathrm{cat}}[\mathrm{S}]}{K_{\mathrm{m}}^{0} + [\mathrm{S}]} = \frac{k_{\mathrm{cat}}}{1 + \frac{K_{\mathrm{m}}^{0}}{[\mathrm{S}]}} = \frac{k_{\mathrm{cat}}}{1 + K_{\mathrm{m}}^{0,\mathrm{niche}}}. \tag{3.49}$$

The superscript *niche* in (3.44), (3.47) and (3.49) denotes the Michaelis constant divided by the actual substrate concentration experienced by the enzyme as explained in Sect. 1.5. Thus, $K_{\mathrm{m}}^{0,\mathrm{niche}}$ and $K_{\mathrm{m}}^{\mathrm{app,niche}}$ are dimensionless. Kinetic barrier diagrams can now be constructed with (3.45) and (3.47)–(3.49) using the method outlined in Sect. 1.5 and the actual modifier concentration. Thus, the equivalents of (1.63)–(1.65) are:

for the reaction in the absence of modifier

$$\Delta_{\mathrm{S}}^{0} = -\log\frac{1}{K_{\mathrm{m}}^{0,\mathrm{niche}}} \tag{3.50}$$

$$\Delta^0_{\mathrm{ES}} = \log \frac{k_{\mathrm{B}}T}{h} - \log k_{\mathrm{cat}} \tag{3.51}$$

$$\Delta^0_{\mathrm{E}} = \log \frac{k_{\mathrm{B}}T}{h} - \log \frac{k_{\mathrm{cat}}}{K_{\mathrm{m}}^{0,\mathrm{niche}}} , \tag{3.52}$$

and for the reaction when the modifier is present

$$\Delta^{\mathrm{X}}_{\mathrm{S}} = -\log \frac{1}{K_{\mathrm{m}}^{\mathrm{app,niche}}} \tag{3.53}$$

$$\Delta^{\mathrm{X}}_{\mathrm{ES}} = \log \frac{k_{\mathrm{B}}T}{h} - \log k_{\mathrm{cat}}^{\mathrm{app}} \tag{3.54}$$

$$\Delta^{\mathrm{X}}_{\mathrm{E}} = \log \frac{k_{\mathrm{B}}T}{h} - \log \frac{k_{\mathrm{cat}}^{\mathrm{app}}}{K_{\mathrm{m}}^{\mathrm{app,niche}}} . \tag{3.55}$$

Two additional thermodynamic differences are necessary, namely those of the EX complex relative to E_0 and of ESX relative to ES_0, which are calculated from the respective equilibrium constants:

$$\Delta_{\mathrm{EX}} = \log K_{\mathrm{X}} \tag{3.56}$$

$$\Delta_{\mathrm{ESX}} = \log \alpha K_{\mathrm{X}} . \tag{3.57}$$

The kinetic barrier diagrams can now be represented analogously to those of the Gibbs energy profiles described by Bearne [3, Fig. 2]. Besides dimensionless ordinates, the barrier diagrams make use of local substrate and modifier concentrations, which allow representing the full potential of the 17 basic mechanisms in Fig. 3.2. In particular, the dual nature of the HMxD(I/A) and HMxD(A/I) mechanisms as inhibitors or activators depending on substrate concentration remains masked when considering standard concentrations $[\mathrm{S}] = [\mathrm{X}] = 1\,\mathrm{M}$. While barrier diagrams for each mechanism will be shown in Chap. 5 in dedicated sections, the method is illustrated here taking as example the hyperbolic mixed, predominantly catalytic inhibition mechanism, HMx(Sp<Ca)I, characterized by $\alpha < 1$, $\beta < 1$ and $\alpha > \beta$ (Fig. 3.14).

The Δ-values are first calculated with (3.50)–(3.55) and the parameters defined by (3.45)–(3.49). The left part of Fig. 3.14 shows the construction of the barriers for the reaction in the absence of modifier (black thick lines), which is performed as shown in Sect. 1.5. The absolute position of the various states on the vertical axis of the diagram is not important and what matters are the *differences* between levels. Also in this case the ground state E + S in the absence of modifier is arbitrarily set to zero and all other levels refer to this point that is labeled E_0. The position of ES_0 is then drawn with the value of Δ^0_{S} and that of the transition state, labeled here $0^{\ddagger}$ (zero for the *absence of modifier*), with the value of Δ^0_{E}. The barrier Δ^0_{ES} corresponds to $(\Delta^0_{\mathrm{E}} - \Delta^0_{\mathrm{S}})$ considering that the sign of Δ^0_{S} can be positive or negative.

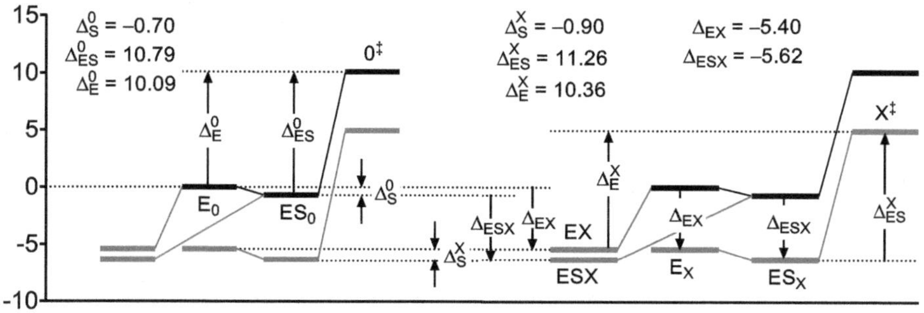

Fig. 3.14 Construction of the kinetic barrier diagram for the general modifier mechanism and linear inhibition mechanisms. This example features a hyperbolic mixed, predominantly catalytic inhibition mechanism with the following parameters and concentrations: $k_{cat} = 100\,s^{-1}$, $K_m = 2.0 \times 10^{-4}$ M, $K_X = 4.0 \times 10^{-6}$ M, $\alpha = 0.6$, $\beta = 0.3$, [S] $= 1.0 \times 10^{-3}$M, [X] $= 4.0 \times 10^{-5}$M

The right part of Fig. 3.14 shows the changes of the states E_0, ES_0 and $0^{\ddagger}$ when the modifier is added (gray thick lines). According to Bearne [3], the position of EX is drawn relative to E_0 using Δ_{EX} from (3.56). Analogously, the position of ESX relative to ES_0 is drawn with Δ_{ESX} from (3.57). Bearne considered four classical linear inhibition mechanisms, in which EX and ESX lie off the unique catalytic pathway, and therefore put them to the left side of the diagram. In the general modifier mechanism, both EX and ESX *lie on* the pathway with catalytic constant βk_{cat} and could be omitted since their positions coincide with those of E_X and ES_X, respectively. Since the system treated here can be equally used for hyperbolic and linear enzyme modification, the method of Bearne is maintained. The new states that originate from the changes Δ_{EX} and Δ_{ESX} are labeled E_X and ES_X, respectively. This is just one possible notation to represent the composite, rapidly equilibrating species A and B in Scheme 3.5.

The final step in the construction of the barriers is to add the transition state denoted $X^{\ddagger}$ (*in the presence* of modifier) with the calculated value Δ^X_{ES}. The complete diagram results from the overlap of the two halves of Fig. 3.14 after omitting unnecessary labels to avoid overload at the benefit of clarity (Fig. 3.15a). The relevant changes in the states of enzyme-containing species caused by the modifier are explicitly shown in Fig. 3.15b. From these values we can recalculate the apparent kinetic parameters and compute the rate equation. It is obvious that if we start from a diagram that we constructed with known parameters, the Δ-values will give back the known parameters. On the other hand, it is instructive to apply this procedure to diagrams sketched by hand and to make guesses on the underlying mechanism. Equations (3.58)–(3.61) show how to do it:

$$k_{cat}^{app} = 10^{\left(12.79 - \Delta^X_{ES}\right)} \tag{3.58}$$

$$K_m^{app,niche} = 10^{\left(\Delta^X_E - \Delta^X_{ES}\right)} \tag{3.59}$$

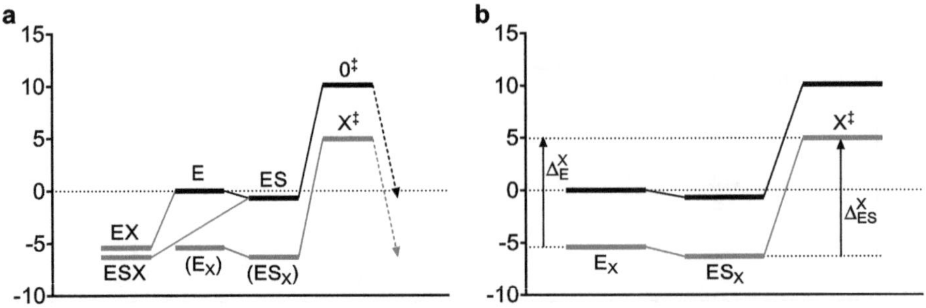

Fig. 3.15 The kinetic barrier diagram resulting from the construction in Fig. 3.14. (**a**) The diagram showing the essential details, where the states in parentheses can be omitted if their significance is not in doubt. (**b**) The *arrows* indicate the changes that can be used to calculate the apparent kinetic parameters and the reaction rate

$$\frac{k_{\text{cat}}^{\text{app}}}{K_{\text{m}}^{\text{app,niche}}} = 10^{\left(12.79-\Delta_{\text{E}}^{\text{X}}\right)} \tag{3.60}$$

$$\frac{v_{\text{X}}}{[\text{E}]_{\text{t}}} = \frac{10^{\left(12.79-\Delta_{\text{ES}}^{\text{X}}\right)}}{1+10^{\left(\Delta_{\text{E}}^{\text{X}}-\Delta_{\text{ES}}^{\text{X}}\right)}}\,. \tag{3.61}$$

The constant 12.79 (more exactly 12.79315851) in these equations represents $\log\left(k_{\text{B}}T/h\right)$. To appreciate the effect of the modifier on the rate, the result obtained from (3.61) can be compared with the rate in the absence of modifier calculated as

$$\frac{v_0}{[\text{E}]_{\text{t}}} = \frac{10^{\left(12.79-\Delta_{\text{ES}}^{0}\right)}}{1+10^{\left(\Delta_{\text{E}}^{0}-\Delta_{\text{ES}}^{0}\right)}}\,. \tag{3.62}$$

With the values in Fig. 3.14, $v_0/v_{\text{X}} = 2.76$, which corresponds to 63.8 % inhibition. The direct calculation using (3.44)–(3.46) gives $v_0/[\text{E}]_{\text{t}} = 83.3\,\text{s}^{-1}$ and $v_{\text{X}}/[\text{E}]_{\text{t}} = 30.2\,\text{s}^{-1}$, i.e., $v_0/v_{\text{X}} = 2.76$.

References

1. Alberty RA (2011) Enzyme kinetics: rapid-equilibrium applications of Mathematica. Wiley, Hoboken
2. Baici A (1981) The specific velocity plot. A graphical method for determining inhibition parameters for both linear and hyperbolic enzyme inhibitors. Eur J Biochem 119:9–14. doi:10.1111/j.1432-1033.1981.tb05570.x
3. Bearne SL (2012) Illustrating enzyme inhibition using Gibbs energy profiles. J Chem Educ 89:732–737. doi:10.1021/ed200395n
4. Bernhard S (1968) The structure and function of enzymes. W. A. Benjamin, New York
5. Cárdenas ML, Cornish-Bowden A (1989) Characteristics necessary for an interconvertible enzyme cascade to generate a highly sensitive response to an effector. Biochem J 257:339–345

6. Cornish-Bowden A (2006) The IUBMB recommendations of symbolism and terminology in enzyme kinetics. In: Hicks MG, Kettner C (eds) Proceedings of the 2nd international Beilstein symposium on experimental standard conditions on enzyme characterizations. Logos Verlag, Berlin, pp 35–50
7. Cornish-Bowden A (2012) Fundamentals of enzyme kinetics, 4th edn. Wiley, Weinheim
8. Dixon M, Webb EC (1979) Enzymes, 3rd edn. Longman, London
9. Frieden C (1964) Treatment of enzyme kinetic data. I. The effect of modifiers on the kinetic parameters of single substrate enzymes. J Biol Chem 239:3522–3531
10. Laidler KJ (1983) Kinetic equations and mechanisms for activation and inhibition in enzyme systems. Can J Biochem Cell Biol 61:1208–1218
11. Monod J, Wyman J, Changeux JP (1965) On the nature of allosteric transitions: a plausible model. J Mol Biol 12:88–118
12. Nomenclature Committee of the International Union of Biochemistry (1982) Symbolism and terminology in enzyme kinetics. Recommendations 1981. Eur J Biochem 128:281–291
13. Segal HL, Kachmar JF, Boyer PD (1952) Kinetic analysis of enzyme reactions. I. Further considerations of enzyme inhibition and analysis of enzyme activation. Enzymologia 15:187–198
14. Segel IH (1975) Enzyme kinetics. Behavior and analysis of rapid equilibrium and steady-state enzyme systems. Wiley, New York
15. Shikita M, Fahey JW, Golden TR, Holtzclaw WD, Talalay P (1999) An unusual case of 'uncompetitive activation' by ascorbic acid: purification and kinetic properties of a myrosinase from Raphanus sativus seedlings. Biochem J 341:725–732
16. Talbot PHB (1971) Principles of fungal taxonomy. Macmillan, London
17. Tian WX, Tsou CL (1982) Determination of the rate constant of enzyme modification by measuring the substrate reaction in the presence of the modifier. Biochemistry 21:1028–1032
18. Topham CM (1990) A generalized theoretical treatment of the kinetics of an enzyme-catalysed reaction in the presence of an unstable irreversible modifier. J Theor Biol 145:547–572
19. Volkenstein MV, Goldstein BN (1966) A new method for solving the problems of the stationary kinetics of enzymological reactions. Biochim Biophys Acta 115:471–477
20. Whitehead EP (1976) Simplifications of the derivations and forms of steady-state equations for non-equilibrium random substrate-modifier and allosteric enzyme mechanisms. Biochem J 159:449–456
21. Woods MJ, Findlater JD, Orsi BA (1979) Kinetic mechanism of the aliphatic amidase from Pseudomonas aeruginosa. Biochim Biophys Acta 567:225–237

Chapter 4
Complements to Enzyme–Modifier Interactions

··· readers who understand the methods used to discriminate between the simple cases are well equipped to adapt them to special experimental circumstances and to understand the more detailed discussions found elsewhere.
A. Cornish-Bowden (2012) Fundamentals of Enzyme Kinetics, Wiley-Blackwell. p. 190

Abstract Starting from the definitions of the basic modifier mechanisms, extensions to more complex systems are not always straightforward but can be achieved if the underlying theories are well understood and used correctly. The mathematical aspects can become more complex but the labor is compensated by rewarding results. This chapter fulfills a double purpose by showing the way for managing system of increasing complexity illustrating the theory with practical examples, and by adding methods that complement those described in the preceding chapters.

4.1 Introduction

The introductory motto to this chapter is taken from a comment by Athel Cornish-Bowden on the impossibility of writing an encyclopedic, exhaustive treatment of enzyme kinetics, a task appropriately defined self-defeating. A large number of specific cases that constitute variations on the theme *enzyme modification* have been published. In many cases these represent known basic mechanisms to which reaction paths have been appended thus adding also complexity to the rate equation.

In the preceding chapters, priority has been given to the theories that support the basic mechanisms of enzyme modification for unireactant enzymes. However, the same concepts can be extended to more complex systems provided the limits of the analytical methods and the criteria for choosing simplifying assumptions are well understood. To illustrate this idea, the first section will deal with the effects exerted by modifiers on proteolysis. We wish to prove the legitimation for treating these nontrivial systems with methodologies that render them manageable.

Additional concepts, methods used in the analysis of enzyme–modifier interactions and corollaries to Chaps. 2 and 3 are placed in this chapter. In combination with theoretical aspects, worked examples aim at showing that every experiment

A. Baici, *Kinetics of Enzyme-Modifier Interactions*,
DOI 10.1007/978-3-7091-1402-5_4

performed in the laboratory is best considered in its individuality rather than being assigned to a stereotype without further, dedicated analysis.

4.2 The Action of Modifiers on Peptide Bond Hydrolysis

An example of a two-substrate/two-product reaction, in which one of the substrates is naturally kept constant at a saturating concentration, is peptide-bond hydrolysis catalyzed by peptidases. The substrate whose concentration does not change is in this case water. The peptide bond hydrolases or peptidases (EC 3.4.x.y) catalyze the hydrolytic cleavage of peptide bonds in proteins and peptides of various sizes down to dipeptides. Despite using different strategies at the molecular level, a catalytic strategy shared by the seven classes of peptidases consists in delivering a proton from a nucleophile to the leaving group at the cleavage site. As shown in Scheme 4.1a, the substrate (S = peptide) reacts with water to generate two products corresponding to the peptides at the N-terminal side (P_N) and at the C-terminal side (P_C) of the bond being cleaved. For brevity, only the side chains R_1 and R_2 of the amino acids joined by the cleaved peptide bond are shown. In the nomenclature of Schechter and Berger [32], R_1 and R_2 correspond to the P_1 and P'_1 amino acid side chains of the substrate that bind to the S_1 and S'_1 pockets of the enzyme, respectively.

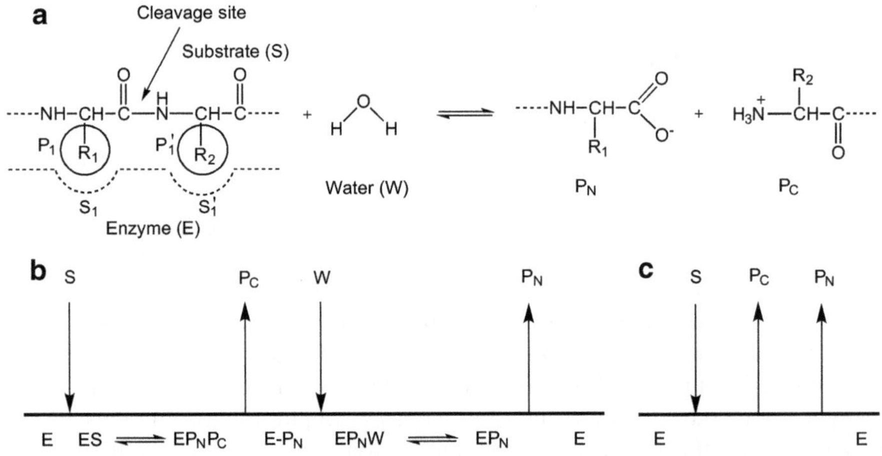

Scheme 4.1 Hydrolytic cleavage of the peptide bond catalyzed by peptidases. (**a**) Polypeptide substrate bound to the enzyme showing the specificity subsites in the nomenclature of Schechter and Berger [32]. Only the first specificity subsite is shown on either side of the scissile bond. The *nonprimed* sites of the enzyme ($S_1 \cdots$) bind substrate residues at the N-terminal side of the scissile bond and the *primed* subsites $S'_1 \cdots$) are those that bind residues of the substrate at the C-terminal side of the cleavage site. The corresponding amino acid side chain residues of the substrate are called ($P_1 \cdots$) and ($P'_1 \cdots$). (**b**) Steps in the mechanism of peptide bond hydrolysis catalyzed by serine, cysteine, and threonine peptidases that corresponds to the substituted-enzyme mechanism in Cleland's notation. (**c**) Formal reduction of the mechanism in (**b**) to an Uni-Bi ordered sequence

Scheme 4.2 The acyl-enzyme mechanism (**a**) and its *squeezed* form (**b**)

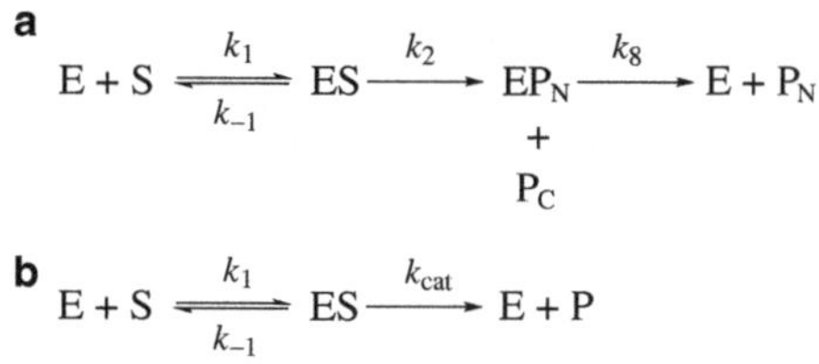

For peptide bond hydrolysis catalyzed by the serine, cysteine, and threonine peptidases, the sequence of kinetic events can be sketched in Cleland's notation as shown in Scheme 4.1b [6]. This is the sequence of reactions of the substituted-enzyme mechanism (also named double-displacement or Ping Pong Bi Bi mechanism) that involves two substrates and two-products. Since water is present in excess, the mechanisms in Scheme 4.1b can be reduced to that in panel (c) that corresponds to an ordered Uni Bi mechanism in Cleland's nomenclature, in which water is only *formally* omitted.[1]

In the customary representation of enzymatic reactions, the ordered Uni-Bi sequence in Scheme 4.1c can be represented as shown in Scheme 4.2a known as the *acyl-enzyme mechanism*.[2] Besides having to do with the initial rate, the unidirectional arrows denote irreversibility as in other hydrolytic reactions without infringing microscopic reversibility. This is justified by the highly exergonic character of peptide bond hydrolysis with a high energy barrier of the reverse reaction due to water in excess. Confirming the general notion that all enzymic reactions are reversible, the reaction is in fact reversible and peptides can be synthesized using peptidases in nonaqueous media as catalysts, e.g., [41]. By measuring only steady-state parameters without the support of pre-steady-state kinetics, viscosity-dependence of the reaction steps and isotope effects, just k_{cat} and K_m are accessible, not the individual constants. The acyl-enzyme mechanism is thus often represented in the *squeezed* Briggs–Haldane form shown in Scheme 4.2b. However, this representation implies that the rate equation and the kinetic parameters therein contain the whole information present in Scheme 4.2a:

$$k_{cat}^{0} = \frac{k_2 k_8}{k_2 + k_8}, \quad K_m^{0} = \frac{k_8 (k_{-1} + k_2)}{k_1 (k_2 + k_8)}, \quad k^{0} = \frac{k_1 k_2}{k_{-1} + k_2}. \tag{4.1}$$

In the presence of a modifier, the acyl-enzyme mechanism must consider the addition of the modifier not only to free enzyme and to the ES complex, but also to EP_N. This results in an extended Botts–Morales mechanism that includes the deacylation step as shown in Scheme 4.3a. The steady-state rate equation is quite

[1]It is important to remember that this is mere formalism because whenever a reagent is present in excess, its concentration is part of the equilibrium constant of the reaction and can by no means be omitted.

[2]To avoid confusion, the microscopic constants are numbered respecting the topology of reaction schemes already used in the preceding chapters by adding the new constant k_8.

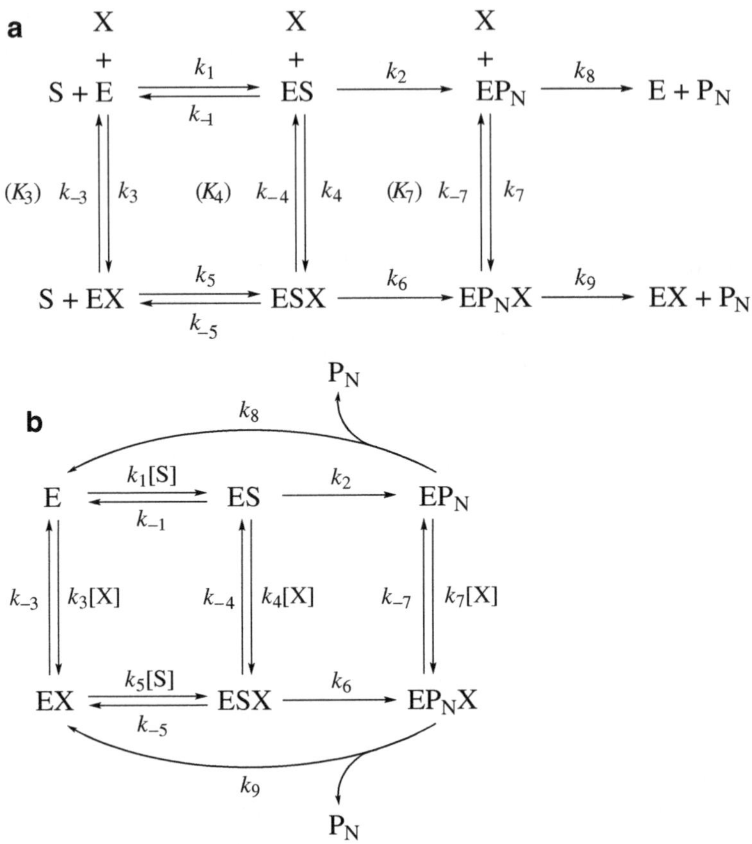

Scheme 4.3 The acyl-enzyme mechanism in the presence of a modifier. The product EP_C that dissociates first is omitted to avoid unnecessary overload of the scheme. The representation in (**a**), which also indicates in parentheses the equilibrium constants of the *vertical steps* is equivalent to mechanism (**b**) that helps understanding the relationships (4.6)–(4.10)

complex and can be derived with the same procedure used in Chap. 2 for the general modifier mechanism [13, 14].

With this mechanism, Di Cera and collaborators investigated the kinetics of thrombin in the presence of modifiers starting from the simplest but nevertheless vital modifier, *the proton* [11, 12]. The pH-dependence of the kinetic parameters in enzyme-catalyzed reactions represents in fact modification reactions in which H^+ is the modifier that behaves as activator or as inhibitor depending on its concentration, i.e., from the pH-value. With the same mechanism, the Di Cera group performed detailed studies on the effects exerted on thrombin by the sodium ion [40], by the natural inhibitor hirudin [2] and by hirudin$^{55-65}$, the C-terminal fragment of hirudin [13] that behaved as activator. These studies are ideal examples to illustrate the extension of the basic principles of the general modifier mechanism to a more complex system. Increase in *complexity* does not mean *complication* as long as the fundamental concepts are clearly understood.

To the thermodynamic cycle of the general modifier mechanism, an extension has now been added (Scheme 4.3) and the linked functions therein can be better appreciated from the equivalent representation shown in Scheme 4.3b.[3] In the impossibility of using the complex steady-state rate equation for data analysis, simplification is desirable. This is possible by considering the vertical steps to equilibrate rapidly with respect to the horizontal steps in a steady-state regime. This assumption is logical for modifiers that are expected to bind rapidly, like the proton or simple ions like Na^+. For other modifiers, the validity of the quasi-equilibrium assumption can be ascertained with the methods discussed in Chap. 2.

At saturating modifier concentration, the kinetic parameters (4.1) become

$$k_{cat}^{\infty} = \frac{k_6 k_9}{k_6 + k_9}, \quad K_m^{\infty} = \frac{k_9 (k_{-5} + k_6)}{k_5 (k_6 + k_9)}, \quad k^{app} = \frac{k_5 k_6}{k_{-5} + k_6} . \tag{4.2}$$

The steady-state rate equation of the acyl-enzyme mechanism in the presence of a modifier can be expressed in condensed form as done analogously in Chap. 2 with (2.10) and (2.11) for the general modifier mechanism:

$$v = k_{cat}[E]_t \frac{a\,[S] + [S]^2}{b + c\,[S] + [S]^2} , \tag{4.3}$$

where a, b, and c represent combinations of all rate constants of the system and contain the modifier concentrations as [X], $[X]^2$ and $[X]^3$ (not shown here but given in full elsewhere [13]). It results that the cumbersome steady-state equation simplifies to

$$v = \frac{k_{cat}^{app}[E]_t\,[S]}{K_m^{app} + [S]} , \tag{4.4}$$

which does not contain higher-power reactant concentrations if the following relationships between rate constants in Scheme 4.3b are true *simultaneously*:

$$k_1 k_4 k_{-5} k_{-3} = k_{-1} k_3 k_5 k_{-4} \tag{4.5}$$

$$k_1 k_7 k_9 k_{-3} = k_8 k_3 k_5 k_{-7} \tag{4.6}$$

$$k_2 k_7 k_9 k_{-4} = k_8 k_4 k_6 k_{-7} , \tag{4.7}$$

where (4.5) is the same expression for microscopic reversibility of the general modifier mechanism (2.6).

[3] Readers interested in the cited papers published by Di Cera and collaborators should note that the authors use association constants in place of dissociation constants as in this book.

With the vertical steps in equilibrium, the corresponding microscopic constants can be substituted by equilibrium constants. Thus, the whole system *operates at equilibrium* if:

$$k_1 k_{-5} K_3 = k_{-1} k_5 K_4 \tag{4.8}$$

$$k_1 k_9 K_3 = k_8 k_5 K_7 \tag{4.9}$$

$$k_2 k_9 K_4 = k_8 k_6 K_7 \ . \tag{4.10}$$

Extracting the ratio k_6/k_2 from (4.10), the ratio k_{-5}/k_{-1} from (4.8) and $k_1 K_3$ from (4.9), the algebraic manipulation results in

$$k_6 k_{-1} = k_2 k_{-5} \ , \tag{4.11}$$

which is the expression of the principle of *generalized microscopic reversibility* discussed in Sect. 2.4.2. This principle depends on microscopic reversibility, i.e., from the fulfilment of (4.8). If all relevant rate constants of the acyl-enzyme mechanism are accessible experimentally, the test for the reduction of the steady-state rate equation to the equilibrium form (4.4) depends on whether the K_3/K_4–ratio calculated by the two following expressions is the same within experimental error.

$$\frac{K_3}{K_4} = \frac{k_{-1}k_5}{k_1 k_{-5}} \quad \text{and} \quad \frac{K_3}{K_4} = \frac{k_2 k_5}{k_1 k_6} \tag{4.12}$$

The left expression in (4.12) describes microscopic reversibility [from (4.8)] and the right expression describes generalized microscopic reversibility [from (4.9) and (4.10)]. While unraveling the condition for the simplification of the acyl-enzyme mechanism to the Michaelis–Menten form in the presence of a modifier, the solution was deduced with the inverse logic made in Sect. 2.4.2.

The concept expressed by (4.5)–(4.11) can be visualized graphically as shown in Scheme 4.4 that shows elements of Scheme 4.3b. Only if the conditions represented by (a–c) are simultaneously satisfied, then condition (d) is also satisfied. In this case, the apparent parameters to be introduced in (4.4) are given by

$$k_{cat}^{app} = \frac{k_2 + k_6 \dfrac{[X]}{K_4}}{\left(\dfrac{k_2 + k_8}{k_8}\right) + \left(\dfrac{k_6 + k_9}{k_9}\right)\dfrac{[X]}{K_4}} \tag{4.13}$$

$$K_m^{app} = \frac{\left(\dfrac{k_{-1} + k_2}{k_1}\right) + \left(\dfrac{k_{-5} + k_6}{k_5}\right)\dfrac{[X]}{K_4}}{\left(\dfrac{k_2 + k_8}{k_8}\right) + \left(\dfrac{k_6 + k_9}{k_9}\right)\dfrac{[X]}{K_4}} \ . \tag{4.14}$$

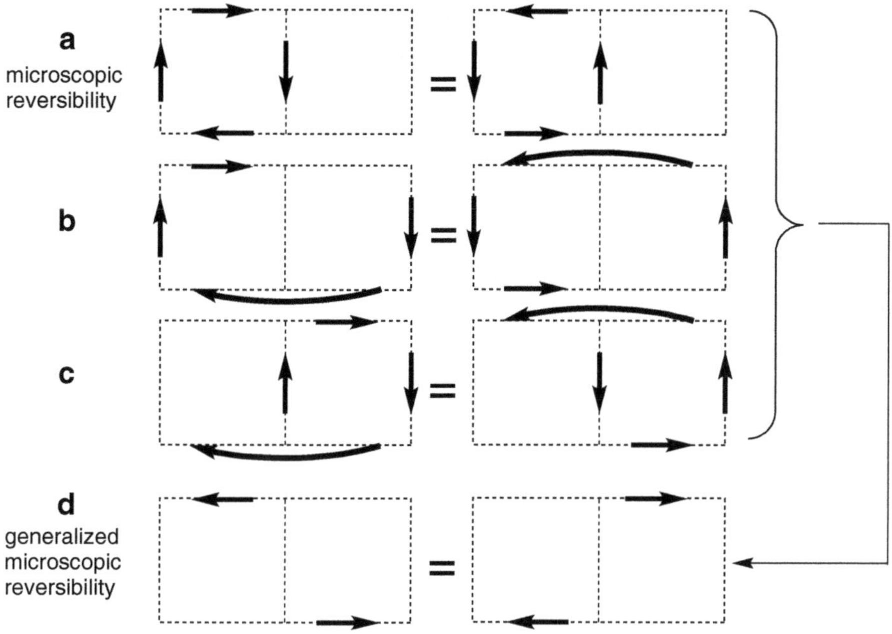

Scheme 4.4 Graphical representation of the equalities (4.5)–(4.7) as (**a**)–(**c**), respectively, and (4.11) as (**d**). The rate constants represented by the *arrows* in each diagram are multiplied together. If (**a**)–(**c**) are simultaneously true, then (**d**) is also true

Without detailing the derivation method (the same described in Sect. 2.4.3), we see that (4.13)–(4.14), which define the parameters at finite modifier concentrations, give the expressions (4.1) when $[X] = 0$ and (4.2) when $[X] = \infty$.

This elegant solution to an otherwise hard problem in the interpretation of kinetic measurements with systems that obey the acyl-enzyme mechanism requires validation. This has been provided by the Di Cera team, who measured the individual kinetic constants of the horizontal steps and the equilibrium constants of the vertical steps in Scheme 4.3 exploiting the sensitivity of the rate constants that involve diffusion to the viscosity of the reaction medium. In the acyl-enzyme mechanism, k_1, k_{-1}, and k_8 depend on diffusion and are thus affected by viscosity [40]. Considering the values of the constants with their standard errors for Na^+ and hirudin$^{55-65}$ listed in Table 4.1 and calculating error propagation, we can verify if the constraints between rate constants (4.12) are satisfied by comparing K_3/K_4 with the two other ratios:

$$\frac{K_3}{K_4} = 2.26 \pm 0.21 \qquad (4.15)$$

$$\text{Na}^+ \qquad \frac{k_{-1}k_5}{k_1k_{-5}} = 1.98 \pm 0.28 \qquad (4.16)$$

Table 4.1 Parameters for the amidase activity of thrombin

Parameters	Units	Without modifier	
k_1	$M^{-1}\,s^{-1}$	$(1.10 \pm 0.10) \times 10^7$	
k_{-1}	s^{-1}	47 ± 4	
k_2	s^{-1}	78 ± 1	
k_8	s^{-1}	59 ± 1	
		Na^+	Hirudin$^{55-65}$
k_5	$M^{-1}\,s^{-1}$	$(1.22 \pm 0.10) \times 10^8$	$(3.6 \pm 0.6) \times 10^7$
k_{-5}	s^{-1}	263 ± 29	53 ± 7
k_6	s^{-1}	94 ± 1	91 ± 2
k_9	s^{-1}	541 ± 22	107 ± 2
K_3	M	$(2.17 \pm 0.16) \times 10^{-2}$	$(4.17 \pm 0.69) \times 10^{-6}$
K_4	M	$(9.62 \pm 0.55) \times 10^{-3}$	$(1.41 \pm 0.10) \times 10^{-6}$
K_7	M	$(1.72 \pm 0.12) \times 10^{-2}$	$(2.17 \pm 0.14) \times 10^{-6}$

The original data [13, Table I] have been transformed to express all concentrations as M and dissociation in place of association constants

$$\frac{k_2 k_5}{k_1 k_6} = 9.20 \pm 1.14 \tag{4.17}$$

$$\frac{K_3}{K_4} = 2.96 \pm 0.54 \tag{4.18}$$

hirudin$^{55-65}$

$$\frac{k_{-1} k_5}{k_1 k_{-5}} = 2.90 \pm 0.72 \tag{4.19}$$

$$\frac{k_2 k_5}{k_1 k_6} = 2.81 \pm 0.54\,. \tag{4.20}$$

For Na^+, (4.15) and (4.16) give the same result within experimental error, while the result in (4.17) is decidedly larger than that in (4.15). Conversely, (4.18)–(4.20) yield the same value. Thus, the binding of Na^+ to thrombin does not adhere to generalized microscopic reversibility while that of hirudin$^{55-65}$ does.

These interactions are good examples for putting in practice the criteria of acceptance of simplifying conditions in performing calculations of kinetic parameters from raw data discussed in Chap. 2 (Sects. 2.6–2.8). Since all relevant constants have been measured, catalytic, Michaelis, and specificity constants in the absence and in the presence of modifiers, as well as other useful information on their properties can be calculated as shown in Table 4.2.

The substrate used is moderately sticky (k_2/k_{-1} =1.7) but this has no consequences for calculating the kinetic parameters in the presence of hirudin$^{55-65}$ because the system is governed by generalized microscopic reversibility. Thus, a very good correspondence between true and measured values of the parameters is expected and the quality of their estimates depends exclusively on experimental error. The situation can be considered satisfactory also for Na^+ as modifier because

Table 4.2 Calculation of the kinetic parameters for the amidase activity of thrombin and the effect of modifiers

Parameters/Properties	Units	Without modifier	
k_{cat}^{0}	s^{-1}	33.6 ± 0.8	
K_{m}^{0}	M	$(4.9 \pm 0.5) \times 10^{-6}$	
k^{0}	$M^{-1}\,s^{-1}$	$(6.9 \pm 0.7) \times 10^{6}$	
		Na^{+}	Hirudin$^{55-65}$
k_{cat}^{∞}	s^{-1}	80.1 ± 4.4	49.2 ± 1.6
K_{m}^{∞}	M	$(2.5 \pm 0.3) \times 10^{-6}$	$(2.2 \pm 0.4) \times 10^{-6}$
k^{∞}	$M^{-1}\,s^{-1}$	$(3.2 \pm 0.4) \times 10^{7}$	$(2.3 \pm 0.4) \times 10^{7}$
$k_{cat}^{0}/k_{cat}^{\infty} = \beta$	–	2.38 ± 0.14	1.46 ± 0.06
$K_3/K_4 = 1/\alpha$ [a]	–	2.16 ± 0.21	2.96 ± 0.54
α	–	0.44 ± 0.04	0.34 ± 0.06
MR[b] obeyed?		Yes	Yes
GMR[c] obeyed?		No	Yes
Expected quality of the		Satisfactory	Very good
Measured parameters			

[a] Allosteric coupling constant
[b] Microscopic reversibility
[c] Generalized microscopic reversibility

substrate stickiness in the presence and absence of modifier is in the same range of the theoretical examples analyzed in Sect. 2.7. Thus, the differences between the calculated parameters and their true values are expected to lie within a few percent and are very likely masked by experimental error.

In Table 4.2, $\beta > 1$ and $\alpha < 1$ for both modifiers identifies both of them as nonessential activators and, to conclude this section, we apply the specific velocity plot to determine the mechanism using the original data of Di Cera and colleagues (Table 4.1). Initial rates, simulated at five substrate and five modifier concentrations using (4.4) and the apparent parameters (4.13), (4.14), were reorganized as v_0/v_X and plotted against $\sigma/(1+\sigma)$ (Fig. 4.1). The specific velocity plots and their replots are characteristic of the hyperbolic mixed, predominantly catalytic activation mechanism [HMx(Sp<Ca)A].

4.3 Dedicated to the Lovers of IC_{50}

IC_{50} corresponds to the inhibitor concentration that decreases the rate of an enzyme-catalyzed reaction from v_0, in the absence of inhibitor, to $0.5 \times v_0$. Under specified conditions of pH, ionic strength, and temperature, at variance with the inhibitor dissociation *constant*, IC_{50} *is not* a constant.

Several authors dealt with the properties and the practical use of this parameter. Cheng and Prusoff pointed out that IC_{50} differs from the inhibition constant if the

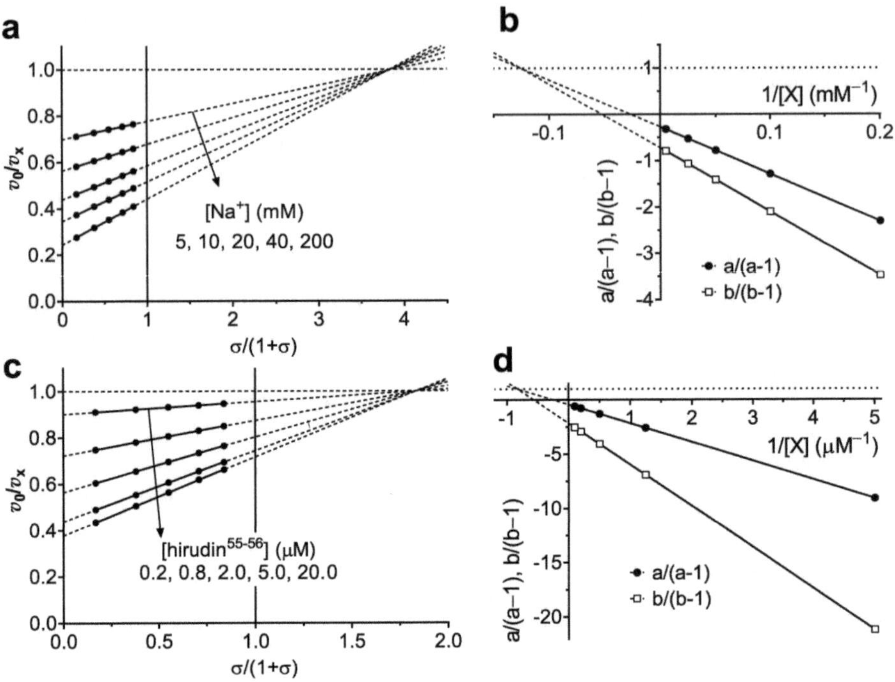

Fig. 4.1 Specific velocity plot for the activation of human thrombin by the sodium ion and the hirudin C-terminal fragment hir$^{55-65}$. Data were simulated with the parameters in Table 4.1 and (4.4). (**a**), (**b**) Results for Na^+, (**c**), (**d**) results for hirudin$^{55-65}$. (**a**) and (**c**) primary plots, (**b**) and (**d**) secondary plots. Both modifiers belong to the mechanism HMx(Sp<Ca)A

mechanism has a competitive character [5]. The authors also advised against its indiscriminate use to compare results from different laboratories if the experimental conditions do not match perfectly.[4] Cortés and coworkers [9] observed that, for linear inhibitors, IC_{50} can be conveniently measured by exploiting the geometric properties of the plot $[S]/v$ against the concentration of inhibitor, the method developed by Cornish-Bowden for determining the inhibition constant of classical inhibitors [7]. In the same paper the authors pointed out that the IC_{50} concept can be equally applied to reactions in the presence of activators and cooperativity. While commenting the misuse of IC_{50}, Naqui remarked its inadequacy for characterizing hyperbolic inhibitors because IC_{50} can attain physically meaningless negative values [30].

This section is aimed at describing the properties of IC_{50} and at proposing an alternative to this much discussed parameter that applies equally well to linear and hyperbolic inhibition and activation.

[4]This sound advice seems to have been largely ignored.

An expression of IC_{50} can be derived starting from the rate equation of the general modifier mechanism under equilibrium conditions (v_X) and the Michaelis–Menten equation ($v_0 = (k_2[E]_t[S])/(K_m + [S])$) that define the velocities in the presence and in the absence of inhibitor, respectively. To make easier deducing the IC_{50} expression for particular mechanisms, two equivalent forms will be shown below (4.21): one that uses the symbolism in Scheme 3.1a (with α and K_X) and the other that uses the symbols in Scheme 3.1b (with K_{Sp} and K_{Ca}):

$$v_X = \frac{k_2\left(1 + \frac{\beta[X]}{\alpha K_X}\right)[E]_t[S]}{K_m\left(1 + \frac{[X]}{K_X}\right) + [S]\left(1 + \frac{[X]}{\alpha K_X}\right)} \equiv \frac{k_2\left(1 + \frac{\beta[X]}{K_{Ca}}\right)[E]_t[S]}{K_m\left(1 + \frac{[X]}{K_{Sp}}\right) + [S]\left(1 + \frac{[X]}{K_{Ca}}\right)} . \tag{4.21}$$

When the rate in the presence of inhibitor is reduced to one half of v_0, then

$$\frac{v_0}{v_X} = \frac{K_m\left(1 + \frac{[X]}{K_{Sp}}\right) + [S]\left(1 + \frac{[X]}{K_{Ca}}\right)}{(K_m + [S])\left(1 + \frac{\beta[X]}{K_{Ca}}\right)} = 2 \tag{4.22}$$

$$\frac{v_0}{v_X} = \frac{K_m\left(1 + \frac{[X]}{K_X}\right) + [S]\left(1 + \frac{[X]}{\alpha K_X}\right)}{(K_m + [S])\left(1 + \frac{\beta[X]}{\alpha K_X}\right)} = 2 , \tag{4.23}$$

and solving for [X] gives the sought IC_{50}:

$$IC_{50} = \frac{K_{Ca}(K_m + [S])}{K_m\left(\frac{K_{Ca}}{K_{Sp}} - 2\beta\right) + [S](1 - 2\beta)} \tag{4.24}$$

$$IC_{50} = K_X\left\{\frac{\alpha(K_m + [S])}{K_m(\alpha - 2\beta) + [S](1 - 2\beta)}\right\} . \tag{4.25}$$

Since all parameters in (4.24) and (4.25) are constant under specified conditions, IC_{50} depends on [S] in a way that is characteristic of the individual mechanisms and thus represents an apparent constant.

The definition of IC_{50} contains the implicit statement that this parameter is unambiguous only for linear inhibitors, i.e., those expected to reduce the rate to zero at saturating concentration. Transferring this concept to hyperbolic modifiers does not make sense because inspection of (4.24) and its equivalent (4.25) reveals that it can take positive or negative values or even become infinite. Moreover, the nonessential activators act by increasing, not by reducing the rate. Therefore, IC_{50}

can be used in a physically meaningful sense *only* for the five linear inhibition mechanisms in Fig. 3.2.

To deduce the expression of IC_{50} for the linear inhibition mechanisms ($\beta = 0$), (4.24) is conveniently rearranged in the two equivalent forms

$$IC_{50} = \frac{K_{\mathrm{Sp}}\left(K_{\mathrm{m}} + [\mathrm{S}]\right)}{K_{\mathrm{m}} + [\mathrm{S}]\dfrac{K_{\mathrm{Sp}}}{K_{\mathrm{Ca}}}} \equiv \frac{K_{\mathrm{Ca}}\left(K_{\mathrm{m}} + [\mathrm{S}]\right)}{K_{\mathrm{m}}\dfrac{K_{\mathrm{Ca}}}{K_{\mathrm{Sp}}} + [\mathrm{S}]}, \tag{4.26}$$

and the explicit expressions are as follows:

$$\text{LSpI} \quad (K_{\mathrm{Ca}} = \infty): \quad IC_{50} = K_{\mathrm{Sp}}\left(1 + \frac{[\mathrm{S}]}{K_{\mathrm{m}}}\right) \tag{4.27}$$

$$\text{LCaI} \quad \left(K_{\mathrm{Sp}} = \infty\right): \quad IC_{50} = K_{\mathrm{Ca}}\left(1 + \frac{K_{\mathrm{m}}}{[\mathrm{S}]}\right) \tag{4.28}$$

$$\text{LMx(Sp} > \text{Ca)I} \quad \left(K_{\mathrm{Sp}} < K_{\mathrm{Ca}}\right): \quad IC_{50} = K_{\mathrm{Sp}}\left(\frac{K_{\mathrm{m}} + [\mathrm{S}]}{K_{\mathrm{m}} + [\mathrm{S}]\dfrac{K_{\mathrm{Sp}}}{K_{\mathrm{Ca}}}}\right) \tag{4.29}$$

$$\text{LMx(Sp} < \text{Ca)I} \quad \left(K_{\mathrm{Sp}} > K_{\mathrm{Ca}}\right): \quad IC_{50} = K_{\mathrm{Ca}}\left(\frac{K_{\mathrm{m}} + [\mathrm{S}]}{K_{\mathrm{m}}\dfrac{K_{\mathrm{Ca}}}{K_{\mathrm{Sp}}} + [\mathrm{S}]}\right) \tag{4.30}$$

$$\text{LMx(Sp} = \text{Ca)I} \quad \left(K_{\mathrm{Sp}} = K_{\mathrm{Ca}}\right): \quad IC_{50} = K_{\mathrm{Sp}} = K_{\mathrm{Ca}} \tag{4.31}$$

The tables containing countless IC_{50} values in the very rich pharmacological literature should only intend to represent either one of the cases in (4.27)–(4.31). However, very often neither the method for calculating IC_{50} nor the experimental conditions are specified.

If IC_{50} fails in its scope with the hyperbolic modifier mechanisms, an analogous parameter, MC_{50}, removes this obstacle by considering the limiting effect produced on the rate at saturating modifier concentration in comparison with the rate in the absence of modifier:

> MC_{50} is defined as the concentration of a modifier, either linear or hyperbolic inhibitor or nonessential activator, which produces an effect on the rate of an enzyme-catalyzed reaction corresponding to one half the limiting value attainable at saturating modifier concentration.

If v_0 and v_∞ are the rates in the absence of and at saturating modifier, the rate between these two extremes is given by

$$v_{0.5} = \frac{v_0 + v_\infty}{2}, \tag{4.32}$$

and MC_{50} can be calculated using the expressions of v_X in (4.21) by setting $v_X = v_{0.5}$ given by (4.32). Solving for [X] yields a choice of four equivalent expressions that are useful to deduce MC_{50} for *all* 17 mechanisms in Fig. 3.2:

$$MC_{50} = \frac{K_{Sp}(K_m + [S])}{K_m + [S]\dfrac{K_{Sp}}{K_{Ca}}} \equiv \frac{K_{Ca}(K_m + [S])}{K_m\dfrac{K_{Ca}}{K_{Sp}} + [S]}, \tag{4.33}$$

$$MC_{50} = \frac{K_X(K_m + [S])}{K_m + \dfrac{[S]}{\alpha}} \equiv \frac{\alpha K_X(K_m + [S])}{\alpha K_m + [S]}. \tag{4.34}$$

The enhanced performance of MC_{50} versus IC_{50} can be appreciated from the fact that MC_{50} implicitly contains the factor by which the modifier affects the catalytic constant (β), which does not appear in (4.33) and (4.34). It is always positive and cannot attain physically meaningless negative values or infinity.

When $\beta = 0$, MC_{50} is equivalent to IC_{50} and the five linear inhibition mechanisms in (4.27)–(4.31) can be equally characterized with either parameter. Considering the raising interest for allosteric modifiers, this is a tentative suggestion to drop IC_{50} in favor of MC_{50}, which provides the concentration of modifier that causes a decrease or an increase of the rate by 50 %. However, also this parameter says very little if not accompanied by adequate descriptions of the experimental conditions.

During labor-intensive screenings of synthetic compounds as potential enzyme inhibitors (but why not also as activators?), as typically seen in pharmacologically oriented publications, MC_{50} can perform better than IC_{50} in interpreting *anomalous* results, i.e., those that yield negative IC_{50}-values. Again, the question arises if specifying the mechanism, any allosteric coupling between substrate and modifier and the modifier dissociation constants is still a better choice.

4.4 Tightly Bound Modifiers

The reactant concentrations that appear in the rate equations of enzyme-catalyzed reactions represent the concentrations of *free* (unbound) reactants. For reactions in the presence of a modifier this is accurate as long as $[X]_{free} \approx [X]_t$ because depletion of X for its participation in the formation of complexes is negligible. It can however happen that sizable inhibition or activation is observed at comparable modifier and enzyme concentrations. This behavior is a purely experimental, not a mechanistic issue, and if it were possible to lower $[E]_t$ with respect to $[X]_t$ the tight-binding condition would disappear. This intention is fine but may contrasts with the detection limits of the measuring instruments because, for any system, a threshold is found by lowering the concentration of the catalyst, under which the measured signal disappears within the background noise.

Suppose that an experiment is set up with enzyme and modifier in the millimolar range and that almost full inhibition is observed because K_X is two orders of magnitude lower. Repeating the same experiment at $[X]_t$ and $[E]_t$ two orders of magnitude lower, the degree of inhibition will be much lower. Considering the relationships between $[X]_t$, $[E]_t$ and their dissociation constant K_X, the tight-binding condition becomes certainly appreciable when $[X]_t \approx [E]_t$ *and* K_X. If K_X is much smaller with respect to $[X]_t$ and $[E]_t$ the tight-binding condition is enhanced.

The quantitative analysis of tight-binding linear inhibition has been considered early by Straus and Goldstein [37] and by Ackermann and Potter [1]. Further important elaborations have been provided by Morrison [29], who extended the treatment to reactions with more than one substrate, and Henderson [15]. Cha considered tight-binding in the frame of both rapid [4] and slow-onset enzyme-inhibitor interactions. These methods are fine until the inhibitors are linear but cannot be used for hyperbolic inhibition mechanisms and nonessential activation. Therefore, Baici attempted the derivation of an equation aimed at covering the tight-binding condition of both linear and hyperbolic modifiers [3]. Unfortunately, a deplorable mathematical mistake produced an equation that did not exactly fulfil its purposes. However, Stefan Szedlacksek recognized the error and amended the equation [38]. I wish to thank Stefan once again for repairing my mistake.

The tight-binding equation of the general modifier mechanism is written as follows using the familiar symbols used so far (the derivation is performed in the Appendix):

$$v_X = \frac{v_0}{2}\left[1 - \frac{\beta(1+\sigma)}{\alpha+\sigma}\right]\left\{\begin{array}{l}\sqrt{\left[\left(\frac{1+\sigma}{\alpha+\sigma}\frac{\alpha K_X}{[E]_t} + \frac{[X]_t}{[E]_t} - 1\right)^2 + 4\frac{1+\sigma}{\alpha+\sigma}\frac{\alpha K_X}{[E]_t}\right]} \\ + \frac{\alpha+\sigma+\beta(1+\sigma)}{\alpha+\sigma-\beta(1+\sigma)} - \frac{1+\sigma}{\alpha+\sigma}\frac{\alpha K_X}{[E]_t} - \frac{[X]_t}{[E]_t}\end{array}\right\}. \tag{4.35}$$

This equation can be used for the analysis of all basic mechanisms in Fig. 3.2 and its structure is similar to the customary expression (4.36) for calculating the proportion of bound ligand in the generic reaction A + B $\rightleftharpoons$ AB, with dissociation constant K_d. This expression, derived below, occurs in the equations of ligand binding and enzyme inhibition whenever a sizable proportion of a ligand, inhibitor, or activator (B) with total concentration $[B]_t$ is bound to a receptor, protein or enzyme (A) with total concentration $[A]_t$.

$$\mathrm{A} + \mathrm{B} \rightleftharpoons \mathrm{AB}$$

$$K_d = \frac{[A][B]}{[AB]}$$

$$[A] = [A]_t - [AB]$$

$$
\begin{aligned}
[\mathrm{B}] &= [\mathrm{B}]_t - [\mathrm{AB}] \\
K_d\,[\mathrm{AB}] &= ([\mathrm{A}]_t - [\mathrm{AB}])\,([\mathrm{B}]_t - [\mathrm{AB}]) \\
[\mathrm{AB}]^2 &- [\mathrm{AB}]\,([\mathrm{A}]_t + [\mathrm{B}]_t + K_d) + [\mathrm{A}]_t[\mathrm{B}]_t = 0 \\
[\mathrm{AB}] &= \frac{([\mathrm{A}]_t + [\mathrm{B}]_t + K_d) \pm \sqrt{([\mathrm{A}]_t + [\mathrm{B}]_t + K_d)^2 - 4[\mathrm{A}]_t[\mathrm{B}]_t}}{2}\,. \qquad (4.36)
\end{aligned}
$$

The quadratic equation has two roots but only the root with the *minus sign* before the square root is physically significant.

4.4.1 *Determination of the Active Site Concentration of Enzymes by Titration*

Equation (4.35) can be exploited for determining the concentration of enzyme active sites. Knowing the exact concentration of active enzyme is important in kinetic experiments and, having a suitable method at hand, enzyme stability over time can be checked periodically. Titrations of enzymes can be performed with various approaches. One of these consists in incubating a small volume of an enzyme solution with increasing known concentrations of an irreversible inhibitor and assaying the residual activity. If an irreversible inhibitor is not available, the titration can also be performed with a reversible inhibitor, whose mechanism and inhibition constant is known. In this case, the experimental conditions do not necessarily need to be optimized for attaining enzyme-inhibitor binding that is as tight as possible, also with the purpose of using little amounts of enzyme and/or titrant if the reagents are expensive or available only in small amounts.

The plots in Fig. 4.2a show the changing shape of titration curves with increasing affinity between enzyme and modifier that is a hyperbolic mixed, predominantly specific activator in this example. The numbers in correspondence of the horizontal dashed lines indicate the concentrations of free modifier (the total concentration of X is 10 nM along the vertical dashed line for the five curves). The curve labeled $K_X = 10$ nM illustrates the threshold condition for tight-binding mentioned above, i.e., $[\mathrm{X}]_t \approx [\mathrm{E}]_t$ *and* K_X, a point for which the assumption $[\mathrm{X}] \approx [\mathrm{X}]_t$ does not hold and even less when $K_X < [\mathrm{X}]_t, [\mathrm{E}]_t$. The curve for $K_X = 0.1$ nM looks like a typical titration curve that can be exploited for measuring the concentration of enzyme active sites. Such titrations are customarily performed with irreversible or linear tight-binding inhibitors but the idea of using a *nonessential activator* in place of inhibitors should not be discarded if a substance of this type is available.

In Fig. 4.2b,a titration of enzyme active sites is shown using a hyperbolic specific inhibitor. This procedure supposes that the mechanism of inhibition for the considered enzyme is known and so α, β, and K_X. Instead of using pencil and ruler,

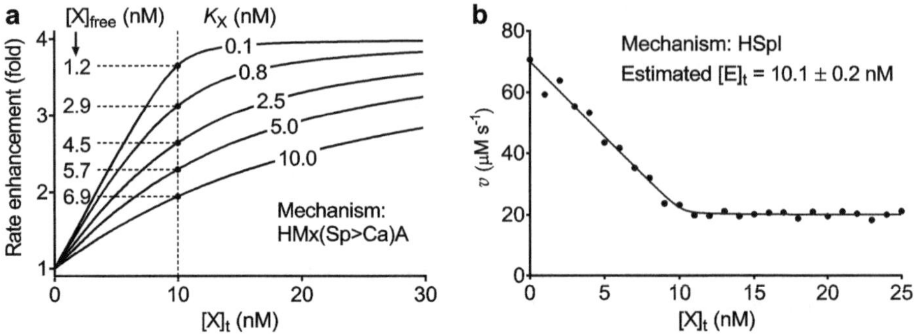

Fig. 4.2 Tightly bound modifiers. The *curves* in panel (**a**) were plotted with (4.35) for the hyperbolic mixed, predominantly specific activation mechanism with the following parameters: $\alpha = 3.0$, $\beta = 8.0$, $\sigma = 1.0$, $[\mathrm{E}]_t = 10\,\mathrm{nM}$, $v_0 = 1$ (arbitrary units to show the fold increase in activity) and five values of K_X (nM) as indicated. The numbers aligned with the *black dots* represent the free modifier concentrations calculated for the corresponding K_X-values. In panel (**b**), the *black circles* were simulated with (4.35) for the hyperbolic specific inhibition mechanism and the following parameters: $\alpha = 6.0$, $\beta = 1.0$, $\sigma = 1.0$, $[\mathrm{E}]_t = 10\,\mathrm{nM}$, $K_X = 0.01\,\mathrm{nM}$, $v_0 = 70\,\mu\mathrm{M\,s^{-1}}$; Gaussian relative error (5 % SD) was added. The *solid line* was generated by fitting (4.35) to the fake data that supplied an estimated value of $[\mathrm{E}]_t = 10.1 \pm 0.2\,\mathrm{nM}$

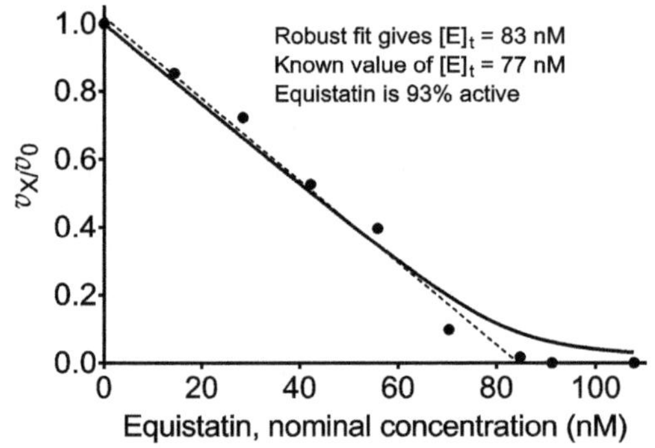

Fig. 4.3 Titration of a known amount of human cathepsin D (77 nM in the assay) with a preparation of native equistatin from *Actinia equina* whose concentration was known only from protein content. The *solid line* is the result of robust fit of (4.35) to data with all parameters in the equation treated as known constants with the exception of $[\mathrm{E}]_t$ that was fitted. Equistatin is a linear specific inhibitor, i.e., $\alpha = \infty$ and $\beta = 0$, $K_X = 0.3\,\mathrm{nM}$, $\sigma = 1.9$. The *dashed line* is as extrapolated by inspection in the original publication, which gives 91 % active equistatin [21]. Data and details of the experiment were kindly provided by Brigita Lenarčič, Faculty of Chemistry and Chemical Technology, University of Ljubljana, Slovenia

the active enzyme concentration can be estimated by fitting (4.35) to data, with the total enzyme concentration as the sole variable to be estimated.

Two real applications are analyzed in Figs. 4.3 and 4.4. The first example features the inhibition of human cathepsin D (EC 3.4.23.5) by equistatin, a protein isolated from the sea anemone *Actinia equina* by Brigita Lenarčič and coworkers [22].

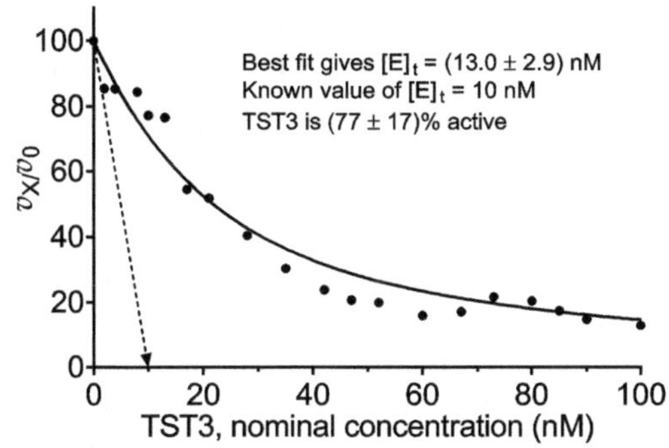

Fig. 4.4 Titration of a known amount of human cathepsin B (10 nM in the assay) with a sample of the human thyroglobulin type-1 domain of testican 3 (TST3), a linear specific inhibitor. The *solid line* is a best fit of (4.35) to data with all parameters in the equation treated as known constants with the exception of $[E]_t$ that was left free to float: $\alpha = \infty$ and $\beta = 0$, $K_X = 13.6$ nM, $\sigma = 0.12$. The *arrowed dashed line* represents the theoretical titration curve that had been obtained using an irreversible inhibitor. The original data were kindly provided by Tilen Vidmar and Brigita Lenarčič, Faculty of Chemistry and Chemical Technology, University of Ljubljana, Slovenia

Equistatin, a protein composed of three thyroglobulin-type 1 domains, inhibits the cysteine peptidases papain, cathepsin L, cathepsin B and cruzipain, and the aspartic peptidase cathepsin D. Only the protein concentration of equistatin was known after purification to homogeneity and its concentration as active inhibitor had therefore to be measured by titration using an enzyme of known active site concentration. Human cathepsin D was first titrated with the hexapeptide pepstatin, a potent inhibitor of aspartic peptidases. A known amount of cathepsin D was then titrated with equistatin as shown in Fig. 4.3 and (4.35) was fitted to data with all parameters constrained as known constants while allowing the enzyme concentration to float. Due to the inherent difficulties in measuring exactly the initial rates near the equivalence point, robust nonlinear regression fit was performed for its larger tolerance to non-Gaussian error distribution and outliers. This fitting procedure does not provide standard errors or confidence intervals. The best fit value for $[E]_t$ was 83 nM that compared with the known value of 77 nM shows that equistatin was 93 % active. The dashed line in Fig. 4.3, drawn by inspection in the original publication [21], gives 91 % active inhibitor. Although this result is the same as that obtained by nonlinear fit, the shape of the fitted curve suggests the importance of considering, in general, the underlying modifier mechanism. In this case the substrate competes with the linear specific inhibitor, an effect that is taken into account by Eq. (4.35).

In the equistatin example (Fig. 4.3) it was possible to adjust the experimental conditions to obtain a steep titration curve, from which the unknown concentration of the titrant could be appreciated also by inspection. The second example (Fig. 4.4) shows a titration of human cathepsin B (EC 3.4.22.1) with a sample of the human thyroglobulin type-1 domain of testican 3 that inhibits the enzyme with a linear specific mechanism. The biochemistry of the thyroglobulin type-1 domain is an attractive research topic for the many functions it exerts in unrelated proteins. One of these functions is as inhibitor of several peptidases, for which the name *thyropins* (acronym of thyroglobulin type-1 domain proteinase inhibitors) has been coined by Lenarčič and Bevec [20].

The titrant was not available at a concentration sufficient to produce a steep titration profile. Concentrating a precious sample just for this purpose had been meaningless for the inevitable losses of protein and activity. Therefore, the experiment was performed under the condition $[E]_t \approx K_X$. The method is the same as that in Fig. 4.3 for equistatin but there is nothing that can be extrapolated *by hand* from Fig. 4.4, even if the substrate concentration was kept low enough not to cause excessive competition. However, fitting (4.35) to data with the enzyme as the sole parameter to be optimized afforded the sought information on the activity status of the inhibitor that was 77 % of the protein content.

4.5 The Hill Equation and the Cooperativity Index

The successful equation proposed by Archibald V. Hill to explain oxygen binding to hemoglobin [16] can be written in a form that is suitable for representing the degree of cooperativity in enzyme-catalyzed reactions [8, p. 286]:

$$v = \frac{V[\mathrm{S}]^h}{K_{0.5}^h + [\mathrm{S}]^h}, \tag{4.37}$$

where h is the Hill coefficient, V the limiting rate, and $K_{0.5}$ the substrate concentration at which $v = 0.5V$. As pointed out by Cornish-Bowden [8, p. 286], $K_{0.5}$ should not be called Michaelis constant and is best represented elevated to the power h in line with $[\mathrm{S}]^h$.

Equation (4.37) considers cooperativity due to binding of substrate: $h > 1$ indicates positive cooperativity, $h < 1$ negative cooperativity, and $h = 1$ no cooperativity, in which case the expression is identical to the Michaelis–Menten equation.

For enzyme-catalyzed reactions in the presence of modifiers, even if some heroic kineticist were able to create cooperative models using full steady-state equations without simplifying assumptions, the final result would be unusable in practice. Assuming that modifier binding occurs at equilibrium mitigates the complications but only for relatively simple systems and for calculating the relevant parameters from experimental data an empirical rather than a mechanistic approach is better suited.

Deviations from hyperbolic behavior in enzyme kinetics, both in the presence of substrates alone and with added modifiers, can be ascertained by calculating a characteristic ratio of reactant concentrations that elicit responses at two selected levels. Taketa and Pogell, who first descried this method, suggested *the ratio of concentrations necessary to change activity from* 90 *to* 10 % *of maximal* [39]. This ratio is called today the *cooperativity index*. The choice of the ratio of concentrations that produce 90 and 10 % of the maximal response (briefly 90/10) is not imperative and other ranges can be used as well, especially when the conditions for achieving the 90/10 effect are experimentally problematic, e.g., poor precision at low reactant concentrations or limited solubility in the high concentration range. Taketa and Pogell showed a nomogram with alternative ratios of concentrations for responses such as 70/30 and 60/40 [39].

In the context of cooperative substrate binding, the 90/10 ratio has been called R_S [17] or R_a [8, pp. 289–290]. Similarly, $[I]_{0.9}/[I]_{0.1}$ has been used with the same meaning in the framework of enzyme inhibition [34, p. 106]. Here, dealing with *modifiers* that embrace inhibitors and activators, the symbol R_M will be employed. To avoid the same problem that arises with the use of IC_{50} with inhibitors that are unable to drive the initial velocity to zero at saturation (Sect. 4.3), it is important to underline that the cooperativity index refers to activity changes as percent of the *maximal* (= *limiting*) response, as clearly stated in the definition above by Takeda and Pogell. The same observation has been made by Segel regarding hyperbolic inhibitors [34, p. 164].

The principle of this method relies on the properties of the rectangular hyperbola. This function, written in explicit form as

$$Y = \frac{a \times X}{b + X}, \tag{4.38}$$

where a represents the limiting value of Y for $X \to \infty$ and b is the value of X for which $Y = a/2$, can be converted to an implicit equation:

$$Yb + YX - aX = 0\,. \tag{4.39}$$

The values of X that correspond to 90 and 10 % of a can be obtained by solving (4.39) for X after setting $Y = 0.9a$ and $Y = 0.1a$. The solutions are $X_{0.9} = 9b$ and $X_{0.1} = b/9$, respectively, whose ratio $X_{0.9}/X_{0.1} = 81$.

A relationship between the cooperativity index and the Hill coefficient in (4.37) can be deduced by substituting $v = 0.9$ or 0.1 in this equation and solving for [S]:

$$[S]_{0.9} = 9^{1/h} K_{0.5}\,, \quad [S]_{0.1} = \frac{9^{1/h}}{K_{0.5}}\,. \tag{4.40}$$

from which the cooperativity index results to be

$$R_S = [S]_{0.9} \Big/ [S]_{0.1} = 81^{1/h}\,, \tag{4.41}$$

and in the inverse operation, h can be calculated from R_M as

$$h = \frac{\log(81)}{\log(R_M)}\,. \tag{4.42}$$

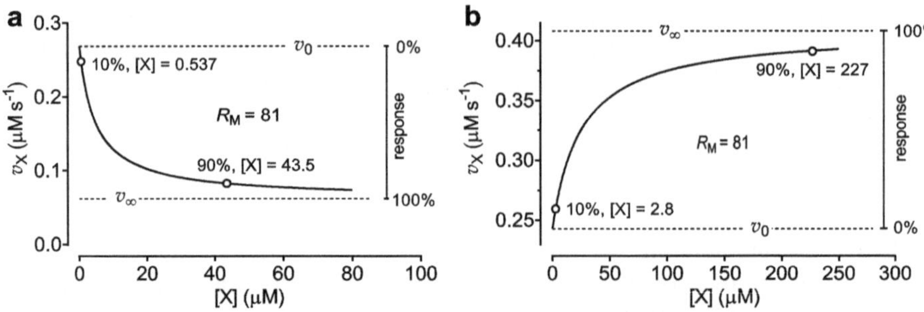

Fig. 4.5 Examples of the calculation of the cooperativity index R_M. The *solid lines* represent plots of (3.8). (**a**) Hyperbolic mixed, predominantly catalytic inhibition mechanism, [HMx(Sp<Ca)I]: $\alpha = 0.6$, $\beta = 0.2$, [S] = 45 μM, K_m = 22 μM, $k_2 = 40\,s^{-1}$, $[E]_t$ = 10 nM, K_X = 7 μM. (**b**) Hyperbolic catalytic activation, HCaA: $\alpha = 3.0$, $\beta = 3.0$, [S] = 34 μM, K_m = 22 μM, $k_2 = 40\,s^{-1}$, $[E]_t$ = 10 nM, K_X = 15 μM

Compared with the Hill coefficient, R_S represents an inverse measure of cooperative effects: if $h > 1$ $R_S < 81$, if $h < 1$ $R_S > 81$ and if $h = 1$ $R_S = 81$.

The function (4.38) is of the same type of the Michaelis–Menten equation and nonadherence to Michaelis–Menten behavior can be tested by evaluating any deviation of the cooperativity index from 81. This test can be applied to all basic modifier mechanisms discussed in Chaps. 3 and 5.

Primary data in the form of initial rates as a function of modifier concentration at different fixed concentrations of the substrate should be collected in a sufficiently broad concentration range and plotted as illustrated in Fig. 4.5 for a hyperbolic inhibitor and a nonessential activator. After making allowance for experimental error, values of R_M that significantly deviate from 81 represent an invitation to look more deeply into the reaction investigated.

A popular method for analyzing experimental results that yields an estimate of the Hill coefficient, consists in plotting data against [X] using a logarithmic scale for the latter and fitting the following function to data:

$$v_X = v_0 + \frac{v_\infty - v_0}{1 + 10^{\log((MC_{50} - [X]) \times h)}}, \tag{4.43}$$

where MC_{50} is the modifier concentration that produces an effect that lies midway between v_0 and v_∞ and h is the Hill coefficient. Equation (4.43), which is empirical and mechanism-independent, is known as the *four-parameter dose–response* or four-parameter logistic equation. Though, the term *logistic* has been used for unrelated purposes in statistics and is therefore discouraged in this context. As illustrated in Fig. 4.6, the Hill coefficient is positive for activation curves ($v_\infty > v_0$) and negative for inhibition curves ($v_\infty < v_0$).

A Note on Cooperativity In the context of enzyme inhibition and activation, $R_M < 81$ or $h > 1$ does not necessarily indicate positive cooperativity in the

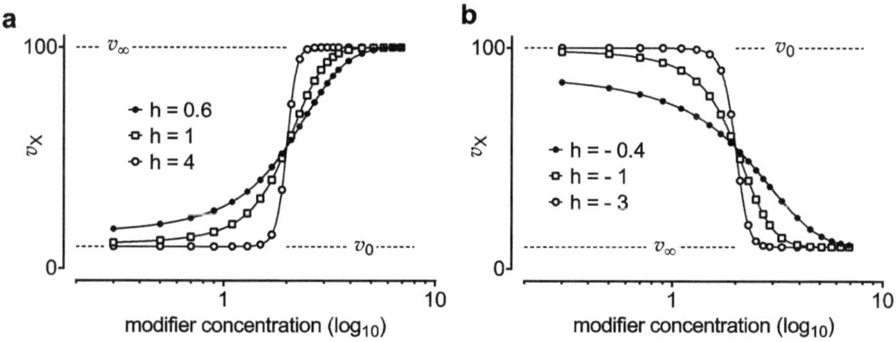

Fig. 4.6 Four-parameter dose–response curves: (**a**) activation, (**b**) inhibition. Data points and *solid lines* were drawn with (4.43) using arbitrary values of 10 and 100 for v_0 and v_∞, arbitrary concentration units to give $\log(MC_{50}) = 2$, and the displayed values of the Hill coefficient

classical sense [27, 28] without further investigation of the properties of the enzyme. In fact, $R_M < 81$ or $h > 1$ can be measured in a monomeric enzyme with only one binding site for the substrate(s) reflecting a genuine mechanistic property that *is not* related to cooperativity if two or more modifier molecules bind to the enzyme, sequentially or at random, without affecting the equilibrium constant of the vacant site(s). The apparent cooperativity that might be judged from a Hill coefficient greater than 1 stems from the existence of multiple enzyme forms in the mechanism, all of which are represented in the rate equation and introduce higher order terms of the modifier concentration in the denominator. Furthermore, artificial deviations from $R_M = 81$ or $h = 1$ may also be elicited by other, nonmechanistic phenomena. These points are discussed in Sects. 4.6 and 4.7.

4.6 The Enzyme Binds More Than One Molecule of Modifier

Two or more molecules of the same modifier can bind an enzyme molecule at two or more sites either sequentially or at random. The reactions shown in Scheme 4.5 represent a choice of mechanisms for an enzyme possessing two binding sites for the same modifier but the treatment can be extended to multiple sites. A new symbol, εK_X, indicates the equilibrium constant for the binding of the second modifier molecule. The coefficient ε has been chosen in place of a used by Segel [34, p. 466] and Schenker et al. [33] to avoid confusion with the typographic representation of the reciprocal allosteric coupling constant α when used in the same context. The nature of the mechanisms is explained in the legend of Scheme 4.5, in which ε can either be 1, i.e., binding of the first modifier molecule does not change the affinity of the vacant binding site, or $\neq 1$ if binding of the first molecule of X changes the affinity of the vacant site by the factor ε that can be greater or smaller than 1.

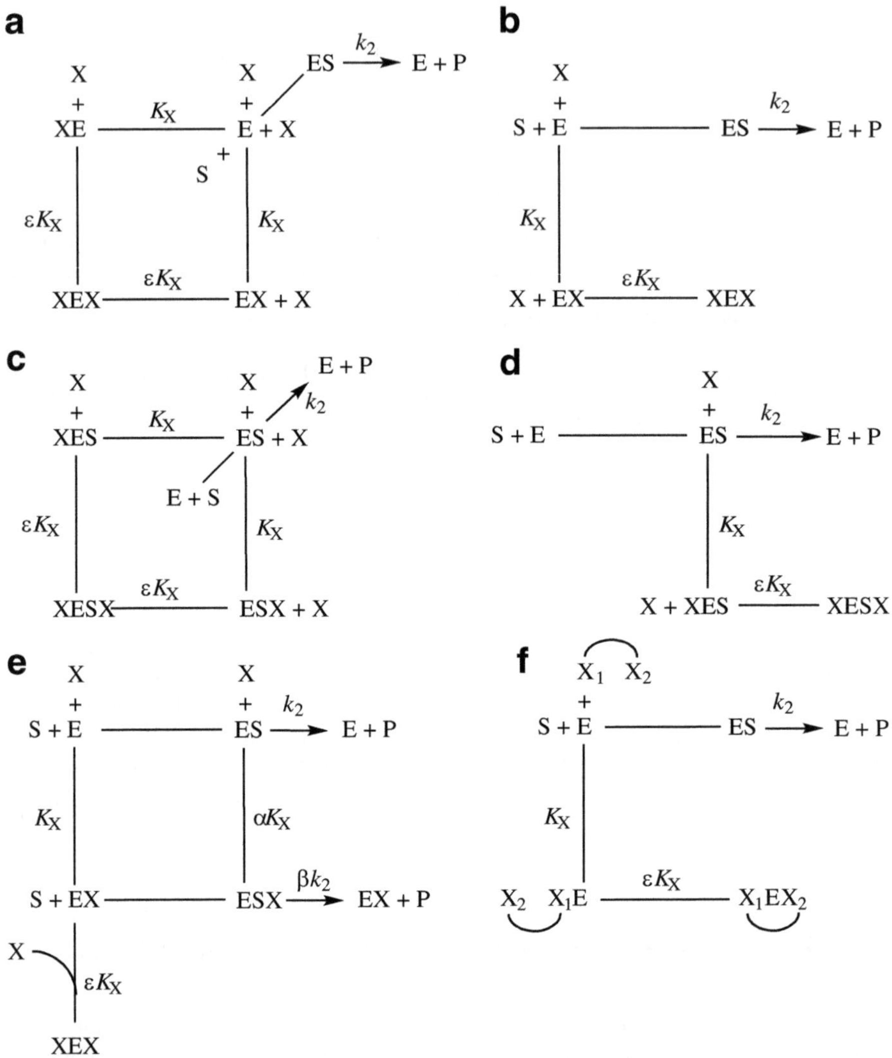

Scheme 4.5 Two molecules of the same modifier bind at different sites of the enzyme. The two sites are represented with X attached left and/or right of E. (**a**) Linear specific inhibition, random binding. (**b**) Linear specific inhibition, sequential binding. (**c**) Linear catalytic inhibition, random binding. (**d**) Linear catalytic inhibition, sequential binding. (**e**) Hyperbolic modification with a dead-end complex. (**f**) Linear mixed inhibition by a double-headed inhibitor that possesses two distinct inhibitory moieties joined by a flexible spacer. To avoid overloading of the reaction schemes all segments without arrows symbolize *reversible* steps in place of double arrows

The rate equations were derived for the mechanisms in Scheme 4.5 assuming equilibrium binding and purposely rearranged with v_X/v_0 as dependent variable. What matters in the expression of R_M is the ratio $[S]/K_m$, and therefore the symbol $\sigma = [S]/K_m$ will be used to make the equations slender. The cooperativity index was calculated by solving $v_X/v_0 = 0.1$ and $v_X/v_0 = 0.9$ for [X]. There are two roots of the resulting quadratic equations, of which only the positive solutions have physical meaning and the ratio of the two significant roots corresponds to R_M.

4.6.1 Two-Sites Linear Specific Inhibition

When two molecules of the same inhibitor can bind to the same enzyme molecule, the binding process can occur at random or in an ordered fashion as shown in Scheme 4.5a, b, respectively. The rate equations and the expressions that relate the cooperativity index to the substrate concentration and to the interaction factor between the bound inhibitor molecules are given in (4.44) and (4.45), respectively:

random

$$\frac{v_X}{v_0} = \frac{1+\sigma}{1+\sigma+\dfrac{2\,[X]}{K_X}+\dfrac{[X]^2}{\varepsilon K_X^2}}; \quad R_M = \frac{\varepsilon-\sqrt{\varepsilon^2+9\varepsilon\,(1+\sigma)}}{\varepsilon-\sqrt{\varepsilon^2+\dfrac{1}{9}\varepsilon\,(1+\sigma)}}. \tag{4.44}$$

sequential

$$\frac{v_X}{v_0} = \frac{1+\sigma}{1+\sigma+\dfrac{[X]}{K_X}+\dfrac{[X]^2}{\varepsilon K_X^2}}; \quad R_M = \frac{\varepsilon-\sqrt{\varepsilon^2+36\varepsilon\,(1+\sigma)}}{\varepsilon-\sqrt{\varepsilon^2+\dfrac{4}{9}\varepsilon\,(1+\sigma)}}. \tag{4.45}$$

R_M is always less than 81 at any value of σ and ε, which should not be immediately considered a proof of positive cooperativity. If $\varepsilon = 1$ this is certainly not the case because binding of X at either site does not change the affinity of the vacant site, but the presence of $[X]^2$ in the denominator of the rate equation affects the apparent Michaelis constant. A characteristic marker of the random mechanism and of its mate in Scheme 4.5c is the value taken by R_M at a substrate concentration identical to K_m: if $\varepsilon = 1$ then $R_M = 31.8$. If after binding of the first molecule of X the affinity for the vacant site is changed, ε can be greater or smaller than 1 but $R_M < 81$ in any case. Thus, testing whether or not ε significantly differs from 1 consists in finely tuning measurements at a substrate concentration fixed at the value of K_m and in measuring R_M: if $\varepsilon = 1$, $R_M = 31.8$. Otherwise if $\varepsilon < 1$, $R_M < 31.8$ and $R_M > 31.8$ if $\varepsilon > 1$. As discussed by Schenker et al. [33], the successful application of this criterion depends on an accurate determination of the substrate concentration and of K_m. The discriminating criterion against the corresponding random catalytic mechanism in Scheme 4.5c is that in the specific inhibition mechanism R_M *decreases* with increasing σ at any value of ε while it *increases* in catalytic inhibition.

The sequential mechanism (b) in Scheme 4.5 has a rate equation similar to that of the random mechanism (a), differing only for the factor 2 that multiplies [X] in the denominator (4.45). $R_M < 81$ irrespective of substrate concentration and

Table 4.3 Values assumed by the cooperativity index (R_M) and by the Hill coefficient (h) at a substrate concentration equal to K_m ($\sigma = 1$) for three values of the interaction constant between binding sites (ε)

	Random mechanisms		Sequential mechanisms	
ε	R_M	h	R_M	h
0.1	15.7	1.60	12.1	1.76
1	31.8	1.27	20.1	1.46
10	60.9	1.07	42.8	1.17

The values equally apply to the two random and the two sequential inhibition mechanisms in Scheme 4.5a, c and b, d, respectively

effects exerted on the affinity of the vacant site after binding of the first inhibitor molecule. A characteristic marker of this and of mechanism (d) in Scheme 4.5 is the value assumed by R_M at a fixed substrate concentration to give $\sigma = 1$. If there is no interaction between the binding sites, i.e., $\varepsilon = 1$, at $\sigma = 1$ $R_M = 20.1$, thus significantly smaller than under the corresponding conditions in the random mechanism (a). These properties are collected with those of the two catalytic mechanisms in Scheme 4.5c, d in Table 4.3 that shows the characteristic values assumed by the cooperativity index and by the Hill coefficient under specified conditions.

4.6.2 Two-Sites Linear Catalytic Inhibition

Catalytic inhibition with two molecules of inhibitor binding at random (Scheme 4.5c) is characterized by a different dependence of the rate and of the cooperativity index on substrate concentration with respect to the specific counterpart:

random

$$\frac{v_X}{v_0} = \frac{1+\sigma}{1+\sigma\left(1+\frac{2\,[\mathrm{X}]}{K_X}+\frac{[\mathrm{X}]^2}{\varepsilon K_X^2}\right)};\quad R_M = \frac{\varepsilon - \sqrt{\varepsilon^2 + 9\varepsilon\left(1+\frac{1}{\sigma}\right)}}{\varepsilon - \sqrt{\varepsilon^2 + \frac{1}{9}\varepsilon\left(1+\frac{1}{\sigma}\right)}}, \tag{4.46}$$

sequential

$$\frac{v_X}{v_0} = \frac{1+\sigma}{1+\sigma\left(1+\frac{[\mathrm{X}]}{K_X}+\frac{[\mathrm{X}]^2}{\varepsilon K_X^2}\right)};\quad R_M = \frac{\varepsilon - \sqrt{\varepsilon^2 + 36\varepsilon\left(1+\frac{1}{\sigma}\right)}}{\varepsilon - \sqrt{\varepsilon^2 + \frac{4}{9}\varepsilon\left(1+\frac{1}{\sigma}\right)}}. \tag{4.47}$$

The catalytic character of this mechanism with respect to the specific mechanism (Scheme 4.5a) can be recognized by the dependence of R_M on substrate concentration, i.e., *increasing* with increasing σ. The cooperativity index is always smaller than 81 and the properties of this parameters are the same as those of the specific inhibition mechanism with inhibitor molecules binding at random (Scheme 4.5a). In the random mechanism, at $\sigma = 1$, if the sites do not interact with each other $\varepsilon = 1$ and $R_M = 31.8$.

Like in the specific sequential mechanism (b), the rate equation of the catalytic counterpart (d) in Scheme 4.5 lacks the factor 2 multiplying [X] in the denominator (4.47). For a given value of ε, R_M *increases* with increasing substrate concentration. At $\sigma = 1$ and $\varepsilon = 1$, $R_M = 20.1$. The properties of the catalytic mechanisms (b) and (d) are summarized in Table 4.3.

4.6.3 Two-Sites Mixed Modification

The mechanism in Scheme 4.5e corresponds to the general modifier mechanism with an added dead-end complex. The rate equation derived under the equilibrium assumption contains an additional term in the denominator, which causes a different dependence of the apparent Michaelis constant on modifier concentration:

$$\frac{v_X}{v_0} = \frac{\left(1 + \beta \frac{[X]}{\alpha K_X}\right)(1+\sigma)}{1 + \frac{[X]}{K_X} + \frac{[X]^2}{\varepsilon K_X{}^2} + \sigma\left(1 + \frac{[X]}{\alpha K_X}\right)} . \tag{4.48}$$

With respect to mechanisms (a)–(d) in Scheme 4.5 the cooperativity index depends now on ε, α, β, and σ in a rater complex way:

$$\begin{aligned} R_M &= \frac{90\beta\varepsilon\,(1+\sigma) - 9\varepsilon\,(\alpha+\sigma) + 9\sqrt{A}}{10\beta\varepsilon\,(1+\sigma) - 9\varepsilon\,(\alpha+\sigma) + \sqrt{B}} \qquad (4.49)\\ A &= (10+10\sigma)^2\beta^2\varepsilon^2 - 20\,[\alpha\,(1+\sigma) + \sigma\,(1+\sigma)]\,\beta\varepsilon^2 \\ &\quad +(\alpha+\sigma)^2\varepsilon^2 + 36\alpha^2\varepsilon\,(1+\sigma) \\ B &= (10+10\sigma)^2\beta^2\varepsilon^2 - 180\,[\alpha\,(1+\sigma) + \sigma\,(1+\sigma)]\,\beta\varepsilon^2 \\ &\quad +(9\sigma+9\alpha)^2\varepsilon^2 + 36\alpha^2\varepsilon\,(1+\sigma)\ . \end{aligned}$$

Complexity arises from allosteric coupling between the substrate and the modifier and from the possibility of a second catalytic path if $\beta \neq 0$. We should not get frightened by this expression, which can be very useful for plotting R_M using alleged values of the parameters and thus getting inspiration for interpreting experimental results. If $\beta = 0$ the complex expression (4.49) simplifies to

$$R_M = \frac{-3\varepsilon\,(\alpha+\sigma) + \sqrt{9\varepsilon^2(\alpha+\sigma)^2 + 324\alpha^2\,(1+\sigma)\,\varepsilon}}{-3\varepsilon\,(\alpha+\sigma) + \sqrt{\varepsilon^2(3\alpha+3\sigma)^2 + 4\alpha^2\,(1+\sigma)\,\varepsilon}}\,. \tag{4.50}$$

The rate law (4.48) suggests that both mixed inhibition and mixed activation may result from modifiers that belong to either one of the basic mechanisms in the taxonomic groups 2 and 3 in Fig. 3.2. However, the XEX dead-end complex interferes with catalysis and the net effect on the rate is a blend of this inhibition and either inhibition or activation via ESX→ EX + P that results in a broad palette of interesting effects. Though, at high modifier concentrations the rate is driven to zero.

The two examples in Fig. 4.7 illustrate this tug-of-war between catalysis and the path that leads to the dead-end complex XEX. Panel (a) features a hyperbolic mixed, predominantly specific activation mechanism [HMx(Sp>Ca)A] in which a further modifier molecule binds to EX. The tree curves were plotted with common parameters and three values of ε, the factor by which the inhibition constant K_X is modified after the first inhibitor molecule has bound: $\varepsilon = 0.1$ means that the second inhibitor molecule binds with a tenfold affinity with respect to the first one, with $\varepsilon = 1$ the affinity remains unchanged and with $\varepsilon = 10$ the affinity decreases by a factor of 10. In the low modifier concentration range activation predominates, the initial rate increases until reaching a maximum and then declines to zero for $[X] \to \infty$ more or less steeply depending on ε.

The example in panel (b) shows hyperbolic balanced inhibition [HMx(Sp=Ca)I] *helped* in its inhibitory intent by the dead-end branch. Also in this case the

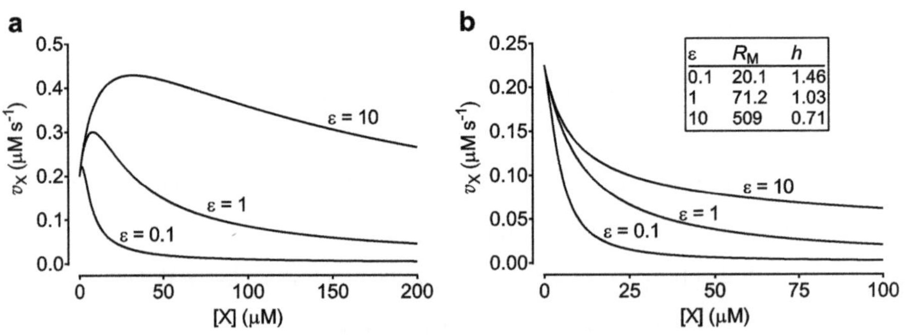

ε	R_M	h
0.1	20.1	1.46
1	71.2	1.03
10	509	0.71

Fig. 4.7 Two examples of the manifold facets of the mechanism in Scheme 4.5e. (**a**) Hyperbolic mixed, predominantly specific activation [HMx(Sp>Ca)A] with an additional dead-end branch. The curves were drawn with (4.48) and: $k_2 = 30\,\text{s}^{-1}$, $[\text{E}]_t = 0.01\,\mu\text{M}$, $K_X = 8\,\mu\text{M}$, $\alpha = 2$, $\beta = 4$, $\sigma = 2$ and the three values of ε indicated in the plot. (**b**) Hyperbolic balanced inhibition [HMx(Sp=Ca)I] with an additional dead-end branch. The curves were drawn with (4.48) and: $k_2 = 30\,\text{s}^{-1}$, $[\text{E}]_t = 0.01\,\mu\text{M}$, $K_X = 8\,\mu\text{M}$, $\alpha = 1$, $\beta = 0.3$, $\sigma = 3$ and the three values of ε indicated in the plot

curves were plotted with the same parameters and three different values of ε. The cooperativity index and the Hill coefficient calculated for each curve are shown in the inset table: for $\varepsilon = 0.1$ and 1 the behavior is similar to that of the four cases discussed above but the combination of parameters when $\varepsilon = 10$ yields a Hill coefficient that is smaller than 1 and the cooperativity index is 509, much greater than 81, which looks very much like negative cooperativity.

The properties discussed in this section are not only theoretical. A practical case has been studied in detail by Küçükkilinç and Özer while screening cationic dyes as inhibitors of cholinesterase (EC 3.1.1.8) and acetylcholinesterase (EC 3.4.1.7) [18, 19]. Cholinesterase, previously reputed to be mainly involved in scavenging xenobiotics, has more recently gained popularity for its participation, together with acetylcholinesterase, in the pathology of Alzhheimer's disease. The phenoxazine dyes meldola blue and nile blue, as well as methylene blue act as inhibitors of both enzymes. The tree dyes inhibit acetylcholinesterase by the hyperbolic mixed, predominantly specific inhibition mechanism [HMx(Sp>Ca)I], while the mechanism of interaction with cholinesterase is more complex. To explain their results, the authors postulated the formation of an additional inhibitor-containing complex in which two molecules of the dye bind to the enzyme according to mechanism (e) in Scheme 4.5.

Figure 4.8 shows the salient points for the inhibition of cholinesterase and acetylcholinesterase by the cationic phenoxazine dye meldola blue (technical details in the figure legend). The relative rates in (a) show a slightly sigmoid course with curves that tend to zero at saturating modifier concentration. The values of R_M and of the Hill coefficient decrease and increase with increasing substrate concentration, respectively (panel d), a characteristic of inhibition mechanisms with predominantly specific character that involve binding of two or more modifier molecules. For a substrate concentration that gives $\sigma = 1$, $R_M = 20.5$, which is very close to 20.1, i.e., the value taken by this parameter in the sequential inhibition mechanisms described above (Table 4.3). At the highest substrate concentration used in the experiments by Küçükkilinç and Özer, $R_M = 15.5$ and $h = 1.6$, the same value reported by the authors [19]. The experimentally determined kinetic parameters, with an allosteric coupling constant of 0.05 (the reciprocal of α, legend of Fig. 4.8) point to a mechanism with a strong prevalence of specific character with a minor function of the catalytic component. This point becomes more evident from the pre-steady-state analysis in panel (c) that shows the distribution of the concentration of the five enzyme-containing species immediately after attainment of the steady-state. The ESX complex is merely 1.3 % of all species and it does not participate to catalysis indicating that the dead-end complex XEX is the predominant species in the steady-state (49.1 %). Thus, the mechanism in Scheme 4.5e almost degenerates to mechanism (b) in the same Scheme.

Meldola blue inhibits also acetylcholinesterase (Fig. 4.8b) but in this case the mechanism is hyperbolic mixed, predominantly specific inhibition [HMx(Sp>Ca)I]. Judging from the inhibition constant $K_X = 35\,\text{nM}$, depending on the enzyme concentration used in the experiment, the tight binding condition may become important. Therefore the curves in panel (b) were plotted with the tight-binding

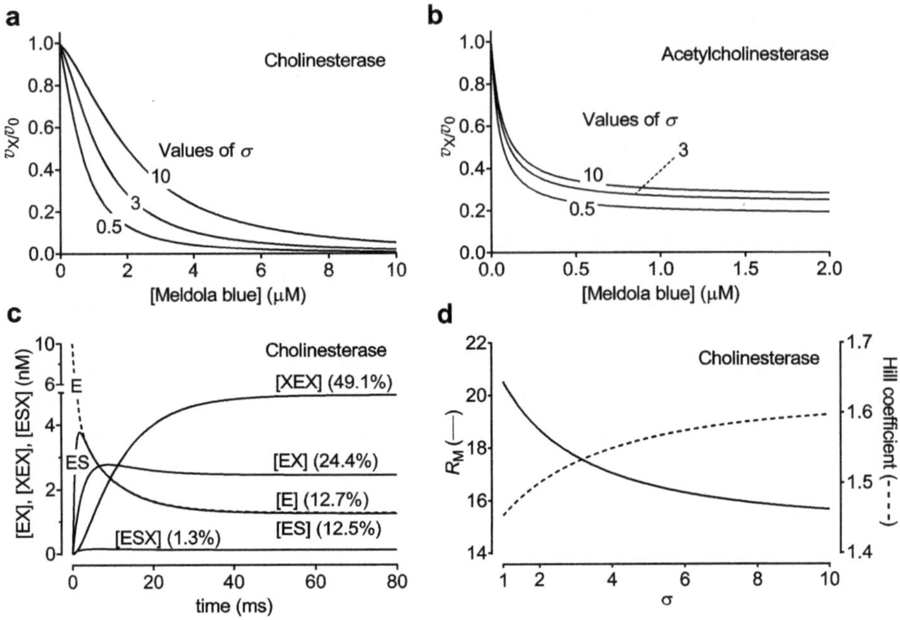

Fig. 4.8 Inhibition of cholinesterase and acetylcholinesterase by meldola blue. All curves were plotted with original data from [19]. Using the symbols adopted in this book, the parameters used to plot the relative rate against [X] for cholinesterase, using (4.48), were: $\alpha = 19.48$, $\beta = 0$, $K_X = 0.77\,\mu M$, $K_m = 38\,\mu M$, $\varepsilon = 0.961$ and substrate concentrations that give the ratios $[S]/K_m = \sigma$ as shown. In (**d**) the cooperativity index R_M was plotted with α and ε as above using (4.50) and the Hill coefficient using (4.42). The pre-steady-state profiles in (**c**) were generated by numerical integration with KinTek Explorer using the same data as in (**a**) for a rapidly equilibrating system, $\sigma = 1$ and $[X] = 1.5\,\mu M$. The *curves* in (**b**) refer to acetylcholinesterase and were plotted considering tight-binding using (4.35) and parameters: $\alpha = 1.9$, $\beta = 0.28$, $K_X = 0.035\,\mu M$, $[E]_t = 10\,nM$, $K_m = 48\,\mu M$ and three [S] values to give the ratios $[S]/K_m = \sigma$ as shown. I am very grateful to Tuba Küçükkilinç, Hacettepe University, Ankara, Turkey, for providing additional information, necessary for plotting these results, which was not present in the original publication [19]

Eq. (4.35) at an enzyme concentration of 10 nM. The hyperbolic nature of the mechanism can be appreciated by the curves leveling-off at a finite value of the rate.

4.6.4 Linear Specific Inhibition by a Double-Headed Inhibitor

Two inhibitory moieties joined by a flexible spacer may be distinct parts of the same molecule and act by the mechanism shown in Scheme 4.5f. It consists of sequential binding of the two inhibitory moieties and the kinetic properties, including the rate

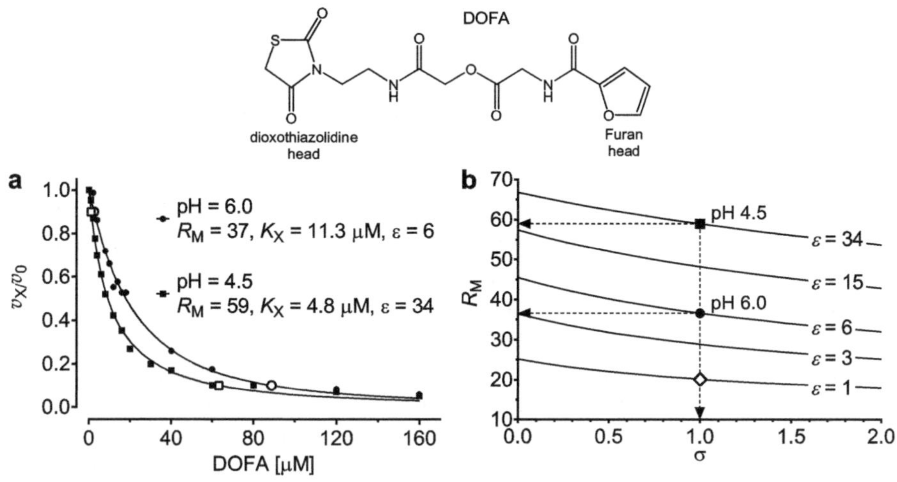

Fig. 4.9 Inhibition of human cathepsin B by DOFA (full name in the footnote). The structure of the inhibitor that owns two *heads* able to bind the enzyme is shown on the *top*. (**a**) Linear inhibition as a function of inhibitor concentration at two pH values as indicated, accompanied by the values of the cooperativity index R_M, the inhibition constant and ε. Equation (4.45) was fitted to data and the *solid lines* represents best fits. The *white symbols* indicate the responses at 10 and 90 % of the limiting value that were interpolated during the fitting procedure. (**b**) Nomogram that relates R_M with ε depending on substrate concentration expressed as its value divided by the Michaelis constant (discussion in the main text)

equation and the expression of the cooperativity index (4.45), are indistinguishable from those of mechanism (b).

Patricia Schenker and coworkers studied a group of inhibitors of the cysteine peptidase cathepsin B (EC 3.4.21.1) that were targeted to a particular structural element of the enzyme know as the *occluding loop*. This is a flexible structural element that can be laid over the primed subsites of the substrate binding cleft (see Fig. 4.1a, binding pocket S_1') to prevent extended binding of polypeptide substrates and their proteolytic cleavage. Using computational methods that included docking of a large group of candidate molecules to the surface of cathepsin B in the occluding loop region, an inhibitor (DOFA)[5] was identified that matched the working hypothesis [33].

Plotting the normalized initial rates measured at pH = 4.5 and 6.0 versus inhibitor concentration as shown in Fig. 4.9a, R_M resulted to be smaller than 81 in both cases. The lower pH-value promotes the closure of the occluding loop of cathepsin B, while at pH 6.0 the open conformation predominates. The two curves were measured at substrate concentrations that were equal to their K_m. The R_M values were 37 and 59 at pH 6.0 and 4.5, respectively. The criteria discussed above

[5]Acronym of [2-[2-(2,4-dioxo-1,3-thiazolidin-3-yl)ethylamino]-2-oxoethyl] 2-(furan-2- carbonylamino)acetate.

and outlined in Table 4.3 indicated that the value of the interaction coefficient ε had to be larger than 1 because in this case R_M should correspond to 20.1 that is the position marked by a white diamond in the nomogram of Fig. 4.9b. After discarding nonmechanistic causes such as poor solubility and aggregation of inhibitor molecules, discrimination was made between candidate mechanisms that included those in Scheme 4.5 coming to the conclusion that the data were not adequately supported by models that predict binding of two individual molecules and that the sequential binding model of the two moieties of the inhibitor according to the double-headed mechanism in Scheme 4.5f was superior.

With the support of molecular modeling, the combined results suggested that the occluding loop is stabilized in its closed conformation by making hydrogen bonds with the inhibitor. Furthermore, the dioxothiazolidine head of the inhibitor, bound away from the catalytic center, plays the part of the protagonist acting as a sterical barrier that hinders binding of the C-terminal residue of peptide substrates. In the double-headed mechanism, the K_X-value of 4.8 μM at pH $=$ 4.5 was associated with binding of the dioxothiazolidine head, while $\varepsilon K_X = 4.8 \times 34 = 163$ μM was ascribed to the looser binding of the furan head of the inhibitor. The increase of pH to 6.0 weakens the affinity of the dioxothiazolidine moiety with $K_X = 11.3$ μM and strengthens binding of the furan head with $\varepsilon K_X = 68$ μM.

4.7 Deviations from Hyperbolic Saturation Kinetics Due to Nonmechanistic Causes

Several kinds of artifacts can affect the outcome of kinetic measurements hampering the correct evaluation of the kinetic parameters. Running experiments with suitable blanks helps avoiding errors by introducing the necessary corrections to photometric or fluorometric readings. Yet, situations exist in which artifacts may produce results that can hardly be distinguished from genuine effects without deeper inspection. Some of these pitfalls are commented below.

Screening in a Spirit Haunted World is the suggestive title of a review by Brian Shoichet on the annoying phenomenon *promiscuous inhibition* that plagues drug hunters screening libraries of organic molecules of potential pharmacological interest: typically 10^5–10^6 compounds at a time [36]. This procedure is called *high-throughput screening*. Promiscuous inhibition refers to the property of organic molecules to form aggregates in the micromolar concentration range, but also less than this. The particles that are formed in the assay buffers are said to *inhibit* the target enzymes, but in reality the cause is sequestration of the catalyst within the colloid-like aggregates. In running automated assays that provide IC_{50} values for whatever is measured, this looks like inhibition.

There is however a difference between a promiscuous and a genuine inhibitor because the former yields inconsistent results from one assay to the other and, generally, it is characterized by a dose–response curve exhibiting a Hill coefficient greater than 1 [35] similar to the curve with open circles in Fig. 4.6b, and R_M values smaller than 81. If repeating the experiment by *increasing* the enzyme concentration, the Hill coefficient decreases or inhibition disappears at all, the molecule under investigation is likely to be a promiscuous inhibitor. This property has been demonstrated in systematic studies by McGovern et al. in the Shoichet laboratory [25, 26], who observed than increasing the enzyme concentration by a factor of 10 strongly reduced or eliminated inhibition even in the presence of a 1,000-fold concentration of the putative inhibitor. The authors established the propensity of the promiscuous modifiers to form aggregates by dynamic light scattering and measured their size by transmission electron microscopy. Typically, the spherical particles have a diameter of 50–400 nm and can either adsorb or absorb the enzymes under investigation, mostly in an unspecific way. If this is the case, another control is to measure the effects on the activity of enzymes unrelated to the target enzyme since the aggregating molecule will possibly capture also these enzymes mocking inhibition.

Therefore, any deviation from hyperbolic saturation curves that yield $R_M = 81$ and $h = 1$ should be looked up by performing the necessary verifications because one is tempted to think at a mechanism involving the binding of two inhibitor molecules by one of the mechanisms described in Sect. 4.6. For instance Liu et al. identified such a phony modifier of phosphomannomutase (EC 5.4.2.8) and characterized its properties in detail [23]. In the pharmaceutical industry, false-positives are the more costly both in terms of money and time [36].

Fixed-Point Methods for evaluating the efficiency of inhibitors are inadequate if the inhibition process is of the slow-onset type (Chap. 8). These methods consist in measuring the change in some optical or other property, which is proportional to product concentration, at a predetermined time after starting the reaction. If the initial rise of product concentration is not linear with time but exponential, as it occurs in slow-onset inhibition, large errors can be made in calculating the inhibited initial rates. Preincubation of the enzyme with various concentrations of the inhibitor may fix the problem but only if the preincubation is sufficiently long to allow equilibration of the enzyme-inhibitor complex. Fig. 4.10 illustrates this concept. The reaction was started and readings of the product formed were made after 200 s. These correspond to the black dots along the dotted line in panel (a), from which the slopes of the dashed lines were calculated for all concentrations of inhibitor (only thee lines are shown to avoid crowding of the graphic). Two more time-points at 60 and 400 s were included and the calculations were performed in the same way. This procedure ignores that the formation of the EX complex can span from milliseconds to hours. The correct (*true*) slopes for the inhibited reaction are those of the straight portions of the progress curves after completion of the slow exponential phase. Therefore, the time-point assay overestimates the inhibited rates and thus underestimates the degree of inhibition at the various inhibitor concentrations.

Wrong and correct slopes from the data in panel (a) were plotted against the logarithm of the inhibitor concentration in panel (b) to which (4.43) was fitted

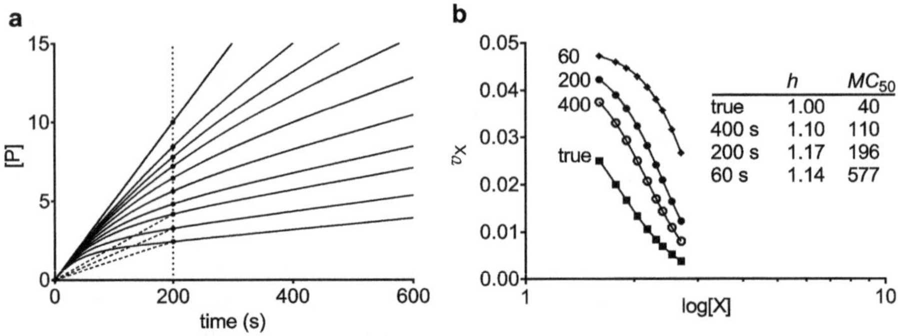

Fig. 4.10 Errors in estimating the inhibited rate in the presence of a slow-onset inhibitor if a fixed-time assay is performed. (**a**) Progress curves consisting of an exponential rise followed by a linear steady-state that yields the *true* slopes (linear part). The *black dots* indicate readings at a fixed point of 200 s to calculate the slopes of the *dashed lines* (only three are shown), which are however *wrong* estimates. (**b**) Four-parameter dose–response curves constructed with the wrong and with the true slopes in (**a**). (4.43) was fitted to data for calculating the Hill coefficient and MC_{50}. Besides results obtained from the time-point assay at 200 s, also the results from the same assay with readings at 60 and 400 s are shown. Concentration units are arbitrary but match in the two panels. The plots in (**a**) are truncated on the *top* to allow expansion and show better separation of the curves

for calculating the Hill coefficient and MC_{50}. The numbers in (b) show that the efficiency of the inhibitor is underestimated by different factors depending on the time-point chosen in the assay by the factors: 14, 5, and 2.8 for the time-points 400, 200, and 60 s, respectively. The fact that $h > 1$ for the three time-points suggests deviation from hyperbolic saturation kinetics but this is not the case because it is an artifact produced by the method used.

This sort of erroneous results stemming from time-point assays is not limited to slow-onset inhibition. Depending on the experimental conditions, the same phenomenon applies to interactions in which the formation of the enzyme-inhibitor complex occurs rapidly but linearity of product formation with time is not always guaranteed (Sect. 1.9). If sizable deviations from linearity are present, the efficiency of the investigated inhibitor will be *overestimated* because at the sampling time the rate is smaller than its true value.

A particular situation is when the time-point assay is performed under experimental conditions that result in tight-binding of the inhibitor. Although the enzyme-inhibitor complex is formed rapidly, what is plotted on the abscissa of the four-parameter dose–response curve is the *total*, not the *free* inhibitor concentration, i.e., on the abscissa of this plot a larger number appears and thus the efficiency of the inhibitor is *underestimated*. Corrections can of course be performed (Sect. 4.4) but this depends on whether high-throughput screening does or does not take into account this important point.

Uncorrected Inner Filter Effects in Fluorometry can be responsible for remarkable artifacts, such as misidentifying a given molecule as efficient inhibitor although the enzyme and the putative inhibitor do not form any kind of complex.

Of course, such a mistake can only occur if blind analyses are run disregarding the spectroscopic properties of the reactants. Inner filter effects are due to two factors: reabsorption of emitted light in the wavelength region of fluorescence emission or high absorption at the excitation wavelength. In the first case the absorption spectrum of one of the components of the reaction mixture overlaps the emission spectrum of the fluorophore: increasing the concentration of the absorbing species is accompanied by a decrease of fluorescence intensity. In the second case any absorbing molecule, including the fluorophore, may exhibit high absorbance at the excitation wavelength and, depending on the light path of the cell in which the solution is contained and on the geometry of the measuring system, the excitation beam loses intensity throughout the solution. For instance, if the cell has a light path of 10 mm and a slit of 0.5 mm is placed to match the center of the cell in direction of the detector, e.g., at a right angle with excitation, before the light beam reaches the center of the cell part of the incident light has been absorbed in proportion to the total absorption coefficient and reactant concentration. Increasing the concentration of one or more of the absorbing components causes attenuated fluorescence emission that simulates enzyme inhibition. The resulting putative titration curve is unrelated to hyperbolic saturation behavior and any attempt to fit customary equations to such data will be unsuccessful. To ascertain and in favorable cases eliminate this problem, the excitation and emission wavelengths can be changed until possibly finding a combination for which the inner filter effect disappears.

Prevention of the inner filter effect has been described by Schenker et al. [33]. For assaying cathepsin B in the presence of putative inhibitors, the internally quenched, peptide fluorogenic substrate Abz-GIVR↓YAK(Dnp)-OH[6] was used for monitoring enzyme activity by Förster resonance energy transfer. The change in fluorescence that accompanies the cleavage of Abz-GIVR↓YAK(Dnp)-OH is typically detected by exciting the sample at 320 nm and measuring fluorescence at 420 nm [10]. One of the compounds tested had a strong absorbance at the excitation wavelength of 320 nm that simulated enzyme inhibition. However, by changing the excitation wavelength to 370 nm and monitoring fluorescence at 430 nm fixed the problem revealing that the compound did not interfere with enzyme activity.

If the inner filter effect cannot be eliminated by appropriate choice of the excitation and emission wavelengths, corrections can be made as described in several reports, e.g., [31]. The use of quenched fluorescent substrates for determining the efficiency of enzyme inhibitors has been considered in detail by Ludewig and coworkers [24]. The authors discussed the sources of errors in fluorometric measurements, considered the inner filter effect, and provided guidelines for the validation of assays in the screening of peptidase inhibitors.

[6] Abz = ortho-aminobenzoyl; Dnp = Nε-2,4-dinitrophenyl; the arrow indicates the scissile bond.

Appendix

Derivation of the Tight-Binding Rate Equation for the General Modifier Mechanism

Redirected from Sect. 4.4. The reactant concentrations in a rate equation represent free concentrations. However, in the presence of tight-binding, disambiguation between free and total concentrations is made by appending a 't' (total), and where necessary an 'f' (free) to the symbols of concentration. We start the derivation of the tight-binding rate equation of the general modifier mechanism by putting in evidence the free modifier concentration while rearranging (3.35) discussed in Chap. 3

$$v_X = \frac{(1+\sigma)\left(1+\beta\dfrac{[X]_f}{\alpha K_X}\right)}{1+\sigma+\dfrac{[X]_f}{K_X}+\dfrac{\sigma[X]_f}{\alpha K_X}}v_0\,, \tag{4.51}$$

from which $[X]_f$ can be extracted as:

$$[X]_f = \frac{(1+\sigma)\,(v_0 - v_X)}{v_X\left(\dfrac{1}{K_X}+\dfrac{\sigma}{\alpha K_X}\right) - v_0\dfrac{\beta}{\alpha K_X}\,(1+\sigma)}\,. \tag{4.52}$$

To proceed, an expression of the total modifier concentration is needed as the sum [EX] + [ESX] + $[X]_f$, and [EX] + [ESX] can be obtained considering the rate (4.51) at saturating modifier, in which case all enzyme molecules are present in these two complexes with product generation occurring only by decomposition of ESX into EX + P:

$$v_\infty = \frac{(1+\sigma)\,\dfrac{\beta}{\alpha K_X}}{\dfrac{1}{K_X}+\dfrac{\sigma}{\alpha K_X}}v_0\,. \tag{4.53}$$

The denominator of (4.53) represents now EX + ESX, while the denominator of (4.51) represents E + ES + EX + ESX. According to the distribution equations of Cleland [6], dividing the first by the second of these denominators and multiplying by $[X]_f[E]_t$ yields [EX] + [ESX]:

$$[EX]+[ESX] = \frac{\left(\dfrac{[X]_f}{K_X}+\dfrac{\sigma[X]_f}{\alpha K_X}\right)[E]_t}{1+\sigma+\dfrac{[X]_f}{K_X}+\dfrac{\sigma[X]_f}{\alpha K_X}}\,. \tag{4.54}$$

Experimentally, rates are measured which contain the concentrations of the complexes. Thus, such concentrations must be deduced from the differences between the rate in the absence (v_0) of modifier, at a given modifier concentration (v_X) and at saturating modifier (v_∞), i.e., ($v_0 - v_X$) and ($v_0 - v_\infty$) using the explicit expression $v_0 = V\sigma/(1+\sigma)$, (4.51) and (4.53):

$$v_0 - v_X = \frac{V\sigma\,[X]_f\,(\alpha + \sigma - \beta - \beta\sigma)}{(1+\sigma)\,(\alpha K_X + \alpha\sigma\,K_X + [X]_f\sigma + [X]_f\alpha)}$$

$$v_0 - v_\infty = \frac{V\sigma\;(\alpha + \sigma - \beta - \beta\sigma)}{(1+\sigma)\,(\alpha+\sigma)}$$

$$\frac{v_0 - v_X}{v_0 - v_\infty} = \frac{(\alpha+\sigma)\,[X]_f}{\alpha K_X + \sigma\alpha K_X + [X]_f\sigma + [X]_f\alpha} = \frac{\dfrac{[X]_f}{K_X} + \dfrac{\sigma[X]_f}{\alpha K_X}}{1 + \sigma + \dfrac{[X]_f}{K_X} + \dfrac{\sigma[X]_f}{\alpha K_X}}\,. \tag{4.55}$$

After multiplication by $[E]_t$, the ratio in (4.55) is identical with (4.54) and the sought conservation equation for the modifier concentration becomes:

$$[X]_t = [EX] + [ESX] + [X]_f = \frac{v_0 - v_X}{v_0 - v_\infty}[E]_t + [X]_f\,. \tag{4.56}$$

We have now all elements for writing the rate equation of the general modifier mechanism under tight-binding conditions but it would contain v_∞, which can be usefully replaced by other parameters to produce an expression that contains only familiar elements present in (4.51) and in the specific velocity equation (3.36). Inspection of (4.53) suggest the following, equivalent rearrangement obtained by multiplying the numerator and the denominator by αK_X:

$$v_\infty = v_0\beta\frac{1+\sigma}{\alpha+\sigma}\,. \tag{4.57}$$

Furthermore, purposeful rearrangement of (4.52) gives

$$\begin{aligned}[X]_f &= \frac{v_0 - v_X}{v_X - v_0\dfrac{\beta}{\alpha K_X}\left(\dfrac{1+\sigma}{\dfrac{1}{K_X} + \dfrac{\sigma}{\alpha K_X}}\right)} \times \frac{1+\sigma}{\dfrac{1}{K_X} + \dfrac{\sigma}{\alpha K_X}} \\ &= \frac{v_0 - v_X}{v_X - \underbrace{v_0\beta\left(\dfrac{1+\sigma}{\alpha+\sigma}\right)}_{v_\infty}} \times \frac{1+\sigma}{\dfrac{1}{K_X} + \dfrac{\sigma}{\alpha K_X}}\,,\end{aligned} \tag{4.58}$$

where the term underscored by the curled brace represents v_∞ in (4.57).

The conservation equation for the modifier's concentration is finally obtained introducing (4.51), (4.57), and (4.58) in (4.56). The resulting expression is then used to extract v_X, a quadratic equation, whose physically significant root is displayed as (4.35) in the main text.

References

1. Ackermann WW, Potter VR (1949) Enzyme inhibition in relation to chemotherapy. Proc Soc Exp Biol Med 72:1–9
2. Ayala Y, Di Cera E (1994) Molecular recognition by thrombin. Role of the slow to fast transition, site-specific ion binding energetics and thermodynamic mapping of structural components. J Mol Biol 235:733–746. doi:10.1006/jmbi.1994.1024
3. Baici A (1987) Graphical and statistical analysis of hyperbolic, tight-binding inhibition [This paper contains a mathematical mistake: see correction in Szedlacsek, S. et al. (1988) Biochem. J. 254:311–312]. Biochem J 244:793–796
4. Cha S (1975) Tight-binding inhibitors - I. Kinetic behavior [See corrections, ibid. 25 (1976) 1561]. Biochem Pharmacol 24:2177–2185
5. Cheng Y, Prusoff WH (1973) Relationship between the inhibition constant (Ki) and the concentration of inhibitor which causes 50 per cent inhibition (I50) of an enzymatic reaction. Biochem Pharmacol 22:3099–3108
6. Cleland WW (1963) The kinetics of enzyme-catalyzed reactions with two or more substrates or products. I. Nomenclature and rate equations. Biochim Biophys Acta 67:104–136
7. Cornish-Bowden A (1974) A simple graphical method for determining the inhibition constants of mixed, uncompetitive and non-competitive inhibitors. Biochem J 137:143–144
8. Cornish-Bowden A (2012) Fundamentals of enzyme kinetics, 4th edn. Wiley, Weinheim
9. Cortés A, Cascante M, Cárdenas ML, Cornish-Bowden A (2001) Relationships between inhibition constants, inhibitor concentrations for 50% inhibition and types of inhibition: new ways of analysing data. Biochem J 357:263–268
10. Cotrin SS, Puzer L, Judice WAD, Juliano L, Carmona AK, Juliano MA (2004) Positional-scanning combinatorial libraries of fluorescence resonance energy transfer peptides to define substrate specificity of carboxydipeptidases: assays with human cathepsin B. Anal Biochem 335:244–252
11. Cristofaro RD, Di Cera E (1990) Effect of protons on the amidase activity of human α-thrombin: analysis in terms of a general linkage scheme. J Mol Biol 216:1077–1085. http://dx.doi.org/10.1016/S0022-2836(99)80021-7
12. Di Cera E, Cristofaro RD, Albright DJ, Fenton JW (1991) Linkage between proton binding and amidase activity in human.alpha.-thrombin: effect of ions and temperature. Biochemistry 30:7913–7924. doi:10.1021/bi00246a007
13. Di Cera E, Dang QD, Ayala Y, Vindigni A (1995) Linkage at steady state: allosteric transitions of thrombin. Meth Enzymol 259:127–144
14. Di Cera E, Hopfner KP, Dang QD (1996) Theory of allosteric effects in serine proteases. Biophys J 70:174–181
15. Henderson PJF (1972) A linear equation that describes the steady-state kinetics of enzymes and subcellular particles interacting with tightly bound inhibitors. Biochem J 127:321–333
16. Hill AV (1910) The possible effect of the aggregation of the molecules of haemoglobin on its dissociation curves. J Physiol 40:4–7
17. Koshland DE, Némethy G, Filmer D (1966) Comparison of experimental binding data and theoretical models in proteins containing subunits. Biochemistry 5:365–385
18. Küçükkilinç T, Özer I (2005) Inhibition of human plasma cholinesterase by malachite green and related triarylmethane dyes: mechanistic implications. Arch Biochem Biophys 440:118–122. http://dx.doi.org/10.1016/j.abb.2005.06.003

19. Küçükkilinç T, Özer I (2007) Multi-site inhibition of human plasma cholinesterase by cationic phenoxazine and phenothiazine dyes. Arch Biochem Biophys 461:294–298
20. Lenarčič B, Bevec T (1998) Thyropins: new structurally related proteinase inhibitors. Biol Chem 379:105–111
21. Lenarčič B, Turk V (1999) Thyroglobulin type-1 domains in equistatin inhibit both papain-like cysteine proteinases and cathepsin D. J Biol Chem 274:563–566
22. Lenarčič B, Ritonja A, Strukelj B, Turk B, Turk V (1997) Equistatin, a new inhibitor of cysteine proteinases from Actinia equina, is structurally related to thyroglobulin type-1 domain. J Biol Chem 272:13899–13903
23. Liu HY, Wang Z, Regni C, Zou X, Tipton PA (2004) Detailed kinetic studies of an aggregating inhibitor; inhibition of phosphomannomutase/phosphoglucomutase by disperse blue 56. Biochemistry 43:8662–8669. doi:10.1021/bi0491907
24. Ludewig S, Kossner M, Schiller M, Baumann K, Schirmeister T (2010) Enzyme kinetics and hit validation in fluorimetric protease assays. Curr Top Med Chem 10:368–382
25. McGovern SL, Caselli E, Grigorieff N, Shoichet BK (2002) A common mechanism underlying promiscuous inhibitors from virtual and high-throughput screening. J Med Chem 45:1712–1722
26. McGovern SL, Helfand BT, Feng B, Shoichet BK (2003) A specific mechanism of nonspecific inhibition. J Med Chem 46:4265–4272
27. Monod J, Changeux JP, Jacob F (1963) Allosteric proteins and cellular control systems. J Mol Biol 6:306–329
28. Monod J, Wyman J, Changeux JP (1965) On the nature of allosteric transitions: a plausible model. J Mol Biol 12:88–118
29. Morrison JF (1969) Kinetics of the reversible inhibition of enzyme-catalysed reactions by tight-binding inhibitors. Biochim Biophys Acta 185:269–286
30. Naqui A (1983) What does I50 mean? Biochem J 215:429–430
31. Palmier MO, Doren SRV (2007) Rapid determination of enzyme kinetics from fluorescence: Overcoming the inner filter effect. Anal Biochem 371:43–51
32. Schechter I, Berger A (1967) On the size of the active sites in proteases. I. Papain. Biochem Biophys Res Commun 27:157–162
33. Schenker P, Alfarano P, Kolb P, Caflisch A, Baici A (2008) A double-headed cathepsin B inhibitor devoid of warhead. Protein Sci 17:2145–2155
34. Segel IH (1975) Enzyme kinetics. Behavior and analysis of rapid equilibrium and steady-state enzyme systems. Wiley, New York
35. Shoichet BK (2006) Interpreting steep dose-response curves in early inhibitor discovery. J Med Chem 49:7274–7277
36. Shoichet BK (2006) Screening in a spirit haunted world. Drug Discov Today 11:607–615
37. Straus OH, Goldstein A (1943) Zone behavior of enzymes. Illustrated by the effect of dissociation constant and dilution on the system cholinesterase-physostigmine. J Gen Physiol 26:559–585
38. Szedlacsek SE, Ostafe V, Serban M, Vlad MO (1988) A re-evaluation of the kinetic equations for hyperbolic tight-binding inhibition. Biochem J 254:311–312
39. Taketa K, Pogell BM (1965) Allosteric inhibition of rat liver fructose 1,6-diphosphatase by adenosine 5'-monophosphate. J Biol Chem 240:651–662
40. Wells CM, Di Cera E (1992) Thrombin is a Na+-activated enzyme. Biochemistry 31:11721–11730. doi:10.1021/bi00162a008
41. Xing GW, Li XW, Tian GL, Ye YH (2000) Enzymatic peptide synthesis in organic solvent with different zeolites as immobilization matrixes. Tetrahedron 56:3517–3522. doi:10.1016/S0040-4020(00)00261-1

Chapter 5
The Basic Mechanisms of Inhibition and Nonessential Activation

The vast majority of textbooks (even the most recent ones) continue the 'romance' with the double-reciprocal plot, in spite of the severe way it is affected by experimental errors.
Adams KAH, Storer AC, Cornish-Bowden A (1984). J Chem Educ 61:527

Abstract The 17 enzyme–modifier interaction mechanisms identified by taxonomic criteria are by no means only theoretical concepts. They are all represented in thousands of reports, though their real identity may not be recognized at a glance. One factor is inconsistent nomenclature and another is collecting mechanisms in pools of undifferentiated special cases. Representative examples that span several branches of the biological sciences are discussed in this chapter for the 17 basic mechanisms highlighting the methods used by the authors in data interpretation. Substrates and reaction products in the role of modifiers are discussed within the mechanisms to which they belong.

5.1 Introduction

As mentioned in the Preface, this book focusses on the interpretation of enzyme modification in vitro, a knowledge that can be exploited in the study of systems in vivo, where modulating the activity of a particular enzyme involves the flux control coefficient of the same enzyme in the whole system [21, Chap. 13]. Rather than being conceived as a stand-alone learning unit, the present chapter is intended as reference point for readers engaged in analyzing their own data. After having designed an experiment and performed all necessary measurements, the goal is to ascribe the results to the most probable mechanism compatible with the data. The primary purpose of this chapter is to provide readers with adequate information about the individual mechanisms by summarizing their properties and discussing practical examples.

The biochemical literature is very rich of studies that deal with enzyme inhibition and (less) with activation, in which the characterization of the mechanisms of action has been performed with methods that vary from elementary to very sophisticated. The effects elicited by substrates and reaction products in the role of modifiers will be discussed in the sections to which they belong since there are no

A. Baici, *Kinetics of Enzyme-Modifier Interactions*,
DOI 10.1007/978-3-7091-1402-5_5

Table 5.1 The basic mechanisms of inhibition and nonessential activation

No.	Mechanism	Acronym	α	β	Other	Page
1	Linear specific inhibition	LSpI	∞[a]	0	$K_{Ca} = \infty$	212
2	Linear catalytic inhibition	LCaI	0[b]	0	$K_{Sp} = \infty$	217
3	Linear mixed, predominantly specific inhibition	LMx(Sp > Ca)I	(1,∞)	0		223
4	Linear mixed, predominantly catalytic inhibition	LMx(Sp < Ca)I	(0,1)	0		227
5	Linear mixed, balanced inhibition	LMx(Sp = Ca)I	1	0		230
6	Hyperbolic specific inhibition	HSpI	(1,∞)	1		235
7	Hyperbolic mixed, predominantly specific inhibition	HMx(Sp > Ca)I	(1,∞)	(0,1)		238
8	Hyperbolic catalytic inhibition	HCaI	(0,1)	(0,1)	$\alpha = \beta$	244
9	Hyperbolic mixed, predominantly catalytic inhibition	HMx(Sp < Ca)I	(0,1)	(0,1)	$\alpha > \beta$	249
10	Hyperbolic mixed, balanced inhibition	HMx(Sp = Ca)I	1	(0,1)		254
11	Hyperbolic mixed, dual modification (inhibition → activation)	HMxD(I/A)	(1,∞)	(1,∞)	$\alpha > \beta$	257
12	Hyperbolic catalytic activation	HCaA	(1,∞)	(1,∞)		263
13	Hyperbolic mixed, predominantly specific activation	HMx(Sp > Ca)A	(1,∞)	(1,∞)	$\alpha < \beta$	265
14	Hyperbolic mixed, dual modification (activation → inhibition)	HMxD(A/I)	(0,1)	(0,1)	$\alpha < \beta$	268
15	Hyperbolic specific activation	HSpA	(0,1)	1		276
16	Hyperbolic mixed, predominantly catalytic activation	HMx(Sp < Ca)A	(0,1)	> 1		279
17	Hyperbolic mixed, balanced activation	HMx(Sp = Ca)A	1	> 1		281

The 17 mechanisms are those in Fig. 3.2. The two columns on the right specify additional requirements and the page where a mechanism is discussed

[a]Purely formal, a better definition is $K_{Ca} = \infty$

[b]Formal for $\alpha \to 0$ and $K_{Sp} \to \infty$ at the same time

special mechanisms for substrates and inhibitors as modifiers. Products more than substrates have been shown to obey a variety of mechanisms that go beyond linear specific, mixed or catalytic inhibition and can behave for instance as hyperbolic mixed, predominantly specific inhibitors or hyperbolic mixed, dual modifiers acting as either activators or inhibitors depending on substrate concentration.

The individual mechanisms will be described in the same order they are displayed in Fig. 3.2 grouped in their own taxonomic level. For practical reasons, the 17 mechanisms are listed in the synoptic Table 5.1 with full names, acronyms, properties, and the page number where they are described. Primary and secondary specific velocity plots will be used as a guide to mechanism diagnosis using error-free simulations. The discussion will be accompanied by examples from various fields of biochemistry that have been selected from the literature. The only criterion for the choice of the examples was clear documentation containing either graphical or tabular details to allow unambiguous identification. When necessary, original data have been reinterpreted by plotting existing graphics in a different way.

Some examples will be discussed in more detail than others with a double aim. First, to assure beginners in the field of enzyme inhibition and activation that theories and rate equations are not abstract concepts but that they find direct application in disparate fields of pharmacology, physiology, animal and plant biology, and biotechnology. Second, the examples illustrate how authors used known methods or developed their own strategies in front of the kinetic puzzles that had to be solved.

To complement the ensemble of characters that identify a mechanism, kinetic barrier diagrams will be shown beneath the specific velocity plot. The method for constructing these diagrams, the significance of the symbols therein and the relationships between barriers, kinetic parameters, and reaction rates have been discussed in Sect. 3.8. For every mechanism three kinetic barrier diagrams will be shown at the same fixed modifier concentration ($[X] = 10K_X$) and $[S] = 0.2K_m$, K_m and $5K_m$ to cover a typical spectrum of substrate concentrations used for studies in vitro. A fourth figure will show *titrations* in barcode-like form at fixed modifier concentration and four substrate concentrations different from those already shown in the kinetic barrier diagrams. This approach aims at highlighting the regulatory strategies of the individual mechanisms.

If the barrier diagrams are examined cursory they look all the same or almost so. This is not the case because none of the kinetic barriers of the 17 featured mechanisms is superimposable to that of another mechanism. I encourage in particular the neophytes in enzyme kinetics to dedicate some time in examining these diagrams with the necessary inquisitiveness (tutorial comments will be provided to every mechanism). After learning the subtle differences between the various mechanisms, kinetic barrier diagrams represent a valid companion for interpreting the behavior of a system in its niche of action.

Essential activation will not be treated here because the theoretical aspects discussed in Sect. 3.4 should be sufficient to understand and interpret experimental results.

5.2 Mechanisms in Taxonomic Level 1: Linear Inhibitors

Linear inhibition is undoubtedly the most studied topic in the biochemical literature and *competitive inhibition* is mastered without problems. However, the other mechanisms of this group are sometimes placed in a doubtful position with inconsistent nomenclature. This section collects the properties of the five linear mechanisms listed in the first taxonomic level of Fig. 3.2.

5.2.1 *Linear Specific Inhibition*

Assigned acronym: LSpI. The modifier binds exclusively the free enzyme. Since EX is only capable to dissociate back to E + X and is thus not connected to other paths in the mechanism Scheme 3.2, it is a dead-end complex present in a regime of equilibrium when the system is at steady-state. The reaction paths of LSpI can be formally deduced from the general modifier mechanism (Scheme 3.1) by setting $\alpha = \infty$ that excludes the catalytic step through ESX. The rate equation and the expressions of the apparent kinetic parameters can also be obtained directly from (3.2)–(3.5) for $\alpha = \infty$. More elegantly, the rate equation can be derived using the King–Altman method and applying Cha's rule for steps in quasi-equilibrium (see Appendix). Relevant equations, definitions of constants, and their dependence on modifier concentration for linear specific inhibition are collected in (5.1)–(5.6) with expressions of the specific velocity equation in (5.6).

$$v_{\mathrm{X}} = \frac{k_2[\mathrm{E}]_{\mathrm{t}}\,[\mathrm{S}]}{K_{\mathrm{m}}^{0}\left(1 + \dfrac{[\mathrm{X}]}{K_{\mathrm{Sp}}}\right) + [\mathrm{S}]} \tag{5.1}$$

$$V = V^{\mathrm{app}} = k_2[\mathrm{E}]_{\mathrm{t}}; \quad k_{\mathrm{cat}} = k_{\mathrm{cat}}^{\mathrm{app}} = k_2 \tag{5.2}$$

$$K_{\mathrm{m}}^{\mathrm{app}} = \frac{k_{-1} + k_2}{k_1}\left(1 + \frac{[\mathrm{X}]}{K_{\mathrm{Sp}}}\right) \tag{5.3}$$

$$k^{\mathrm{app}} = \frac{k_1 k_2}{k_{-1} + k_2} \times \frac{1}{1 + \dfrac{[\mathrm{X}]}{K_{\mathrm{Sp}}}}; \quad \frac{1}{k^{\mathrm{app}}} = \frac{k_{-1} + k_2}{k_1 k_2}\left(1 + \frac{[\mathrm{X}]}{K_{\mathrm{Sp}}}\right) \tag{5.4}$$

$$\frac{v_{\mathrm{X}}}{v_0} = \frac{1 + \sigma}{1 + \dfrac{[\mathrm{X}]}{K_{\mathrm{Sp}}} + \sigma} \tag{5.5}$$

$$\frac{v_0}{v_{\mathrm{X}}} = -\frac{[\mathrm{X}]}{K_{\mathrm{Sp}}}\frac{\sigma}{1 + \sigma} + \left(1 + \frac{[\mathrm{X}]}{K_{\mathrm{Sp}}}\right); \quad \frac{\mathrm{a}}{\mathrm{a} - 1} = \frac{K_{\mathrm{Sp}}}{[\mathrm{X}]} + 1 \tag{5.6}$$

Here, the dissociation constant of the EX complex is called K_{Sp} to explicitly state that it refers to a specific inhibitor but, when the specific character of inhibition is not in doubt, it is simply named K_{i} (inhibition constant). To avoid ambiguous interpretation, inhibition constants should always be accompanied by the name of the mechanism to which they refer. The limiting rate V (5.2) (left) and therefore k_{cat} (5.2) (right) are unaffected by the inhibitor, whereas the apparent Michaelis constant increases linearly for increasing [X] by the factor $(1 + [\mathrm{X}]/K_{\mathrm{Sp}})$ (5.3). The definition *specific inhibition* derives from the effect the modifier exerts on the specificity constant (5.4) (left), which decreases hyperbolically as [X] increases, and the adjective linear applies because the reciprocal of the specificity constant

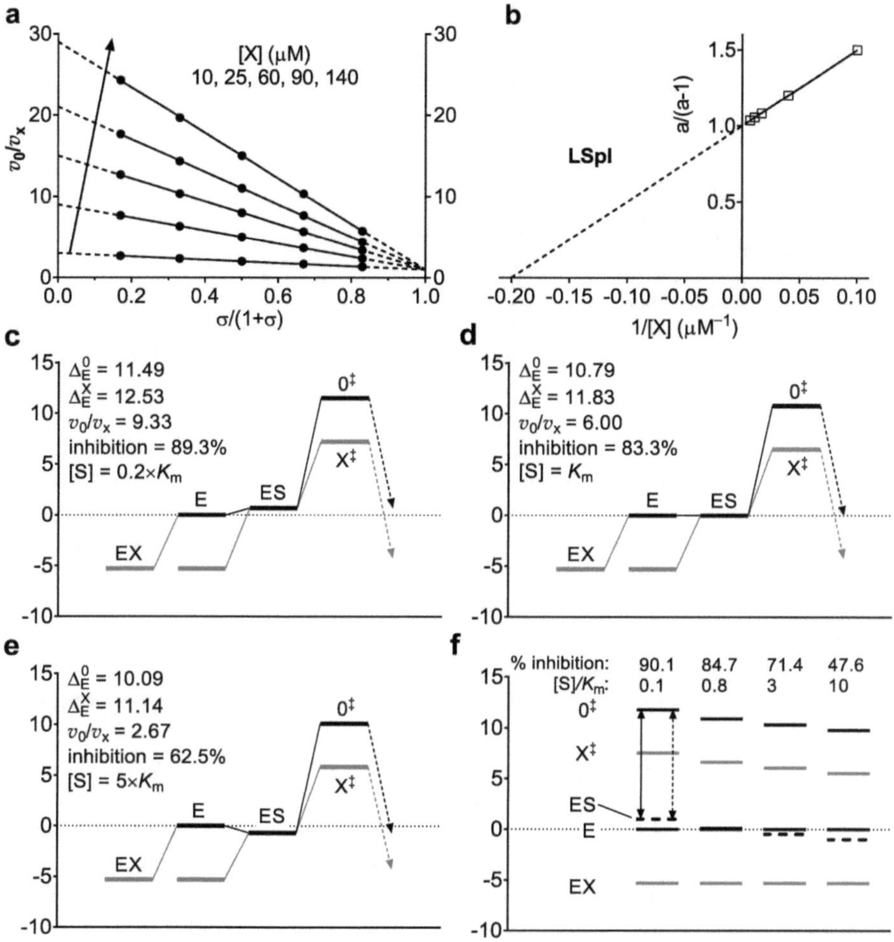

Fig. 5.1 Linear specific inhibition: LSpI. Primary (**a**) and secondary (**b**) specific velocity plot. The *arrow* in (**a**) indicates the increasing direction of [X]. (**c**)–(**e**) Kinetic barrier diagrams, (**f**) barcode-like representation of the various states at four substrate concentrations as indicated. $k_{cat} = 100\,s^{-1}$, $K_m = 200\,\mu M$, $K_{Sp} = 5\,\mu M$, in (**c**)–(**f**) $[X] = 10 \times K_{Sp}$. In (**f**) the *continuous double arrow* is Δ_{ES}^{X} and the *dashed double arrow* is Δ_{ES}^{0}. At any substrate concentration $\Delta_{ES}^{X} = \Delta_{ES}^{0}$

increases linearly with [X] (5.4) (right). These properties are summarized graphically in Fig. 3.3.

The diagnosis of linear specific inhibition is generally straightforward. However, plotting data as specific velocity plot as shown in Fig. 5.1 is recommended for differentiating against hyperbolic specific inhibition and the hyperbolic mixed mechanism with predominantly specific character. Only linear specific inhibition exhibits at the same time a primary specific velocity plot with negative slope, intersecting at a common point with coordinates (1,1), and a secondary plot

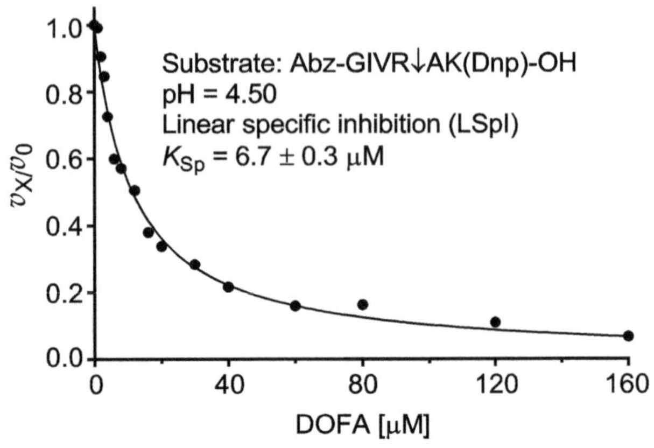

Fig. 5.2 Linear specific inhibition of human cathepsin B by DOFA. In the formula of the Förster resonance energy transfer-labeled substrate used in fluorometric measurements Abs = ortho-aminobenzoyl, Dnp = Nϵ-2,4-dinitrophenyl, and the *arrow* indicates the scissile bond. Florescence readings were corrected for inner filter effects [83]

intersecting the ordinate at the value of 1 (Fig. 5.1). K_{Sp} can be estimated as the slope of the secondary plot or from the reciprocal of the intersection point of the straight line on the abscissa (Eq. (5.6), right). With this preliminary value of K_{Sp} and knowing the mechanism of inhibition, calculations can then be refined on primary data as rate against [S] at several modifiers concentrations by globally fitting (5.1) to data.

Alternatively, experience has shown that a series of measurements at fixed substrate and several inhibitor concentrations to cover the range 10–90 % inhibition is suitable for determining the inhibition constant. Data are plotted as v_X/v_0 and (5.5) is fitted to them with K_{Sp} as the sole parameter. An example is shown in Fig. 5.2 for the inhibition of cathepsin B (EC 3.4.21.1) by DOFA (for the discussion of this case see Sect. 4.6.4). This substrate occupies the primed and the nonprimed subsites of the enzyme (Fig. 4.1) and acts as a linear specific inhibitor.

As shown by the kinetic barrier diagrams in Fig. 5.1c–f, the ES complex is not affected by the presence of the modifier because the ESX complex does not exist. Thus, the kinetic barrier Δ^0_{ES} (the distance between ES and the transition state $0^{\ddagger}$ in the absence of modifier) remains the same in the presence of X: Δ^0_{ES} and Δ^X_{ES}, which depend on the reciprocal k_{cat} and k^{app}_{cat}, respectively, coincide . Δ^X_E decreases with increasing [S] because this barrier, calculated with (3.55), depends on the reciprocal specificity constant given by (3.48), which in the kinetic barrier diagrams contains the concentrations of the substrate and of the modifier. The inhibitory effect is due to sequestration of enzyme in the dead-end EX complex and the efficiency of linear specific inhibitors decreases with increasing substrate concentration as shown by the decreasing percent inhibition in Fig. 5.1c–f. This property must be taken into account in pharmacological applications as clearly remarked by Cornish-Bowden [20, p. 91]

> An essential point that is obvious but is nearly always forgotten in efforts to design drugs is that anything that can compete with the substrate for the active site of an enzyme is something that the substrate can compete with.

Table 5.2 Examples of linear specific inhibition, LSpI

Enzymes	EC no.	Inhibitors	Ref.
Oligo-1,6-glucosidase[a]	3.2.1.10	Glucose	[64]
β-Fructofuranosidase	3.2.1.26	Fructose	[63]
Dihydropteroate synthase	2.5.1.15	Sulfonamides	[108]
Cathepsin H	3.4.22.16	Chicken egg cystatin	[80]
Thrombin	3.4.21.5	Hirudin and hirudin fragments	[85]

[a]The authors experimented with an yeast extract, not with purified enzyme. Oligo-1,6-glucosidase is likely to be the main activity they dealt with

In all enzyme classes, linear competitive inhibition is a common type of interaction between enzymes and modifiers, either natural or synthetic. This is the inhibition mechanism that suffers less than others from misapplied nomenclature and wrong diagnosis. Three classical examples of linear specific inhibition are listed in Table 5.2 together with two more recent examples of the application of the specific velocity plot.

Michaelis and Rona studied the inhibition by products of the reaction catalyzed by *maltase* (oligo-1,6-glucosidase) and derived the following equation to explain their results [64, Eq. A, p. 65]:

$$\frac{k_1}{k} = \frac{[R]}{([S] + k)\left(\dfrac{v}{v_1} - 1\right)} . \tag{5.7}$$

Translated 1:1 in the terminology of this book (terms in the same place as in the original equation) and rearranging, (5.7) becomes

$$\frac{K_{\mathrm{Sp}}}{K_{\mathrm{m}}^{0}} = \frac{[\mathrm{X}]}{([\mathrm{S}] + K_{\mathrm{m}}^{0})\left(\dfrac{v_0}{v_{\mathrm{X}}} - 1\right)}$$

$$\frac{v_0}{v_{\mathrm{X}}} = 1 + \frac{[\mathrm{X}]\,K_{\mathrm{m}}^{0}}{K_{\mathrm{Sp}}\left([\mathrm{S}] + K_{\mathrm{m}}^{0}\right)} = \frac{K_{\mathrm{Sp}}\left([\mathrm{S}] + K_{\mathrm{m}}^{0}\right) + [\mathrm{X}]\,K_{\mathrm{m}}^{0}}{K_{\mathrm{Sp}}\left([\mathrm{S}] + K_{\mathrm{m}}^{0}\right)}$$

$$\frac{v_0}{v_{\mathrm{X}}} = \frac{1 + \dfrac{[\mathrm{X}]}{K_{\mathrm{Sp}}} + \dfrac{[\mathrm{S}]}{K_{\mathrm{m}}^{0}}}{1 + \dfrac{[\mathrm{S}]}{K_{\mathrm{m}}^{0}}} \equiv \frac{1 + \dfrac{[\mathrm{X}]}{K_{\mathrm{Sp}}} + \sigma}{1 + \sigma} . \tag{5.8}$$

Thus, nothing changed in 100 years because (5.8) is the reciprocal of (5.5). This is the original rate equation for competitive inhibition, proposed for inhibition by a product of the reaction, but valid in general. With the same equation Michaelis and

Pechstein studied in parallel the inhibition of *invertase* (β-fructofuranosidase) by fructose [63].

Sulfonamides represent a great discovery as the first drugs used to contrast bacterial attack. In 1932 Gerhard Domagk demonstrated the effect 4-sulfonamide-2′,4′-diaminoazobenzene in protecting mice from infection by *Streptococcus pyogenes* [27]. In 1940 Woods explained the mechanism of action of sulphanilamide in a qualitative but convincing way [108]:

> As there was no possibility of a direct molecular reaction between sulphanilamide and the factor, this constant quantitative relationship between inhibitor and active substance was reminiscent of the competitive inhibition of enzyme reactions by substances chemically related to the substrate or product.

In this paper Wood identified the *factor* as 4-aminobenzoic acid. In Wood's clever approach, the inhibitor (sulphanilamide) has been identified thanks to the competing action of the substrate (4-aminobenzoic acid) in an enzymatic reaction. Today we know that the target enzyme was *dihydropteroate synthase*. This enzyme catalyzes the condensation of 4-aminobenzoic acid and (2-amino-4-hydroxy-7,8-dihydropteridin-6-yl)methyl diphosphate to form dihydropteroate. This is the biosynthetic step preceding the formation of 7,8-dihydrofolate in microorganisms. Since mammals lack dihydropteroate synthase, sulfonamides have been the first drug candidates against bacterial infection. Due to bacterial resistance they have been progressively dismissed and are seldom used today.

The Specific Velocity Plot has been invoked to differentiate between linear specific inhibition and other mechanisms in several cases. For instance, wild type chicken egg white cystatin and two recombinant variants were classified by Rothe et al. as linear specific inhibitors of *aleurain* from barley, a plant homologous of mammalian cathepsin H now put in synonymy with this enzyme [80].

Schmitz et al. [85] studied the inhibition mechanism of thrombin by hirudin and two hirudin-derived fragments, an N-terminal hirudin(1–47) and a C-terminal hirudin(45–65) fragment. Using a synthetic tripeptide substrate, full-length hirudin and hirudin(1–47) behaved as linear specific inhibitors, while hirudin(45–65) was a hyperbolic catalytic inhibitor (Sect. 5.3.3). Both hirudin fragments behave also as linear specific inhibitors of the enzyme in the presence of fibrinogen A_2 as substrate [85].

5.2.1.1 Nonproductive Binding

Nonproductive substrate binding is discussed here because of the analogy with specific inhibition. This is a *phantom effect*, meaning that we cannot always ascertain its presence. Nonproductive binding is a property of hydrolytic enzymes that act in the extracellular space on macromolecular substrates such as the peptidases and the nucleases. As schematically shown in Fig. 4.1a, the peptidases own an extended region for binding their polypeptide substrates. Only two of the subsites

Scheme 5.1 Nonproductive substrate binding

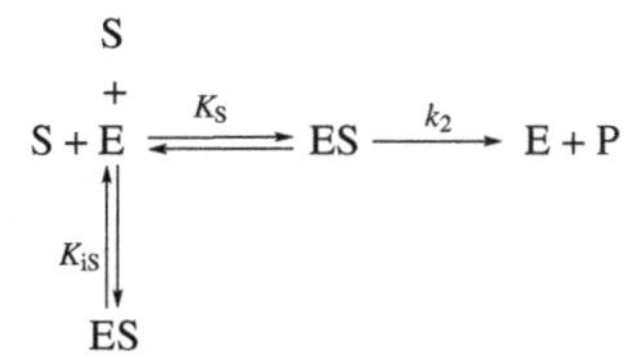

are shown in the figure but they can be up to four on each side of the position marked with the cleaved bond. Cleavage of a peptide bond takes place only if the substrate binds productively, i.e., if the amino acids side chains accommodated in the enzyme pockets S and S′ correctly align the susceptible peptide bond. However, (poly)peptides can also bind unspecifically by occupying either only the nonprimed (S) or the primed subsites (S′). This is similar to adding a specific inhibitor, as shown in Scheme 5.1, for which the rate equation is

$$v = \frac{V\,[\mathrm{S}]}{K_\mathrm{m}\left(1 + \dfrac{[\mathrm{S}]}{K_\mathrm{iS}}\right) + [\mathrm{S}]}. \tag{5.9}$$

Expanding the denominator of (5.9) and rearranging, the rate equation can be recast as

$$v = \frac{\dfrac{V}{1 + K_\mathrm{m}/K_\mathrm{iS}}[\mathrm{S}]}{\dfrac{K_\mathrm{m}}{1 + K_\mathrm{m}/K_\mathrm{iS}} + [\mathrm{S}]}, \tag{5.10}$$

from which it is evident that the limiting rate and the Michaelis constant are diminished by the same factor. Since the modified V- and K_m-values cannot be compared with the corresponding values in the absence of the inhibitory effect, the phenomenon remains unobserved. As a drawback, the parameters expected from undisturbed substrate binding will contain an unknown bias. The specificity of the peptidases is usually studied by mapping the specificity subsites with synthetic oligopeptides. The doubt can remain if comparing the specificity constants determined as the ratio $k_\mathrm{cat}/K_\mathrm{m}$ for a series of peptides in reactions catalyzed by peptidases does reflect the true specificity of the enzymes investigated.

5.2.2 Linear Catalytic Inhibition

Assigned acronym: LCaI. The formation of a dead-end ESX complex exclusively through binding of X to ES is characteristic of catalytic inhibition, i.e., reactant addition to enzyme is accomplished in compulsory order with substrate binding first

(Scheme 3.2). The rate equation for linear catalytic inhibition and the dependence from modifier concentration of the parameters therein can be deduced from (3.2)–(3.5) as shown in the equation array (5.11)–(5.15). A more elegant and formally correct derivation using the King–Altman method is given in the Appendix.

In linear catalytic inhibition, the apparent specificity constant, k^{app}, does not depend on inhibitor concentration (5.14) because k_{cat}^{app} and K_m^{app} are affected by the same factor $(1 + [X]/K_{Ca})$, which cancels out when taking the ratio of the two parameters. The plots of the dependences of the parameters on modifier concentration are summarized in Fig. 3.3. Since in this mechanism the specificity constant is not affected by the modifier but catalysis is affected, *catalytic inhibition* is an appropriate name.

In the specific velocity plot, linear catalytic inhibition is characterized by a family of straight lines with positive slope converging to a common point with coordinates (0,1) and a secondary plot intersecting the ordinate at the value 1 (Fig. 5.3). The inhibition constant K_{Ca} can be obtained from the slope or from the abscissa intercept of the secondary plot according to (5.15) (right). Final refinement is best performed by globally fitting (5.11) to rates against [S] using the estimate of K_{Ca} as initial guess in nonlinear regression and/or using the method shown in Fig. 5.2 for linear specific inhibition.

The EX complex does not exist in catalytic inhibition and the modifier exerts its effect by withdrawing enzyme from the catalytic path. Contrary to specific inhibition, catalytic inhibition becomes more efficient when the concentration of the substrate increases because this promotes the formation of the ESX complex as shown by the kinetic barrier diagrams in Fig. 5.3c–f. The value of Δ_E^0 remains unaltered in the presence of the modifier and so $\Delta_E^X = \Delta_E^0$ at any modifier and substrate concentration, which is equivalent to say that the specificity constant does not change with [X].

Linear catalytic inhibition is a rare mechanism for unireactant enzymes, but occurs in multireactant systems, were one of the products is often a catalytic inhibitor with respect to one of the substrates, a property exploited for mechanistic diagnostic purposes [89, Chap. 9]. Cornish-Bowden put forward the important concept that, in metabolic pathways, natural metabolites acting as catalytic inhibitors have been disfavored by evolution against specific inhibitors because of their severe toxic effects. However, synthetic catalytic inhibitors may prove useful in drug design and pest control, provided any interaction with enzymes of the organism to be treated/protected can be excluded [19]. This concept was strengthened by Westley and Westley, who showed the superiority of catalytic inhibitors toward specific inhibitors in open systems, i.e., in metabolic pathways with constant input of substrates and removal of products [105].

$$v = \frac{\dfrac{k_2[E]_t}{1 + [X]/K_{Ca}}[S]}{\dfrac{K_m^0}{1 + [X]/K_{Ca}} + [S]} \tag{5.11}$$

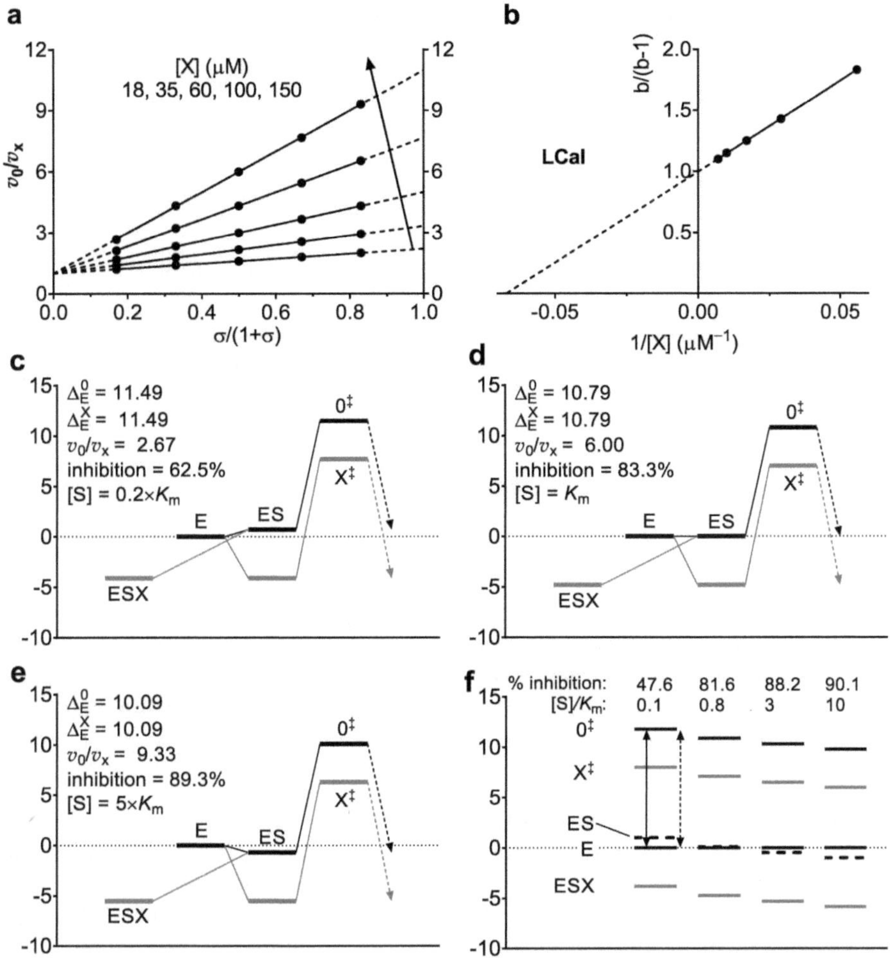

Fig. 5.3 Linear catalytic inhibition: LCaI. Primary (**a**) and secondary (**b**) specific velocity plot. The *arrow* in (**a**) indicates the increasing direction of [X]. (**c**)–(**e**) Kinetic barrier diagrams, (**f**) barcode-like representation of the various states at four substrate concentrations as indicated. $k_{cat} = 100\,s^{-1}$, $K_m = 200\,\mu M$, $K_{Ca} = 15\,\mu M$, in (**c**)–(**f**) $[X] = 10 \times K_{Ca}$. In (**f**) the *solid double arrow* is Δ_E^X and the *dashed double arrow* is Δ_E^0. At any substrate concentration $\Delta_E^X = \Delta_E^0$. Common to all diagrams: $\Delta_{ES}^0 = 10.79$, $\Delta_{ESX} = -4.82$

$$V^{app} = \frac{k_2[E]_t}{1 + [X]/K_{Ca}}; \quad k_{cat}^{app} = \frac{k_2}{1 + [X]/K_{Ca}} \tag{5.12}$$

$$K_m^{app} = \frac{k_{-1} + k_2}{k_1} \times \frac{1}{1 + [X]/K_{Ca}} \tag{5.13}$$

Table 5.3 Examples of linear catalytic inhibition, LCaI

Enzymes	EC no.	Inhibitors	Ref.
Inositol-phosphate phosphatase	3.1.3.25	Li^{+}[a]	[40, 56, 100]
3-oxo-5α-steroid 4-dehydrogenase ($NADP^{+}$)	1.3.1.22	A-ring aryl steroids[b]	[16]
γ-Glutamyl transferase	2.3.2.2	OU749[c]	[50, 107]
Elastase-2	3.4.21.27	Silanediol peptide isosters	[60]
Acinetobacter baylyi cultures	1.13.11.-	Aniline, 4-chlorocatecol	[43]

[a]Only at low [Li^{+}]
[b]E.g. 17β-(*N*,*N*-diisopropylcarboxamide)-estra-1,3,5(10)-triene-3-carboxylic acid and analogues
[c]*N*-[5-(4-methoxybenzyl)-1,3,4-thiadiazol-2-yl]benzenesulfonamide and analogues

$$k^{\text{app}} = \frac{k_1 k_2}{k_{-1} + k_2} \tag{5.14}$$

$$\frac{v_0}{v_X} = \frac{[X]}{K_{Ca}} \frac{\sigma}{1+\sigma} + 1; \qquad \frac{b}{b-1} = \frac{K_{Ca}}{[X]} + 1 \tag{5.15}$$

Pharmacologically relevant examples of catalytic inhibition elicited by molecules and ions unrelated to substrates or products of the reaction are known, particularly for multireactant enzymes. The examples commented below and summarized in Table 5.3 serve to illustrate the importance of knowing accurately the inhibition mechanism.

Bipolar Disorder, or Manic-Depressive Disorder occupies the sixth place among the causes of disability in the human population. Patients affected by this disorder manifest episodes of mania, characterized by very high mood, interchanging with depression. At low concentrations, Li^{+} ions inhibit the inositol-phosphate phosphatase with a linear catalytic mechanism, resulting in modulation of the phosphatidylinositol signaling response and efficient alleviation of the manic-depressive symptoms [40, 56, 100].

Testosterone is converted to 5α-dihydrotestosterone by the $NADP_{red}$-dependent 3-oxo-5α-steroid 4-dehydrogenase (acceptor). This metabolite of testosterone, which is a more potent androgen, is responsible for benign prostatic hyperplasia, male pattern baldness, and acne. Modern therapies are directed toward both the androgen receptor, with the use of antagonists, and inhibition of the 3-oxo-5a-steroid 4-dehydrogenase. Finasteride is successfully used for the treatment of the mentioned disorders. Research on further inhibitors of the 3-oxo-5a-steroid 4-dehydrogenase lead to the discovery of aryl steroids that behave as linear catalytic inhibitors [16].

γ-Glutamyl Transferase cleaves the gamma-glutamyl bond of glutathione and transfers this residue to an acceptor, either water, an amino acid, or a peptide. Making extracellular glutathione available for intracellular use, γ-glutamyl transferase is a key enzyme in the homeostasis of glutathione and in turn for drug detoxification. However, γ-Glutamyl transferase in tumor cells plays a role in resistance to radiation and chemotherapy, suggesting this enzyme as a target for clinically efficient inhibition. OU749 (full name in Table 5.3) and derivatives of this molecule have proven effective inhibitors acting with a linear catalytic mechanism [50, 107].

Elastase-2 is a hot pharmacological target for the many pathological states associated with its uncontrolled activity. This enzyme is implicated in the development of chronic obstructive pulmonary disease, estimated at the fifth place under the leading causes of death. Among many hundreds of synthetic inhibitors synthesized against elastase-2, belonging to a plethora of chemical classes, silanediol peptide isosters have been found to behave as linear catalytic inhibitors of the enzyme [60]. This mechanism is advantageous compared to other modes of action for the increased elastase selectivity in the presence of related serine peptidases.

Bioremediation strategies of sites contaminated by halogenated anilines originated from the biotransformation of herbicides have been investigated kinetically with several combinations of substrates and inhibitors [43] (last row in Table 5.3). Cultures of *Acinetobacter baylyi* degrade 4-chlorocathecol, 4-chloroaniline, and 3,4-dichloroaniline and are inhibited by reaction products. Two major enzymes are involved in the biotransformation reactions, cathecol 1,2-dioxygenase (EC 1.13.11.1) and chlorocathecol 1,2-dioxygenase (EC 1.13.11.-). In this system, the authors demonstrated inhibition of enzymatic activity by a linear catalytic mechanism with 4-chloroaniline as substrate and aniline or 4-chlorocathecol as inhibitor.

5.2.2.1 Catalytic Substrate Inhibition

In its genuine sense substrate inhibition is analogous to catalytic inhibition because an additional substrate molecule is added to the ES complex in compulsory order (Fig. 5.4). Owing to the sequential addition of substrate to enzyme, the rate equation contains a squared substrate concentration term:

$$v = \frac{V' [\mathrm{S}]}{K'_\mathrm{m} + [\mathrm{S}] + \dfrac{[\mathrm{S}]^2}{K_\mathrm{Si}}} \tag{5.16}$$

where the inhibition constant is called K_Si as distinct from K_iS in nonproductive binding (p. 217).

When performing initial rate measurements with increasing substrate concentrations, substrate inhibition causes deviations from hyperbolic saturation showing instead a maximum followed by declining velocities (Fig. 5.4, right). The inhibition caused by substrate in this way is linear, i.e., the initial rate tends to zero when $[\mathrm{S}] \rightarrow \infty$. The position of the maximum in the plot of Fig. 5.4 is obtained

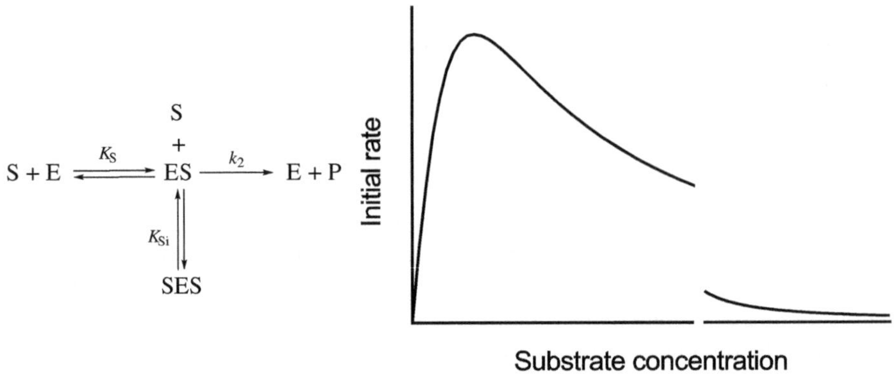

Fig. 5.4 Inhibition by substrate. The analogy to catalytic inhibition is evident from the scheme on the *left*. The initial rate as a function of substrate concentration shows a maximum

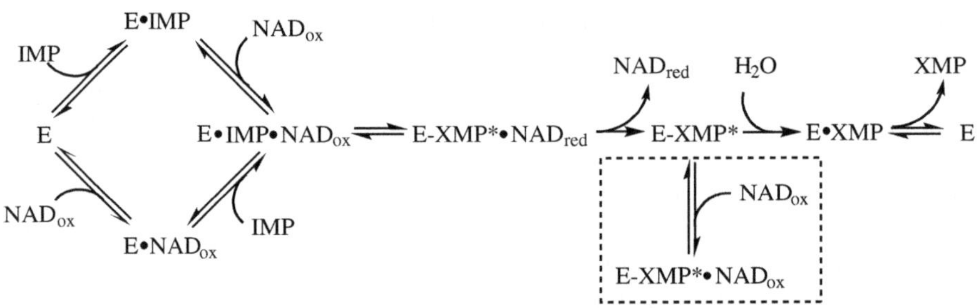

Scheme 5.2 Kinetic mechanism of human IMP dehydrogenase (from [104]). The substrates are inosine monophosphate and NAD_{ox} (coenzyme) and the products are xanthosine monophosphate and NAD_{red}. E-XMP* is a covalent intermediate that is converted into XMP by hydrolysis. The *boxed branch* represents substrate inhibition at high concentrations of NAD_{ox}

by differentiating (5.16), equating the result to 0 and solving for [S]. Thus, the maximum occurs at $[S] = \sqrt{K'_m K_{Si}}$. The primed parameters in (5.16) indicate that they should not be confused with the corresponding parameters of the Michaelis–Menten equation because the rate has a limiting value of zero when $[S] \to \infty$ and thus there is no Michaelis constant that can correspond to a substrate concentration for which the rate is one half the limiting rate.

Equation (5.16), shown here for a single-substrate single-product reaction, is equally valid for more complex systems like a reaction with two substrates and two products. For example, the reaction catalyzed by IMP dehydrogenase (EC 1.1.1.205) is illustrated in Scheme 5.2. Early believed to follow an ordered bi-bi reaction mechanism, the IMP dehydrogenase has been shown to obey a mechanism in which the substrates IMP and NAD_{ox} bind at random, while the products are released in compulsory order with the reduced coenzyme being released first. This information and much more on the properties of the IMP dehydrogenase have been reviewed by Lizbeth Hedstrom [41].

At high concentrations of NAD_{ox}, the covalent intermediate E-XMP* is trapped in a ternary, dead-end E-XMP*·NAD_{ox} complex resulting in substrate inhibition. If measurements are performed at a fixed, saturating concentration of IMP and variable [NAD_{ox}], the rate equation is the same as (5.16) with [NAD_{ox}] replacing [S]. The very reasons of this behavior can be interpreted considering that IMP dehydrogenase catalyzes two chemical transformations, the first being the reaction as dehydrogenase with formation of the reduced coenzyme and the second being hydrolysis of the covalent intermediate E-XMP*. How this mechanism has been unraveled cannot be detailed here and the reader is invited to read the mentioned review by Hedstrom, where the mechanism is described at the structural level [41, Sect. 6].

5.2.3 Linear Mixed, Predominantly Specific Inhibition

Assigned acronym: LMx(Sp>Ca)I. Linear mixed inhibition exhibits traits of both specific and catalytic character, where either one can dominate over the other or be equally represented. The reaction paths are shown in Scheme 3.2, in which the ESX complex is catalytically inactive and α can be less than, equal to, or larger than one. The rate equation, the dependences of the parameters on [X], and the specific velocity equation are shown in (5.17)–(5.22). When this mechanism has a predominantly specific character ($\alpha > 1$), k_{cat}^{app} depends only on the catalytic inhibition constant and decreases for increasing [X] by the factor $(1 + [X]/\alpha K_X)$ according to (5.18), while K_m^{app} depends on both the catalytic and the specific inhibition constant (5.19). Since in this mechanism $\alpha > 1$, the right fraction in (5.19) is greater than 1 at any [X] and K_m^{app} increases with increasing [X] accordingly. The apparent specificity constant, however, decreases hyperbolically when [X] increases (5.20) (left). The linearity of the mechanism can be ascertained by plotting $1/k^{app}$ against [X], which gives a straight line according to (5.20) (right). These properties are represented graphically in Fig. 3.3.

$$v = \frac{\dfrac{k_2[E]_t}{1 + \dfrac{[X]}{\alpha K_X}}[S]}{K_m^0 \dfrac{1 + \dfrac{[X]}{K_X}}{1 + \dfrac{[X]}{\alpha K_X}} + [S]} \tag{5.17}$$

$$V^{app} = \frac{k_2[E]_t}{1 + \dfrac{[X]}{\alpha K_X}}; \quad k_{cat}^{app} = \frac{k_2}{1 + \dfrac{[X]}{\alpha K_X}} \tag{5.18}$$

$$K_m^{app} = \frac{k_{-1}+k_2}{k_1} \times \frac{1+\dfrac{[X]}{K_X}}{1+\dfrac{[X]}{\alpha K_X}} \tag{5.19}$$

$$k^{app} = \frac{k_1 k_2}{k_{-1}+k_2} \times \frac{1}{1+\dfrac{[X]}{K_X}}; \quad \frac{1}{k^{app}} = \frac{k_{-1}+k_2}{k_1 k_2}\left(1+\frac{[X]}{K_X}\right) \tag{5.20}$$

$$\frac{v_0}{v_X} = \left(\frac{[X]}{\alpha K_X} - \frac{[X]}{K_X}\right)\frac{\sigma}{1+\sigma} + \left(1+\frac{[X]}{K_X}\right) \tag{5.21}$$

$$\frac{a}{a-1} = \frac{K_X}{[X]} + 1; \quad \frac{b}{b-1} = \frac{\alpha K_X}{[X]} + 1 \tag{5.22}$$

The primary specific velocity plot in Fig. 5.5a consists of straight lines with negative slopes because in (5.21), with $\alpha > 1$, $(1/\alpha K_X) < (1/K_X)$. As illustrated in Fig. 3.9, the abscissa coordinate of the intersection point of the straight lines in panel (a) corresponds to $(\alpha - \beta)/(\alpha - 1)$, which together with intersections and slopes of the secondary plots (Fig. 5.5b) allow to calculate α, β and K_X.

The kinetic barrier diagrams in Fig. 5.5c–f reveal the mixed nature with predominantly specific character of this linear inhibitor because its inhibitory power is moderately influenced by substrate concentration. In fact, in the same range of substrate concentrations and for the same $[X]/K_X$ ratio, the percent inhibition of LMx(Sp>Ca)I decreases much less than in specific inhibition (Fig. 5.1c–f). This property, together with the fact that all Δ-values are different from one another at any substrate concentration, is characteristic of LMx(Sp>Ca)I.

Among naturally occurring and man-made inhibitors, those that act by a linear mixed, predominantly specific mechanism are relatively frequent. Many cases have been properly characterized and a selected choice of these is listed in Table 5.4 of which the first four entries will not be commented.

Aminopeptidase N Inactivates Enkephalins and Other Bioactive Peptides by cleaving the N-terminal amino acid. A study by Nakanishi and coworkers [69] on the inhibition of aminopeptidase N by bile acids is didactically instructive because it shows the importance of considering in detail the kinetic mechanism of action of the modifiers. This information has been exploited by the authors for making hypotheses on the spatial arrangement of reactants in the ESX complex and the allosteric coupling constant has been used to calculate the free energy change caused by the substrate-modifier interaction on the enzyme. The authors performed measurements with Leu-Met, Leu-Gly, and Leu-*p*-nitroanilide as substrates, and cholic acid, deoxycholic acid, and chenodeoxycholic acid as inhibitors. Four mixed mechanisms either linear or hyperbolic could be characterized: LMx(Sp>Ca)I, LMx(Sp<Ca)I, LMx(Sp=Ca)I, and HMx(Sp>Ca)I, which will be shortly commented in the respective sections that will follow. With cholic acid as inhibitor, measuring α for a series of Leu-Y dipeptides as substrates, allowed to establish that

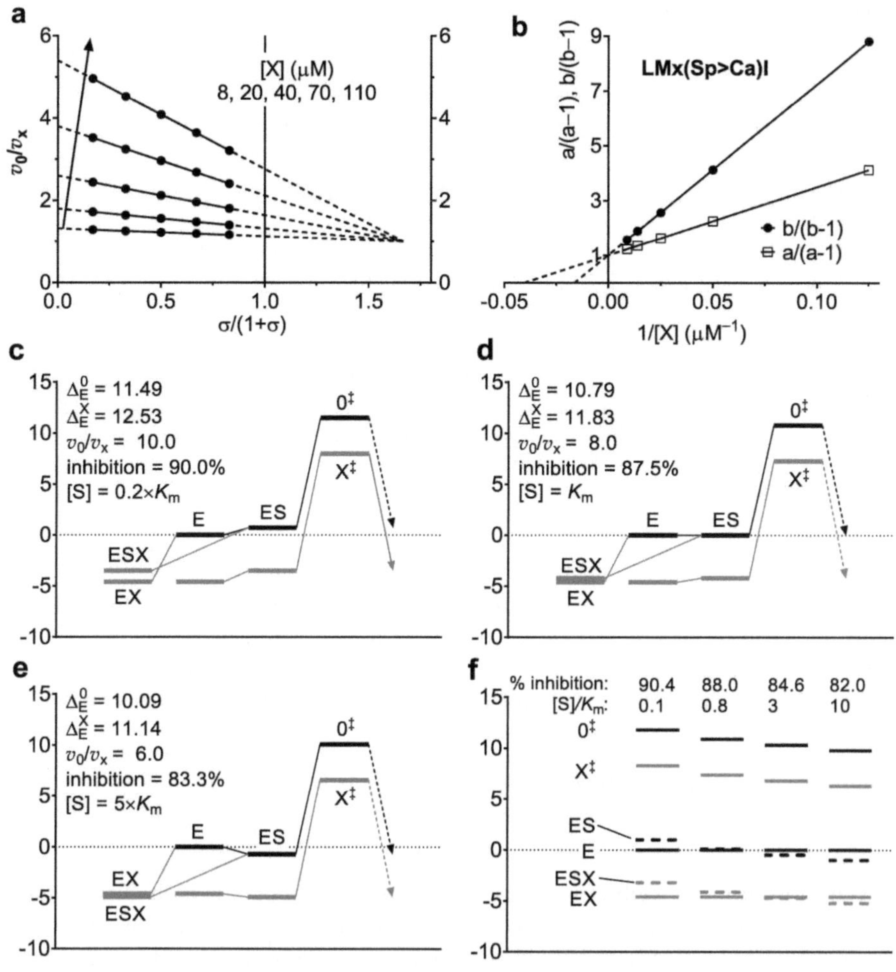

Fig. 5.5 Linear mixed, predominantly specific inhibition: LMx(Sp>Ca)I. Primary (**a**) and secondary (**b**) specific velocity plot. The *arrow* in (**a**) indicates the increasing direction of [X]. (**c**)–(**e**) Kinetic barrier diagrams, (**f**) barcode-like representation of the various states at four substrate concentrations as indicated. $k_{cat} = 100\,\text{s}^{-1}$, $K_m = 200\,\mu\text{M}$, $K_X = 25\,\mu\text{M}$, $\alpha = 2.5$, in (**c**)–(**f**) $[X] = 10 \times K_X$. Constant values common to (**c**)–(**f**): $\Delta_{ES}^0 = 10.79$, $\Delta_{ES}^X = 11.49$, $\Delta_{EX} = -4.60$, $\Delta_{ESX} = -4.20$. The degree of inhibition decreases for increasing [S] and there are no Δ-values that differ from one another in a special way at any [S]

α increased with the increasing length of the side chain of the C-terminal amino acid (Y = Gly, Ser, Val, Leu, Met), i.e., the specific character increased with respect to the catalytic component.

Human Liver Aldehyde Oxidase is responsible for the oxidation of aldehydes to carboxylic acids and of nitrogen-containing heterocyclic molecules, playing a role in drug metabolism. Implications on drug–drug interactions involving the

Table 5.4 Examples of linear mixed, predominantly specific inhibition, LMx(Sp>Ca)I

Enzymes	EC no.	Inhibitors	Ref.
Peptidyltransferase	2.3.2.12	Chloramphenicol	[28]
Dipeptidyl-peptidase IV	3.4.14.5	Oligopeptides from the HIV-1 Tat protein	[58]
β-Lactamase	3.5.2.6	*N*-arylsulfonyl hydrazones	[93]
γ-Secretase[a]	3.4.23.–	Pepstatin A methylester	[101]
Aminopeptidase N	3.4.11.2	Bile acids[b]	[69]
Aldehyde oxidase	1.2.3.1	Domperidone, chlorpromazine, clozapine	[10]
Linoleate 13S-lipoxygenase	1.13.11.12	(Z)-9-palmitoleyl sulfate	[81]
NADH:ubiquinone reductase[c]	1.6.5.3	Zn^{++}	[86]
Urease	3.5.1.5	Hydroxamic acids	[112]

[a]Part of a membrane-bound molecular organization involving presenilins 1 and 2 (see p. 227)
[b]Inhibitors: cholic, deoxycholic, and chenodeoxycholic acids with Leu-Met as substrate; cholic and chenodeoxycholic acids with Leu-Gly as substrate
[c]H^+-translocating

aldehyde oxidase-mediated pathway have been investigated by Barr and Jones [10]. With kinetic measurements, the authors demonstrated that structurally different molecules inhibit aldehyde oxidase by linear mixed mechanisms. Domperidone, chlorpromazine, and clozapine obey the linear mixed, predominantly specific mechanism with α values of 2.6, 5.3 and 15, respectively, while four other compounds act via a predominantly catalytic mechanism (Sect. 5.2.4).

Linoleate 13S-Lipoxygenase from soybean binds (Z)-9-palmitoleyl sulfate at an allosteric site resulting in linear mixed, predominantly specific mechanism with $\alpha = 10.2$ (see more in Sect. 5.3.2 for hyperbolic inhibition by sulfated oleic acid).

NADH:ubiquinone Reductase (H^+-translocating, complex I) is the first component of the respiratory chain in both eukaryotes and bacteria. Its function is to couple the transfer of two electrons from NAD_{red} to ubiquinone which results in the transfer of four protons across the membrane. Schulte and coworkers exploited the inhibition of complex I by Zn^{2+} to clarify the function of two components of the complex responsible for electron transfer and proton translocation. The inhibition mechanism could be identified as linear mixed, predominantly specific.

Inhibition of Urease from *Helicobacter pilori* is aimed at alleviating gastric ulcer symptoms. Apparently, the authors of a study on hydroxamic acids as potential drugs directed to *Helicobacter*-urease were surprised in finding deviations from the well-known competitive and noncompetitive inhibition mechanisms [112]. For the clearly diagnosable linear mixed, predominantly specific inhibition mechanism the authors proposed the new name *apparent mixed competitive mechanism*. The rate equation they proposed is correct but the reaction scheme in [112, Scheme 5] is unfortunately incomplete for lacking the connection between the EI and ESI complexes, a violation of the principle of microscopic reversibility and the equation for noncompetitive inhibition has a wrong denominator [112, Eq. 2].

5.2.4 Linear Mixed, Predominantly Catalytic Inhibition

Assigned acronym: LMx(Sp<Ca)I. The mechanism is that in Scheme 3.2 with $0 < \alpha < 1$. The rate equation and the expressions for the dependences of the parameters on modifier concentration are the same as for the preceding mechanism, (5.17)–(5.22). Since $\alpha < 1$, $K_{\mathrm{m}}^{\mathrm{app}}$ increases with increasing [X], while $k_{\mathrm{cat}}^{\mathrm{app}}$ and the specificity constant decrease with increasing [X] because this concentration appears only in the denominators of the respective equations. The plots of the dependences of the parameters on modifier concentration are summarized in Fig. 3.3.

Since $\alpha < 1$, the specific velocity plot (Fig. 5.6) is the specular image of that in Fig. 5.5 and the straight lines have a positive slope. α, β, and K_{X} can be calculated from the primary and secondary plots as instructed in Fig. 3.9. Some examples of this mechanism that have been interpreted correctly are summarized in Table 5.5 and briefly commented below.

From the kinetic barrier diagrams in Fig. 5.6c–f the prevalence of the catalytic character of this linear mixed inhibitor can be appreciated by the inhibitory power, which increases moderately when the substrate concentration is increased considerably. Moreover, all Δ-values differ from one another irrespective of substrate concentration.

Water Decontamination is an Imperative Duty in the relationship between human mankind and nature. Chlorinated hydrocarbons as groundwater contaminants can be eliminated using bacteria that like such non-appetizing molecules as food. To stimulate the appropriate bacteria, growth substrate must be added, which is butane in this case. Aerobic cometabolism of the chlorinated hydrocarbons that should be removed can be hampered by their competition for the growth substrate butane and optimization of the water-treatment process requires the knowledge of the inhibition mechanism. Notwithstanding measurements conducted with mixed cultures of gram-positive and gram-negative bacteria grown on butane, not with purified enzymes, Kim and coworkers could quantitatively characterize the transformation of three chlorinated hydrocarbons often found as contaminants of groundwater by means of detailed kinetic characterization of the inhibition mechanisms [49]. The diagnosis of the inhibition types was performed using the *direct linear plot* [29]. With 1,1,1-trichloroethane as substrate, butane acted as a linear mixed inhibitor with predominantly catalytic character, while butane was an inhibitor of the LMx(Sp>Ca)I type with 1,1-dichloroethylene and 1,1-dichloroethane as substrates. These results can be exploited for the optimization of aerobic bacterial growth conditions in the plants of water decontamination.

The Study on γ-Secretase mentioned in Table 5.4 dealt with inhibitors aimed at contrasting cerebral deposition of amyloid β-peptides, which is a determinant process in the development of Alzheimer's disease. Although not yet completely characterized, the membrane-bound complex containing the γ-secretase activity,

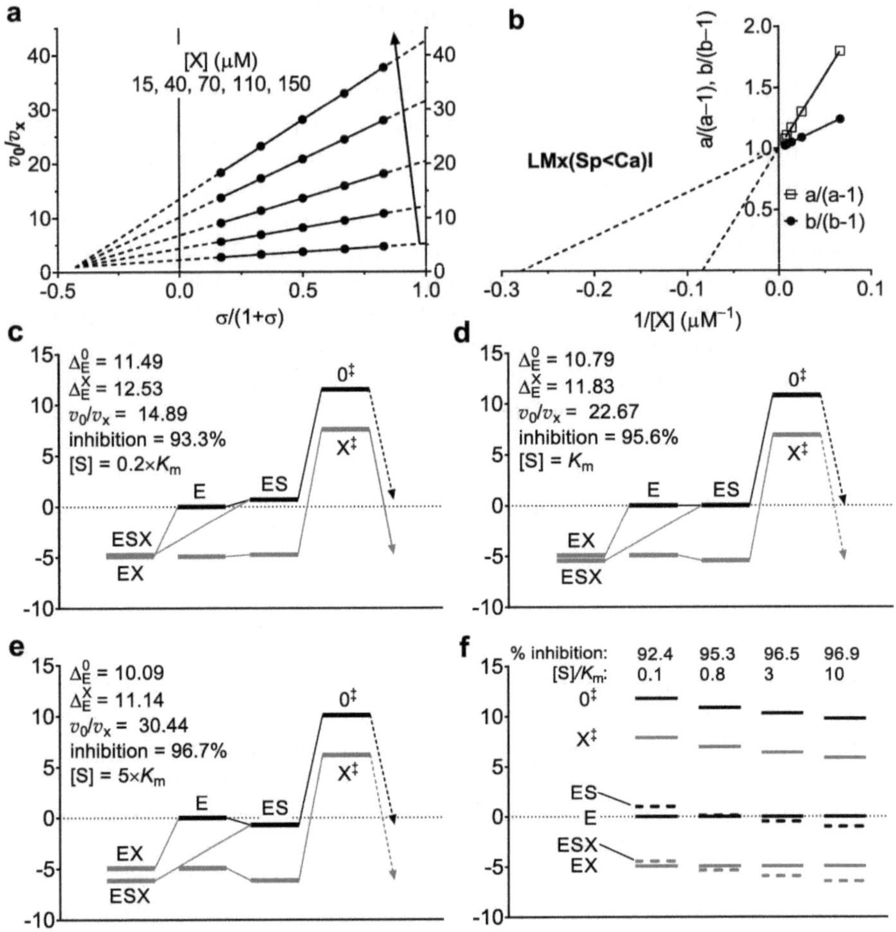

Fig. 5.6 Linear mixed, predominantly catalytic inhibition: LMx(Sp<Ca)I. Primary (**a**) and secondary (**b**) specific velocity plot. The *arrow* in (**a**) indicates the increasing direction of [X]. (**c**)–(**e**) Kinetic barrier diagrams, (**f**) barcode-like representation of the various states at four substrate concentrations as indicated. $k_{cat} = 100\,s^{-1}$, $K_m = 200\,\mu M$, $K_X = 12\,\mu M$, $\alpha = 0.3$, in (**c**)–(**f**) $[X] = 10 \times K_X$. Constant values common to (**c**)–(**f**): $\Delta_{ES}^{0} = 10.79$, $\Delta_{ES}^{X} = 12.33$, $\Delta_{EX} = -4.92$, $\Delta_{ESX} = -5.44$. The degree of inhibition increases for increasing [S]. The Δ-values do not share particular relationships with one another when the substrate concentration is increased

an aspartic peptidase, is inhibited by pepstatin derivatives, L-685,458,[1] benzodiazepines and sulfonamides [101]. Pepstatin A methylester behaves as a linear mixed, predominantly specific inhibitor (Table 5.4), while a sulfonamide-based molecule is still a linear mixed inhibitor, but with predominant catalytic character ($\alpha = 0.54$)

[1] [(2R,4R,5S)-2-Benzyl-5-(Boc-amino)-4-hydroxy-6-phenyl-hexanoyl]-Leu-Phe-NH_2.

Table 5.5 Examples of linear mixed, predominantly catalytic inhibition, LMx(Sp<Ca)I

Enzymes	EC no.	Inhibitors	Ref.
Bacterial cultures	–	Butane[a]	[49]
γ-Secretase[b]	3.4.23.–	F- and Cl-containing sulfonamide	[101]
Aminopeptidase N	3.4.11.2	Cholic acid[c]	[69]
Aldehyde oxidase	1.2.3.1	DCPIP[d], β-estradiol, menadione, ethinyl estradiol	[10]
Adenosine kinase	2.7.1.20.	Compound 1a in the paper[e]	[94]
Tyrosinase	1.14.18.1	4-Acetylphenylthiourea	[98]

[a]Substrate: 1,1,1- trichloroethane
[b]Part of a complex membrane-bound molecular organization involving presenilins 1 and 2
[c]Substrate: Leu-*p*-nitroanilide amino acids
[d]2,6-Dichlorophenolindophenol
[e]A 7-(het)aryl-7-deazaadenine ribonucleoside derivative

(Table 5.5). The remaining compounds investigated in the same study, considering the standard deviations of the kinetic constants, were characterized by α-values close to one. All compounds were classified as linear noncompetitive inhibitors but the authors provided figures of good quality as well as a detailed table of kinetic constants with their statistical significance. This allowed identification and naming of all mechanisms.

Aminopeptidase N is inhibited by cholic acid with Leu-pNA as substrate conforming to the LMx(Sp<Ca)I mechanism with $\alpha = 0.6$ and $\beta = 0$ [69]. As seen in Table 5.4, the same inhibitor obeys the LMx(Sp>Ca)I mechanism in the presence of Leu-Met. Substitution of the bulky methionine side chain with p-nitroaniline drops the level of specific inhibition with respect to the catalytic component as predictable from the experiment described in the preceding section with a series of Leu-Y amino acids.

The Aldehyde Oxidase-Mediated Pathway of drug metabolism has been already mentioned in Sect. 5.2.3. Barr and Jones also identified four further compounds that inhibit aldehyde oxidase with a linear mixed, predominantly catalytic mechanism [10]. 2,6-Dichlorophenolindophenol ($\alpha = 0.13$) was the compound with the mostly pronounced catalytic character, followed by β-estradiol ($\alpha = 0.14$), menadione ($\alpha = 0.16$) and ethinyl estradiol ($\alpha = 0.21$).

Adenosine Kinase is a pharmacological target for treating tuberculosis, which is unfortunately still a leading disease with increasing therapeutic challenges because of drug-resistant *Mycobacterium tuberculosis* strains. Out of several compounds investigated by Snášel and coworkers as inhibitors of adenosine kinase and characterized by their IC_{50} values, four nucleosides have been examined under the magnifying glass by determining their kinetic mechanism of inhibition. Three of these compounds behaved as LMx(Sp>Ca)I and one (name in Table 5.5) was a linear mixed, predominantly catalytic inhibitor [94, Fig. 1].

A Technically Impeccable Study is commented to conclude this section. Sun and coworkers investigated the mechanisms of inhibition of tyrosinase with the aim at suggesting a suitable approach to the determination of mechanisms and inhibition parameters. The authors globally fitted the equation of the general modifier mechanism by Botts and Morales to a whole ensemble of data and illustrated their results graphically. As a mean for discriminating among models, the correlation coefficient and the Akaike information criterion have been used. Importantly, all inhibition mechanisms have been named correctly! Among linear specific inhibition and linear mixed, predominantly specific inhibition, this study reported an inhibitor of tyrosinase that belongs to LMx(Sp<Ca)I discussed in this section, namely 4-acetylphenylthiourea, characterized by a reciprocal allosteric coupling constant $\alpha = 0.41$.

5.2.5 Linear Mixed, Balanced Inhibition

Assigned acronym: LMx(Sp=Ca)I. When the scales in Scheme 3.1 neither tip in favor of specific nor of catalytic inhibition, the inhibitory character of a modifier is *balanced* between the two extremes, a property described in Scheme 3.2 by $\alpha = 1$. The rate equation and the expressions of the kinetic parameters simplify considerably with respect to the two preceding cases as shown in the equation array (5.23)–(5.28).

The modifier affects the apparent catalytic constant, which decreases for increasing [X] (5.24) (right), while the Michaelis constant is independent of [X] (5.25). Accordingly, also the specificity constant decreases hyperbolically with increasing [X] (5.26) (left) and its reciprocal is a linear function of [X] (5.26) (right). The plots of the dependences of the parameters on modifier concentration are summarized in Fig. 3.3. The primary specific velocity plot consists of lines parallel to the abscissa (Fig. 5.7a), whose intersection points, plotted as secondary plot, describe a straight line with ordinate intersection = 1 and slope = K_X (Fig. 5.7b).

A characteristic property of the kinetic barrier diagrams in balanced inhibition is $\left(\Delta_E^0 - \Delta_{ES}^0\right) = \left(\Delta_E^X - \Delta_{ES}^X\right)$. These differences (see p. 165) correspond to Δ_S^0 and Δ_S^X, respectively, are the same at a given substrate concentration and can be positive or negative as illustrated by the facing arrowheads in Fig. 5.7f. Δ_S^0 and Δ_S^X are inversely proportional to the Michaelis constant in the absence and presence of modifier and, if they have the same value, the Michaelis constant is independent of [X]. Concurrently, $\Delta_{EX} = \Delta_{ESX}$ with the same value at any substrate concentration, indicated by double arrows in panel (f), because the specific and the catalytic components are equally represented ($\alpha = 1$). The outcome is that the degree of inhibition does not depend on substrate concentration.

This is the only mechanism to which the epithet *noncompetitive* applies as *pure noncompetitive inhibition* [70]. As discussed in Sect. 3.1, the term *balanced* will however be used consistently to avoid nomenclature conflicts with the two other mechanisms characterized by $\alpha = 1$ in the taxonomic tree of Fig. 3.2.

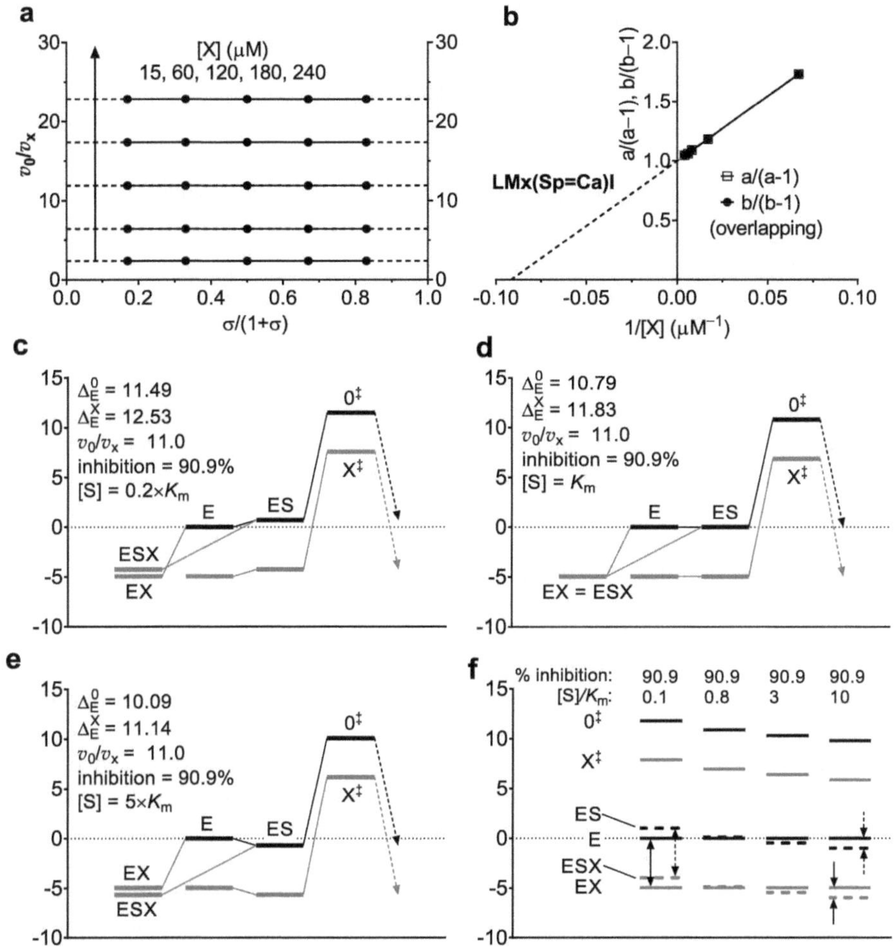

Fig. 5.7 Linear mixed, balanced inhibition: LMx(Sp=Ca)I. Primary (**a**) and secondary (**b**) specific velocity plot. The *arrow* in (**a**) indicates the increasing direction of [X]. (**c**)–(**e**) Kinetic barrier diagrams, (**f**) barcode-like representation of the various states at four substrate concentrations as indicated. $k_{cat} = 100\,s^{-1}$, $K_m = 200\,\mu M$, $K_X = 11\,\mu M$, $\alpha = 1.0$, in (**c**)–(**f**) $[X] = 10 \times K_X$. Common to panels (**c**)–(**f**): $\Delta_{EX} = \Delta_{ESX} = -4.96$ (*solid* and *dashed double arrows* in (**f**), respectively), $\Delta_{ES}^0 = 10.79$, $\Delta_{ES}^X = 11.83$. $\left(\Delta_E^0 - \Delta_{ES}^0\right)$ and $\left(\Delta_E^X - \Delta_{ES}^X\right)$ (the amplitudes between the *dashed* and the *solid facing arrows*, respectively), change with changing substrate concentrations but are equal at a given substrate concentration

In its strict definition, linear mixed, balanced inhibition is very rare and innumerable cases of *noncompetitive* inhibition in the literature correspond to one of the mixed mechanisms described above. Though, a few cases have been appropriately characterized as LMx(Sp=Ca)I; three of them are listed in Table 5.6 and will be commented below.

Table 5.6 Examples of linear balanced inhibition, LMx(Sp=Ca)I

Enzymes	EC no.	Inhibitors	Ref.
γ-Secretase[a]	3.4.23.–	L-685,458[b], a sulfonamide and two benzodiazepines	[101]
Aminopeptidase N	3.4.11.2	Deoxycholic and chenodeoxycholic acid[c]	[69]
Caspases[d]	3.4.22.–	Copper-containing compounds[e]	[30]

[a]Part of a complex membrane-bound molecular organization involving presenilins 1 and 2
[b]See Sect. 5.2.4
[c]Leu-*p*-nitroanilide as substrate
[d]Caspase-2 (EC 3.4.22.55), caspase-7 (EC 3.4.22.60), caspase-9 (EC 3.4.22.62)
[e]Codes NSC321205, NSC277584, NSC321206, and NSC310547 in the Chemical Repository of the Developmental Therapeutics Program of the National Cancer Institute

The fact that balanced inhibitors affect the limiting rate but not the Michaelis constant (5.24) and (5.25) has erroneously inspired the name *noncompetitive* for molecules that irreversibly inactivate enzymes. When in an enzymatic assay catalysis is blocked by an irreversibly binding substance, part of the enzyme is sequestered and only the remaining enzyme molecules can turn over substrates. Of course the limiting rate decreases in this case because $V = k_{cat}[E]_t$, but it is depletion of enzyme from the system that causes the apparent limiting rate to decrease, not any effect on the catalytic constant [21, p. 137]. In any case, before arriving at interpreting results of this type, one has first to collect data by measuring either substrate consumption or product appearance using an appropriate detection method. To do this, either enzyme and the modifier under investigation are preincubated for a given time and reaction is started by adding substrate, or enzyme is added to a solution containing substrate and modifier. Confounding enzyme inactivation with reversible balanced inhibition is likely to occur using the first method but not using the second. In fact, enzyme inactivation is usually a time-dependent process that is seen in progress curves as a single or double exponential followed by a straight line or a plateau, depending on modifier concentration [9].

$$v_X = \frac{\dfrac{k_2[E]_t}{1+\dfrac{[X]}{K_X}}[S]}{K_m^0 + [S]} \tag{5.23}$$

$$V^{app} = \frac{k_2[E]_t}{1+\dfrac{[X]}{K_X}}; \quad k_{cat}^{app} = \frac{k_2}{1+\dfrac{[X]}{K_X}} \tag{5.24}$$

$$K_m^{app} = K_m^0 = \frac{k_{-1}+k_2}{k_1} \tag{5.25}$$

$$k^{app} = \frac{k_1 k_2}{k_{-1}+k_2} \times \frac{1}{1+\dfrac{[X]}{K_X}}; \quad \frac{1}{k^{app}} = \frac{k_{-1}+k_2}{k_1 k_2}\left(1+\frac{[X]}{K_X}\right) \tag{5.26}$$

$$\frac{v_0}{v_X} = 1 + \frac{[X]}{K_X} \tag{5.27}$$

$$\frac{a}{a-1} = \frac{b}{b-1} = 1 + \frac{K_X}{[X]} \tag{5.28}$$

The Study on γ-Secretase by Tian et al. [101] mentioned in the two preceding sections included also occurrences of linear mixed, balanced inhibition. The inhibitors considered in this investigation shared structural similarities with similar albeit not identical modes of binding that may influence the coupling energy between substrate and inhibitor. Having established for all substances a mixed inhibitory character that varied between predominantly specific and predominantly catalytic, the finding of intermediate cases in which the two characters were equally represented is not surprising.

The Same Reasoning applies to the investigation by Nakanishi et al. on aminopeptidase N [69] acting on substrates with variable size of the C-terminal amino acid. After establishing the linear mixed, either predominantly specific or catalytic inhibition mechanisms (Sects. 5.2.3 and 5.2.4), the authors found that two bile acids behaved as linear mixed, balanced inhibitors when the substrate was Leu-*p*-nitroanilide.

Caspases are Inhibited by Copper-Containing Compounds as demonstrated in a study by Feldman et al. [30]. Four related, pyridinyl-based, copper-containing molecules were characterized as reversible inhibitors of the caspases-1,-2,-3, -7, -8, and -9. The crystal structure of caspase-7 in complex with compound NSC 321205 (Table 5.6 and Fig. 5.8b) was determined revealing a caspase-7 dimer per crystallographic asymmetric unit with two molecules of inhibitor bound to the solvent-exposed dimerization interface of the enzyme (Fig. 5.8a). Caspase-7 is still a dimer in the presence of the inhibitor while the same compound disrupts the caspase-8 dimer [30]. The active site Cys186, shown in Fig. 5.8a as spheres in both monomeric units, does not interfere directly with the bound inhibitor molecules. However, inhibitor binding provokes several distortions of the enzyme structure, which is kept in a nonproductive conformation.

The kinetic mechanism of inhibition of each caspase with each of the four compounds was analyzed using the specific velocity plot and was found to be either linear mixed or hyperbolic mixed balanced for the six caspases with the four inhibitors. For caspases-2, -7, and -9 the mechanism was LMx(Sp=Ca)I. For diagnosing the mechanism and calculating the kinetic parameters the authors used the specific velocity plot shown in Fig. 5.8c,d for caspase-7. The primary plot shown in panel (c) consists of a family of lines with slope zero or very close to zero, i.e., the degree of inhibition does not depend on substrate concentration, which is a criterion for mixed, balanced inhibition. The secondary plot, constructed with the intersections at abscissa 0 or 1 in the primary plot, intersects the ordinate at a value indistinguishable from 1 when the 95 % confidence intervals are considered. K_X, estimated from the slopes of the secondary plot was 0.37 μM or 0.40 μM

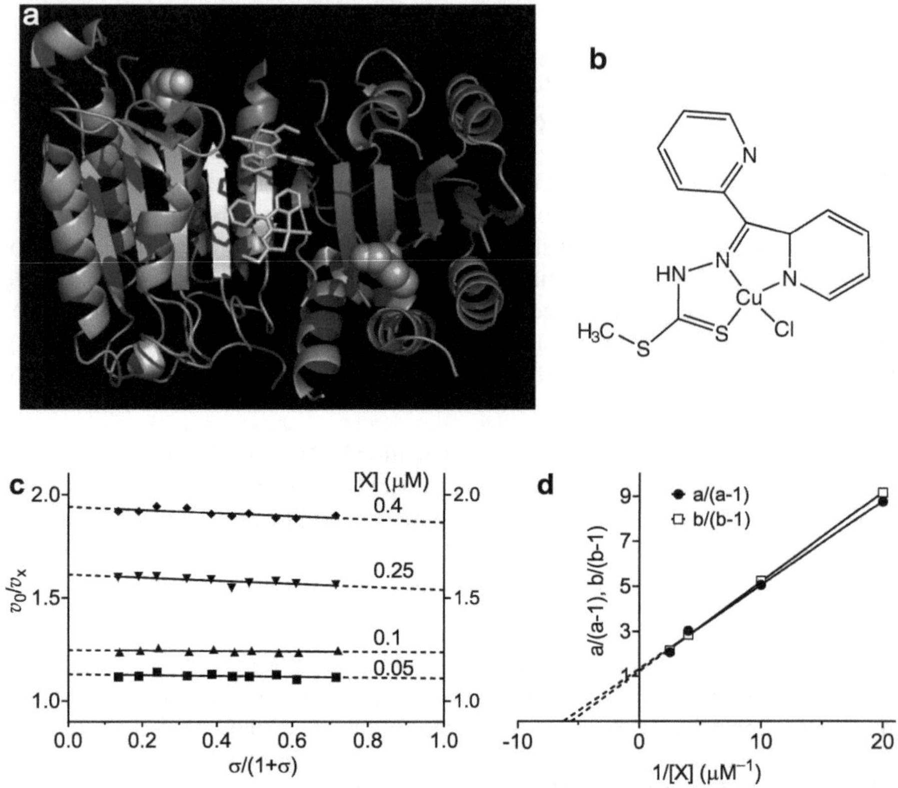

Fig. 5.8 Caspase-7 in complex with the allosteric inhibitor NSC 321205 (chemical repository of the developmental therapeutics program of the national cancer institute). (**a**) Structure of the complex with two molecules of inhibitor, represented as sticks, binding at the dimerization interface of the enzyme. The active site cysteines of the two caspase-7 monomers are shown as spheres. Accession number of the crystal structure of the complex: PDB ID code 4FEA. (**b**) Chemical formula of the inhibitor. (**c**) and (**d**) Primary and secondary specific velocity plots, respectively, for the inhibition of caspase-7 by NSC 32205. Original data were kindly provided by Taya Feldman and Xuejun Jiang, Memorial Sloan-Kettering Cancer Center, New York, U.S.A. [30]

from the $a/(a-1)$ and $b/(a-1)$ replots, respectively. The abscissa intercepts gave $K_X = 0.28$ and $0.32\,\mu M$. The authors estimated $K_X = 0.34\,\mu M$, which is the mean of the four values calculated from Fig. 5.8d.

To confirm that NSC 321205 binds caspase-7 equally well in the presence and absence of substrate, a property expected for the balanced inhibition mechanism, a separate experiment was performed by incubating the enzyme (10 nM) with an excess inhibitor (2 μM) in the presence and absence of the hexapeptide substrate DEVDGA (2 μM). Inhibitor and substrate in excess were removed by dialysis and enzyme activity was measured using the tetrapeptide fluorogenic substrate based on the sequence DEVD. The progress curves of the preincubates with/without the DEVDGA substrate were practically superimposable.

Linear balanced inhibition is certainly a very rare mechanism but this study shows, with a marriage between structure and kinetics, that it can really exist in particular systems. In the same study [30], using cell cultures Feldman et al. demonstrated that the copper-containing compounds inhibit apoptosis induced by multiple stimuli by arresting cellular caspase activation. With this result one can think that balanced inhibitors are better candidates than specific or predominantly specific inhibitors since their affinity for the target enzymes is not affected by the substrates.

5.3 Mechanisms in Taxonomic Level 2: Hyperbolic Inhibitors

The general kinetic equations for all mechanisms in this section correspond to the full expressions (3.2)–(3.5) and the equations of the specific velocity plot with its replots are (3.36)–(3.38). The diagnostic plots with the dependences of $k_{\mathrm{cat}}^{\mathrm{app}}$, $K_{\mathrm{m}}^{\mathrm{app}}$ and k^{app} are collected in Fig. 3.4 for the five members of this group. The reaction paths valid for this section and Sect. 5.4 are reproduced again here as Scheme 5.3.

5.3.1 Hyperbolic Specific Inhibition

Assigned acronym: HSpI. Hyperbolic specific inhibition has been recognized *as another plausible mechanism* by Segal, Kachmar and Boyer who proposed for it the name *apparent competitive* [88, p. 190]. Numerous examples of this mechanism are known, whose reaction paths are those of Scheme 5.3 with $1 < \alpha < \infty$ and $\beta = 1$. The primary specific velocity plot (Fig. 5.9a) looks very similar to that of linear specific inhibition (Fig. 5.1), but the difference is made by the slopes, which do not increase proportionally to modifier concentration leveling off to a constant value. This effect is better appreciated in the secondary plot (Fig. 5.9b), where the ordinate intercept is larger than 1. Since the combination of plot characteristics in Fig. 5.9 is unique for HSpI, the inhibition constant and α can be calculated (with

Scheme 5.3 Reaction paths common to all mechanisms in Sects. 5.3 and 5.4

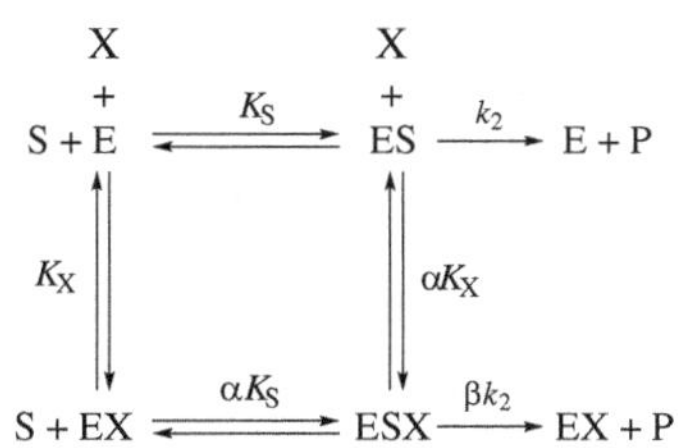

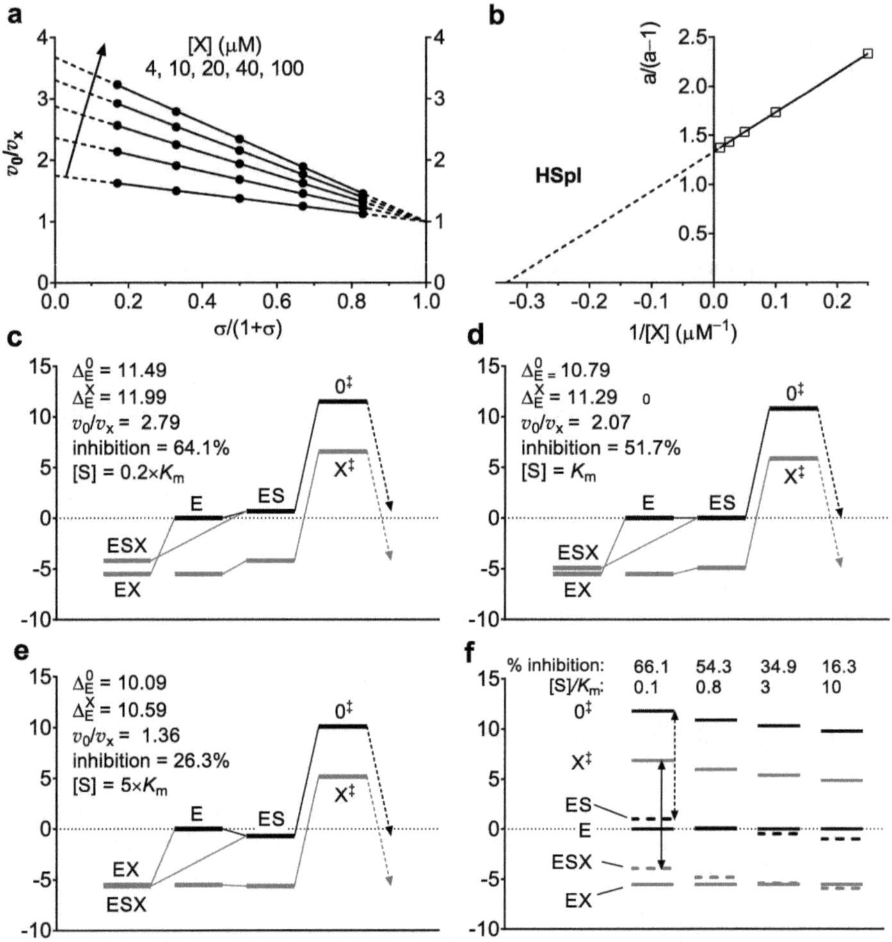

Fig. 5.9 Hyperbolic specific inhibition: HSpI. Primary (**a**) and secondary (**b**) specific velocity plot. The *arrow* in (**a**) indicates the increasing direction of [X]. (**c**)–(**e**) Kinetic barrier diagrams, (**f**) barcode-like representation of the various states at four substrate concentrations as indicated. $k_{cat} = 100\,s^{-1}$, $K_m = 200\,\mu M$, $K_X = 3\,\mu M$, $\alpha = 4.0$, $\beta = 1.0$, in (**c**)–(**f**) $[X] = 10 \times K_X$. Constant values common to (**c**)–(**f**): $\Delta_{EX} = -5.52$, $\Delta_{ESX} = -4.92$. In (f) the *solid double arrow* represents Δ_{ES}^{X} and the *dashed double arrow* is Δ_{ES}^{0}. At any substrate concentration $\Delta_{ES}^{X} = \Delta_{ES}^{0}$

$\beta = 1$) from the slope and abscissa intercept of the secondary plot as shown in Fig. 3.9b.

At a fixed modifier concentration, the degree of inhibition decreases with increasing [S] (Fig. 5.9c–f). The kinetic barrier diagrams differ from those of linear specific inhibition for the presence of the ESX complex, which is catalytically active but with the characteristic property of transforming substrate into product at the same rate as the ES complex. Accordingly, $\Delta_{ES}^{X} = \Delta_{ES}^{0}$, i.e., the catalytic constant is the same in the presence and absence of modifier and does not depend on [X]. The

Table 5.7 Examples of hyperbolic specific inhibition, HSpI

Enzymes	EC no.	Inhibitors	Ref.
NAD(P)H oxidase	1.6.3.1	NAD_{ox}	[111]
Sucrose-phosphate synthase	2.4.1.14	Inorganic phosphate	[26]
Thrombin	3.4.21.15	Hirudin fragments	[22, 99]
DNA helicase	3.6.4.12	Biphenylsulfonacetic acid derivatives	[106]
Acetylcholinesterase	3.1.1.7	Propidium iodide	[11]
Cathepsin K	3.4.22.38	Methylamine derivative[a]	[73]

[a] 1-[1-(2-Phenylethyl)-4-piperidinyl]-methanamine

net effect of the inhibitor is to reduce the fraction of enzyme available for catalysis by trapping it in the EX complex that is however not a dead-end complex as in LSpI.

Hyperbolic specific inhibition can be misidentified as linear specific inhibition if the effects of sufficiently large modifier concentrations are not explored. It can thus not be excluded that hyperbolic specific inhibition is more frequent than realized. In view of the possibility of interacting on the surface of the same enzyme molecule with other allosteric modifiers, a closer look at *competitive* inhibitors is worthwhile. The multifaceted interactions that can occur between two hyperbolic modifiers acting simultaneously on the same enzyme will be treated in Chap. 6.

Some examples of HSpI are summarized in Table 5.7 and discussed below.

Allosteric Interactions in the NAD(P)H Oxidase of *Mycobacterium tuberculosis* have been studied shortly after the publication of Monod, Changeux, and Jacob on allosteric proteins [66] by Worcel and coworkers (including W. W. Cleland). AMP was identified as allosteric activator [110] and NAD_{ox} as inhibitor [111]. The allosteric nature of the NAD_{ox}-interaction was proposed on the basis of the kinetic mechanism, hyperbolic specific inhibition.

Sucrose-Phosphate Synthase catalyzes the synthesis of sucrose from fructose-6-phosphate and uridine diphosphoglucose. While the reaction is allosterically activated by glucose-6-phosphate, activation is contrasted by inorganic phosphate [25]. In this two-substrate reaction, inorganic phosphate behaves as a hyperbolic specific inhibitor independently of whether fructose-6-phosphate or uridine diphosphoglucose is kept constant while varying the other substrate in the presence of a constant concentration of the activator glucose-6-phosphate [26]. From this and other results the authors concluded that inorganic phosphate interacts with both the enzyme's catalytic site and with an allosteric site. However, when glucose-6-phosphate is bound at its activation site, phosphate interacts only with the catalytic site.

The Hirudin Variant N^{α}-acetyl desulfo hirudin$^{45-65}$ inhibits thrombin with a hyperbolic specific mechanism ad demonstrated by DiMaio et al. [22]. Szewczuk and coworkers found the same mechanism in further variants of the 45–65 hirudin segment containing peptide spacers of different length and composition [99].

Infection by Papillomaviruses of squamous and mucosal epithelial cells can have consequences spanning from benign plantar warts, to anogenital warts and cervical neoplasia. Papillomaviruses use their small circular double-stranded DNA for encoding eight proteins responsible for viral pathogenesis. One of these proteins, the E1 DNA helicase, utilizes the energy from ATP hydrolysis to unwind double-stranded DNA, a necessary process for viral replication and a target for antiviral therapy. White et al. characterized DNA helicase inhibitors based on biphenyl-sulfonacetic acid that are particularly efficient against the DNA helicase of the human papillomavirus HPV6 [106]. The authors studied the kinetic mechanism of inhibition of nine compounds against the helicases from HPV6, HPV11, and HPV18 and determined the common mechanism as hyperbolic specific inhibition.

Propidium Iodide Inhibits Acetylcholinesterase While analyzing the modes of interaction between acetylcholinesterase and several cationic ligands, Berman and Leonad observed that propidium iodide behaved as linear mixed, predominantly specific inhibitor when the substrate was acetylthiocholine (a charged molecule), but inhibition changed to hyperbolic specific in the presence of the uncharged substrate 7-hydroxy-4-methylcoumarin [11]. Propidium binds to a peripheral anionic site of the enzyme acting as allosteric modifier located at more than 20 Å from the active site. Occupation of this site results in linear inhibition of the enzymatic acylation step with ionic substrates and hyperbolic inhibition with uncharged acetyl ester substrates.

Several Modification Mechanisms were identified by Marko Novinec and collaborators in a project aimed at targeting allosteric sites at the surface of cathepsin K using computational methods. A methylamine derivative, one of several investigated compounds, behaved as hyperbolic specific inhibitor [73]. See also Sects. 5.3.2, 5.3.3 and the discussion in Sect. 5.4.1.

5.3.2 Hyperbolic Mixed, Predominantly Specific Inhibition

Assigned acronym: HMx(Sp>Ca)I. In Scheme 5.3, this mechanism is characterized by $1 < \alpha < \infty$ and $0 < \beta < 1$ with a specific velocity plot resembling that of linear mixed, predominantly specific inhibition. However, the secondary plots intersect the ordinate at values larger than one, which is diagnostic of the hyperbolic character (Fig. 5.10). In a number of published studies, this hyperbolic inhibition mechanism, in which the specific component dominates, is often disguised under the generic epithet noncompetitive. However, the identification is possible when graphical proofs or lists of kinetic parameters are provided.

In the barrier diagrams of HMx(Sp>Ca)I, Fig. 5.10c–f, the distinguishing features versus other hyperbolic inhibition mechanisms are Δ-values that differ from one another at any [S] and the degree of inhibition decreasing with increasing [S].

A choice of examples is listed in Table 5.8 and briefly commented below in the same order of appearance in the table. HMx(Sp>Ca)I appears to be a widespread

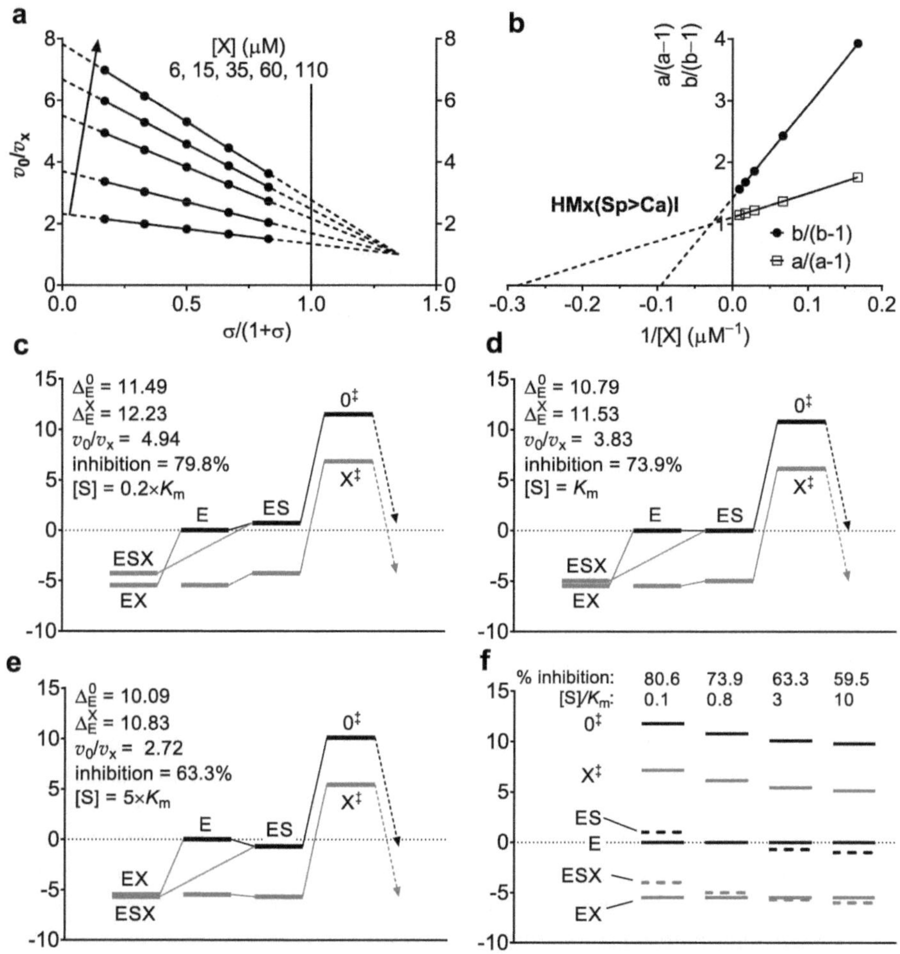

Fig. 5.10 Hyperbolic mixed, predominantly specific inhibition: HMx(Sp>Ca)I. Primary (**a**) and secondary (**b**) specific velocity plot. The *arrow* in panel (**a**) indicates the increasing direction of [X]. (**c**)–(**e**) Kinetic barrier diagrams, (**f**) barcode-like representation of the various states at four substrate concentrations as indicated. $k_{cat} = 100\,\mathrm{s}^{-1}$, $K_m = 200\,\mu\mathrm{M}$, $K_X = 3.5\,\mu\mathrm{M}$, $\alpha = 3.0$, $\beta = 0.3$, in (**c**)–(**f**) $[X] = 10 \times K_X$. Constant values common to (**c**)–(**f**): $\Delta_{ES}^0 = 10.79$, $\Delta_{ES}^X = 11.13$, $\Delta_{EX} = -5.46$, $\Delta_{ESX} = -4.98$. The degree of inhibition decreases for increasing [S] and there are no Δ-values that differ from one another in a special way at any [S]

mechanism in enzyme systems from several branches of the biological sciences. Further examples, not listed in Table 5.8, are for instance product inhibition of cellulose hydrolysis catalyzed by cellulose 1,4-beta-cellobiosidase (EC 3.2.1.91) [12], inhibition of ethanolamine-phosphate phospho-lyase (EC 4.2.3.2) by the reaction product phosphate [84], inhibition of electric eel acetylcholinesterase (EC 3.1.1.7) by triarylmethane dyes [54], inhibition of elastase-2 (EC 3.4.21.37), cathepsin G (EC 3.4.21.20) and myeloblastin (EC 3.4.21.76) by suramin (a hexasulfonated

Table 5.8 Examples of hyperbolic mixed, predominantly specific inhibition, HMx(Sp>Ca)I

Enzymes	EC no.	Inhibitors	Ref.
Sterol esterase	3.1.1.13	5-benzylidenerhodanine derivative	[42]
Acetylcholinesterase	3.1.1.7	Tetracyclic thienopyrimidines	[38]
ADAMTS13 endopeptidase[a]	3.4.24.87	cVWF63[b]	[37]
Linoleate 13S-lipoxygenase	1.13.11.12	Oleyl sulfate[c]	[65]
Elastase-2	3.4.21.37	Sulfated glycosaminoglycans	[52]
Kallikrein-related peptidase 7[d]	3.4.21.117	Antileukoproteinase	[34]
Aminopeptidase N	3.4.11.2	Cholic and chenodeoxycholic acid[e]	[69]
Cathepsin K	3.4.22.38	Three aromatic compounds[f,g,h]	[73]
Caspase-2	3.4.22.55	Designed ankyrin repeat AR_F8	[87]

[a]Recombinant ADAMTS13 truncated after the spacer domain
[b]C-terminal cleavage product of Von Willebrand factor
[c](Z)-9-octadecenyl sulfate
[d]Synonym: stratum corneum chymotryptic enzyme
[e]Leu-Gly as substrate
[f]3-Amino-4-[(tetrahydro-2-furanylmethyl)amino] benzoic acid
[g]Piperidin-2-yl-[2-(trifluoromethyl)-6-[4-(trifluoromethyl)phenyl]pyridin-4-yl] methanol
[h]*N*-Benzoyl-*p*-aminosalicylic acid

naphtylurea) [17], and aminopyrine N-demethylase[2] by the antimalarial agent primaquine [68].

Atherosclerosis and Coronary Heart Disease Sterol esterase catalyzes the hydrolysis of cholesterol esters into cholesterol and represents a pharmacological target in the treatment of hypercholesterolemia, a factor of atherosclerosis and coronary heart disease. Among several inhibitors developed against this enzyme, a group of rhodanine- and thiazolidinedione-based molecules have been characterized by Heng et al. [42]. Besides listing IC_{50} values, the authors performed a detailed analysis of the inhibition mechanism of a 5-benzylidenerhodanine derivative that resulted to be a hyperbolic mixed, predominantly specific inhibitor. One more rhodanine- and two thiazolidinedione-based derivatives were found to be hyperbolic inhibitors with inhibition constants in the low micromolar range.

The Cholinergic Hypothesis preceded the amyloid and other pathogenetic hypotheses of Alzheimer's disease. It is based on reduced synthesis of the neurotransmitter acetylcholine, for which a treatment based on inhibition of cholinesterase received approval for human use [33]. Among eight reversible inhibitors of acetylcholinesterase by tetracyclic thienopyrimidine-derivatives investigated by González-Tanarro and Gütschow [38], four substances could be identified as belonging to HMx(Sp>Ca)I. The predominantly specific character of the four inhibitors obeying the HMx(Sp>Ca)I mechanism with electric eel acetylcholinesterase was characterized by α-values between 1.5 and 2.0 while β

[2]A microsomal cytochrome P-450-dependent activity.

ranged from 0.12 and 0.17. In the opinion of the authors these compounds were weak modifiers with respect to other cholinesterase inhibitors. In general, however, low values of α suggest less competition with substrates, a property that may prove useful in a natural environment provided this behavior is maintained in the presence of natural substrates.

Platelet Adhesion to injured blood vessels mediated by the Von Willebrand factor is controlled by proteolytic intervention of the metallopeptidase ADAMTS13 on the factor. Proteolysis at a single locus is regulated by interaction between specific domains of the two proteins: a spacer domain of ADAMTS13 binds to an exosite of the factor and deletion of either one of these domains drastically reduces reaction rate. The proteolytic event is promoted by mechanical forces originating from flowing blood that triggers unfolding of the susceptible von Willebrand factor exosite exposing the scissile bond to the action of ADAMTS13. A polypeptide fragment of the von Willebrand factor containing the exosite behaves as a hyperbolic mixed inhibitor with predominantly specific character ($\alpha = 5.4$, $\beta = 0.12$). This result of *product inhibition* confirms the role of the extended interaction moieties of ADAMTS13 and von Willebrand factor in the feedback control mechanism to prevent microvascular thrombosis during blood coagulation [37].

The Lipoxygenases contain nonheme iron to catalyze the dioxygenation of polyunsaturated fatty acids with a cis,cis-1,4-pentadiene structure. These enzymes are pharmacological targets for their implication in inflammatory states, atherosclerosis, asthma, and cancer. In a study that explored the inhibition of two lipoxygenases by (Z)-9-octadecyl sulfate, an oleic acid derivative, Mogul et al. [65] observed hyperbolic mixed, predominantly specific inhibition ($\alpha = 4.6$, $\beta = 0.85$) and proposed the existence of an allosteric site in linoleate 13S-lipoxygenase and possibly also in arachidonate 15-lipoxygenase (EC 1.13.11.33).

Enzymes with Positively-Charged Residues at their Surface can interact with negatively charged molecules via electrostatic interactions. Owing to a high isoelectric point and to a characteristic distribution of positively-charged residues at the surface of the enzyme, the serine peptidase elastase-2 interacts with negatively-charged molecules such as glycosaminoglycans. This property, shared with two related enzymes of human polymorphonuclear leukocytes, cathepsin G and myeloblastin, determines interactions spanning from activation, inhibition, or inhibition followed by reactivation as a function of modifier concentration. Elastase-2 binds glycosaminoglycans resulting in hyperbolic mixed inhibition with either predominating specific [52] or balanced character [6, 7].

Desquamation of the Skin is accomplished by the migration of corneocytes toward the skin surface followed by cell shedding. The process requires proteolysis by the kallikrein-related peptidase, which is regulated in both physiological and pathological conditions by the inhibitor antileukoproteinase. In vitro, using a synthetic tripeptide substrate, Franzke and coworkers demonstrated that the mechanism of inhibition was hyperbolic mixed with predominantly specific character ($\alpha = 5.6$, $\beta = 0.16$) [34]. Refinement of a preliminary K_X-value estimated with the specific velocity plot was performed by taking into account tight-binding, due to experimental conditions, as described in Sect. 4.4.

Aminopeptidase N commented above for being inhibited by bile acids through the mechanisms LMx(Sp>Ca)I, LMx(Sp<Ca)I, and LMx(Sp=Ca)I is inhibited by cholic and chenodeoxycholic acids in the presence of the Leu-Gly dipeptide as substrate with HMx(Sp>Ca)I as mechanism [69]. The authors exploited the pooled kinetic information to formulate a hypothesis about the relative positions occupied by substrates and bile acids on the enzyme.

Human Cathepsin K in the project by Novinec et al. mentioned above, dedicated to targeting allosteric sites of human cathepsin K [73], was inhibited by three compounds via the hyperbolic mixed, predominantly specific inhibition mechanism (compounds 1–3, names in Table 5.8). The goal of this study was not optimizing the efficiency of the compounds targeted to the enzyme but at demonstrating the *existence* of allosteric sites in cathepsin K. Despite the modest efficiency of compounds 1–3 ($K_X = 100 - 190\ \mu$M) the goal was reached showing that modifiers can bind at defined patches located at the surface of the enzyme, which communicate with the active site through an evolutionarily-conserved network of residues. More details on this project are in Sect. 5.4.1.

Designed Ankyrin Repeat Proteins have been developed at the University of Zurich in the laboratory of Andreas Plückthun by a team of talented young researchers who I wish to mention: Patrik Forrer, Kaspar Binz, Michael Stumpp, Patrick Amstutz, and Christian Zahnd [3, 14, 15, 32, 51].

The modular architecture of the ankyrin repeat proteins has been exploited as a strategy for the production of specifically targeted modifiers in protein–protein interactions. Stacked units containing usually 33 amino acids form the domains of ankyrin repeats. A β-turn in each unit is followed by two antiparallel helices and a loop that joins the β-turn of the next repeat. After generating combinatorial libraries, ankyrin repeat modules are produced that contain fixed residues and randomly mutated amino acids that constitute binding contacts. Two-three or more of these fundamental blocks are cloned between N-terminal and C-terminal repeats as capping elements. Randomization of mutated amino acids in adjacent repeats makes it possible to create interaction surfaces with any target molecule. Ankyrin repeat proteins are very stable, do not contain disulfide bridges representing an alternative strategy to antibodies whose stability depends on intact Cys–Cys bridges. Notably, ankyrin repeats are ideal binders in reducing environments as the cytosol.

A project between groups of the University of Zurich was dedicated to the production of designed ankyrin repeats specifically targeted to *caspases*, acronym for cysteine-dependent aspartyl-specific peptidases. These enzymes play a role in initiating and executing cell death by apoptosis and have been implicated in inflammation and cancer. The crystal structure of caspase-2 in complex with an ankyrin repeat protein, identified by the code AR_F8, has been solved and the kinetic mechanism of the caspase-2/AR_F8 interaction complemented the structural results [87]. Preliminary kinetic surface plasmon resonance measurements with

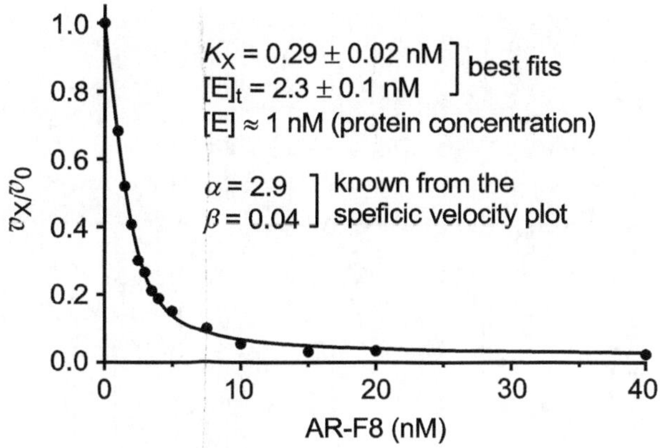

Fig. 5.11 Tight-binding inhibition of caspase-2 by the ankyrin repeat protein AR_F8. Initial rates measured at a fixed substrate concentration and variable modifier concentrations (*black dots*) to which (4.35) was fitted using α and β as constant parameters, while K_X and $[E]_t$ were treated as parameters to be fitted [87]

immobilized caspase-2 yielded a dissociation constant of 4.1 nM for AR_F8. Thus, the tight-binding condition was circumvented while investigating the kinetic mechanism with the specific velocity plot by keeping the enzyme concentration at 5 nM and using AR_F8 concentrations in the range 25–400 nM. AR_F8 inhibited caspase-2 by the hyperbolic mixed, predominantly specific mechanism with $\alpha = 2.9 \pm 0.8$ and $\beta = 0.04 \pm 0.01$, values that were reproducible in several independent experiments.

Refinement of the kinetic data was completed by measurements at constant substrate concentration and variable modifier concentrations and fitting the tight-binding equation (4.35) to data (Fig. 5.11). Knowing the mechanism, α and β, these two parameters were considered as known constants while K_X and $[E]_t$ were left free to float because a suitable titrant for caspase-2 was not available. The best-fit concentration was 2.3 ± 0.1 nM of enzyme active sites and the estimated concentration from protein content was $\approx$ 1 nM (as dimer). This means that the stoichiometry of the interaction between caspase-2 and AR_F8 is one molecule of modifier per enzyme monomer.

When fitting equations to data, particular attention should be paid with the number of fitted parameters compared with the amount of data available. Constraining one or more parameters while running least-squares minimization is not only advisable but also necessary in some cases. With (4.35) it is impossible to consider *all parameters* as unknowns. Notably, the two inhibition constants $K_X \equiv K_{Sp}$ and $\alpha K_X \equiv K_{Ca}$ are *intertwined* and fitted curves neatly overlapping the data can be obtained with different combinations of the two constants. The way out of the problem is to use the specific velocity plot to calculate *the ratio* of the two constants, i.e., $K_{Ca}/K_{Sp} = \alpha$, and to use this value as a constant in (4.35) to obtain statistically sound estimates of the sought parameters.

The kinetic analysis was made more than 1 year prior the determination of the three-dimensional structure of the caspase-2/AR_F8 complex, which confirmed the allosteric nature of the interaction and the stoichiometry [87] (Fig. 5.12).

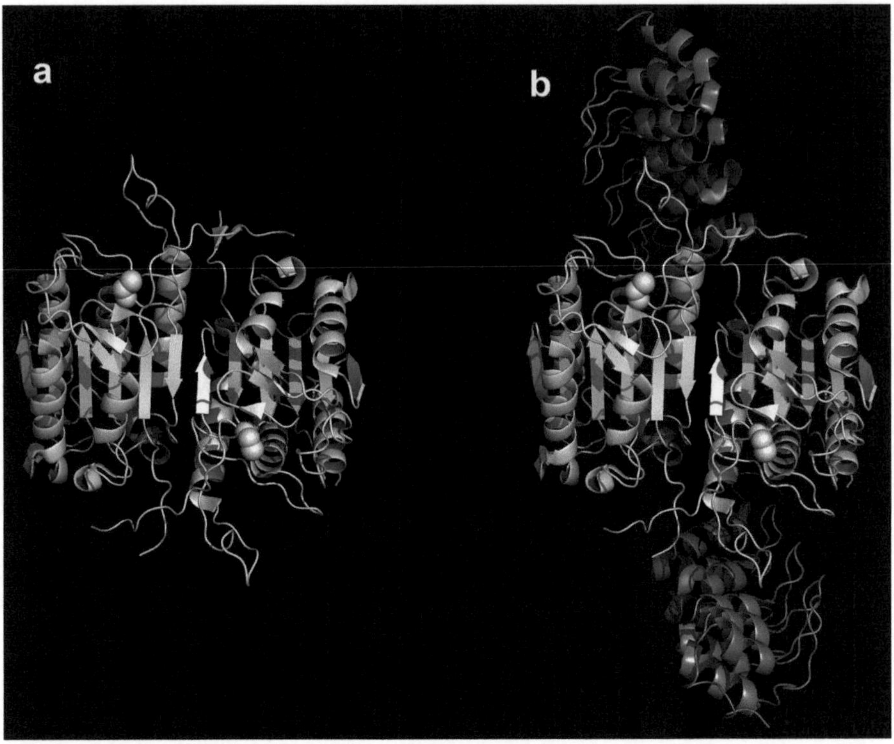

Fig. 5.12 Structure of human caspase-2. (**a**) The caspase-2 dimer represented as *ribbons* and the active site cysteines represented as *spheres*. (**b**) Caspase-2 in complex with the ankyrin repeat protein AR_F8 showing one molecule of the modifier (*dark ribbons*) bound per enzyme monomer. AR_F8 binds with high specificity to two flexible regions of caspase-2 located at the enzyme surface without involving the catalytic cysteine residues [87]

5.3.3 Hyperbolic Catalytic Inhibition

Assigned acronym: HCaI. The name of this mechanism derives from the plots of the apparent parameters against modifier concentration and the rate equation that are similar to those of linear catalytic inhibition. In the general modifier mechanism (Scheme 5.3), HCaI is characterized as $0 < \alpha,\ \beta < 1$, and $\alpha = \beta$. As in linear catalytic inhibition, the specific velocity plot consists of straight lines with positive slope intersecting at a common point when the substrate concentration is extrapolated to zero (Fig. 5.13a). If a sufficient range of modifier concentrations is explored, the slopes of this plot will change very little by increasing [X] till reaching an upper limit. This property is also evident in the secondary plot (Fig. 5.13b), where the ordinate intercept is larger than one and from which $\alpha = \beta$ can be calculated.

Noteworthy properties of the kinetic barrier diagrams are the values of $\Delta_{\mathrm{E}}^{\mathrm{X}}$ and Δ_{E}^{0}, which change for every value of [S] but always $\Delta_{\mathrm{E}}^{\mathrm{X}} = \Delta_{\mathrm{E}}^{0}$. This feature is shown in Fig. 5.13f as dashed double arrow and solid double arrow, respectively.

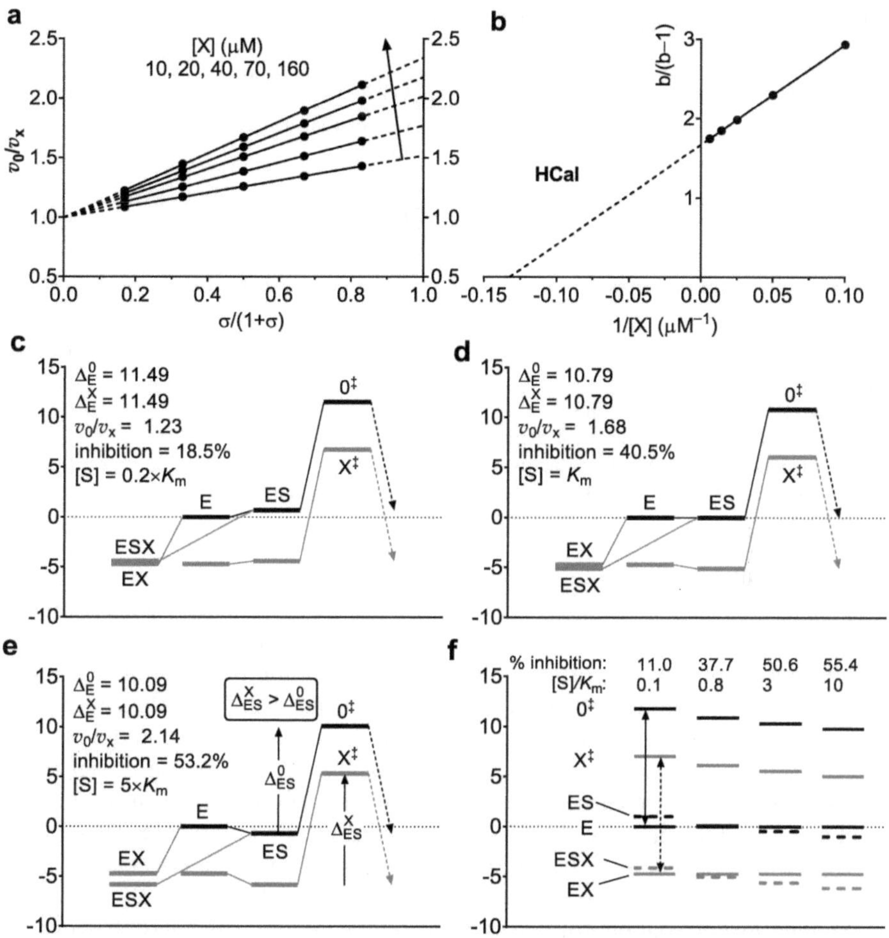

Fig. 5.13 Hyperbolic catalytic inhibition: HCaI. Primary (**a**) and secondary (**b**) specific velocity plot. The *arrow* in (**a**) indicates the increasing direction of [X]. (**c**)–(**e**) Kinetic barrier diagrams, (**f**) barcode-like representation of the various states at four substrate concentrations as indicated. $k_{cat} = 100\,s^{-1}$, $K_m = 200\,\mu M$, $K_X = 19\,\mu M$, $\alpha = 0.4$, $\beta = 0.4$, in (**c**)–(**f**) $[X] = 10 \times K_X$. Constant values common to (**c**)–(**f**): $\Delta_{ES}^0 = 10.79$, $\Delta_{ES}^X = 11.17$, $\Delta_{EX} = -4.72$, $\Delta_{ESX} = -5.12$. In panel (**f**) the *solid double arrow* represents Δ_E^0 and the *dashed double arrow* is Δ_E^X; their values change with substrate concentration but always is $\Delta_E^0 = \Delta_E^X$. Note that Δ_{ES}^X is larger than Δ_{ES}^0 at any substrate concentration (shown as example in panel **e**)

That inhibition should result from the kinetic barrier diagrams can be deduced from the relative heights of the kinetic barriers for the transition states $0^\ddagger$ and $X^\ddagger$, which maintain a constant difference at any substrate concentration with $\Delta_{ES}^X > \Delta_{ES}^0$. Δ_{ES}^X depends on k_{cat}^{app} that contains β, which is < 1 in HCaI: a higher kinetic barrier in the presence of modifier corresponds to a greater activation energy that means inhibition.

Table 5.9 Examples of hyperbolic catalytic inhibition, HCaI

Enzymes	EC no.	Inhibitors	Ref.
Glutamate dehydrogenase (NAD(P)$^+$)	1.4.1.3	GTP	[35]
Cathepsin B	3.4.22.1	Cyclopalladated complex	[13]
4-Hydroxy-tetrahydrodipicolinate synthase	4.3.3.7	Lysine	[24]
NADH:ubiquinone reductase[a]	1.6.5.9	Flavone[b]	[103]
Thrombin	3.4.21.15	Hirudin(45–65) fragment	[85]
Glutathione-disulfide reductase	1.8.1.7	Safranin & menadione derivatives[c]	[59]
Penicillin amidase	3.5.1.11	Polyethylene glycol	[48]
Cathepsin K	3.4.22.38	Two compounds[d,e]	[73]

[a]Nonelectrogenic
[b]2-phenyl-4H-1-benzopyran-4-one
[c]Safranin O, thionin O, 2-anilino-1,4-naphtoquinone, LY-83583
[d]2-[(3-Nitrophenyl)carbamoyl]benzoic acid
[e]Methyl N-[1-(4-methoxyphenyl)-2,5-dioxopyrrolidin-3-yl]glycinate

Occasionally the concept of hyperbolic catalytic inhibition, with the characteristics discussed in this section, has been applied inappropriately, with confusion deriving from nomenclature and ambiguity in the reaction scheme. A cause of misconception is naming *partial uncompetitive inhibition* the reaction scheme of linear catalytic inhibition (Scheme 3.2) in which the ESX complex is catalytically competent. Without further specification, this mechanism should neither be addressed to as *inhibition*, nor be called *uncompetitive*. The reason is that this is nothing else than the hyperbolic mixed, dual modification mechanism in the third level of Fig. 3.2 labeled HMxD(A/I). This is characterized by $0 < \alpha,\ \beta < 1$ and $\alpha < \beta$, which results in activation below the critical substrate concentration and inhibition above this value as will be discussed in Sect. 5.4.4.1.

Among several examples of hyperbolic catalytic inhibition that have been appropriately characterized, the selection summarized in Table 5.9 will be commented.

Glutamate Dehydrogenase catalyzes the oxidative deamination of L-glutamate into 2-oxoglutarate and ammonia (5.29). In 1963 Frieden published a study with emphasis on the correlation between molecular and kinetic properties of beef liver glutamate dehydrogenase [35]. In the reaction scheme used by Frieden, coenzyme and modifier are added first, followed by substrate addition. The reaction paths recast from the six individual reactions in the original publication are shown in Scheme 5.4 using symbols of this book. The scheme is reminiscent of the general modifier mechanism, in this case extended to a reaction with two substrates and two products. This way of representing the reaction communicates directly the idea of allosteric coupling between coenzyme and modifier, which is necessary to interpret the behavior of GTP and ADP as modifiers that were investigated by Frieden.

$$\text{L-glutamate} + H_2O + \text{NADP}_{ox} \rightleftarrows \text{2- oxoglutarate} + \text{ammonia} + \text{NADP}_{red} \quad (5.29)$$

Scheme 5.4 Reaction paths for the glutamate dehydrogenase reaction. Adapted from Frieden [35], C = coenzyme

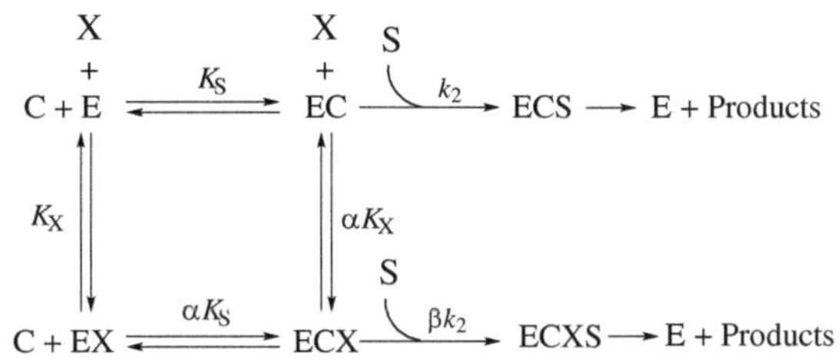

The reaction was measured in the direction of $NADP_{red}$ oxidation at saturating concentrations of 2- oxoglutarate and ammonia and variable $NADP_{red}$. GTP behaved as a hyperbolic catalytic inhibitor and ADP as a hyperbolic catalytic activator, which is discussed here in place of Sect. 5.4.2. The rates in the presence of modifiers were measured at only one concentration of the modifiers and five concentrations of the reduced coenzyme. The estimated inhibition constants for ADP were $K_X = 2\text{–}4\ \mu M$ and $\alpha K_X = 10\text{–}20\ \mu M$. Frieden observed that the condition for obtaining *uncompetitive behavior* is $\alpha = \beta$ (in the original: $k_2/k_1 = K_3/K_2$) [35, Eq. 1a]). Thus, from K_X and αK_X we can estimate for the activation of glutamate dehydrogenase by ADP $\alpha = \beta \approx 5$, in line with the character of HCaA.

The Cysteine Peptidase Cathepsin B has been implicated in several pathological states including tumor invasion. Bincoletto and coworkers synthesized organometallic compounds containing palladium and demonstrated that administering one of the cyclopalladates to Walker-tumor bearing rats reduced by 90 % tumor growth [13]. The authors also performed inhibition experiments of a putative agent of tumor invasion, cathepsin B, observing that the specificity constant was the same in the presence and absence of the inhibitor. Basing their calculations on the general modifier mechanism they calculated $\alpha = 0.19 \pm 0.02$ and $\beta = 0.18 \pm 0.02$. The authors did not name the mechanism but from their clear description their results can be assigned to hyperbolic catalytic inhibition.

4-Hydroxy-Tetrahydrodipicolinate Synthase catalyzes, in bacteria and plants, the first committed step of the lysine biosynthesis pathway by condensation of (S)- aspartate-4-semialdehyde with pyruvate. In bacteria, this step is necessary in the process of cross-linking elements of the cell wall and, since the synthase is not expressed in animals, inhibition of the enzyme is interesting as a target for antibiotic therapy and for the development of herbicides. The reaction follows a substituted-enzyme mechanism and lysine behaves as allosteric feedback inhibitor with respect to both substrates. Feedback inhibition of 4-hydroxy-tetrahydrodipicolinate synthase occurs in Gram-negative bacteria and plants but not in Gram-positive bacteria.

With the enzyme from *Escherichia coli* and pyruvate as the varied substrate, lysine was found by Dobson et al. to obey the hyperbolic catalytic inhibition mechanism [24], called by the authors *mixed partial inhibition*, whereas the enzyme from *Vitis vinifera* displayed a hyperbolic mixed inhibition mechanism with predominantly specific character [4]. That the same enzyme from different species shows different inhibition mechanisms by lysine is not surprising after considering that the bacterial and plant synthases have alternative architectures, as seen in the crystal structures (commented in [4]).

Electrons from $\mathbf{NAD_{red}}$ are transferred to ubiquinone in complex I in most eukaryotes, whereby protons are translocated across the inner mitochondrial membrane to generate an electrochemical potential used to synthesize ATP. However, the mitochondria of the yeast *Saccharomyces cerevisiae* do not possess complex I and electrons are transferred from NAD_{red} to ubiquinone in a process catalyzed by the NADH:ubiquinone reductase without proton translocation across the membrane. The enzyme is therefore labeled as *nonelectrogenic*.[3] Since the NADH:ubiquinone reductase from yeast can reestablish NAD_{red} oxidation in mammalian cells deficient in complex I, knowing the properties of the enzyme is relevant to gene therapy. In this frame, Velásquez and Pardo [103] explored the effects of reaction products on the rate as well as the effects of flavone while investigating the substituted-enzyme kinetic mechanism of NADH:ubiquinone reductase. Flavone behaved as hyperbolic catalytic inhibitor, HCaI, with respect to the first substrate NAD_{red} and as hyperbolic balanced inhibitor, HMx(Sp=Ca)I, with respect to the second substrate (either a quinone or the artificial electron acceptor 2,6-dichlorophenolindophenol, Sect. 5.3.5).

The Inhibition of Thrombin by hirudin fragments was commented in Sect. 5.3.1. In the same study Schmitz et al. found that the C-terminal fragment, hirudin(45–65), is a hyperbolic catalytic inhibitor, contrary to the N-terminal fragment which is a linear specific inhibitor [85]. Differences in this behavior in the presence of a synthetic tripeptide substrate or fibrinogen led the authors to formulate hypotheses about the binding mode of hirudin to thrombin: the N-terminal part of hirudin interacts with the active center and blocks catalysis, whereas the C-terminal part interferes with parts of the thrombin molecule containing an anion-binding locus used to interact with thrombomodulin and the blood clotting factors V and VII.

Antimalarial Strategies include the inhibition of glutathione-disulfide reductase. Since glutathione reductase is a dimer, the contact interface between the two subunits has been explored as a site for binding allosteric modifiers that might affect enzyme activity. Indeed safranin O and menadione bind in a cavity of this interface in glutathione-disulfide reductase from human erythrocytes. Lüönd et al. [59], investigating the mechanism of inhibition of the enzyme from yeast by several analogues of safranin O and menadione, observed that safranin O, thionin O as well as the menadione analogues 2-anilino-1,4-naphtoquinone and LY-83583 behaved as hyperbolic catalytic inhibitors. In these four cases the mechanism could clearly be diagnosed ($\alpha = \beta$ in the range 014–0.19) and thionin O was the most efficient inhibitor with $K_X = 0.4\,\mu$M.

[3]This is EC 1.6.5.9, whereas the proton-translocating enzyme is EC 1.6.5.3.

Fermentation at High Temperatures offers some advantages in biotechnology including better solubility of reactants, increased reaction rates, and decreased viscosity. Among other reactions, penicillin amidase from *Escherichia coli* catalyzes the hydrolysis of penicillin G to phenylacetic acid and 6-aminopenicillanate, which is important in the chemical synthesis of penicillins. While studying the protective effects of polyethylene glycol against thermal denaturation of penicillin amidase, Kazan and Erarslan [48], besides finding enhanced thermostability of the enzyme, observed that polyethylene glycol inhibited penicillin oxidase by a hyperbolic catalytic mechanism with $\alpha = \beta$ in the range 0.17–0.34 depending on the molecular size of the inhibitor. The partial loss of enzyme catalytic power by inhibitors and stabilizers of covalently immobilized enzymes for industrial applications may be compensated by the gain in their thermal stability.

The Allosteric Modifiers of Cathepsin K investigated by Novinec et al. [73] (see Tables 5.7 and 5.8 for other cases) included also two examples of hyperbolic catalytic inhibition. The names of the two compounds are listed in Table 5.9. Thus, three members of the taxonomic level 2 in Fig. 3.2 are represented in this group of novel allosteric effectors of cathepsin K. The finding is not surprising in view of the sophisticated system of binding sites predicted by computational methods and the poses of the modifiers suggested by docking. The many colored figures necessary to illustrate these properties cannot be rendered here but interested readers can refer to the original publication [73].

5.3.4 Hyperbolic Mixed, Predominantly Catalytic Inhibition

Assigned acronym: HMx(Sp<Ca)I. The specific velocity plot (Fig. 5.14), with the straight lines extrapolating to the negative side of the abscissa in the primary plot and both ordinate intercepts larger than one in the secondary plot, discriminates this mechanism toward HCaI described in the preceding section and LMx(Sp<Ca)I (Sect. 5.2.4). The mechanistic hallmark is $0 < \alpha, \beta < 1$ and $\alpha > \beta$.

In the kinetic barrier diagrams of HMx(Sp<Ca)I there are no special relationships between barriers and/or the thermodynamic differences listed in Fig. 5.14. However, together with the degree of inhibition that increases for increasing [S], this is a sufficiently distinctive criterion among the hyperbolic inhibitors. Recognizing this mechanism in the available literature is not always immediate, particularly when either appropriate graphical representations or values of the kinetic constants are not provided. Many examples may have been overlooked and filed in the large pool of insufficiently characterized *noncompetitive* cases. Five examples are listed in Table 5.10. Although two of these studies contain little experimental information, the inhibition mechanism can be deduced from the material presented.

Acetylcholinesterase Inhibition by Thienopyrimidine Derivatives was discussed in Sect. 5.3.2. This study included also two compounds that manifested hyperbolic mixed, predominantly catalytic inhibition [38]. The α/β parameters for

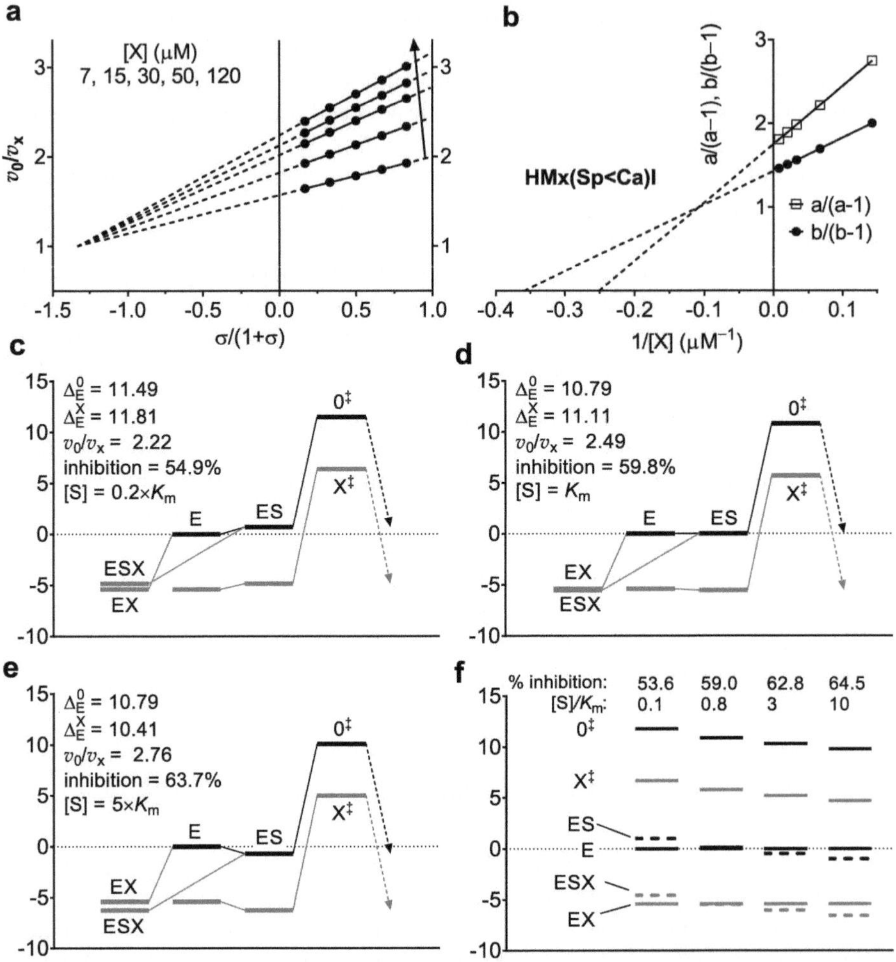

Fig. 5.14 Hyperbolic mixed, predominantly catalytic inhibition: HMx(Sp<Ca)I. Primary (**a**) and secondary (**b**) specific velocity plot. The *arrow* in (**a**) indicates the increasing direction of [X]. (**c**)–(**e**) Kinetic barrier diagrams, (**f**) barcode-like representation of the various states at four substrate concentrations as indicated. $k_{cat} = 100\,s^{-1}$, $K_m = 200\,\mu M$, $K_X = 4\,\mu M$, $\alpha = 0.7$, $\beta = 0.3$, in (**c**)–(**f**) [X] $= 10 \times K_X$. Constant values common to (**c**)–(**f**): $\Delta_{ES}^0 = 10.79$, $\Delta_{ES}^X = 11.25$, $\Delta_{EX} = -5.40$, $\Delta_{ESX} = -5.55$. The degree of inhibition increases for increasing [S] and there are no Δ-values that differ from one another in a special way at any [S]

acetylcholinesterase from electric eel were 0.70/0.08 and 0.47/0.24, respectively, for the molecules numbered 9 and 10 in the original publication.

Blood-Sucking Animals use anticoagulants to prevent blood clotting in the host. This strategy has suggested their application as anticoagulants for human use, for instance hirudin from leeches as inhibitor of thrombin (mentioned in Sect. 5.3.1) and nitrophorin-2, a ferriheme protein from the insect *Rhodnius prolixus* investigated by

Table 5.10 Examples of hyperbolic mixed, predominantly catalytic inhibition, HMx(Sp<Ca)I

Enzymes	EC no.	Inhibitors	Ref.
Acetylcholinesterase	3.1.1.7	Tetracyclic thienopyrimidines	[38]
Coagulation factor IXa	3.4.21.22	Nitrophorin-2	[114]
Inositol-phosphate phosphatase	3.1.3.25	Li^{+}	[56]
"Cytochrome P450"[a]	–	4-Nitrobenzoate	[97]
Glycerol kinase	2.7.1.30	IIA^{Glc} [b]	[77]
β-Lactamase	3.5.2.6	Single-domain antibody fragment	[95]

[a] An enzyme activity from rat liver microsomes called by the authors zoxazolamine oxidase, possibly EC 1.14.14.1 (unspecific monooxygenase)

[b] Glucose-specific phosphocarrier protein of the phosphoenolpyruvate:glycose phosphotransferase system

Zhang et al. as inhibitor of the coagulation factor IXa [114]. The last interaction is highly specific to inhibit the FIXa-catalyzed activation of FX since neither thrombin nor FVIIa, FXIIa, FXIa, FXa, and kallikrein activities change in the presence of nitrophorin-2. The authors measured the kinetic parameters with several combinations of the reacting species in a complex mixture that included, besides catalyst, substrate, and inhibitor, an artificial membrane surface consisting of a mixture of L-α-phosphatidyl-L-serine and L-α-dioleoylphosphatidylcholine, and FVIIIa. Either the surface or FVIIIa or both together were necessary for the inhibitory action, which was in all cases hyperbolic mixed, predominantly catalytic [HMx(Sp<Ca)I] with $\alpha = 0.07$–0.57, $\beta = 0.035$–0.21 and $K_X = 13$–23 nM).

Lithium Administration to patients with manic-depressive disorder has been commented in Sect. 5.2.2 for the property of Li^{+} to inhibit inositol-phosphate phosphatase with a lineal catalytic inhibition mechanism at low [Li^{+}]. Increasing the lithium concentration, Leech at al. reported a change in the inhibition mechanism to *nonlinear noncompetitive* [56]. Closer inspection of the original graphics reveals a hyperbolic mixed, predominantly catalytic mechanism despite the paucity of data.

A Microsomal Rat Liver Enzyme Activity that was reported by Sternson and Gammans as *zoxazolamine oxidase*, is inhibited by aromatic nitro compounds [97]. This enzyme, possibly an unspecific, cytochrome P450-associated monooxygenase, was inhibited by 4-nitrobenzoate. The publication contains little data to allow quantitative attempts at calculating kinetic parameters. However, a graphical representation of the results [97, Fig. 1] strongly suggest that the inhibition mechanism is hyperbolic mixed with predominantly catalytic character that was named by the authors *S,I-hyperbolic noncompetitive*.

Transplantation of Allosteric Control into a Non-Allosteric Enzyme Allosteric modulation is not limited to the action of low molecular mass modifiers on macromolecules. An instructive system, in whose complex biochemistry the interaction between modifier and enzyme can be reduced to one of the basic mechanisms in taxonomic level 2 (this section), is the inhibition of glycerol kinase by IIA^{Glc} described by Donald Pettigrew [77]. The modifier (IIA^{Glc}) is an enzyme that transplants allosteric control into an otherwise non-allosteric enzyme (glycerol kinase) [76]. IIA^{Glc} is the acronym for the 18-kDa glucose-specific phosphocarrier protein of the phospho*enol*pyruvate:glycose phosphotransferase system. In its unphosphorylated form, IIA^{Glc} acts as inhibitor of the glycerol kinase of *Escherichia coli*, a member of the sugar kinase/heat shock protein 70/actin superfamily. The kinetic mechanism of inhibition is fully supported by the crystal structure of IIA^{Glc} in complex with glycerol kinase [76, 77].

The interaction between IIA^{Glc} and glycerol kinase was reported as K-type activation and V-type inhibition with an allosteric coupling constant of about 3 (i.e., $\alpha \approx 0.3$) and $\beta \approx 0.1$[4] [77], which can be unmistakably identified as the mechanism HMx(Sp<Ca)I.

***β*-Lactamases** are responsible for bacterial resistance against β-lactam antibiotics for cleaving the β-lactam ring characteristic of these drugs. Sohier et al. developed a single-domain antibody fragment, the camelid nanobody cherry-NbVIM_38, capable of inhibiting the clinically relevant Verona integron-encoded metallo-β-lactamase [95]. Single-domain antibodies have a molecular mass of 12–15 kDa, consist of a single monomeric variable domain, bind antigens with high specificity, and share properties with low molecular mass drugs avoiding some limitations of the conventional antibodies. The camelid nanobody cherry-NbVIM_38 raised against the β-lactamase behaved as a hyperbolic mixed, predominantly catalytic inhibitor acting as allosteric modifier [95].

The system analyzed by the authors required an additional path in the general modifier mechanism to account for a special form of substrate inhibition, i.e., a second molecule of substrate bound by the β-lactamase in the ES complex. As it is conceivable that this pattern of substrate inhibition might be represented also in other cases of hyperbolic inhibition in which the catalytic component predominates, it is discussed below.

5.3.4.1 Substrate Inhibition in a Quaternary Complex

The reaction paths, rewritten with symbols used in this book, are shown in Scheme 5.5 and the corresponding rate equation, written in the form of (3.32) is given by (5.30)

$$\frac{v_X}{V} = \frac{[S] + \beta \frac{[X]}{\alpha K_X}[S]}{K_m^0\left(1 + \frac{[X]}{K_X}\right) + [S]\left(1 + \frac{[X]}{\alpha K_X} + \frac{[S][X]}{\alpha K_X K_{SX}}\right)}. \tag{5.30}$$

[4] Reported as Q ≈ 3 and W ≈ 0.1.

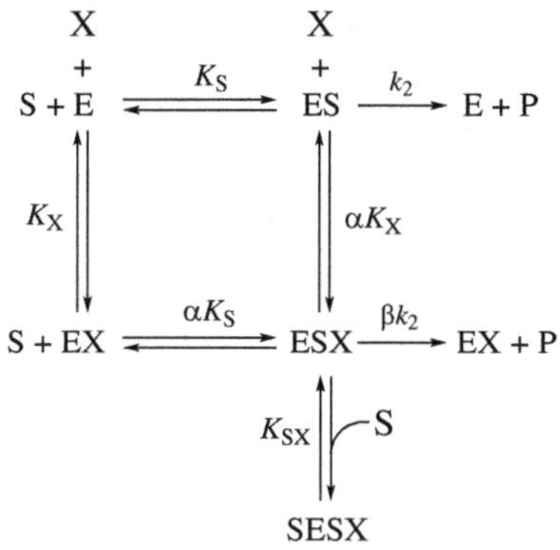

Scheme 5.5 Extension of the general modifier mechanism by adding a further substrate molecule to the ternary complex [95]

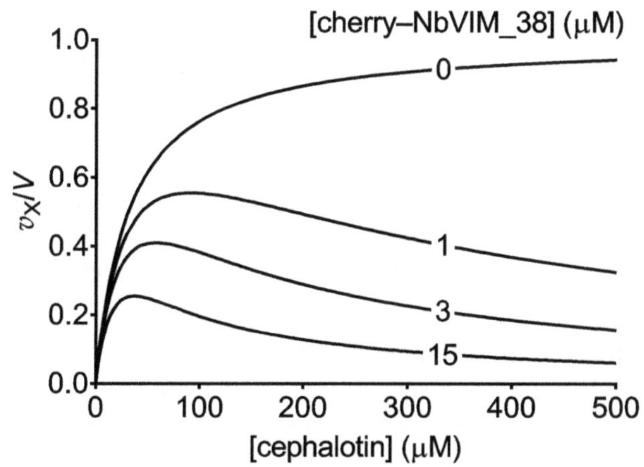

Fig. 5.15 Substrate inhibition in a quaternary complex. Plots of the function (5.30) using the parameters of Sohier et al. [95]. The enzyme is β-lactamase, the substrate is the cephalosporin antibiotic cephalothin, and the modifier is the camelid nanobody cherry-NbVIM_38

The additional term rightmost in the denominator of (5.30) represents the quaternary SESX complex, showing that this additional branch enhances the inhibitory effect of the modifier. The addition of the second substrate molecule to the ESX complex occurs in compulsory order to preformed ESX since no substrate inhibition could be detected in the absence of modifier (Fig. 5.15, curve with label 0). The parameters calculated for the β-lactamase reaction in the presence of the camelid nanobody in the study of Sohier et al. [95] were $\alpha = 0.6$, $\beta = 0.4$, $K_X = 9\ \mu M$ and $K_{SX} = 45\ \mu M$.

5.3.5 *Hyperbolic Mixed, Balanced Inhibition*

Assigned acronym: HMx(Sp=Ca)I. The survey of the modifier mechanisms in taxonomic level 2 ends with the hyperbolic counterpart of linear mixed balanced inhibition. No problems should arise in diagnosing these mechanisms since the degree of inhibition does not depend on substrate concentration, a property defined in Scheme 5.3 by $\alpha = 1$ and $0 < \beta < 1$. The primary specific velocity plot consists of a family of lines parallel to the abscissa (Fig. 5.16a) whose distance from one another becomes narrower as the modifier concentration increases until reaching an upper limit. This property is confirmed by the ordinate intercept of the secondary plot, which is larger than 1 (Fig. 5.16b). Hyperbolic balanced inhibition has been documented in numerous studies with some examples listed in Table 5.11 and commented below.

As in linear balanced inhibition, the kinetic barrier diagrams of the hyperbolic counterpart show that $\left(\Delta_{\mathrm{E}}^{0} - \Delta_{\mathrm{ES}}^{0}\right) = \left(\Delta_{\mathrm{E}}^{\mathrm{X}} - \Delta_{\mathrm{ES}}^{\mathrm{X}}\right)$. These differences (see p. 165) correspond to Δ_{S}^{0} and $\Delta_{\mathrm{S}}^{\mathrm{X}}$, respectively, have the same value at a given substrate concentration and can be positive or negative (facing arrowheads in Fig. 5.16f). This property derives from the fact that Δ_{S}^{0} and $\Delta_{\mathrm{S}}^{\mathrm{X}}$ depend on the Michaelis constant. If the two values in the absence and presence of modifier are identical the Michaelis constant does not depend on [X]. $\Delta_{\mathrm{EX}} = \Delta_{\mathrm{ESX}}$ have the same value shared at any substrate concentration as indicated by double arrows in panel (f). This derives from the equal participation of the catalytic and the specific components to inhibition ($\alpha = 1$). The resulting hyperbolic inhibitory effect is independent of substrate concentration.

4-Hydroxy-Tetrahydrodipicolinate Synthase from *Vitis vinifera* catalyzes the condensation of (S)-aspartate-4-semialdehyde with pyruvate. In Sect. 5.3.3 we have seen that the enzyme is inhibited by lysine via a hyperbolic mixed mechanism with predominantly specific character with respect to the substrate pyruvate. Lysine behaves instead as a hyperbolic balanced inhibitor toward the substrate (S)-aspartate-4-semialdehyde, [4].

Mimetics of Heparin stimulate tissue repair in models of skin, bone, colon, cornea, and muscle injury by protecting heparin-binding growth factors from degradation catalyzed by plasmin. Dextrans modified by addition of defined proportions of carboxymethyl, sulfate and benzylamide groups have been studied as plasmin inhibitors by Ledoux et al. [55], who demonstrated inhibition of the enzyme by polymers of different molecular mass. Considering the general modifier mechanism and a variable stoichiometry of binding between plasmin and the polymers, a hyperbolic balanced inhibition mechanism was found to fit experimental data optimally. The most efficient modified dextran polymer had an average M_{r} of 140,000 and a linear chain able to bind 20 enzyme molecules. The kinetic parameters were: $\alpha = 1$, $\beta = 0.2$ and $K_{\mathrm{X}} = 28$ nM.

Spermidine Synthase is involved in the biosynthesis of polyamines in eukaryotes. Raina and coworkers studied the kinetic mechanism of the reaction catalyzed by

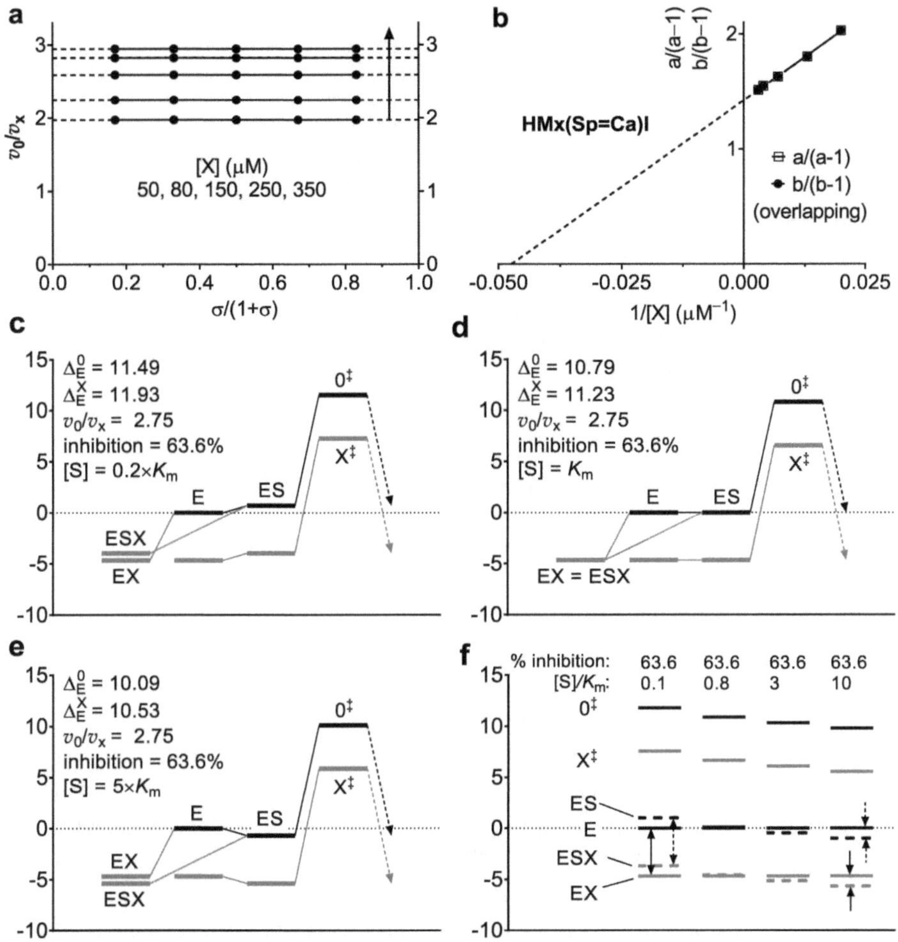

Fig. 5.16 Hyperbolic mixed, balanced inhibition: HMx(Sp=Ca)I. Primary (**a**) and secondary (**b**) specific velocity plot. The *arrow* in (**a**) indicates the increasing direction of [X]. (**c**)–(**e**) Kinetic barrier diagrams, (**f**) barcode-like representation of the various states at four substrate concentrations as indicated. $k_{cat} = 100\,s^{-1}$, $K_m = 200\,\mu M$, $K_X = 21\,\mu M$, $\alpha = 1.0$, $\beta = 0.3$, in (**c**)–(**f**) $[X] = 10 \times K_X$. The characteristic properties are $\Delta_{EX} = \Delta_{ESX}$, evidenced by the *solid* and the *dashed double arrows* in (**f**), respectively, and $\left(\Delta_E^0 - \Delta_{ES}^0\right) = \left(\Delta_E^X - \Delta_{ES}^X\right)$ represented by the amplitudes between the *dashed* and the *solid facing arrows*, respectively. Constant values common to (**c**)–(**f**): $\Delta_{ES}^0 = 10.79$, $\Delta_{ES}^X = 11.23$, $\Delta_{EX} = \Delta_{ESX} = -4.68$. The degree of inhibition is the same at any substrate concentration and is thus determined solely by the inhibitor concentration

spermidine synthase [79]. The authors performed measurements to determine the inhibition of reaction products versus both substrates in the reaction

$$\text{decarboxylated adenosylmethionine} + \text{putrescine}$$
$$\rightleftharpoons \text{5}'\text{-methylthioadenosine} + \text{spermidine}\,.$$

Table 5.11 Examples of hyperbolic mixed, balanced inhibition, HMx(Sp=Ca)I

Enzymes	EC no.	Inhibitors	Ref.
4-Hydroxy-tetrahydrodipicolinate synthase	4.3.3.7	Lysine	[4]
Plasmin	3.4.21.7	Modified dextran	[55]
Spermidine synthase	2.5.1.16	5′-methylthioadenosine[a]	[79]
Caspases[b]	3.4.22.–	Copper-containing compounds[c]	[30]
Acetylcholinesterase	3.1.1.7	*d*-Tubocurarine	[11]
NADH:ubiquinone reductase[d]	1.6.5.9	Flavone	[103]

[a]A reaction product, acts as HMx(Sp=Ca)I toward both substrates of the reaction
[b]Caspase-1 (EC 3.4.22.36), caspase-3 (EC 3.4.22.56), caspase-8 (EC 3.4.22.61)
[c]Codes NSC321205, NSC277584, NSC321206, and NSC310547 in the Chemical Repository of the Developmental Therapeutics Program of the National Cancer Institute
[d]Nonelectrogenic

5′-Methylthioadenosine was a hyperbolic mixed, balanced inhibitor with respect to both substrates, while spermidine behaved as a hyperbolic mixed, predominantly specific inhibitor [HMx(Sp>Ca)I] with respect to both substrates. From this and other results, the authors excluded a substituted-enzyme mechanism in favor of a more complex one including abortive ternary complexes.

The Inhibition of Caspases by pyridinyl-based copper-containing molecules discussed in Sect. 5.2.5 comprises also cases of HMx(Sp=Ca)I. Feldman et al. [30] established that the four compounds listed in Table 5.11 inhibit the caspases 1, 3, and 8 by a hyperbolic mixed, balanced mechanism. For all compounds and enzymes tested, α was equal to or very close to 1. Contrary to caspases 2, 7, and 9, for which the parameter β was very close to zero, for caspases 1, 3, and 8 this value was between 0.11 and 0.60, which is diagnostic for HMx(Sp=Ca)I.

Acetylcholinesterase interacts with a multitude of compounds. Among other modes of interaction between this enzyme and reversible cationic ligands investigated by Berman and Leonad [11], *d*-tubocurarine was found to inhibit the enzyme with a hyperbolic mixed, balanced mechanism. Tubocurarine was formerly used to complement anesthesia as a skeletal muscle relaxant during surgery. This molecule binds at the same allosteric site as propidium iodide, which behaves as a hyperbolic specific inhibitor (Table 5.7).

Inhibition by Flavone of the reaction catalyzed by NADH:ubiquinone reductase with respect to the second substrate, which is naturally a quinone but commonly replaced by 2,6-dichlorophenolindophenol for in vitro measurements, occurs by the hyperbolic mixed, balanced mechanism. In the same reaction, inhibition with respect to the first substrate, NAD_{red}, is hyperbolic catalytic (Sect. 5.3.3).

5.4 Mechanisms in Taxonomic Level 3: Nonessential Modifiers

The reaction paths common to all members of this group are those in Scheme 5.3, the rate equation and the dependences of the apparent parameters on modifier concentration are (3.2)–(3.5), and the equations of the specific velocity plot with its replots are (3.36)–(3.38). The graphical representations of the dependences of k_{cat}^{app}, K_{m}^{app} and k^{app} are collected in Fig. 3.5 for the whole group.

Five of the seven mechanisms in this taxonomic level, shown in Fig. 3.2 at level 3 with a completely gray background, belong to *nonessential activation* [70]. The two other mechanisms have a dual character as inhibitors or activators depending on substrate concentration, and to show this property the background of their boxes in level 3 of Fig. 3.2 is half white and half gray.

5.4.1 Hyperbolic Mixed, Dual Modification (Inhibition to Activation)

Assigned acronym: HMxD(I/A). In Fig. 3.2, the branches of the taxonomic tree descending from LMx(Sp>Ca)I in level 1 with the dominant character $1 < \alpha < \infty$, end up in level 3 with three mechanisms in which the ESX complex is catalytically more active than ES, i.e., $\beta > 1$. At first sight, this may insinuate activation under all circumstances if it were not for the relationships between α and β, which embraces also inhibition when $\alpha > \beta$. This is the case of the first mechanism in this group, HMxD(I/A), for which this property can be appreciated from Fig. 5.17 that shows two zones of behavior determined by substrate concentration. Left to the vertical dashed line the modifier behaves as inhibitor, at the critical substrate concentration there is no difference between the rate in presence and absence of modifier, while activation ensues at higher substrate concentrations.

In the specific velocity plot, the family of straight lines converges to a point within the left and right vertical axes and the resulting sheafs of curves represent inhibition left to the intersection point and activation on the right of it (Fig. 5.18a). The secondary plot unmistakably shows two straight lines, one with positive and one with negative slope that, together with the primary plot, are diagnostic for the mechanism. A similar behavior is manifested by the related mechanism HMxD(A/I) described in Sect. 5.4.4, which can be differentiated from HMxD(I/A) from the opposite inhibition and activation effects depending on substrate concentration.

The salient aspect of this mechanism is its dual action as inhibitor or activator depending on substrate concentration. Guessing this property by inspection of the kinetic barrier diagrams is not immediate (Fig. 5.18c–f) but it becomes evident when the rate is calculated from the barriers Δ_{E}^{X} and Δ_{ES}^{X} using (3.61) and (3.62). Namely, there is only one substrate concentration, the critical substrate concentration

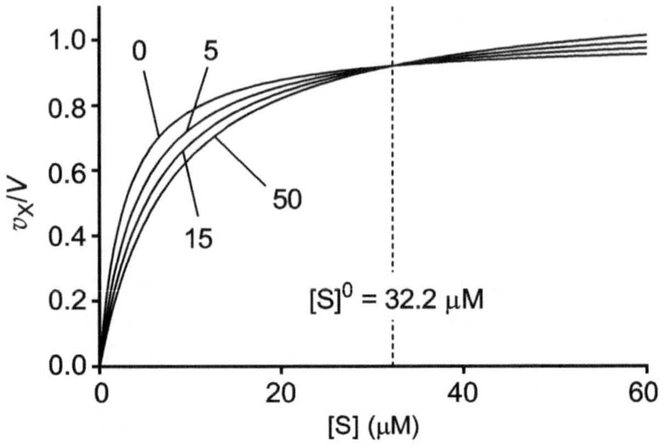

Fig. 5.17 Hyperbolic mixed, predominantly specific dual modification: HMxD(I/A). Normalized rate against [S] at various [X] (*numbers labeling the lines*). The critical substrate concentration ($[S]^0$, *dashed line*) demarcates the zones of inhibition and activation. The *curves* were plotted with (3.32) and the parameters of Joshi et al. for the oxidation of γ-linolenic acid (the substrate) catalyzed by arachidonate 15-lipoxygenase [47, Fig. 3]: $\alpha = 3.5$, $\beta = 1.2$, $K_X = 4.9\,\mu M$, $K_m = 2.8\,\mu M$. The modifier was 13(*S*)-HOTrE(γ)

(Sect. 3.7) for which the rate is the same in the absence and in the presence of modifier at *any* concentration.

The dual nature as inhibitor and activator of modifiers obeying this mechanism can escape attention because activation may become manifest only at prohibitively high substrate concentrations. There are indeed cases in the literature, in which this interesting mechanism has been published under the name *partial mixed inhibition* because the inversion of behavior and the critical point with no effect illustrated in Figs. 5.17 and 5.18 could not be observed for practical reasons. This has been the case of some published examples discussed below (summary in Table 5.12).

Arachidonate 15-Lipoxygenase Inhibition by Reaction Products has been investigated by Joshi et al. [47]. Arachidonate 15-lipoxygenase, besides arachidonic acid, uses also γ-linolenic acid and linoleic acid as substrates that have the following reactions (on the left of the arrow the substrates, on the right the acronym of the hydroperoxide products with full names given in the footnotes of Table 5.12):

$$\gamma\text{-linolenic acid} + O_2 \rightarrow 13(S)\text{-HOTrE}(\gamma) \tag{5.31}$$

$$\text{linoleic acid} + O_2 \rightarrow 13(S)\text{-HODE}\,. \tag{5.32}$$

The authors measured the effects of the two hydroperoxides on the arachidonate 15-lipoxygenase reaction performing a fine analysis based on plotting the apparent Michaelis constant and specificity constant against modifier concentration. For 13(*S*)-HOTrE(γ) the reaction was characterized by $\alpha = 3.5$ and $\beta = 1.2$, which agree with the HMxD(I/A) mechanism. This behavior was ascribed by the authors

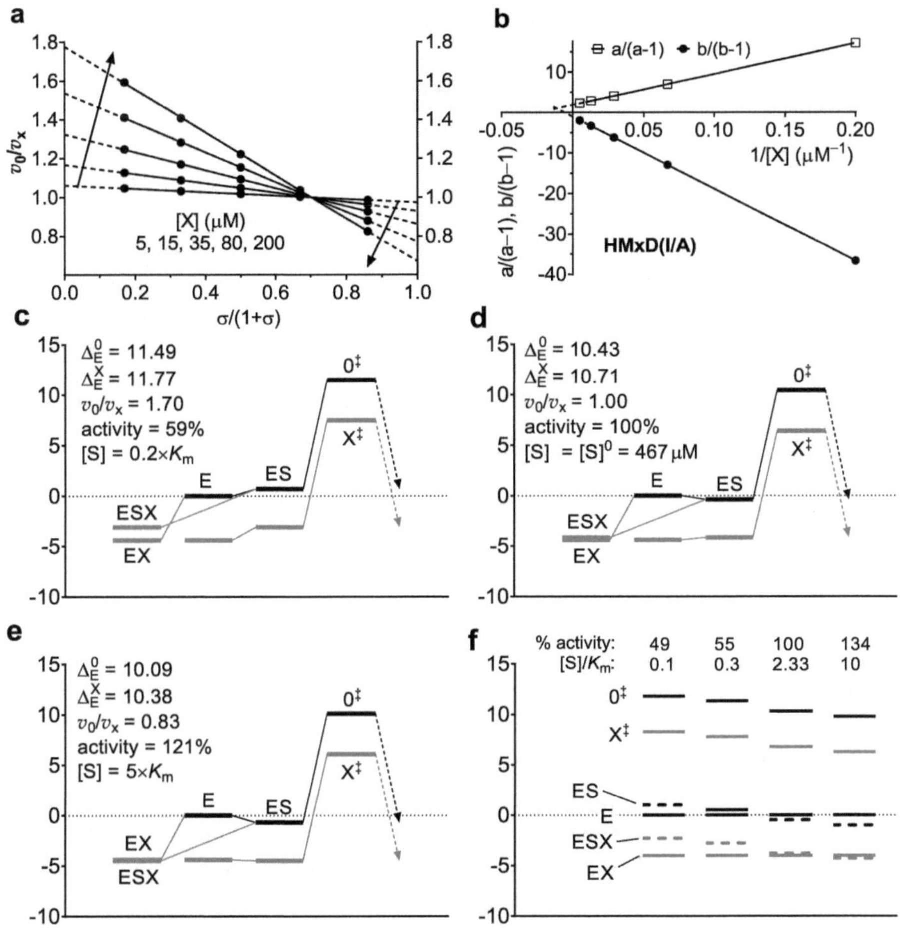

Fig. 5.18 Hyperbolic mixed, dual modification (inhibition→activation: HMxD(I/A). Primary (**a**) and secondary (**b**) specific velocity plot. The *arrows* in (**a**) indicate the increasing direction of [X]. (**c**)–(**e**) Kinetic barrier diagrams, (**f**) barcode-like representation of the various states at four substrate concentrations as indicated. $k_{cat} = 100\,s^{-1}$, $K_m = 200\,\mu M$, $K_X = 40\,\mu M$, $\alpha = 4.0$, $\beta = 1.9$, in (**c**)–(**f**) $[X] = 10 \times K_X$. Constant values common to (**c**)–(**f**): $\Delta_{ES}^0 = 10.79$, $\Delta_{ES}^X = 11.58$, $\Delta_{EX} = -4.40$, $\Delta_{ESX} = -3.80$. The critical substrate concentration is 467 μM, corresponding to $2.33 \times K_m$. The numerical values of the activation in panels (**c**)–(**f**) refer to 100 % activity in the absence of modifier

to *mixed hyperbolic allostery*: $\alpha > 1$ (K-type inhibition) and $\beta > 1$ (V-type activation) resulting in k_{cat}/K_m *allosteric inhibition.* It is irrelevant that the dual character of the modifier could not be seen experimentally because the critical substrate concentration for this system (Sect. 3.7) is 32.2 μM. The highest substrate concentration used in the experiment was possibly 20 μM and thus the annihilation of inhibition at the critical substrate concentration and the switch from inhibition to activation could not be perceived. Moreover, in the region of the critical substrate

Table 5.12 Examples of the hyperbolic mixed, dual modification (inhibition→activation) mechanism, HMxD(I/A)

Enzymes	EC no.	Modifiers	Ref.
Arachidonate 15-lipoxygenase	1.13.11.33	13(S)-HOTrE(γ)[a], 13(S)-HODE[b]	[47]
Cholinesterase	3.1.1.8	Benzalkonium	[62]
Acetylcholinesterase	3.1.1.7	Decamethonium	[11]
Chymotrypsin	3.4.21.11	α-Naphtylamine and phenol[c]	[61]
Glutathione transferase	2.5.1.18	ethacrinic acid	[78]
Cathepsin K	3.4.22.38	NSC13345[d]	[72]
Cathepsin K	3.4.22.38	Two compounds[e]	[73]

[a] 13S-hydroxy-6Z,9Z,11E-octadecatrienoic acid, a product of the γ-linolenic acid reaction catalyzed by arachidonate 15-lipoxygenase
[b] 13S-hydroxy-9Z,11E-octadecanoic acid, a product of the linoleic acid reaction catalyzed by arachidonate 15-lipoxygenase
[c] The substrate was *N*-acetylglycine ethyl ester for both modifiers
[d] (2-[(2- carbamoylsulphanylacetyl)-amino]benzoic acid)
[e] (2-Biphenylylmethyl) malonic acid and 1, 4, 7, 9B-Tetraazaphenalene

concentration, there is no significant difference between the rates measured in the presence or absence of modifier because of masking by experimental errors.

The same mechanism, HMxD(I/A), was observed also when the substrate was again γ-linolenic acid but the modifier was 13(*S*)-HODE, the product of linoleic acid peroxidation. With arachidonic acid as substrate, the behavior of 13(*S*)-HODE as modifier corresponded to the HMxD(A/I) mechanism (Sect. 5.4.4).

The Effects of Benzalkonium on Cholinesterase have been studied by Masson et al. [62], who assigned the interaction to the HMxD(I/A) modification mechanisms. Benzalkonium chloride is a detergent commonly used as preservative in pharmaceutical preparations, e.g., eye and nasal drops. The authors identified the dual nature of the modifier as inhibitor and activator providing the parameters $\alpha = 5.7$, $\beta = 3.1$ and $K_X = 0.05\,\mu$M.

The Interaction of Acetylcholinesterase with Decamethonium has been reported in the same study discussed in Sects. 5.3.1 and 5.3.5. Decamethonium (syncurine) has been used in anesthesia to induce paralysis but this procedure has now been dismissed. Berman and Leonad [11] were surprised of the anomalous behavior of this modifier because at high substrate concentrations the rate in the presence or absence of inhibitor was the same. Binding of a second molecule of substrate that may preclude the access of inhibitor was suggested as a possible interpretation. However, the annihilation of inhibition by increasing [S] and the data in [11, Fig. 1C] strongly suggest that this mechanism is HMxD(I/A). The dual character of the mechanism is also in line with similar results obtained with cholinesterase by Masson et al. (discussed above [62]).

The Chymotrypsin-Catalyzed Hydrolysis of *N*-acetylglycine ethyl ester in the presence of α-naphthylamine and phenol as modifiers has been studied by Martinek and coworkers [61]. The patterns of inhibition revealed two clearly distinct ranges of substrate concentration for which inhibition (at low concentrations) or activation (at high concentrations) was observed. The authors did not discuss the mechanistic origin of this finding and simply suggested to name the two modifiers *effectors* in place of *inhibitors*.

Glutathione Transferase and Occam's Razor would be an appropriate subtitle for a paper by Phillips and Mantle, who described an *anomalous* behavior in the reaction catalyzed by this enzyme in the presence of ethacrynic acid, a diuretic drug, as modifier [78]. The observed pattern of the interaction [78, Fig. 3] matches with that of the mechanism discussed in this section, HMxD(I/A). For the interpretation of their results the authors proposed a model consisting of two connected thermodynamic cycles, which includes twelve complexes and two catalytic paths. With the help of a computer programm, the system could be analyzed and values of the kinetic parameters calculated. The statistical significance of such calculations with 12 complexes, 15 parameters, and 25 experimental points is unclear.

The most intriguing argument is why one-two simpler models, with less complexes and much less parameters, have not been invoked as alternatives: the general modifier mechanism can fit the data under discussion neatly. This is a case in which William of Ockham had used his razor without hesitation.

The Rich Repertoire of Allosteric Cathepsin K Modifiers in the studies performed by Novinec and collaborators [72, 73] includes also three examples of the hyperbolic mixed, dual modification mechanism HMxD(I/A). Besides determining the kinetic mechanism of modification by NSC13345 (full name in Table 5.12) using a steady-state approach, the crystal structure of this compound in complex with human cathepsin K was resolved. NSC13345 was one of several compounds identified as potential allosteric modifiers of cathepsin K by computational methods.

In collaboration with Rama Ranganathan, who developed successful algorithms aimed at detecting possible allosteric sites in entire protein families, statistical coupling analysis was performed on the peptidase subfamily C1A in the MEROPS database (the papain-like cysteine peptidases). Multiple sequence alignments of 1,239 catalytic domains were constructed and well-conserved residues were identified from positional conservation. After additional computational procedures, conserved networks of residues that represent the essence of allosteric communication in proteins and called *protein sectors* could be identified [72]. Since protein sectors can mediate allosteric communication, the next step was searching for their expression at the surface of cathepsin K in the form of binding cavities. Eight such structures were identified as candidates for the role of allosteric sites and one of them was already known as a chondroitin sulfate-binding site from structural studies of cathepsin K [57].

The validity of these findings needed experimental validation, which was accomplished in collaboration with Amedeo Caflisch, who developed original algorithms for computer-aided drug design based on structure. Docking procedures led to the identification of more than 200 compounds that could potentially interact with the predicted eight allosteric sites of cathepsin K. All compounds were tested

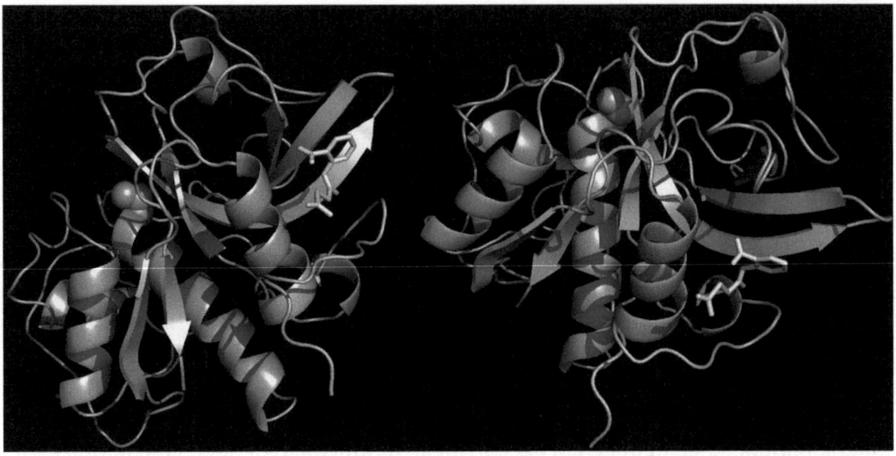

Fig. 5.19 Structure of human cathepsin K in complex with NSC13345. Two orientations in which the modifier is represented as sticks and the active site cysteine as spheres [72]

experimentally with steady-state kinetic methods using synthetic peptide substrates to monitor reactions and, additionally, collagen degradation assays were performed to ascertain their efficiency in the presence of a natural substrate of the enzyme.

Acting on a synthetic substrate, NSC13345 behaves as a hyperbolic mixed, dual modifier that inhibits at substrate concentrations smaller than the critical substrate concentration and activates at higher [S]. The parameters are: $K_X = 28 \pm 8\,\mu M$, $\alpha = 2.4 \pm 0.4$, and $\beta = 1.3 \pm 0.2$. The critical substrate concentration ($K_m = 3.8\,\mu M$) is calculated as $[S]^0 = 13.9\,\mu M$, the relative substrate concentration is $\sigma^0 = 3.7$ and the critical specific velocity $\sigma^0/(1+\sigma^0) = 0.79$ (Sect. 3.7).

Importantly, the compound completely inhibited collagen degradation by cathepsin K. Using a general proteolytic assay with azocasein as substrate, the selectivity of NSC13345 over the related cysteine peptidases cathepsin L, B, S, V was high while papain activity was not affected by the modifier.

The structure of human cathepsin K in complex with NSC13345 is shown in Fig. 5.19 in two different orientations to illustrate the spatial arrangement of the modifier with respect to the active site. The modifier, represented as sticks, is bound at the surface of the cathepsin K molecule at a considerable distance from the active site cysteine represented as spheres and this distance is in agreement with the meaning of the word *allosteric* (Appendix 1 in Chap. 2). At the structural level no major conformational changes are associated with modifier binding and thus the effects exerted on catalysis are likely due to population shifts of preexisting conformational states. At the mechanistic level, allosteric coupling between substrate and modifier is quantifiable by the allosteric coupling constant $1/\alpha = 0.42 \pm 0.07$, which, calculated with (2.4), corresponds to a coupling free energy $\Delta G_c = 2.2\,\mathrm{kJ\,mol^{-1}}$. The positive sign of ΔG_c signifies reciprocal binding hindrance between substrate and modifier.

Two other compounds that obey the HMxD(I/A) mechanism (listed in the last row of Table 5.12) were identified in this project [73]. The diagnosis of the kinetic mechanism clearly points to this dual behavior for the consistent values of α and $\beta > 1$ and $\alpha > \beta$ using a synthetic peptide substrate. However, both modifiers were relatively weak in the presence of azocasein, a customary protein used as unspecific substrate for assaying peptidases. What remains established is the existence of binding pockets at the surface of cathepsin B to which small molecules bind triggering the transmission of a signal toward the active site thus demonstrating the functionality of the conserved network of residues that make up the protein sectors.

5.4.2 Hyperbolic Catalytic Activation

Assigned acronym: HCaA. The characteristic traits of this nonessential activation mechanism are $1 < \alpha, \beta < \infty$, and $\alpha = \beta$. The Michaelis constant and the catalytic constant are equally affected in the presence of modifier resulting in the independence of the specificity constant on modifier concentration (Fig. 3.5). The primary specific velocity plot is characterized by straight lines with negative slopes, all located below the ordinate value 1. There is only one secondary plot with negative calculated values and negative slope.

Although the kinetic barriers of HCaA shown in Fig. 5.20c–f look very much like those of HCaI in Fig. 5.13, the degree of *nonessential activation* increases with increasing [S] in HCaA, while it is the degree of *inhibition* that increases for the same reason in HCaI. In all barrier diagrams this property is identified as $\Delta^{X}_{ES} < \Delta^{0}_{ES}$. Δ^{X}_{ES} depends on k^{app}_{cat} that contains β, which is > 1 in HCaA and < 1 in HCaI. Ultimately, the effect is exerted on the height of the barriers $0^{\ddagger}$ and $X^{\ddagger}$ as shown in Fig. 5.20e.

Well-characterized examples of this activation mechanism in the literature are limited to a few cases or they are *hidden*, conceivably because they might have been confused with other mechanisms and published with insufficient details. Three occurrences of HCaA that have been adequately documented are discussed below and summarized in Table 5.13.

The Glutamate Dehydrogenase reaction scheme proposed by Frieden for explaining hyperbolic catalytic inhibition by GTP (p. 246) has been used by the author also for describing hyperbolic catalytic activation by ADP. The catalytic constant and the Michaelis constant were equally affected in the presence of ADP and the enzyme was catalytically competent also in the absence of modifier. The diagnosis of this nonessential activation mechanism points therefore directly to HCaA.

Amidases catalyze the reaction of an amide with water (substrates) to yield a carboxylate plus ammonia as products (5.33). For studying the kinetic mechanism of the reaction, the concentration of water can obviously not be varied but fortunately

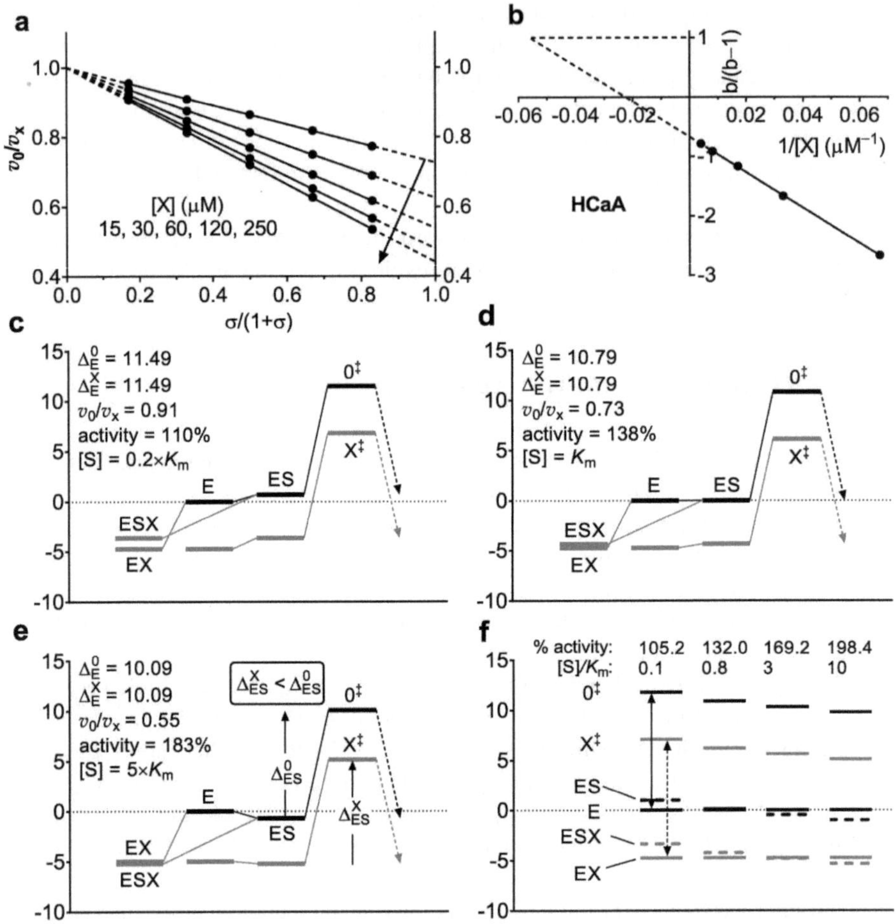

Fig. 5.20 Hyperbolic catalytic activation: HCaA. Primary (a) and secondary (b) specific velocity plot. The *arrow* in (a) indicates the increasing direction of [X]. (c)–(e) Kinetic barrier diagrams, (f) barcode-like representation of the various states at four substrate concentrations as indicated. $k_{cat} = 100\,s^{-1}$, $K_m = 200\,\mu M$, $K_X = 18\,\mu M$, $\alpha = 2.5$, $\beta = 2.5$, in (c)–(f) $[X] = 10 \times K_X$. Constant values common to (c)–(f): $\Delta^0_{ES} = 10.79$, $\Delta^X_{ES} = 10.45$, $\Delta_{EX} = -4.74$, $\Delta_{ESX} = -4.35$. The numerical values of the activation in panels (c)–(f) refer to 100 % activity in the absence of modifier. In panel (f) the *solid double arrow* represents Δ^0_E and the *dashed double arrow* is Δ^X_E; their values change with substrate concentration but always is $\Delta^0_E = \Delta^X_E$

the amidases are also capable of transferring acyl groups to form hydroxamic acids (5.34):

$$RCONH_2 + H_2O \rightarrow RCOOH + NH_3 \tag{5.33}$$

$$RCONH_2 + NH_2OH \rightarrow RCONHOH + NH_3\,. \tag{5.34}$$

Table 5.13 Examples of hyperbolic catalytic activation, HCaA

Enzymes	EC no.	Activators	Ref.
Glutamate dehydrogenase (NAD(P)$^+$)	1.4.1.3	ADP[a]	[35]
Amidase	3.5.1.4	Hydroxylamine	[109]
Thioglucosidase	3.2.1.147	Ascorbate	[92]

[a]$NADP_{red}$ as coenzyme

Woods and coworkers were interested in elucidating the kinetic mechanism of amidase from *Pseudomonas aeruginosa* and, besides other strategies, exploited the amidase transfer reaction (5.34) using acetamide as substrate and hydroxylamine (NH_2OH) as alternate acceptor [109]. The reaction studied in the presence of hydroxylamine agreed with a substituted mechanism termed crypto ping pong [74]. In this reaction hydroxylamine behaved also as a strong activator of the reaction, a mechanism the authors identified as *uncompetitive activation*. The modification patterns were hyperbolic and exhibited all characteristics of HCaA [109, Fig. 3].

White Mustard, Wasabi, Horseradish, and Watercress have something in common with hyperbolic catalytic activation. No panic for the lovers of these delicacies with agreeable, pungent taste to accompany sausages, sushi, and salads! The relationship with kinetics has to do with the strategies used by some plants to persuade in a convincing way herbivores, chiefly insects, to avoid them as food. In fact, the enzyme thioglucosidase, also known under the synonyms myrosinase and sinigrinase, catalyzes the reaction between a tioglucoside and water to give the corresponding sugar and, depending on pH and other factors, an isothiocyanate, a thiocyanate, or a nitrile. These molecules are toxic to insects but Shikita et al. were rather interested in studying the mechanism of action of thioglucosidase [92]. This enzyme was activated by ascorbate but had a finite activity in its absence, as noted by the authors. Adding increasing concentrations of ascorbate to the reaction mixture, V and K_m increased but their ratio remained constant. These properties were attributed to an *unusual and possibly unique example of linear uncompetitive activation* but the criterion for using the adjective *linear* was not explained. From the ensemble of the data presented in this paper, notably the *increase* of both V and K_m at a constant ratio and the rapid saturation by modifier in [92, Fig. 4], the mechanism of thioglucosidase described by Shikita et al. is hyperbolic catalytic activation. For a discussion of *uncompetitive activation* see Sect. 3.4.4 (Table 5.13).

5.4.3 Hyperbolic Mixed, Predominantly Specific Activation

Assigned acronym: HMx(Sp>Ca)A. For this system $1 < \alpha, \beta < \infty$ and $\alpha < \beta$. A characteristic property, which is sufficient for establishing the identity of this nonessential activation mechanism, is the hyperbolic dependence of the apparent values of k_{cat}, K_m and k on modifier concentration, in which all three parameters

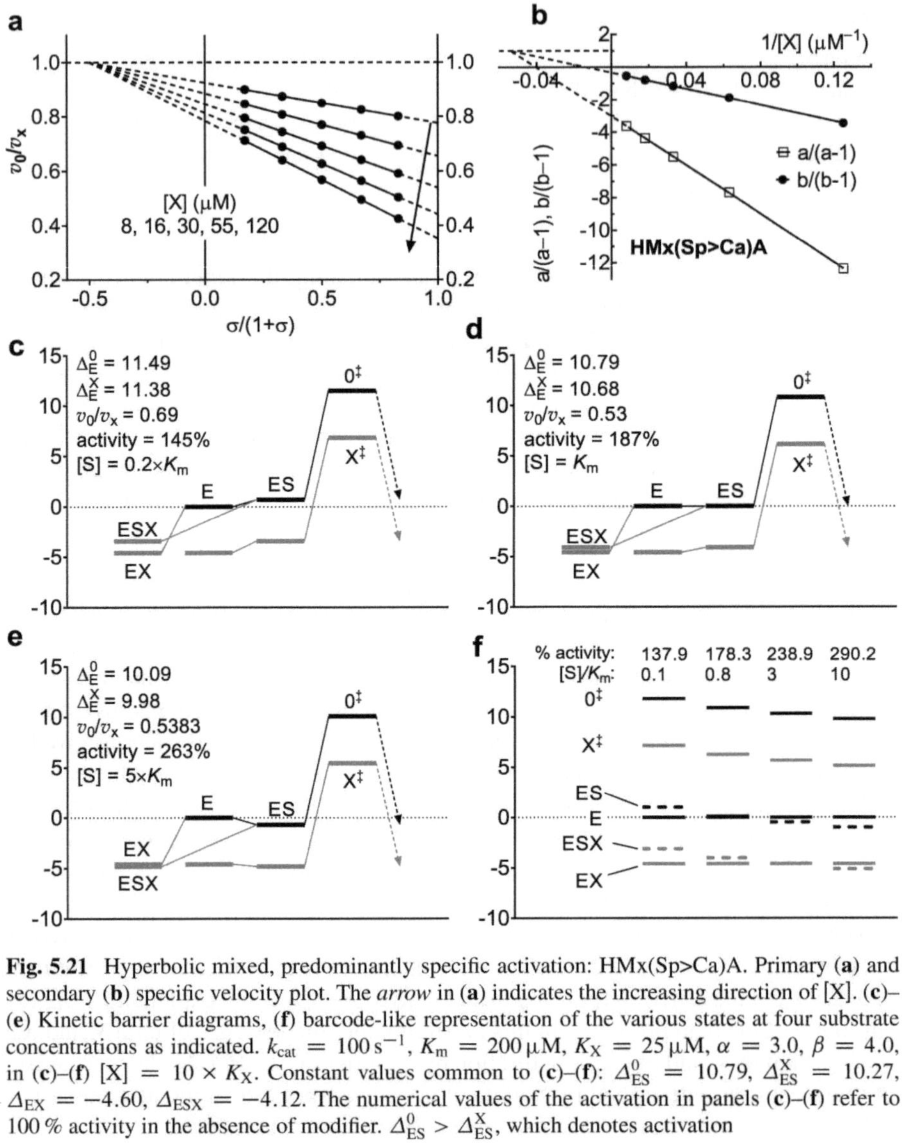

Fig. 5.21 Hyperbolic mixed, predominantly specific activation: HMx(Sp>Ca)A. Primary (**a**) and secondary (**b**) specific velocity plot. The *arrow* in (**a**) indicates the increasing direction of [X]. (**c**)–(**e**) Kinetic barrier diagrams, (**f**) barcode-like representation of the various states at four substrate concentrations as indicated. $k_{cat} = 100\,s^{-1}$, $K_m = 200\,\mu M$, $K_X = 25\,\mu M$, $\alpha = 3.0$, $\beta = 4.0$, in (**c**)–(**f**) $[X] = 10 \times K_X$. Constant values common to (**c**)–(**f**): $\Delta^0_{ES} = 10.79$, $\Delta^X_{ES} = 10.27$, $\Delta_{EX} = -4.60$, $\Delta_{ESX} = -4.12$. The numerical values of the activation in panels (**c**)–(**f**) refer to 100 % activity in the absence of modifier. $\Delta^0_{ES} > \Delta^X_{ES}$, which denotes activation

increase with [X] (Fig. 3.5). The primary specific velocity plot consists of straight lines with negative slopes converging right-to-left to a common point at the ordinate value 1 on the negative side of the abscissa. The two secondary plots have negative slopes.

The kinetic barrier diagrams in Fig. 5.21c–f show efficient activation increasing when the substrate concentration increases. The basis for activation resides in a lower activation energy of the transition state $X^\ddagger$ with respect to $0^\ddagger$ because the barrier Δ_{ES}^{X} is lower than Δ_{ES}^{0} and these are the kinetic barriers governed by k_{cat}^{app} and k_{cat}, respectively.

It is difficult to establish whether this mechanism is rare or merely described with insufficient details for recognizing its identity. A clear case of nonessential activation that has all characteristics of HMx(Sp>Ca)A has been published by Kovačič and collaborators [53].

Snakes Use Venoms containing various components that serve to immobilize, kill, and digest their preys. The venoms consist mostly of polypeptides of various sizes, including enzymes (oxidoreductases, transferases, a large group of hydrolases, and at least one lyase), neurotoxins, myotoxins, and anticoagulants. The most dangerous snake venom in Europe is that of *Vipera ammodytes ammodytes*, which contains *ammodytoxins* in its killer cocktail. The ammodytoxins are phospholipase A_2 hydrolases ranked under the presynaptic neurotoxins of the group IIA, whose biological effect is paralysis at the neuromuscular junction.

Phospholipase A_2 (EC 3.1.1.4) catalyzes the hydrolysis of phosphatidylcholine to give 1-acylglycerophosphocholine and a carboxylate as products. Calmodulin, the calcium-binding messenger expressed in eukaryotes, whose functions as mediator include binding Ca^{2+} for modulating the properties of other proteins, interacts with secreted phospholipase A_2 in the cytosol of nerve cells. In a study dedicated to the interactions between calmodulin and ammodytoxins, Kovačič and collaborators reported two biologically relevant effects: nonessential activation and stabilization of the structure of ammodytoxins [53].

The enzymatic assays were performed by interfacial kinetics in large unilamellar vesicles and reaction progress was monitored fluorometrically. Despite the individual readings did not reach the quality of measurements performed in customary dilute solutions, the activation mechanism could be determined using the specific velocity plot, which consistently exhibited patterns like those shown in Fig. 5.21. The measurements were performed under both nonreducing and reducing conditions, the second of which aimed at representing a cytosol-like environment. The values of α and β determined under nonreducing conditions were 4.4 and 5.9, and under reducing conditions 4.1 and 4.9, respectively, which are characteristic of the hyperbolic mixed, predominantly specific activation mechanism.

The second relevant result of this study was stabilization of the structure of ammodytoxin by calmodulin. In a reducing environment similar to that found in the cytosol, ammodytoxin undergoes a slow transition to a state exhibiting decreased catalytic activity. However, full activity is rapidly restored upon transfer to a nonreducing environment. In the presence of calmodulin, independently of reducing or nonreducing conditions, a complex is formed between ammodytoxin and calmodulin which not only stabilizes the phospholipase but also enhances its catalytic power.

5.4.4 Hyperbolic Mixed, Dual Modification (Activation to Inhibition)

Assigned acronym: HMxD(A/I). A modifier behaves as activator at substrate concentrations smaller than the critical substrate concentration, changing to inhibition when the substrate concentration increases over this threshold, if $0 < \alpha, \beta < 1$ and $\alpha < \beta$. Figure 3.5 shows that the apparent values of k_{cat} and K_m *decrease* hyperbolically and that *k increases* hyperbolically with modifier concentration. The specific velocity plot (Fig. 5.22) is the specular image of that seen in the other dual modification mechanism, HMxD(I/A), with a pivotal point represented by the critical specific velocity (Sect. 3.7).

The dual nature of HMxD(A/I) as activator or inhibitor can be appreciated in the kinetic barrier diagrams (Fig. 5.22c–f), where it is seen that a very low substrate concentration may be necessary for measuring appreciable activation, in this particular example visible only at $[S] < K_m$ ($K_m = 200\,\mu$M and the critical substrate concentration $[S]^0 = 150\mu$M). The switch from activation to inhibition when [S] increases cannot be predicted just by inspection of Fig. 5.22c–f. However, the rate in the presence and absence of modifier can be calculated with (3.61) and (3.62), respectively using the values of the barriers Δ_E^X and Δ_{ES}^X as well the corresponding barriers in the absence of modifier, Δ_E^0 and Δ_{ES}^0. Only at the critical substrate concentration, which is integrated in the expressions of Δ_E^0 and Δ_E^X, the rate is the same at any modifier concentration.

The dual mechanism under discussion has been interpreted precisely with the correct rate equation in some cases discussed below and listed in Table 5.14 despite the use of an ambiguous reaction scheme. HMxD(A/I) is prone to misinterpretation due to either the application of a wrong rate equation or for overlooking pivotal details of the mechanism because of insufficient analysis. These points will be commented on p. 273.

3-Phosphoshikimate 1-Carboxyvinyltransferase is an enzyme responsible for the transfer of an enolpyruvyl group from phosphoenolpyruvate to 3-phosphoshikimate in the sixth of seven steps in the shikimate pathway. Microorganisms, fungi, and plants, but *not by animals*, use the shikimate pathway for synthesizing the aromatic amino acids phenylalanine, tyrosine, and tryptophan (Scheme 5.6). A potent herbicide used to kill weeds is *glyphosate* (*N*-[phosphonomethyl]glycine) whose mechanism of action is to modulate the activity of the carboxyvinyltransferase. The kinetic mechanism of this interaction has been studied by Fischer et al. who demonstrated that glyphosate acts as a linear specific inhibitor with respect to phosphoenolpyruvate and as hyperbolic mixed modifier with mechanism HMxD(A/I) with respect to 3-phosphoshikimate [31]. An additional parameter had to be taken into account by the authors, namely the activation of the enzyme by monovalent cations with effectiveness in the order $NH_4^+ > K^+, Rb^+, Na^+$. The effectiveness of glyphosate as inhibitor of 3-phosphoshikimate 1-carboxyvinyltransferase was attributed to the ability of

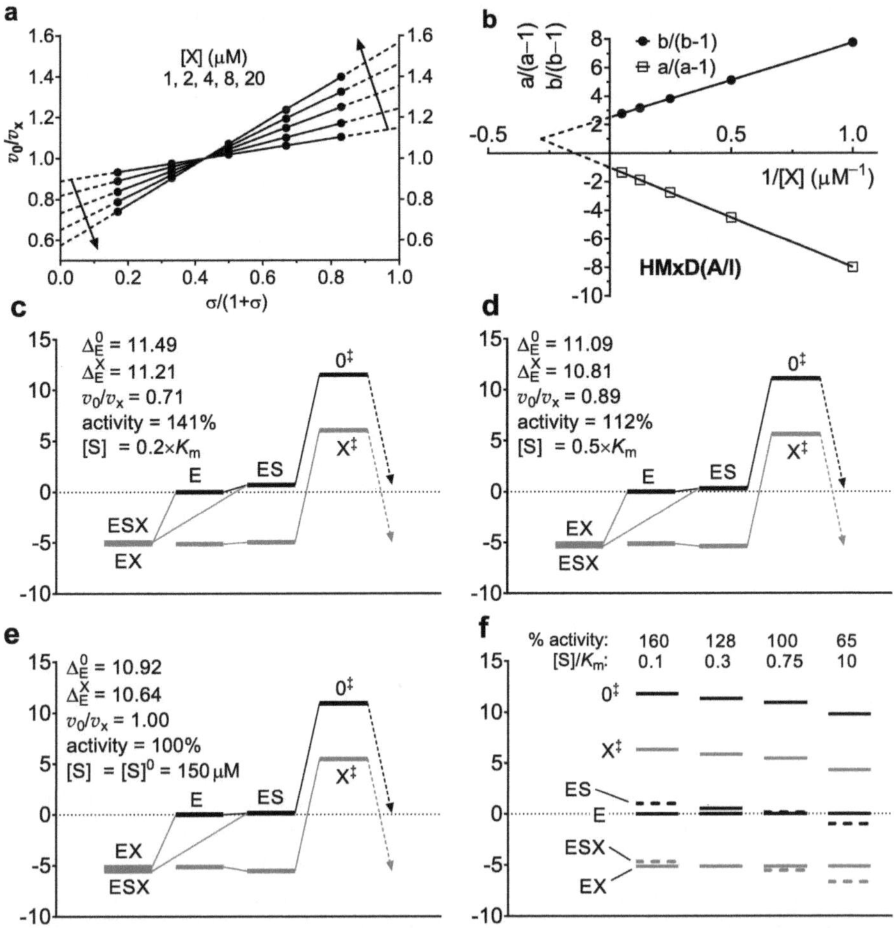

Fig. 5.22 Hyperbolic catalytic activation: HMxD(A/I). Primary (**a**) and secondary (**b**) specific velocity plot. The *arrow* in (**a**) indicates the increasing direction of [X]. (**c**)–(**e**) Kinetic barrier diagrams, (**f**) barcode-like representation of the various states at four substrate concentrations as indicated. $k_{cat} = 100\,s^{-1}$, $K_m = 200\,\mu M$, $K_X = 7.0\,\mu M$, $\alpha = 0.3$, $\beta = 0.6$, in (**c**)–(**f**) $[X] = 10 \times K_X$. Constant values common to (**c**)–(**f**): $\Delta_{ES}^0 = 10.79$, $\Delta_{ES}^X = 11.01$, $\Delta_{EX} = -5.15$, $\Delta_{ESX} = -5.68$. The critical substrate concentration is 150 μM, corresponding to $0.75 \times K_m$. The numerical values of the activation in panels (**c**)–(**f**) refer to 100 % activity in the absence of modifier

ammonium ions to decrease the inhibition constant of glyphosate at physiologically relevant NH_4^+-concentrations.

Intrinsic Tenase is the name given the complex consisting of the active blood coagulation factor IXa, its cofactor VIIIa and the substrate, factor X. The name *tenase* derives from the substrate cleaved by coagulation factor IXa, i.e., factor X: *ten* with the suffix *-ase*. Controlling the activity of the tenase complex represents a possible pharmacological strategy to prevent thrombosis. In a study on the inhibition

Table 5.14 Examples of hyperbolic mixed, dual modification, activation→inhibition, HMxD(A/I)

Enzymes	EC no.	Modifiers	Ref.
3-Phosphoshikimate 1-carboxyvinyltransferase	2.5.1.19	Glyphosate[a]	[31]
Intrinsic tenase complex	3.4.21.22[b]	ISIS 2302[c]	[91]
D-Amino acid oxidase	1.4.3.3	Risperidone	[1]
Cathepsin K	3.4.22.38	Chondroitin sulfate	[71]
Arachidonate 15-lipoxygenase	1.13.11.33	13(S)-HODE	[47]

[a]The herbicide *N*-[phosphonomethyl]glycine
[b]Coagulation factor IXa, the enzyme that activates factor X
[c]20-mer phosphorothioate oligodeoxyribonucleotide 5′-GCCCAAGCTGGCATCCGTCA-3′

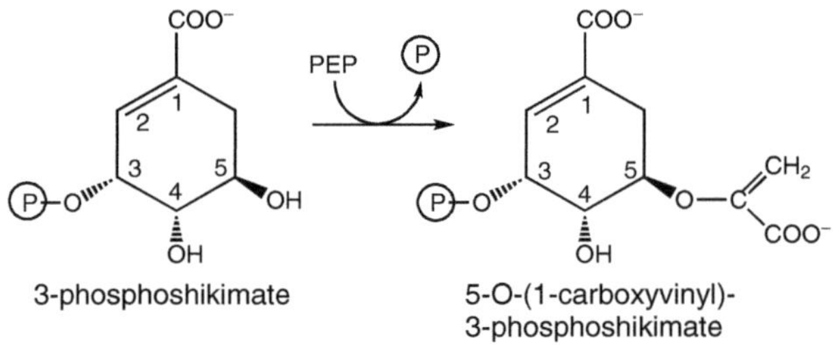

Scheme 5.6 The sixth of seven steps in the shikimate pathway

of coagulation factor IXa by phosphorothioate oligonucleotides, Sheehan and Phan [91] performed measurements to calculate the inhibition constant and fitted the equation for *partial uncompetitive inhibition* of Dixon and Webb [23, Eq. VIII.50] to initial rate against [X] data obtained ar a fixed substrate concentration. Furthermore, the authors performed measurements at variable substrate concentration in the absence and presence of saturating [X] to calculate k^0_{cat}, K^0_m, k^∞_{cat} and K^∞_m. Using the measured values of these parameters, $\alpha = K^\infty_m / K^0_m = 0.17$ and $\beta = k^\infty_{cat} / k^0_{cat} = 0.26$ can be calculated [91, Fig. 8]. The critical relative substrate concentration calculated with (3.41) is thus 0.12. These values are characteristic of the dual modification mechanism HMxD(A/I) discussed in this section. The interpretation given the oligonucleotide as inhibitor, in the substrate concentration ranges investigated by the authors, is correct because detecting activation had meant performing measurements at prohibitively small substrate concentrations, i.e., $[S] < 0.12K^0_m$.

D-Amino Acid Oxidase catalyzes the oxidative transformation of D-amino acids into 2-oxo acids with FAD as coenzyme. The substrates are a D-amino acid, water, and oxygen and the products are a 2-oxo acid, ammonia, and hydrogen peroxide. Abou El-Magd and collaborators determined the kinetic modification mechanism of D-amino acid oxidase by risperidone, an antipsychotic drug used to threat schizophrenia [1]. The diagnosis of the mechanism was performed clearly and the authors quoted Dixon and Webb [23, Eq. VIII.51] as the source of the rate equation

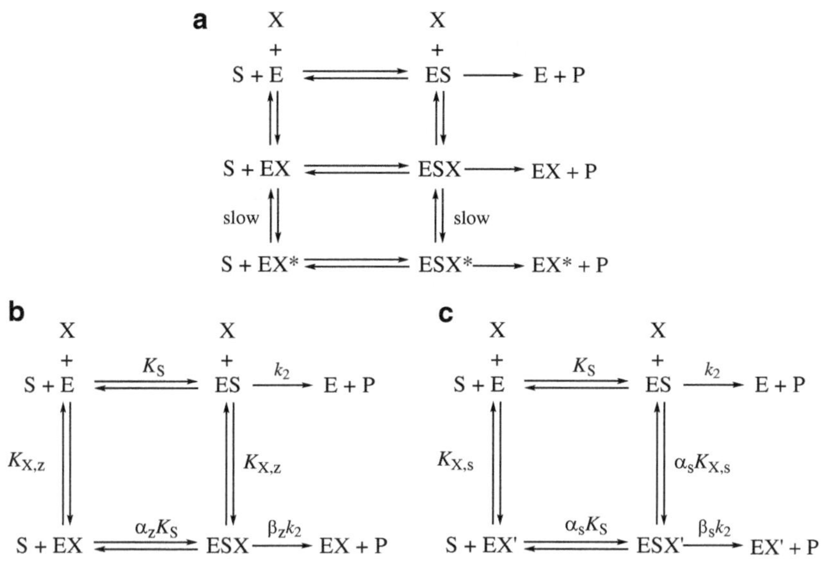

Scheme 5.7 Interaction of cathepsin K with glycosaminoglycans. (**a**) Rapid binding of the modifier to the enzyme followed by a slow reassessment of the EX* and ESX* complexes. (**b**) Simplification of scheme (**a**) at time zero, i.e., when the modifier has rapidly bound as EX and ESX but the transition to EX* and ESX* has not yet taken place. Following a transient, the steady-state is reached and the system can be characterized with the parameters shown in (**c**)

used to perform calculations. The reaction mechanism used by the authors was that in Scheme 5.8a. See also the comments on p. 273.

Cathepsin K interacts with sulphated polysaccharides that modulate its activity. In a study of the conformational flexibility of cathepsin K, Novinec et al. [71] found that chondroitin sulfate modulates enzyme activity by the hyperbolic mixed dual mechanism HMxD(A/I) while dermatan sulfate is a nonessential activator with predominantly catalytic character HMx(Sp<Ca)A (Sect. 5.4.6). These interactions are of the *slow-onset type*, which can be taken into account by the minimal mechanism shown in Scheme 5.7a. An analytical expression of the integrated rate equation for this system is quite complicated and of little practical use. However, for interpreting the net effect of glycosaminoglycans on catalysis by cathepsin K, mechanism (a) can be split to represent the rate at time zero, v_z, and after attainment of the steady-state, v_s, as shown in Scheme 5.7, panels (b) and (c), respectively.

The integrated rate equation of slow-onset interactions (8.10) (Chap. 8) was fitted to progress curves at various modifier concentrations to extract v_z and v_s, whose expressions correspond to (5.36) and (5.36) respectively. Using these relationships, the parameters for the interaction between cathepsin K and chondroitin sulfate could be calculated: $K_{X,z} = 83\ \mu M$, $K_{X,s} = 33\ \mu M$, $\alpha_z = 0.23$, $\alpha_s = 0.13$ and $\beta_z = \beta_s =$

0.85. The values of α and β are significantly smaller than 1 and $\alpha < \beta$, which are characteristics of the HMxD(A/I) mechanism.

$$v_{\mathrm{z}} = \frac{v_0 (1 + \sigma) \left(1 + \beta_{\mathrm{z}} \frac{[\mathrm{X}]}{\alpha_{\mathrm{z}} K_{\mathrm{X,z}}}\right)}{1 + \frac{[\mathrm{X}]}{K_{\mathrm{X,z}}} + \sigma \left(1 + \frac{[\mathrm{X}]}{\alpha_{\mathrm{z}} K_{\mathrm{X,z}}}\right)} \tag{5.35}$$

$$v_{\mathrm{s}} = \frac{v_0 (1 + \sigma) \left(1 + \beta_{\mathrm{s}} \frac{[\mathrm{X}]}{\alpha_{\mathrm{s}} K_{\mathrm{X,s}}}\right)}{1 + \frac{[\mathrm{X}]}{K_{\mathrm{X,s}}} + \sigma \left(1 + \frac{[\mathrm{X}]}{\alpha_{\mathrm{s}} K_{\mathrm{X,s}}}\right)} \tag{5.36}$$

The dual mechanism described here should be placed among the slow-onset interactions in Chap. 8. It is however retained in this chapter because the properties of the slow transient are not exploited for the analysis of its overall behavior.

Arachidonate 15-Lipoxygenase Inhibition by Reaction Products In the study by Joshi et al. [47] mentioned on p. 258, we discussed the modulation of arachidonate 15-lipoxygenase by 13(*S*)-HOTrE(γ) and 13(*S*)-HODE with γ-linolenic acid as substrate (see p. 258 for names and details). The same enzyme in the presence of arachidonic acid as substrate interacts with the hydroperoxide reaction product of linoleic acid, 13(*S*)-HODE, by the HMxD(A/I) mechanism. The parameters, the critical substrate concentration, and the change in rate as a function of modifier concentration for a fixed concentration of the substrate are shown in Fig. 5.23 without further comments.

The examples of product inhibition in the reactions catalyzed by arachidonate 15-lipoxygenase are important to reflect on the significance of enzyme regulation in an environment that has little to do with a measuring cell placed inside a photometer. It suffices to imagine the crowding of molecules in the small intestine after ingesting a meal, where carbohydrates, lipids, and proteins are split in their constituents to allow intestinal absorption. Regarding the fatty acids, once absorption has taken place, some have nutritional purposes while others become the substrates of lipoxygenases, cyclooxygenases, and cytochrome P450. The transformations catalyzed by these enzymes generate a multitude of metabolites that play biological roles either as direct mediators of cellular responses or as precursors of other mediators such as prostaglandins and leukotrienes. In the congested environment of lipid metabolites, traffic lights are necessary to regulate their functions.

Allosteric modulation of the enzymes responsible for fatty acid metabolism is one of the possible regulation mechanisms investigated in the laboratory of Theodore Holman, leader of the group who investigated product inhibition of arachidonate 15-lipoxygenase discussed here and in Sect. 5.4.1. In humans, the lipoxygenase consists of six isozymes that act on seven fatty acids as substrates displaying various degrees of specificity. The finding that reaction products bind to allosteric sites of the isozymes is important for the changes induced in substrate

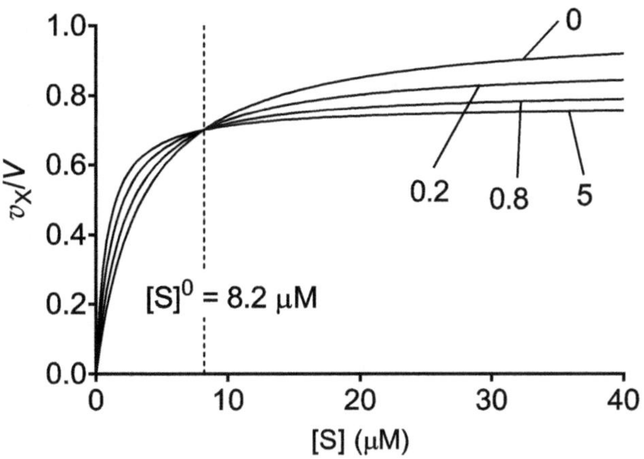

Fig. 5.23 Hyperbolic mixed, dual modification: HMxD(A/I). Normalized rate against [S] at various [X] (*numbers labeling the lines*). The critical substrate concentration ($[S]^0$, *dashed line*) demarcates the zones of activation (*left to the line*) and activation (*right to the line*). The curves were plotted with (3.32) and the parameters of Joshi et al. for the oxidation of arachidonic acid (the substrate) catalyzed by arachidonate 15-lipoxygenase [47, Fig. 7]: $\alpha = 0.2$, $\beta = 0.76$, $K_X = 1.2\,\mu M$, $K_m = 3.5\,\mu M$. The modifier was 13(*S*)-HODE

specificity among the isozymes. Regulation mechanisms can thus become operative at the level of redistribution of reaction products depending on the relative activities of the isozymes and substrate availability.

5.4.4.1 Alternative Schemes and Misinterpretations

The hyperbolic mixed, dual modification mechanism HMxD(A/I) has been recognized and correctly interpreted for a long time. It was considered and commented for its dual property to activate or inhibit enzyme activity depending on substrate concentration by Dixon and Webb [23, Eq. VIII.51], Segel [89, Eq. IV-43] and Tipton [102, p. 133]. A comment that might be possibly felt formal and pedantic is however necessary regarding the reaction schemes used for this mechanism. The criticism to the reaction paths shown in Scheme 5.8 is the same made in Sect. 3.4.4 for essential activation.

Scheme 5.8a has been used by Dixon and Webb [23, VIII.49], by Tipton [102, p. 133] and by Abou El-Magd et al. (with credits to Dixon and Webb) [1]. Since reinventions are common in all branches of science including enzyme kinetics, Scheme 5.8a has been re-published as new in 2005 by Park and coworkers [75]. The problem with Scheme 5.8a is that the catalytic step through ESX considers a termolecular reaction. The convention of writing the catalytic steps with an unidirectional arrow in the direction of product formation is based on the assumption that only the initial rate is measured so that $[P] \approx 0$. If however the reverse reaction is also considered, the collision of three molecules to yield ESX is highly improbable.

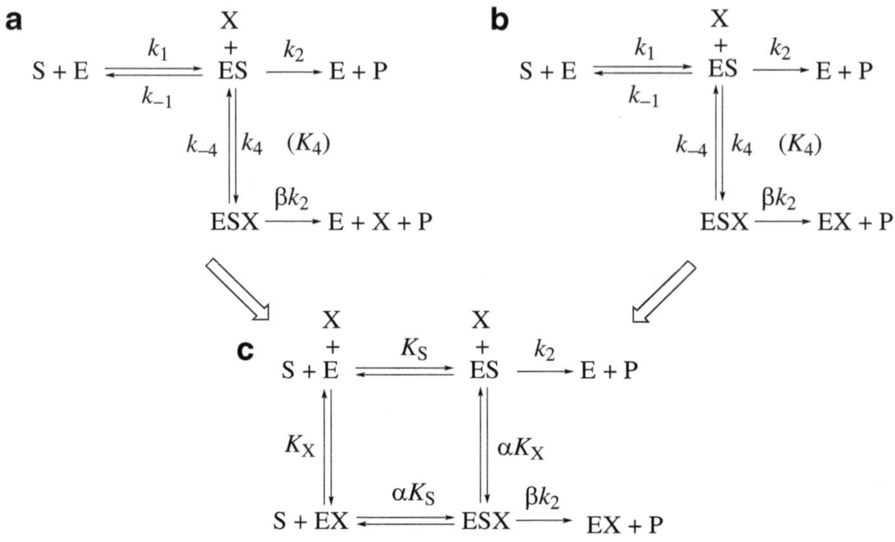

Scheme 5.8 Ambiguous representations of reactions schemes involving catalytic, predominantly catalytic or dual modification mechanisms. (**a**) The ESX complex decays into free enzyme, modifier, and product(s). (**b**) Product(s) and the EX complex are generated but, as written, EX is not further utilized being equivalent to enzyme inactivation. (**c**) Consideration of a thermodynamic cycle in which all enzyme-containing species communicate with each other and are capable of allosteric coupling

The rate equation for the mechanism in Scheme 5.8a has two terms in the numerator (corresponding to the two catalytic steps) and three terms in the denominator (corresponding to E, ES and ESX). For the mechanism in (c) of the same scheme the numerator is the same but the denominator contains a fourth term that represents the EX complex. Both equations used to analyze hyperbolic mixed dual modification HMxD(A/I) are equivalent because the specific component is scarcely or no represented ($\alpha < 1$ or $\alpha \ll 1$). The dependence of the parameters on modifier concentration is the same with both equations meaning that we are faced with a case of homeomorphism. However, the examples commented above in this section, notably the case of arachidonate 15-lipoxygenase, were concerned with HMxD(A/I) in the context of *allosteric regulation*, which *is not* taken into account by Scheme 5.8a because there is no allosteric coupling between substrate and modifier. This property is however intrinsic in Scheme 5.8c.

More problems are associated with Scheme 5.8b, used for instance by Segel [89, Eq. IV-43], in which EX suffers from the fact that it is not further utilized after its formation as decay product of ESX and should therefore be formally treated as an irreversible species. Nevertheless, the equation provided by Segel does not contain the term representative of EX in the denominator and is therefore formally incorrect if referred to this mechanism. Also in this case, allosteric coupling is not taken into account. However, the equation derived by Segel predicts the behavior of the hyperbolic mixed dual mechanism HMxD(A/I) and as such it has been discussed. Though, this was not the case in the two following occurrences.

Data and Fitted Model Do Not Agree Tyrosinase (EC 1.14.18.1) is involved in the synthesis of betalains, e.g., the red color of beets, and melanin, the black pigment of many organisms. Tyrosinase catalyzes the oxidation of L-tyrosine to dopaquinone and of L-dopa to dopaquinone, the precursor of betalains and melanin. Park et al., measuring the effects of chloride ions on the activity of mushroom tyrosinase, diagnosed a new type of *partial hyperbolic uncompetitive inhibition* and provided a new rate equation to evaluate the inhibition constant [75]. This equation, based on Scheme 5.8a, is the same as published by Dixon, Webb, Segel, and Tipton mentioned above who correctly discussed the dual nature of this equation for describing activation or inhibition depending on substrate concentration. The main issue is that this equation does not agree with the experimental findings by Park et al. [75]. In fact, the lines in the plots of Figs. 1, 4, and 5 (panels B) in this paper have been obviously drawn by hand because the employed equation predicts a compelling intersection point in the first quadrant. Despite being seemingly parallel to one another, the lines intersect somewhere in the third quadrant that indicates $\alpha < 1$. Since the catalytic constant at saturating chloride concentration was about one fifth of that in the absence of modifier ($\beta \approx 0.2$), the data of Park et al. [75] belong to the hyperbolic mixed, predominantly catalytic inhibition mechanism HMx(Sp<Ca)I. The putative reason of this disagreement between data and model is that the authors expected an *uncompetitive behavior* in which the catalytic constant and the Michaelis constant are equally affected by the modifier, i.e., the specificity constant does not depend on [X]. However, the considered rate equation predicts that the specificity constant increases with increasing [X].

Graphical Method Yoshino described a graphical method for the determination of inhibition parameters for linear and hyperbolic inhibitors [113]. The method consists in plotting $v_X/(v_0 - v_X)$ against $1/[X]$: the resulting graphic consists of a family of straight lines that converge to a common point on the negative side of the abscissa for hyperbolic inhibitors and at the origin of the axes for linear inhibitors. The author illustrated the method for the common mechanisms, including *partial uncompetitive inhibition*. The rate equation based on Scheme 5.8b was the same as that of Dixon, Webb, Segel, and Tipton mentioned above, rearranged for the purpose of the graphical method and reproduced as follows using the symbols in Scheme 5.8b:

$$\frac{v_X}{v_0 - v_X} = \frac{\left(1 + \frac{[S]}{K_S}\right)\left(\frac{K_4}{[X]} + \beta\right)}{(1-\beta)\frac{[S]}{K_S} - \beta}\,. \tag{5.37}$$

The same comments made above on the reactions in Scheme 5.8 apply also to this case: allosteric coupling exists although this information is not present in (5.37) because the EX complex has been swept under the carpet. Possibly the author was

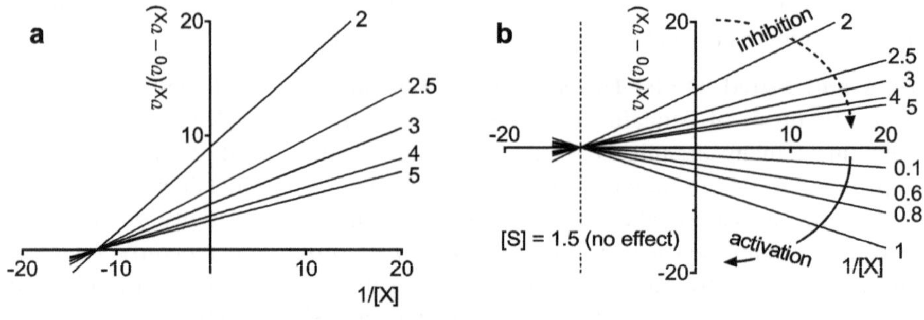

Fig. 5.24 The Yoshino plot for *partial uncompetitive inhibition*. The *lines* were plotted with (5.37) and the parameters $K_4 = 0.05$, $\beta = 0.6$, $K_S = 1$ and variable [S] as indicated by the numbers next to the lines (in arbitrary units). (**a**) Using the [S]-values of the original publication [113, Fig. 4 (a)]. (**b**) Additional [S]-values including the critical substrate concentration (1.5) that marks the boundary between activation (*continuous arched arrow*) and inhibition (*dashed arched arrow*)

interested in discussing only inhibition and did not show the full potential of the employed equation because (5.37) describes the HMxD(A/I) mechanism. The figure shown by Yoshino to illustrate this mechanism is shown in Fig. 5.24 in its original version (panel a) and after adding a few substrate concentrations (panel b). The critical substrate concentration can be calculated from $\beta = 0.6$, $\alpha \approx 0$ (in reality $\alpha < 1$ or $\alpha \ll 1$) and $K_S = 1$ using (3.40), i.e., 1.5 in the arbitrary units chosen. At this substrate concentration the dependent variable in (5.37) is indefinite because $v_0 = v_X$ (division by zero). The vertical dashed line in Fig. 5.24b represents the imaginary demarcation between the zones of activation ([S] < 1.5) and those of inhibition ([S] > 1.5).

In conclusion, the preferred reaction scheme for describing the mechanism HMxD(A/I), as well as related mechanisms with predominantly catalytic character is Scheme 5.8c.

5.4.5 *Hyperbolic Specific Activation*

Assigned acronym: HSpA. The modifier is characterized by $0 < \alpha < 1$ and $\beta = 1$. Similar to hyperbolic specific inhibition (HSpI, Sect. 5.3.1) the modifier does not affect catalysis since $\beta = 1$, while the apparent Michaelis constant decreases with increasing modifier concentration. Activation occurs by promoting substrate binding to the enzyme ($\alpha < 1$) so that the steady-state concentration of ESX is enhanced. Nonessential activation corresponding to this mechanism has been described in 1952 by Segal et al. in their survey of enzyme inhibition and activation

[88, p. 193].[5] It matches pure K-type activation in the nomenclature of Monod et al. [67] (Table 3.1).

The specific velocity plot consisting of lines with positive slopes that intersect at the coordinates 1,1 is sufficient to diagnose this mechanism. If a mirror is placed along the abscissa of the specific velocity plot, HSpA appears as the specular image of HSpI (Fig. 5.25 compared with Fig. 5.9).

The kinetic barriers $\Delta_{\mathrm{ES}}^{\mathrm{X}}$ and Δ_{ES}^{0} are equal and have the same heights at any substrate concentration (Fig. 5.25c–f). Since they depend on $k_{\mathrm{cat}}^{\mathrm{app}}$ and k_{cat}, respectively, this property reflects the independence of the catalytic constant on modifier concentration, a trait common to LSpI, HSpI, and HSpA. Three examples are listed in Table 5.15 and discussed below.

Intestinal Sugar Transport The mechanism of activation of sucrose α-glucosidase by sodium ions varies from species to species. Giorgio Semenza, a pioneer of intestinal sugar transport, was the first to report an allosteric mechanism of the K-type for the activation of hamster and rat sucrose α-glucosidase by Na^+ [90], which was accompanied by cooperative interaction. Kinetic experiments revealed that the apparent Michaelis constant decreased with $[Na^+]$ while the limiting rate was unaffected, a finding that fits the hyperbolic specific activation mechanism, HSpA. The experiments were internally consistent since the ratios of the Michaelis constants in the absence of and at saturating $[Na^+]$ were the same as the ratios of the dissociation constants of $[Na^+]$ at $[S] = 0$ and at saturating [S], thus respecting the principle of microscopic reversibility.

Hyperbolic specific activation was not the mechanism determined for the sucrose α-glucosidase extracted from the small intestine of the Greek turtle. For this species, the dependence of the kinetic parameters on $[Na^+]$ revealed K-type inhibition and V-type activation, which is compatible with either one of the activation mechanisms HMxD(I/A), HCaA or HMx(Sp>Ca)A (see Table 3.1). Since the increase of $K_{\mathrm{m}}^{\mathrm{app}}$ with increasing [X] was modest, the mechanism could also be HMx(Sp=Ca)A. As commented by Semenza, this result could exclude the presence of more than one *sucrase* in the turtle intestine.

Sodium Ion Activation of Sucrose α-Glucosidase depends on pH as demonstrated by Gupta et al. [39]. The authors used the K- and V-type nomenclature and provided adequate numeric information by evaluating the kinetic parameters in the presence and absence of $[Na^+]$. At pH $= 5.0$ the activation mechanism was HMx(Sp=Ca)A and therefore did not depend on allosteric coupling between the binding of substrate and sodium ($\alpha = 1$ and $\beta = 2.6$). At pH $= 7.2$ the mechanism was clearly HSpA because $\alpha \approx 0.8$ and β was statistically indistinguishable from 1. In contrast, at pH $= 8.5$ Na^+ was inhibitory with a mechanism that, judging from the values of the parameters and error propagation, corresponds to hyperbolic balanced inhibition [HMx(Sp=Ca)I].

[5] As an aid to locate the unnumbered rate equation, this is in the middle of p. 193, the case for which $\phi_{\mathrm{s}} = \phi_{\mathrm{a}}$.

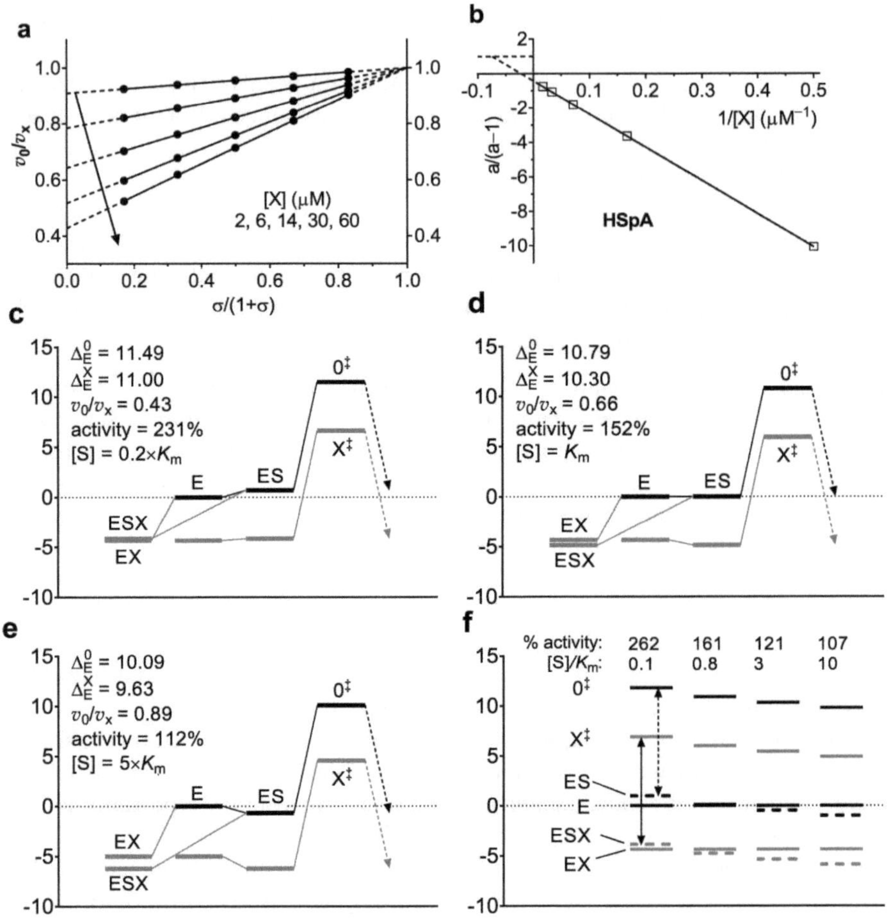

Fig. 5.25 Hyperbolic specific activation: HSpA. Primary (**a**) and secondary (**b**) specific velocity plot. The *arrow* in (**a**) indicates the increasing direction of [X]. (**c**)–(**e**) Kinetic barrier diagrams, (**f**) barcode-like representation of the various states at four substrate concentrations as indicated. $k_{cat} = 100\,s^{-1}$, $K_m = 200\,\mu M$, $K_X = 45\,\mu M$, $\alpha = 0.3$, $\beta = 1.0$, in (**c**)–(**f**) $[X] = 10 \times K_X$. Constant values common to (**c**)–(**f**): $\Delta_{ES}^X = \Delta_{ES}^0 = 10.79$ (the same value at any [S]), $\Delta_{EX} = -4.35$, $\Delta_{ESX} = -4.87$. In (f) the *solid double arrow* represents Δ_{ES}^X and the *dashed double arrow* is Δ_{ES}^0. The numerical values of the activation in panels (**c**)–(**f**) refer to 100 % activity in the absence of modifier

Legionnaires' Disease A case of hyperbolic specific activation has been documented by Srinivasan and coworkers [96] while studying allosteric modulation of the cytosolic nucleotidase II from the Gram-negative bacterium *Legionella pneumophila*. This is the primary pathogenic agent of Legionnaires' disease, a form of pneumonia. The bacterium grows preferentially in water at high temperatures and, in the civilized world, can be an unwanted host in reservoirs for hot water and in the air-conditioning plants of department stores, large office buildings and

Table 5.15 Examples of hyperbolic specific activation, HSpA

Enzymes	EC no.	Activators	Ref.
Sucrose α-glucosidase	3.2.1.48	Na^+ (species-dependent)	[90]
Sucrose α-glucosidase	3.2.1.48	Na^+ (pH-dependent)	[39]
5′-Nucleotidase	3.1.3.5	GMP	[96]

hotels. Nucleotidases catalyze the hydrolytic cleavage of a 5′-ribonucleotide into the corresponding ribonucleoside and inorganic phosphate with a wide specificity for 5′-nucleotides. Srinivasan and colleagues found that the hydrolysis of GMP was activated by GTP, GDP, and by the substrate itself. An activator binding site consisting of residues R86 and Y421 was identified in the crystal structure at the dimer interface of the tetrameric enzyme. Double mutation of R86 and Y421 abolished activation by the allosteric modifiers. The observed cooperative activation by GMP was explained by its simultaneous binding to the active site and to the allosteric site.

With GMP as substrate, the mechanism of activation by GTP was studied by measuring the effects of modifier concentration on the kinetic parameters. GTP elicited activation by promoting the formation of the ESX complex, documented by a decrease of K_m^{app} and by the independence of the limiting rate with increasing [GTP], i.e., the mechanism was hyperbolic specific activation ($\alpha = 0.7$ and $\beta = 1$).

5.4.6 *Hyperbolic Mixed, Predominantly Catalytic Activation*

Assigned acronym: HMx(Sp<Ca)A. As in all other mechanisms with catalytic or predominantly catalytic character, the affinity of the substrate for the enzyme is promoted by the modifier and vice versa, i.e., $0 < \alpha < 1$. Activation is achieved by increasing the catalytic constant by a factor $\beta > 1$. The straight lines in the primary specific velocity plot have positive slopes and converge to a point at an abscissa value > 1, which are properties sufficient for diagnosing the mechanism.

The distinctive trait of the kinetic barrier diagrams of this nonessential activation mechanism is the degree of activation decreasing with increasing substrate concentration (Fig. 5.26c–f). Activation occurs at any [S] because $\Delta_{ES}^{X} < \Delta_{ES}^{0}$ despite the tiny difference between the two barriers. Δ_{ES}^{0} and Δ_{ES}^{X} depend on k_{cat}^{app} and k_{cat}, respectively.

Only one case is discussed here, for which the kinetic mechanism has been reported with the due details.

The Activation of Human Cathepsin K by dermatan sulfate represents a case of HMx(Sp<Ca)A studied by Novinec et al. [71]. The allosteric interaction of this system occurs by slow-onset binding of the modifier and therefore the kinetic parameters were determined with the approach described on p. 271, with reaction paths in Scheme 5.7 and Eqs. (5.35), (5.36). The values of the relevant parameters

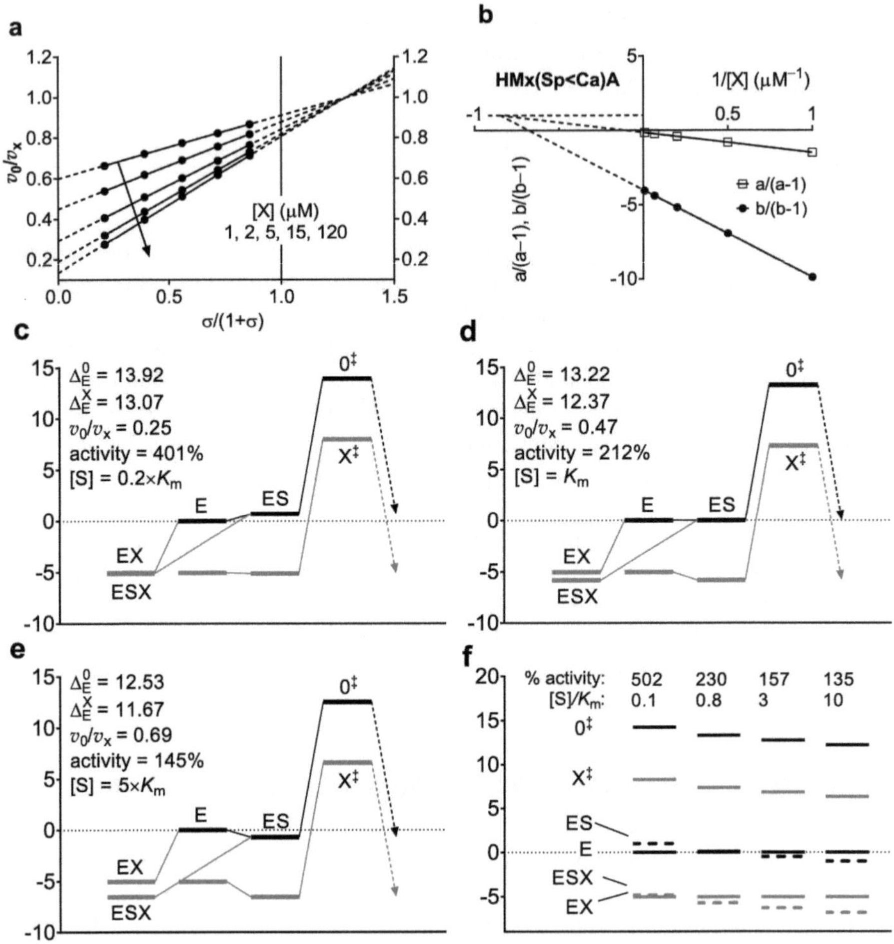

Fig. 5.26 Hyperbolic mixed, predominantly catalytic activation: HMx(Sp<Ca)A. Primary (**a**) and secondary (**b**) specific velocity plot. The *arrow* in (**a**) indicates the increasing direction of [X]. (**c**)–(**e**) Kinetic barrier diagrams, (**f**) barcode-like representation of the various states at four substrate concentrations as indicated. $k_{cat} = 0.37\,\mathrm{s}^{-1}$, $K_m = 7.7\,\mu\mathrm{M}$, $K_X = 9.2\,\mu\mathrm{M}$, $\alpha = 0.16$, $\beta = 1.25$, in (**c**)–(**f**) $[X] = 10 \times K_X$. These are the parameters for the activation of human cathepsin K by dermatan sulfate [71]. Constant values common to (**c**)–(**f**): $\Delta^0_{ES} = 13.22$, $\Delta^X_{ES} = 13.13$, $\Delta_{EX} = -5.04$, $\Delta_{ESX} = -5.83$. The numerical values of the activation in panels (**c**)–(**f**) refer to 100 % activity in the absence of modifier. The degree of activation decreases for increasing [S]

for the interaction between cathepsin K and dermatan sulfate were: $K_{X,z} = 53\,\mu\mathrm{M}$, $K_{X,s} = 9.2\,\mu\mathrm{M}$, $\alpha_z = 0.17$, $\alpha_s = 0.16$ and $\beta_z = \beta_s = 1.25$, which are characteristic of an allosteric, nonessential activator with mechanism HMx(Sp<Ca)A.

On the basis of crystallographic evidence [57], of statistical coupling analysis to identify allosteric sites (p. 261), intrinsic fluorescence changes upon binding [71] and enzyme kinetic results [71], the binding process of glycosaminoglycans

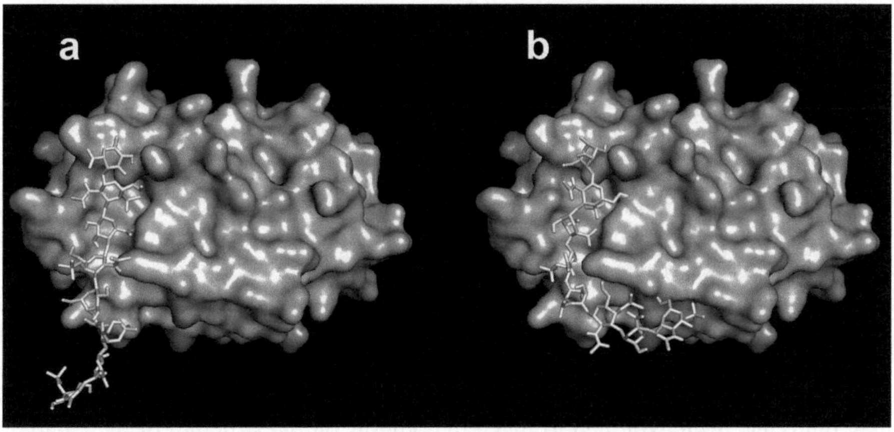

Fig. 5.27 Binding mode of chondroitin sulfate and dermatan sulfate to human cathepsin K exemplified by a fragment of dermatan sulfate. The glycosaminoglycan chain, shown as sticks, consists of four repetitive 4-sulfo-*N*-acetylgalactosamine-iduronic acid disaccharides. (**a**) Rapid recognition of an allosteric site by one part of the glycosaminoglycan molecule followed by a slow rearrangement to form a tighter complex (**b**). Efficient binding to the enzyme is accomplished by a large structural rearrangement of dermatan sulfate with respect to its conformation in solution [71]

to cathepsin K can be visualized as shown in Fig. 5.27 for dermatan sulfate. The slow-onset interaction determined kinetically has been interpreted as rapid binding of one segment of the glycosaminoglycan to a structurally ascertained allosteric site, followed by tightening of binding in a slower step accompanied by a rearrangement of the backbone of the molecule in comparison with its structure in solution [2]. Among the glycosaminoglycans, those containing iduronic acid, i.e., dermatan sulfate, heparin, and heparan sulfate, own additional conformational flexibility with respect to glycosaminoglycans with similar negative charge distribution but containing glucuronic acid [18]. This property explains the stronger interactions of heparin, heparan sulfate and dermatan sulfate with biological targets.

5.4.7 *Hyperbolic Mixed, Balanced Activation*

Assigned acronym: HMx(Sp=Ca)A. A family of parallel straight lines with ordinate intercept decreasing for increasing modifier concentration is sufficient to recognize this mechanism. Allosteric coupling between substrate and modifier is absent as indicated by the dominant character $\alpha = 1$, while nonessential activation results from $\beta > 1$. Another characteristic feature of this mechanism is the independence of K_m on modifier concentration, while the catalytic and the specificity constant increase hyperbolically with [X] (Fig. 3.5). These properties can be deduced from (3.3)–(3.5) after inserting $\alpha = 1$. The activating effect exerted by the modifier is due by stimulating catalysis while not interfering with substrate binding.

In the kinetic barrier diagrams, the traits common to all balanced mechanisms, LMx(Sp=Ca)I, HMx(Sp=Ca)I, and HMx(Sp=Ca)A, are $(\Delta_{E}^{0} - \Delta_{ES}^{0}) = (\Delta_{E}^{X} - \Delta_{ES}^{X})$ and $\Delta_{EX} = \Delta_{ESX}$. As discussed in Sects. 5.2.5 and 5.3.5 these properties are due to the independence of the Michaelis constant on modifier concentration and to the equal participation of the specific and the catalytic component to enzyme modification. The common outcome to the tree balanced mechanisms is that inhibition or activation do not depend on substrate concentration. Figure 5.28c–f shows activation because the kinetic barrier Δ_{ES}^{X} (transition state $X^{\ddagger}$) is smaller than Δ_{ES}^{0} (transition state $0^{\ddagger}$).

This activation mechanism may have a physiological significance as exemplified by the interaction between myeloblastin and a glycosaminoglycan briefly discussed below. Also, the occurrence of hyperbolic mixed, balanced activation in interactions between molecules unrelated to the natural environment of the enzyme to which they bind, has been exploited to make hypotheses on the structure of the catalytic site. The examples discussed below are listed in Table 5.16.

Kinetic Studies with Trypsin and related digestive peptidases, such as the much investigated chymotrypsin, attracted the interest of early researchers in enzyme kinetics chiefly because of their availability in large amounts from animal pancreas. Inagami and Murachi measured the effects of alkylammonium ions on the trypsin-catalyzed hydrolysis of the low molecular mass acetylglycine ethyl ester as substrate [45]. Methyl-, ethyl-, and 1-propylammonium ions acted as nonessential activators, while the larger 1-*n*-butylammonium ion was an inhibitor. Trypsin activation was evaluated by measuring the effects of the modifiers on the Michaelis constant and on the catalytic constant. K_m did not depend on the concentration of the modifiers while the apparent k_{cat} increased with increasing [X]. The authors did not name the activation mechanism but we can safely say that this is hyperbolic mixed, balanced activation.

The significance of these results must be sought in comparison with parallel measurement using the more bulky substrate benzoyl-L-arginine ethyl ester, in the presence of which the same ammonium ions behaved as lineal specific inhibitors [44]. Numerically, the *inhibition* constants with benzoyl-L-arginine ethyl ester as substrate were practically identical to the *activation* constants with acetylglycine ethyl ester as substrate. From this information the authors deduced that the alkylammonium ions occupied the specificity-determining site of trypsin by excluding a bulky substrate that fully occupies the *specificity site* (linear *specific* inhibition) and activating the enzyme when this acts on a small (pseudo)substrate that does not share a relationship with this site. At the time the papers by Inagami were published (1964) the subsites nomenclature of the peptidases of Schechter and Berger (published in 1967) was unknown [82]. Inagami's specificity site can be identified with the subsite S_1 of trypsin (Scheme 4.1a) that accommodates the arginine residue of the bulky substrate.

Methylguanidine and Ethylguanidine activate trypsin with the same mechanism as alkylammonium ions, hyperbolic mixed, balanced activation as demonstrated by Inagami and York [46]. As in the case of the alkylammonium ions (preceding case),

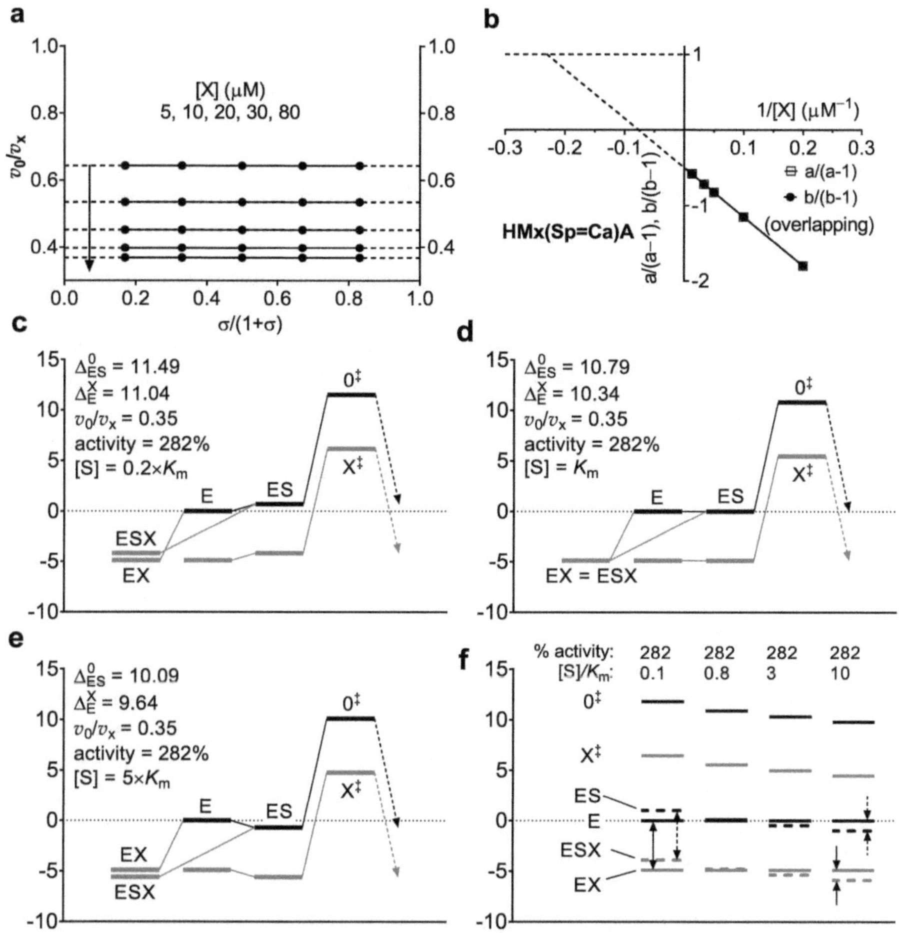

Fig. 5.28 Hyperbolic balanced activation: HMx(Sp=Ca)A. Primary (**a**) and secondary (**b**) specific velocity plot. The *arrow* in (**a**) indicates the increasing direction of [X]. (**c**)–(**e**) Kinetic barrier diagrams, (**f**) barcode-like representation of the various states at four substrate concentrations as indicated. $k_{cat} = 100\,s^{-1}$, $K_m = 200\,\mu M$, $K_X = 13\,\mu M$, $\alpha = 1.0$, $\beta = 3.0$, in (**c**)–(**f**) $[X] = 10 \times K_X$. The salient traits are $\Delta_{EX} = \Delta_{ESX}$, marked by the *solid* and the *dashed double arrows* in (**f**), respectively, and $(\Delta^0_E - \Delta^0_{ES}) = (\Delta^X_E - \Delta^X_{ES})$ represented by the amplitudes between the *dashed* and the *solid facing arrows*, respectively. Constant values common to (**c**)–(**f**): $\Delta^0_{ES} = 10.79$, $\Delta^X_{ES} = 10.34$, $\Delta_{EX} = \Delta_{ESX} = -4.89$. The degree of activation is the same at any substrate concentration and is thus determined by the concentration of the modifier

activation was attributed to the occupancy of the (primary) specificity site of the enzyme. In this study, the authors complemented kinetics with thermodynamics and demonstrated that the activating effect of the alkylguanidine derivatives was due to an increase of the entropy of activation.

Table 5.16 Examples of hyperbolic mixed, balanced activation, HMx(Sp=Ca)A

Enzymes	EC no.	Activators	Ref.
Trypsin	3.4.21.4	Methyl-, ethyl- and 1-propylammonium	[45]
Trypsin	3.4.21.4	Methylguanidine and ethylguanidine	[46]
Myeloblastin	3.4.21.76	Glycosaminoglycans[a]	[36]

[a]Chondroitin sulfate, dermatan sulfate, and a polysulfated chondroitin

Myeloblastin is a serine peptidase of the azurophil granules of neutrophilic polymorphonuclear leukocytes (shortly *neutrophils*) that impenitent scientists continue to call proteinase 3.[6] Myeloblastin is a protein that exerts a double physiological role as endopeptidase and antibiotic and a pathological role as autoantigen in Wegener's granulomatosis [5].

In synergy with cathepsin G, elastase-2 and azurocidin, myeloblastin is an efficient anti-bactericidal agent. In this function, myeloblastin *does not act* as a peptidase but as a protein devoid of catalytic properties. This fact has been put in relation with a very slow transition from an inactive state at low pH, as that found in resting lysosomes, to an active state as peptidase by raising the pH to a level found in phagolysosomes. Damping the transition from inactive to active means that the unwanted proteolytic activity of myeloblastin is greatly retarded in favor of microbicidal activity during the pH excursion from acidic to neutral and back during the phagocytic process [8]. This fits the notion that the microbicidal proteins of the neutrophils have preferred pH values to exert their function and for myeloblastin this is pH $= 5.5$. The rate constants measured for the hysteretic transition of myeloblastin from inactive to active are compatible with the phagocytic times in human phagolysosomes [8].

Regulation of myeloblastin activity can be accomplished by pH-excursions within organelles, through inhibition by plasma peptidase inhibitors in the extracellular space (α-1-peptidase inhibitor, α-2-macroglobulin and elafin) and possibly through modulation by electrostatic interaction with glycosaminoglycans at the interface with connective tissues or within intracellular organelles, as proposed for elastase-2 that shares many properties with myeloblastin [52].

In virtue of their high isoelectric point and a distribution of positively-charged arginine residues at the surface of the molecules, myeloblastin, cathepsin G and elastase-2 interact with polyanions. While for cathepsin G and elastase-2 this interaction results in inhibition of the enzymatic activity, myeloblastin is activated. This effect has been demonstrated by Hans Früh et al. for chondroitin sulfate, dermatan sulfate and a chondroitin polysulfate [36]. The last substance was produced by introducing additional sulfates into chondroitin for an average of 3.25 sulfate groups per disaccharide unit: one each on C-4 and C-6 of *N*-acetylgalactosamine and

[6]The reluctance in accepting rules and recommendations released by International Organizations in the interest of the scientific community does not help the dissemination of scientific information using a common language.

one sulfate (occasionally two) on glucuronic acid. Steady-state measurement performed at variable concentrations of a synthetic substrate and glycosaminoglycans yielded families of parallel lines in the specific velocity plot with intersections on the ordinate decreasing for increasing modifier concentration. The interaction was interpreted as nonessential activation and, using the nomenclature proposed in this book, the mechanism can now be ascribed to hyperbolic mixed, balanced activation. However, when the substrate was radiolabeled elastin coated to a surface, the interaction was hyperbolic mixed, balanced inhibition. The switch from activation to inhibition passing from a low-molecular mass substrate to a polymeric substrate is not surprising in view of a more extended binding mode of the polymer that is likely to occupy all specificity subsites of the peptidase.

Appendix

Rate Equations for Linear Specific Inhibition and Linear Specific Activation

In linear specific inhibition, the association between enzyme and inhibitor (X) is in equilibrium because EX is a dead-end complex. Therefore, using the Cha simplification of the King–Altman method for paths in equilibrium, this step can be represented as node (A), while the second node (B) is constituted by the single species ES as shown in Scheme 5.9a.

In deriving the rate equation for essential activation the association of E and X is also considered to be in rapid equilibrium. This assumption is partly justified by the fact that essential activators are often either rapidly reacting protons or small ions. The values of the nodes are as shown in Scheme 5.9b. The numbering of constants for linear specific activation is maintained coherent with the topology of the general modifier mechanism. Values of the nodes:

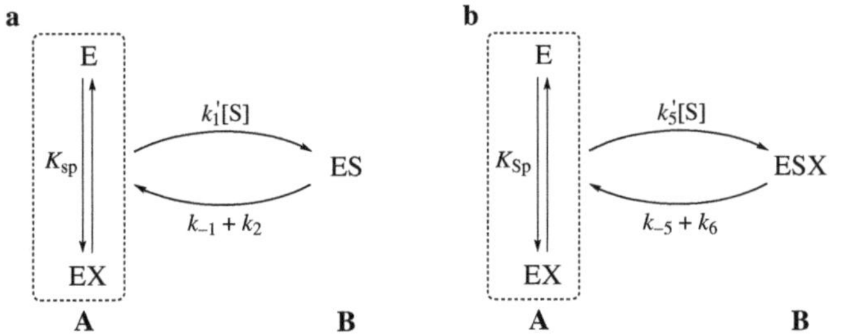

Scheme 5.9 Linear specific inhibition (**a**) and essential activation (**b**). A and B represent the nodes of the mechanisms

$$\text{LSpI}: \quad E_A = k_{-1} + k_2\,, \quad E_B = k_1'\,[\text{S}]$$
$$\text{LSpA}: \quad E_A = k_{-5} + k_6\,, \quad E_B = k_5'\,[\text{S}]\;.$$

Fractions of species in the nodes:

$$\text{LSpI}: \quad f_E = \frac{1}{1 + [\text{X}]/K_{\text{Sp}}}\,, \quad f_{EX} = \frac{[\text{X}]/K_{\text{Sp}}}{1 + [\text{X}]/K_{\text{Sp}}}\,, \quad f_{ES} = 1$$
$$\text{LSpA}: \quad f_E = \frac{1}{1 + [\text{X}]/K_{\text{Sp}}}\,, \quad f_{EX} = \frac{[\text{X}]/K_{\text{Sp}}}{1 + [\text{X}]/K_{\text{Sp}}}\,, \quad f_{ESX} = 1\;.$$

Values of the nodes corrected for fractions of the species within the nodes:

$$\text{LSpI}: \; E_A = (k_{-1} + k_2) f_{ES} = k_{-1} + k_2\,, \quad E_B = k_1 f_E\,[\text{S}] = \frac{k_1\,[\text{S}]}{1 + [\text{X}]/K_{\text{Sp}}}$$
$$\text{LSpA}: \; E_A = (k_{-5} + k_6) f_{ESX} = k_{-5} + k_6\,, \quad E_B = k_5 f_{EX}\,[\text{S}] = \frac{k_5\left([\text{X}]/K_{\text{Sp}}\right)[\text{S}]}{1 + [\text{X}]/K_{\text{Sp}}}\;.$$

Fractions of nodes B, needed for calculating the velocity; the fractions in nodes A are not needed for this purpose:

$$\text{LSpI}: \quad f_B = \frac{E_B}{E_A + E_B} = \frac{\dfrac{k_1}{1 + [\text{X}]/K_{\text{Sp}}}\,[\text{S}]}{k_{-1} + k_2 + \dfrac{k_1}{1 + [\text{X}]/K_{\text{Sp}}}\,[\text{S}]}$$

$$\text{LSpA}: \quad f_B = \frac{E_B}{E_A + E_B} = \frac{\dfrac{k_5\left([\text{X}]/K_{\text{Sp}}\right)[\text{S}]}{1 + [\text{X}]/K_{\text{Sp}}}}{k_{-5} + k_6 + \dfrac{k_1\left([\text{X}]/K_{\text{Sp}}\right)[\text{S}]}{1 + [\text{X}]/K_{\text{Sp}}}}\;.$$

Rate equations:

$$\text{LSpI}: \quad v = k_2 f_{ES} f_B [\text{E}]_\text{t} = \frac{k_2[\text{E}]_\text{t}\dfrac{k_1}{1 + [\text{X}]/K_{\text{Sp}}}\,[\text{S}]}{k_{-1} + k_2 + \dfrac{k_1}{1 + [\text{X}]/K_{\text{Sp}}}\,[\text{S}]}$$

$$= \frac{k_2[\text{E}]_\text{t}\,[\text{S}]}{\dfrac{k_{-1} + k_2}{k_1}\left(1 + \dfrac{[\text{X}]}{K_{\text{Sp}}}\right) + [\text{S}]} \qquad (5.38)$$

$$\text{LSpA}: \qquad v = k_6 f_{ESX} f_B [\text{E}]_t = \frac{k_6[\text{E}]_t \dfrac{k_5\left([\text{X}]/K_{\text{Sp}}\right)}{1+[\text{X}]/K_{\text{Sp}}}[\text{S}]}{k_{-5}+k_6+\dfrac{k_5\left([\text{X}]/K_{\text{Sp}}\right)}{1+[\text{X}]/K_{\text{Sp}}}[\text{S}]}$$

$$= \frac{k_6[\text{E}]_t\,[\text{S}]}{\dfrac{k_{-5}+k_6}{k_5}\left(1+\dfrac{K_{\text{Sp}}}{[\text{X}]}\right)+[\text{S}]}\,. \tag{5.39}$$

The rate equations for LSpI and LSpA differ for the term in the denominator that multiplies the Michaelis constant, i.e., $1 + [\text{X}]/K_{\text{Sp}}$ and $1 + K_{\text{Sp}}/[\text{X}]$, respectively. Thus the apparent Michaelis constant of LSpI depends linearly on [X], while it is linear with $1/[\text{X}]$ in LSpA.

Rate Equation for Linear Catalytic Inhibition

In linear catalytic inhibition, ESX is a dead-end complex that operates at equilibrium when the system is in steady-state (Scheme 5.10). Using the reasoning above for linear specific inhibition and activation, the steps to obtain the rate equation are shown below without comments.

$$E_A = k'_{-1} + k'_2\,, \quad E_B = k_1\,[\text{S}]$$

$$f_E = 1, \quad f_{ES} = \frac{1}{1+[\text{X}]/K_{\text{Ca}}}\,, \quad f_{ESX} = \frac{[\text{X}]/K_{\text{Ca}}}{1+[\text{X}]/K_{\text{Ca}}}$$

$$E_A = \left(k'_{-1}+k'_2\right) f_{ES} = \left(k_{-1}+k_2\right)\frac{1}{1+[\text{X}]/K_{\text{Ca}}}\,, \quad E_B = k_1 f_E\,[\text{S}] = k_1\,[\text{S}]$$

$$f_B = \frac{E_B}{E_A+E_B} = \frac{k_1\,[\text{S}]}{\left(k_{-1}+k_2\right)\dfrac{1}{1+[\text{X}]/K_{\text{Ca}}}+k_1\,[\text{S}]}$$

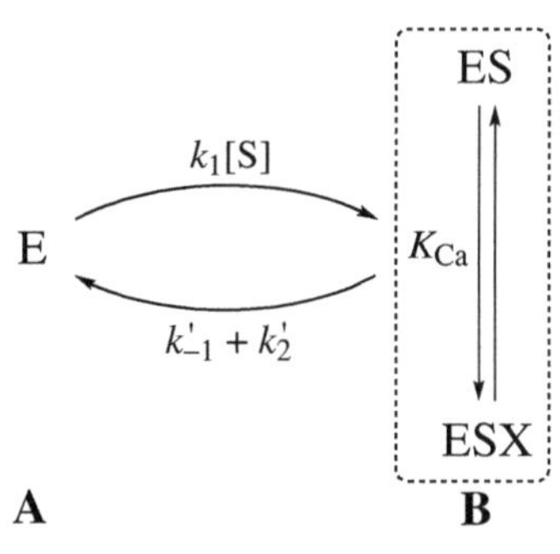

Scheme 5.10 Linear catalytic inhibition. A and B represent the nodes of the mechanism

$$v = k_2 f_{ES} f_B [\mathrm{E}]_\mathrm{t} = \frac{k_2[\mathrm{E}]_\mathrm{t} \dfrac{k_1}{1 + [\mathrm{X}]/K_{\mathrm{Ca}}} [\mathrm{S}]}{(k_{-1} + k_2) \dfrac{1}{1 + [\mathrm{X}]/K_{\mathrm{Ca}}} + k_1 [\mathrm{S}]}$$

$$= \frac{\dfrac{k_2[\mathrm{E}]_\mathrm{t}}{1 + [\mathrm{X}]/K_{\mathrm{Ca}}} [\mathrm{S}]}{\dfrac{k_{-1} + k_2}{k_1} \left(\dfrac{1}{1 + [\mathrm{X}]/K_{\mathrm{Ca}}} \right) + [\mathrm{S}]} = \frac{\dfrac{k_2[\mathrm{E}]_\mathrm{t}}{1 + [\mathrm{X}]/K_{\mathrm{Ca}}} [\mathrm{S}]}{\dfrac{K_\mathrm{m}^0}{1 + [\mathrm{X}]/K_{\mathrm{Ca}}} + [\mathrm{S}]}. \quad (5.40)$$

References

1. Abou El-Magd RM, Park HK, Kawazoe T, Iwana S, Ono K, Chung SP, Miyano M, Yorita K, Sakai T, Fukui K (2010) The effect of risperidone on D-amino acid oxidase activity as a hypothesis for a novel mechanism of action in the treatment of schizophrenia. J Psychopharmacol 24:1055–1067. doi:10.1177/0269881109102644
2. Almond A, Sheehan JK (2000) Glycosaminoglycan conformation: do aqueous molecular dynamics simulations agree with x-ray fiber diffraction? Glycobiology 10:329–338. doi:10.1093/glycob/10.3.329
3. Amstutz P, Binz HK, Parizek P, Stumpp MT, Kohl A, Grütter MG, Forrer P, Plückthun A (2005) Intracellular kinase inhibitors selected from combinatorial libraries of designed ankyrin repeat proteins. J Biol Chem 280:24715–24722
4. Atkinson SC, Dogovski C, Downton MT, Czabotar PE, Dobson RCJ, Gerrard JA, Wagner J, Perugini MA (2013) Structural, kinetic and computational investigation of Vitis vinifera DHDPS reveals new insight into the mechanism of lysine-mediated allosteric inhibition. Plant Mol Biol 81:431–446. doi:10.1007/s11103-013-0014-7
5. Baici A (2002) Myeloblastin. In: Kazazian HH (ed) Wiley encyclopedia of molecular medicine, vol 3. Wiley, New York, pp 2181–2183
6. Baici A, Bradamante P (1984) Interaction between human leukocyte elastase and chondroitin sulfate. Chem-Biol Interact 51:1–11
7. Baici A, Diczházi C, Neszmélyi A, Móczár E, Hornebeck W (1993) Inhibition of the human leukocyte endopeptidases elastase and cathepsin G and of porcine pancreatic elastase by N-oleoyl derivatives of heparin. Biochem Pharmacol 46:1545–1549. doi:10.1016/0006-2952(93)90321-M
8. Baici A, Szedlacsek SE, Früh H, Michel BA (1996) pH-dependent hysteretic behaviour of human myeloblastin (leucocyte proteinase 3). Biochem J 317:901–905
9. Baici A, Schenker P, Wächter M, Rüedi P (2009) 3-Fluoro-2,4-dioxa-3-phosphadecalins as inhibitors of acetylcholinesterase. A reappraisal of kinetic mechanisms and diagnostic methods. Chem Biodivers 6:261–282
10. Barr JT, Jones JP (2011) Inhibition of human liver aldehyde oxidase: implications for potential drug-drug interactions. Drug Metab Disp 39:2381–2386. doi:10.1124/dmd.111.041806
11. Berman HA, Leonard K (1990) Ligand exclusion on acetylcholinesterase. Biochemistry 29:10640–10649. doi:10.1021/bi00499a010
12. Bezerra RM, Dias AA, Fraga I, Pereira AN (2011) Cellulose hydrolysis by cellobiohydrolase Cel7A shows mixed hyperbolic product inhibition. Appl Biochem Biotechnol 165:178–189. doi:10.1007/s12010-011-9242-y
13. Bincoletto C, Tersariol ILS, Oliveira CR, Dreher S, Fausto DM, Soufen MA, Nascimento FD, Caires ACF (2005) Chiral cyclopalladated complexes derived from N,N-dimethyl-1-phenethylamine with bridging bis(diphenylphosphine)ferrocene ligand as inhibitors of

the cathepsin B activity and as antitumoral agents. Bioorg Med Chem 13:3047–3055. doi:10.1016/j.bmc.2005.01.057
14. Binz HK, Stumpp MT, Forrer P, Amstutz P, Plückthun A (2003) Designing repeat proteins: well-expressed, soluble and stable proteins from combinatorial libraries of consensus ankyrin repeat proteins. J Mol Biol 332:489–503
15. Binz HK, Amstutz P, Kohl A, Stumpp MT, Briand C, Forrer P, Grütter MG, Plückthun A (2004) High-affinity binders selected from designed ankyrin repeat protein libraries. Nat Biotechnol 22:575–582
16. Brandt M, Greway AT, Holt DA, Metcalf BW, Levy MA (1990) Studies on the mechanism of steroid 5-a-reductase inhibition by 3-carboxy A-ring aryl steroids. J Steroid Biochem Mol Biol 37:575–579. doi:10.1016/0960-0760(90)90403-8
17. Cadène M, Duranton J, North A, Sitahar M, Chignard M, Bieth JG (1997) Inhibition of neutrophil serine proteinases by suramin. J Biol Chem 272:9950–9955
18. Casu B, Petitou M, Provasoli M, Sinaÿ P (1988) Conformational flexibility: a new concept for explaining binding and biological properties of iduronic acid-containing glycosaminoglycans. Trends Biochem Sci 13:221–225
19. Cornish-Bowden A (1986) Why is uncompetitive inhibition so rare? A possible explanation, with implications for the design of drugs and pesticides. FEBS Lett 203:3–6. doi:10.1016/0014-5793(86)81424-7
20. Cornish-Bowden A (1998) Two centuries of catalysis. J Biosci 23:87–92
21. Cornish-Bowden A (2012) Fundamentals of enzyme kinetics, 4th edn. Wiley, Weinheim
22. DiMaio J, Gibbs B, Munn D, Lefebvre J, Ni F, Konishi Y (1990) Bifunctional thrombin inhibitors based on the sequence of hirudin45-65. J Biol Chem 265:21698–21703
23. Dixon M, Webb EC (1979) Enzymes, 3rd edn. Longman, London
24. Dobson RCJ, Griffin MDW, Roberts SJ, Gerrard JA (2004) Dihydrodipicolinate synthase (DHDPS) from Escherichia coli displays partial mixed inhibition with respect to its first substrate, pyruvate. Biochimie 86:311–315. doi:10.1016/j.biochi.2004.03.008
25. Doehlert DC, Huber SC (1983) Spinach leaf sucrose phosphate synthase: activation by glucose 6-phosphate and interaction with inorganic phosphate. FEBS Lett 153:293–297. doi:10.1016/0014-5793(83)80627-9
26. Doehlert DC, Huber SC (1984) Phosphate inhibition of spinach leaf sucrose phosphate synthase as affected by glucose-6-phosphate and phosphoglucoisomerase. Plant Physiol 76:250–253
27. Domagk G (1935) Ein Beitrag zur Chemotherapie der bakteriellen Infektionen. Deut Med Wochenschr 61:250–253
28. Drainas D, Kalpaxis DL, Coutsogeorgopoulos C (1987) Inhibition of ribosomal peptidyltransferase by chloramphenicol. Kinetic studies. Eur J Biochem 164:53–58
29. Eisenthal R, Cornish-Bowden A (1974) The direct linear plot. A new graphical procedure for estimating enzyme kinetic parameters. Biochem J 139:715–720
30. Feldman T, Kabaleeswaran V, Jang SB, Antczak C, Djaballah H, Wu H, Jiang X (2012) A class of allosteric caspase inhibitors identified by high-throughput screening. Mol Cell 47:585–595. doi:10.1016/j.molcel.2012.06.007
31. Fischer RS, Rubin JL, Gaines CG, Jensen RA (1987) Glyphosate sensitivity of 5-enolpyruvylshikimate-3-phosphate synthase from Bacillus subtilis depends upon state of activation induced by monovalent cations. Arch Biochem Biophys 256:325–334
32. Forrer P, Stumpp MT, Binz HK, Plückthun A (2003) A novel strategy to design binding molecules harnessing the modular nature of repeat proteins. FEBS Lett 539:2–6
33. Francis PT, Palmer AM, Snape M, Wilcock GK (1999) The cholinergic hypothesis of Alzheimer's disease: a review of progress. J Neurol Neurosurg Psychiatry 66:137–147. doi:10.1136/jnnp.66.2.137
34. Franzke CW, Baici A, Bartels J, Christophers E, Wiedow O (1996) Antileukoprotease inhibits stratum corneum chymotryptic enzyme. Evidence for a regulative function in desquamation. J Biol Chem 271:21,886–21,890

35. Frieden C (1963) Glutamate dehydrogenase. V. The relation of enzyme structure to the catalytic function. J Biol Chem 238:3286–3299
36. Früh H, Kostoulas G, Michel BA, Baici A (1996) Human myeloblastin (leukocyte proteinase 3): Reactions with substrates, inactivators and activators in comparison with leukocyte elastase. Biol Chem 377:579–586
37. Gao WQ, Anderson PJ, Majerus EM, Tuley EA, Sadler JE (2006) Exosite interactions contribute to tension-induced cleavage of von Willebrand factor by the antithrombotic ADAMTS13 metalloprotease. Proc Natl Acad Sci USA 103:19099–19104
38. González-Tanarro CM, Gütschow M (2011) Hyperbolic mixed-type inhibition of acetylcholinesterase by tetracyclic thienopyrimidines. J Enzyme Inhib Med Chem 26:350–358. doi:10.3109/14756366.2010.504674
39. Gupta S, Mahmood S, Mahmood A (2009) Kinetic characteristics of brush border sucrase activation by Na+ ions in mice intestine. Indian J Exp Biol 47:811–815
40. Hallcher LM, Sherman WR (1980) The effects of lithium ion and other agents on the activity of myo-inositol-1-phosphatase from bovine brain. J Biol Chem 255:10896–10901
41. Hedstrom L (2009) IMP dehydrogenase: structure, mechanism, and inhibition. Chem Rev 109:2903–2928. doi:10.1021/cr900021w
42. Heng S, Tieu W, Hautmann S, Kuan K, Pedersen DS, Pietsch M, Gütschow M, Abell AD (2011) New cholesterol esterase inhibitors based on rhodanine and thiazolidinedione scaffolds. Bioorg Med Chem 19:7453–7463. doi:10.1016/j.bmc.2011.10.042
43. Hongsawat P, Vangnai AS (2011) Biodegradation pathways of chloroanilines by Acinetobacter baylyi strain GFJ2. J Hazard Mater 186:1300–1307. doi:10.1016/j.jhazmat.2010.12.002
44. Inagami T (1964) The mechanism of the specificity of trypsin catalysis. I. Inhibition by alkyl ammonium ions. J Biol Chem 239:787–791
45. Inagami T, Murachi T (1964) The mechanism of the specificity of trypsin catalysis. III. Activation of the catalytic site of trypsin by alkylammonium ions in the hydrolysis of acetylglycine ethyl ester. J Biol Chem 239:1395–1401
46. Inagami T, York SS (1968) Effect of alkyl guanidines and alkyl amines on trypsin catalysis. Biochemistry 7:4045–4052. doi:10.1021/bi00851a036
47. Joshi N, Hoobler EK, Perry S, Diaz G, Fox B, Holman TR (2013) Kinetic and structural investigations into the allosteric and pH effect on the substrate specificity of human epithelial 15-lipoxygenase-2. Biochemistry 52:8026–8035. doi:10.1021/bi4010649
48. Kazan D, Erarslan A (1997) Stabilization of Escherichia coli penicillin G acylase by polyethylene glycols against thermal inactivation. Appl Biochem Biotechnol 62:1–13
49. Kim Y, Arp DJ, Semprini L (2002) Kinetic and inhibition studies for the aerobic cometabolism of 1,1,1-trichloroethane, 1,1-dichloroethylene, and 1,1-dichloroethane by a butane-grown mixed culture. Biotechnol Bioeng 80:498–508
50. King JB, West MB, Cook PF, Hanigan MH (2009) A novel, species-specific class of uncompetitive inhibitors of g-glutamyl transpeptidase. J Biol Chem 284:9059–9065
51. Kohl A, Binz HK, Forrer P, Stumpp MT, Plückthun A, Grütter MG (2003) Designed to be stable: crystal structure of a consensus ankyrin repeat protein. Proc Natl Acad Sci USA 100:1700–1705
52. Kostoulas G, Hörler D, Naggi A, Casu B, Baici A (1997) Electrostatic interactions between human leukocyte elastase and sulfated glycosaminoglycans: physiological implications. Biol Chem 378:1481–1489
53. Kovačič L, Novinec M, Petan T, Baici A, Križaj I (2009) Calmodulin is a nonessential activator of secretory phospholipase A(2). Biochemistry 48:11,319–11,328
54. Küçükkilinç TT, Ozer I (2008) Inhibition of electric eel acetylcholinesterase by triarylmethane dyes. Chem-Biol Interact 175:309–311. doi:10.1016/j.cbi.2008.05.008
55. Ledoux D, Papy-Garcia D, Escartin Q, Sagot MA, Cao Y, Barritault D, Courtois J, Hornebeck W, Caruelle JP (2000) Human plasmin enzymatic activity is inhibited by chemically modified dextrans. J Biol Chem 275:29383–29390. doi:10.1074/jbc.M000837200
56. Leech AP, Baker GR, Shute JK, Cohen MA, Gani D (1993) Chemical and kinetic mechanism of the inositol monophosphatase reaction and its inhibition by Li+. Eur J Biochem 212:693–704

57. Li Z, Kienetz M, Cherney MM, James MN, Brömme D (2008) The crystal and molecular structures of a cathepsin K:chondroitin sulfate complex. J Mol Biol 383:78–91. doi:S0022-2836(08)00880-2 [pii] 10.1016/j.jmb.2008.07.038
58. Lorey S, Stöckel-Maschek A, Faust J, Brandt W, Stiebitz B, Gorrell MD, Kähne T, Mrestani-Klaus C, Wrenger S, Reinhold D, Ansorge S, Neubert K (2003) Different modes of dipeptidyl peptidase IV (CD26) inhibition by oligopeptides derived from the N-terminus of HIV-1 Tat indicate at least two inhibitor binding sites. Eur J Biochem 270:2147–2156
59. Lüönd RM, McKie JH, Douglas KT, Dascombe MJ, Vale J (1998) Inhibitors of glutathione reductase as potential antimalarial drugs. Kinetic cooperativity and effect of dimethyl sulphoxide on inhibition kinetics. J Enzyme Inhib 13:327–345
60. Madsen JL, Andersen TL, Santamaria S, Nagase H, Enghild JJ, Skrydstrup T (2012) Synthesis and evaluation of silanediols as highly selective uncompetitive inhibitors of human neutrophil elastase. J Med Chem 55:7900–7908. doi:10.1021/jm301000k
61. Martinek K, Varfolomeev SD, Levashov AV, Berezin IV (1971) Kinetic manifestations of the structure of the active center of a-chymotrypsin on interacting with fragments of specific substrates. Mol Biol Engl Transl 5:278–286
62. Masson P, Froment MT, Gillon E, Nachon F, Lockridge O, Schopfer LM (2008) Kinetic analysis of effector modulation of butyrylcholinesterase-catalysed hydrolysis of acetanilides and homologous esters. FEBS J 275:2617–2631. doi:10.1111/j.1742-4658.2008.06409.x
63. Michaelis L, Pechstein H (1914) Über die verschiedenartige Natur der Hemmungen der Invertasewirkung. Biochem Z 60:79–90
64. Michaelis L, Rona P (1914) Die Wirkungsbedingungen der Maltase aus Bierhefe. III. Über die Natur der verschiedenenartigen Hemmungen der Fermentwirkungen. Biochem Z 60:62–78
65. Mogul R, Johansen E, Holman TR (2000) Oleyl sulfate reveals allosteric inhibition of soybean lipoxygenase-1 and human 15-lipoxygenase. Biochemistry 39:4801–4807
66. Monod J, Changeux JP, Jacob F (1963) Allosteric proteins and cellular control systems. J Mol Biol 6:306–329
67. Monod J, Wyman J, Changeux JP (1965) On the nature of allosteric transitions: a plausible model. J Mol Biol 12:88–118
68. Murray M, Farrell GC (1986) Mechanistic aspects of the inhibition of microsomal drug oxidation by primaquine. Biochem Pharmacol 35:2149–2155
69. Nakanishi M, Moriyama A, Narita Y, Sasaki M (1989) Aminopeptidase-M from human liver. II. Kinetic analysis of inhibition of the enzyme by bile acids. J Biochem 106:826–830
70. Nomenclature Committee of the International Union of Biochemistry (1982) Symbolism and terminology in enzyme kinetics. Recommendations 1981. Eur J Biochem 128:281–291
71. Novinec M, Kovačič L, Lenarčič B, Baici A (2010) Conformational flexibility and allosteric regulation of cathepsin K. Biochem J 429:379–389. doi:10.1042/BJ20100337
72. Novinec M, Korenč M, Caflisch A, Ranganathan R, Lenarčič B, Baici A (2014) A novel allosteric mechanism in the cysteine peptidase cathepsin K discovered by computational methods. Nat Commun 5:3287. doi:10.1038/ncomms4287
73. Novinec M, Lenarčič B, Baici A (2014) Probing the activity modification space of the cysteine peptidase cathepsin K with novel allosteric modifiers. PLoS One 9:e106642. doi:10.1371/journal.pone.0106642
74. Orsi BA, McFerran N, Hill A, Bingham A (1972) Kinetics and the mechanism of action of adenosine aminohydrolase. Biochemistry 11:3386–3392
75. Park YD, Kim SY, Lyou YJ, Lee JY, Yang JM (2005) A new type of uncompetitive inhibition of tyrosinase induced by Cl- binding. Biochimie 87:931–937. doi:10.1016/j.biochi.2005.06.006
76. Pawlyk AC, Pettigrew DW (2002) Transplanting allosteric control of enzyme activity by protein-protein interactions: Coupling a regulatory site to the conserved catalytic core. Proc Natl Acad Sci USA 99:11115–11120. doi:10.1073/pnas.132393599
77. Pettigrew DW (2009) Amino acid substitutions in the sugar kinase/hsp70/actin superfamily conserved ATPase core of E. coli glycerol kinase modulate allosteric ligand affinity but do not alter allosteric coupling. Arch Biochem Biophys 481:151–156. doi:10.1016/j.abb.2008.11.020

78. Phillips MF, Mantle TJ (1991) The initial-rate kinetics of mouse glutathione S-transferase YfYf. Evidence for an allosteric site for ethacrynic acid. Biochem J 275:703–709
79. Raina A, Hyvönen T, Eloranta T, Voutilainen M, Samejima K, Yamanoha B (1984) Polyamine synthesis in mammalian tissues. Isolation and characterization of spermidine synthase from bovine brain. Biochem J 219:991–1000
80. Rothe M, Zichner A, Auerswald EA, Dodt J (1994) Structure/function implications for the aminopeptidase specificity of aleurain. Eur J Biochem 224:559–565. doi:10.1111/j.1432-1033.1994.00559.x
81. Ruddat VC, Whitman S, Holman TR, Bernasconi CF (2003) Stopped-flow kinetic investigations of the activation of soybean lipoxygenase-1 and the influence of inhibitors on the allosteric site. Biochemistry 42:4172–4178. doi:10.1021/bi020698o
82. Schechter I, Berger A (1967) On the size of the active sites in proteases. I. Papain. Biochem Biophys Res Commun 27:157–162
83. Schenker P, Alfarano P, Kolb P, Caflisch A, Baici A (2008) A double-headed cathepsin B inhibitor devoid of warhead. Protein Sci 17:2145–2155
84. Schiroli D, Ronda L, Peracchi A (2015) Kinetic characterization of the human O-phosphoethanolamine phospho-lyase reveals unconventional features of this specialized pyridoxal phosphate-dependent lyase. FEBS J 282:183–199. doi:10.1111/febs.13122
85. Schmitz T, Rothe M, Dodt J (1991) Mechanism of the inhibition of a-thrombin by hirudin-derived fragments hirudin(1–47) and hirudin(45–65). Eur J Biochem 195:251–256
86. Schulte M, Mattay D, Kriegel S, Hellwig P, Friedrich T (2014) Inhibition of Escherichia coli respiratory complex I by Zn2+. Biochemistry 53:6332–6339. doi:10.1021/bi5009276
87. Schweizer A, Roschitzki-Voser H, Amstutz P, Briand C, Gulotti-Georgieva M, Prenosil E, Binz HK, Capitani G, Baici A, Plückthun A, Grütter MG (2007) Inhibition of caspase-2 by a designed ankyrin repeat protein: specificity, structure, and inhibition mechanism. Structure 15:625–636. doi:10.1016/j.str.2007.03.014
88. Segal HL, Kachmar JF, Boyer PD (1952) Kinetic analysis of enzyme reactions. I. Further considerations of enzyme inhibition and analysis of enzyme activation. Enzymologia 15:187–198
89. Segel IH (1975) Enzyme kinetics. Behavior and analysis of rapid equilibrium and steady-state enzyme systems. Wiley, New York
90. Semenza G (1969) A kinetic investigation on the allosteric effects in intestinal sucrase. Eur J Biochem 8:518–529. doi:10.1111/j.1432-1033.1969.tb00557.x
91. Sheehan JP, Phan TM (2001) Phosphorothioate oligonucleotides inhibit the intrinsic tenase complex by an allosteric mechanism. Biochemistry 40:4980–4989
92. Shikita M, Fahey JW, Golden TR, Holtzclaw WD, Talalay P (1999) An unusual case of 'uncompetitive activation' by ascorbic acid: purification and kinetic properties of a myrosinase from Raphanus sativus seedlings. Biochem J 341:725–732
93. Siemann S, Evanoff DP, Marrone L, Clarke AJ, Viswanatha T, Dmitrienko GI (2002) N-arylsulfonyl hydrazones as inhibitors of IMP-1 metallo-beta-lactamase. Antimicrob Agents Chemother 46:2450–2457
94. Snášel J, Nauš P, Dostál J, Hnízda A, Fanfrlík J, Brynda J, Bourderioux A, Dušek M, Dvoráková H, Stolaríková J, Zábranská H, Pohl R, Konečný P, Džubák P, Votruba I, Hajdúch M, Rezáčová P, Veverka V, Hocek M, Pichová I (2014) Structural basis for inhibition of mycobacterial and human adenosine kinase by 7-substituted 7-(Het)aryl-7-deazaadenine ribonucleosides. J Med Chem 57:8268–8279. doi:10.1021/jm500497v
95. Sohier JS, Laurent C, Chevigné A, Pardon E, Srinivasan V, Wernery U, Lassaux P, Steyaert J, Galleni M (2013) Allosteric inhibition of VIM metallo-beta-lactamases by a camelid nanobody. Biochem J 450:477–486. doi:10.1042/bj20121305
96. Srinivasan B, Forouhar F, Shukla A, Sampangi C, Kulkarni S, Abashidze M, Seetharaman J, Lew S, Mao L, Acton TB, Xiao R, Everett JK, Montelione GT, Tong L, Balaram H (2014) Allosteric regulation and substrate activation in cytosolic nucleotidase II from Legionella pneumophila. FEBS J 281:1613–1628. doi:10.1111/febs.12727
97. Sternson LA, Gammans RE (1976) Effect of aromatic nitro compounds on oxidative metabolism by cytochrome P-450 dependent enzymes. J Med Chem 19:174–177

98. Sun W, Wendt M, Klebe G, Röhm KH (2014) On the interpretation of tyrosinase inhibition kinetics. J Enzyme Inhib Med Chem 29:92–99. doi:10.3109/14756366.2012.755621
99. Szewczuk Z, Gibbs BF, Yue SY, Purisima E, Zdanov A, Cygler M, Konishi Y (1993) Design of a linker for trivalent thrombin inhibitors: interaction of the main chain of the linker with thrombin. Biochemistry 32:3396–3404
100. Takimoto K, Okada M, Matsuda Y, Nakagawa H (1985) Purification and properties of myo-inositol-1-phosphatase from rat brain. J Biochem (Tokyo) 98:363–370
101. Tian GC, Sobotka-Briner CD, Zysk J, Liu XD, Birr C, Sylvester MA, Edwards PD, Scott CD, Greenberg BD (2002) Linear non-competitive inhibition of solubilized human gamma-secretase by pepstatin a methylester, L685458, sulfonamides, and benzodiazepines. J Biol Chem 277:31499–31505
102. Tipton KF (1996) Patterns of enzyme inhibition. In: Engel PC (ed) Enzymology LabFax. Academic Press Inc., San Diego, pp 115–174
103. Velazquez I, Pardo JP (2001) Kinetic characterization of the rotenone-insensitive internal NADH: ubiquinone oxidoreductase of mitochondria from Saccharomyces cerevisiae. Arch Biochem Biophys 389:7–14. doi:10.1006/abbi.2001.2293
104. Wang W, Hedstrom L (1997) Kinetic mechanism of human inosine 5'-monophosphate dehydrogenase type II: random addition of substrates and ordered release of products. Biochemistry 36:8479–8483. doi:10.1021/bi970226n
105. Westley AM, Westley J (1996) Enzyme inhibition in open systems. Superiority of uncompetitive agents. J Biol Chem 271:5347–5352
106. White PW, Faucher AM, Massariol MJ, Welchner E, Rancourt J, Cartier M, Archambault J (2005) Biphenylsulfonacetic acid inhibitors of the human papillomavirus type 6 E1 helicase inhibit ATP hydrolysis by an allosteric mechanism involving tyrosine 486. Antimicrobial agents and chemotherapy 49:4834–4842. doi:10.1128/AAC.49.12.4834-4842.2005
107. Wickham S, Regan N, West MB, Thai J, Cook PF, Terzyan SS, Li PK, Hanigan MH (2013) Inhibition of human gamma-glutamyl transpeptidase: development of more potent, physiologically relevant, uncompetitive inhibitors. Biochem J 450:547–557. doi:10.1042/bj20121435
108. Woods DD (1940) The relation of p-aminobenzoic acid to the mechanism of the action of sulphanilamide. Brit J Exp Pathol 21:74–90
109. Woods MJ, Findlater JD, Orsi BA (1979) Kinetic mechanism of the aliphatic amidase from Pseudomonas aeruginosa. Biochim Biophys Acta 567:225–237
110. Worcel A, Goldman DS (1964) Activation by AMP of the NADH oxidase of Mycobacterium tuberculosis. Biochem Biophys Res Commun 17:559–564
111. Worcel A, Goldman DS, Cleland WW (1965) An allosteric reduced nicotinamide adenine dinucleotide oxidase from Mycobacterium tuberculosis. J Biol Chem 240:3399–3407
112. Xiao ZP, Peng ZY, Dong JJ, Deng RC, Wang XD, Ouyang H, Yang P, He J, Wang YF, Zhu M, Peng XC, Peng WX, Zhu HL (2013) Synthesis, molecular docking and kinetic properties of beta-hydroxy-beta-phenylpropionyl-hydroxamic acids as Helicobacter pylori urease inhibitors. Eur J Med Chem 68:212–221. doi:10.1016/j.ejmech.2013.07.047
113. Yoshino M (1987) A graphical method for determining inhibition parameters for partial and complete inhibitors. Biochem J 248:815–820
114. Zhang Y, Ribeiro JMC, Guimarães JA, Walsh PN (1998) Nitrophorin-2, a novel mixed-type reversible specific inhibitor of the intrinsic factor-X activating complex. Biochemistry 37:10681–10690. doi:10.1021/bi973050y

Chapter 6
Multiple Enzyme-Modifier Interactions

> *··· the simultaneous action of three ligands on one enzyme may result in unexpected effects. These effects may change in the function of the concentrations of the ligands and the liberation or inhibition may turn into activation.*
> Fajszi Cs, Keleti T (1974) Symp Biol Hung 18:105–119 (p. 116)

Abstract Binding of two or more modifiers to an enzyme can alter their affinities for the enzyme and synergy or antagonism may result in the inhibition or activation processes. If binding of the modifiers does not alter their affinities for the enzyme the condition is called zero-interaction. Knowing the kinetic mechanism of the modifiers acting individually, a rate equation is provided that accounts for simultaneous interactions and allows to represent graphically the resulting effects as isoboles, lines that describe equieffective concentrations of the modifiers. Apparently odd effects, such as activation resulting from the combination of two inhibitors and efficient inhibition ensuing from the concomitant action of two nonessential activators, have a defined mechanistic meaning and embody sophisticated regulatory strategies.

6.1 Introduction

Knowing the details of multiple enzyme-modifier interactions is fundamental for several reasons. Enzyme inhibitors used as drugs may interact with endogenous inhibitors by reducing or enhancing their activity. The diversity of interactions than can be expected is considerable because these can occur between inhibitors and/or activators operating with different kinetic mechanisms. Multiple inhibition studies offer the possibility of exploring and mapping the active centers of enzymes, including their conformational states, thus allowing the measurement of individual binding constants for defined elements of substrates, coenzymes, and inhibitors [10, 13, 33]. This information can in turn be exploited for the rational design of pharmacologically relevant modifiers. An example is the intensive research around inosine 5′-monophosphate dehydrogenase that is amplified in neoplastic transformation and, more generally, in rapidly proliferating tissues. Lizbeth Hedstrom reviewed

A. Baici, *Kinetics of Enzyme-Modifier Interactions*,
DOI 10.1007/978-3-7091-1402-5_6

the properties of the enzyme and commented the use of inhibitors as drugs in immunosuppressive and antiviral chemotherapy as well as antimicrobial agents [12].

In vitro, the study of the interactions between enzymes and modifiers usually begins by characterizing one modifier at a time in the presence of suitable substrates. This procedure is necessary to gather every possible mechanistic information regarding a particular modifier. However, when two or more modifiers act simultaneously on an enzyme, their combined action may generate effects that do not necessarily correspond to the simple sum of their individual effects. Slater and Bonner observed such a behavior while studying the action of fluoride and phosphate, two inhibitors of *succinic oxidase* [35]. Yagi and Ozawa measured the combined effects of inhibitors of D-amino acid oxidase and devised a graphical method for assessing the interaction occurring between two simultaneously acting inhibitors [40, 41]. An extension of the mentioned studies was brought up by Yonetani and Theorell, who investigated the double inhibition of liver alcohol dehydrogenase by six combinations in pairs of the NAD_{ox}-competitive inhibitors ADP, AMP, adenosine diphosphate ribose, and *o*-phenanthroline [42]. With a newly designed plot, Yonetani and Theorell could establish if the inhibitors interacted with the enzyme at a common site or at different sites.

Unsuspected double enzyme-modifier interactions can occur in vitro if for assaying the enzyme under investigation a racemic substrate is used. It is not rare that in this case only one of the optical isomers is the true substrate while the other isomer behaves as inhibitor. If such a system is used to study the effect of an inhibitor or an activator, the enzyme will possibly bind two modifiers resulting in biased results. This was the case in a study by Mares-Guia and Shaw who investigated the reaction of trypsin using the racemate N^{α}-benzoyl-DL-arginine-4-nitroanilide as substrate of which the L-isomer was the substrate whereas the D-isomer behaved as linear specific inhibitor [23].

Combinations of modifiers other than linear specific inhibitors have been considered by other authors who expanded the theoretical treatment of multiple modifier interactions. Important contributions are those of Webb [38, Chap. 10], Keleti and Fajszi [16], Fajszi [6] and Segel [34, Chap. 8]. Significant contributions to the kinetic theory of multiple enzyme-modifier interactions are those by Keleti and Fajszi. Despite a few shortcomings that will be mentioned below, the authors created the quantitative basis of our present understanding in this field. The concept of *liberator* was published in 1964 [15], a thorough study of *double inhibition* dates back to 1971 [16], and the kinetic basis of *enzyme regulation* by allosteric interactions in double modifier interactions and the *inhibition paradox* concept also appeared in 1971 [7]. Mazat and coworkers considered the same topics and obtained the same results in three papers published between 1977 and 1986 [24–26] with marginal or no credits to the mentioned work by Keleti and Fajszi.

Patricia Schenker and the author of the present book scrutinized the literature of multiple enzyme-modifier interactions and of the theories published to predict the combined effects of modifiers [32]. This survey revealed conflicting results, misinterpretations, and occasional mathematical mistakes. Therefore, the entire topic was reconsidered by comparing existing theories, by merging established concepts and by proposing a general model for the simultaneous interaction of two modifiers on the same enzyme. This work constitutes the theoretical basis of the present chapter, where further mathematical and practical aspects of multiple interactions between enzymes, substrates, and modifiers will be described. Avoiding an extensive review of the literature of multiple enzyme-modifier interactions in this place, the reader is referred to the cited paper [32], where contributions to this field not cited here have been discussed and appropriately acknowledged.

A Journey in an Uncharted Territory A classical example of multiple enzyme-modifier interactions is the regulation of mammalian phosphofructokinases, which evolved to protein sizes more than twice those of the prokaryotic counterparts. The gain in regulatory properties of mammalian phosphofructokinases is due to the presence of activator allosteric binding sites for AMP (or ADP) and fructose 2,6-bisphosphate and of inhibitory allosteric sites for ATP and citrate [17].

In Chap. 3, the endorsement of hyperbolic mixed modifiers to the level of distinct mechanisms instead of considering them as special cases of two undifferentiated pools of hyperbolic mixed inhibitors and nonessential activators, had also in mind applications to the systematics of multiple enzyme-modifier interactions. It is hoped that this arrangement may help future studies in this field. In fact, with the exception of some well-investigated systems, such as the mentioned phosphofructokinase, the theoretical contributions mentioned above have not yet been paralleled by experimental evidence on a large scale. The gaps, mostly positioned around the interactions occurring between hyperbolic mixed modifiers, are by no means due to negligence or disrelish of complicate-looking mathematics, but to the fact that this is a field that started to emerge *recently* after the discovery that even monomeric enzymes can possess more than one, often multiple conserved allosteric sites. The work performed by Ranganathan and his team is certainly worth of being read [11, 18, 20, 31, 37], and see also the discussion in Sect. 5.4.1, p. 261, on cathepsin K regulation based on the studies by Novinec and coworkers [28, 29].

The *uncharted territory* is therefore the large, unexplored area of interactions that can occur between at least four molecules: an enzyme, a substrate, and two modifiers. I believe to have computed and analyzed graphically most of the possible double interactions between the 17 basic mechanisms in Fig. 3.2 and in addition combinations with essential activators and liberators. In this chapter only a minute selection of representative cases will be shown as a clue dedicated to all zealous explorers who wish to map the uncharted territory.

6.2 Multiple Enzyme-Modifier Interactions

The formation of complexes between an enzyme and two modifiers will be discussed considering catalysis involving a single substrate. The treatment of systems with more than one substrate implies analogous rate equations, in which

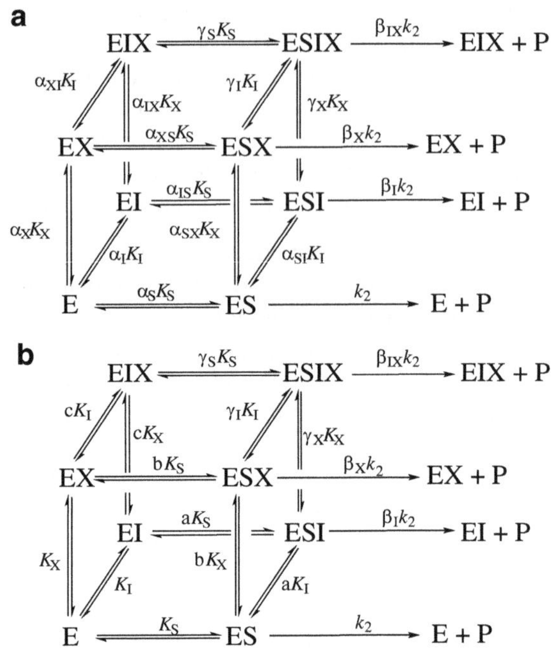

Scheme 6.1 General mechanism for two modifiers, I and X, interacting simultaneously with an enzyme. (**a**) Complete representation with all individual dissociation constants and coefficients. (**b**) Simplified representation of the same mechanism with simpler notation resulting from the equalities (6.1). In the most general case, as illustrated here, two hyperbolic mixed modifiers interact on E and ES and with four catalytic steps, i.e., all possible reaction paths exist

one substrate at a time is set to a known, constant value. With the inclusion of substrate(s), the minimum number of molecules that bind an enzyme to be considered in such interactions is three and the adjective *multiple* in the title of this section suggests that the mathematical treatment can in principle be extended to more modifiers concomitantly acting on an enzyme. Obviously, the algebraic formalism may become cumbersome and hardly usable in practice. The interactions between an enzyme and two modifiers, herein called I and X, can be schematically described as an extension of the general modifier mechanism (Chap. 2). The regulatory potential of multiple interactions dramatically increases when the two modifiers, besides acting independently, also interact with each other. The taxonomy and nomenclature of the basic modifier mechanisms used in this chapter have been introduced in Chap. 3.

The reaction scheme for two modifiers that act on the same enzyme contains paths leading to a ternary EIX and to a quaternary ESIX complex. The resulting reactions are best visualized in three dimensions as a cube in which the enzyme-containing species are located at the eight corners. In the general case, all four substrate-containing complexes are catalytically competent (Scheme 6.1). In order to improve legibility of the reaction scheme, the 12 reactant-binding occurrences, such as S + EIX $\rightleftharpoons$ ESIX, are not shown explicitly but these are provided in Table 6.1 together with the definitions of the corresponding equilibrium constants.

In Scheme 6.1a, each of the 12 dissociation constants at the edges of the cube consists of a K-term, a dissociation constant, multiplied by a dimensionless coefficient, α or γ, bearing a subscript that unequivocally identifies its position in the reaction scheme. The apparently odd coefficients α_S, α_I and α_X might look redundant without further explanation: these coefficients can only take the values 1 if the equilibrium they represent exists, or ∞ if the path does not exist. This notation is necessary to avoid ambiguities when the reactants bind in *compulsory order* as in *catalytic inhibition* and *essential activation*. To avoid confusion due to the presence of many coefficients with/without subscripts, the symbol "α" used in the general modifier mechanism (Scheme 3.1) has been renamed in the case of multiple interactions as "a" and "b" and applied to the modifiers I and X, respectively.[1]

When all enzyme-containing complexes exist, the notation in Scheme 6.1a can be rendered less cumbersome as shown in Scheme 6.1b because parallel pathways and microscopic reversibility imply the equalities (6.1). It is needless to say that microscopic reversibility (Sect. 1.3) must be granted for all cycles defined by the six faces of the cubes in Scheme 6.1. For mechanisms involving ordered reactant binding, the equalities (6.1) must be interpreted in one of two ways depending on the path that really exists (6.2). For practical reasons, in displaying reaction mechanisms of double enzyme-modifier interactions, a simpler version than Scheme 6.1 will be used later on in this chapter, in which double arrows and non-necessary paths will be omitted, e.g., as shown in Scheme 6.2.

$$\begin{aligned} \mathrm{a} &= \alpha_I \alpha_{IS} = \alpha_S \alpha_{SI} \\ \mathrm{b} &= \alpha_X \alpha_{XS} = \alpha_S \alpha_{SX} \\ \mathrm{c} &= \alpha_X \alpha_{XI} = \alpha_X \alpha_{IX} , \end{aligned} \tag{6.1}$$

$$\begin{aligned} \mathrm{a} &= \alpha_I \alpha_{IS} \quad \text{or} \quad \mathrm{a} = \alpha_S \alpha_{SI} \\ \mathrm{b} &= \alpha_X \alpha_{XS} \quad \text{or} \quad \mathrm{b} = \alpha_S \alpha_{SX} \\ \mathrm{c} &= \alpha_X \alpha_{XI} \quad \text{or} \quad \mathrm{c} = \alpha_I \alpha_{IX} . \end{aligned} \tag{6.2}$$

There are three paths leading to the quaternary ESIX complex in Scheme 6.1, with γ-factors multiplying K_S, K_I and K_X. These are used for defining a new constant, e, which is a measure of the interaction of the modifiers I and X on the ES complex based on microscopic reversibility of the system:

$$\mathrm{e} = \mathrm{a}\gamma_X = \mathrm{b}\gamma_I = \mathrm{c}\gamma_S . \tag{6.3}$$

When e = ∞, I and X exclude reciprocally on the ES complex and ESIX does not exist. This interaction constant, which is not shown explicitly in Scheme 6.1, is

[1] a and b are set in upright font instead italics to improve readability toward *alpha* symbols in the same rate equation.

Table 6.1 Reaction steps and dissociation constants for the general case of the interaction of two modifiers with the same enzyme shown in Scheme 6.1

#	Reaction	Equilibrium	#	Reaction	Equilibrium
1	$E + S \rightleftharpoons ES$	$\alpha_S K_S = \frac{[E]\,[S]}{[ES]}$	7	$EX + S \rightleftharpoons ESX$	$\alpha_{XS} K_S = \frac{[EX]\,[S]}{[ESX]}$
2	$E + I \rightleftharpoons EI$	$\alpha_I K_I = \frac{[E]\,[I]}{[EI]}$	8	$EI + X \rightleftharpoons EIX$	$\alpha_{IX} K_X = \frac{[EI]\,[X]}{[EIX]}$
3	$E + X \rightleftharpoons EX$	$\alpha_X K_X = \frac{[E]\,[X]}{[EX]}$	9	$EX + I \rightleftharpoons EIX$	$\alpha_{XI} K_I = \frac{[EX]\,[I]}{[EIX]}$
4	$ES + I \rightleftharpoons ESI$	$\alpha_{SI} K_I = \frac{[ES]\,[I]}{[ESI]}$	10	$ESI + X \rightleftharpoons ESIX$	$\gamma_X K_X = \frac{[ESI]\,[X]}{[ESIX]}$
5	$ES + X \rightleftharpoons ESX$	$\alpha_{SX} K_X = \frac{[ES]\,[X]}{[ESX]}$	11	$ESX + I \rightleftharpoons ESIX$	$\gamma_I K_I = \frac{[ESX]\,[I]}{[ESIX]}$
6	$EI + S \rightleftharpoons ESI$	$\alpha_{IS} K_S = \frac{[EI]\,[S]}{[ESI]}$	12	$EIX + S \rightleftharpoons ESIX$	$\gamma_S K_S = \frac{[EIX]\,[S]}{[ESIX]}$

necessary for deriving the rate equation of double enzyme modification in the most general way. For particular situations in which only one of the paths leading to ESIX exists, the interaction constant e corresponds to the appropriate term in (6.3).

After dealing with all necessary definitions, the rate equation (6.4) for the double-modifier mechanisms in Scheme 6.1 can be derived assuming all steps to operate at equilibrium. Considering all steps in a steady-state regime, the same equation is overly complicated and not usable in practice. Furthermore, we assume that the relationships between [E], [I], [X] and the dissociation constants K_I and K_X are such that depletion of [I] and [X] due to enzyme binding can be neglected:

$$v_{IX} = \frac{v_0\,(1+\sigma)\left(\frac{1}{\alpha_S} + \beta_I \frac{[I]}{aK_I} + \beta_X \frac{[X]}{bK_X} + \beta_{IX} \frac{[I]\,[X]}{eK_I K_X}\right)}{1 + \frac{[I]}{\alpha_I K_I} + \frac{[X]}{\alpha_X K_X} + \frac{[I]\,[X]}{cK_I K_X} + \sigma\left(\frac{1}{\alpha_S} + \frac{[I]}{aK_I} + \frac{[X]}{bK_X} + \frac{[I]\,[X]}{eK_I K_X}\right)}. \tag{6.4}$$

This rate equation has *four* terms in the numerator corresponding to the four catalytic steps and *eight* terms in the denominator that correspond to the eight enzyme-containing species at the corners of the cubes in Scheme 6.1, with 1 representing free enzyme.

Depending on the values of the interaction constants c and e in (6.4), as defined in (6.5) and illustrated in Scheme 6.2, four *types of interaction*, originally defined by Keleti and Fajszi [16], can occur between the modifiers I and X on free enzyme (formation of EIX) and on the ES complex (formation of ESIX):

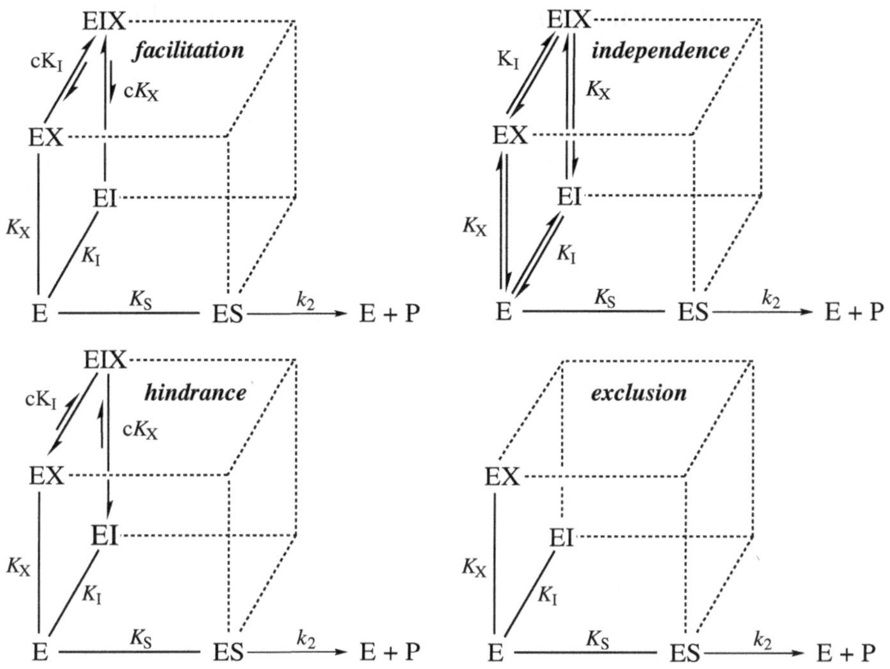

Scheme 6.2 The four types of interaction of two simultaneously acting modifiers I and X on free enzyme as determined by the interaction constant c (6.5). The *size of the arrows* indicates how the complex EIX is favored or disfavored with respect to EI and EX. Other paths and constants are omitted for clarity

$$
\begin{array}{llll}
 & \text{EIX} & & \text{ESIX} \\
0 < c < 1 & \textit{facilitation}\,; & 0 < e < 1 & \textit{facilitation} \\
c = 1 & \textit{independence}\,; & e = 1 & \textit{independence} \\
1 < c < \infty & \textit{hindrance}\,; & 1 < e < \infty & \textit{hindrance} \\
c = \infty & \textit{exclusion}\,; & e = \infty & \textit{exclusion}\,,
\end{array}
\qquad (6.5)
$$

where the short notations $(0,1)$ and $(1,\infty)$ will replace $0 < c < 1$ and $1 < c < \infty$, elsewhere in this book, respectively, and the same for e. The notation of the interaction constants as $(1,\infty)$ indicates that they can take relatively large positive values but not sufficiently large to cause exclusion of the modifiers from E and ES, i.e., so large to preclude the formation of EIX (interaction constant c) or ESIX (interaction constant e). This condition is expressed as $c = \infty$ and $e = \infty$. It is important to emphasize that the definitions of the four types of interaction (6.5) as facilitation, independence, hindrance, and exclusion are based on the validity of the rate equation (6.4). These *model-based* definitions should neither be confused with nor put in synonymy with the *phenomenological* effects of synergy, zero-interaction, and antagonism (6.6) discussed below.

6.3 Effects Resulting from the Combination of Modifiers

The theories discussed in this section have been inherited from pharmacology but their application to enzyme kinetics is less straightforward than originally realized. Therefore, reference will be made to a few concepts in need of more precise definitions to set the limits of validity of published theories and to avoid their misinterpretation.

Very much like the elevation above mean sea level of geographic locations, represented in topographic maps as contour lines joining points having the same altitude (isopleths), an effect resulting from the combined administration of drugs is represented in pharmacology as *isobolograms*, which consist of lines joining dose combinations of different drugs that produce the same effect and called *isoboles*. In 1872, Fraser represented antagonism using this graphical method without giving it a specific name [8, 9] and demonstrated that

> This result appears a very remarkable one, when it is considered that successful counteraction is produced by much smaller doses of atropia against the poisonous action of doses of physostigma greatly in excess of the minimum-lethal, ⋯ [9, p. 486]

Experiments consisted in giving rabbits various doses of the alkaloids physostigmine (a reversible acetylcholinesterase inhibitor) and atropine (a competitive antagonist of the muscarinic acetylcholine receptor) to study their mutual effects. Half a century later, Loewe and Muischnek reconsidered the definitions of antagonism, synergy and related concepts in the context of drug interactions and coined the term isobole [21].

After noticing *fallacious criteria while determining the nature of drug interactions*, Berenbaum conducted a meticulous survey of the literature, reconciled nomenclature conflicts [1] and corroborated the mathematical tools for correctly describing the interactions between drugs [1, 3].

The definitions of synergy, antagonism, and zero-interaction are based on the following relationships for two substances A and B and their resulting effect E:

$$\frac{d_a}{D_a} + \frac{d_b}{D_b} = E \qquad \begin{array}{ll} E < 1 & \textit{synergy} \\ E = 1 & \textit{zero-interaction} \\ E > 1 & \textit{antagonism}\ , \end{array} \tag{6.6}$$

where d_a, d_b, D_a, and D_b represent doses or concentrations of A and B: d_a and d_b refer to A and B used in combination, D_a and D_b represent the isoeffective doses of A and B, i.e., the doses that produce the same effect as their combination when used separately. *Summation*, *additivism* an *addition* have been used as synonyms of zero-interaction with inconsistent meanings among different authors, for which reason this nomenclature has been discouraged [3].

The representation of synergy, antagonism, and zero-interaction by means of isoboles is illustrated in Fig. 6.1. Synergy and antagonism can be appreciated by making reference to the zero-interaction isobole represented by the dashed straight line in Fig. 6.1, in which the sum of the ratios $d_a/D_a + d_b/D_b$ is always 1, e.g., for the point Z, $E = (10/20 + 5/10) = 1.0$. On the other hand, for the point labeled S on the synergy isobole $E = (2.75/20 + 5/10) = 0.64$, which is less than 1, and for point A on the antagonism isobole $E = (17.4/20 + 5/10) = 1.37$, a value larger than 1.

The application of the pharmacological concepts of synergy, antagonism, and zero-interaction to enzyme inhibition and their representation as isobolograms has been introduced by Webb [38, Chap. 10]. Although apparently straightforward, important conceptual differences exist between the pharmacological and the enzymological applications of these notions. In pharmacology and toxicology, synergy and antagonism are often determined empirically by measuring effects, e.g., giving rats various doses of a poison and of an antidote and counting the number of surviving animals. Such phenomenological observations in vivo are generally not related to a fundamental theory based on a mathematical model.

Regarding the interactions than can occur between administered substances, the term *zero-interaction* is used in pharmacology and toxicology in a broader context as in enzymology. Considering two chemicals A and B, these may interact directly in a chemical reaction, or A and B may reciprocally affect their affinity for target sites, or either one of the chemicals may influence the metabolism of the other. Bosgra and coworkers [4] stated that

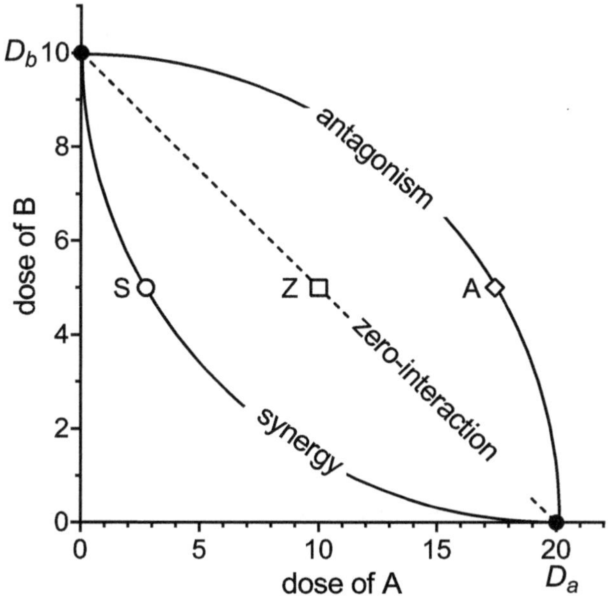

Fig. 6.1 Graphical representation of equieffective dose combinations of two drugs A and B by means of isoboles. For the zero-interaction isobole, the sum of all fractions of D_a and D_b that form the combination is 1

Zero interaction may be defined as the absence of all such interactions: The properties of chemical B remain unaltered, regardless of the presence or concentration of chemical A.

Any pharmacological or toxicological mechanistic approach in living organisms is subordinated to the knowledge of the individual mechanisms of action of the investigated compounds. These mechanisms must then be expressed in mathematical terms in the form of a model ⋯ *which is not a trivial undertaking* [4].

In enzymology instead, all reactions in the presence of substrates, cofactors, and modifiers can be conducted in vitro and reaction rates measured in the presence and absence of modifiers can be used to characterize their mutual interactions using rate equations such as (6.4), i.e., the analysis is performed *quantitatively* on the basis of an underlying theory formulated as a mathematical model. Under these premises, the definition of zero-interaction assumes just one meaning that can be unambiguously defined in mathematical terms. Hence, while criticism to Berenbaum's criteria [2, 3] on the application of the isobole method to interactions between concomitantly acting compounds may apply to systems in vivo [4], the same criticism does not concern the present treatment of multiple enzyme-modifier interactions based on a defined mathematical model.

The goal of multiple inhibition experiments is to evaluate the interaction constants c and e in (6.4) and to ascertain any synergy, zero-interaction, or antagonism between the modifiers. The interaction modes facilitation, independence, hindrance, and exclusion (6.5) are defined by the constants c and e, based on the validity of model (6.4). They are a measure of the *change of the dissociation constant* of a modifier in complex with E or ES when a second modifier is bound, whereas the interaction types synergy, zero and antagonism depend on *how much* the reaction rate changes in the presence of both modifiers.

The following relationships have been proposed by Keleti and Fajszi for differentiating between synergy, zero-interaction, and antagonism on the basis of reaction rates measured in double-modifier experiments [16]:

$$v_{1,2} < \frac{v_1 v_2}{v_0} \qquad \textit{synergy} \tag{6.7}$$

$$v_{1,2} = \frac{v_1 v_2}{v_0} \qquad \textit{zero-interaction} \tag{6.8}$$

$$v_{1,2} > \frac{v_1 v_2}{v_0} \qquad \textit{antagonism}\,, \tag{6.9}$$

where $v_{1,2}$ is the rate in the presence of both modifiers, v_1 and v_2 are the individual rates and v_0 is the rate in the absence of modifiers.

Lienhard and coworkers proposed an equation for describing the combined effects of a linear specific and a linear balanced inhibitor that do not interact when present concomitantly [19]

$$\frac{1}{v_{1,2}} = \frac{1}{v_1} + \frac{1}{v_2} - \frac{1}{v_0}\,, \tag{6.10}$$

and Chou and Talalay [5] extended this expression to n inhibitors

$$\frac{1}{v_{1,2,3,\cdots n}} = \sum_{i=1}^{n} \frac{1}{v_i} - \frac{n-1}{v_0}. \tag{6.11}$$

The same authors provided also a generalization of (6.8) for the simultaneous action of n inhibitors

$$\frac{v_{1,2,3,\cdots n}}{v_0} = \frac{v_1}{v_0} \times \frac{v_2}{v_0} \times \frac{v_3}{v_0} \cdots \frac{v_n}{v_0}. \tag{6.12}$$

All relationships between rates in (6.7)–(6.12) are valid under a given set of reactant concentrations and can be used for evaluating experimental data but should be used cautiously in making *generalizations* and *predictions*. A measurement can agree with either (6.7) or (6.9) but changing the concentrations of the modifiers may convert synergy into antagonism or vice versa. Similarly, (6.11) does not warrant the claim of *universal validity* for the analysis of multiple inhibitions [5]. Zero-interaction between mutually exclusive inhibitors can be diagnosed irrespective of their individual mechanisms but if the equality in (6.11) is not satisfied, the sole possible judgment is that either one or both of the EIX and ESIX complexes is/are present in the mechanism and/or that at least one of the inhibitors has a hyperbolic character.

The *prediction* of synergy or antagonism on the basis of the equations under discussion is wrong in a number of published cases because of mathematical mistakes, disregard of the individual kinetic mechanisms of the two modifiers or misinterpretation of the interactions that can occur between them. In a few cases, linguistic inaccuracy also caused confusion, e.g., facilitation ($0 < c < 1$) used as a synonym of synergy and exclusion ($c = \infty$) as a synonym of antagonism. Further misapplied terms are *positive cooperativity* for synergy and *negative cooperativity* for antagonism.

Most predictions of synergy and antagonism for combinations of inhibitors in a paper by Fajszi and Keleti [7, Table 2] are unfortunately incorrect. As an example, the interaction type between two mutually exclusive, linear specific inhibitors ($c = e = \infty$) was filed as antagonism. Without mathematics, mutual exclusion of two inhibitors on E and ES means that they are bound to two different enzyme molecules, so neither EIX nor ESIX exist precluding any possibility of synergy or antagonism. This condition is called zero-interaction.

Equation (6.4) offers a straightforward way with respect to (6.7)–(6.12) for predicting synergy, zero-interaction, or antagonism. Modeling the effects resulting from any combination of modifiers and interaction constants and creating a three-dimensional plot of the results can be accomplished using software for mathematical calculations and graphical rendering. Predictions of the effects that can result from physically meaningful combinations of the five linear and the five hyperbolic basic inhibition mechanisms are globally summarized in Table 6.2, while selected single cases among the large choice of 918 theoretically possible combinations between inhibitors and nonessential activators will be discussed in the following sections. In the same sections, examples of synergy, zero-interaction, and antagonism will be explained with numerical examples and illustrated by three-dimensional surface plots and isoboles.

For modeling, predicting, and visualizing the interactions that occur between two modifiers, a rearranged form of (6.4) is particularly useful. This consists in separating the concentrations of the modifiers as variables, whereby either one of [I] or [X] is considered as the dependent variable. In order to assure uniform descriptions of mechanisms throughout this chapter, [X] will be treated as the dependent variable. With some algebraic manipulations, (6.4) becomes

$$[\mathrm{X}] = \frac{K_\mathrm{X}(1+\sigma)\left(\dfrac{v_0}{v_\mathrm{IX}}-1\right) - \dfrac{K_\mathrm{X}}{K_\mathrm{I}}\left(\dfrac{1}{\alpha_\mathrm{I}}+\dfrac{\sigma}{\mathrm{a}}-\dfrac{v_0(1+\sigma)}{v_\mathrm{IX}}\dfrac{\beta_\mathrm{I}}{\mathrm{a}}\right)[\mathrm{I}]}{\left(\dfrac{1}{\alpha_\mathrm{X}}+\dfrac{\sigma}{\mathrm{b}}-\dfrac{v_0(1+\sigma)}{v_\mathrm{IX}}\dfrac{\beta_\mathrm{X}}{\mathrm{b}}\right)+\dfrac{1}{K_\mathrm{I}}\left(\dfrac{1}{\mathrm{c}}+\dfrac{\sigma}{\mathrm{e}}-\dfrac{v_0(1+\sigma)}{v_\mathrm{IX}}\dfrac{\beta_\mathrm{IX}}{\mathrm{e}}\right)[\mathrm{I}]}\,, \tag{6.13}$$

which represents the *general equation of the isoboles* in multiple enzyme-modifier interactions. The participation of an essential activator in multiple enzyme-modifier interactions will be treated separately from other interactions in Chap. 7. For this reason, α_S that appears in (6.4) has been set equal to 1 in (6.13), which means that the direct combination path of substrate with enzyme exist. However, α_I and α_X are retained for taking into account binding of catalytic inhibitors in compulsory-order. Equation (6.13) represents a hyperbola that can take different shapes: concave-up or concave-down curves in the entire inhibitor concentration range, or both effects can coexist when the concentration of the modifiers is increased. However, (6.13) becomes a straight line if

$$K_\mathrm{X}\left(\frac{1}{\alpha_\mathrm{I}}+\frac{\sigma}{\mathrm{a}}-\frac{v_0(1+\sigma)}{v_\mathrm{IX}}\frac{\beta_\mathrm{I}}{\mathrm{a}}\right) \gg [\mathrm{I}]\left(\frac{1}{\mathrm{c}}+\frac{\sigma}{\mathrm{e}}-\frac{v_0(1+\sigma)}{v_\mathrm{IX}}\frac{\beta_\mathrm{IX}}{\mathrm{e}}\right). \tag{6.14}$$

An obvious situation for this behavior is when c = e = ∞, i.e., when the modifiers exclude each other on E and ES.

How many combinations are possible for two modifiers acting on the same enzyme? In the mentioned survey of double enzyme-modifier interactions, Schenker and Baici [32] focussed on combinations between linear and hyperbolic inhibitors. Nevertheless, the classification of the basic modifier mechanisms in Chap. 3 recognizes the existence of 17 entities, which are all candidates in the context of multiple interactions. Considering the participation of *two modifiers*, the order of their binding to the enzyme is not important and, after making a combination, the same mechanism can be considered again. In combinatorial mathematical language this represents an r-combination of a multiset M of size r chosen from a set of n elements: $M(n+r-1, r)$, also known as *r-combination with repetition allowed*:

$$\left\langle {n \atop r} \right\rangle = \binom{n+r-1}{r} = \frac{(n+r-1)!}{r!\,(n-1)!}\,. \tag{6.15}$$

The theoretically possible number of such combinations between 17 mechanisms is thus 153:

$$M\,(17 + 2 - 1, 2) = \frac{18!}{2!16!} = 153\,. \tag{6.16}$$

Additionally, we must consider whether or not the two modifiers interact with each other on the free enzyme, on the ES complex or on both, properties defined by the interaction constants c and e. Moreover, ESIX can or cannot be catalytically competent. There are thus six modes of interaction between I and X when bound to the enzyme, which enlarge the palette of the 153 combinations to 918 theoretically possible effects. Without enumerating explicitly the 153 combinations, they are grouped concisely in Table 6.2 subdividing the effects in six columns numbered as Case (1–6) together with the predicted effects that result from the combinations. Herein, the effects elicited by the mixed inhibitors are similar for the predominantly specific, predominantly catalytic or balanced mechanisms (acronyms in italics). The theoretical combinations that are physically meaningless are indicated by the symbol –.

6.3.1 Notes for the Following Sections

The effects resulting from the concomitant binding of two modifiers to the same enzyme will be discussed in six groups as *cases* 1–6 corresponding to columns 1–6 in Table 6.2. These will include combinations between two inhibitors, one inhibitor, and one nonessential activator, and between two nonessential activators. The dual modifier HMxD(A/I), with its property to behave as activator or inhibitor depending on substrate concentration, represents a flexible mechanism to illustrate combinations between two inhibitors or between an inhibitor and a nonessential activator. The dual mechanism HMxD(I/A) behaves in a similar way with inverted roles as activator or inhibitor depending on substrate concentration. Combinations that involve the action of an essential activator and a liberator together with any modifier will be discussed separately in Chap. 7.

Because of the large number of possible combinations, only representative examples will be considered. The plots will be generated with (6.4) and appropriate parameters, respecting microscopic reversibility and avoiding the tight-binding condition. The results of the combinations will be displayed as three-dimensional activity surfaces with the concentrations of the two modifiers [I] and [X] on two axes and the rate, v_{IX}, on the third axis. The three-dimensional activity surfaces shown for explanatory purposes should not suggest that experimental data may be treated in the same way, e.g., to represent best fits from regression analysis.

Table 6.2 Overview of the interactions between two modifiers I and X bound to an enzyme

		Case 1	Case 2	Case 3	Case 4	Case 5	Case 6
		~~EIX~~	EIX	~~EIX~~	~~EIX~~	EIX	EIX
		~~ESIX~~	~~ESIX~~	ESIX	ESIX	ESIX	ESIX
	c:	∞	$(0,\infty)$	∞	∞	$(0,\infty)$	$(0,\infty)$
	e:	∞	∞	$(0,\infty)$	$(0,\infty)$	$(0,\infty)$	$(0,\infty)$
	β_{IX}:	0	0	0	(0,1]	0	(0,1]
I	X						
LSpI	LSpI	Zi	Sy	–	–	–	–
LSpI	*LMxI*	Zi	Sy	–	–	–	–
LSpI	LCaI	Zi	–	–	–	–	–
LMxI	*LMxI*	Zi	Sy	Sy	–	Sy	–
LMxI	LCaI	Zi	–	Sy	–	–	–
LCaI	LCaI	Zi	–	Sy	–	–	–
LSpI	HSpI	Zi	Sy	–	–	–	–
LSpI	*HMxI*	Zi	Variable	–	–	–	–
LSpI	HCaI	Zi	Sy,Sy/An	–	–	–	–
LMxI	HSpI	Zi	Sy,Sy/An	Sy,Sy/An	–	Sy	–
LMxI	*HMxI*	Zi	Variable	Variable	–	Variable	–
LMxI	HCaI	Zi	Sy/An	Sy/An	–	Sy,Sy/An	–
LCaI	HSpI	Zi	–	Sy	–	–	–
LCaI	*HMxI*	Zi	–	Variable	–	–	–
LCaI	HCaI	Zi	–	Sy,Sy/An	–	–	–
HSpI	HSpI	Zi	Sy	Sy	An/Sy	Sy	Variable
HSpI	*HMxI*	Zi	Variable	Variable	An/Sy	Variable	Variable
HSpI	HCaI	Zi	Sy	Variable	An/Sy	Sy	Variable
HMxI	*HMxI*	Zi	Variable	Variable	Variable	Sy,An/Sy	Sy,An/Sy
HMxI	HCaI	Zi	Variable	Variable	An,An/Sy	Variable	Sy,An/Sy
HCaI	HCaI	Zi	Sy,Sy/An	Sy	An,Sy/An	Sy	Sy,Sy/An
HMxI	*HMxA*	Zi	Variable	Variable	Variable	Variable	Variable
HMxA	*HMxA*	Zi	Variable	Variable	Variable	Variable	Variable

See the classification of the mechanisms in Fig. 3.2. The modifiers can or cannot be mutually exclusive on E and/or ES and the ESIX complex can or cannot be catalytically competent. ~~EIX~~ and ~~ESIX~~ mean that these complexes are not formed. To allow a synoptic representation, LMx(Sp>Ca)I, LMx(Sp<Ca)I and LMx(Sp=Ca)I are pooled as *LMxI*. Similarly, HMx(Sp>Ca)I, HMx(Sp<Ca)I, HMx(Sp=Ca)I, HMxD(I/A) and HMxD(A/I) (the latter two in the role of inhibitors) are pooled as *HMxI*. The five nonessential activators plus HMxD(I/A) and HMxD(A/I) in the role of activators are pooled as *HMxA*. The combinations that are physically meaningless are marked –. The typical *effects* produced by the combinations, calculated with (6.4) using appropriate parameters and verified by graphical analysis, are shown as Sy (synergy), Zi (zero-interaction) and An (antagonism) or *variable*. Sy/An and An/Sy indicate that on increasing [I] and [X] the observed effect changes from synergy to antagonism or vice versa. *Variable* means that the effects may be only synergy, only antagonism or their combinations depending on [I], [X], [S] as well as on the values of the interaction constants c and e

The isoboles will be plotted from equieffective concentrations of the modifiers to illustrate synergy, zero-interaction, and antagonism by projecting the effects on the [I]-[X] plane. The unlabeled isoboles drawn beneath the surface plots do not necessarily correspond to those shown in some figures as a separate two-dimensional plots to illustrate the most representative curves labeled with the value of v_{IX}. The rate in the absence of modifiers, v_0, will be normalized to the value 100 and that in the presence of modifiers, v_{IX}, will be expressed as $\%v_0$. In the isobole plots, [X] will be formally considered as the dependent and [I] as the independent variable according to (6.13).

6.4 Case 1: Modifiers Exclusive on E and ES

A feature common to all modifier combinations in Table 6.2 as case 1, are the values $c = e = \infty$, which also imply $\beta_{IX} = 0$ in (6.4). For all these mechanisms zero-interaction is predicted at any reactant concentration. This is due to mutual exclusion of the modifiers on the free enzyme and on the ES complex, which precludes any synergy or antagonism since the two modifiers are bound to two different enzyme molecules (Schemes 6.3, 6.4 and 6.5). The combinations between linear modifiers are characterized by parallel isoboles, the combinations between a linear and a hyperbolic modifier have intersecting isoboles, and the combinations between two hyperbolic modifiers may have either parallel or intersecting isoboles.

6.4.1 *Case 1.1: Two Linear Inhibitors*

Examples of reaction paths for combinations between linear inhibitors are shown in Scheme 6.3 and the rate equation for this group can be obtained from (6.4) by setting $c = e = \infty$ and $\beta_I = \beta_X = \beta_{IX} = 0$:

$$v_{IX} = \frac{v_0\,(1+\sigma)}{1 + \dfrac{[I]}{\alpha_I K_I} + \dfrac{[X]}{\alpha_X K_X} + \sigma\left(1 + \dfrac{[I]}{aK_I} + \dfrac{[X]}{bK_X}\right)}. \tag{6.17}$$

In (6.17), the coefficient α_S that appears in (6.4) has been set to 1 because the other alternative, $\alpha_S = \infty$, applies to essential activators that have nothing to do here, while α_I and α_X are retained for combinations that include linear catalytic inhibitors. Setting α_I and/or α_X to infinite causes deletion of the second and/or of the third term in the denominator, while retaining all terms in the expression within parentheses in the denominator, where the [I]- and [X]-containing terms represent the ESI and ESX complexes, respectively. The equation for the isoboles obtained

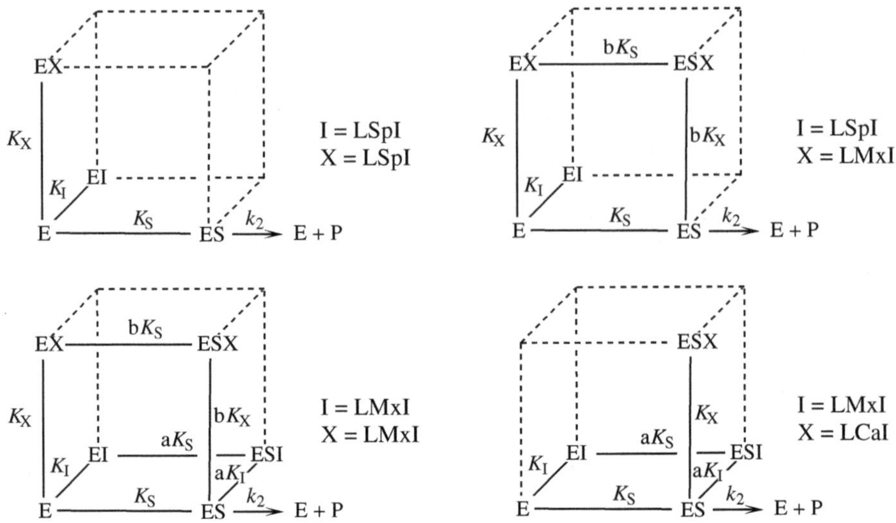

Scheme 6.3 Modifiers exclusive on E and ES: four examples of combinations between two linear inhibitors. Remark to the notation used in multiple interactions: a and b correspond to α in the general modifier mechanism, applied to I and X, respectively

from (6.18), considering [X] as the dependent and [I] as the independent variable, describes a straight line:

$$[X] = \frac{K_X(1+\sigma)}{\dfrac{1}{\alpha_X}+\dfrac{\sigma}{b}}\left(\frac{v_0}{v_{IX}}-1\right) - \frac{K_X\left(\dfrac{1}{\alpha_I}+\dfrac{\sigma}{a}\right)}{K_I\left(\dfrac{1}{\alpha_X}+\dfrac{\sigma}{b}\right)}[I]\,. \tag{6.18}$$

The slope of (6.18) contains only constants and σ, which is also kept constant during an experiment. Thus, varying [I] and [X] has no influence on the slope of the isoboles that run parallel to one another. At fixed σ, the intercepts on the ordinate depend on v_{IX} and increase with increasing [I] and [X], i.e with decreasing v_{IX}. An example featuring the combination between two linear specific inhibitors is shown in Fig. 6.2, while all other combinations between linear inhibitors in case 1 behave in a similar way. The smooth appearance of the inhibition surface in Fig. 6.2 a intuitively suggests that neither synergy nor antagonism occurs between I and X, as confirmed by the parallel isoboles. The equieffective concentrations of the two inhibitors can be read out directly from panel b. For instance, from the isobole labeled 20, i.e., 80 % inhibition, this effect is brought about by 8 μM of I alone or by 40 μM of X alone, or by 6 μM I plus 10 μM X, or by 2 μM I plus 30 μM of modifier X.

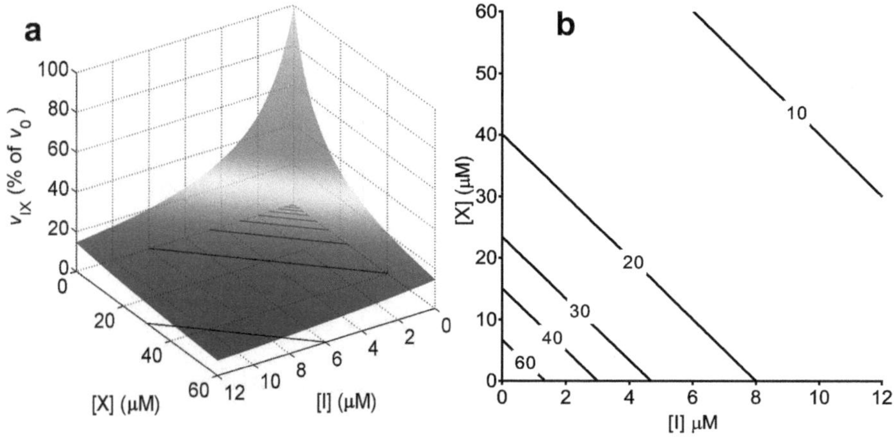

Fig. 6.2 Case 1, modifiers exclusive on E and ES: combination of two linear specific inhibitors (LSpI). The inhibition surface plot (**a**) and the isoboles (**b**) were generated with (6.17) and (6.18), respectively, and the parameters: a = b = ∞, $\alpha_I = \alpha_X = 1$, $\sigma = 1$, $K_I = 1\,\mu$M, $K_X = 5\,\mu$M. The isoboles are labeled with the values of v_{IX}

6.4.2 *Case 1.2: Linear Inhibitor and Hyperbolic Modifier*

In the examples shown in Scheme 6.4, the linear inhibitors are assigned to I and the hyperbolic modifiers to X. The rate equation is (6.19), in which the numerator contains the β_X coefficient that accounts for the catalytic competence of the ESX complex.

$$v_{IX} = \frac{v_0\,(1+\sigma)\left(1 + \beta_X \dfrac{[X]}{bK_X}\right)}{1 + \dfrac{[I]}{\alpha_I K_I} + \dfrac{[X]}{K_X} + \sigma\left(1 + \dfrac{[I]}{aK_I} + \dfrac{[X]}{bK_X}\right)}. \tag{6.19}$$

From (6.19) the equation of the isoboles is obtained by rearrangement and separation of [X] and [I] as variables:

$$[X] = \frac{K_X\,(1+\sigma)\left(1 - \dfrac{v_0}{v_{IX}}\right)}{\dfrac{1}{b}\left(\dfrac{\beta_X v_0\,(1+\sigma)}{v_{IX}} - \sigma\right) - 1} + \frac{K_X\left(\dfrac{1}{\alpha_I} + \dfrac{\sigma}{a}\right)}{K_I\left[\dfrac{1}{b}\left(\dfrac{\beta_X v_0\,(1+\sigma)}{v_{IX}} - \sigma\right) - 1\right]}\,[I]\,. \tag{6.20}$$

Besides constants, both the slope and the intercept in (6.20) contain v_{IX} and thus they depend on the concentration of the modifiers. As a result, the isoboles consist of lines that intersect at a common point when [X] is plotted versus [I]. The coordinates

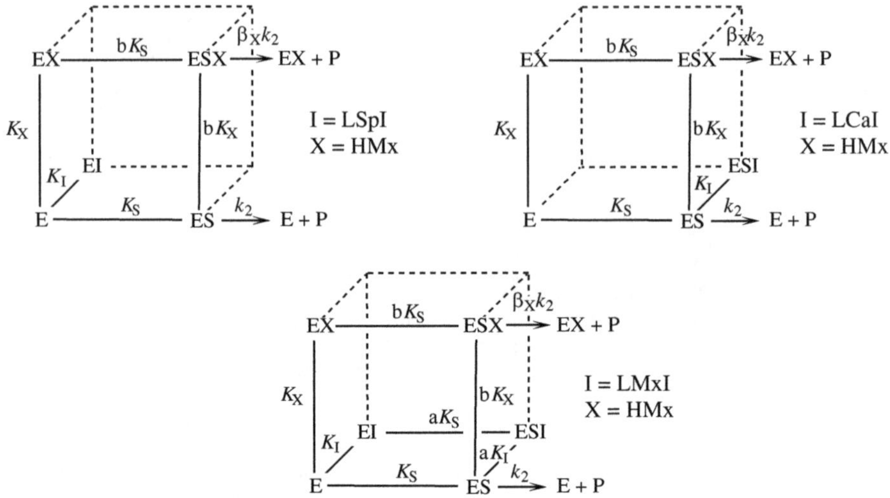

Scheme 6.4 Modifiers exclusive on E and ES: combinations between a linear inhibitor (I) and one of twelve hyperbolic modifiers (X), generically indicated as HMx

of the intersection point, $[\mathrm{I}]^0$ and $[\mathrm{X}]^0$, are

$$[\mathrm{I}]^0 = \frac{\left(\dfrac{\sigma + \mathrm{b}}{\beta_{\mathrm{X}}} - 1 - \sigma\right) K_{\mathrm{I}}}{1 + \dfrac{\sigma}{\mathrm{a}}}; \quad [\mathrm{X}]^0 = -\frac{\mathrm{b}K_{\mathrm{X}}}{\beta_{\mathrm{X}}}, \tag{6.21}$$

i.e., the intersection occurs in the fourth quadrant because we assigned the linear modifier to I. More generally, for a combination between a linear and a hyperbolic modifier, the linear isoboles converge toward the axis that represents the linear modifier. Inhibition profiles and isobole plots are very similar for all mechanisms of this group, just the slope and the position of the isoboles changes according to the values of the constants but the straight lines always converge to a common point.

Zero-interaction, which represents the unique effect observable in all instances of case 1, applies to modifier combinations that involve two inhibitors, one inhibitor and one activator or two activators. Examples of two inhibitors are shown in Fig. 6.3 for mechanisms in the combinations LSpI-HMx(Sp<Ca)I (panels a, b) and LMx(Sp>Ca)I-HCaI (panels c,d).

The rate that results from combinations of different concentrations of the modifiers such as shown in panel (c) should not be confused with antagonism. Instead, for any set of [I] and [X], the rate simply results as sum of the individual contributions. In this example, the rate in the presence of I alone at $[\mathrm{I}] = 20\,\mu\mathrm{M}$ is $v_{\mathrm{I}} = 7.7$ (92.3 % inhibition), and for X alone at $[\mathrm{X}] = 50\,\mu\mathrm{M}$ $v_{\mathrm{X}} = 59.5$ (40.5 % inhibition). The combination $[\mathrm{I}] = 20\,\mu\mathrm{M} + [\mathrm{X}] = 50\,\mu\mathrm{M}$ gives $v_{\mathrm{IX}} = 36.1$ (63.9 % inhibition) not because of antagonisms exerted by X on I but because the modifiers,

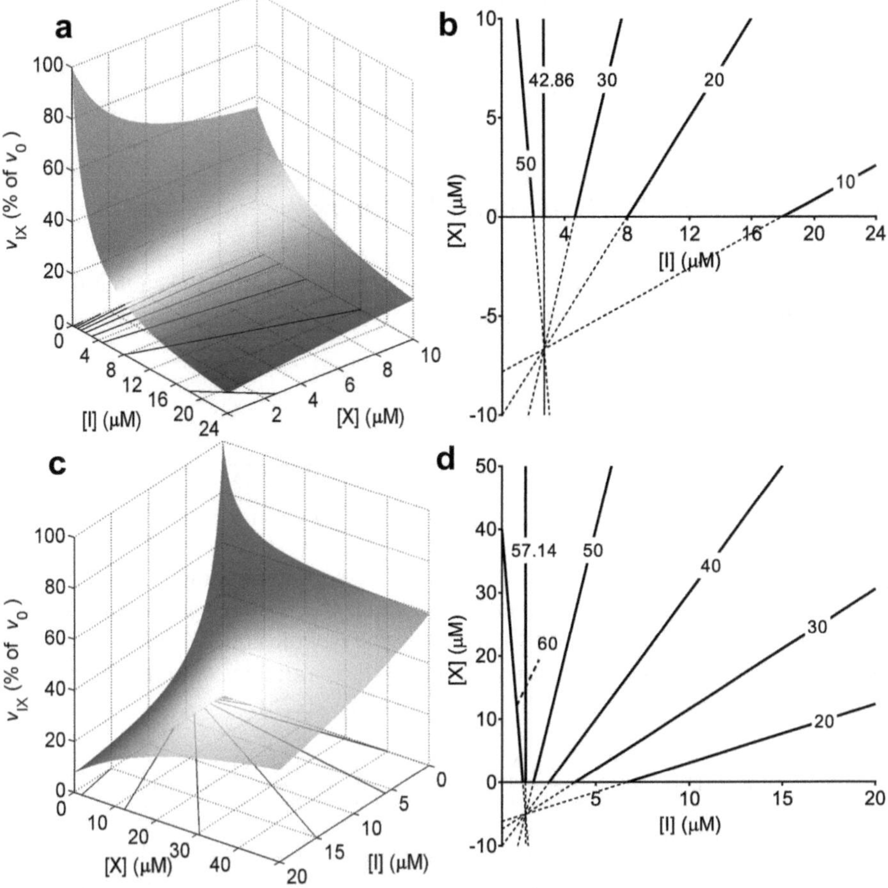

Fig. 6.3 Case 1, modifiers exclusive on E and ES: combinations between a linear (I) and a hyperbolic modifier (X). (**a, b**) LSpI combined with HMx(Sp<Ca)I: a = ∞, b = 0.4, $\beta_X = 0.3$, $[I]^0 = 2.7\,\mu M$, $[X]^0 = -6.7\,\mu M$. (**c, d**) LMx(Sp>Ca)I combined with HCaI: a = 5.0, b = 0.4, $\beta_X = 0.4$, $[I]^0 = 1.25\,\mu M$, $[X]^0 = -5.0\,\mu M$. Constants common to both mechanisms: $\alpha_I = \alpha_X = 1$, $\beta_I = 0$, $\sigma = 1$, $K_I = 1\,\mu M$, $K_X = 5\,\mu M$. The activity surfaces and the isoboles were generated with (6.19) and (6.20), respectively. The isoboles in (**b**) and (**d**) are labeled with the values of v_{IX}

when acting together, have the same amount of enzyme available for binding as if they would act individually. Thus, I and X independently form the EI and EX complexes according to their own kinetic constants. In both panels (b) and (d) of Fig. 6.3 one of the isoboles runs parallel to the ordinate and intersects the abscissa at [I] = $[I]^0$: at this concentration of modifier I (the linear modifier), v_{IX} remains the same at any value of [X]. The value of v_{IX} for [I] = $[I]^0$ can be calculated by substituting $[I]^0$ from (6.21) in (6.19):

$$v_{IX,[I]^0} = \frac{v_0 (1+\sigma)\,\beta_X}{b+\sigma} . \tag{6.22}$$

6.4.3 Case 1.3: Two Hyperbolic Modifiers

In the presence of two mutually exclusive hyperbolic modifiers (Scheme 6.5), the rate equation contains three terms in the numerator since the mechanism comprises three product-generating paths

$$v_{IX} = \frac{v_0 (1+\sigma)\left(1 + \beta_I \frac{[I]}{aK_I} + \beta_X \frac{[X]}{bK_X}\right)}{1 + \frac{[I]}{K_I} + \frac{[X]}{K_X} + \frac{[I]\,[X]}{cK_I K_X} + \sigma\left(1 + \frac{[I]}{aK_I} + \frac{[X]}{bK_X}\right)} . \tag{6.23}$$

The isoboles for this group intersect either in the second quadrant or in the fourth quadrant but there is a special situation in which the isoboles are parallel. These properties are better understood by examining the isobole equation (6.24), which is obtained from (6.23) by separation of [X] and [I] as variables, and the coordinates of the intersection point of all isoboles, $[I]^0$ and $[X]^0$ (6.25):

$$[X] = \frac{K_X (1+\sigma)\left(1 - \frac{v_0}{v_{IX}}\right)}{\frac{1}{b}\left[\frac{v_0 (1+\sigma)}{v_{IX}}\beta_X - \sigma\right] - 1} - \frac{K_X\left[\frac{1}{a}\left(\frac{v_0 (1+\sigma)}{v_{IX}}\beta_I - \sigma\right) - 1\right]}{K_I\left[\frac{1}{b}\left(\frac{v_0 (1+\sigma)}{v_{IX}}\beta_X - \sigma\right) - 1\right]}[I] , \tag{6.24}$$

$$[I]^0 = \frac{[(b+\sigma) - (1+\sigma)\,\beta_X]\,aK_I}{(a+\sigma)\,\beta_X - (b+\sigma)\,\beta_I}; \qquad [X]^0 = \frac{[(a+\sigma) - (1+\sigma)\,\beta_I]\,bK_X}{(b+\sigma)\,\beta_I - (a+\sigma)\,\beta_X} . \tag{6.25}$$

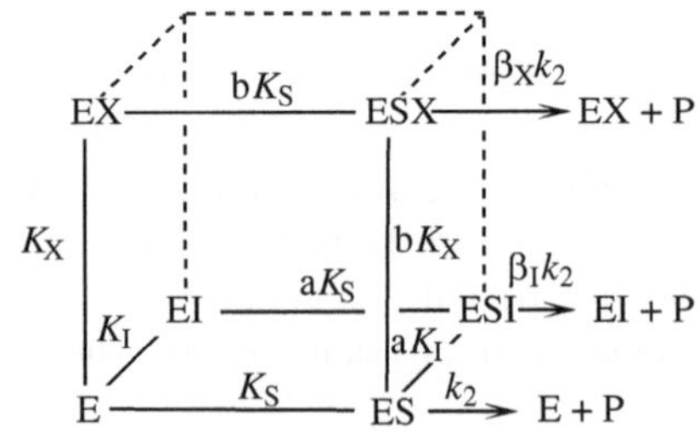

Scheme 6.5 Modifiers exclusive on E and ES: combinations between two hyperbolic modifiers. Any two of the twelve modifiers in Fig. 3.2 (levels 2 and 3) can combine in this way

The slope of the straight line (6.24) contains v_{IX}: when this value decreases by increasing [I] and [X] at fixed [S], the numerator and the denominator will vary according to the values of a, b, K_I, K_X, β_I and β_X and the slope can take any value. The following three situations can be observed.

1. If a $\neq$ b and $\beta_I \neq \beta_X$, the isoboles either intersect in the second quadrant with $[I]^0$ negative and $[X]^0$ positive (Fig. 6.4a,b) or in the fourth quadrant, in which case $[I]^0$ is positive and $[X]^0$ negative (Fig. 6.4c,d). If the combination of parameters is such that the isoboles intersect in the second quadrant, increasing [I] and [X] causes the slopes to change from negative to positive passing through slope zero for $[X] = [X]^0$ (6.25). At this concentration of X, v_{IX} is insensitive to changes in [I] (Fig. 6.4b). The value of v_{IX} for $[X] = [X]^0$ can be calculated by substituting $[X]^0$ from (6.25) in (6.23):

$$v_{IX,[X]^0} = \frac{v_0 (1 + \sigma) \beta_I}{a + \sigma}. \tag{6.26}$$

   In the specific example of Fig. 6.4a,b, $[I]^0 = -0.35\,\mu M$, $[X]^0 = 4.05\,\mu M$ and $v_{IX,[X]^0} = 51.6\,\%$ of v_0. In this example the modifier I obeys the dual mechanism HMxD(A/I): only inhibition is observed because the relative critical substrate concentration, $\sigma^0 = 0.5$, is smaller than the relative substrate concentration employed in drawing the plots ($\sigma = 3.$)
2. When the intersection of the isoboles occurs in the fourth quadrant, increasing [I] and [X] promotes inhibition, v_{IX} decreases and the slope changes from positive to negative. In between, at the characteristic concentration $[I]^0$ in (6.25), the slope is infinite with the isobole running perpendicular to the abscissa and, at this concentration of I, v_{IX} is insensitive to changes in [I] (Fig. 6.4c,d). This characteristic value of v_{IX} can be calculated from (6.23) by setting $[I] = [I]^0$, with $[I]^0$ from (6.25):

$$v_{IX,[I]^0} = \frac{v_0 (1 + \sigma) \beta_X}{b + \sigma}. \tag{6.27}$$

   In the example shown in Fig. 6.4c,d, $[I]^0 = 5.68\,\mu M$, $[X]^0 = -4.09\,\mu M$ and $v_{IX,[I]^0} = 37.5\,\%$ of v_0. The modifier X obeys the dual mechanism HMxD(A/I): only inhibition is observed because the relative critical substrate concentration, $\sigma^0 = 0.14$, is smaller than the relative substrate concentration used for drawing the plots ($\sigma = 3$).

   For the analysis of multiple modifier interactions, measurements are typically performed at a fixed substrate concentration and at a fixed concentration of one of the modifiers while the concentration of the second modifier is varied. The measurements are then repeated inverting the role of the fixed modifier. If the chosen conditions are such that the concentration of the fixed modifier corresponds to either $[I]^0$ or $[X]^0$ as appropriate, or if the fixed [I] or [X] is close to these values, there is a *risk* of getting frustrated because no changes in the rate

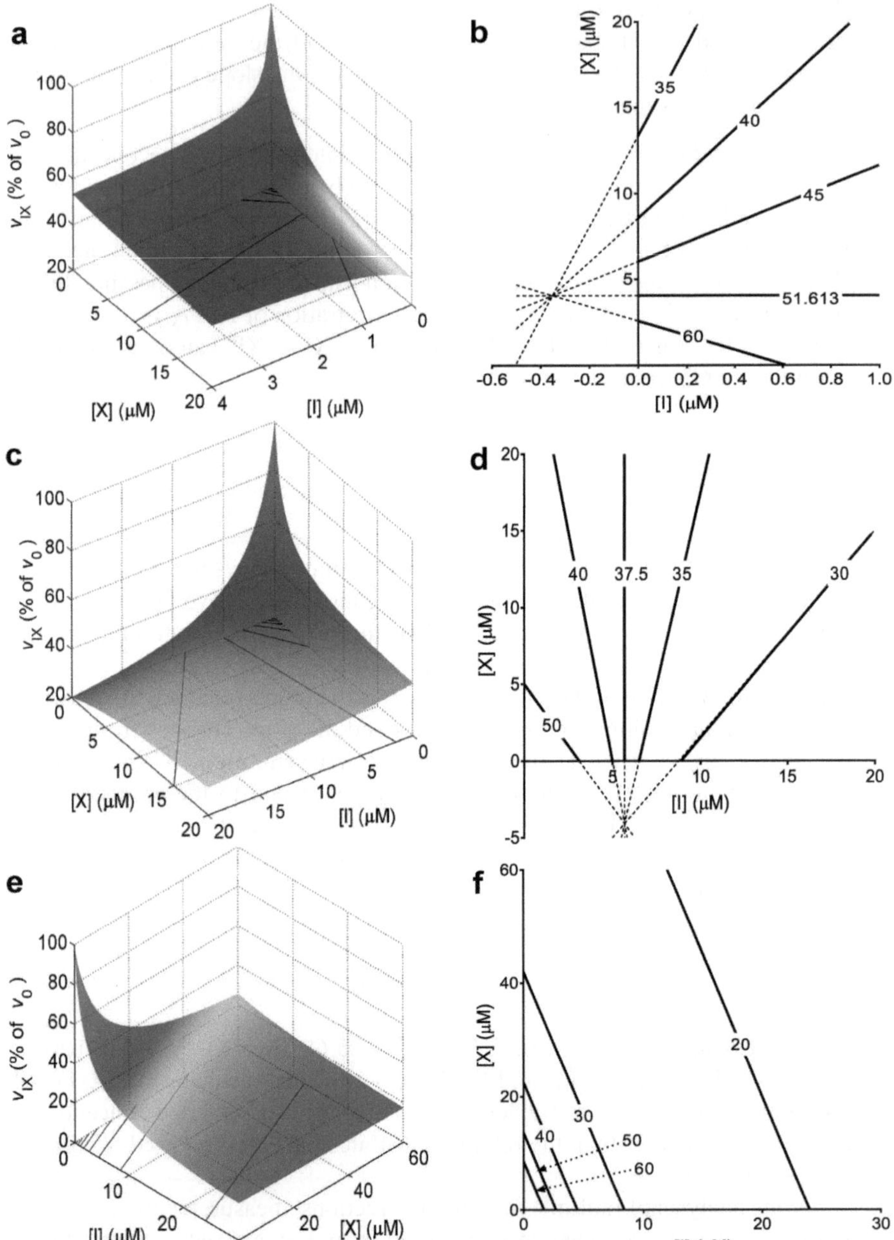

Fig. 6.4 Case 1, modifiers exclusive on E and ES: combinations between two hyperbolic mixed modifiers. (**a, b**) HMxD(A/I) with HMx(Sp<Ca)I: a = 0.1, $\beta_I = 0.4$, b = 0.4, $\beta_X = 0.2$. (**c, d**) HMx(Sp>Ca)I with HMxD(A/I): a = 5, $\beta_I = 0.2$, b = 0.2, $\beta_X = 0.3$. (**e, f**) HMx(Sp>Ca)I with HMx(Sp>Ca)I: a = b = 3, $\beta_I = \beta_X = 0.2$. Constants common to (**a**)–(**f**): $\alpha_I = \alpha_X = 1$, $\sigma = 3$, $K_I = 1\,\mu M$ and $K_X = 5\,\mu M$. The activity surfaces and the isoboles were plotted with (6.23) and (6.24), respectively. The isoboles in (**b**), (**d**) and (**f**) are labeled with the values of v_{IX}

will be noticed. However, it suffices to repeat the same experiment by changing the substrate concentration while maintaining the other conditions and the rate will change proportionally to the concentration of the varied inhibitor [see (6.27) and (6.26)].

3. Two molecules I and X that behave as hyperbolic inhibitors toward an enzyme may be chemically similar. For instance, they can share a basic structure in which a given position is occupied by similar residues. Such inhibitors may possess different inhibition constants, K_I and K_X, while sharing the same inhibitory character and residual activities, i.e., a = b and $\beta_I = \beta_X$. These conditions result in parallel isoboles with slope $-(K_X/K_I)$ because all terms in the numerator and in the denominator of the slope in (6.24) cancel out with the exception of K_X and K_I. An example is shown in Fig. 6.4e,f.

It is obvious that the intersection points and the characteristic v_{IX} around which the slopes of the isoboles pivot by changing the concentrations of the modifiers depend on which mechanism is assigned to I and which one is assigned to X. To avoid confusion, we just remember that in this chapter a and β_I refer to I, while b and β_X refer to X.

From the account given in this section, the double interactions between two hyperbolic modifiers that exclude reciprocally on E and ES, emerge for the diversity of the effects that can arise, as exemplified in the example illustrated in Fig. 6.4.

If we consider five hyperbolic inhibitors (Fig. 3.2, level 2) and five nonessential activators plus two dual modifiers (Fig. 3.2, level 3), the palette of possible effects becomes considerable even in the absence of synergistic and antagonistic effects. The most significant aspect for regulation is that the activity of an enzyme in the presence of a constant concentration of a given modifier can be increased or decreased by changing the concentration of a second modifier.

Figure 6.5 illustrates combinations of two activators (a) and of an activator with one inhibitor (b), in which the modifiers reciprocally exclude on E and ES and therefore the isoboles consist of straight lines (zero-interaction). In panel (a), a hyperbolic catalytic activator (HCaA) is combined with a hyperbolic balanced activator [HMx(Sp=Ca)A]. These modifiers behave as nonessential activators at any concentration of the reactants. The resulting degree of enzyme activation takes a value that lies between the activations produced by either modifier acting individually.

Figure 6.5b features the combination between the dual modifier HMxD(A/I) (as I) and the predominantly catalytic inhibitor HMx(Sp<Ca)I (as X). From the parameters in the legend of Fig. 6.5, $\sigma_I^0 = 0.29$ and thus I behaves as activator since $\sigma = 0.15$. When I acts alone the activity of the enzyme is raised to 120 % at [I] $= 20\,\mu$M, while X alone inhibits the enzyme by 71 % at [X] $= 50\,\mu$M. In this example, the combination of I and X results in very modest or imperceptible inhibition. Such effects have more than a theoretical significance and can explain apparently odd results, for instance when measuring the effects of an inhibitor on an enzyme that co-purifies with a nonessential activator.

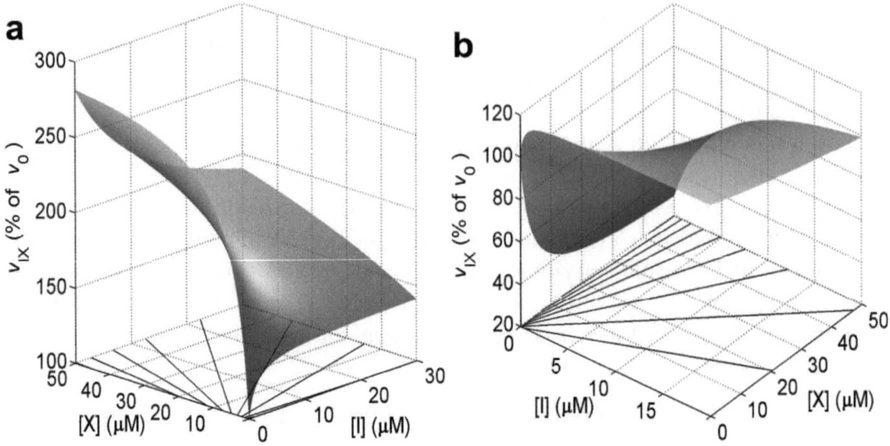

Fig. 6.5 Case 1, modifiers exclusive on E and ES: combinations between two hyperbolic mixed modifiers. (**a**): HCaA (I) combined with HMx(Sp=Ca)A (X): a = 2.0, β_I = 2.0, b = 1.0, β_X = 3.0, σ = 1.5. (**b**): HMxD(A/I) (I) combined with HMx(Sp<Ca)I (X): a = 0.1, β_I = 0.3, b = 0.3, β_X = 0.1, σ = 0.20, σ_I^0 = 0.29. The surface plots were generated with (6.23). Constants common to both mechanisms: K_I = 1 μM, K_X = 5 μM

6.5 Case 2: Modifiers Exclusive on ES

When two modifiers do not form the ESIX complex but form EIX (exclusion on ES), all their combinations have in common the properties $0 < c < \infty$ and $e = \infty$, which imply $\beta_{IX} = 0$. Since I and X can form the EIX complex, depending on the mechanisms of the two modifiers, the effects may result in synergy at any [S], [I], and [X] or in complex effects that depend on the concentrations of reactants. Unlike case 1, case 2 comprises nonsensical combinations when linear catalytic inhibitors are involved. In fact, it is hard to imagine that a catalytic inhibitor, which does not exist as EI, can form the EIX complex when another modifier is present. Nevertheless there are 136 modes of combination between any couple of the remaining 16 mechanisms.

The interaction of the two modifiers on the free enzyme is quantified by the constant c and defined as facilitation, independence, or hindrance in (6.5). Although the definition of hindrance as $1 < c < \infty$ suggests that the constant c may attain high values, a *modestly large* value with respect to $[I]\,[X]\,/K_IK_X$, say in the range 10–50, is sufficient to make the $[I]\,[X]\,/cK_IK_X$ term in the denominator of (6.4) small enough to be insignificant.

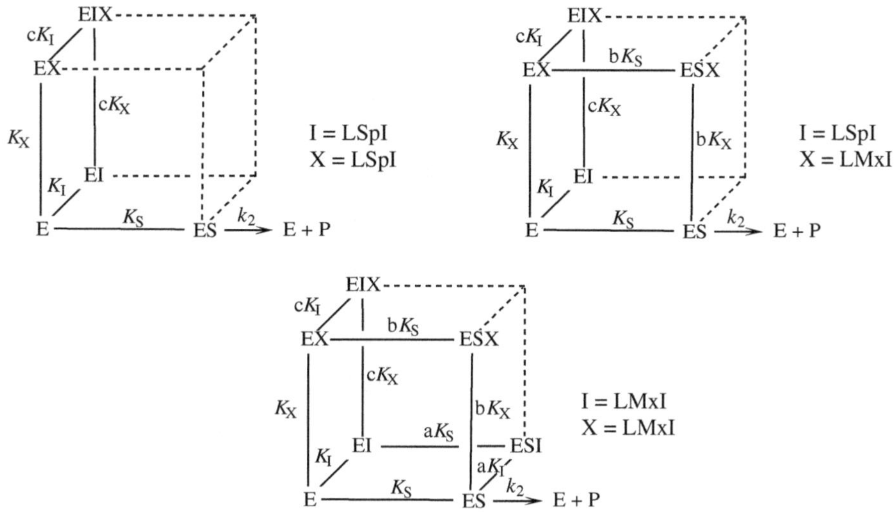

Scheme 6.6 Modifiers exclusive on ES: combinations between two linear inhibitors. Depending on c, the interaction between the two inhibitors can be facilitation ($0 < c < 1$), independence ($c = 1$), and hindrance ($1 < c < \infty$)

6.5.1 *Case 2.1: Two Linear Inhibitors*

Logical combinations between two linear inhibitors that reciprocally exclude on the ES complex are shown in Scheme 6.6. Further diversity in the combinations arises from the nature of the linear mixed mechanism that can be predominantly specific, predominantly catalytic, or balanced. The rate and the isobole equations correspond to (6.28) and (6.29), respectively:

$$v_{IX} = \frac{v_0\,(1+\sigma)}{1 + \dfrac{[I]}{K_I} + \dfrac{[X]}{K_X} + \dfrac{[I]\,[X]}{cK_IK_X} + \sigma\left(1 + \dfrac{[I]}{aK_I} + \dfrac{[X]}{bK_X}\right)}, \tag{6.28}$$

$$[X] = \frac{K_X\,(1+\sigma)\left(\dfrac{v_0}{v_{IX}} - 1\right) - K_X\left(1 + \dfrac{\sigma}{a}\right)\dfrac{[I]}{K_I}}{1 + \dfrac{\sigma}{b} + \dfrac{[I]}{cK_I}}. \tag{6.29}$$

Since linear catalytic inhibitors do not participate in the combinations described in this section, in (6.28) $\alpha_I = \alpha_X = 1$. The resulting effect of the combinations is synergy unless $K_X\,(1 + \sigma/a) \gg [I]/c$, in which case the effect is zero-interaction because (6.29) becomes a straight line and the system degenerates to case 1. In the

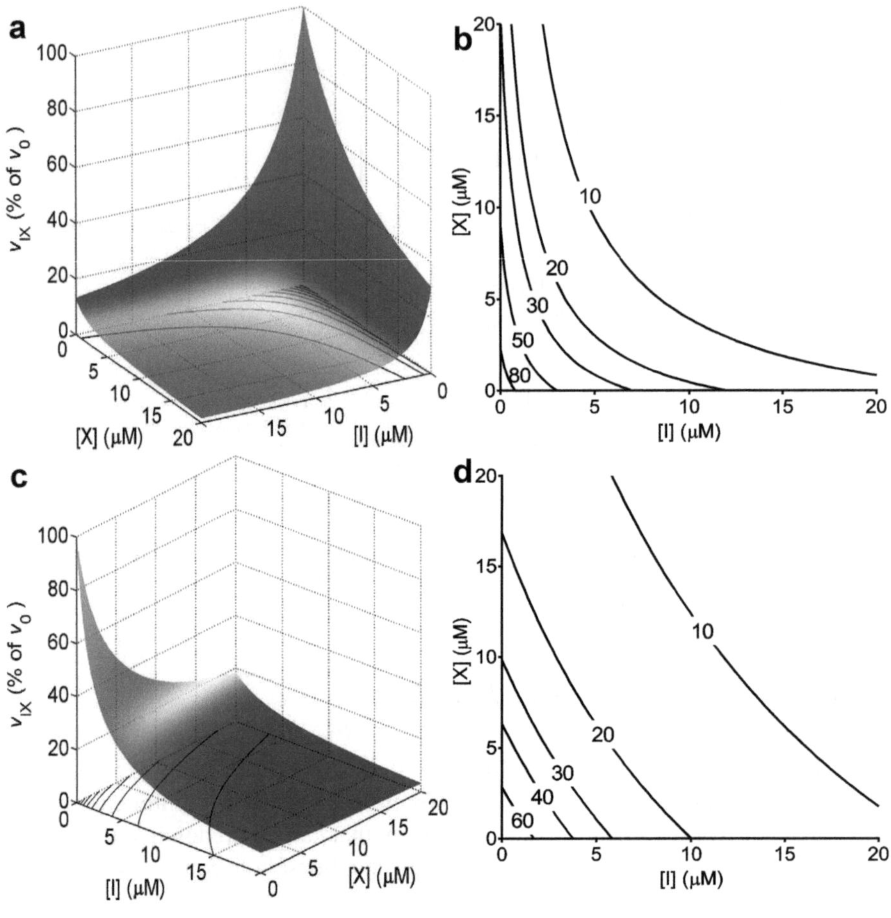

Fig. 6.6 Case 2, modifiers exclusive on ES: combinations of two linear inhibitors. (**a, b**) LSpI (I) combined with LMx(Sp>Ca)I (X): a $= \infty$, b $= 3$, c $= 0.5$, $\sigma = 2$. (**c, d**) LMx(Sp>Ca)I (I) combined with LMx(Sp<Ca)I (X): a $= 5$, b $= 0.8$, c $= 3$, $\sigma = 3$. The inhibition surfaces and the isoboles were generated with (6.28) and (6.29), respectively. Common parameters: $K_I = 1\,\mu M$, $K_X = 5\,\mu M$. The isoboles in (**b**) and (**d**) are labeled with v_{IX}

combinations LSpI-LSpI and LSpI-LMxI, the comparison is between K_X and [I]/c because a $= \infty$ and consequently $1 + \sigma/a = 1$.

The combination of two linear mixed inhibitors produces manifold nuances of synergy depending on the individual prevalence of the specific and of the catalytic character. The examples shown in Fig. 6.6 feature a combination between a linear specific inhibitor and a linear mixed, predominantly specific inhibitor, LSpI-LMx(Sp>Ca)I (a, b), and the effects arising when LMx(Sp>Ca)I is combined with LMx(Sp<Ca)I (c, d). In (a, b) the synergy is strong because c $= 0.5$ *facilitates* the formation of the EIX complex, while in (c, d) there is *hindrance* in forming EIX because c $= 3$.

The concomitant action of two inhibitors and their possible interaction in complex with an enzyme has been treated by Yonetani and Theorell, whose dedicated plot to establish an interaction between the inhibitors or mutual exclusion has been used successfully in a number of reports [42]. The rate equation on which this method is based contains the interaction constant that corresponds to c in (6.28). An example in which c = ∞ (exclusion) is the inhibition of choline-sulfatase (EC 3.1.6.6) by the reaction product choline in combination with either trimethylamine, tetramethylamine, or dimethylaminoethanol, where all modifiers behave as linear catalytic inhibitors [22]. Interaction in the EIX complex occurs for instance in the concomitant inhibition of inositol-phosphate phosphatase (EC 3.1.3.25) by Li^+ and inorganic phosphate [36].

6.5.2 Case 2.2: Linear Inhibitor and Hyperbolic Modifier

Combinations with linear catalytic inhibitors are senseless in this group but there are no restrictions for other combinations. The rate and the isobole equations are given by (6.30) and (6.31), respectively:

$$v_{\mathrm{IX}} = \frac{v_0 (1+\sigma)\left(1+\beta_{\mathrm{X}}\dfrac{[\mathrm{X}]}{\mathrm{b}K_{\mathrm{X}}}\right)}{1+\dfrac{[\mathrm{I}]}{K_{\mathrm{I}}}+\dfrac{[\mathrm{X}]}{K_{\mathrm{X}}}+\dfrac{[\mathrm{I}]\,[\mathrm{X}]}{\mathrm{c}K_{\mathrm{I}}K_{\mathrm{X}}}+\sigma\left(1+\dfrac{[\mathrm{I}]}{\mathrm{a}K_{\mathrm{I}}}+\dfrac{[\mathrm{X}]}{\mathrm{b}K_{\mathrm{X}}}\right)}, \tag{6.30}$$

$$[\mathrm{X}] = \frac{K_{\mathrm{X}}(1+\sigma)\left(\dfrac{v_0}{v_{\mathrm{IX}}}-1\right) - K_{\mathrm{X}}\left(1+\dfrac{\sigma}{\mathrm{a}}\right)\dfrac{[\mathrm{I}]}{K_{\mathrm{I}}}}{1+\dfrac{\sigma}{\mathrm{b}}-\dfrac{v_0(1+\sigma)}{v_{\mathrm{IX}}}\dfrac{\beta_{\mathrm{X}}}{\mathrm{b}}+\dfrac{[\mathrm{I}]}{\mathrm{c}K_{\mathrm{I}}}}. \tag{6.31}$$

Depending on the constants in (6.31), the isoboles, which consist of hyperbolae, intersect at a common point in the third or in the fourth quadrant with coordinates of the intersection point given by

$$[\mathrm{I}]^0 = \frac{[(\mathrm{b}+\sigma)-\beta_{\mathrm{X}}(1+\sigma)]\,K_{\mathrm{I}}}{\beta_{\mathrm{X}}\left(1+\dfrac{\sigma}{\mathrm{a}}\right)-\dfrac{\mathrm{b}}{\mathrm{c}}}; \qquad [\mathrm{X}]^0 = -\frac{\mathrm{b}K_{\mathrm{X}}}{\beta_{\mathrm{X}}}. \tag{6.32}$$

The combinations (Scheme 6.7) involve at least one mixed modifier which can manifest predominantly specific, predominantly catalytic, or balanced character. By interaction with another modifier in the EIX complex, a large palette of effects can arise that may be relevant for allosteric regulation of enzyme activity. A few selected

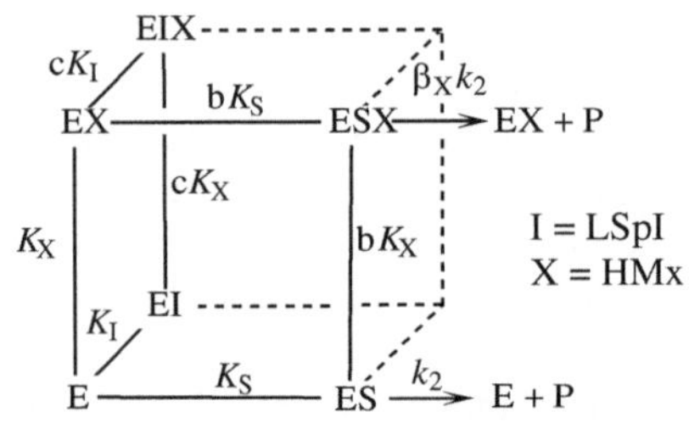

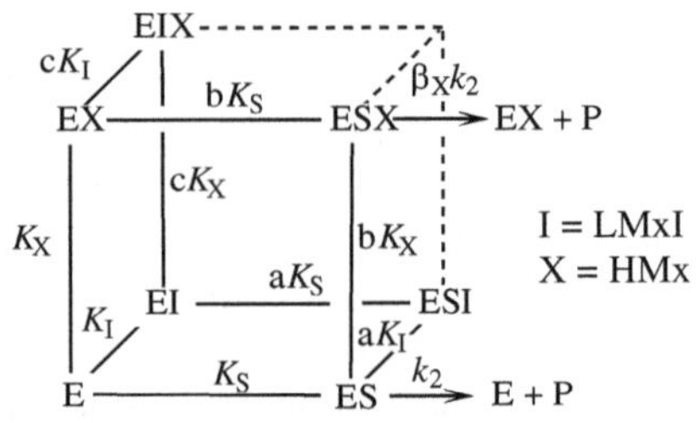

Scheme 6.7 Modifiers exclusive on ES: combinations between a linear inhibitor and a hyperbolic modifier, generically called HMx, which can be any one of twelve mechanisms in Fig. 3.2 (levels 2 and 3)

combinations of modifiers will now be commented and, when appropriate, their properties and effects summarized in tabular form.

6.5.2.1 Combinations Between a Linear Inhibitor and a Hyperbolic Inhibitor or Activator

When a linear specific (LSpI) and a hyperbolic specific inhibitor (HSpI) bind simultaneously to free enzyme but not to ES, the relevant parameters are a $= \infty$, b > 1 and $\beta_X = 1$. Inspection of (6.31) shows that synergy can be predicted unless $K_X(1 + \sigma/\text{a}) \gg [\text{I}]/\text{c}$, in which case we have zero-interaction, linear isoboles, and the same considerations as in Sect. 6.5.1. The isoboles in case of synergy are of the same type as in Fig. 6.6.

The combination with mutual exclusion on ES between a linear specific inhibitor, LSpI (I), and a modifier (X) that obeys the HMxD(A/I) mechanism is illustrated in Fig. 6.7. Panels (a, b) and (c, d) illustrate different effects of the same system obtained by calculating the activity surface with the same kinetic constants but at two different substrate concentrations. For the modifier X, b $= 0.2$ and $\beta_X = 0.5$ yield $\sigma_X^0 = 0.60$ (Sect. 3.7). In panels (a) and (b), $\sigma = 0.3$, which is less than σ_X^0 and thus X *activates* the enzyme as can be seen in the activity profile of X in the absence of I (panel a). The activator X antagonizes the inhibitor by I with concave-down isoboles (panel b). In panels (c) and (d), $\sigma = 5$, which is larger than σ_X^0 and X behaves as inhibitor. When $[\text{I}] < [\text{I}]^0$, there is synergy between I and X (concave-up isoboles); when $[\text{I}] = [\text{I}]^0$, $v_{\text{IX}} = 54.17$ and maintains this value at any [X]; when $[\text{I}] > [\text{I}]^0$, X and I antagonize (concave-down isoboles). Thus, [X] acts as a *brake* that counteracts the action of I resulting in antagonism. The properties discussed in this paragraph are summarized in Table 6.3, row 1.

Figure 6.8 illustrates the combination of LSpI with HMx(Sp<Ca)I (the same behavior is found in the combination of LSpI with HMx(Sp=Ca)I). If X is predominantly catalytic or if the hyperbolic inhibition is balanced, properties defined by $0 < \text{b} \leq 1$ and $\text{b} > \beta_X$, the effect observed by increasing [I] and [X] depends on the interaction constant c. With $0 < \text{c} \leq 1$, i.e., facilitation or independence, only synergy is observed at any combination of reactant concentrations as illustrated in

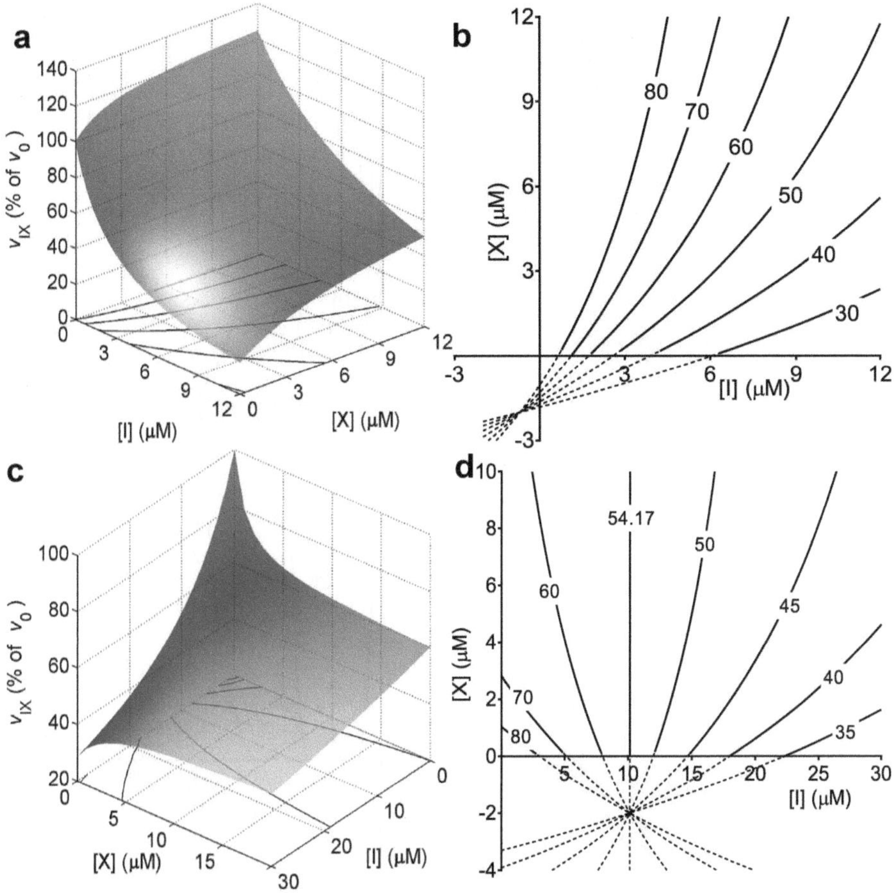

Fig. 6.7 Case 2, modifiers exclusive on ES: combination between a linear inhibitor and a dual modifier to show the effects of substrate concentration. LSpI (I) is combined with HMxD(A/I) (X) using the following constants in (**a**)–(**d**): a $= \infty$, b $= 0.2$, c $= 3.0$, $\beta_I = \beta_{IX} = 0$, $\beta_X = 0.5$, $K_I = 2\,\mu$M, $K_X = 5\,\mu$M, $\sigma_X^0 = 0.6$. (**a, b**): $\sigma = 0.3$; (**c, d**): $\sigma = 5.0$. In (**b**), $[I]^0 = -0.69\,\mu$M, $[X]^0 = -2.0\,\mu$M; in (**d**), $[I]^0 = 10.15\,\mu$M, $[X]^0 = -2.0\,\mu$M. Summary of the effects in Table 6.3, row 1. The activity surface and the isobole plots were generated with (6.30) and (6.31), respectively, and the isoboles are labeled with the value of v_{IX}

Fig. 6.8a,b and the synergistic effect fades out with increasing [X]. With c > 1 (hindrance), synergy or antagonism is manifested depending on reactant concentrations as shown in Fig. 6.8c,d. However, at high substrate concentration antagonism can be observed only at very high values of $[I]/K_I$. The behavior shown in Fig. 6.8 can be explained intuitively. If c ≤ 1, the formation of the catalytically inactive EIX complex is favored because the fourth term in the denominator of (6.30) increases and the formation of EX that leads to ESX and catalysis is disfavored (the second term in the numerator of (6.30) decreases). Thus, increasing $[I]/K_I$ at any [S], lower

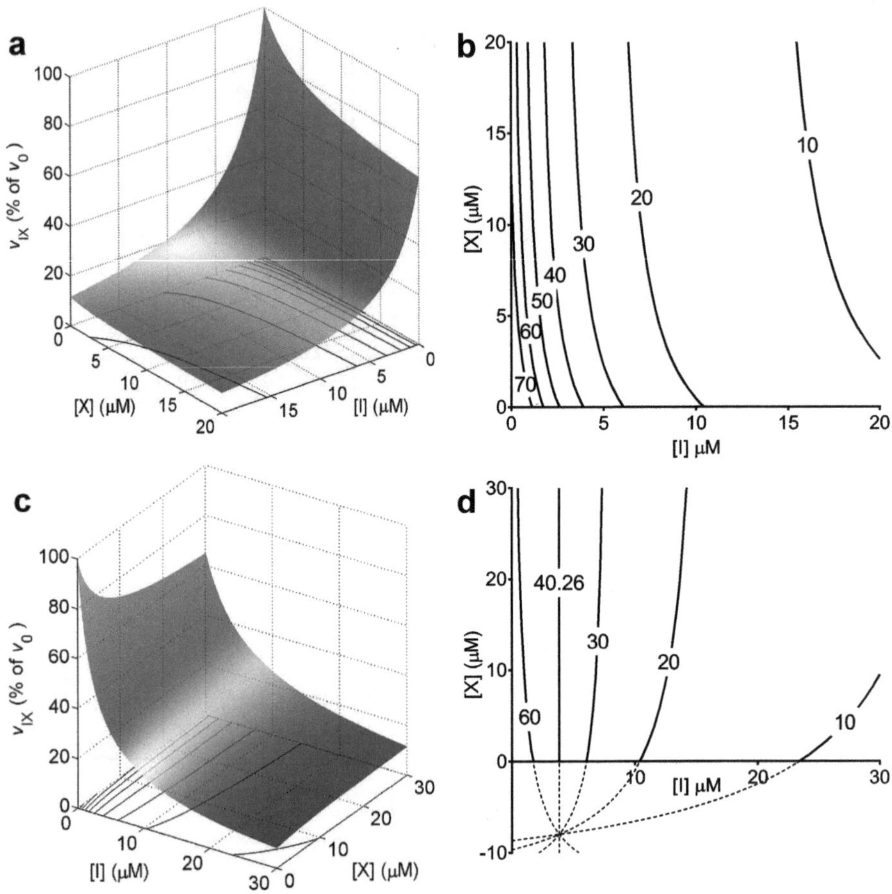

Fig. 6.8 Case 2, modifiers exclusive on ES: combination between LSpI and HMx(Sp<Ca)I to show the effects of the interaction constant c. The following constants are common to all plots: $\sigma = 0.3$, $a = \infty$, $b = 0.8$, $\beta_I = \beta_{IX} = 0$, $\beta_X = 0.5$, $K_I = 2\,\mu M$, $K_X = 5\,\mu M$. (**a, b**): $c = 1.0$; (**c, d**): $c = 3.0$. In (**d**), $[I]^0 = 3.85\,\mu M$, $[X]^0 = -8.0\,\mu M$. The activity surface and the isobole plots were generated with (6.30) and (6.31), respectively and the isoboles are labeled with the value of v_{IX}

concentrations of X are needed to reach the value of v_{IX} that would be obtained if the two modifiers were mutually exclusive on both free enzyme and on the ES complex. If $c > 1$, EI and EX are favored with respect to EIX and the antagonism that arises when [I] increases is due to progressive sequestration of enzyme in the EI complex. These cases are summarized in Table 6.3, rows 2 and 3.

When I is a linear specific inhibitor (LSpI), and X a hyperbolic, predominantly specific inhibitor [HMx(Sp>Ca)I], only synergy results independently of c, [S], [I] and [X]. This is row 4 in Table 6.3. Figures are not shown but they are of the same type as those in Fig. 6.8a,b.

Table 6.3 Modifiers exclusive on ES, summary of the effects arising from combinations of LSpI, which is represented by I in all cases, with hyperbolic mixed modifiers

#	Mechanism of X	b	c	↑ [I] and [X] $\sigma \leq \sigma^0$	↑ [I] and [X] $\sigma > \sigma^0$	↑ [I] and [X] at any σ
1	HMxD(A/I)	(0,1), $b < \beta_X$	$(0,\infty)$	An	Sy→An	–
2	HMx(Sp<Ca)I	(0,1], $b > \beta_X$	(0,1]	–	–	Sy
	or HMx(Sp=Ca)I					
3	HMx(Sp<Ca)I	(0,1], $b > \beta_X$	$(1,\infty)$	–	–	Sy→An
	or HMx(Sp=Ca)I					
4	HMx(Sp>Ca)I	$(1,\infty)$, $b > \beta_X$	$(0,\infty)$	–	–	Sy

Sy synergy, *An* antagonism, σ_X^0 = critical substrate concentration for the modifier X. The symbol ↑ means increasing

Combining a linear specific (LSpI) with a hyperbolic catalytic inhibitor HCaI ($a = \infty$ and $b = \beta_X$) leads to a simplification of (6.31): in the numerator, $\sigma/a = 0$ and in the denominator $\beta_X/b = 1$. When $0 < c \leq 1$ (facilitation or independence) synergy is observed at any reactant concentration with isoboles like those in Fig. 6.8b. When $1 < c < \infty$ (hindrance), progressive increase of the concentration of inhibitors yields first synergy and then antagonism with isoboles similar to those in Fig. 6.8d. However, at high [S] only synergy can be observed unless $[I] \gg K_I$.

The interaction in the EIX complex between a linear mixed inhibitor with predominantly specific character [LMx(Sp>Ca)I] and a hyperbolic specific inhibitor (HSpI) results is synergy that, depending on the constants and substrate concentration, can be weak or very strong. Such an example of pronounced synergy is shown in Fig. 6.9a,b. Here, the conditions that favor synergy between the two inhibitors with specific character are the relatively low substrate concentration ($\sigma = 1$) and the interaction constant $c = 0.3$ that facilitates the formation of the EIX complex.

The substrate concentration affects the interaction in the combination between LMx(Sp<Ca)I and HSpI in a different way with respect to the preceding case because I has, in this case, a predominantly catalytic character. As shown in Fig. 6.9c,d, when acting alone, I is very efficient because $\sigma = 3$ promotes the formation of ESI. However, the formation of ESX is contrasted by the relatively high substrate concentration. Together with the constant $c = 3$, which hinders the formation of EIX, the system shows moderate synergy when [I] increases up to about $6 \times K_I$. Synergy changes to antagonism at concentrations of I larger than 8.4 μM that corresponds to $[I]^0$, the concentration of I for which the rate remains constant at any concentration of the second modifier, X. In this example, at $[I]^0 = 8.4\,\mu M$, $v_{IX} = 24.1\,\%$ of v_0 as shown in Fig. 6.9d.

The palette of effects that can be achieved by combining a linear inhibitor with any of 12 hyperbolic mixed modifiers is by no means described in full by the examples above, both qualitatively and quantitatively. Though, the expected effects are similar to those shown in Figs. 6.7, 6.8, and 6.9.

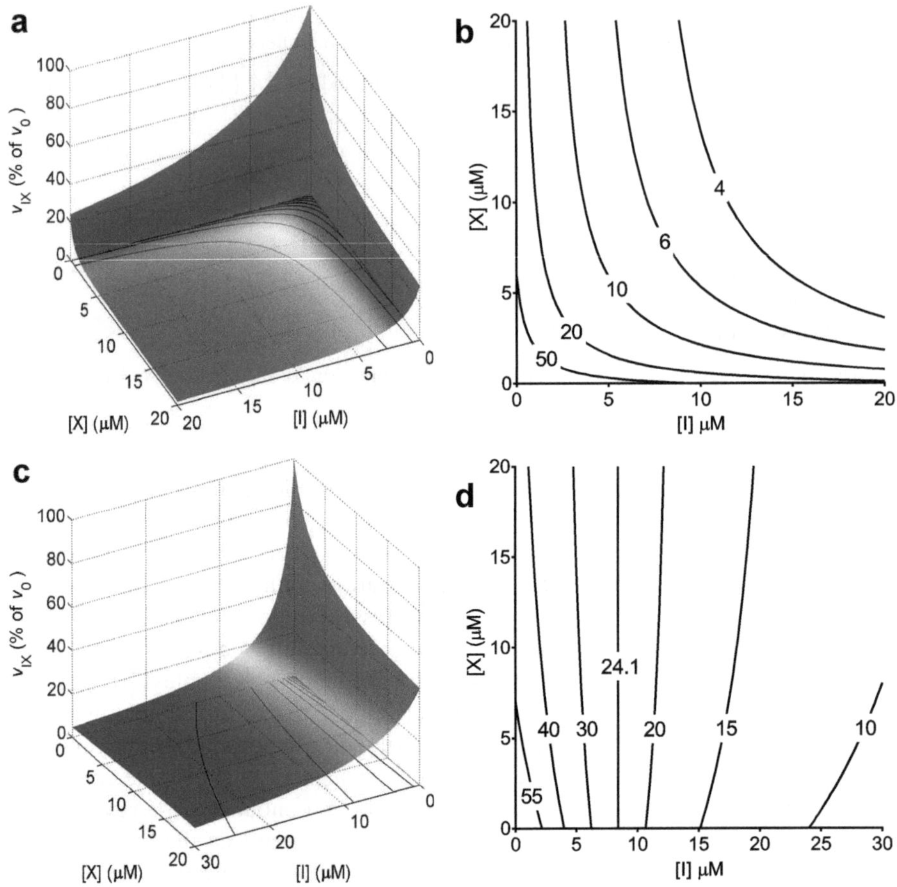

Fig. 6.9 Case 2, modifiers exclusive on ES: combination between a linear mixed inhibitor (I) and a hyperbolic mixed inhibitor (X). (**a, b**) LMx(Sp>Ca)I combined with HSpI: a $=$ 4, $\beta_I = 0$, b $=$ 8, $\beta_X = 1$, c $=$ 0.3 (facilitation), $\sigma = 1$. (**c, d**) LMx(Sp<Ca)I combined with HSpI: a $=$ 0.6, $\beta_I = 0$, b $=$ 8, $\beta_X = 1$, c $=$ 3 (hindrance), $\sigma = 3$. Inhibition constants common to all plots: $K_I = 4\,\mu M$, $K_X = 1\,\mu M$. In (**d**), $[I]^0 = 8.4\,\mu M$, $[X]^0 = -8.0\,\mu M$. The activity surface and the isobole plots were generated with (6.30) and (6.31), respectively. The isoboles in (**b**) and (**d**), labeled with the values of v_{IX}, are not the same as in the corresponding panels (**a**) and (**c**)

6.5.3 *Case 2.3: Two Hyperbolic Modifiers*

The way two hyperbolic modifiers combine by excluding reciprocally on the ES complex is shown in Scheme 6.8. The rate equation is given by (6.33) and the corresponding isobole equation is (6.34). The combinations HSpI-HSpI and HSpI-HCaI result in synergistic interaction at any value of the interaction constant c and substrate concentration whose graphical representation is the same as that in Fig. 6.9b. The combination HCaI-HCaI can result in synergy or synergy followed by antagonism by increasing the concentration of the two inhibitors. Other combinations between a hyperbolic inhibitor and a nonessential activator result in synergy or antagonism depending on modifier concentration.

Some of the 78 possible combinations between 12 hyperbolic modifier mechanisms give rise to complex effects that depend on substrate concentration and on the interaction constant c. This is the case of the dual mechanisms HMxD(I/A) and HMxD(A/I), whose existence in practice is not in doubt (Sects. 5.4.1 and 5.4.4). It cannot be excluded that the subtle effects connected with the dual mechanisms may play an important role in enzyme regulation in the niches of action of some enzymes because of combinations between the dual mechanisms themselves or between one of them and either a hyperbolic inhibitor or a nonessential activator. The analytical expressions of the coordinates of the intersection points of the isoboles as well as the characteristic modifier concentrations for which v_{IX} does not depend on the concentration of the other modifier are very complex. Nevertheless, they can be calculated after introducing the appropriate constants in (6.33).

$$v_{\mathrm{IX}} = \frac{v_0 (1+\sigma)\left(1 + \beta_{\mathrm{I}}\dfrac{[\mathrm{I}]}{\mathrm{a}K_{\mathrm{I}}} + \beta_{\mathrm{X}}\dfrac{[\mathrm{X}]}{\mathrm{b}K_{\mathrm{X}}}\right)}{1 + \dfrac{[\mathrm{I}]}{K_{\mathrm{I}}} + \dfrac{[\mathrm{X}]}{K_{\mathrm{X}}} + \dfrac{[\mathrm{I}]\,[\mathrm{X}]}{\mathrm{c}K_{\mathrm{I}}K_{\mathrm{X}}} + \sigma\left(1 + \dfrac{[\mathrm{I}]}{\mathrm{a}K_{\mathrm{I}}} + \dfrac{[\mathrm{X}]}{\mathrm{b}K_{\mathrm{X}}}\right)}, \tag{6.33}$$

$$[\mathrm{X}] = \frac{K_{\mathrm{X}}(1+\sigma)\left(\dfrac{v_0}{v_{\mathrm{IX}}} - 1\right) - K_{\mathrm{X}}\left(1 + \dfrac{\sigma}{\mathrm{a}} - \dfrac{v_0(1+\sigma)}{v_{\mathrm{IX}}}\dfrac{\beta_{\mathrm{I}}}{\mathrm{a}}\right)\dfrac{[\mathrm{I}]}{K_{\mathrm{I}}}}{1 + \dfrac{\sigma}{\mathrm{b}} - \dfrac{v_0(1+\sigma)}{v_{\mathrm{IX}}}\dfrac{\beta_{\mathrm{X}}}{\mathrm{b}} + \dfrac{[\mathrm{I}]}{\mathrm{c}K_{\mathrm{I}}}}. \tag{6.34}$$

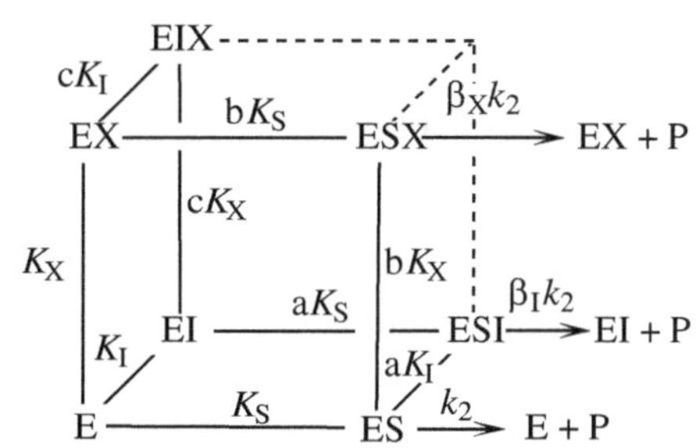

Scheme 6.8 Modifiers exclusive on ES: combinations between two hyperbolic modifiers. Any two of the twelve modifiers in Fig. 3.2 (levels 2 and 3) can combine in this way

Table 6.4 Modifiers exclusive on ES: combinations between HSpI (I) and HMxD(A/I) as modifier (X)

#	Mechanism of X	a	b	c	↑ [I] and [X] $\sigma \leq \sigma_X^0$	↑ [I] and [X] $\sigma > \sigma_X^0$
1	HMxD(A/I)	(1,∞)	(0,1), b < β_X	(0,1]	Sy→An	Sy
2	HMxD(A/I)	(1,∞)	(0,1), b < β_X	(1,∞)	An→Sy [a]	Sy

Sy synergy, *An* antagonism, σ_X^0 = relative critical substrate concentration for the modifier X. ↑ means increasing

[a]Saturation by I is required to see all effects

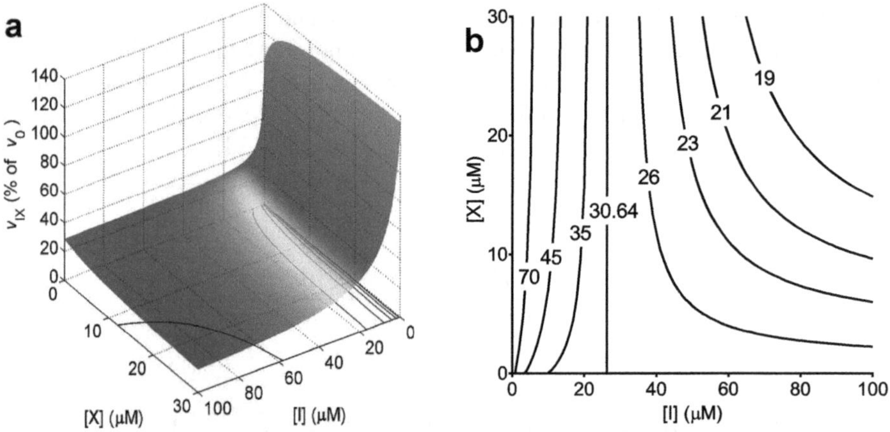

Fig. 6.10 Case 2, modifiers exclusive on ES: combination between an inhibitor and a nonessential activator. Here, HSpI (I) is combined with HMxD(A/I) (X). Constants used to draw the plots: a = 4, $\beta_I = 1.0$, b = 0.1, $\beta_X = 0.3$, c = 3.0 (hindrance), $\sigma = 0.15$, $\sigma_X^0 = 0.29$, $K_I = 1\,\mu M$, $K_X = 5\,\mu M$. The activity surface and the isobole plots were generated with (6.33) and (6.34), respectively. The chosen isoboles in (**b**), labeled with the value of v_{IX}, are not the same as in (**a**)

6.5.3.1 Combination Between a Hyperbolic Inhibitor and a Nonessential Activator

The combination of HSpI with HMxD(A/I) is considered here. By increasing the concentration of the two modifiers, the dual character of HMxD(A/I) as activator or inhibitor elicit effects that span from synergy alone, synergy followed by antagonism, or antagonism followed by synergy (synopsis in Table 6.4). This changing behavior, which can be inferred from (6.33) and (6.34), depends on substrate concentration, i.e., σ smaller or greater than the critical relative substrate concentration σ_X^0, as well as on the interaction constant c. The graphical representation of the combination, with HMxD(A/I) in the role of activator because $\sigma < \sigma_X^0$, is shown in Fig. 6.10. The interaction constant c > 1 means some hindrance in the formation of EIX (Table 6.4, row 2) In Fig. 6.10b the change from antagonism to synergy is delimited by a straight line parallel to the ordinate and intersecting the abscissa at [I] = 26.28 μM, for which $v_{IX} = 30.64\,\%$ of v_0.

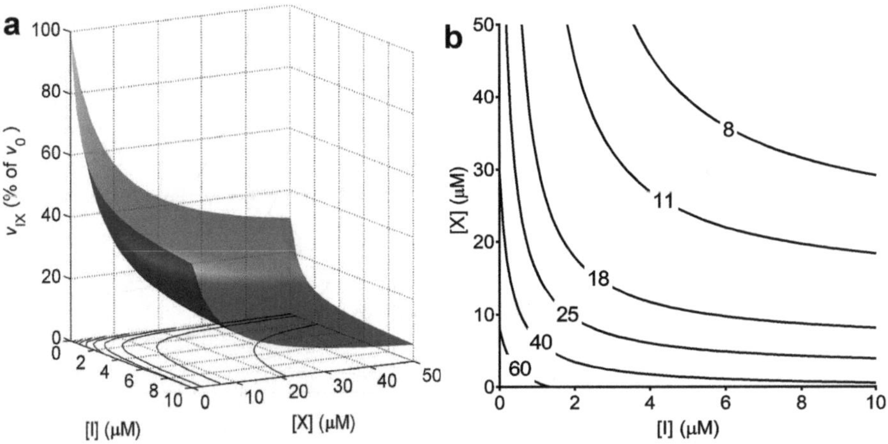

Fig. 6.11 Case 2, modifiers exclusive on ES: combinations between two hyperbolic inhibitors. Combination between the inhibitors HMx(Sp<Ca)I as I and HMx(Sp>Ca)I as X: a = 0.7, β_I = 0.3; b = 4.0, β_X = 0.5; c = 0.5, σ = 1.0, K_I = 1 μM, K_X = 5 μM. The activity surface (**a**) and isobole plots (**b**) were generated with (6.33) and (6.34), respectively. The isoboles in (**b**) are labeled with v_{IX}

6.5.3.2 Combination Between Two Hyperbolic Inhibitors

HSpI combined with another HSpI, HMx(Sp>Ca)I, HMx(Sp<Ca)I, or HMx(Sp=Ca)I produces the effect of synergy independently of the interaction constant c and reactant concentrations. The activity surfaces and the isobole plots are of the same type as in Fig. 6.9a,b.

Figure 6.11 illustrates the combination of HMx(Sp<Ca)I with HMx(Sp>Ca)I under conditions of facilitation in the formation of ESX (c = 0.5). The two modifiers behave as inhibitors in the whole concentration range of the reactants, either alone or in combination. When present together, the resulting inhibitory effect is larger than that obtained by either modifier alone at the same concentration in the combination (strong synergy) as can be appreciated from the concave-up inhibition surface. Despite the predominantly catalytic character of one of the modifiers, there is considerable accumulation of the EIX complex that withdraws enzyme from the three catalytic paths (Scheme 6.8).

6.5.3.3 Combination Between Two Nonessential Activators

The formation of the EIX complex between two nonessential activators implies an interplay of antagonism and synergy between the modifiers depending on their concentrations. Enzyme activation observed in the presence of either one of

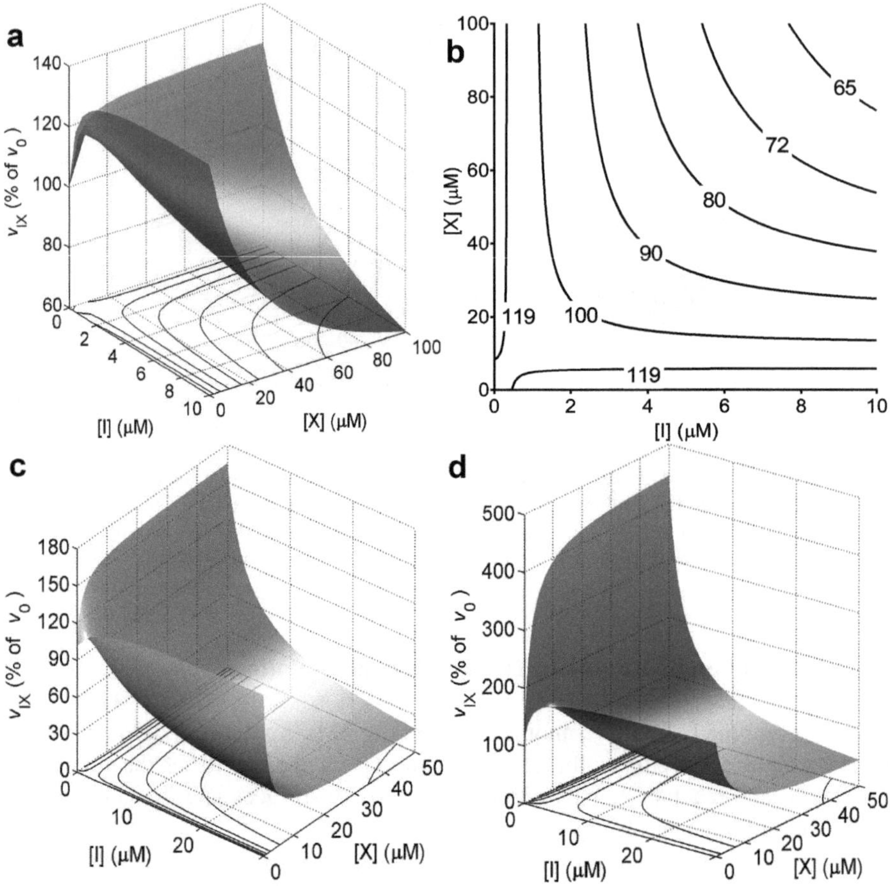

Fig. 6.12 Case 2, modifiers exclusive on ES: combinations between two nonessential activators. (**a, b**) Combination between two dual HMxD(A/I) modifiers with constants: a = 0.1, $\beta_I = 0.3$ ($\sigma_I^0 = 0.29$); b = 0.3, $\beta_X = 0.5$ ($\sigma_X^0 = 0.40$); c = 3.0, $\sigma = 0.15$. (**c**) Combination between HSpA (I) and HCaA (X): a = 0.5, $\beta_I = 1.0$; b = 5.0, $\beta_X = 5.0$; c = 0.5, $\sigma = 1.0$, $K_I = 5\,\mu$M, $K_X = 1\,\mu$M. (**d**) Combination of HMx(Sp=Ca)A with HMx(Sp<Ca)A: a = 1.0, $\beta_I = 2.0$; b = 0.3, $\beta_X = 3.0$; c = 0.5, $\sigma = 1.0$, $K_I = 1\,\mu$M, $K_X = 5\,\mu$M. The activity surface and isobole plots were generated with (6.33) and (6.34), respectively. The isoboles in (**b**) are labeled with the values of v_{IX}

the modifiers acting individually turns into inhibition because of sequestration of enzyme in the EIX complex (three examples in Fig. 6.12). The changing shape of the activity surface does not vary considerably when the interaction constant c is one of facilitation (c < 1), independence (c = 1), or hindrance (c > 1).

The effects expected by combining two hyperbolic dual modifiers HMxD(A/I) are listed in Table 6.5. The interesting situation in the first row of this table is displayed graphically in Fig. 6.12a,b. The modifiers act in a niche where the substrate concentration is lower than the relative critical substrate concentration of both I and X. Acting individually on the enzyme, both modifiers behave as activators in their whole concentration range as can be ascertained by following the upper edges of the activity surface for [I] = 0 and [X] = 0. Despite some hindrance in forming the EIX complex (in this case c = 3), I and X are present together on the enzyme: antagonism ensues by progressively increasing [I] and [X], i.e., the activating effects become weaker. Arrived at the threshold delimited by the antagonism isoboles labeled 119 in Fig. 6.12b, activation turns back to the level 100 that means no effect whatsoever on the enzyme. By further increasing [I] and [X], antagonism turns into strong synergy but in this case the resulting effect is inhibition. The reason for this behavior can be deduced intuitively from Scheme 6.8: ESI as well as ESX are catalytically competent and with the interaction constant c = 3 there is hindrance in the formation of EIX, whose steady-state concentration is progressively reduced as [I] and [X] increase. At low substrate concentration the paths leading to EI and EX are thus favored withdrawing enzyme from all three catalytic paths.

If σ lies between the lower and the highest of σ_I^0 and σ_X^0 (rows 2 and 3 in Table 6.5), I behaves as inhibitor and X as activator. If σ is larger than both σ_I^0 and σ_X^0, I and X behave as inhibitors (rows 4 and 5 in Table 6.5). The interaction constant c governs synergy and antagonism that result from increasing the concentration of the modifiers.

Grasping at a glance the combined effects of modifiers in Table 6.5 and the other tables in this chapter may not be easy. Nevertheless the systems therein are worth of attention for their remarkable potential to respond in different ways to changes in substrate and modifier concentrations. Yet another boring piece of pure theory? This is certainly not the case if we look back at Sects. 5.4.1 and 5.4.4, where real cases of allosteric interactions by the mechanisms HMxD(I/A) and HMxD(A/I) are discussed. In particular, inhibition of arachidonate 15-lipoxygenase by products of lipid metabolism is brought about by these mechanisms (pp. 258 and 272). Allosteric regulation of human lipoxygenase is receiving much attention in physiology and for its implication in inflammation and cancer [14, 39]. The

Table 6.5 Modifiers exclusive on ES: combinations between two HMxD(A/I) modifiers for different values of the interaction constant c and of the relative substrate concentration

#	I and X	a	b	c	Effects by increasing [I] and [X] $\sigma \le \sigma_I^0$ $\sigma \le \sigma_X^0$	 $\sigma > \sigma_I^0$ $\sigma \le \sigma_X^0$	 $\sigma > \sigma_I^0$ $\sigma > \sigma_X^0$
1	HMxD(A/I)	(0,1)	(0,1)	(0,∞)	An→Sy	–	–
2	HMxD(A/I)	(0,1)	(0,1)	(0,1]	–	Sy	–
3	HMxD(A/I)	(0,1)	(0,1)	(1,∞)	–	An→Sy	–
4	HMxD(A/I)	(0,1)	(0,1)	(0,1]	–	–	Sy [a]
5	HMxD(A/I)	(0,1)	(0,1)	(1,∞)	–	–	Sy→An

Sy synergy, *An* antagonism. σ_I^0 and σ_X^0 are the relative critical substrate concentrations of the modifiers I and X, respectively. In all combinations a < β_I and b < β_X

[a]Sy→An for high values of σ

possibility of multiple interactions between reaction products as well as between reaction products and other effectors such as ATP and Ca^{2+} cannot be excluded and is definitely worthwhile of further investigation.

Finally, panels (c) and (d) in Fig. 6.12 show the combinations HSpA-HCaA and HMx(Sp=Ca)A-HMx(Sp<Ca)A, respectively. All involved modifiers behave as nonessential activators independently of reactant concentrations when acting individually. Though, in the combinations with a second activator, antagonism at low [I] and [X] changing to synergy as the modifier concentrations are increased turns the original activation into inhibition. Noteworthy is the range of modulation of enzyme activity in the space depicted in Fig. 6.12d, which spans between 50 % inhibition and 450 % activation.

6.6 Case 3: Modifiers Exclusive on E, ESIX Catalytically Inactive

In case 3 the ESIX complex is formed but not EIX. Any choice of pairs among the basic modifiers is possible with the exception of linear specific inhibitors because it is unlikely that a specific inhibitor forms a complex with an enzyme, which is already in complex with a second modifier and substrate. The common characteristics of the 136 possible combinations in case 3 (see Table 6.2) are $c = \infty$, $0 < e < \infty$ and $\beta_{IX} = 0$. Instead of interacting in the EIX complex as in case 2, the two modifiers interact in the quaternary ESIX complex, which is catalytically inert. The concepts of facilitation, independence or hindrance in the formation of ESIX, when $0 < e < 1$, $e = 1$ and $1 < e < \infty$, respectively, apply in the same sense of the interaction constant c in case 2.

The *effects* of synergy and antagonism originating from interactions in the ESIX complex differ from those in the EIX complex in the way they are modulated by substrate concentration. Comparing the pairs of Eqs. (6.36)-(6.29), (6.38)-(6.31) and (6.40)-(6.34) we see that the isobole equations contain in their denominators the interaction constant e in case 3 and c in case 2. What matters in these pairs of equations is their difference in the multiplicative term for $[I]/K_I$ in the denominator, which is $1/c$ in case 2 and σ/e in case 3. This suggests that substrate concentration has a different impact in the combinations of case 3 compared with those in case 2. Essentially, this establishes whether only synergy or synergy followed by antagonism are the effects that result from increasing the concentration of the modifiers. Graphical examples with nuances of synergy and antagonism, as well as combinations in case 3 will not be detailed in this section as has been done in Sect. 6.5 for case 2 because the effects are similar, but short summaries and references to previous figures will be provided.

6.6.1 Case 3.1: Two Linear Inhibitors

The plausible combinations between two linear inhibitors that interact in a catalytically inert ESIX complex are shown in Scheme 6.9, with rate and isobole equations given by (6.35) and (6.36), respectively. These equations contain the coefficients α_I and α_X to take into account the participation of catalytic inhibitors. The effect predicted for these three combinations is always synergy, regardless of the interaction constant e and reactant concentrations with shapes as those shown in Fig. 6.6. The synergistic inhibitory action results from segregation of enzyme in the ESI, ESX and ESIX complexes resulting in less catalyst available for substrate turnover. Formation of the ESIX complex, and consequently stronger synergy in enzyme inhibition, is promoted by low values of the interaction constant e (e < 1, facilitation).

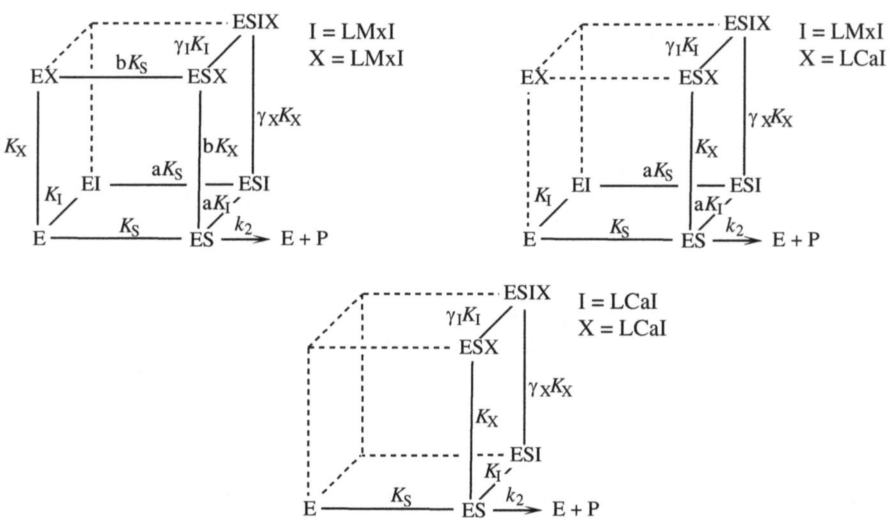

Scheme 6.9 Modifiers exclusive on E and forming a catalytically inactive ESIX complex. Three plausible combinations between two linear inhibitors

Diversity with fine modulation of the elicited effects derives from the linear mixed inhibitor engaged in the formation of the ESIX complex, which can have a predominantly specific or catalytic character or be a balanced inhibitor.

$$v_{\mathrm{IX}} = \frac{v_0\,(1+\sigma)}{1+\dfrac{[\mathrm{I}]}{\alpha_{\mathrm{I}}K_{\mathrm{I}}}+\dfrac{[\mathrm{X}]}{\alpha_{\mathrm{X}}K_{\mathrm{X}}}+\sigma\left(1+\dfrac{[\mathrm{I}]}{\mathrm{a}K_{\mathrm{I}}}+\dfrac{[\mathrm{X}]}{\mathrm{b}K_{\mathrm{X}}}+\dfrac{[\mathrm{I}]\,[\mathrm{X}]}{\mathrm{e}K_{\mathrm{I}}K_{\mathrm{X}}}\right)}, \tag{6.35}$$

$$[\mathrm{X}] = \frac{K_{\mathrm{X}}\,(1+\sigma)\left(\dfrac{v_0}{v_{\mathrm{IX}}}-1\right)-\dfrac{K_{\mathrm{X}}}{K_{\mathrm{I}}}\left(\dfrac{1}{\alpha_{\mathrm{I}}}+\dfrac{\sigma}{\mathrm{a}}\right)[\mathrm{I}]}{\left(\dfrac{1}{\alpha_{\mathrm{X}}}+\dfrac{\sigma}{\mathrm{b}}\right)+\dfrac{\sigma}{\mathrm{e}}\dfrac{[\mathrm{I}]}{K_{\mathrm{I}}}}. \tag{6.36}$$

6.6.2 Case 3.2: Linear Inhibitor and Hyperbolic Modifier

Either a linear mixed or a linear catalytic inhibitor combines in this case with any hyperbolic modifier according to the reaction paths in Scheme 6.10. The rate and isobole equations are given by (6.37) and (6.38), respectively, where α_{X} has been set equal to 1 because of the convention used in this chapter that in combinations between linear and hyperbolic modifiers, the hyperbolic partner is assigned to X.

$$v_{\mathrm{IX}} = \frac{v_0\,(1+\sigma)\left(1+\beta_{\mathrm{X}}\dfrac{[\mathrm{X}]}{\mathrm{b}K_{\mathrm{X}}}\right)}{1+\dfrac{[\mathrm{I}]}{\alpha_{\mathrm{I}}K_{\mathrm{I}}}+\dfrac{[\mathrm{X}]}{K_{\mathrm{X}}}+\sigma\left(1+\dfrac{[\mathrm{I}]}{\mathrm{a}K_{\mathrm{I}}}+\dfrac{[\mathrm{X}]}{\mathrm{b}K_{\mathrm{X}}}+\dfrac{[\mathrm{I}]\,[\mathrm{X}]}{\mathrm{e}K_{\mathrm{I}}K_{\mathrm{X}}}\right)}, \tag{6.37}$$

$$[\mathrm{X}] = \frac{K_{\mathrm{X}}\,(1+\sigma)\left(\dfrac{v_0}{v_{\mathrm{IX}}}-1\right)-\dfrac{K_{\mathrm{X}}}{K_{\mathrm{I}}}\left(\dfrac{1}{\alpha_{\mathrm{I}}}+\dfrac{\sigma}{\mathrm{a}}\right)[\mathrm{I}]}{\left(1+\dfrac{\sigma}{\mathrm{b}}-\dfrac{v_0(1+\sigma)}{v_{\mathrm{IX}}}\dfrac{\beta_{\mathrm{X}}}{\mathrm{b}}\right)+\dfrac{\sigma}{\mathrm{e}}\dfrac{[\mathrm{I}]}{K_{\mathrm{I}}}}. \tag{6.38}$$

The most finely tunable effects in this group are those arising from the combinations of a linear mixed inhibitor and one of the two hyperbolic dual modifiers: the interactions depend on the concentrations of I, X, and S and give rise to antagonism, synergy changing to antagonism as I and X are increased or synergy alone. Otherwise, the prevalent effect between linear and hyperbolic inhibitors is synergy which may turn to antagonism at high concentrations of the inhibitors. A selection of such effects is summarized in Table 6.6 for some combinations of modifiers and facilitation, independence or hindrance in the formation of the ESIX complex.

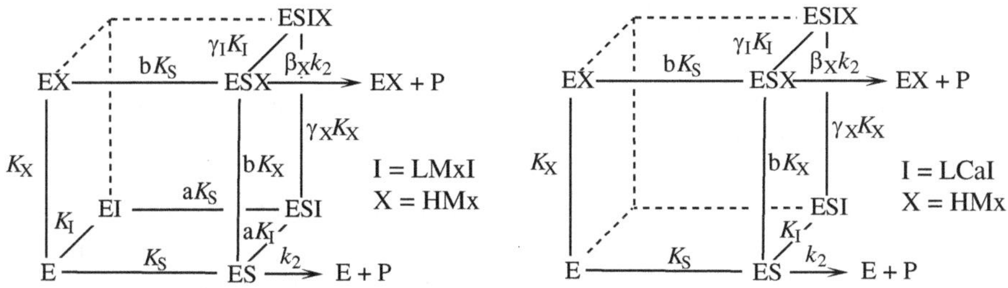

Scheme 6.10 Modifiers exclusive on E and forming a catalytically inactive ESIX complex. Meaningful combinations between a linear inhibitor and any hyperbolic mixed modifier (HMx)

Table 6.6 Modifiers exclusive on E and forming a catalytically inactive ESIX complex

#	I	X	e	Effects by increasing [I] and [X]		
				$\sigma \le \sigma_X^0$	$\sigma > \sigma_X^0$	any σ
1	LMx(Sp⋛Ca)I	HMxD(A/I)	(0,∞)	An	–	–
2	LMx(Sp≤Ca)I	HMxD(A/I)	(0,1]	–	Sy→An	–
3	LMx(Sp>Ca)I	HMxD(A/I)	(0,1]	–	Sy	–
4	LMx(Sp>Ca)I	HMxD(A/I)	(1,∞)	–	Sy→An	–
5	LMx(Sp≤Ca)I	HMx(Sp⋛Ca)I	(1,∞)	–	–	Sy→An
6	LMx(Sp<Ca)I	HSpI	(0,∞)	–	–	Sy

Combinations between a linear inhibitor and a hyperbolic modifier (activator or inhibitor)
Sy synergy, *An* antagonism, σ^0 = critical substrate concentration for modifier X

The surface activity plot and the isoboles of the combinations in Table 6.6 (row 1) are of the same type shown in Fig. 6.7a,b that shows how the linear mixed inhibitor (I) antagonizes nonspecific activation by the dual modifier (X) at $\sigma \le \sigma_X^0$. The net effect is reduced inhibition when both modifier concentrations are increased. However, if the local substrate concentration satisfies the inequality $\sigma > \sigma_X^0$ the resulting effects are no longer of sole antagonism but of synergy that may change into antagonism depending on the relative concentrations of the modifiers and the either predominantly specific, catalytic of balanced character of the linear inhibitor. Therefore, Table 6.6 shows only selected effects while leaving empty other positions that would require further subdivision.

The effects that result only in synergy undisturbed by antagonism are elicited by several combinations of modifiers, such as the example in Table 6.6 (row 6), which shows a linear, predominantly catalytic inhibitor acting together with a hyperbolic specific inhibitor. When the interaction constant is one of facilitation ($e < 1$), very strong synergy in the inhibition process is predicted to occur at much lower combined modifier concentrations with respect to the individual effects. The reason for strong inhibition in this case is sequestration of enzyme in the quaternary complex. The activity surface and the isobole plots are of the type shown in Fig. 6.9a,b.

6.6.3 Case 3.3: Two Hyperbolic Modifiers

The way two hyperbolic modifiers can simultaneously bind to the same enzyme, represented in Scheme 6.11, is analogous to that in Scheme 6.8 with the difference that the catalytically inactive ESIX complex in case 3 replaces EIX in case 2. The presence of ESIX introduces changes in the denominators of the equations for the rate (6.39) and for the isoboles (6.40). In particular, the term σ/e in the denominators of (6.39) and (6.40) is responsible for subtle effects on the isoboles depending on substrate concentration.

$$v_{\mathrm{IX}} = \frac{v_0\,(1+\sigma)\left(1+\beta_{\mathrm{I}}\dfrac{[\mathrm{I}]}{\mathrm{a}K_{\mathrm{I}}}+\beta_{\mathrm{X}}\dfrac{[\mathrm{X}]}{\mathrm{b}K_{\mathrm{X}}}\right)}{1+\dfrac{[\mathrm{I}]}{K_{\mathrm{I}}}+\dfrac{[\mathrm{X}]}{K_{\mathrm{X}}}+\sigma\left(1+\dfrac{[\mathrm{I}]}{\mathrm{a}K_{\mathrm{I}}}+\dfrac{[\mathrm{X}]}{\mathrm{b}K_{\mathrm{X}}}+\dfrac{[\mathrm{I}]\,[\mathrm{X}]}{\mathrm{e}K_{\mathrm{I}}K_{\mathrm{X}}}\right)} \tag{6.39}$$

$$[\mathrm{X}] = \frac{K_{\mathrm{X}}\,(1+\sigma)\left(\dfrac{v_0}{v_{\mathrm{IX}}}-1\right)-\dfrac{K_{\mathrm{X}}}{K_{\mathrm{I}}}\left(1+\dfrac{\sigma}{\mathrm{a}}-\dfrac{v_0(1+\sigma)\,\beta_{\mathrm{I}}}{v_{\mathrm{IX}}\quad\mathrm{a}}\right)[\mathrm{I}]}{\left(1+\dfrac{\sigma}{\mathrm{b}}-\dfrac{v_0(1+\sigma)\,\beta_{\mathrm{X}}}{v_{\mathrm{IX}}\quad\mathrm{b}}\right)+\dfrac{\sigma}{\mathrm{e}}\dfrac{[\mathrm{I}]}{K_{\mathrm{I}}}} \tag{6.40}$$

Representative effects that result from combinations between modifiers in this category are illustrated by four examples in Fig. 6.13. In panel (a) a hyperbolic catalytic inhibitor (HCaI as I) acts together with a hyperbolic catalytic activator (HCaA as X). Inhibition and activation that would be observed from the modifiers acting individually can be appreciated along the [I] and [X] axes. Although not all isoboles can be discerned in this picture, at low concentrations of both modifiers antagonism ensues lowering activation by modifier X, which changes to synergy at higher concentrations to result in net inhibition.

Two hyperbolic inhibitors present concomitantly (Fig. 6.13b) act in synergy by increasing the inhibitory power of the system. Here, a predominantly catalytic inhibitor [HMx(Sp<Ca)I as I] is combined with a hyperbolic specific inhibitor (HSpI as X). As in the analogous situation in which the EIX complex is formed instead of ESIX (Sect. 6.5.3), the combination between two hyperbolic inhibitors

Scheme 6.11 Modifiers exclusive on E and forming a catalytically inactive ESIX complex. Seventy-eight combinations with three catalytic paths are possible between two hyperbolic modifiers, either inhibitors or nonessential activators

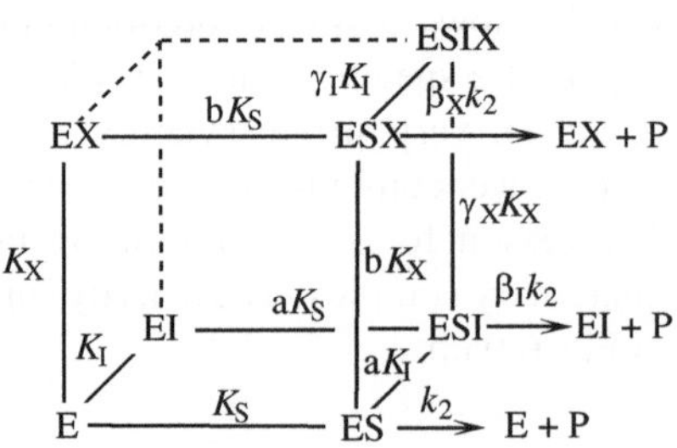

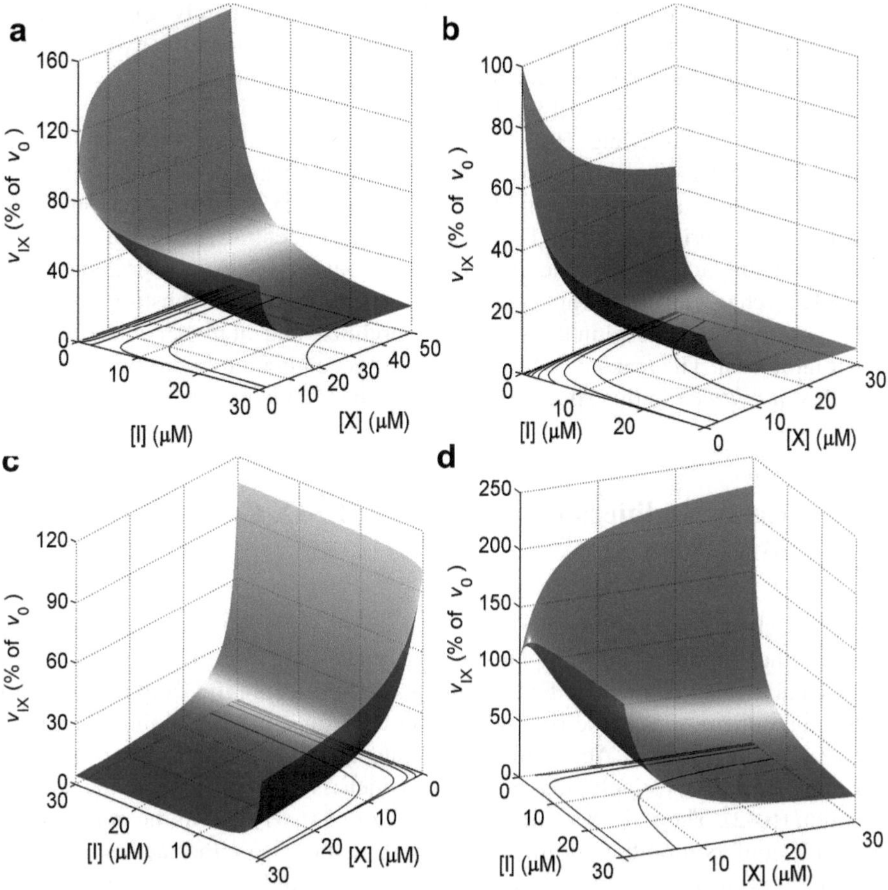

Fig. 6.13 Case 3, modifiers exclusive on E, ESIX is catalytically inactive. Combinations between two hyperbolic modifiers. (**a**) HCaI combined with HCaA: a = 0.4, $\beta_I = 0.4$, b = 4.0, $\beta_X = 4.0$, $\sigma = 1.0$, e = 0.5, $K_I = 4\,\mu M$, $K_X = 2\,\mu M$. (**b**) HMx(Sp<Ca)I combined with HSpI: a = 0.5, $\beta_I = 0.2$, b = 6.0, $\beta_X = 1.0$, $\sigma = 1.0$, e = 0.3, $K_I = 2\,\mu M$, $K_X = 6\,\mu M$. (**c**) Combination between HMxD(I/A) and HMx(Sp=Ca)I: a = 4.0, $\beta_I = 2.0$, b = 1.0, $\beta_X = 0.3$, $\sigma = 2.5$, e = 0.4, $K_I = 1\,\mu M$, $K_X = 4\,\mu M$. (**d**) Combination between HCaA and HMx(Sp>Ca)A: a = 5.0, $\beta_I = 5.0$, b = 2.0, $\beta_X = 4.0$, $\sigma = 0.5$, e = 0.5, $K_I = 1\,\mu M$, $K_X = 3\,\mu M$. The activity surface and the isobole plots were generated with (6.39) and (6.40), respectively

is advantageous in terms of the inhibition degree that can be obtained with either modifier alone.

A dual modifier HMxD(I/A) (I) is combined in Fig. 6.13c with a hyperbolic balanced inhibitor (HMx(Sp=Ca)I as X). The relative substrate concentration used to draw the plot ($\sigma = 2.5\,\mu M$) is slightly greater than the relative critical substrate concentration ($\sigma_I^0 = 2.0\,\mu M$) so that the dual modifier acts as a very weak activator. These conditions are close to the situation $\sigma = \sigma_I^0$ for which the dual modifier has neither inhibitory nor activating effects toward the enzyme thus remaining *invisible* (Sects. 3.7 and 5.4.1). In the example of Fig. 6.13c the resulting effect at very low modifier concentrations is antagonism, which is however barely appreciable for the reason just discussed. Instead, strong synergy that yields effective inhibition is observed in almost the entire concentration range of [I] and [X].

The fourth example, Fig. 6.13d, deals with the combination of two nonessential activators: HCaA as I and HMx(Sp>Ca)A as X. This figure displays a result analogous to that shown in the two examples of Fig. 6.12, namely inhibition ensuing from the concomitant presence of two activators. Such a baffling outcome may undoubtedly surprise anybody. The effect at low concentrations of the modifiers is antagonism, not appreciable in this figure but of the same type shown in Fig. 6.12b, changing to strong synergy that results first in reconstitution of the enzyme activity to its level in the absence of modifiers and, at higher concentrations in net inhibition. The reason for this behavior is accumulation of the inactive ESIX complex, *facilitated* in this specific example by the interaction constant e < 1.

6.7 Case 4: Modifiers Exclusive on E, ESIX Catalytically Active

Table 6.2 shows that the combinations that involve at least one linear inhibitor with another modifier are not plausible candidates for this group characterized by a catalytically competent ESIX complex. The meaningful combinations are those that adhere to Scheme 6.12 where the four substrate-containing complexes are catalytically active. Accordingly, the rate equation contains four terms in the numerator (6.41) including now β_{IX}, which is also present in the isobole equation (6.42). The ratio β_{IX}/e in the numerator of (6.41) and in the denominator of (6.42) confers modifiers in this group the possibility to either act as activators or inhibitors at any modifier concentration, or to exhibit a dual behavior. The best way to appreciate the changes produced on enzyme activity by combinations of hyperbolic modifiers in case 4 is to examine a few characteristic examples (Fig. 6.14), which are represented as surface plots that include unlabeled isoboles without the accompanying detailed isobole plots that were shown in most of the preceding figures.

Scheme 6.12 Combination of modifiers exclusive on E and forming a catalytically active ESIX complex. In the turnover of ESIX the complex is transformed into product and EI + EX in equal parts since EIX does not exist

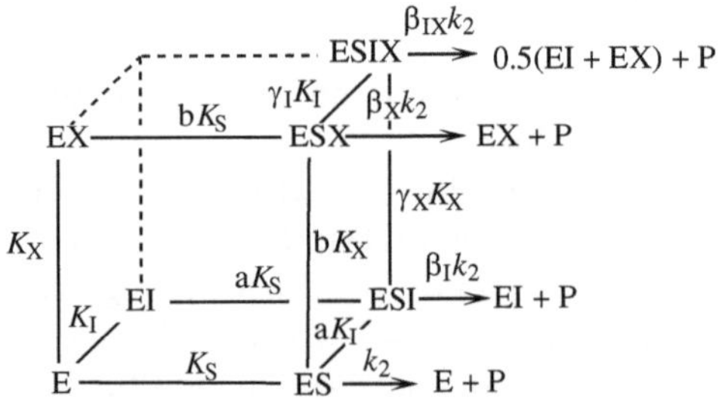

$$v_{IX} = \frac{v_0\,(1+\sigma)\left(1+\beta_I\dfrac{[I]}{aK_I}+\beta_X\dfrac{[X]}{bK_X}+\beta_{IX}\dfrac{[I]\,[X]}{eK_IK_X}\right)}{1+\dfrac{[I]}{\alpha_IK_I}+\dfrac{[X]}{\alpha_XK_X}+\sigma\left(1+\dfrac{[I]}{aK_I}+\dfrac{[X]}{bK_X}+\dfrac{[I]\,[X]}{eK_IK_X}\right)}, \tag{6.41}$$

$$[X] = \frac{K_X\,(1+\sigma)\left(\dfrac{v_0}{v_{IX}}-1\right)-\dfrac{K_X}{K_I}\left(1+\dfrac{\sigma}{a}-\dfrac{v_0(1+\sigma)}{v_{IX}}\dfrac{\beta_I}{a}\right)[I]}{\left(1+\dfrac{\sigma}{b}-\dfrac{v_0(1+\sigma)}{v_{IX}}\dfrac{\beta_X}{b}\right)+\left(\dfrac{\sigma}{e}-\dfrac{v_0(1+\sigma)}{v_{IX}}\dfrac{\beta_{IX}}{e}\right)\dfrac{[I]}{K_I}}. \tag{6.42}$$

In the combination between two hyperbolic specific inhibitors (HSpI) shown in Fig. 6.14a, the two modifiers behave as inhibitors when acting separately, as seen from the rates that decrease hyperbolically in the [I]-v_{IX} and [X]-v_{IX} planes. When present concomitantly, increasing [I] and [X] results first in antagonism between the two modifiers at low concentrations, but the interaction changes to synergy with an opposite effect on enzyme activity, i.e., reactivation, which does not level off at the original value of 100 % v_0 but increases beyond this threshold resulting in net activation. This phenomenon, named *inhibition paradox*, will be discussed in Sect. 6.9.1. Enzyme reactivation by increasing [I] and [X], following inhibition at low modifier concentrations, depends on the magnitude of the coefficient β_{IX} that multiplies the catalytic constant for the turnover of ESIX and on the interaction constant e (6.41).

An example similar to the preceding one is shown in Fig. 6.14b to illustrate that the observable degree of enzyme reactivation upon increasing [I] and [X] depends on the availability of precise measurements at high modifier concentrations. In this combination between two hyperbolic inhibitors, HSpI (I) and HMx(Sp>Ca)I (X), v_{IX} levels off to about 110 % of v_0 at [I] and [X] concentrations exceeding 200 μM. This effect at saturation is however not fully appreciable in the range of concentrations shown in the figure.

The example in Fig. 6.14c shows once more the richness of effects that can arise from the combination of modifiers in dependence of the constants in (6.41). Here, two dual HMxD(A/I) modifiers behave as nonessential activators in the entire range of their concentrations because the substrate concentration is smaller than either one of their critical substrate concentrations. Activation is not only the effect exerted by the modifiers acting individually but also when they are present together. In fact, at saturating [I] and [X], enzyme activity rises by 2.9-fold the activity in the absence of modifiers (activation up to a factor of about 2.5 is visible for the concentrations in Fig. 6.14c). The effect is that of a strong synergy as indicated by the concave-up isoboles visible in the [I]-[X] plane.

Two inhibitors with contrasting character, such as the combination between HMx(Sp<Ca)I and HMx(Sp>Ca)I shown in Fig. 6.14d as I and X, respectively, induce the enzyme to regain activity albeit the net effect remains inhibition. Here, the predominantly catalytic character of I and the predominantly specific character of X, together with e = 0.5 that means facilitation in the formation of ESIX, consent an appreciable steady-state concentration of the catalytically competent quaternary complex, which increases when [I] and [X] increase. At low values of [I] and [X] the antagonism between the two modifiers is very strong but this behavior changes to synergy that leads to reactivation by further increasing [I] and [X].

In Fig. 6.14a,b,d the isoboles are organized in distinct zones of synergy and antagonism in the [I]-[X] plane, which are delimited by threshold lines at character-

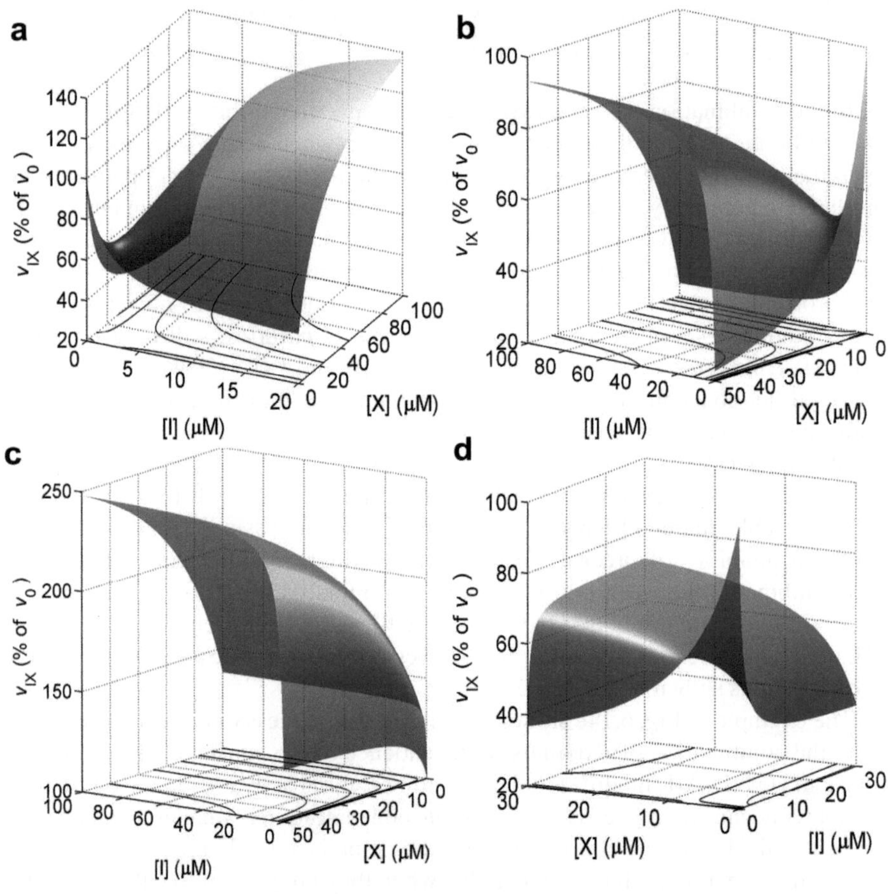

Fig. 6.14 Case 4, modifiers exclusive on E and forming a catalytically active ESIX complex. The surface plots were generated with (6.41). The inhibition constants $K_I = 1\,\mu$M and $K_X = 5\,\mu$M are common to all plots. (**a**) Combination between two hyperbolic specific inhibitors (HSpI): a = 4.0, β_I = 1.0, b = 8.0, β_X = 1.0, β_{IX} = 1.0, e = 5.0, σ = 1.0. (**b**) HSpI (I) combined with HMx(Sp>Ca)I (X): a = 8.0, β_I = 1.0, b = 5.0, β_X = 0.3, β_{IX} = 0.6, e = 3.0, σ = 1.0. (**c**) Combination between two hyperbolic mixed, dual modifiers HMxD(A/I): a = 0.1, β_I = 0.3, b = 0.3, β_X = 0.5, β_{IX} = 0.4, e = 0.3, σ = 0.15, σ_I^0 = 0.29, σ_X^0 = 0.40. (**d**) HMx(Sp<Ca)I (I) combined with HMx(Sp>Ca)I (X): a = 0.7, β_I = 0.3, b = 4.0, β_X = 0.5, β_{IX} = 0.4, e = 0.5, σ = 1.0

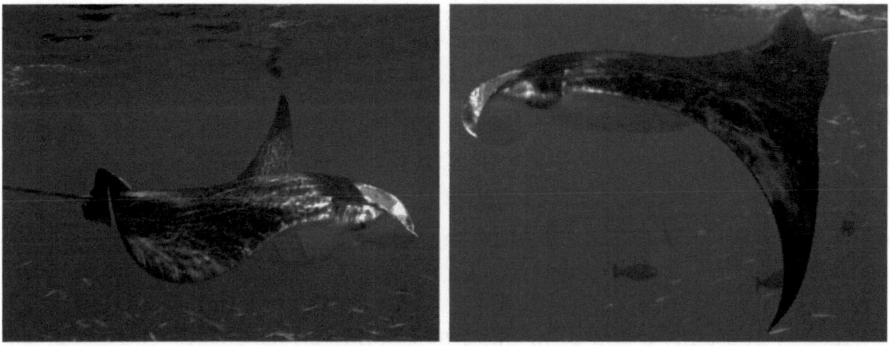

Fig. 6.15 Reef manta rays (*Manta alfredi*) swimming in the Southern Ari Atoll, Maldives. Original photos by courtesy of Federico Baici, 2008

istic concentrations of the modifiers. These zones of demarcation will be discussed for analogous examples in Sect. 6.9 and illustrated in Figs. 6.19, 6.20, and 6.21. The characteristic shapes of the inhibition-reactivation profiles in Fig. 6.14a,b,d recall a swimming *manta ray* (Fig. 6.15). The tip of the tail corresponds to 100 % activity, the reference value in the absence of modifiers ([I] = [X] = 0) but in some images, depending on the orientation of the three-dimensional graphs, the tip of the tail may not be revealed as originating from the vertical axis at the value 100 for perspective reasons. The tips of the manta wing-like pectoral fins represent v_I and v_X at the maximum values of [I] and [X] used to draw the graph, while the tip of the head (opposite to the tail tip) represents the maximum (or minimum, e.g., Fig. 6.18a) value of the combination.

Unexpected effects possibly observed by mixing two modifiers in an enzyme assay should not be quickly discarded as anomalous and attributed to impurities or to inexplicable causes. Enigmatic results are worthwhile of further investigation as any other result. Thus, if two modifiers known as inhibitors happen to activate the enzyme when present concomitantly, this is likely to be a case of interaction of the type discussed in this section. If not already known, the mechanisms of the two modifiers and their kinetic parameters should then be determined and used to simulate plausible interactions with the equations provided in this chapter.

6.8 Case 5: Modifiers Interacting on E and ES, ESIX Catalytically Inactive

This case merges the properties of cases 2 and 3 since two modifiers interact on both the free enzyme and the ES complex, the quaternary ESIX complex being catalytically inactive. In case 5 there are thus two complexes, EIX and ESIX, in which the modifiers can be sequestered in an unproductive way depending on the

Scheme 6.13 Modifiers interacting on E and ES, ESIX is catalytically inactive. Combinations between two linear mixed inhibitors

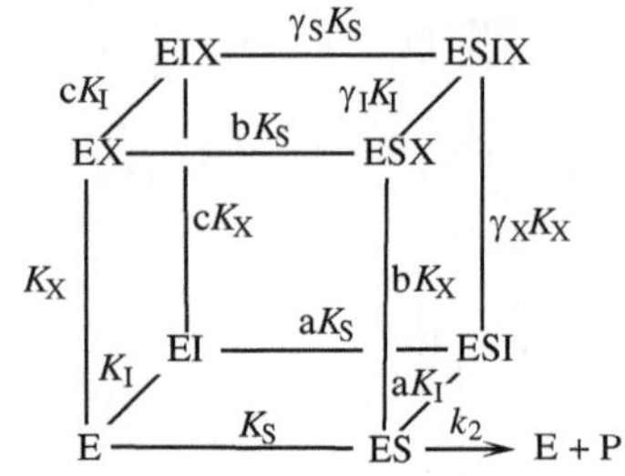

interaction constants c and e. In this framework, various shades of synergy and antagonism depending on the system constants and reactant concentrations are possible. As seen in Table 6.2, not all theoretically possible combinations are likely to occur in practice.

6.8.1 Case 5.1: Two Linear Mixed Inhibitors

In case 5 the only likely combination between linear inhibitors is that between two mixed inhibitors, while combinations involving specific and catalytic inhibitors make little sense because a specific inhibitor will not form the ESIX complex and a catalytic inhibitor will not participate in the formation of EIX (Scheme 6.13). The rate and isobole equations of this combination, with the possibility of interaction on the free enzyme and the ES complex, are given by (6.43) and (6.44), respectively. Independently of the characters of the two mixed inhibitors, i.e., predominantly specific, catalytic, or balanced, and of the interaction constants c and e, synergy is observed for this system at any substrate and modifier concentration. The isoboles have the same aspect as those in Fig. 6.9a,b and when both interaction constants c and e are less than one, the EIX and ESIX complexes are richly populated, withdrawing a considerable proportion of the enzyme from the catalytic path and resulting in efficient, synergistic inhibition.

$$v_{IX} = \frac{v_0(1+\sigma)}{1+\frac{[I]}{K_I}+\frac{[X]}{K_X}+\frac{[I][X]}{cK_IK_X}+\sigma\left(1+\frac{[I]}{aK_I}+\frac{[X]}{bK_X}+\frac{[I][X]}{eK_IK_X}\right)} \tag{6.43}$$

$$[X] = \frac{K_X(1+\sigma)\left(\frac{v_0}{v_{IX}}-1\right)-\frac{K_X}{K_I}\left(1+\frac{\sigma}{a}\right)[I]}{1+\frac{\sigma}{b}+\left(\frac{1}{c}+\frac{\sigma}{e}\right)\frac{[I]}{K_I}} \tag{6.44}$$

6.8.2 *Case 5.2: Linear Mixed Inhibitor and Hyperbolic Modifier*

For the same reason discussed in Sect. 6.8.1, in combinations involving a linear and a hyperbolic modifier, only linear mixed inhibitors are likely to interact with hyperbolic partners by the mechanism sketched in Scheme 6.14. The numerator of the rate equation contains two terms (6.45) and the coefficient β_X appears both in this and in the isobole Eq. (6.46).

The combinations between a linear mixed inhibitor with either predominantly specific, predominantly catalytic or balanced character with HSpI or HCaI result always in synergy between the interacting partners independently of the constants of the system and reactant concentrations. The graphical representations are of the type shown in Fig. 6.9a,b. Other combinations between linear mixed inhibitors with hyperbolic inhibitors or activators give rise to synergy, antagonism or both effects depending on the concentrations of substrate and modifiers. Figure 6.16a illustrates the combination between LMx(Sp=Ca)I as I and HMx(Sp=Ca)I as X. The two modifiers act in synergy in their entire concentration range. The example in Fig. 6.16b shows a linear, predominantly specific inhibitor, LMx(Sp>Ca)I (I) flanked in its action by the nonessential activator HMx(Sp>Ca)A (X). Antagonism is the effect when the modifiers are present at low concentrations. The linear inhibitor takes the lead by weakening the activating effect of X until antagonism switches to synergy producing net inhibition at high modifier concentrations.

$$v_{IX} = \frac{v_0 (1+\sigma)\left(1+\beta_X \frac{[X]}{bK_X}\right)}{1+\frac{[I]}{K_I}+\frac{[X]}{K_X}+\frac{[I][X]}{cK_IK_X}+\sigma\left(1+\frac{[I]}{aK_I}+\frac{[X]}{bK_X}+\frac{[I][X]}{eK_IK_X}\right)} \qquad (6.45)$$

$$[X] = \frac{K_X (1+\sigma)\left(\frac{v_0}{v_{IX}}-1\right)-\frac{K_X}{K_I}\left(1+\frac{\sigma}{a}\right)[I]}{1+\frac{\sigma}{b}-\frac{v_0(1+\sigma)}{v_{IX}}\frac{\beta_X}{b}+\left(\frac{1}{c}+\frac{\sigma}{e}\right)\frac{[I]}{K_I}} \qquad (6.46)$$

6.8.3 *Case 5.3: Two Hyperbolic Modifiers*

When both participants in a combination are of the hyperbolic type the rate (6.47) and isobole equations (6.48) contain the coefficients β_I and β_X that take into account the contributions of ESI and ESX to catalysis (Scheme 6.15). The predicted interaction mode for the combinations HSpI-HSpI, HSpI-HCaI, and HCaI-HCaI is synergy of inhibition under all circumstances, which does not need additional

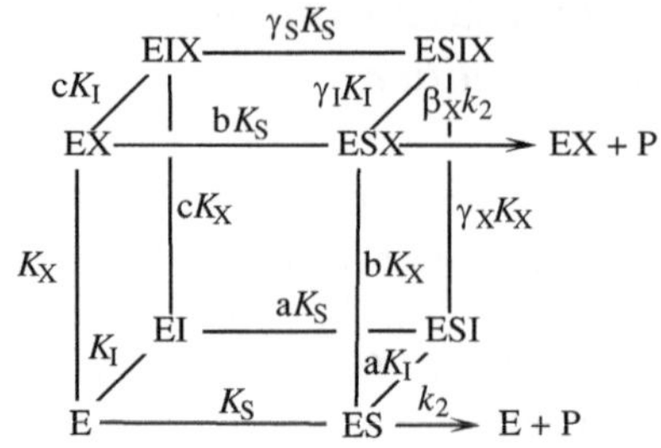

Scheme 6.14 Modifiers interacting on E and ES, ESIX is catalytically inactive. Combinations between a linear inhibitor and a hyperbolic modifier

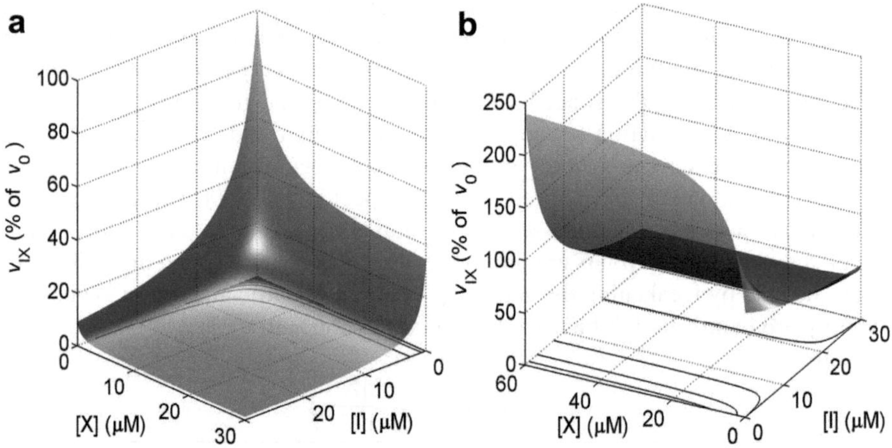

Fig. 6.16 Case 5, modifiers that interact on E and AS, ESIX is catalytically inactive. (**a**) Combination between LMx(Sp=Ca)I and HMx(Sp=Ca)I: a = 1.0, $\beta_I = 0$, b = 1.0, $\beta_X = 0.3$, c = 0.5, e = 0.3, $\sigma = 1.0$, $K_I = 3\,\mu M$, $K_X = 2\,\mu M$. (**b**) Combination between LMx(Sp>Ca)I and HMx(Sp>Ca)A: a = 4.0, $\beta_I = 0$, b = 3.0, $\beta_X = 5.0$, c = 0.5, e = 0.3, $\sigma = 1.0$, $K_I = 20\,\mu M$, $K_X = 3\,\mu M$

comments. Enzyme activities and isobole plots for these mechanisms are analogous to those shown in Fig. 6.9a,b.

The combinations that involve the remaining mechanisms are those expected to give manyfold effects in dependence of concentrations and interaction constants by exhibiting the merged properties of the corresponding combinations in cases 2 and 3. Noteworthy is the strength of the synergistic interactions between inhibitors, which result from the presence of all eight enzyme-containing complexes and from the catalytically inert ESIX complex. This implies lowering of the ES concentration, which accounts for the highest catalytic activity, by forcing more enzyme to form the catalytically less efficient ESI and ESX complexes as well in the inert species EIX and ESIX.

Scheme 6.15 Modifiers interacting on E and ES, ESIX is catalytically inactive. Combinations between two hyperbolic modifiers

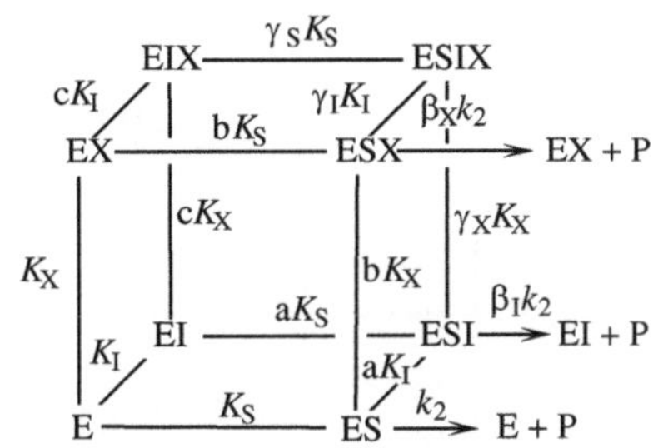

$$v_{\mathrm{IX}} = \frac{v_0\,(1+\sigma)\left(1+\beta_{\mathrm{I}}\dfrac{[\mathrm{I}]}{\mathrm{a}K_{\mathrm{I}}}+\beta_{\mathrm{X}}\dfrac{[\mathrm{X}]}{\mathrm{b}K_{\mathrm{X}}}\right)}{1+\dfrac{[\mathrm{I}]}{K_{\mathrm{I}}}+\dfrac{[\mathrm{X}]}{K_{\mathrm{X}}}+\dfrac{[\mathrm{I}]\,[\mathrm{X}]}{\mathrm{c}K_{\mathrm{I}}K_{\mathrm{X}}}+\sigma\left(1+\dfrac{[\mathrm{I}]}{\mathrm{a}K_{\mathrm{I}}}+\dfrac{[\mathrm{X}]}{\mathrm{b}K_{\mathrm{X}}}+\dfrac{[\mathrm{I}]\,[\mathrm{X}]}{\mathrm{e}K_{\mathrm{I}}K_{\mathrm{X}}}\right)} \tag{6.47}$$

$$[\mathrm{X}] = \frac{K_{\mathrm{X}}\,(1+\sigma)\left(\dfrac{v_0}{v_{\mathrm{IX}}}-1\right)-\dfrac{K_{\mathrm{X}}}{K_{\mathrm{I}}}\left(1+\dfrac{\sigma}{\mathrm{a}}-\dfrac{v_0(1+\sigma)}{v_{\mathrm{IX}}}\dfrac{\beta_{\mathrm{I}}}{\mathrm{a}}\right)[\mathrm{I}]}{1+\dfrac{\sigma}{\mathrm{b}}-\dfrac{v_0(1+\sigma)}{v_{\mathrm{IX}}}\dfrac{\beta_{\mathrm{X}}}{\mathrm{b}}+\left(\dfrac{1}{\mathrm{c}}+\dfrac{\sigma}{\mathrm{e}}\right)\dfrac{[\mathrm{I}]}{K_{\mathrm{I}}}} \tag{6.48}$$

An example is shown in Fig. 6.17 to illustrate the great advantage that can result from combining two hyperbolic inhibitors that interact on both E and ES: here HSpI is combined with HMx(Sp>Ca)I. To facilitate understanding of the synergy arising from the interaction, the isobole plot in Fig. 6.17b was drawn using different v_{IX} levels and axes scaling with respect to panel (a). Table 6.7 lists the inhibitory effects that can be obtained at given concentrations of the modifiers acting individually or in combination at the same concentrations. The two hyperbolic inhibitors I and X, acting separately, can maximally inhibit the enzyme by 45.5 and 80 %, respectively. These are however asymptotic values that can possibly not be exploited in practice. Already at modest concentrations, e.g., [I] = 4 μM and [X] = 2 μM with individual inhibitions of 25 and 10 %, respectively, synergy in the combination gives 67 % inhibition. At 30 μM each of I and X the individual inhibitions are 41 and 55 %, respectively, while the combination gives 98 % inhibition.

As last example in this section, Fig. 6.18 illustrates the effects of two dual modifiers HMxD(A/I) present simultaneously. When I and X act individually as activators because the substrate concentration is lower than both critical substrate concentrations of the two modifiers (Fig. 6.18a), antagonism between the two activators occurs at low [I] and [X], but with increasing modifier concentrations antagonism changes to synergy and activation switches to inhibition. When both I and X behave as inhibitors because σ_{I}^0, $\sigma_{\mathrm{X}}^0 < \sigma$at (Fig. 6.18b), only synergy with efficient inhibition is manifested in the whole concentration range of the modifiers.

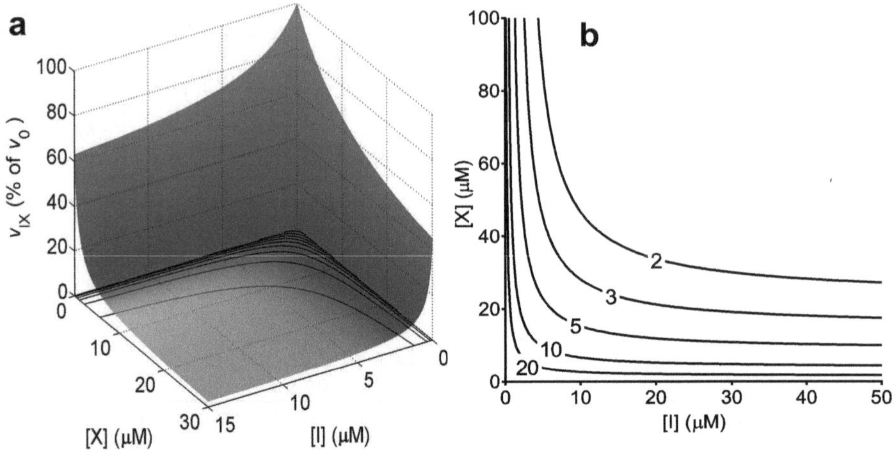

Fig. 6.17 Case 5, modifiers interacting on E and ES, ESIX is catalytically inactive: combination between HSpI (I) and HMx(Sp>Ca)I (X). The surface (**a**) and isobole (**b**) plots were generated with (6.47) and (6.48), respectively. The constants were: a = 6.0, $\beta_I = 1.0$, b = 4.0, $\beta_X = 0.3$, c = 3.0, e = 0.5, $K_I = 1\,\mu$M, $K_X = 5\,\mu$M, $\sigma = 5.0$. The isoboles in (**b**), labeled with the values of v_{IX}, are not the same as in (**a**) and are drawn with different scaling of the [I] and [X] axes

Table 6.7 Explanation of the synergistic effects in the combination between a hyperbolic specific inhibitor (HSpI as I) and a hyperbolic mixed, predominantly specific inhibitor [HMx(Sp>Ca)I as X], the same illustrated in Fig. 6.17

Modifier I alone			Modifier X alone			I + X	
[I]	v_I	% inh.	[X]	v_X	% inh.	v_{IX}	% inh.
2	82.76	17.24	1	94.42	5.58	56.77	43.23
4	75.00	25.00	2	89.57	10.43	33.09	66.91
6	70.59	29.41	3	85.31	14.69	22.09	77.91
8	67.74	32.26	4	81.54	18.46	16.21	83.79
10	65.75	34.25	5	78.18	21.82	12.66	87.34
12	64.29	35.71	6	75.17	24.83	10.33	89.67
15	62.69	37.31	10	65.71	34.29	6.29	93.71
20	60.94	39.06	20	52.00	48.00	3.17	96.83
30	59.02	40.98	30	44.62	55.38	2.00	98.00
50	57.34	42.66	90	30.32	69.68	0.68	99.32
100	55.99	44.01	120	28.00	72.00	0.47	99.53
∞	54.55	45.45	∞	20.00	80.00	0	100

[I] and [X] are in μM units, v_I, v_X and v_{IX} are expressed as % of v_0. [I] and [X] in the rows showing combinations of the two modifiers are the same as the individual modifiers in the same row. The calculations were performed with the same constants of Fig. 6.17

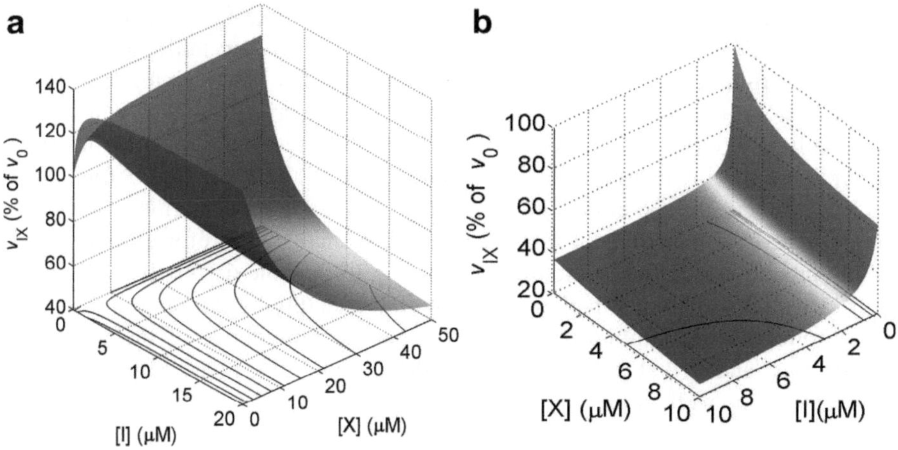

Fig. 6.18 Case 5, modifiers interacting on E and ES with ESIX catalytically inactive: combination between two hyperbolic mixed, dual modifiers HMxD(A/I). (**a**) Both modifiers in the role of activators ($\sigma = 0.15$), (**b**) the same modifiers in the role of inhibitors ($\sigma = 5.0$). Both plots have the following common constants: $a = 0.1$, $\beta_I = 0.3$, $b = 0.3\beta_X = 0.5$, $c = 0.5$, $e = 0.5$, $K_I = 1\,\mu M$, $K_X = 5\,\mu M$, $\sigma_I^0 = 0.29$, $\sigma_X^0 = 0.40$. The activity surfaces were generated with (6.47)

As discussed above, the reason for synergy in the inhibitory process must be sought in the promotion of EIX and ESIX formation, which withdraws enzyme from the catalytic paths.

The two examples in Fig. 6.18 can be taken as reference points for other combinations between two nonessential activators (panel a) and two hyperbolic mixed inhibitors (panel b). In this respect, the dual mechanisms HMxD(A/I) and (HMcD(I/A) are the most responsive among all 17 members of the taxonomic tree in Fig. 3.2 in terms of sensitivity to changes in reactant concentrations.

6.9 Case 6: Modifiers Interacting on E and ES, ESIX is Catalytically Active

In this case all paths for the combination of two modifiers exist and four complexes are catalytically active (Scheme 6.16). The physically meaningful mechanisms in case 6 do not include linear inhibitors because they cannot form the catalytically active ESIX complex. The rate and the isobole equations are not reproduced here again because they correspond to the expressions discussed at the beginning of this chapter, i.e. (6.4) and (6.13), respectively.

The salient property of the combinations between hyperbolic modifiers that obey the mechanism in Scheme 6.16 is the high sensitivity to changes in substrate and modifier concentrations, which can result in restoration of enzyme activity after inhibition has occurred. By progressively increasing the concentrations of the two modifiers, the enzyme activity profiles may take three forms: just inhibition, inhibition followed by reactivation with plateau $\leq v_0$, or inhibition followed by reactivation to the level of v_0 and surpassing this value giving rise to net activation. While making predictions of the effects that arise from the combinations of modifiers in cases 1–5 discussed in the preceding sections was straightforward in terms of the interaction constants c and e and of the catalytic coefficients β_I and β_X, forecasting the effects is not equally obvious in case 6. This is *uncharted territory*, as commented in the introduction to this chapter. Therefore, the properties of the interactions between modifiers according to Scheme 6.16 will be discussed considering just the combination of two modifiers acting by the same mechanism, hyperbolic specific inhibition (HSpI) to show the importance of the interaction constants c and e that govern the formation of the EIX and ESIX complexes, respectively, as well as the concentration of substrate and modifiers in regulating enzyme activity.

The graphics in Figs. 6.19 and 6.20 were purposely drawn with the same kinetic constants and $\sigma = 1.0$, but using four different sets of the interaction constants c and e: c = e = 0.5 (Fig. 6.19a,b), c = 0.5, e = 3.0 (Fig. 6.19c,d), c = e = 3.0 (Fig. 6.20a,b), and c = 3.0, e = 0.5 (Fig. 6.20c,d). For appreciating the impact of substrate concentration on the activity profiles of the four examples, Fig. 6.21 shows activity profiles calculated with the same kinetic constants as in Figs. 6.19 and 6.20 but using four values of σ: 0.5, 1.0, 3.0 and 6.0. In Fig. 6.21, the two-dimensional plots display v_{IX} as a function of [I] and [X] increasing in equimolar concentrations, i.e., the curves correspond to v_{IX} on the diagonal over the whole interval of [I] and [X] in panels (a) and (c) of Figs. 6.19 and 6.20. The modifier concentrations were extended enough to reach saturation of the effects and, in Fig. 6.21, they are shown on a logarithmic scale for better visualization.

In Figs. 6.19a,b and 6.21a the formation of both EIX and ESIX is facilitated (c < 1, e < 1). In a narrow range of low [I]- and [X]-values, inhibition results from antagonism between the modifiers shown in the hatched area of Fig. 6.19b. In the two shaded areas, antagonism between the modifiers results in reactivation of enzyme activity. The two perpendicular dashed lines, defined by [I] and [X] and corresponding to $v_{IX} = 79.89\,\%$ of v_0 in this particular example, delimit zones of inhibition/reactivation (hatched area/shaded areas) characterized by antagonism between I and X, and antagonistic/synergistic reactivation (shaded

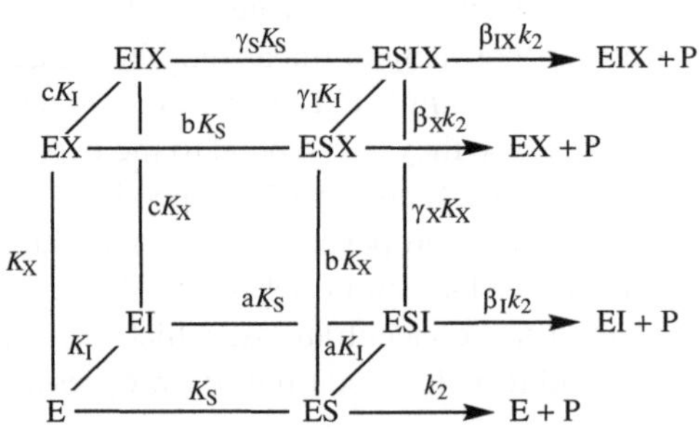

Scheme 6.16 Two hyperbolic modifiers interact on E and ES and ESIX is catalytically active. Nomenclature and overview in Table 6.2

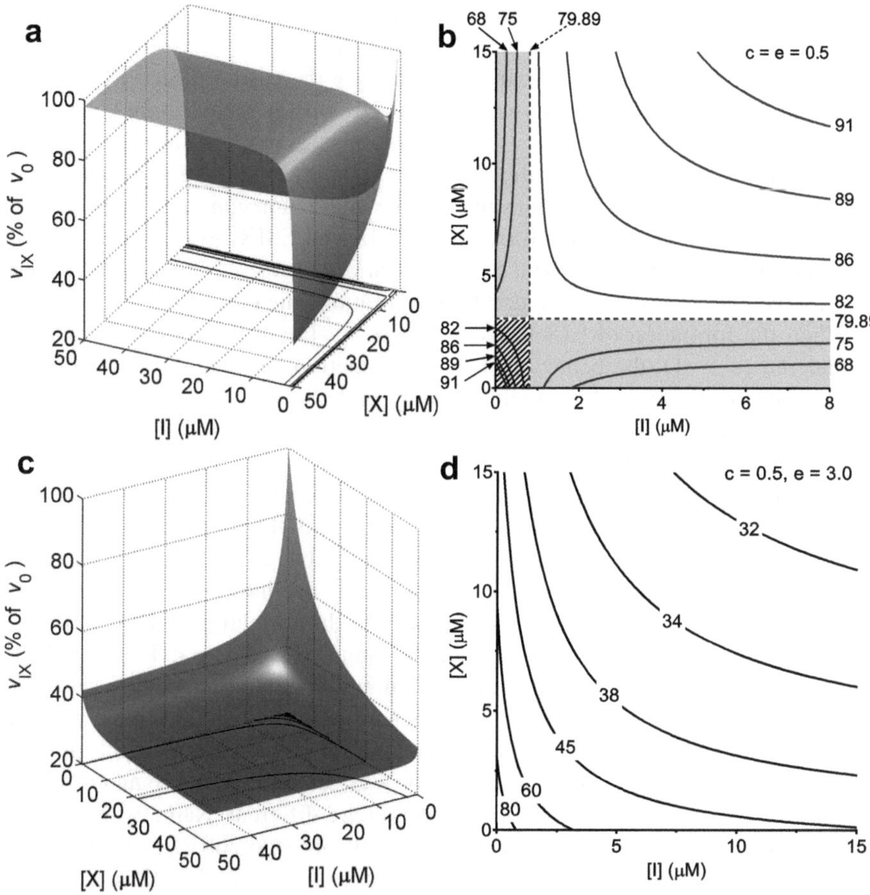

Fig. 6.19 Case 6, modifiers that interact on E and ES, ESIX is catalytically active. All panels represent the combination between two hyperbolic specific inhibitors (HSpI). The detailed isobole plots (**b** and **d**) are shown in a narrower range of modifier concentrations with respect to the corresponding surface plots to emphasize the demarcation between zones of synergy and antagonism. Common parameters for drawing the plots were: a = 4.0, b = 8.0, $\beta_I = \beta_X = \beta_{IX} = 1.0$, $K_I = 1.0\,\mu M$, $K_X = 5.0\,\mu M$ and $\sigma = 1.0$. The individual parameters were c = e = 0.5 in (**a, b**) and c = 0.5, e = 3.0 in (**c, d**). For the combination in panels (**a, b**), the only zone in which inhibition occurs is evidenced in panel (**b**) as *narrow dashed area*, while the *two shaded areas* and the *white area* represent zones of steep reactivation from inhibition occurring on the v_{IX} − [I] and v_{IX} − [X] faces of the cube

areas/white area). The dashed lines cross the axes at [I] = 0.81 μM and [X] = 3.1 μM. At these concentrations of [I] and [X], v_{IX} does not depend on [X], or [I], respectively. In Fig. 6.19a, the steep increase of enzyme activity along both *wings of the manta* (p. 341) corresponds to the shaded antagonism areas in panel (b), in which reactivation occurs rapidly while increasing the concentration of one of the modifiers in the presence of just low concentrations of the second modifier. At the top of the manta body an almost flat surface characterizes complete enzyme reactivation to control levels at saturating [I] and [X].

The degree of inhibition that can be reached by increasing [I] and [X] depends on substrate concentration as shown in Fig. 6.21a with best inhibition achieved at low values of the $[S]/K_m$ ratio, while inhibition is annihilated at saturating [S]. The reason for the contrasting effect exerted on inhibition by the substrate can be deduced from Scheme 6.17a′. The values of the interaction constants c and e, being both less than one, facilitate the formation of EIX and ESIX. At the same time, the substrate *pushes* the four catalytic paths to turn over substrate making the system progressively less sensitive to modifier concentration when [S] increases.

When the formation of EIX is facilitated ($c < 1$) but the formation of ESIX is hindered ($e > 1$), the behavior of the combination between the two modifiers changes drastically to inhibition with synergy in the whole modifier concentration range (Figs. 6.19c,d and 6.21b). Reactivation does not occur as in Fig. 6.19a because the favored EIX complex withdraws enzyme from the catalytic path via ESIX, as can be judged from Scheme 6.17b′. The substrate modulates the inhibitory process as shown in Fig. 6.21b: at low [S] the enzyme is preferentially trapped in EIX resulting in increased inhibition at high [I] and [X], while at high [S] the catalytic paths become more involved and inhibition concomitantly declines.

The behavior of the system with hindrance in the formation of EIX and ESIX ($c > 1$, $e > 1$) is analogous to that of the situation $c < 1$, $e < 1$ in Fig. 6.19a. The effects of inhibition and reactivation as a function of modifier concentration are delimited by two linear isoboles crossing each other at right angles. In the example of Fig. 6.20b these isoboles occur at $v_{IX} = 65.68\,\%$ of v_0 with intercepts at $[I] = 2.14\,\mu M$ and $[X] = 7.02\,\mu M$. While the global effects in Figs. 6.19b and 6.20b are similar, the mechanistic reasons for their origin are different. Namely, the favored complexes are EI, EX, ESI and ESX in the example of Fig. 6.20b with explanation in Scheme 6.17c′, while the most populated complexes are EIX and ESIX in Fig. 6.19b as explained in Scheme 6.17a′. At saturating concentrations of I and X, reactivation reaches a plateau corresponding to the rate in the absence of modifiers.

6.9.1 *The Inhibition Paradox*

The fourth example is characterized by interaction coefficients $c > 1$ and $e < 1$ that imply hindrance in the formation of EIX and facilitation in the formation of ESIX: Figs. 6.20c,d and 6.21d with explanation in Scheme 6.17d′. The activity surface plot in Fig. 6.20c reveals an apparently eccentric effect: by increasing the concentration of the two *inhibitors* the activity of the enzyme passes from inhibition to reactivation and ends up in net *activation*. The isobole plot in panel (d), drawn in a lower modifier concentration range with respect to (c), serves to illustrate five regions of interaction. The narrow, densely hatched area represents the inhibition zone in which the modifiers operate in antagonism. This zone is flanked by two shaded regions of antagonism between the modifiers but in this case the enzyme is reactivated in a very

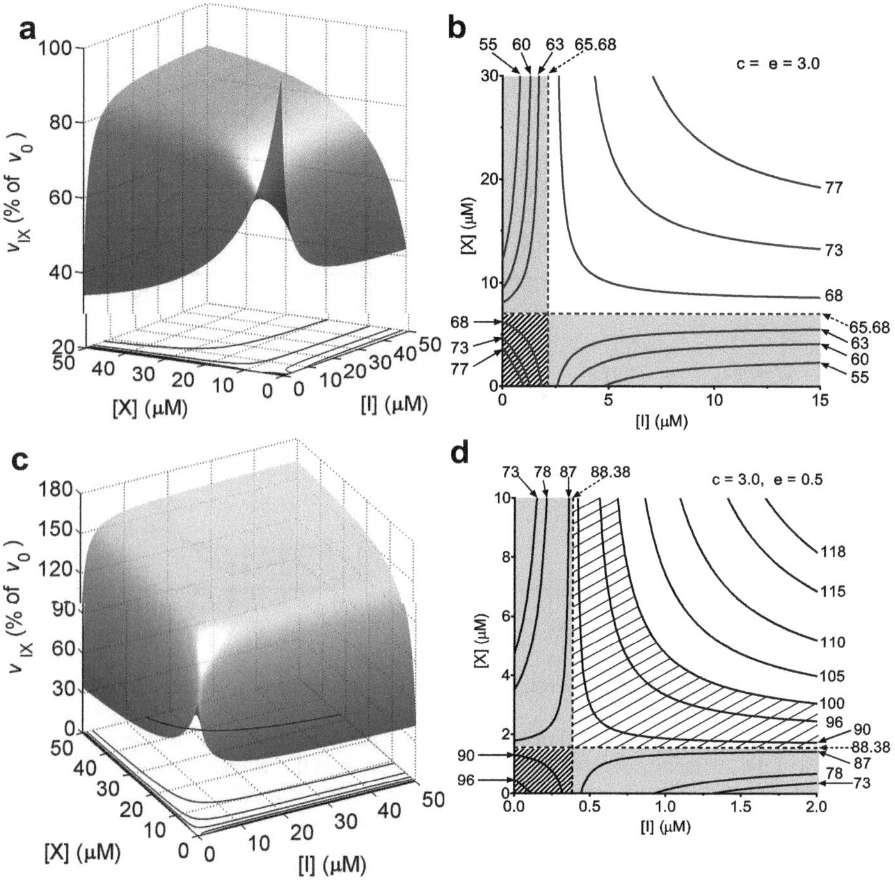

Fig. 6.20 Case 6, modifiers that interact on E and ES, ESIX is catalytically active. All panels represent the combination between two hyperbolic specific inhibitors (HSpI). (**a**) and (**c**) Surface plots with isoboles plotted in (**b** and **d**) in a narrow concentration range to reveal the zones of demarcation between antagonism and synergy. Both mechanisms were simulated with the same parameters with the exception of the interaction constants c and e that are shown in panels (**b** and **d**). The common parameters were: a = 4.0, b = 8.0, $\beta_I = \beta_X = \beta_{IX} = 1.0$, $K_I = 1.0\,\mu$M, $K_X = 5.0\,\mu$M, $\sigma = 1.0$

sensitive concentration-dependent manner evidenced by the steep rise in activity *along the wings of the manta* in panel (c). At I and X concentrations for which the rate does not depend on [X] and [I], respectively, (dashed perpendicular lines crossing the axes at [I] = 0.38 μM and [X] = 1.56 μM with $v_{IX} = 88.38\,\%$ of v_0 in this example) there is a switch from antagonism to synergy that leads to reactivation until reaching the rate in the absence of modifiers (broadly dashed area delimited by the dashed straight lines and the isobole $v_{IX} = 100\,\%$ of v_0), which then proceeds to net activation. With the parameters used to draw the activity surface in panel (c) and $\sigma = 1.0$, the activation levels off to $v_{IX} = 171\,\%$ of v_0.

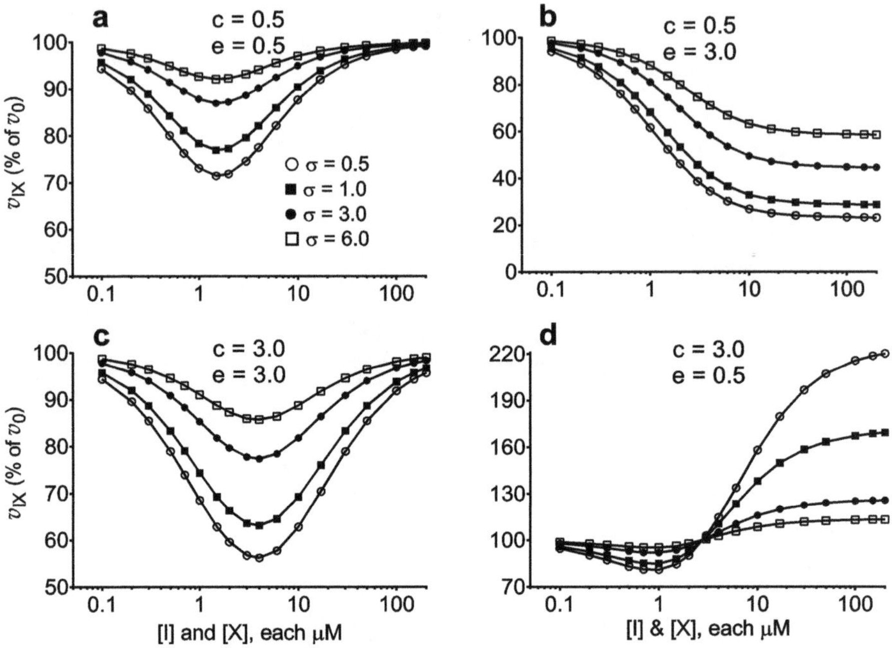

Fig. 6.21 Case 6, combination between two hyperbolic specific inhibitors. Inhibition, reactivation, and the inhibition paradox. The curves were drawn at four substrate concentrations each, represented as $\sigma = [S]/K_m$ in panel (**a**), using the same parameters for the four examples, with the exception of the interaction constants c and e that are shown in the individual panels. The reaction schemes corresponding to the four inhibition profiles are depicted in Scheme 6.17a′–d′, where the complexes in boldface are those that are particularly populated for given combinations of c and e. The parameters common to (**a–d**) were: a = 4.0, b = 8.0, $\beta_I = \beta_X = \beta_{IX} = 1.0$, $K_I = 1.0\,\mu M$, $K_X = 5.0\,\mu M$

The substrate-concentration dependence of the effects just described is shown in Fig. 6.21d, where the switch point that leads to net activation is clearly visible. Large [S]-values tend to annihilate all effects while low substrate concentrations result in activation that, for instance at $\sigma = 0.5$, reaches the plateau of $v_{IX} =$ 225 % of v_0, i.e., the rate at saturating concentrations of both inhibitors is more than doubled with respect to the rate in their absence. The mechanistic reason for this effect is explained in Scheme 6.17d′ that shows how ESI is disfavored while ESIX is favored. Thus, at high [I] and [X], the ESIX complex contributes substantially to increasing the rate of reaction giving rise to net activation. Another example of this type of enzyme-modifier interaction was shown in Fig. 6.14a. In general, this effect is expected to occur in combinations between two hyperbolic modifiers, for which the kinetic constants lead to the formation of appreciable concentrations of the catalytically competent ESIX complex.

The effect described in this section has been called by Fajszi and Keleti *the inhibition paradox* in the frame of a theory named by the authors *the triple-faced*

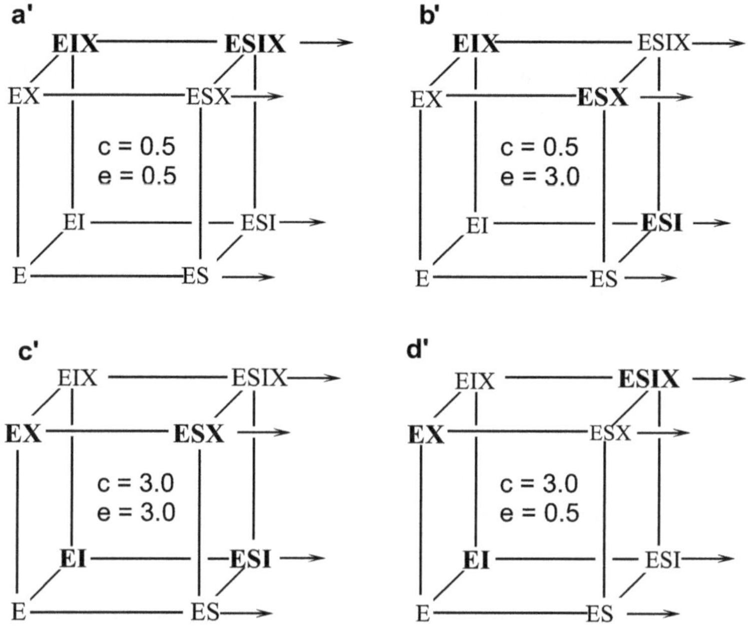

Scheme 6.17 Reaction schemes of the four conditions in Figs. 6.19 and 6.20 and in the four panels in Fig. 6.21. Panels (**a′**–**d′**) refer to the corresponding inhibition profiles (a–d) in Fig. 6.21. The complexes in boldface represent those that are more populated with respect to the other complexes indicated by smaller, plain letters

enzyme-inhibitor relation [7]. Although this theory contains unfortunately a few mistakes that compromise its validity, the main conclusions and the inhibition paradox concept retain their legitimacy [32]. This phenomenon, which represents a mechanism distinct from enzyme activation, still awaits clear-cut experimental verification *with sufficiently precise measurements* in real systems. It cannot be excluded that its occurrence escaped attention either because of confusion with enzyme activation of by insufficient screening under appropriate sets of reagent concentrations.

Musdal et al. observed that inhibition of glutathione transferase by ethacrynic acid was partially restored by another inhibitor, chlorophyllide [27]. The authors interpreted this fact as inhibition paradox. This was however *not the case* since the inhibition was only partially attenuated but neither eliminated nor the effect was inverted into net activation. As discussed on p. 261 in Chap. 5, the mechanism of inhibition of glutathione transferase by ethacrynic acid is a dual one, HMxD(I/A) [30]. Glutathione transferase is a two-substrate enzyme that is assayed in vitro with reduced glutathione and 1-chloro-2,4-dinitrobenzene as substrates. Although insufficiently documented, according to [27, Fig. 2] chlorophyllide behaves as linear specific inhibitor with respect to 1-chloro-2,4-dinitrobenzene and as *mixed-type noncompetitive inhibitor* with respect to reduced glutathione. The latter looks like

LMx(Sp>Ca)I. In the reactivation experiment, both 1-chloro-2,4-dinitrobenzene and reduced glutathione were fixed at saturating concentrations. Under these conditions ethacrynic acid behaves as activator and it is therefore logical that reactivation is the observed effect. This is of the same type shown in Fig. 6.7a that features a linear inhibitor in conjunction with a nonessential activator. In this figure, HMxD(A/I) is the hyperbolic modifier that behaves as activator at the considered (low) substrate concentration. Exactly the same picture is obtained with LMx(Sp>Ca)I as inhibitor coupled to a HMxD(I/A) modifier at saturating substrate concentration, under which condition it behaves as activator. The interaction type, antagonism with partial relief from inhibition, explains the results by Musdal et al. [27].

References

1. Berenbaum MC (1977) Synergy, additivism and antagonism in immunosuppression. A critical review. Clin Exp Immunol 28:1–18
2. Berenbaum MC (1985) The expected effect of a combination of agents: the general solution. J Theor Biol 114:413–431
3. Berenbaum MC (1989) What is synergy? Pharmacol Rev 41:93–141
4. Bosgra S, van Eijkeren JC, Slob W (2009) Dose addition and the isobole method as approaches for predicting the cumulative effect of non-interacting chemicals: a critical evaluation. Crit Rev Toxicol 39:418–426. doi:10.1080/10408440902787592
5. Chou TC, Talalay P (1977) A simple generalized equation for the analysis of multiple inhibitions of Michaelis-Menten kinetic systems. J Biol Chem 252:6438–6442
6. Fajszi C (1974) Methods of analysis of double inhibition experiments. Symp Biol Hung 18:77–103
7. Fajszi C, Keleti T (1974) Kinetic basis of enzyme regulation. The triple-faced enzyme-inhibitor relation and the inhibition paradox. Symp Biol Hung 18:105–119
8. Fraser TR (1870–1871) An experimental research on the antagonism between the actions of physostigma and atropia. Proc R Soc Edin 7:506–511
9. Fraser TR (1872) The antagonism between the actions of active substances. Brit Med J 2:485–487
10. Guillén-Schlippe YV, Riera TV, Seyedsayamdost MR, Hedstrom L (2004) Substitution of the conserved Arg-Tyr dyad selectively disrupts the hydrolysis phase of the IMP dehydrogenase reaction. Biochemistry 43:4511–4521. doi:10.1021/bi035823q
11. Halabi N, Rivoire O, Leibler S, Ranganathan R (2009) Protein sectors: evolutionary units of three-dimensional structure. Cell 138:774–786. doi:10.1016/j.cell.2009.07.038
12. Hedstrom L (2009) IMP dehydrogenase: structure, mechanism, and inhibition. Chem Rev 109:2903–2928. doi:10.1021/cr900021w
13. Hedstrom L, Wang CC (1990) Mycophenolic acid and thiazole adenine dinucleotide inhibition of Tritrichomonas foetus inosine 5'-monophosphate dehydrogenase: implications on enzyme mechanism. Biochemistry 29:849–854
14. Joshi N, Hoobler EK, Perry S, Diaz G, Fox B, Holman TR (2013) Kinetic and structural investigations into the allosteric and pH effect on the substrate specificity of human epithelial 15-lipoxygenase-2. Biochemistry 52:8026–8035. doi:10.1021/bi4010649
15. Keleti T (1967) The liberator. J Theor Biol 16:337–355. doi:10.1016/0022-5193(67)90060-4
16. Keleti T, Fajszi C (1971) The system of double inhibitions. Math Biosci 12:197–215. doi:10.1016/0025-5564(71)90016-2
17. Kemp RG, Gunasekera D (2002) Evolution of the allosteric ligand sites of mammalian phosphofructo-1-kinase. Biochemistry 41:9426–9430

18. Lee J, Natarajan M, Nashine VC, Socolich M, Vo T, Russ WP, Benkovic SJ, Ranganathan R (2008) Surface sites for engineering allosteric control in proteins. Science 322:438–442. doi:10.1126/science.1159052
19. Lienhard GE (1971) Enzymatic catalysis and the transition state theory of reaction rates: transition state analogs. Cold Spring Harbor Symp Quant Biol 36:45–51
20. Lockless SW, Ranganathan R (1999) Evolutionarily conserved pathways of energetic connectivity in protein families. Science 286:295–299
21. Loewe S, Muischnek H (1926) Über Kombinationswirkungen. I. Mitteilung: Hilfsmittel der Fragestellung. Arch Exp Pathol Pharmakol 114:313–326
22. Lucas JJ, Burchiel SW, Segel IH (1972) Choline sulfatase of Pseudomonas aeruginosa. Arch Biochem Biophys 153:664–672
23. Mares-Guia M, Shaw E (1965) Studies on active center of trypsin. The binding of amidines and guanidines as models of the substrate side chain. J Biol Chem 240:1579–1585
24. Mazat JP, Mazat F (1986) Double site enzyme and squatting: where one regulatory ligand is also a substrate of the reaction. J Theor Biol 121:89–103
25. Mazat JP, Langla J, Mazat F (1977) Double-site enzymes and squatting. A study of the regulation by one or several ligands binding at two different classes of site. J Theor Biol 68:365–383
26. Mazat F, Langla J, Mazat JP (1981) The measure of synergy in enzymatic regulation. A general coefficient. Biochimie 63:107–111. doi:10.1016/S0300-9084(81)80173-3
27. Musdal Y, Hegazy UM, Aksoy Y, Mannervik B (2013) FDA-approved drugs and other compounds tested as inhibitors of human glutathione transferase P1-1. Chem-Biol Interact 205:53–62. doi:10.1016/j.cbi.2013.06.003
28. Novinec M, Korenč M, Caflisch A, Ranganathan R, Lenarčič B, Baici A (2014) A novel allosteric mechanism in the cysteine peptidase cathepsin K discovered by computational methods. Nature Commun 5. doi:10.1038/ncomms4287
29. Novinec M, Lenarčič B, Baici A (2014) Probing the activity modification space of the cysteine peptidase cathepsin K with novel allosteric modifiers. PLoS One 9:e106,642. doi:10.1371/journal.pone.0106642
30. Phillips MF, Mantle TJ (1991) The initial-rate kinetics of mouse glutathione S-transferase YfYf. Evidence for an allosteric site for ethacrynic acid. Biochem J 275:703–709
31. Reynolds KA, McLaughlin RN, Ranganathan R (2011) Hot spots for allosteric regulation on protein surfaces. Cell 147:1564–1575. doi:10.1016/j.cell.2011.10.049
32. Schenker P, Baici A (2009) Simultaneous interaction of enzymes with two modifiers: reappraisal of kinetic models and new paradigms. J Theor Biol 261:318–329. doi:10.1016/j.jtbi.2009.07.033
33. Schimerlik MI, Cleland WW (1977) Inhibition and alternate-substrate studies on the mechanism of malic enzyme. Biochemistry 16:565–570
34. Segel IH (1975) Enzyme kinetics. Behavior and analysis of rapid equilibrium and steady-state enzyme systems. Wiley, New York
35. Slater EC, Bonner, Jun WD (1952) The effect of fluoride on the succinic oxidase system. Biochem J 52:185–196
36. Strasser F, Pelton PD, Ganzhorn AJ (1995) Kinetic characterization of enzyme forms involved in metal ion activation and inhibition of myo-inositol monophosphatase. Biochem J 307:585–593
37. Süel GM, Lockless SW, Wall MA, Ranganathan R (2003) Evolutionarily conserved networks of residues mediate allosteric communication in proteins. Nat Struct Biol 10:59–69
38. Webb JL (1963) Enzyme and metabolic inhibitors. Volume 1, General principles of inhibition. Enzyme and metabolic inhibitors. Academic Press, New York
39. Wecksler AT, Kenyon V, Garcia NK, Deschamps JD, van der Donk WA, Holman TR (2009) Kinetic and structural investigations of the allosteric site in human epithelial 15-lipoxygenase-2. Biochemistry 48:8721–8730. doi:10.1021/bi9009242

40. Yagi K, Ozawa T (1960) Complex formation of apo-enzyme, coenzyme and substrate of D-amino acid oxidase. I. Kinetic analysis using indicators. Biochim Biophys Acta 42:381–387. doi:10.1016/0006-3002(60)90167-0
41. Yagi K, Ozawa T (1960) Mechanism of inhibition of D-amino acid oxidase. III. Kinetic analysis of the behaviour of chloramphenicol, streptomycin and penicillin in their competition with flavin-adenine dinucleotide. Biochim Biophys Acta 39:304–310. doi:10.1016/0006-3002(60)90815-5
42. Yonetani T, Theorell H (1964) Studies on liver alcohol dehydrogenase complexes. III. Multiple inhibition kinetics in the presence of two competitive inhibitors. Arch Biochem Biophys 106:243–251. doi:10.1016/0003-9861(64)90184-5

Chapter 7
Multiple Interactions: Essential Activation and Liberation

··· the antibody does not inhibit but, rather, protects the enzyme molecule against both inhibition and activation induced by a variety of nucleotides.

Corman C, Kaplan NO (1965) Biochemistry 4:2175–2179

Abstract An account is given of the participation of essential activators acting jointly with any other modifier that may be an inhibitor or a nonessential activator. The mechanisms of action of the liberator, a substance that reverts inhibition or activation by any modifier, can optimally be understood and interpreted using the formalism of multiple enzyme-modifier interactions.

Multiple enzyme-modifier interactions that involve essential activation and liberation from activation or inhibition are placed in this chapter, not in Chap. 6. The separation aims at underlining that essential activation belongs to a group that differs from the taxonomic lineage of linear inhibitors and mixed modifiers discussed in Chap. 3. The kinetic properties of liberation are also discussed here because a liberator can theoretically revert any effect produced by modifiers on enzymes and therefore is logically placed after the presentation of the various modification mechanisms and their mutual interactions.

7.1 Interactions Involving Essential Activators

For the nomenclature and taxonomy of essential activation see Sect. 3.4. Once an enzyme has been put in action from his dormant state after binding an essential activator, the active enzyme may interact with either an inhibitor or a nonessential activator. Assigning the nonessential activator to X, valid combinations with another modifier I are shown in Scheme 7.1. In panel (a) X is a linear mixed activator and I a hyperbolic mixed inhibitor, while panel (b) features the combination between a linear specific activator (X) and a linear specific inhibitor (I). The activator in Scheme 7.1a can be predominantly specific ($b > 1$), balanced ($b = 1$), or predominantly catalytic ($b < 1$) (Fig. 3.7) and the modifier I can belong to any of the 17 mechanisms in Fig. 3.2.

A. Baici, *Kinetics of Enzyme-Modifier Interactions*,
DOI 10.1007/978-3-7091-1402-5_7

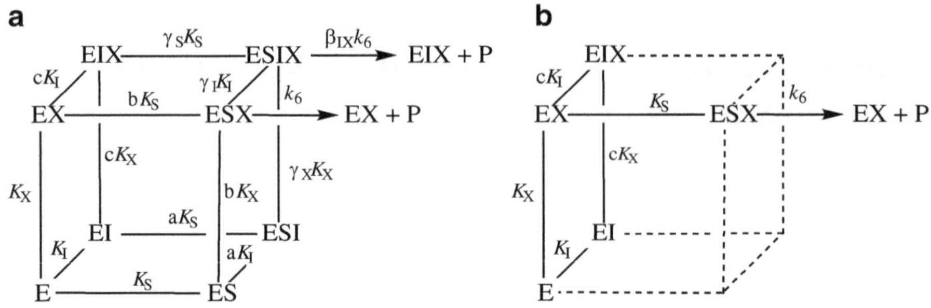

Scheme 7.1 An essential activator (X) and a modifier (I) act concomitantly on the same enzyme. (**a**) X is a linear mixed activator (either predominantly specific, predominantly catalytic, or balanced) and I is a hyperbolic mixed modifier. (**b**) X is a linear specific activator and I a linear specific inhibitor. To maintain uniform symbolism the catalytic constant is numbered after Scheme 3.3

The topology of Scheme 7.1a is preserved to match that of Scheme 6.1b with $k_2 = 0$ and $\beta_X k_2$ renamed k_6 in accord with the notation of essential activation in Sect. 3.4. Unlike (6.4), the rate equation has only two terms in the numerator that correspond to the to catalytic paths and v_0 must be redefined as the rate in the obligate presence of X and absence of I. To maintain an uniform symbolism with reaction schemes and rate constants adopted for essential activation in Sect. 3.4, the rate equation can be represented as:

$$v_{IX} = \frac{k_6[\mathrm{E}]_t\sigma'\left(\dfrac{1}{\mathrm{b}} + \beta_{IX}\dfrac{[\mathrm{I}]}{\mathrm{e}K_I}\right)}{1 + \dfrac{[\mathrm{I}]}{\mathrm{c}K_I} + \dfrac{K_X}{[\mathrm{X}]} + \dfrac{[\mathrm{I}]}{K_I}\dfrac{K_X}{[\mathrm{X}]} + \sigma'\left(\dfrac{K_X}{[\mathrm{X}]} + \dfrac{[\mathrm{I}]}{\mathrm{a}K_I}\dfrac{K_X}{[\mathrm{X}]} + \dfrac{1}{\mathrm{b}} + \dfrac{[\mathrm{I}]}{\mathrm{e}K_I}\right)}. \tag{7.1}$$

The coefficients c and e in (7.1) determine facilitation, independence, hindrance or exclusion in the formation of the EIX and ESIX complexes, respectively, in the same way described in the preceding sections of this chapter. In the mechanism shown in Scheme 7.1b, with rate equation (7.2), a linear specific activator enables the enzyme to catalysis and a linear specific inhibitor present concomitantly counteracts the activation process:

$$v_{IX} = \frac{k_6[\mathrm{E}]_t\sigma'}{1 + \dfrac{[\mathrm{I}]}{\mathrm{c}K_I} + \dfrac{K_X}{[\mathrm{X}]} + \dfrac{[\mathrm{I}]}{K_I}\dfrac{K_X}{[\mathrm{X}]} + \sigma'}. \tag{7.2}$$

In (7.1) and (7.2), σ' is the ratio $[\mathrm{S}]/K'_m$, with $K'_m = (k_{-5} + k_6)/k_5$ (p. 146). The activity surfaces plotted in three dimensions for combinations that involve an essential activator according to the mechanisms in Scheme 7.1 differ from those seen hitherto in this chapter because there is no activity in the absence of the essential activator (here always represented as X). Thus, the surfaces consist of a series of

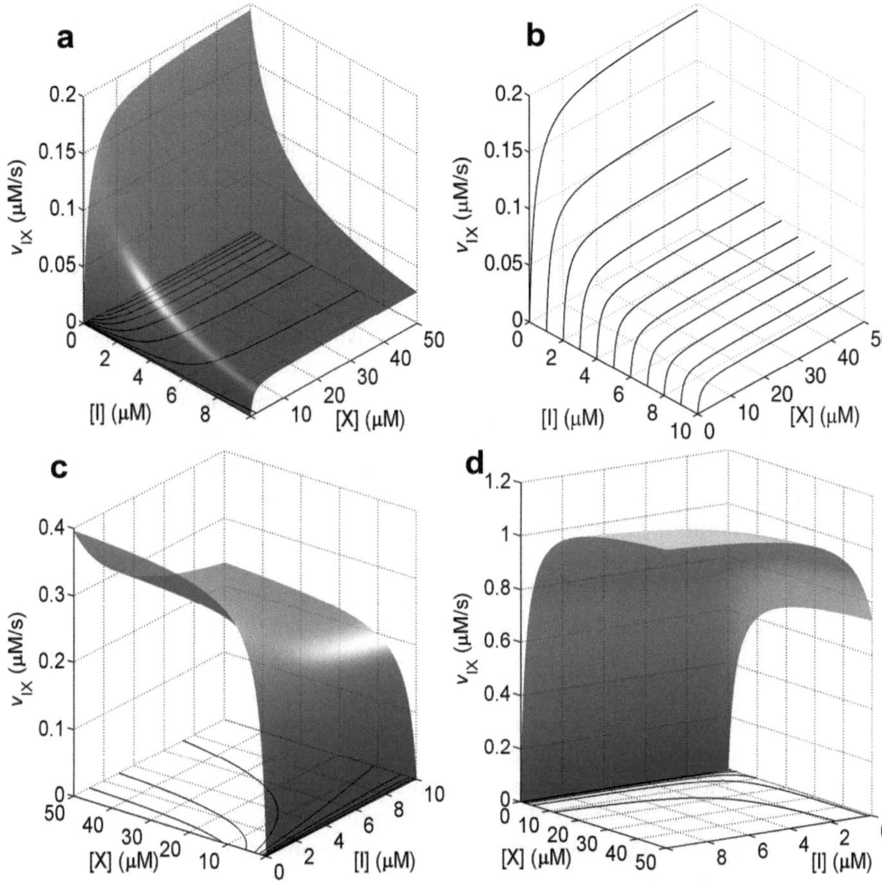

Fig. 7.1 Combinations of an essential mixed activator with another modifier (either inhibitor or nonessential activator). The plots were drawn with (7.1) and parameters common to (**a–c**) were k_6 = 100 s^{-1}, $[E]_t$ = 10 nM. (**a**) LMx(Sp<Ca)I combined with the essential activator LMx(Sp>Ca)A: a = 0.5, b = 4.0, c = 0.5, e = 0.3, $\beta_{IX} = 0$, $\sigma = 1.0$, $K_I = 7\,\mu$M, $K_X = 1\,\mu$M. (**b**) Single line plot with the same parameters as in (**a**) that shows the change of v_{IX} depending on the concentration of the essential activator at 10 fixed concentrations of I. (**c**) HCaI combined with the essential balanced activator LMx(Sp=Ca)A: a = 0.3, b = 1.0, c = 1.0, e = 0.5, $\beta_{IX} = 0.3$, $\sigma = 0.7$, $K_I = 6\,\mu$M, $K_X = 2\,\mu$M. (**d**) A nonessential activator HMx(Sp>Ca)A combined with an essential mixed activator LMx(Sp<Ca)A: a = 5.0, b = 0.3, c = 0.4, e = 0.6, $\beta_{IX} = 3.0$, $\sigma = 1.0$, $K_I = 1\,\mu$M, $K_X = 6\,\mu$M

stacked curves of rate against the concentration of the binding partner (I) that start at $v_{IX} = 0$ as illustrated in Fig. 7.1b, which represents a series of slices parallel to the v_{IX}-[X] plane of the surface plot in panel (a).

The combination in Fig. 7.1a,b is made between a mixed essential activator with predominantly specific character [LMx(Sp>Ca)A as X] and a linear mixed inhibitor with predominantly catalytic character [LMx(Sp<Ca)I as I]. The curve in panel

(b) on the v_{IX}-[X] face represents essential enzyme activation in the absence of modifier I.

In Fig. 7.1c a mixed balanced essential activator [LMx(Sp=Ca)A as X] is present together with a hyperbolic catalytic inhibitor (HCaI as I) and panel (d) features the combination of LMx(Sp<Ca)A as X and the nonessential activator HMx(Sp>Ca)A. In all examples of Fig. 7.1 the mixed essential activator acts in antagonism with its binding partner irrespective of whether this is an inhibitor or an activator. Antagonism is not a synonym of abolishment of activation by the essential activator. It is just one of the modes of interaction between two ligands present concomitantly on the same enzyme, where the strongest takes the lead. Antagonism between two activators of this type can result in additional activation as seen in the combination of 7.1d.

The theoretically possible combinations with a linear specific activator are more restrictive than those with a mixed linear activator because the ES complex does not exist and thus the ESI complex is also not expected. A plausible combination is shown in Scheme 7.1b between LSpA and LSpI. The interaction constant c can lead to facilitation in the formation of the ternary complex EIX ($c < 1$), independence ($c = 1$), hindrance ($c > 1$) or exclusion ($c = \infty$). In Fig. 7.2 the four graphics were generated with the same parameters varying only the interaction constant c. As expected, increasing c from 0.5 to 5, antagonism between the specific activator and the specific inhibitor results in less populated EI and EIX complexes thus favoring catalysis.

The isoboles beneath the surfaces in panels (a–d) are concave downward and their curvature becomes less pronounced as c increases. The diminished efficiency of the inhibitor with the increase of c can be appreciated from the edge of the activity surface along the v_{IX}-[I] plane. The extreme case is exclusion ($c = \infty$) shown in panel (d), so that the EIX complex is not formed. The observation angle of Fig. 7.2d has been set to appreciate the linearity of the isoboles that indicates zero interaction.

The Range of Multiple Interactions between an essential activator and another modifier in living cells does not possibly extend to the wide range of concentration of the essential activator illustrated in this section. In a natural environment, an essential activator can be expected to be present at a saturating concentration and, under this condition, a second modifier in the role of inhibitor or nonessential activator is expected to behave as in a system with just one modifier that acts on an enzyme that does not need to be activated compulsorily. An inhibitor will thus behave as inhibitor and a nonessential activator will do its job irrespective of any antagonism possibly present depending on the interaction constant c.

7.2 The Liberator

The *liberator* has been mentioned on p. 153 in the synoptic representation of allosteric interactions (Fig. 3.8), where it is represented as black diamond at coordinates (1,1). The concept of liberator has received only modest appreciation in

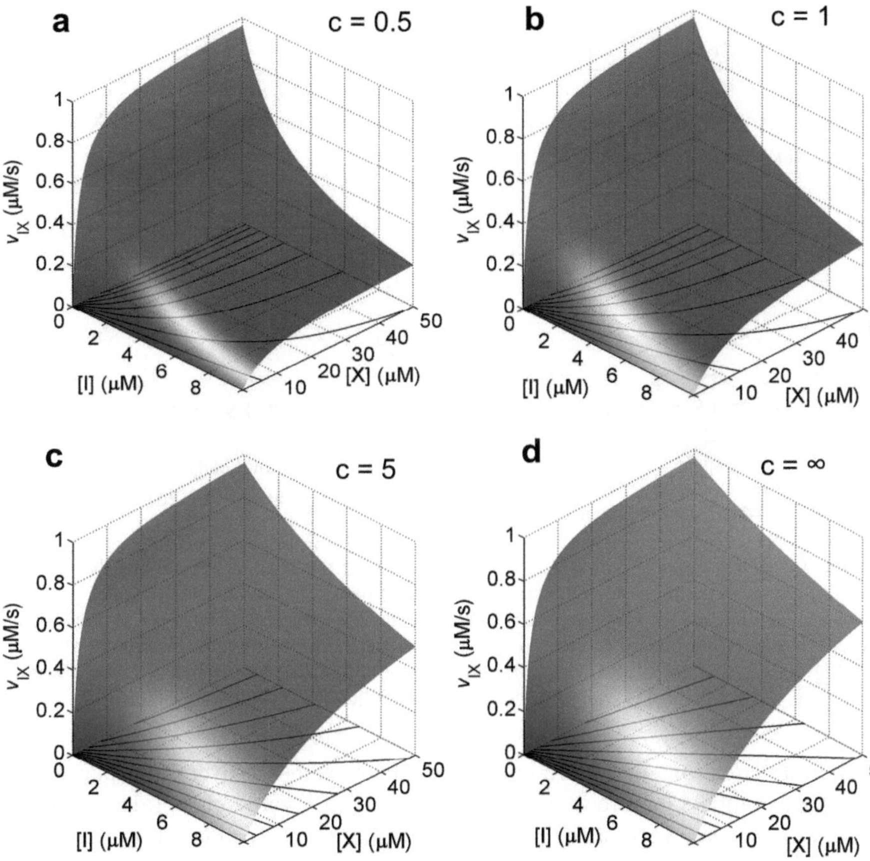

Fig. 7.2 Combination of a linear specific activator (LSpA as X) with a linear specific inhibitor (LSpI as I). The plots were drawn with (7.2). The parameters common to (**a–d**) were $k_6 = 100\,\mathrm{s}^{-1}$, $[E]_t = 10\,\mathrm{nM}$, $\sigma = 1.0$, $K_I = 6\,\mu\mathrm{M}$, $K_X = 2\,\mu\mathrm{M}$. Each of the four plots was generated with a different value of the interaction constant c as indicated in panels (**a–d**)

the biochemical literature since its definition in 1967 by Tamás Keleti [3]. Keleti's original definition is

> The liberator is a substance which may liberate the enzyme from the action of an inhibitor without affecting the enzymic activity alone. ··· The liberator alone is supposed to have no effect on enzymic activity, i.e., it is not an activator.

and in the same paper [3], the concept of liberator has been extended to embrace liberation from activation. Keleti formulated his concept by deriving dedicated rate equations to all possible combinations between a liberator and another modifier considering that a liberator may be *competitive*, *noncompetitive*, or *uncompetitive* with the second modifier [3] (note that Keleti's *noncompetitive* corresponds to mixed modification in this book).

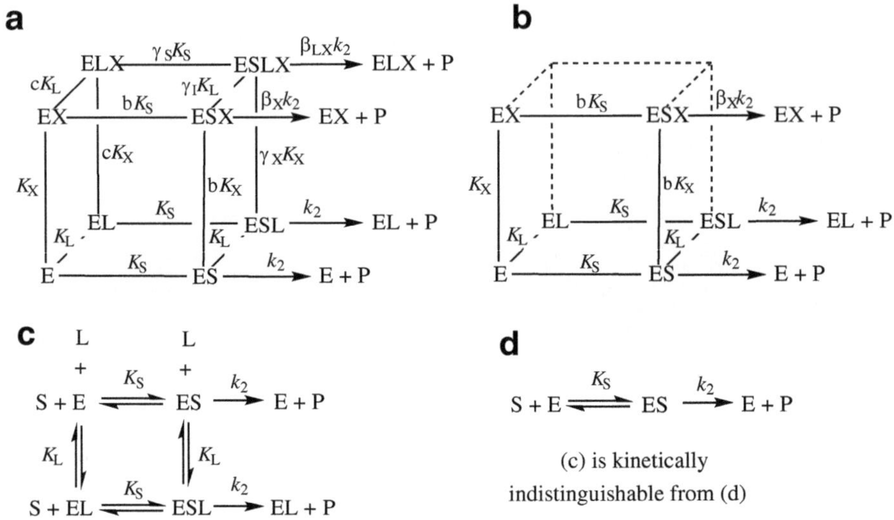

Scheme 7.2 Mechanism of action of a liberator (L) in the presence of any one of the 17 basic modifiers (X). (**a**) General case with all possible reaction paths. (**b**) L and X are mutually exclusive on the free enzyme and the ES complex. (**c**) The liberator interacts with the enzyme in the absence of other modifiers; the reactions are those of the general modifier mechanism without allosteric coupling between S and L and with two catalytic paths that have the same catalytic constant. Kinetically, mechanism (**c**) is equivalent to (**d**), i.e., the Michaelis–Menten mechanism

The rate equations in [3] are correct but their extension over eight pages does not offer a compact picture of an otherwise straightforward concept. However, the notation of Scheme 6.1b accommodates all combinations proposed by Keleti in a synoptic way. Substituting the modifier I with a liberator L, the complete set of relationships between the liberator and another modifier X can thus be represented as in Scheme 7.2a. Here it can be noted that on the face at the bottom of the cube, which contains the properties of the liberator, the constants are not multiplied by coefficients because these are equal to one, a fact shown explicitly using the complete paths in Scheme 7.2c. The initial rate of the system in panel (a) is given by (6.4) after substituting L for I, i.e.

$$v_{\mathrm{LX}} = \frac{v_0\,(1+\sigma)\left(1+\dfrac{[\mathrm{L}]}{K_{\mathrm{L}}}+\beta_{\mathrm{X}}\dfrac{[\mathrm{X}]}{\mathrm{b}K_{\mathrm{X}}}+\beta_{\mathrm{LX}}\dfrac{[\mathrm{L}]\,[\mathrm{X}]}{\mathrm{e}K_{\mathrm{L}}K_{\mathrm{X}}}\right)}{1+\dfrac{[\mathrm{L}]}{K_{\mathrm{L}}}+\dfrac{[\mathrm{X}]}{\mathrm{b}K_{\mathrm{X}}}+\dfrac{[\mathrm{L}]\,[\mathrm{X}]}{\mathrm{c}K_{\mathrm{L}}K_{\mathrm{X}}}+\sigma\left(1+\dfrac{[\mathrm{L}]}{K_{\mathrm{L}}}+\dfrac{[\mathrm{X}]}{\mathrm{b}K_{\mathrm{X}}}+\dfrac{[\mathrm{L}]\,[\mathrm{X}]}{\mathrm{e}K_{\mathrm{L}}K_{\mathrm{X}}}\right)}. \qquad (7.3)$$

Compared with (6.4), the presence of the liberator and the path E + S $\rightleftharpoons$ ES in (7.3) are given for granted and thus $\alpha_{\mathrm{S}} = \alpha_{\mathrm{L}} = 1$. Also, a = β_{L} = 1, equivalent to α and β in the general modifier mechanism (Scheme 3.1a), do not appear explicitly in the equation. In fact, a liberator that binds an enzyme without the intervention of other modifiers obeys the general modifier mechanism with $\alpha = \beta = 1$, whose rate is given by the Michaelis–Menten equation. The kinetic properties of the mechanism in Scheme 7.2c are indistinguishable from those of the mechanism in panel (d). A liberator remains invisible as a sort of *guardian angel* ready to intervene when necessary. Therefore, this tutelary molecule, which is unable to establish allosteric interactions with the substrate in an ESL and/or ESLX complex, becomes evident only in the presence of either an inhibitor or an activator.

The complete mechanism of the liberator, depicted in Scheme 7.2a, considers the presence of the ELX and ESLX complexes. The concept of liberation states that a molecule (L) is able to *relieve* an enzyme from inhibition or activation by a modifier (X). In the absence of the liberator, X operates by the mechanism depicted in the front face of the cube in Scheme 7.2a with catalysis via k_2 and $\beta_X k_2$. It is thus expected that *these* reactions are targeted by the liberator, which seeks to promote the formation of EL and ESL while hindering the formation of ES, EX and ESX. Indeed, liberation from inhibition or activation can be accomplished also if ELX and ESLX do not exist, i.e., if c = e = ∞. In this context, the ELX and ESLX complexes, including catalysis via $\beta_{IX} k_2$, can be responsible for something more than just *liberating* an enzyme from inhibition or activation. Namely, sequestration of enzyme in ELX and ESLX, with or without generation of product from ESLX, can provoke net inhibition after attenuating the action of a nonessential activator as well as net activation after relieving inhibition. Keleti recognized this fact and in a footnote [3, p. 341] stated

> In this case it would be inappropriate to say that *L* was liberating; it would be more appropriate to say it was 'aggravating'.

In the light of these considerations, depending on the concentration of the liberator and its dissociation constant K_L, the reactions in Scheme 7.2a may also induce effects as those elicited by modifiers that interact in the EIX and ESIX complexes. This can occur with or without catalysis via ESIX as discussed under cases 2–6 in the preceding sections when an inhibitor, or activator, is entitled to form the EIX and/or ESIX complexes. See for instance Figs. 6.16a and 6.19a, where the enzyme is relieved from activation or inhibition, respectively, in systems that have nothing to do with the concept of liberation.

The hypothesis that a liberator does not affect enzyme activity per se might be not verifiable experimentally with the necessary precision if such effects are small and possibly partly masked by an unfavorable signal-to-noise ratio. Should the demonstration that a substance has no activity whatsoever toward an enzyme be difficult using a kinetic approach, alternative methods can be invoked to ascertain that binding to the enzyme takes place.

A mechanism of action that harmonizes with the concept of *liberation* from inhibition or activation without overrunning this effect is shown in Scheme 7.2b. This is a system with exclusion of two modifiers on the free enzyme and the ES complex (c = e = ∞) discussed in Sect. 6.4. The isoboles resulting from such a combination are straight lines that indicate zero-interaction. Although being in accord with the definition of liberation, the limitation of this system is that it may result unspecific because there is no reciprocal recognition between liberator and inhibitor or activator. Conversely, the combination with facilitation, independence, and even hindrance in the formation of ELX and or ESLX is possibly a sign that the combination may be specific. The rate equation for the mechanism Scheme 7.2b is

$$v_{\mathrm{LX}} = \frac{v_0\,(1+\sigma)\left(1+\dfrac{[\mathrm{L}]}{K_{\mathrm{L}}}+\beta_{\mathrm{X}}\dfrac{[\mathrm{X}]}{\mathrm{b}K_{\mathrm{X}}}\right)}{1+\dfrac{[\mathrm{L}]}{K_{\mathrm{L}}}+\dfrac{[\mathrm{X}]}{K_{\mathrm{X}}}+\sigma\left(1+\dfrac{[\mathrm{L}]}{K_{\mathrm{L}}}+\dfrac{[\mathrm{X}]}{\mathrm{b}K_{\mathrm{X}}}\right)}\,. \tag{7.4}$$

In place of useless listings of possible interactions that may be realized when a liberator is present together with an inhibitor or activator, the effect of liberation is illustrated with representative examples in Fig. 7.3. Panel (a) shows liberation after inhibition by the linear, predominantly specific inhibitor LMx(Sp>Ca)I. In the absence of liberator, inhibition corresponds to the hyperbola on the $v_{\mathrm{LX}} - [\mathrm{X}]$ face of the cube (97 % inhibition at [X] = 60 μM) but when [L] increases, there is partial liberation from inhibition to reach 54 % inhibition at [L] = 30 μM. The hallmark of the liberator to be itself inactive toward the enzyme is evidenced by the linear trace on the top of the surface plot, parallel to the [L]-axis at level 100. Zero interaction between the liberator and the inhibitor is evident from the linear isoboles, which indicate that neither the ELX nor the ESLX complex exists (c = e = ∞).

Liberation from activation by the nonessential activator HMx(Sp<Ca)A is shown in Fig. 7.3b. Here the surface plot lies over the 100 % activity level and the modifier increases the activity of the enzyme to 420 % at [X] = 100 μM. The increasing liberator concentration progressively counteracts activation by antagonism, and brings back activation to the level of 120 % at [L] = 30 μM. The interaction is unmistakably antagonism as seen from the concave-down isoboles.

In Fig. 7.3c, the hyperbolic specific inhibitor HSpI at a concentration of 100 μM causes 67 % inhibition as seen on the $v_{\mathrm{LX}} - [\mathrm{X}]$ face of the cube. Antagonism by the liberator is very efficient under the conditions of this example and enzyme activity is restored almost in full with a steep liberation curve as the concentration of the liberator increases.

Another liberation from inhibition is shown in Fig. 7.3d. Here, hyperbolic inhibition by HMx(Sp<Ca)I is 74 % at [X] = 100 μM and the liberation results in fine modulation restoring some activity to reach the level of 46 % inhibition at [L] = 40 μM. Liberation is accomplished in this case by a combination of antagonism and synergy that depends on the relative concentrations of L and X.

7.2.1 *Experimental Evidence of Liberation*

The mechanism of liberation from inhibition or activation is not a popular topic in biochemistry and it is only thanks to Keleti [3] if this regulatory mechanism of enzyme action has been put in evidence. Though, Keleti's paper has been cited very

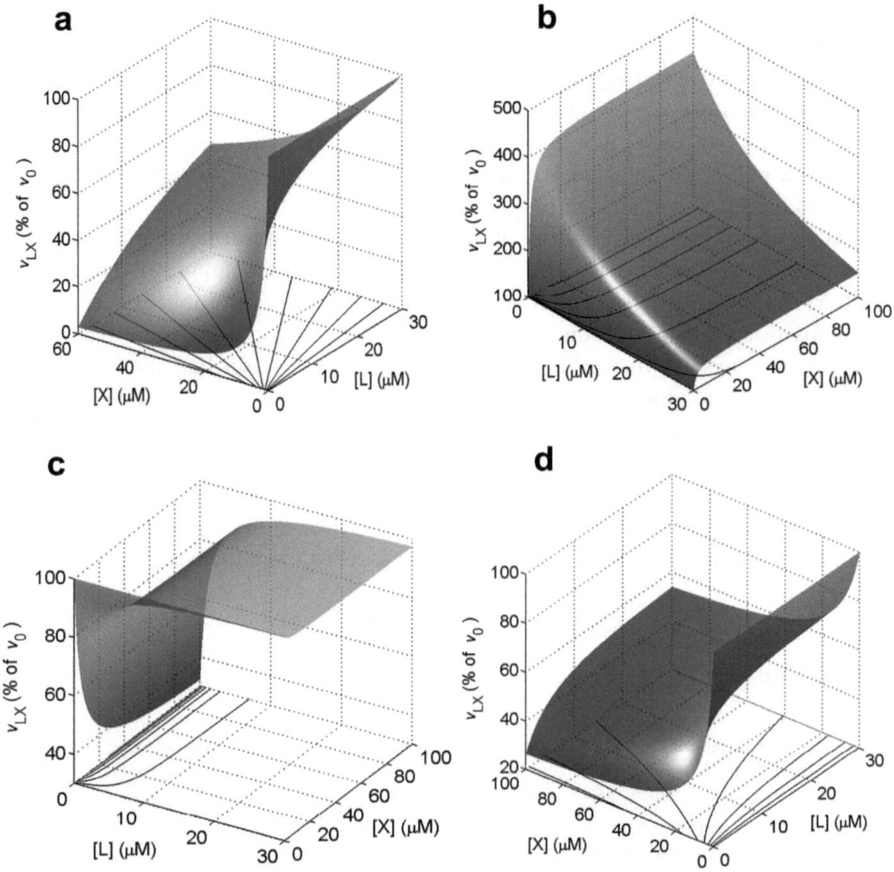

Fig. 7.3 The liberator action. All surface plots were generated with (7.3). (**a**) Liberation from inhibition by the linear inhibitor LMx(Sp>Ca)I: b = 5, $\beta_X = 0$, c = e = ∞ (exclusion of L from both EX and ESX), $\beta_{LX} = 0$, $\sigma = 1$, $K_L = 1\ \mu M$, $K_X = 1\ \mu M$. (**b**) Liberation from activation by the nonessential activator HMx(Sp<Ca)A: b = 0.4, $\beta_X = 3.0$, c = 1, e = 2 (ELX and ESLX exist), $\beta_{LX} = 0$, $\sigma = 1$, $K_L = 5\ \mu M$, $K_X = 8\ \mu M$. (**c**) Liberation from hyperbolic inhibition by HSpI: b = 6, $\beta_X = 1$, c = e = 1, $\beta_{LX} = 1$, $\sigma = 1$, $K_L = 2\ \mu M$, $K_X = 3\ \mu M$. (**d**) Liberation from hyperbolic inhibition by HMx(Sp<Ca)I: b = 0.8, $\beta_X = 0.2$, c = ∞, e = 1, $\beta_{LX} = 0.2$, $\sigma = 0.8$, $K_L = 2\ \mu M$, $K_X = 5\ \mu M$

little. In some reports of relief from inhibition or activation that appeared in the literature, experimental evidence was in many cases insufficient to distinguish the effect of true liberation from double-interaction between inhibitors or between an inhibitor and an activator. However, a clean demonstration of the liberator action is the recovery from inhibition of acetylglutamate kinase (EC 2.7.2.8) by L-arginine when urea is present concomitantly. Faragó and Dénes provided evidence of a true liberation mechanism showing that urea, at the low concentration needed for its liberator activity, had no influence per se on the enzyme [2].

A case of liberation from inhibition and activation of chicken and beef glutamate dehydrogenase (EC 1.4.1.3) by nucleotides has been characterized by Corman and Kaplan [1] (motto at the beginning of this chapter). Antisera raised in rabbits against glutamate dehydrogenase *prevented* inhibition of the enzyme by NAD_{red} and GTP. At pH 8, ATP inhibited glutamate dehydrogenase, while the enzyme was activated by ATP at pH 7: the antisera counteracted both effects. Importantly, the antisera alone did not affect enzyme activity.

Bathophenanthroline inhibition of electron transport in mitochondria is relieved by uncouplers. Measuring the effects of bathophenanthroline on purified mitochondrial ATPase (F_1), Phelps and coworkers demonstrated inhibition of the enzyme and liberation from inhibition by 2,4,-dinitrophenol and other typical uncouplers of electron transport from ATP synthesis. The authors ascribed the relief from inhibition of F_1 as interaction of the uncouplers with the F_1-bathophenanthroline complex [4]. Despite the lack of extensive kinetic characterization, this result agrees with the concept of liberator.

Because of the slow-onset character of the reaction, the liberation of cathepsin K from substrate-inhibition by clusterin will be described in Sect. 8.13.

These few examples show that the liberator is not a mere theoretical concept.

References

1. Corman L, Kaplan NO (1965) The influence of antibody on the interaction between nucleotides and glutamic dehydrogenase. Biochemistry 4:2175–2179. doi:10.1021/bi00886a034
2. Faragó A, Dénes G (1967) Mechanism of arginine biosynthesis in Chlamydomonas reindardti. II. Purification and properties of N-acetylglutamate 5-phosphotransferase, the allosteric enzyme of the pathway. Biochim Biophys Acta 136:6–18
3. Keleti T (1967) The liberator. J Theor Biol 16:337–355. doi:10.1016/0022-5193(67)90060-4
4. Phelps DC, Nordenbrand K, Nelson BD, Ernster L (1975) Inhibition of purified mitochondrial ATPase (F1) by bathophenanthroline and relief of the inhibition by uncouplers. Biochem Biophys Res Commun 63:1005–1012. doi:10.1016/0006-291X(75)90669-5

Chapter 8
Slow-Onset Enzyme Inhibition

Balancing the kinetic properties of an inhibitor between the Scylla of too fast a dissociation rate and the Charybdis of too slow an association rate is one of the current challenges facing those trying to design enzyme inhibitors with commercial utility via this route.
John V. Schloss (1988). Account Chem Res 21:348–357 (p. 353)

Abstract Slow-onset interactions between modifiers and their target enzymes are common in living cells and physiological fluids. This chapter aims at supporting the analysis of such interactions in the test tube by discussing the properties of various mechanisms and providing diagnostic methods to discriminate between them. That this is not a mere academic exercise is demonstrated by the importance of several slow-onset enzyme inhibitors that have been approved as drugs. Knowing their properties in vitro is the logical starting point for interpreting their mode of action at target sites in cells and body fluids.

8.1 Introduction

Not all interactions between enzymes and modifiers are rapid, diffusion-controlled events that can be treated with the methods described in the preceding chapters. To explain slow enzyme responses to rapid changes in substrate or modifier concentrations in the context of enzyme regulation in metabolic processes, Frieden coined the term *hysteretic enzymes*

> The term *hysteretic enzyme* is introduced to refer to enzymes which respond (in terms of configurational or other changes) slowly to rapid changes in ligand concentration.

and derived an integrated rate equation that will be discussed in Sect. 8.3 [12]. This equation has been shown to apply to a variety of *slow* enzymatic mechanisms, including slow-onset inhibition whether or not enzyme hysteresis is involved.

In physiology, slow-onset modification may serve as a fine-tuning mechanism to dampen cellular responses to rapid changes in metabolite concentrations instead of switching them off instantly. On the other hand, the pharmacological use of slow-onset inhibitors may pose problems of efficiency if binding to the target enzyme is insufficiently rapid.

A. Baici, *Kinetics of Enzyme-Modifier Interactions*,
DOI 10.1007/978-3-7091-1402-5_8

The name *slow-binding inhibition* proposed by John Morrison [19] suggests inhibitor binding to the enzyme as the step responsible for the sluggishness of the process. However, in some mechanisms enzyme and inhibitor bind rapidly and the process may even be controlled by diffusion, whereas other events determine the overall slow behavior. Therefore, the more general term *slow-onset* will be consistently used in place of slow-binding. With regard to the time-scale, *slow-onset* is a qualitative definition since the time ranges from a few seconds, to minutes, or even hours. Accordingly, such slow changes must be interpreted in relative terms of their rate compared to that of catalysis.

The kinetic properties of slow-onset enzyme inhibition have been scrutinized in studies authored or co-authored by Sungman Cha, specifically addressed to *tight-binding* inhibitors [4, 6–9], and by Morrison and co-workers [19–21, 28, 29, 35]. These studies were motivated by molecules of pharmacological interest: allopurinol as inhibitor of xanthine oxidase to treat hyperuricemia [8], the antineoplastic synergist and immunosuppressant coformycin used as inhibitor of adenosine deaminase [6, 8], and the antineoplastic agent methotrexate as inhibitor of dihydrofolate reductase [36]. Particularly two papers by Cha [4] and by Morrison [19] contributed to the rapid increase of the interest for this branch of enzyme kinetics.

The first section in this chapter is devoted to slow-onset, linear specific inhibition, in which the enzyme and the inhibitor equilibrate in two steps. Although apparently simple, in its most general case this system is not manageable in practice using an integrated rate equation but it can be analyzed by numerical integration. Ample space will then be dedicated to the analysis of progress curves of slow-onset modification using integrated rate equations. Mechanisms will be described that have in common the rate equation but distinct expressions of three variables, whose dependence on modifier concentration can be exploited for model discrimination and for calculating kinetic parameters.

Slow-onset linear specific inhibition is the basic mechanism most frequently represented in the literature. Besides linear specific inhibition, this chapter will explicitly discuss slow-onset linear catalytic and mixed inhibition. Nevertheless, *all basic modifier mechanisms* described in Chaps. 3 and 5 can be of the slow-onset type. The two slow-onset mechanisms based on HMxD(A/I) (p. 271) and HMx(Sp<Ca)A (p. 279) have not be included in the present chapter only because the properties of the transient in these two known examples could not be exploited for data analysis. All cases of reversible slow-onset modification represent *particular conditions* of the *basic mechanism* rather than representing taxonomically independent entities. For practical reason, however, the cases of slow-onset modification will be referred to as *mechanisms* since they appear in a characteristic, experimentally unmistakable way.

Similarly to multiple enzyme–modifier interactions (Chaps. 6 and 7), slow-onset modification is waiting to be explored as a yet *uncharted territory* and it is hoped that the present chapter will stimulate all the Junior (and Senior!) Woodchucks and Chickadee Patrols with a passion for scouting enzyme kinetics.

8.2 Linear Specific, Slow-Onset Inhibition: Not So Simple

Assigned acronym: SO-LSpI, where SO means slow-onset. We consider here an enzyme that interacts *slowly* with a linear specific inhibitor in a two-step reaction that leads to a non-covalent complex E·I via an intermediate adsorptive complex EI (Scheme 8.1). By accepting restrictive assumptions that include $[S] \approx [S]_t$ and $[I] \approx [I]_t$, Kuzmič succeeded in deriving an approximate, closed-form integrated rate equation that releases the restriction of only one slow step in mechanism SO-LSpI (Scheme 8.1) [14]. If *all* assumptions are denied, it is impossible to derive an analytical solution of the type (8.10) that will be discussed in Sect. 8.3.

In this section, assumptions are made neither on the relative rates of equilibration of the two steps that involve the binding of inhibitor, nor on the relative concentrations enzyme/inhibitor and enzyme/substrate. Computational obstacles that may arise from inhibitor depletion as a consequence of tight-binding to the enzyme and substrate consumption during the reaction can be circumvented by numerical integration methods.

The mechanism in Scheme 8.1 is defined by a set of four differential equations (8.1)–(8.4) and by the mass balance for enzyme, substrate and inhibitor (8.5)–(8.7). The progress curves for this mechanism consist of *two exponential phases* followed by a linear increase of product concentration. Knowing the properties of this system is a good starting point for interpreting all other mechanisms discussed below as well as for understanding the limits of assumptions and approximations for optimizing experimental design.

$$\frac{\mathrm{d}\,[\mathrm{ES}]}{\mathrm{d}t} = k_1\,[\mathrm{E}]\,[\mathrm{S}] - (k_1 + k_2)\,[\mathrm{ES}] \tag{8.1}$$

$$\frac{\mathrm{d}\,[\mathrm{EI}]}{\mathrm{d}t} = k_3\,[\mathrm{E}]\,[\mathrm{I}] - (k_{-3} + k_4)\,[\mathrm{EI}] + k_{-4}\,[\mathrm{E{\cdot}I}] \tag{8.2}$$

$$\frac{\mathrm{d}\,[\mathrm{E{\cdot}I}]}{\mathrm{d}t} = k_4\,[\mathrm{EI}] - k_{-4}\,[\mathrm{E{\cdot}I}] \tag{8.3}$$

$$\frac{\mathrm{d}\,[\mathrm{P}]}{\mathrm{d}t} = k_2\,[\mathrm{ES}] \tag{8.4}$$

Scheme 8.1 Slow-onset, linear specific inhibition. Generic case with no hypotheses about the relative rates of equilibration of the two inhibitory steps

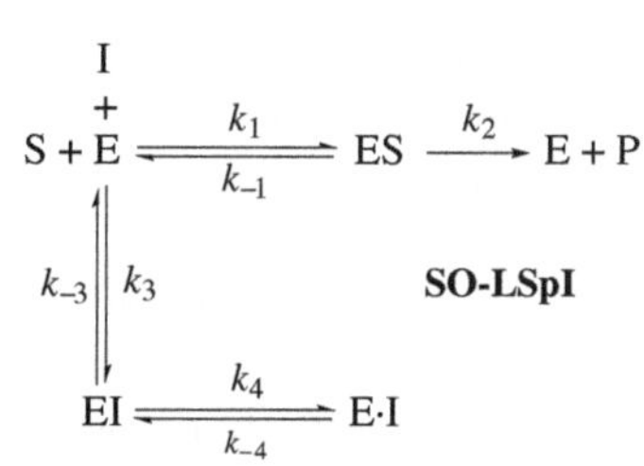

$$[E]_t = [E] + [ES] + [EI] + [E{\cdot}I] \tag{8.5}$$

$$[S]_t = [S] + [ES] + [P] \tag{8.6}$$

$$[I]_t = [I] + [EI] + [E{\cdot}I] \ . \tag{8.7}$$

The first task is to examine the interplay of the relaxation times of the two exponential phases to establish if these can be resolved and calculated individually. There are circumstances under which one of the relaxation times is sufficiently rapid with respect to the other for the system to degenerate to a single step. In relaxation kinetics, following a perturbation of a system in equilibrium, e.g., by a rapid jump in temperature, the system establishes a new equilibrium, from which it *relaxes* exponentially to the original equilibrium with a rate that is proportional to the extent of the deviation. The time needed by this system to return $1/e = 0.368$ of its value at equilibrium is called *relaxation time*, where $e = 2.718$ is the base of natural logarithms [3, pp. 6–8]. This is a law characteristic of first-order kinetic processes that holds also in slow-onset inhibition, but in this case the relaxation time applies to the approach to steady-state after the system has been *perturbed* by initiating the enzymatic reaction. In the integrated rate equations for slow-onset modification, the relaxation time corresponds to the reciprocal of λ in the exponential term of equations (8.8)–(8.10) below.

Whether the pre-steady-state phase of the progress curves for [P], such as those in Fig. 8.1a, consist of one or two exponentials followed by a linear steady-state, can be ascertained by fitting the generic equations (8.8) and (8.9) to data and determining which one provides the best fit

$$[P] = A_1\left(1 - e^{-\lambda_1 t}\right) + kt + d \tag{8.8}$$

$$[P] = A_1\left(1 - e^{-\lambda_1 t}\right) + A_2\left(1 - e^{-\lambda_2 t}\right) + kt + d\,. \tag{8.9}$$

The A-factors in these two equations represent the *amplitudes* of the exponential phases and d is a displacement factor which takes into account any nonzero value of the ordinate at time zero. λ_1 and λ_2 are *frequency constants* that correspond to reciprocal relaxation times and describe the approach to steady-state, i.e., to the straight line represented by the term kt in (8.8) and (8.9). For understanding the development of the reaction, it is instructive to examine not only the time dependence of [P] that is typically measured experimentally but also the calculated concentration changes with time of [ES], [EI] and [E·I] (Fig. 8.1), which are not ordinarily measurable.

From where the two exponentials arise is shown by the time dependences of [EI] and [E·I] shown in Fig. 8.1b,d. In this example, [EI] is formed slowly but anyway faster than [E·I], which then progressively increases at the expenses of the first adsorptive complex until both species reach a plateau and the system enters in the steady-state. The time necessary to attain the steady-state, which depends on inhibitor concentration, can also be appreciated from the progress curves of [ES] in Fig. 8.1c: when the plateau is reached, then $d[ES]/dt = 0$. The low values of [ES]

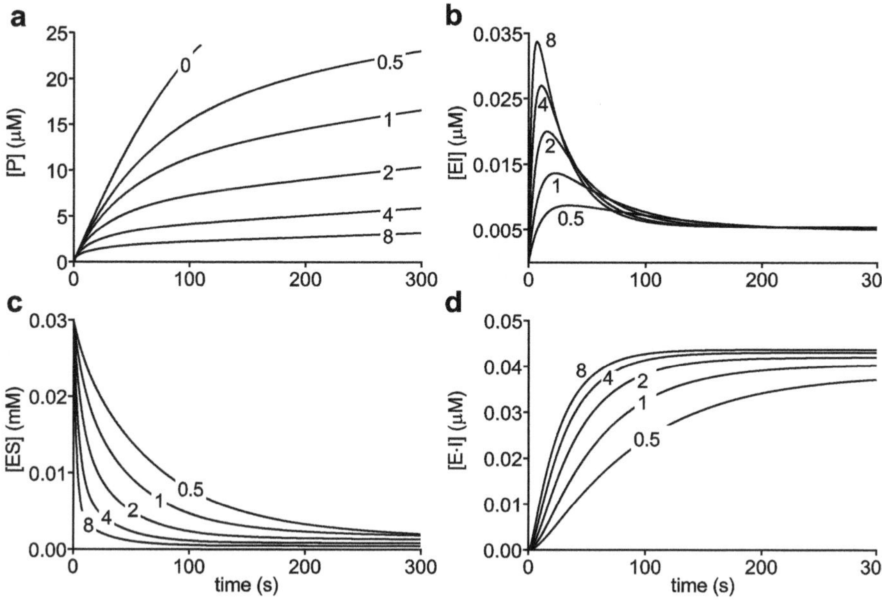

Fig. 8.1 Progress curves for the slow-onset linear specific inhibition mechanism in Scheme 8.1. Typically, only [P] is experimentally observable (**a**), while the time-dependences of [EI] (**b**), [E·I] (**d**) and [ES] (**c**) are usually not accessible to direct measurement but are shown for illustrating the properties of this mechanism. The progress curves were generated by numerical integration with the following constants and concentrations: $k_1 = 100\,\mu\text{M}^{-1}\,\text{s}^{-1}$, $k_{-1} = 1{,}990\,\text{s}^{-1}$, $k_2 = 10\,\text{s}^{-1}$, $k_3 = 0.1\,\mu\text{M}^{-1}\,\text{s}^{-1}$, $k_{-3} = 0.05\,\text{s}^{-1}$, $k_4 = 0.04\,\text{s}^{-1}$, $k_{-4} = 0.005\,\text{s}^{-1}$, $[\text{E}]_t = 0.05\,\mu\text{M}$, $[\text{S}]_t = 30\,\mu\text{M}$, $[\text{I}]_t$ (μM) as indicated by numbers in the panels, $K_m = (k_{-1} + k_2)/k_1 = 20\,\mu\text{M}$

during the steady-state with respect to [ES] at time zero explain enzyme inhibition proportional to [I].

We consider now the progress curves in Fig. 8.1a with the aim at extracting information on the mechanism. We start by fitting (8.8) and (8.9) to data as illustrated in Fig. 8.2. Goodness-of-fit analysis reveals that the double exponential model is superior to the single exponential model with a significance $p < 0.0001$. The analysis of residuals confirms this result by showing systematic deviation from the single exponential model.

Attention: the nice result of fitting (8.9) to data serves *only* to establish that the system approaches steady-state with two exponential phases but the underlying mechanism remains unknown and unknowable because the agnostic equation (8.9) is not related to a particular mechanism. Excessive substrate turnover during the time of the experiment (Fig. 8.1a) and the sequestration a non-negligible amount of inhibitor in the E · I complex at low inhibitor concentration, imply that the conventional method based on fitting an integrated rate equation to data with the aim at extracting mechanistic information would be inadequate.

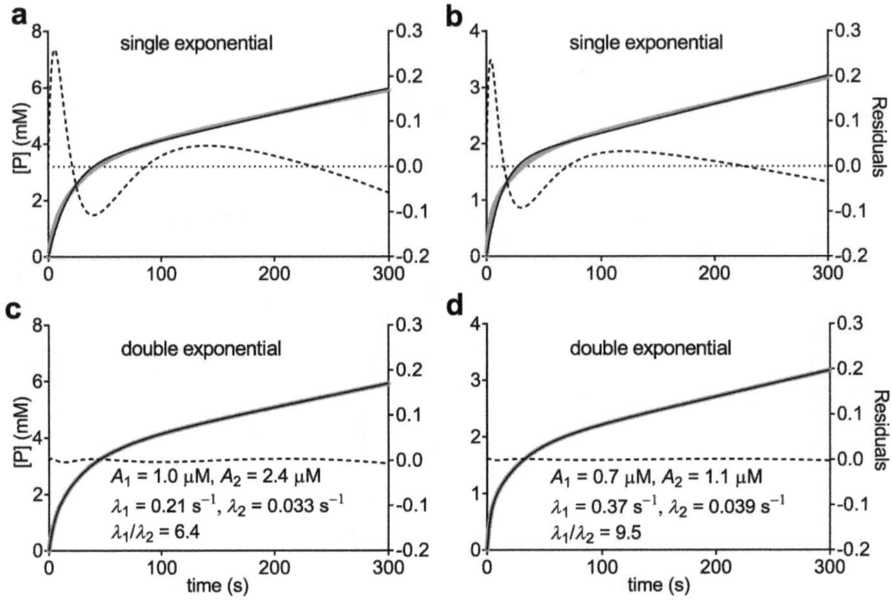

Fig. 8.2 Analysis of two progress curves from Fig. 8.1a, i.e., the curves for [I] = 4 μM (**a, c**) and [I] = 8 μM (**b, d**). The data to be analyzed are shown as *thick gray lines* and the best-fit curves as *thin black lines*. The *dashed lines* represent the residuals around the zero-deviation line drawn as *dotted lines*. In (**a**) and (**b**) a single exponential followed by steady-state was fit to data (8.8), while in (**c**) and (**d**) a double exponential plus steady-state (8.9) function was used. Goodness-of-fit analysis suggests that the double-exponential model (8.9) is superior to the single-exponential one (8.8) with a significance $p < 0.0001$. The calculated amplitudes of the two exponential phases in (8.9), the frequency constants, and their ratio are shown in (**c**) and (**d**)

An appropriate method of analysis is *global numerical integration* of all progress curves in Fig. 8.1a, including the control curve in the absence of inhibitor, based on Eqs. (8.1)–(8.4). However, even for this set of theoretical progress curves generated without introducing simulated errors, the fit aimed at evaluating the four constants k_3, k_{-3}, k_4 and k_{-4} (Scheme 8.1) failed to produce parameter estimates well constrained by data. This is due to the impossibility of extracting all mechanistic information from just one set of data. Therefore, a further set of progress curves corresponding to reactions started by adding substrate to preincubated enzyme and inhibitor was added to improve the previous analysis (Fig. 8.3). The progress curves in panel (a) are the same as in Fig. 8.1a with enlarged ordinate and those in panel (b) correspond to reactions started by adding the substrate. Random scatter was added to mimic instrumental noise.

The curves in Fig. 8.3a contain information about the association between enzyme and inhibitor, while those in panel (b) refer to the dissociation process. Also for the ensemble of curves in this figure, global fitting allowing all parameters to float freely during the optimization procedure did not produce acceptable results. However, splitting the procedure in two stages afforded satisfactory estimates of

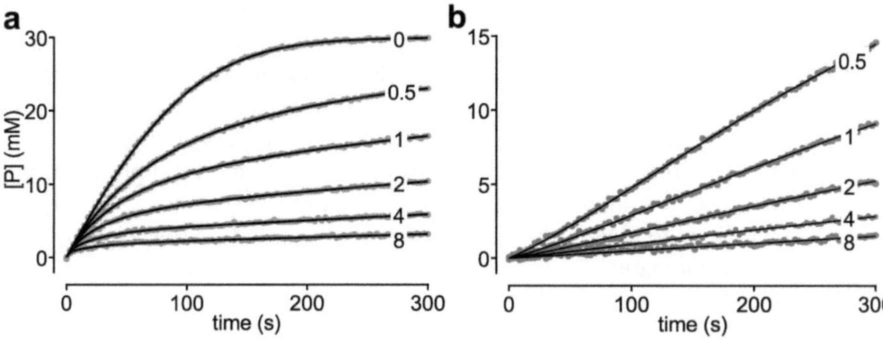

Fig. 8.3 Global fit of progress curves for the slow-onset, linear specific inhibition mechanism in Scheme 8.1. (**a**) Reaction started by adding enzyme to a solution containing substrate and inhibitor; (**b**) reaction started by adding substrate to a solution containing enzyme and inhibitor that had been preincubated for 300 s. Fake data (*gray dots*) generated with random scatter to simulate instrumental noise using the same constants and concentration as in Fig. 8.1. The *solid lines* represent best fit curves calculated by numerical integration coupled to nonlinear regression using the Global Kinetic Explorer software (p. 13)

Table 8.1 Results of the fit in Fig. 8.3

	k_3 ($\mu M^{-1} s^{-1}$)	k_{-3} (s^{-1})	k_4 (s^{-1})	k_{-4} (s^{-1})
Best fit	0.0975	0.0467	0.0388	0.0051
Standard error	0.0012	0.0006	0.0007	0.0002
Lower bound	0.0878	n.c.[a]	0.0341	0.0045
Upper bound	0.111	n.c.[a]	0.0453	0.0060
Theoretical values	0.1000	0.0500	0.0400	0.0050

The exaggerated number of significant digits has the only scope of illustrating the method

[a]Not calculated because k_3 and k_{-3} were constrained to a constant ratio

the parameters. First, a fit was run only with the progress curves in panel (a) until stable values of k_3 and k_{-3} were obtained as indicated by a minimum of χ^2. k_3 and k_{-3} were then constrained to vary at a constant ratio while performing a second round of optimization that included also the data in panel (b). Visual inspection of simulated data and best-fit curves in Fig. 8.3 suggests a good result, but this is not enough to guarantee statistically meaningful estimates of the parameters because they could not be unique. The FitSpace Explorer algorithm (p. 13) was then run to demonstrate that all parameters were well constrained by data. The results are summarized in Table 8.1, where the deviations of the best-fit parameters from the theoretical values used to create the artificial data are due to the scatter introduced to mimic instrumental noise. In fact, the deviations between theoretical and best-fit parameters depend only on the signal-to-noise ratio and vanish when the curves do not contain noise.

The key issue in calculating the rate constants for the two-step inhibition reaction in Scheme 8.1 is the *interplay* between the frequency constants λ_1 and λ_2. If their values differ sufficiently from one another, e.g., by one order of magnitude or almost so, or if they have similar values but the *amplitudes* of the exponential phases differ considerably from one another, then λ_1 and λ_2 can be extracted unambiguously from data. This is why numerical integration was successful in the example discussed in Figs. 8.2 and 8.3. If this is not the case, the frequency constants are said to be *coupled*, their individual values cannot be resolved, and getting clues on the mechanism is problematic.

8.3 The Integrated Rate Equation of Slow-Onset Inhibition

The vast majority of studies on slow-onset inhibition published so far made use of integrated rate equations for extracting kinetic information from data. The popularity of this method grew in parallel with the appearance of helpful commercial software offering regression analysis, statistical tests, and publication-quality graphics in the same package. The limitations associated with the use of analytical solutions of rate equations are balanced by practicality provided the user is aware of the underlying assumptions and approximations. An integrated rate equation can be derived for the mechanisms discussed in this chapter if the following assumptions 1–5 are satisfied.

Assumptions

1. In the system for which the integrated equation is derived, the formation of the enzyme-inhibitor complex involves *only one slow step*, i.e., there may be more than one step on the way to the final inhibited complex but only one of these steps is much slower than the others.
2. During the time of measurement there is no loss of enzyme activity for reasons other than the action of the inhibitor. Although the Selwyn test is invalid for slow-onset modification (Sect. 1.8), enzyme stability can be ascertaining by performing this test on the reaction in the presence of substrate without added modifier.
3. The concentration of free substrate does not change appreciably during the observation time, i.e., $[\mathrm{S}] \approx [\mathrm{S}]_\mathrm{t}$. This condition can be realized if the inequality $[\mathrm{S}]_\mathrm{t} \gg [\mathrm{E}]_\mathrm{t}$ holds for the whole reaction time at all concentrations of reactants.
4. Free inhibitor is not appreciably depleted by sequestration in the complex(es) with enzyme, i.e., $[\mathrm{I}] \approx [\mathrm{I}]_\mathrm{t}$. This condition can be met if $[\mathrm{I}]_\mathrm{t} \gg [\mathrm{E}]_\mathrm{t}$, i.e., if tight-binding can be circumvented by appropriate experimental design.
5. The catalytic step can be considered as being practically irreversible and product inhibition is insignificant. This statement should by no means suggest that the back reaction rate is zero because the corresponding rate constant is zero, but because the concentration of product(s) is approximated to zero.

Information Needed

a. Reaction scheme with all molecules participating in the reaction and associated rate constants.
b. Differential equations for the change in concentration with time of all enzyme-containing complexes and equation for the overall reaction rate.
c. Expressions of the concentrations of species in steady-state or in rapid-equilibrium derived from steady-state or from equilibrium relationships.
d. Mass conservation equations.
e. Boundary conditions for calculating the integration constants, which depend on whether the reaction is started by adding enzyme or substrate.

The integration of rate equations can either be performed with the Laplace transform method or using the long-winded (but intuitive) procedure illustrated in Appendix 2, where the derivation is made for representative mechanisms. The integrated rate equation (8.65) in Appendix 2 is reproduced here by adding the term *d* (displacement) that takes into account any nonzero value of the measured signal at time zero:

$$[\mathrm{P}] = v_{\mathrm{s}}t + \frac{v_{\mathrm{z}} - v_{\mathrm{s}}}{\lambda}\left(1 - e^{-\lambda t}\right) + d\,. \tag{8.10}$$

Equation (8.10) has been first published by Frieden in his study of hysteretic enzymes [12]. While (8.10) applies to all slow-onset inhibition mechanisms, a particular system is characterized by the expressions of v_z, v_s, and λ that represent the rate at time zero, the rate at steady-state, and the frequency constant, respectively.

8.3.1 *Geometric Properties of the Progress Curves for Slow-Onset Inhibition*

In enzyme-catalyzed reactions, slow responses due to the presence of slow-onset inhibitors result in progress curves that show either a *burst* like that in Fig. 8.4b or a *lag* as in curve (c). The shape of the progress curves, with burst or lag preceding the steady-state, depends on the difference $(v_{\mathrm{s}} - v_{\mathrm{z}})$ in (8.10).

In vitro, the reaction can be started by adding the enzyme to a mixture containing the substrate(s) and the inhibitor. Assuming that the ES complex is set up rapidly, the slope of the progress curve at time zero is represented by the dotted line (a) in Fig. 8.4, which is named $v_{\mathrm{z,E}}$ to specify that the reaction is started by adding the enzyme. Though, curve (a) cannot be observed because the enzyme is slowly and progressively sequestered by the inhibitor. The linear steady-state with slope v_{s} is attained after a certain time as explained below. In this experimental arrangement $v_{\mathrm{z,E}} > v_{\mathrm{s}}$.

A lag phase as in Fig. 8.4c is observed when the enzyme is preincubated with the inhibitor for a time sufficient to allow equilibration of the E·I complex and the

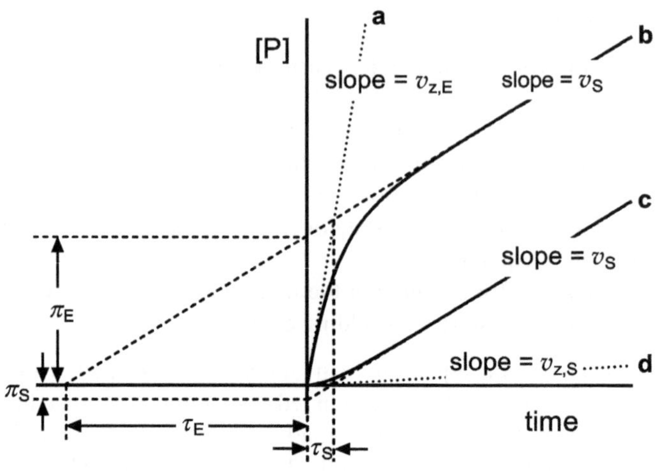

Fig. 8.4 Progress curves according to (8.10) for a generic slow-onset inhibitor. (*a*), Velocity at time zero when a reaction is started by adding enzyme; (*b*), substrate and modifier are preincubated and the reaction is started by adding enzyme; (*c*), the enzyme is preincubated with the modifier for a sufficiently long time to allow complex formation between the two molecules and the reaction is started by adding substrate; (*d*), velocity at time zero when the reaction is started by adding substrate

reaction is then started by adding the substrate. The rate at time zero for this situation is represented by the dotted line (d) and $v_{z,S} < v_S$.

In Fig. 8.4, the straight segments of the progress curves (b) and (c), whose common slope corresponds to the steady-state rate v_s, are extrapolated (dashed lines) to show their intersections with the ordinate at the characteristic points π_E and π_S, and their intersections with the time axis at points τ_E and τ_S, where the subscripts E and S indicate that the extrapolated values are obtained from a curve measured by starting the reaction by adding enzyme or substrate, respectively. Under the aforementioned assumptions (p. 374), preliminary information can be deduced from the geometric properties of the progress curves and their extrapolated steady-state portions, as shown in Fig. 8.4. For π and τ the following relationships hold [1, 7]:

$$\pi = \frac{v_z - v_s}{\lambda}; \quad \tau = \frac{1 - v_z/v_s}{\lambda} \tag{8.11}$$

$$\lambda = \frac{v_z - v_s}{\pi}; \quad \lambda = \frac{1 - v_z/v_s}{\tau}, \tag{8.12}$$

where π and τ may be either negative or positive depending on the relative values of v_s and v_z, namely

$$\text{reaction started with E}: \quad v_z > v_s \quad \therefore \quad \pi_E > 0, \ \tau_E < 0$$

$$\text{reaction started with S}: \quad v_z < v_s \quad \therefore \quad \pi_S < 0, \ \tau_S > 0\,.$$

Estimating graphically λ from the relationships (8.12) depends on data quality and this parameter can also be calculated from τ_S since

$$\tau_S = \frac{1}{\lambda} . \tag{8.13}$$

The value of the overall dissociation rate constant of the inhibitor can be calculated from the velocities v_s, v_z and λ as

$$k_{off} = \lambda \frac{v_s}{v_{z,E}} . \tag{8.14}$$

k_{off} is a first-order constant whose expression depends on the mechanism, e.g., it corresponds to k_{-4} in mechanism SO-LSpI-2 and to $(k_{-4} + k_5)$ in SO-LSpI-2Temp described in Sects. 8.4 and 8.6, respectively. If the results from fitting (8.10) to data can be demonstrated to be statistically valid, this is certainly the most reliable and straightforward method for determining k_{off} from measurable parameters.

8.3.2 *The Quality of the Determined Slow-Onset Inhibition Parameters Depends on Compliance with the Assumptions*

The progress curves in Fig. 8.4 represent the ideal outcome of a perfect experiment, in which all assumptions on p. 374 are respected without compromises. This objective is however more theoretical than practical and, as in other fields of enzyme kinetics, also slow-onset inhibition results must be accepted with some approximation. This policy is supported by the fact that experimental error, comprising instrumental noise and sample manipulation, overwhelms in most cases deviations of the measured parameters from their true values for the nonobservance of the assumptions made in deriving the rate equations.

In slow-onset inhibition, an obstacle is overcoming the problem of substrate consumption during the relatively long time often necessary to measure the exponential phase and a sufficiently long linear segment following it. The rule of thumb accepted among researchers is that no more than 10 % of the total substrate should be turned over during the time of the experiment. The problem is more pronounced in reactions started with the enzyme as in those initiated by adding the substrate (compare curves b and c Fig. 8.4). This comment on the experimental acquisition of data is directed also to teachers and students who wish to visualize graphically the function (8.10) using either a *plot function* approach or *simulation* by numerical integration.

To *plot a function*, v_z, v_s and λ are calculated from a set of rate constants and concentrations and then used to draw the plot of rate against time using (8.10). The linear part of the progress curve increases in a perennial, perfectly linear steady-state because, during derivation, the function has been *instructed* that the substrate concentration is constant. Substrate consumption signifies decrease of the rate, which declines to zero when the substrate has been completely used up. The dashed lines in Fig. 8.5a,b show that product is released linearly even after total depletion of substrate (40 μM in this example).

Simulation by numerical integration (Sect. 1.2.7) is performed starting from rate constants and concentrations using the differential equations of the species in the system. At each integration step, the free substrate and inhibitor concentrations are updated by subtracting the portions that have been withdrawn to form the product and the enzyme-inhibitor complex(es). Substrate depletion causes the rate to decrease independently of the inhibition mechanism and considering the depletion of free inhibitor is important for integrating the rate equation with the actual value of [I], which is particularly important in case of tight-binding of I to E. Any deviation from the assumptions $[S] \approx [S]_t$ and $[I] \approx [I]_t$ is thus taken into account. Provided the enzyme does not loose activity for causes independent of the inhibitor, simulations closely represent real experiments, while plotted functions adhere to a real system only in a narrow time range, during which substrate depletion is negligible. These concepts are illustrated by the comparison of the continuous and the dashed progress curves in Fig. 8.5a. Because of substrate consumption and the progressive lowering of the reaction velocity, the *pseudo-linear* portions of the continuous curves are not parallel with one another since the slope of the curve obtained by starting the reaction with substrate is greater than that of the curve corresponding to a reaction initiated by adding enzyme. *Pseudo-linear* in the preceding sentence underlines the fact that, in addition, the progress curves are bent down in the presence of even modest substrate depletion. This topic has been discussed in Sect. 1.9 in the frame of initial velocities and is illustrated in Fig. 8.6 for the time range 500–800 s of the curves obtained by starting the reaction with enzyme.

Panels (c–f) in Fig. 8.5 aim at exploring the conditions that characterize the steady-state of substrate turnover by showing the changes of concentration with time of E, ES, EI, and E·I. These properties are not ordinarily measurable but are very instructive for visualizing all components of a reaction by simulation. Mimicking an experiment performed with a stopped-flow apparatus, the time interval 0–200 s corresponds to a preincubation time after filling both syringes with the appropriate reagents that should reach the desired temperature in the thermostatted chamber of the instrument. Such an exact time is certainly not necessary and the 200 s chosen here are because of the simulation performed with KinTek software as described for didactic purposes in Appendix 1.

During the preincubation time, nothing happens in panels (c) and (d) because, when the reaction is started with enzyme, the catalyst is put in one syringe and substrate plus inhibitor in the other. Conversely, when the goal is to measure a reaction by adding substrate, enzyme and inhibitor are put together in one of the syringes. According to the rate constants, an equilibrium is set up between E, EI, and E·I, where the last is the predominant species in this particular example (panels e and f). After rapidly mixing the reagents at the time labeled 200 s,

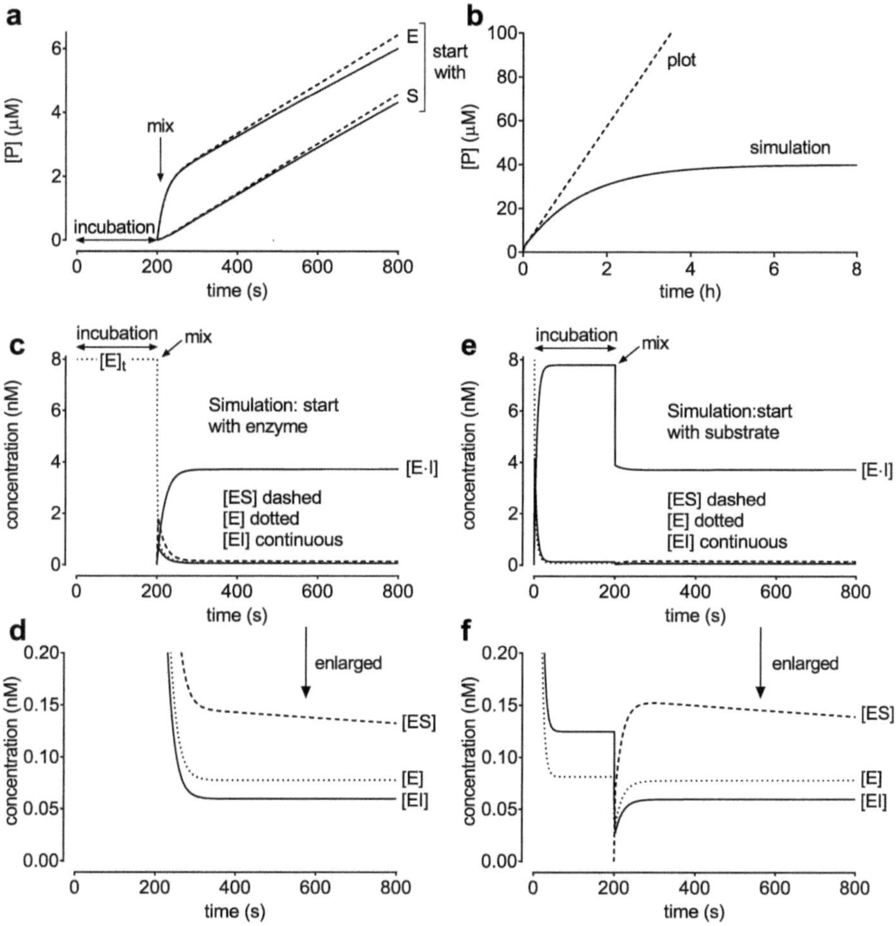

Fig. 8.5 Slow-onset inhibition. (**a, b**) Different progress curves plotted as function or simulated by numerical integration. The *dashed lines* represent progress curves *plotted* with the function (8.10), while the continuous lines were *simulated* by numerical integration using KinTek software. The mechanism considered in this example is SO-LSpI-2 described in Sect. 8.4, based on Scheme 8.2, with the following rate constants and concentrations: $k_1 = 100\,\mu M^{-1}\,s^{-1}$, $k_{-1} = 1{,}950\,s^{-1}$, $k_2 = 50\,s^{-1}$, $k_3 = 2.3\,\mu M^{-1}\,s^{-1}$, $k_{-3} = 12\,s^{-1}$, $k_4 = 0.25\,s^{-1}$, $k_{-4} = 0.004\,s^{-1}$, $[E]_t = 4\,nM$, $[S]_t = 40\,\mu M$, $[I]_t = 4\,\mu M$. (**c–f**) Excessive substrate consumption and non-adherence to the assumptions for using the integrated rate equation. The simulation mimics an experiment performed with a stopped-flow apparatus equipped with two syringes that deliver reactants to the observation chamber in a volume ratio of 1:1 (Fig. 8.29 in Appendix 1). (**c, d**) Variation of [E], [ES], [EI] and [E · I] in a reaction started by adding enzyme. (**e, f**) Reaction started by adding substrate to a preincubated solution containing enzyme and inhibitor. Panels (**d**) and (**f**) show enlarged portions, at low concentrations, of the corresponding curves in (**c**) and (**e**). During the time of the measurement, [E] and [EI] do not change, while [E · I] increases (barely appreciable in the scale of panels **c** and **e**) while [ES] decreases

Scheme 8.2 Slow-onset linear specific inhibition with rapidly equilibrating first step

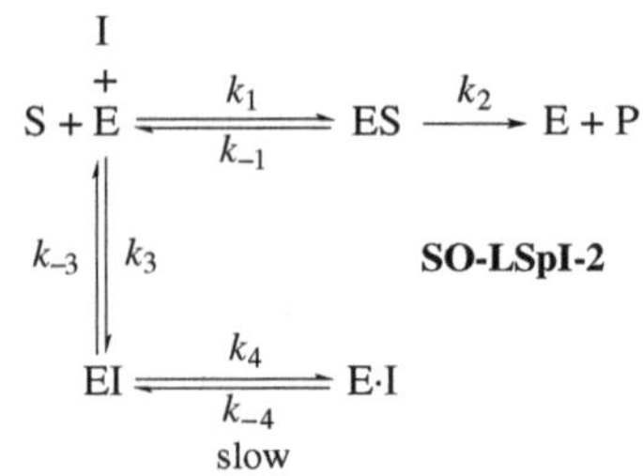

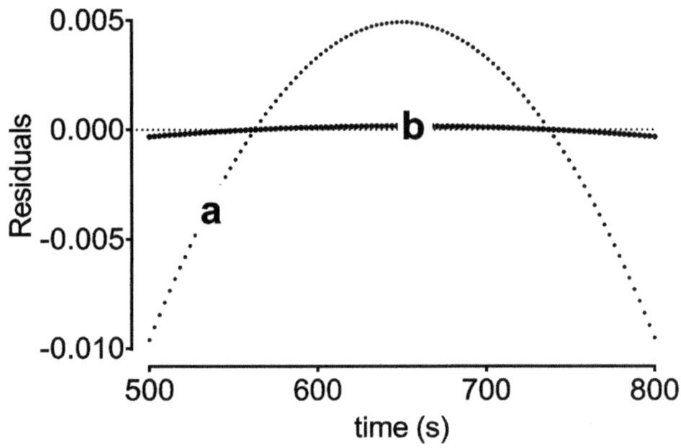

Fig. 8.6 Linear regression of the progress curves in Figs. 8.5a and 8.7a for reactions started by adding enzyme. Residuals of the regression line in the time range 500–800 s : (*a*) from Fig. 8.5, (*b*) from Fig. 8.7

following an exponential phase [E], [EI] and [E · I] remain practically constant, while [ES] slowly decreases as a consequence of substrate depletion. Thus, this portion of the progress curve does not properly conform to the concept of steady-state. The outcome is that the values of v_s and λ will not be exactly the same in reactions started with enzyme or substrate as it should be in the absence of substrate depletion.

The problems outlined above should be taken into account in designing experiments, being aware that at fixed substrate concentration and variable inhibitor concentration, the parameters gathered at the lowest values of [I] will not be as precise as those obtained at higher [I]-values.

The most frequent slow-onset modification cases in the literature are those based on linear specific inhibition, in which the substrate is supposed to *compete* with the inhibitor for the same binding site in the active center. Alleged competition can however result from other causes, including binding of the modifier to only one form of the enzyme in equilibrium between two isomerizing species as will be discussed in Sects. 8.7 and 8.8. There is no competition whatsoever in this case although the kinetic outcome is the same, but in any case this effect can be exploited in experiment design by using substrate concentrations that are relatively high, such as three to four times K_m but anyway only in the absence of substrate inhibition. High substrate concentrations also contribute to partly contrast any depletion of free inhibitor in case of tight-binding. As confirmed by the simulations in Fig. 8.7, this strategy reduces the problem of substrate depletion at the benefit of closer

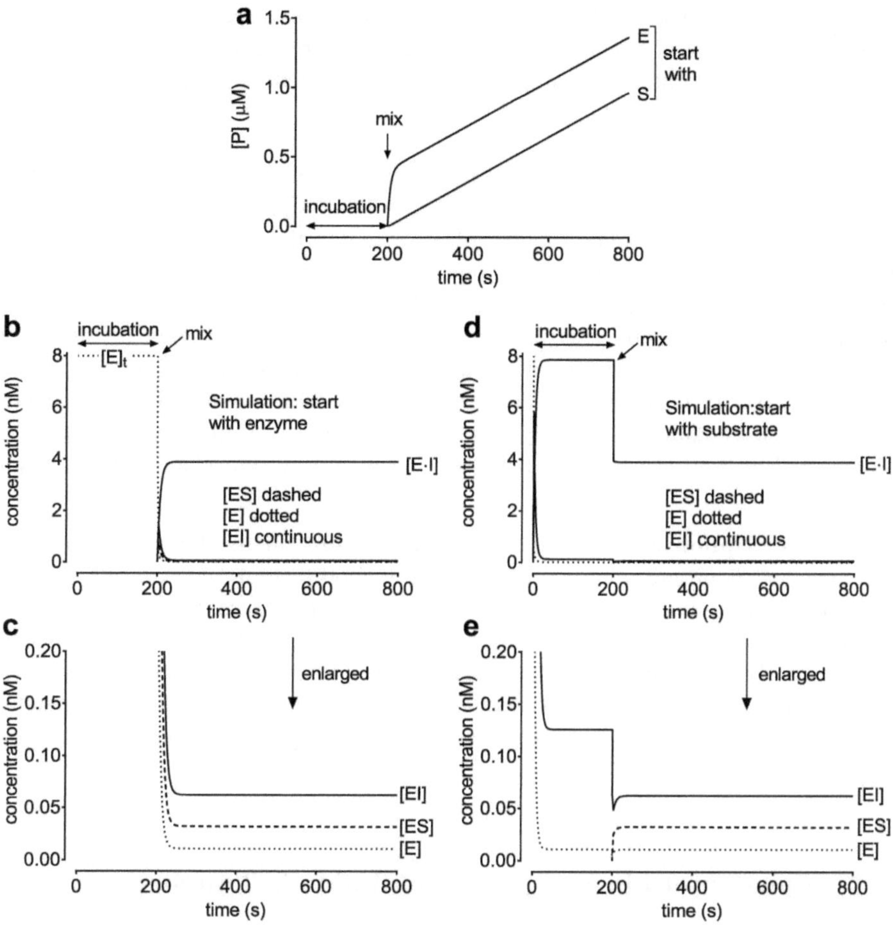

Fig. 8.7 Slow-onset inhibition: adherence to the assumptions for using the integrated rate equation. (**a**) Progress curves obtained by starting the reaction with enzyme or substrate: the linear portions of the two curves are parallel. The mechanism and the constants used in the simulation are the same as in Fig. 8.5 but $[S]_t = 60\,\mu M$, $[I]_t = 30\,\mu M$. (**b, c**) Variation of [E], [ES], [EI] and [E · I] in a reaction started by adding enzyme. (**d, e**) Reaction started by adding substrate to a preincubated solution containing enzyme and inhibitor. Panels (**c**) and (**e**) show enlarged portions, at low concentrations, of the corresponding curves in (**b**) and (**d**). During the time of the measurement, [E] and [EI] do not change and the changes of [ES] and [E · I] are imperceptible

adherence to the assumptions on p. 374. Without commenting all panels, note in this figure that the dependence of [ES] with time is still not perfectly linear but, as shown in the residuals analysis of Fig. 8.6, reasonably acceptable as representative of a steady-state in comparison with the accompanying curve based on the data in Fig. 8.5.

Planning slow-onset inhibition experiments comprises the evaluation of compromises in the choice of reactant concentrations while trying to minimize deviations from the assumptions made in deriving the integrated rate equation. A rapid-mixing instrument for data collection minimizes the time that elapses between reagent mixing and recording the reaction trace. Conventional spectrophotometry or fluorometry with manual reaction start may lead to inconsistent results and progress curves for reactions started with the substrate may miss precious information in the exponential phase. The major source of error is that the exact time at which the reaction is initiated may differ from one progress curve to another in a series of measurements. A stopped-flow apparatus allows data collection by accumulating and averaging several progress curves with the aim at minimizing instrumental noise. A further improvement in data acquisition is the use of a logarithmic timebase. For enzymes that do not tolerate excessive dilutions, the use of a stopped-flow apparatus equipped with syringes of different sizes may be exploited as shown in Fig. 8.30 (Appendix 1), where the enzyme is introduced in the small syringe.

In the Description of the Individual Mechanisms in the following sections, the analysis of progress curves of slow-onset modification mechanisms will be made by showing dedicated equations grouped for practical reasons, which will be complemented by graphical illustration of the procedures. These will be provided in an extensive form for the mechanisms discussed first with the aim at introducing the basic methodology. For other mechanisms, instead of showing the entire palette of illustrations, analogous to those shown in detail, only relevant and characteristic properties will be discussed.

8.4 Slow-Onset Linear Specific Inhibition in Two Steps

The assigned acronym is SO-LSpI-2 (where 2 indicates a two-step mechanism). This is the celebrated *Mechanism B* of Morrison [19], the most frequent one in the literature of slow-onset inhibition (Scheme 8.2).

Managing the mechanism SO-LSpI in Scheme 8.1 is greatly simplified when the first inhibitory step equilibrates much faster than the second one because, in the presence of only one slow step, the rate equation can be integrated under the assumptions made on p. 374. A procedure for the integration is demonstrated for SO-LSpI-2 in parallel with the temporary mechanism SO-LSpI-2Temp (Sect. 8.6) in Appendix 2. The expressions of v_s, v_z, and λ in (8.10), the inhibition constants and other useful diagnostic expressions are collected in the group of equations (8.15).

Complementing the measurements of progress curves for reactions started by adding enzyme at fixed substrate and variable inhibitor concentrations with data collected by starting the reaction with the substrate may be useful as discussed in Sect. 8.2. Another strategy is to measure progress curves at fixed inhibitor

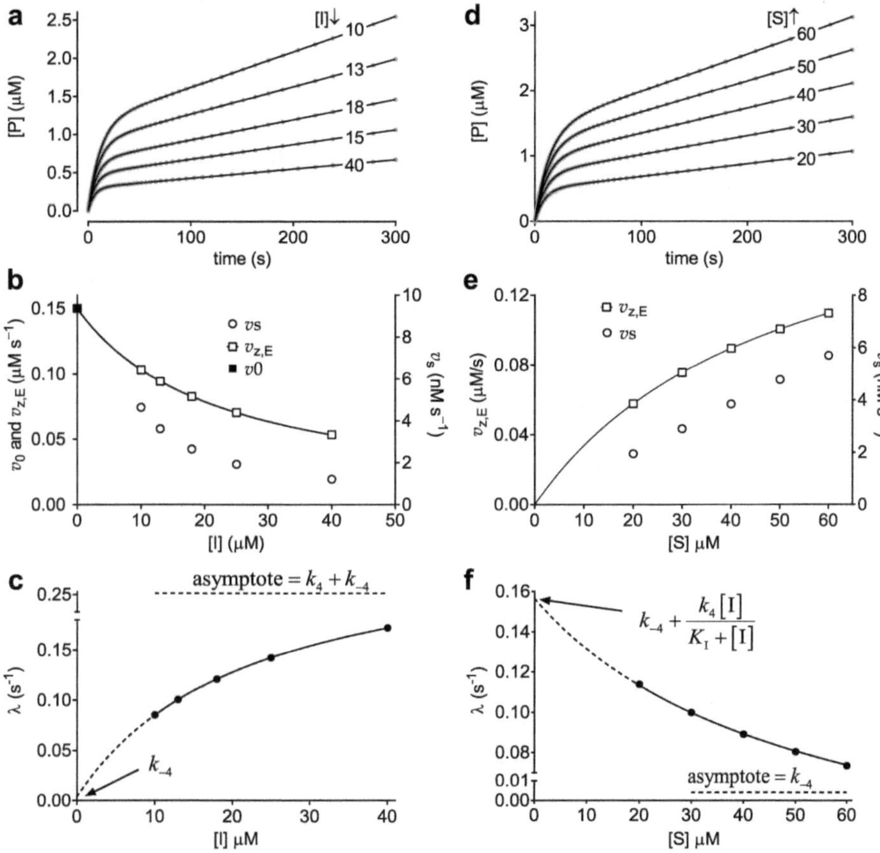

Fig. 8.8 Slow-onset inhibition mechanism SO-LSpI-2. Basing on Scheme 8.2, the error-free curves in (**a**) and (**d**) were generated with KinTek software on a logarithmic time scale but are shown here on a linear scale. The constants were: $k_1 = 100\,\mu\text{M}^{-1}\,\text{s}^{-1}$, $k_{-1} = 1{,}950\,\text{s}^{-1}$, $k_2 = 50\,\text{s}^{-1}$, $k_3 = 2.3\,\mu\text{M}^{-1}\,\text{s}^{-1}$, $k_{-3} = 12\,\text{s}^{-1}$, $k_4 = 0.25\,\text{s}^{-1}$, $k_{-4} = 0.004\,\text{s}^{-1}$, $[\text{E}]_t = 4\,\text{nM}$. Derived rates and parameters: $K_m^0 = 20\,\mu\text{M}$, $V^0 = 0.2\,\mu\text{M}\,\text{s}^{-1}$, $K_I = 5.2\,\mu\text{M}$, $K_I^* = 0.082\,\mu\text{M}$. (**a–c**) Reactions started by adding enzyme at constant [S] = 60 μM and variable [I] (μM, numbers on the curves in panel (**a**)). (**d–f**) Reactions started with enzyme at fixed [I] = 8 μM and variable [S] (μM) as indicated by the numbers in (**d**). The *solid lines* in (**a**) and (**d**) represent best-fits of (8.10) to data, which afforded $v_{z,E}$, v_s and λ, whose dependences on [I] are shown in (**b, c**) while the dependences on [S] are those in (**e, f**). The curves through the values of $v_{z,E}$ and λ represent fits of the corresponding equations in (8.15)

and variable substrate concentrations that can be chosen to minimize the drawback of substrate depletion. Figure 8.8 shows how information about the mechanism can be extracted from raw data exploiting the dependences of the parameters in (8.10) on the concentrations of inhibitor and substrate. (a–c) Describe measurements obtained at fixed substrate and variable inhibitor, while panels (d–f) refer to measurements obtained at fixed inhibitor and variable substrate.

8.4.1 Progress Curves Obtained at Fixed Substrate and Variable Inhibitor Concentration

Figure 8.8a–c. All progress curves were generated with KinTek software using a logarithmic timebase. Certain systems can greatly profit from a logarithmic timebase for the acquisition of data, especially when using a rapid-mixing device such as a stopped-flow apparatus [34]. The advantage for progress curves that contain one or more exponential phases is that samplings are more frequent at the beginning of the reaction, where the signal increases exponentially and more rapidly than in the linear portion of the curve. Samplings taken at larger time intervals in linear segments of progress curves are sufficient for an accurate analysis. At the same time, the superior resolution of the exponential phase(s) allows more accurate analysis of the relaxation time(s).

Progress curves of reactions started by adding the substrate to preincubated mixtures of enzyme and inhibitor are not shown here (but see Figs. 8.5 and 8.7 that contain this information). The best use of these progress curves is in numerical integration of whole sets of data that include reactions started with enzyme and with substrate (Sect. 8.2). Independently of whether the reaction is started by adding enzyme or substrate, the expressions of λ and of the rate at steady-state, v_s, are the same, while the rates at time zero are different. In the reaction started by adding enzyme (a), $v_{z,E} > v_{z,S}$ whereas the contrary is valid when the reaction is started by adding substrate. The coefficient x that multiplies [I] in the expression of $v_{z,S}$ in (8.15) represents a dilution factor that accounts for the increase in volume after adding substrate to a solution in which enzyme and inhibitor have been preincubated.

For SO-LSpI-2, as for other slow-onset inhibition mechanisms, extracting useful information from the velocities $v_{z,S}$ and v_s calculated from progress curves depends very much on their magnitude. In some instances, at high inhibitor concentrations, the experimentally measurable steady-state rates are small. In Fig. 8.8b v_s is plotted on the scale of the right ordinate as nM s^{-1}, i.e., all five values of v_s represent more than 95 % inhibition (v_0 = 150 nM). To obtain larger values of v_s and $v_{z,S}$ very low inhibitor concentrations and long measuring times are required, but this results in excessive substrate consumption, a violation of the third assumption on p. 374 with the consequences discussed in Sect. 8.3.2.

$$\textbf{SO}-\textbf{LSpI}-\textbf{2} \tag{8.15}$$

$$\text{required}: \quad K_{\text{m}} \gg K_{\text{I}}\,, \quad k_{-1} \gg k_4\,, \quad k_{-1} \gg k_{-4}$$

$$\lambda = k_{-4} + \frac{k_4\,[\text{I}]}{K_{\text{I}}\left(1+\dfrac{[\text{S}]}{K_{\text{m}}}\right)+[\text{I}]}\,, \qquad v_{\text{s}} = \frac{V\,[\text{S}]}{K_{\text{m}}\left(1+\dfrac{[\text{I}]}{K_{\text{I}}^*}\right)+[\text{S}]}$$

$$k_{\text{off}} = k_{-4} = \lambda\frac{v_{\text{s}}}{v_{\text{z,E}}}\,, \quad K_{\text{I}} = \frac{k_{-3}}{k_3}\,, \quad K_{\text{I}}^* = K_{\text{I}}\left(\frac{k_{-4}}{k_4+k_{-4}}\right)$$

$$v_{\text{z,E}} = \frac{V\,[\text{S}]}{K_{\text{m}}\left(1+\dfrac{[\text{I}]}{K_{\text{I}}}\right)+[\text{S}]}\,, \quad v_{\text{z,S}} = \frac{V\,[\text{S}]}{K_{\text{m}}\left(1+\dfrac{[\text{I}]}{K_{\text{I}}}\right)+[\text{S}]} \times \frac{1+x\dfrac{[\text{I}]}{K_{\text{I}}}}{1+x\dfrac{[\text{I}]}{K_{\text{I}}^*}}$$

Unlike $v_{\text{z,S}}$, the dependence of $v_{\text{z,E}}$ on [I] can be measured in a broader range of inhibition, which serves to calculate K_{I} by nonlinear regression using the equation for $v_{\text{z,E}}$ in (8.15). This expression corresponds to the rate in the presence of a linear specific (fast-acting) inhibitor, (5.1) in Sect. 5.2.1, which decreases as hyperbola when the inhibitor concentration is increased, a property that distinguishes SO-LSpI-2 toward other slow-onset inhibition mechanisms, such as SO-LSpI-1, for which $v_{\text{z,E}}$ is independent of [I]. For SO-LSpI-2, the reason for the dependence of $v_{\text{z,E}}$ on [I] is that ES and EI are immediately in place after starting the reaction by adding the enzyme, while E·I is formed at a slower rate.

If k_{-4} is very small with respect to k_4, determining its value as the ordinate intercept of the λ against [I] plot (Fig. 8.8c) or as asymptote (Fig. 8.8f) is very difficult or even meaningless because the extrapolated intersection or asymptote may not be statistically distinguishable from zero, especially in the presence of large experimental scatter. This shortcoming can be circumvented using the displacement method described below in Sect. 8.4.3, which in parallel serves to demonstrate the reversibility of the inhibition process, particularly in the presence of tightly-bound inhibitors. The other method for determining k_{-4} consists in its calculation from $v_{\text{z,E}}$, v_{s} and λ according to (8.14). The dependence of λ on [I], which increases hyperbolically with [I] (Fig. 8.8c), is part of the diagnostic criteria for discriminating among slow-onset inhibition mechanisms.

Two inhibition constants characterize this slow-onset, specific inhibition mechanism: K_{I}, the dissociation constant of the rapidly equilibrating EI complex, and K_{I}^*, the overall inhibition constant. Although the latter could be determined from the dependence of v_s on [I] as suggested by its expression in (8.15), this rate can usually be measured only in a narrow range as discussed above. Nevertheless, K_I^* can be calculated from K_I, k_4 and k_{-4} determined separately or by fitting the equation of λ in (8.15), which can be rewritten in the equivalent form (8.17) discussed on p. 391.

8.4.2 *Progress Curves Obtained at Fixed Inhibitor and Variable Substrate Concentration*

This approach, illustrated in Fig. 8.8d–f, complements the set in panels (a–c) by assessing internal consistency of the calculated parameters. The fixed concentration of the inhibitor should allow measuring a sufficiently broad range of inhibition degrees at various substrate concentrations (panel d). After fitting (8.10) to progress curves to determine $v_{z,E}$, v_s and λ for each progress curve, their dependence on [S] is shown in panels (e) and (f), respectively. λ *decreases* exponentially with [S] and the useful information is represented by the intercept of the curve, the best fit of the expression of λ in (8.15), with the ordinate. This number can be compared with the value of the expression shown in panel (f) calculated with the parameters determined from the data in Fig. 8.8a–c.

Useful information is contained in panel (e), where the dependence of $v_{z,E}$ on [S] can be exploited to assess the basic mechanism of the system. The curve through the void squares in panel (e) represents the best fit of the Michaelis–Menten equation that gives $V^{app} = 0.2\,\mu M\,s^{-1}$, identical to V^0 (see the legend of Fig. 8.8), and $K_m^{app} = 50.7\,\mu M$. The linearity of the inhibition mechanism is confirmed by the asymptotic decline to zero of $v_{z,S}$. Since $K_m^0 = 20\ \mu M$, the larger value of the apparent Michaelis constant in the presence of inhibitor demonstrates that the basic mechanism is linear specific inhibition. This is in fact the only linear inhibition mechanism for which the catalytic constant (and thus the limiting rate) is independent of [I] *and* the Michaelis constant *increases* with [I] (Fig. 3.4).

The ensemble of all characteristic traits of SO-LSpI-2 and other mechanisms will be used in the dichotomous keys in Chap. 10 in a guided procedure aimed at discriminating between mechanisms and at deducing the most likely hit from data.

8.4.3 *The Displacement Method to Determine the Off-Rate Constant*

The rate constant for the dissociation of a slow-onset inhibitor, which corresponds to k_{-4} in mechanism SO-LSpI-2 but we can generically call it k_{off} for this and

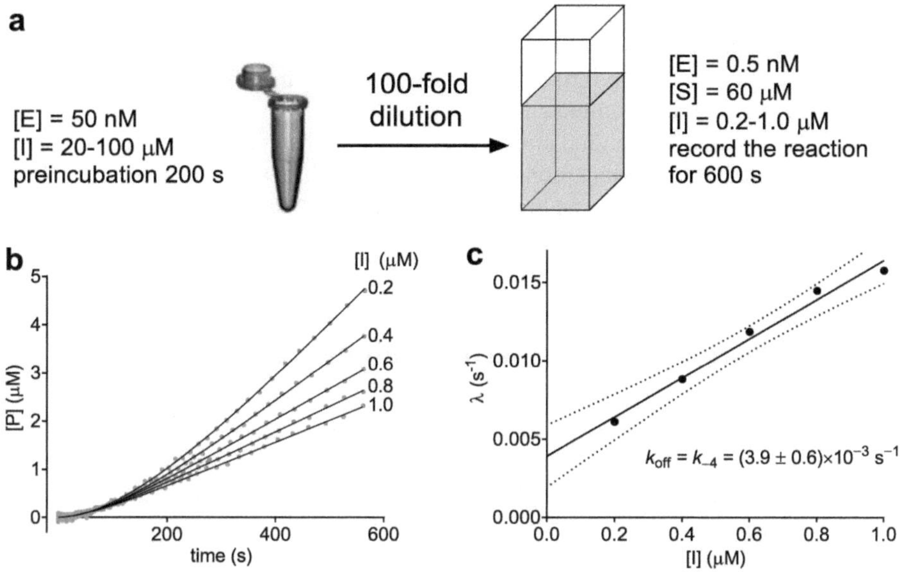

Fig. 8.9 Determination of the dissociation rate constant in slow-onset inhibition. (**a**) Explanation of the procedure. (**b**) Progress curves on logarithmic timebase with added error simulated by numerical integration using the parameters in the legend of Fig. 8.8 for mechanism SO-LSpI-2. The *curves* represent reaction progress after diluting the preincubated mixture 100-fold in buffer containing substrate. To simulated data, shown as *gray dots*, (8.10) was fitted to give the best estimates shown as *black lines*. (**c**) The frequency constants for each inhibitor concentration, determined from the fitting procedure in (**b**), were plotted against [I] and linear regression was run to draw a straight line. The *dotted lines* around the best-fit represent the 95 % confidence band

other slow-onset mechanisms, cannot be determined accurately from plots such as that in Fig. 8.8c. However, when the enzyme is preincubated with the inhibitor for a sufficiently long time to allow equilibration and then rapidly diluted into a solution containing substrate, dissociation of the enzyme-inhibitor complex(es) occurs and the liberated enzyme can turn over the substrate. The progress curves are characterized by a lag phase that is usually pronounced, followed by a linear increase of product concentration.

This procedure is sketched in Fig. 8.9a. In the vial on the left, enzyme and inhibitor were preincubated for a time sufficient to allow equilibration. Then, an aliquot was rapidly diluted 100-fold into assay buffer containing substrate and the reaction was monitored continuously as shown in Fig. 8.9b. Equation (8.10) fitted to progress curves afforded the black lines superimposed to data points shown in light gray. The values of λ obtained from the fitting procedure were then plotted against the actual inhibitor concentration after dilution as shown in Fig. 8.9c. In the assay mixture, the inhibitor is very diluted and therefore the increase of λ with [I] is practically linear. This increase corresponds to the very beginning of the extrapolated dashed line in Fig. 8.8c. According to the expression of λ in (8.15), the *off* rate constant corresponds to the intercept of the straight line on the ordinate in Fig. 8.9c, i.e., $k_{-4} = (3.9 \pm 0.6)\times10^{-3}\,\text{s}^{-1}$, which compares well with the true value of $4.0\times10^{-3}\,\text{s}^{-1}$ used to simulate the experiment.

Besides the dilution method aimed at dissociating the enzyme-inhibitor complex for determining the dissociation rate constant, alternative methods consist in capturing the inhibitor by some high-affinity agent. See for instance the application of antibodies against okadaic acid described on p. 414.

8.5 Slow-Onset, Linear Specific Inhibition in One Step

The assigned acronym is SO-LSpI-1 and the reactions are shown in Scheme 8.3, which correspond to linear specific inhibition (Sect. 5.2.1). With the exception of the slow equilibration of the E·I complex, there is no difference between LSpI and SO-LSpI-1. The progress curves can be analyzed by fitting (8.10) to data for determining v_z, v_s and λ from each curve. The dependence of these parameters on [I] is then exploited for calculating the rate constants for the formation and dissociation of the E·I complex and K_I using the equations collected in (8.16). A set of graphics with the same layout used for SO-LSpI-2 is shown in Fig. 8.10.

Mechanism SO-LSpI-1 is characterized by the independence of $v_{z,E}$ on [I]. The reason is that ES is already in place immediately after adding the enzyme to start the reaction, while E·I is built-up slowly and thus, at the very beginning of the reaction, the rate is that of the Michaelis–Menten mechanism. The dependence of $v_{z,S}$ (not shown) contains scarce quantitative information for the same reasons discussed above for mechanism SO-LSpI-2, i.e., in most cases $v_{z,S}$ represents > 90 % inhibition.

Progress curves measured at fixed [I] and variable [S] and the corresponding dependences of λ and $v_{z,E}$ on [S] (Fig. 8.10d–f) are useful for confirming internal consistency with the calculations from data collected at fixed [S] and variable [I]. Hence, the value of (k_3 [I] + k_{-3}) obtained as the ordinate intercept of the plot of λ against [S] can be compared with that calculated with the same constants obtained from the results in (a–c). In panel (b), fitting the Michaelis–Menten

Scheme 8.3 Slow-onset linear specific inhibition in one step

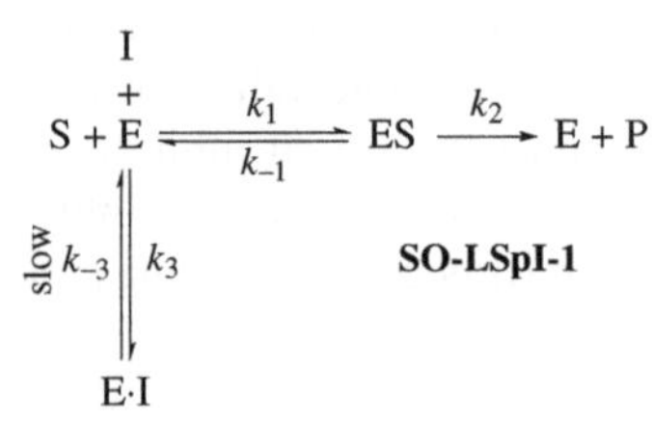

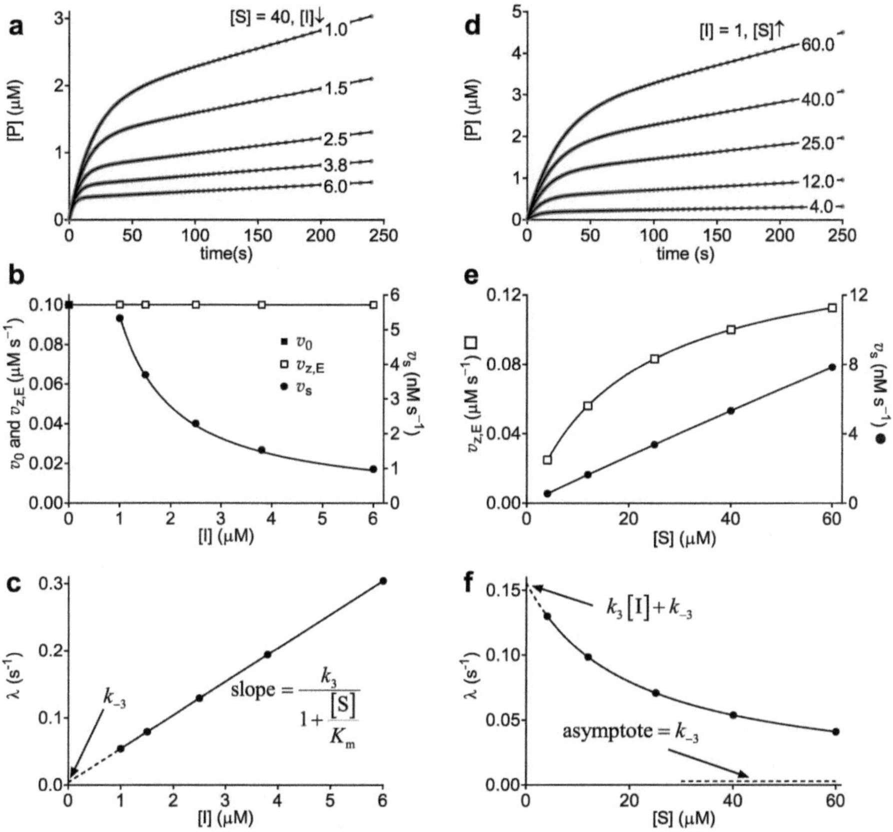

Fig. 8.10 Slow-onset inhibition mechanism SO-LSpI-1. The reactions are those in Scheme 8.3. The curves in (**a**) and (**d**) were generated using KinTek Explorer and a logarithmic timebase. The constants were: $k_1 = 100\,\mu\text{M}^{-1}\,\text{s}^{-1}$, $k_{-1} = 1{,}950\,\text{s}^{-1}$, $k_2 = 50\,\text{s}^{-1}$, $k_3 = 0.15\,\mu\text{M}^{-1}\,\text{s}^{-1}$, $k_{-3} = 0.003\,\text{s}^{-1}$, $[\text{E}]_\text{t} = 3\,\text{nM}$. Derived rates and parameters: $K_\text{m} = 20\,\mu\text{M}$, $v_0 = 0.10\,\mu\text{M}\,\text{s}^{-1}$, $V = 0.15\,\mu\text{M}\,\text{s}^{-1}$, $K_\text{i} = 20\,\text{nM}$. (**a–c**) Reactions started by adding enzyme at constant [S] = 40 μM and variable [I] (μM, numbers on the curves in panel (**a**). (**d–f**) Reactions started by adding enzyme at fixed $[\text{I}]_\text{t} = 1.0\,\mu\text{M}$ and [S] variable as shown by numbers in panel (**d**)

equation to $v_{z,\text{E}}$ against [I] yields $K_\text{m}^\text{app} = K_\text{m}^0 = 20\,\mu\text{M}$ and $V^\text{app} = V^0 = 0.15\,\mu\text{M}\,\text{s}^{-1}$ demonstrating that the inhibitor does not influence the rate at time zero for this mechanism.

$$\textbf{SO} - \textbf{LSpI} - \textbf{1} \tag{8.16}$$

$$\text{required}: \quad K_\mathrm{m} \gg K_\mathrm{I}\,, \quad k_{-1} \gg k_{-3}\,, \quad k_1\,[\mathrm{S}] \gg k_3\,[\mathrm{I}]$$

$$\lambda = k_{-3} + \frac{k_3\,[\mathrm{I}]}{1 + \dfrac{[\mathrm{S}]}{K_\mathrm{m}}}\,, \quad k_\mathrm{off} = k_{-3} = \lambda \frac{v_\mathrm{s}}{v_\mathrm{z,E}}\,, \quad v_\mathrm{s} = \frac{V\,[\mathrm{S}]}{K_\mathrm{m}\left(1 + \dfrac{[\mathrm{I}]}{K_\mathrm{I}}\right) + [\mathrm{S}]}$$

$$K_\mathrm{I} = \frac{k_{-3}}{k_3}\,, \quad v_\mathrm{z,E} = \frac{V\,[\mathrm{S}]}{K_\mathrm{m} + [\mathrm{S}]}\,, \quad v_\mathrm{z,S} = \frac{V\,[\mathrm{S}]}{K_\mathrm{m} + [\mathrm{S}]} \times \frac{1}{1 + x\dfrac{[\mathrm{I}]}{K_\mathrm{I}}}$$

The frequency constant λ depends linearly on [I] for SO-LSpI-1 and hyperbolically for SO-LSpI-2, which represents a good diagnostic tool for distinguishing the two mechanisms as shown in panels (c) of Figs. 8.8 and 8.10. Determining k_3 as slope of the plot of λ against [I] does not pose problems, while k_{-3} cannot always be reliably evaluated from the ordinate intercept unless the extrapolated intersection point can be statistically distinguished from zero. This depends on both the value of k_{-3} and experimental error. However, (8.14) and the dissociation method described in Sect. 8.4.3 provide reliable estimates of k_{-3}.

Shapiro and Riordan proposed a graphical method than can discriminate between mechanism SO-LSpI-1 and SO-LSpI-2 [30]. It consists in plotting $1/(\lambda - k_\mathrm{off})$ against $1/[\mathrm{I}]$. As can be verified from the expressions of λ in (8.15) and (8.16), the plot is a straight line that passes through the origin in mechanism SO-LSpI-1, while the line intersects the ordinate at a value corresponding to $1/k_4$ in mechanism SO-LSpI-2.

The distinction between SO-LSpI-2 and SO-LSpI-1 is based more on handiness than on mechanistic criteria. In fact, SO-LSpI-1 represents a degenerated form of SO-LSpI-2, in which the concentration of the adsorptive EI complex is kinetically insignificant. This fact can be appreciated by rearranging the expressions of λ in (8.15) and (8.16) as shown in (8.17) and (8.18) for SO-LSpI-2 and SO-LSpI-1, respectively. In mechanism SO-LSpI-2, if $K_\mathrm{I} \gg K_\mathrm{I}^*$ and $[\mathrm{I}] \approx K_\mathrm{I}^*$, the term in the denominator of the right form in (8.17) approaches 1 and the expression is formally reduced to the right form of (8.18).

$$\lambda = k_{-4} + \frac{k_4\,[\mathrm{I}]}{K_\mathrm{I}\left(1 + \dfrac{[\mathrm{S}]}{K_\mathrm{m}}\right) + [\mathrm{I}]} \quad \Longrightarrow \quad \lambda = k_{-4}\left\{\frac{1 + \dfrac{[\mathrm{I}]}{K_\mathrm{I}^*\left(1 + \dfrac{[\mathrm{S}]}{K_\mathrm{m}}\right)}}{1 + \dfrac{[\mathrm{I}]}{K_\mathrm{I}\left(1 + \dfrac{[\mathrm{S}]}{K_\mathrm{m}}\right)}}\right\} \tag{8.17}$$

$$\lambda = k_{-3} + \frac{k_3\,[\mathrm{I}]}{1 + \dfrac{[\mathrm{S}]}{K_\mathrm{m}}} \quad \Longrightarrow \quad \lambda = k_{-3}\left\{1 + \frac{[\mathrm{I}]}{K_\mathrm{I}\left(1 + \dfrac{[\mathrm{S}]}{K_\mathrm{m}}\right)}\right\} \tag{8.18}$$

While the criteria described in this section are sufficient to distinguish mechanisms SO-LSpI-2 and SO-LSpI-1 from one another, the problem remains of differentiating them from other mechanisms having similar patterns of the dependences of rates and frequency constants on [I] and [S]. This issue in data analysis and model discrimination is the aim of the dichotomous keys in Chap. 10.

8.6 Slow-Onset, Temporary Inhibition by Competing Substrates

The denomination *temporary inhibition* has been introduced by Laskowski and Wu [15], who observed that inhibition of trypsin by the Kazal pancreatic inhibitor was slowly reversed following proteolytic processing of the modifier. This mechanism is characteristic of several protein inhibitors of peptidases, or mutants thereof, which possess protein domains with inhibitor-like properties, e.g., [16, 17, 32], but has also been reported for synthetic, low-molecular mass inhibitors [10]. The dual behavior of proteins as inhibitors or substrates can be attributed to different binding modes to their target enzymes, to the enzyme/inhibitor molar ratio, and/or to pH and ionic strength variations.

The irreversible segments that lead to I* in both mechanisms of Scheme 8.4 are simplified forms of reactions that may contain more, nonmeasurable intermediates. Owing to the slow-onset inhibitory character often observed in the presence of *temporary inhibitors*, they are placed in this chapter because the mathematical formalism is the same as for mechanisms SO-LSpI-2 and SO-LSpI-1. There-

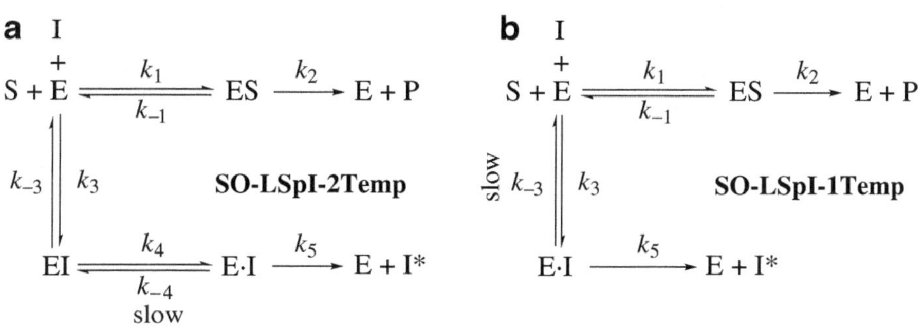

Scheme 8.4 Slow-onset, temporary inhibition by competing substrates. (**a**) Two-step, (**b**) one-step mechanism

fore, slow-onset competing substrates will be formally called *inhibitors* with the acronyms SO-LSpI-2Temp and SO-LSpI-1Temp. The difference from their genuine inhibitory counterparts SO-LSpI-2 and SO-LSpI-1 is the release of a modified form of the competing substrate to a species I* (or more than one such molecules derived from I). I* is assumed to be inert toward the enzyme, i.e., it does neither give rise to product inhibition nor bind unproductively to the enzyme, which is recycled into the system in its native state.

$$\textbf{SO}-\textbf{LSpI}-\textbf{2Temp} \tag{8.19}$$

$$\text{assumption}: \quad \left[\mathrm{I}^*\right] \ll [\mathrm{I}]_t$$

$$\text{required}: \quad K_m \gg K_I\,, \quad k_{-1} \gg k_4\,, \quad k_{-1} \gg k_{-4}$$

$$\lambda = k_{-4} + k_5 + \frac{k_4\,[\mathrm{I}]}{K_I\left(1 + \dfrac{[\mathrm{S}]}{K_m}\right) + [\mathrm{I}]}\,, \qquad v_s = \frac{V\,[\mathrm{S}]}{K_m\left(1 + \dfrac{[\mathrm{I}]}{K^*_{I,\,temp}}\right) + [\mathrm{S}]}$$

$$K^*_{I,\,temp} = K_{m,\,I} = K_I\left(\frac{k_{-4} + k_5}{k_4 + k_{-4} + k_5}\right)\,, \qquad v_{z,E} = \frac{V\,[\mathrm{S}]}{K_m\left(1 + \dfrac{[\mathrm{I}]}{K_I}\right) + [\mathrm{S}]}$$

$$K_I = \frac{k_{-3}}{k_3}\,, \qquad k_{cat,\,I} = \frac{k_4 k_5}{k_4 + k_{-4} + k_5}\,, \qquad \frac{k_{cat,\,I}}{K_{m,\,I}} = \frac{k_3 k_4 k_5}{k_{-3}k_{-4} + k_{-3}k_5}$$

The derivation of the rate equation and associated parameters of SO-LSpI-2Temp have been placed in Appendix 2 for practical reasons, while the rate equation with its parameters of the related mechanism SO-LSpI-1Temp can be obtained in an analogous way. All relevant expressions are collected in the groups of equations (8.19) and (8.20). For integrating the rate equation of the two temporary mechanisms, additional assumptions are necessary besides those mentioned on p. 374, i.e., I* is considered to be inert toward the enzyme and also the condition $[\mathrm{I}^*] \ll [\mathrm{I}]_t$ over the entire measurement time must be satisfied. Unlike reactions related to temporary inhibition by competing substrates that may take place in living organisms (open systems), measurements in vitro, i.e., in closed systems, imply that $[\mathrm{I}]_t$ acts as a sink by delivering fresh molecules of the competing substrate to compensate the withdrawal of I*.

The inhibition constants measured for competing substrates correspond to their Michaelis constants, i.e., $K^*_{I,\,temp}$ and $K_{I,\,temp}$ can also be called $K_{m,\,I}$. Typical of slow-onset inhibition by competing substrates are small catalytic constants, denoted $k_{cat,\,I}$ in the equation groups (8.19) and (8.20). In turn, the values of $k_{cat,\,I}/K_{m,\,I}$ for competing substrates of this type are also small, which explain efficient inhibition and slow turnover. Despite representing simplified models, mechanisms SO-LSpI-

2Temp and SO-LSpI-1Temp are sufficient to describe the kinetics of temporary inhibition, particularly when a limited number of data is available for analysis.

Progress curves and other characteristics of mechanisms SO-LSpI-2Temp and SO-LSpI-1Temp are shown in Figs. 8.11 and 8.12, respectively. Comparing the equation groups (8.19) with (8.15) and (8.20) with (8.16) reveals that the temporary mechanisms cannot be distinguished from their non-temporary counterparts on the basis of the dependence of λ, v_s and $v_{z,E}$ on $[I]_t$ (Figs. 8.11 and 8.12). The net off-constants that can be determined are $(k_{-4} + k_5)$ for SO-LSpI-2Temp and $(k_{-3} + k_5)$ for SO-LSpI-1Temp, not the individual constants. As long as sufficient *inhibitor* is available to satisfy the condition $[I]^* \ll [I]_t$ during the measuring time, the progress curves for mechanisms SO-LSpI-2Temp and SO-LSpI-1Temp as shown in panels (a) do not suggest inhibitor turnover. Thus, the temporary character of slow-onset inhibition elicited by competing substrates may go unnoticed unless measurements are performed for a *sufficiently long time* to allow sizable turnover of I into I* and to reveal any anomalous behavior.

$$\textbf{SO} - \textbf{LSpI} - \textbf{1Temp} \qquad (8.20)$$

$$\text{assumption}: \quad [I^*] \ll [I]_t$$

$$\text{required}: \quad K_m \gg K_I, \quad k_{-1} \gg k_3, \quad k_{-1} \gg k_{-3}$$

$$\lambda = k_{-3} + k_5 + \frac{k_3\,[I]}{1 + \dfrac{[S]}{K_m}}, \quad v_s = \frac{V\,[S]}{K_m\left(1 + \dfrac{[I]}{K_{I,\text{temp}}}\right) + [S]}$$

$$K_{I,\text{temp}} = K_{m,I} = \frac{k_{-3} + k_5}{k_3}, \quad v_{z,E} = \frac{V\,[S]}{K_m + [S]}$$

$$k_{cat,I} = k_5, \quad \frac{k_{cat,I}}{K_{m,I}} = \frac{k_3 k_5}{k_{-3} + k_5}$$

The turnover of a temporary inhibitor is best appreciated by preincubating the enzyme with the alleged competing substrate for a sufficiently long time to allow partial or complete turnover of I to I* and starting reactions by adding the regular substrate. This strategy is shown in panels (g) of Figs. 8.11 and 8.12, where the chosen preincubation time is sufficient to completely transform I to I* in both cases so that the corresponding progress curves have an initial slope corresponding to v_0, the rate in the absence of the competing substrate. How fast the turnover of I occurs is established by the values of $k_{cat,I}$ in (8.19) and (8.20). Relatively low concentrations of the competing substrate speed up this test but long incubation times may be needed to appreciate its turnover. Should the enzyme be unstable during the preincubation time, the slope would decrease, not increase. To exclude

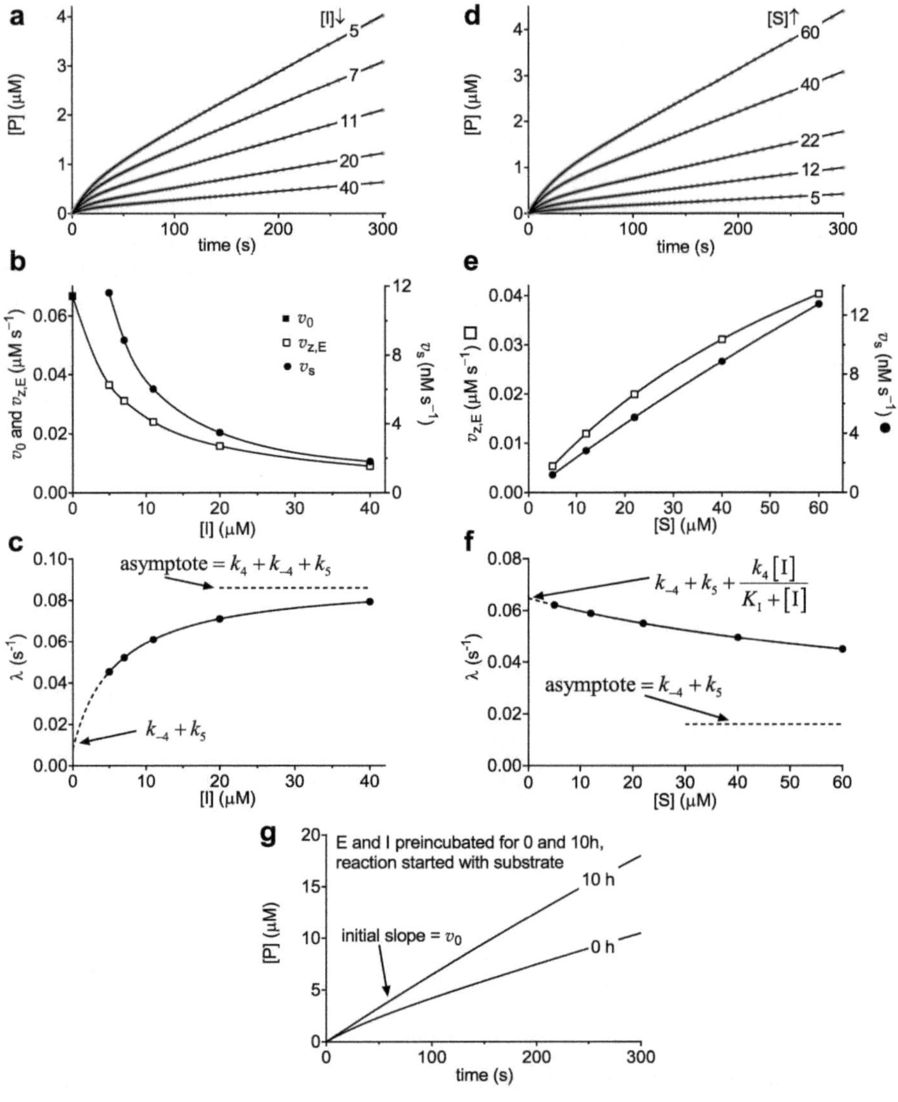

Fig. 8.11 Slow-onset, temporary inhibition mechanism SO-LSpI-2Temp. The model is based on the reactions in Scheme 8.4a. The constants and concentrations used to simulate the progress curves were: $k_1 = 100\,\mu\mathrm{M}^{-1}\,\mathrm{s}^{-1}$, $k_{-1} = 1{,}990\,\mathrm{s}^{-1}$, $k_2 = 10\,\mathrm{s}^{-1}$, $k_3 = 3\,\mu\mathrm{M}^{-1}\,\mathrm{s}^{-1}$, $k_{-3} = 6\,\mathrm{s}^{-1}$, $k_4 = 0.07\,\mathrm{s}^{-1}$, $k_{-4} = 0.004\,\mathrm{s}^{-1}$, $k_5 = 0.012\,\mathrm{s}^{-1}$, $[\mathrm{E}]_\mathrm{t} = 10\,\mathrm{nM}$. Derived rates and parameters: $K_\mathrm{m}^0 = 20\,\mu\mathrm{M}$, $v_0 = 0.067\,\mu\mathrm{M\,s}^{-1}$, $V^0 = 0.1\,\mu\mathrm{M\,s}^{-1}$, $K_\mathrm{I} = 2\,\mu\mathrm{M}$, $K^*_{\mathrm{I,temp}} = 0.372\,\mu\mathrm{M}$. (**a–c**) Reactions started by adding enzyme at constant [S] = 40 μM and variable [I] (μM, numbers on the curves in panel (**a**)). (**d–f**) Reactions started by adding enzyme at fixed $[\mathrm{I}]_\mathrm{t}$ = 7.0 μM and [S] variable as shown by numbers in panel (**e**) as μM. (**g**) Enzyme mixed with inhibitor and immediately assayed by adding substrate (0 h), or preincubated for 10 h with 1.0 μM inhibitor and reaction started by adding substrate at 40 μM final concentration

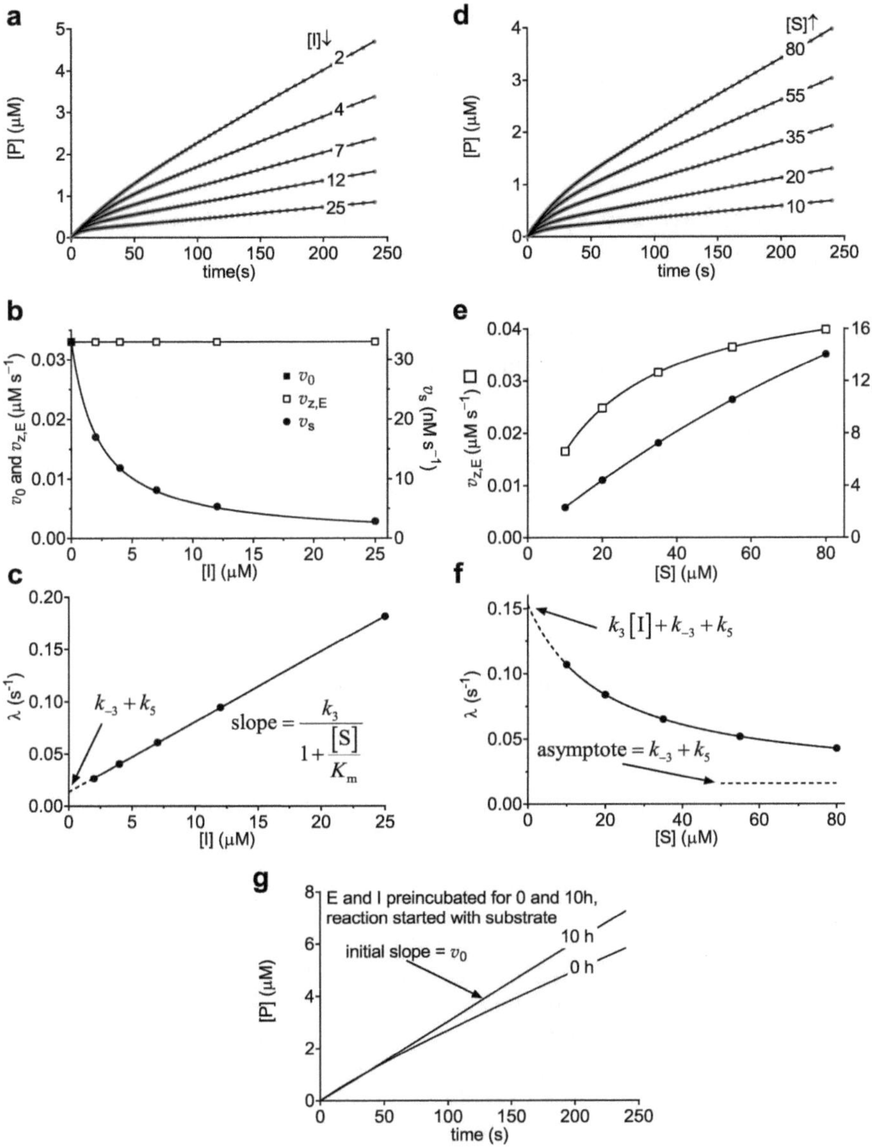

Fig. 8.12 Slow-onset, temporary inhibition mechanism SO-LSpI-1Temp. The model is based on the reactions in Scheme 8.4b. The constants and concentrations used to construct the plots were: $k_1 = 100\,\mu\text{M}^{-1}\,\text{s}^{-1}$, $k_{-1} = 1{,}990\,\text{s}^{-1}$, $k_2 = 10\,\text{s}^{-1}$, $k_3 = 0.02\,\mu\text{M}^{-1}\,\text{s}^{-1}$, $k_{-3} = 0.0007\,\text{s}^{-1}$, $k_5 = 0.015\,\text{s}^{-1}$, $[\text{E}]_t = 5\,\text{nM}$. Derived rates and parameters: $K_m^0 = 20\,\mu\text{M}$, $v_0 = 0.033\,\mu\text{M}\,\text{s}^{-1}$, $V^0 = 0.05\,\mu\text{M}\,\text{s}^{-1}$, $K_{i,temp} = 0.785\,\mu\text{M}$. (**a–c**) Reactions started by adding enzyme at constant [S] = 40 μM and variable [I] (μM, numbers on the curves in panel (**a**)). (**d–f**) Reactions started by adding enzyme at fixed $[\text{I}]_t$ = 7.0 μM and [S] variable as shown by numbers in (**e**) as μM. (**g**) Enzyme mixed with inhibitor and immediately assayed by adding substrate (0 h), or preincubated for 10 h with 1.0 μM inhibitor and reaction started by adding substrate at 40 μM final concentration

the possibility that an unstable competing substrate, preincubated with a stable enzyme, is responsible for apparent temporary inhibition, some analytical method able to reveal chemical changes in the molecule of the competing substrate must be invoked as additional help.

8.7 A Rare Enzyme Species is Responsible for Slow-Onset Inhibition

For a given second-order rate constant, a reaction proceeds at the rate dictated by the concentrations of the involved molecules. The mechanism in Scheme 8.5 shows an enzyme that isomerizes *rapidly* between two forms, of which only E′ binds reversibly a ligand to form the E′·I complex. With these premises, the condition for observing slow-onset inhibition is that the equilibrium between E and E′ lies distinctly in favor of E so that E′ becomes a *rare* species. SO-LSpI-RareE has been chosen as acronym to underline this fact by recalling the origin of the sluggish inhibition process: a very low concentration of E′ that causes the second-order reaction E′ + I to proceed slowly.

$$\textbf{SO}-\textbf{LSpI}-\textbf{RareE} \tag{8.21}$$

$$\text{required}: \quad K_\text{m} \gg K_\text{I}\,, \qquad k_{-1} \gg k_{-7} \gg k_7$$

$$\lambda = k_{-3}\frac{1+\dfrac{[\text{S}]}{K_\text{m}}+\dfrac{[\text{I}]}{K_\text{I}}}{1+\dfrac{[\text{S}]}{K_\text{m}}}\,, \qquad k_\text{off} = \lambda\frac{v_\text{s}}{v_{\text{z,E}}} = k_{-3}$$

$$K_\text{m} = \frac{(k_{-1}+k_2)\,(k_7+k_{-7})}{k_1 k_{-7}}\,, \qquad K_\text{I} = \frac{k_{-3}\,(k_7+k_{-7})}{k_3 k_7}$$

$$v_\text{s} = \frac{V\,[\text{S}]}{K_\text{m}\left(1+\dfrac{[\text{I}]}{K_\text{I}}\right)+[\text{S}]}\,, \qquad v_{\text{z,E}} = \frac{V\,[\text{S}]}{K_\text{m}+[\text{S}]}$$

Equation (8.21) collects the characteristic expressions of this mechanism, which has been considered by Cha in the frame of *tight-binding* inhibition [4] but the tight-binding condition is not a prerequisite for slow-onset inhibition. The case of linear specific inhibition treated in this section corresponds to the simplest model of *heterotropic* interactions mentioned by Monod, Wyman, and Changeux for a monomeric enzyme that exists as two *tautomeric* forms in equilibrium [18, p. 115]. With reference to Scheme 8.5, the tautomeric form E is stabilized by the substrate, while E′ is stabilized by *some other specific ligand* [18] that acts therefore as

Scheme 8.5 Slow-onset inhibition due to enzyme isomerization. The equilibrium is shifted in favor of E and E′ is a rare species

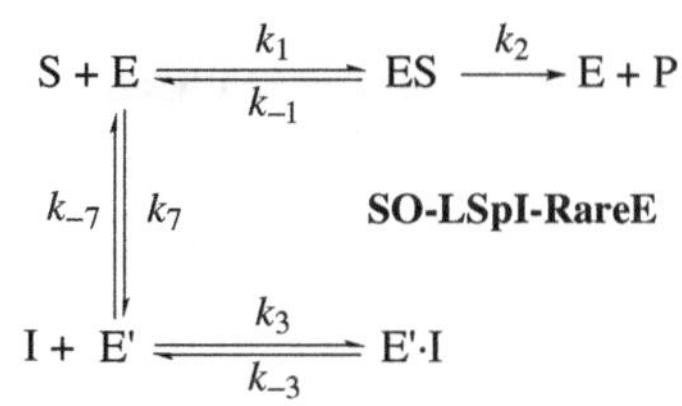

inhibitor. The necessary condition is that the enzyme possesses two different binding sites for the substrate and *the other ligand.*

The requirements for the existence of SO-LSpI-RareE are listed in the first row of (8.21). If the enzyme can catalyze the turnover of different substrates, slow-onset inhibition may be observed with one substrate but not necessarily with another substrate that does not satisfy the condition $k_{-1} \gg k_{-7} \gg k_7$. This restriction enhances the specificity of regulation using a very simple mechanism: a tiny population of enzyme molecules, E′ in Scheme 8.5, is appointed to stand sentinel over unnecessary enzyme activity by damping substrate turnover in a very efficient way, for instance in feedback inhibition by any accumulated reaction product.

In the example illustrated in Figs. 8.13 and 8.14, E′ makes up only 0.5 % of $[E]_t$ in the incubation period, and 0.02 % of $[E]_t$ during the steady-state when [I] = 18 μM. At the same inhibitor concentration, 40 % of the available enzyme is withdrawn from the catalytic path as E′·I showing that the function of E′ is to act as mediator in damping enzyme activity when the concentration of the ligand I increases while returning full activity when $[I] \rightarrow 0$.

The analysis of SO-LSpI-RareE can be performed using the same methods discussed in the preceding sections: primary data are collected by mixing enzyme with a solution containing substrate and inhibitor as shown in Fig. 8.13, followed by fitting (8.10) to data for extracting v_s, $v_{z,E}$ and λ. As predicted from the expressions in (8.21), λ depends linearly on inhibitor concentration, while $v_{z,E}$ does not depend on [I]. That the rate at time zero is insensitive to the concentration of inhibitor can be deduced intuitively after recalling that [E′] is very small. Thus, the steady-state of product release and the equilibrium between E and E′ are already in place before any appreciable amount of E′·I can be detected.

Figure 8.14 illustrates the changes with time of the concentration of the four enzyme-containing species. The phase that precedes mixing of the two reagent solutions has essentially the function of temperature equilibration. It is in fact conceivable that the enzyme, diluted from a stock solution at high concentration in a suitable buffer, reaches rapidly equilibration between E and E′. However, such a rapid equilibration may not be a general property, e.g., with a hysteretic enzyme stored in an inactive state at a given pH and diluted at another pH that turns the enzyme slowly or very slowly to his active state [2]. In the plots shown in Fig. 8.14a,b, E and E′ are immediately in place, their concentrations are halved after mixing (time zero) and, depending on the concentrations of substrate and inhibitor decrease exponentially to lower values until reaching the steady-state (curves running parallel to the time axis). ES is set up immediately after mixing

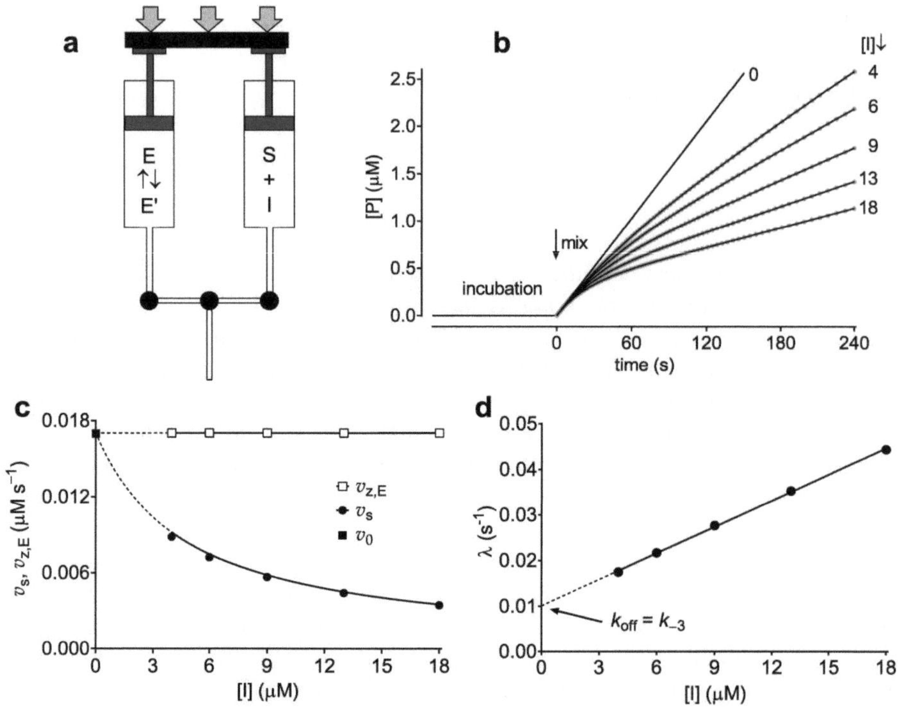

Fig. 8.13 Slow-onset, linear specific inhibition due to a rare enzyme species, SO-LSpI-RareE. The model is based on the reactions in Scheme 8.5. (**a**) Sketch of the experiment using a stopped-flow apparatus. The constants used for simulating the progress curves were: k_1 = 100 μM^{-1} s^{-1}, k_{-1} = 3,000 s^{-1}, k_2 = 10 s^{-1}, k_3 = 1 μM^{-1} s^{-1}, k_{-3} = 0.01 s^{-1}, k_7 = 1 s^{-1}, k_{-7} = 200 s^{-1}. Derived rates and parameters: $v_{z,E} = v_0$ = 0.017 μM s^{-1}, V = 0.03 μM s^{-1}, K_I = 0.01 μM, K_I^* = 2.01 μM. The reagents, left to reach the desired temperature, were rapidly mixed in equal volumes and their concentrations after mixing were: $[S]_t$ = 40 μM, $[I]_t$ (μM) as indicated and $[E]_t$ = 3 nM. (**b**) Progress curves simulated on a logarithmic timebase. Equation (8.10) was fitted to data for determining the dependence on [I] of the rates at time zero and at steady-state (**c**) and of the frequency constant (**d**). *Solid lines* and their extrapolated courses represent best fits using the appropriate equations in (8.21)

and its concentration also decays with time to extents that depend on inhibitor concentration. The most populated enzyme-containing species at steady-state is E′·I, which accounts for slow but efficient inhibition.

8.8 Slow-Onset Inhibition Due to Slow Enzyme Isomerization

If an enzyme isomerizes *slowly* between two forms E and *E, of which only *E binds the inhibitor and only E combines with the substrate, slow-onset inhibition SO-LSpI-IsoE *may be* observed. The emphasized *may be* stresses the fact that

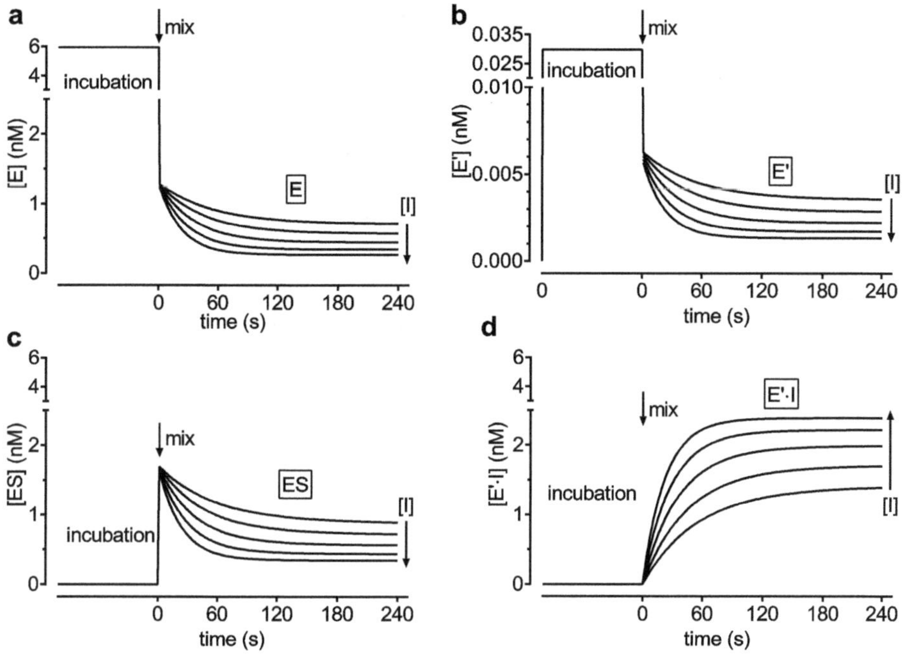

Fig. 8.14 Slow-onset, linear specific inhibition due to a rare enzyme species, SO-LSpI-RareE. The model is based on the reactions in Scheme 8.5, and the curves were constructed with the constants listed in Fig. 8.13. (**a–d**) Variation with time of the concentrations of the enzyme-containing species. The increasing direction of [I] is indicate by *arrows*. Note the different scale of the ordinate in panel (**b**)

progress curves may not show a burst or lag at the beginning of the reaction. The mechanism, with acronym SO-LSpI-IsoE and shown in Scheme 8.6a has its roots in linear specific inhibition because *E·I is a dead-end complex. When enzyme is added to substrate and inhibitor, two reactions start: between E and S and between *E and I, with the assumption of fast binding between *E and I. While the reaction proceeds, the equilibrium between E and *E is adjusted according to the rate constants and to the concentrations of substrate and inhibitor. Notwithstanding a favorable second-order rate constant (k_3), the enzyme necessary to build the *E·I complex is delivered slowly to the inhibitory path.

SO-LSpI-IsoE is a slow-onset inhibition mechanism with variable, complex behavior that has been mentioned only marginally in the literature [11, 20] and might have been overlooked or confused with other systems. Therefore, it is important to learn its properties and the methods of differentiation toward systems with a phenomenologically similar behavior. In SO-LSpI-IsoE the enzyme responds slowly to a sudden change in the concentration of a ligand, thus matching the definition of hysteretic enzyme [12]. Progress curves show a lag phase preceding the steady-state also in the absence of inhibitors and therefore, before describing the properties of the inhibition mechanism, we examine first substrate binding and turnover in the absence of modifier.

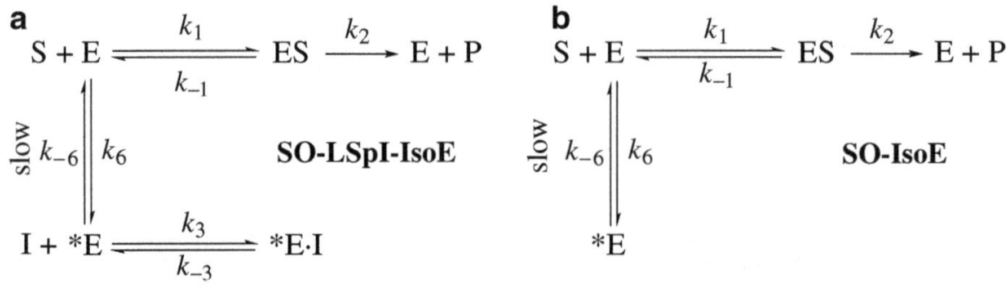

Scheme 8.6 Slow-onset inhibition due to enzyme isomerization, SO-LSpI-IsoE, (**a**). The reaction scheme for slow enzyme isomerization without inhibitor, SO-IsoE, is shown in (**b**)

8.8.1 Enzyme Isomerization in the Absence of Inhibitor

The kinetic mechanism to consider, named SO-IsoE in Scheme 8.6b, is the same as SO-LSpI-IsoE omitting the inhibitor and the characteristic expressions of rates and frequency constant are given in the equation group (8.22).

SO − IsoE (8.22)

$$\lambda = k_{-6} + \frac{k_6}{1 + \dfrac{[\mathrm{S}]}{K'_\mathrm{m}}}, \quad k_\mathrm{off} = \lambda \frac{v_\mathrm{s}}{v_\mathrm{z}} = k_6 + k_{-6}, \quad K_6 = \frac{k_6}{k_{-6}}$$

$$v_\mathrm{s} = \frac{V[\mathrm{S}]}{K_\mathrm{m} + [\mathrm{S}]}, \quad v_\mathrm{z} = \frac{\dfrac{V}{1 + K_6}[\mathrm{S}]}{K'_\mathrm{m} + [\mathrm{S}]}, \quad V = k_2[\mathrm{E}]_\mathrm{t}$$

$$K'_\mathrm{m} = \frac{k_{-1} + k_2}{k_1}, \quad K_\mathrm{m} = \frac{(k_{-1} + k_2)(k_6 + k_{-6})}{k_1 k_{-6}} = K'_\mathrm{m}(1 + K_6)$$

The shape of the progress curves depends very much on the ratio of the isomerization constants k_6 and k_{-6}. In particular, the initial lag may or may not be sufficiently pronounced to be experimentally observable in dependence of substrate concentration. With $K_6 > 1$ the species *E predominates at equilibrium, and the contrary is valid for $K_6 < 1$. In the first case the reaction starts at a low rate, which then increases slowly and exponentially until reaching the steady-state predicted by (8.10). However, with $K_6 \leq 1$, there is enough enzyme in the E-form present at time zero that the values of v_z and v_s are similar, so that the progress curves are nearly linear from the very beginning (Fig. 8.15). This is a handicap when (8.10) is used to globally fit a set of progress curves because those curves that are close to a straight line afford nonsensical values of the frequency constant. v_s can however be estimated from the slopes.

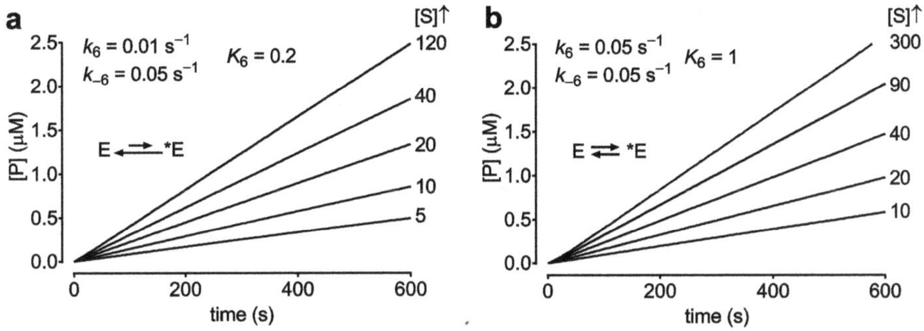

Fig. 8.15 Slow isomerization between two enzyme forms, SO-IsoE, according to Scheme 8.6b. Two examples with different isomerization equilibrium constants between E and *E. Constants common to both plots: $k_1 = 100\,\mu\text{M}^{-1}\,\text{s}^{-1}$, $k_{-1} = 1{,}990\,\text{s}^{-1}$, $k_2 = 10\,\text{s}^{-1}$, $[\text{E}]_t = 0.5\,\text{nM}$, k_6, k_{-6} and [S] as indicated. The *arrows* in (**a**) indicate that the equilibrium is shifted toward E, while in (**b**) E and *E have the same concentration at equilibrium

A further example, in which $K_6 > 1$, is shown in Fig. 8.16. In order to appreciate the effect of time resolution in calculating the frequency constant, simulations of progress curves were purposely performed using a linear and a logarithmic timebase. Panels (b) and (c) show the first 50 s of the curves in (a), generated with a linear and a logarithmic timebase, respectively, illustrating that the logarithmic timebase contains much more data than the linear one at the beginning of the reaction.

The isomerization rate constants k_6 and k_{-6} can be determined by nonlinear regression using the expression of λ in (8.22), i.e., from its dependence on [S] as illustrated in Fig. 8.16d. As seen in panel (d), the triangles (representing data from the linear timebase) deviate more than the squares (logarithmic timebase) from the theoretical curve, drawn as solid line, at low substrate concentrations. At higher substrate concentrations the adherence to the model is perfect. This simulated, error-free example stresses the importance of careful data collection in this type of experiments, which is not a problem using modern rapid-mixing devices. Even small deviations such as those of the triangles in panel (d) produce errors of more than 100 % in determining k_6.

The success in evaluating the asymptote, which corresponds to k_{-6}, depends on the range of substrate concentrations that can be used in measuring the progress curves. The relationship (8.14) can be invoked to calculate the off-rate that corresponds in this case to $k_6 + k_{-6}$. Since the value of $\lambda v_s/v_z$ does not depend on substrate concentration, $k_{\text{off}} = k_6 + k_{-6}$ can be calculated preferentially from v_s, v_z and λ obtained by nonlinear regression or progress curves at high substrate concentration. Using the data from $[\text{S}] = 300\,\mu\text{M}$ in this example, $k_{\text{off}} = 0.06$, identical with the theoretical value. Measuring k_6 and k_{-6} as accurately as possible is useful for data analysis in the presence of an inhibitor (see below).

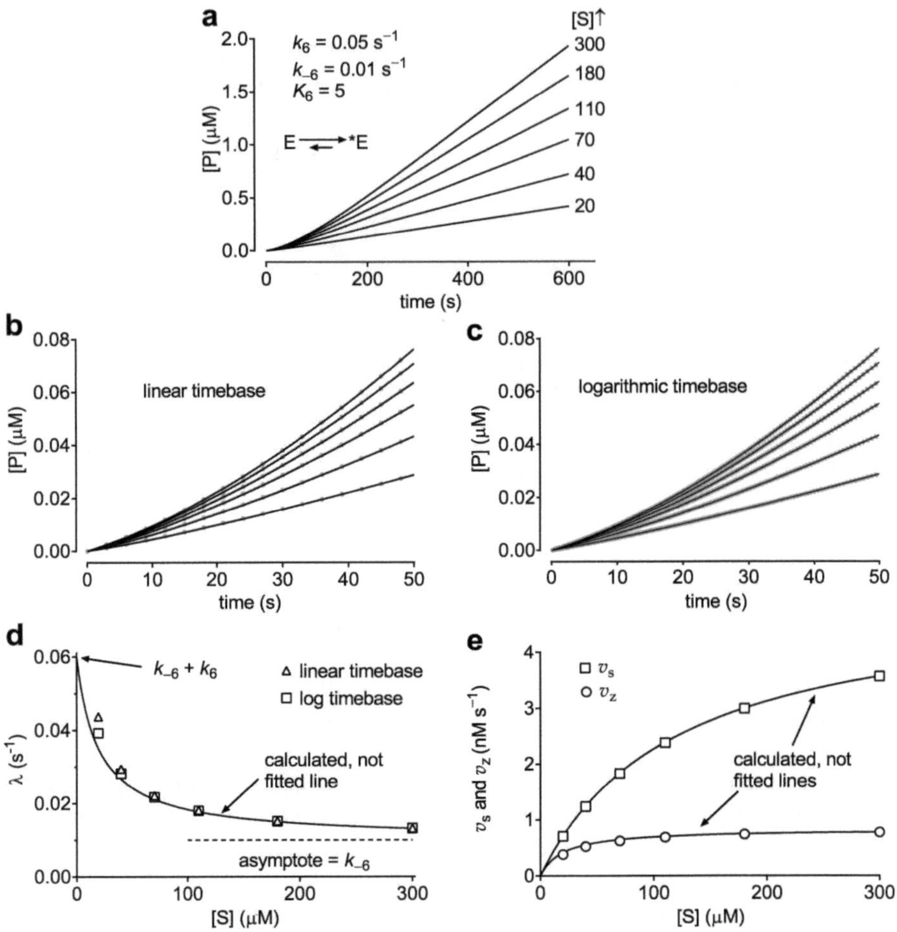

Fig. 8.16 Slow isomerization between two enzyme forms, SO-IsoE, according to Scheme 8.6b. Example in which the equilibrium is in favor of *E. The progress curves were simulated with either logarithmic or linear timebase and the following constants and concentrations: $k_1 = 100\ \mu M^{-1}\ s^{-1}$, $k_{-1} = 1{,}990\ s^{-1}$, $k_2 = 10\ s^{-1}$, $k_6 = 0.05\ s^{-1}$, $k_{-6} = 0.01\ s^{-1}$, $[E]_t = 0.5$ nM and [S] as indicated in panel (**a**). (**b**) and (**c**) Show the first 50 s of the progress curves in (**a**) generated with 200 points on a linear timebase and with 600 points on a logarithmic timebase, respectively. (**d**) Dependence of the frequency constant on substrate concentration: the *symbols* represent data from fitting (8.10) to progress curves on linear or logarithmic timebase, while the *solid line* was plotted with the known constants and the expression of λ in (8.22). (**e**) Dependence of the rates v_s and v_z on [S]: the points are those from linear regression of progress curves, while the *solid lines* were plotted with the corresponding expressions in (8.10)

Regression analysis of progress curves for SO-IsoE yields useful values of v_s, while those of v_z may be too small for practical use. The overlap of v_s data in Fig. 8.16e with the theoretical curve allows to determine the Michaelis constant that is made up by all rate constants of the mechanism as shown in the last row of (8.22). The other constant, K'_m, must be increased in proportion to the equilibrium constant of the isomerization between E and *E, K_6, to take into account the fraction of *E that does not participate to catalysis.

The time that elapses between the start of the reaction and the attainment of steady-state may be too long to ignore depletion of substrate. In simulating the data in Fig. 8.16, this problem was circumvented by decreasing the total enzyme concentration but still to a level that would allow measuring reaction progress with a sensitive method such as fluorometry.

8.8.2 Enzyme Isomerization in the Presence of Inhibitor

The equations for mechanism SO-LSpI-IsoE, whose reactions are shown in Scheme 8.6a, are collected in the group (8.23). The properties of enzyme isomerization described above are maintained when an inhibitor is added to the scenery and are responsible for slow-onset inhibition.

In the presence of inhibitor, the progress curves can exhibit a lag, a burst or be straight lines just by varying the inhibitor concentration at fixed substrate. Figure 8.17 illustrates this fact with progress curves simulated with three values of K_6. It is evident that fitting (8.10) to an ensemble of data such as those in panel (b) is problematic because there is no definition of the frequency constant for linear progress curves.

SO – LSpI – IsoE (8.23)

$$\lambda = \frac{k_6}{1+\dfrac{[\mathrm{S}]}{K_\mathrm{m}}} + \frac{k_{-6}}{1+\dfrac{[\mathrm{I}]}{K_\mathrm{I}}}, \quad k_\mathrm{off} = \lambda\frac{v_\mathrm{s}}{v_\mathrm{z}} = \frac{k_6+k_{-6}}{1+\dfrac{[\mathrm{I}]}{K_\mathrm{I}}}, \quad K_\mathrm{I} = \frac{k_{-3}}{k_3}$$

$$v_\mathrm{s} = \frac{V[\mathrm{S}]}{K_\mathrm{m}^\mathrm{app}+[\mathrm{S}]}, \quad v_\mathrm{z,E} = \frac{\dfrac{V}{1+K_6}[\mathrm{S}]}{K'_\mathrm{m}+S}, \quad V = k_2[\mathrm{E}]_\mathrm{t}, \quad K_6 = \frac{k_6}{k_{-6}}$$

$$K'_\mathrm{m} = \frac{k_{-1}+k_2}{k_1}, \quad K_\mathrm{m}^\mathrm{app} = K'_\mathrm{m}\left[1+K_6\left(1+\frac{[\mathrm{I}]}{K_\mathrm{I}}\right)\right]$$

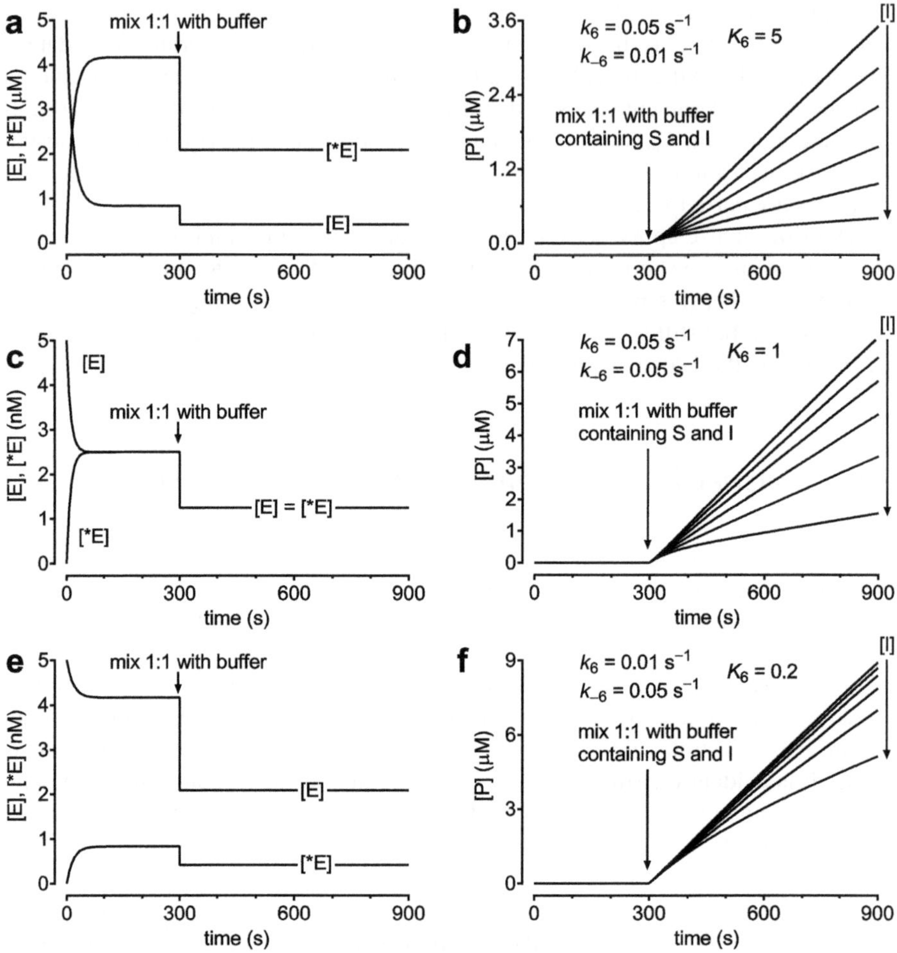

Fig. 8.17 Slow-onset inhibition due to slow enzyme isomerization, SO-LSpI-IsoE. Changing shapes of the progress curves depending on the equilibrium constant for enzyme isomerization. The plots were generated by numerical integration using the following constants and concentrations common to all plots: k_1 = 100 μM^{-1} s^{-1}, k_{-1} = 1,990 s^{-1}, k_2 = 10 s^{-1}, k_3 = 50 μM^{-1} s^{-1}, k_{-3} = 500 s^{-1}. k_6 and k_{-6} varied for the couples of plots (**a, b**), (**c, d**) and (**e, f**) as shown in panels (**b, d, f**). In panels (**a, c, e**) the enzyme was left to equilibrate before rapidly adding an equal volume of buffer solution: dilution 1:1 halves the concentrations of E and *E without affecting their equilibrium. In panels (**b, d, f**) the enzyme was equilibrated for 300 s before adding and equal volume of buffer containing [S] = 40 μM and [I] = 0, 4, 10, 22, 50 or 200 μM (final concentrations)

Two more problems are associated with the analysis of this system: (1) the lag or burst may be too short for producing statistically meaningful fittings; (2) in order to record a sufficiently long steady-state phase, the reactions may proceed too far for the approximation $[S] \approx [S]_t$ to be valid. This is certainly the case of the progress curves in Fig. 8.17d,f and to a certain extent also of the curves in panel (b). All three parameters of (8.10) will be biased for these reasons suggesting that nonlinear regression fitting is not the appropriate method and that numerical integration should be chosen in its place. Refer to Sect. 8.2 for a description of data fitting by numerical integration. In addition to progress curves obtained by starting the reaction with enzyme at constant [S] and variable [I], curves obtained at constant [I] and variable [S] may be included in the analysis, as well as progress curves started by adding substrate to preincubates enzyme and inhibitor.

The panels on the left of Fig. 8.17 are matched with the panels on their right and illustrate what happens when only enzyme is preincubated in one syringe of a stopped-flow apparatus and rapidly mixed with an equal volume of buffer. The preincubation phase of 300 s is an imaginary situation because the enzyme is already in equilibrium in solution: it shows the time that would be required in order for a 5 nM concentration of total enzyme in the E-form to equilibrate between E and *E. Dilution 1:1 by mixing with buffer alone halves the concentration of the two isomers that remain constant with time. At the time of mixing with a solution that contains substrate and inhibitor (panels b, d, and f) the concentrations of the enzyme forms experienced by these reagents are those of the long straight lines in panels (a, c, and e).

8.9 Rare Inhibitor Species Mocking Slow-Onset Inhibition

If an inhibitor molecule is present in solution as two or more species in equilibrium, slow-onset inhibition is observed if only one of the isomers, representing a minor fraction of the total inhibitor concentration, can bind and inhibit the enzyme (mechanism SO-LSpI-RareI, Scheme 8.7). This mechanism is similar to SO-LSpI-RareE with the difference that the rare species is now the inhibitor, I_r. Although k_3, the second-order rate constant for E·I_r formation, may be large, the low concentration of I_r is responsible for a slow second-order association reaction.

Comparison of the kinetic equations of mechanism SO-LSpI-RareI (8.24) and SO-LSpI-1 (8.16) reveals that there are no formal differences in the expressions of rates, equilibrium constants, and frequency constants. Merely, all occurrences in these equations of the inhibitor concentration [I] are multiplied by the constant factor K_r in mechanism SO-LSpI-RareI. This factor, the equilibrium constant of the isomerization reaction of the inhibitor, defines the fraction of inhibitor present as I_r. The smaller the value of K_r, the smaller becomes $[I]_t$, but for large values of K_r SO-LSpI-RareI reduces to slow-onset linear specific inhibition. Progress curves and the dependence of rates and frequency constant on [I] are not shown because they are identical to those in Fig. 8.10. The factor x in the expression of $v_{z,S}$ is the dilution factor that results in adding a non-negligible volume of substrate solution to preincubated enzyme and inhibitor.

Without chemical evidence of the existence of isomers for an inhibitor in solution, kinetic characterization is insufficient for identifying mechanism SO-

Scheme 8.7 Slow-onset inhibition due to inhibitor isomerization. The subscript r denotes a rare inhibitor species

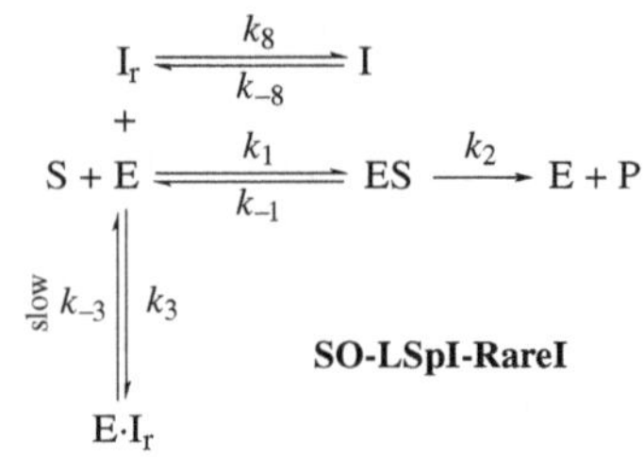

LSpI-RareI, which will be diagnosed as SO-LSpI-1 instead. An example of mechanism SO-LSpI-RareI is the interaction of leupeptin with the cysteine peptidases cathepsin B and papain [27], which had originally been mistaken for a representative of mechanism SO-LSpI-1 [1]. Leupeptins are peptide derivatives isolated from cultures of *Streptomyces* species and identified as mixtures of N-α-acetyl- (or N-α-propionyl)-L-leucyl-L-leucyl-L-argininal [33]. In aqueous solution leupeptin exists in three forms: as cyclic carbinolamine, as free aldehyde, and as aldehyde hydrate, of which only the free aldehyde form (I_A in Fig. 8.18a) acts as inhibitor of cathepsin B and other cysteine- and serine peptidases [27]. Using combined ^{1}H-NMR measurements in D_2O, deuterium isotope effects and semicarbazone trapping of the free aldehyde, Schultz et al. calculated the percent distributions of the three leupeptin forms and the rate constants for the conversions of the carbinolamine and hydrate forms into free aldehyde as shown in Fig. 8.18a [27].

$$\textbf{SO} - \textbf{LSpI} - \textbf{RareI} \tag{8.24}$$

$$\text{assumption}: \quad K_r = \frac{[\mathrm{I}]_r}{[\mathrm{I}]} = \frac{k_{-8}}{k_8} \ll 1$$

$$\lambda = k_{-3} + \frac{k_3 K_r [\mathrm{I}]}{1 + \frac{[\mathrm{S}]}{K_m}}, \quad v_s = \frac{V[\mathrm{S}]}{K_m\left(1 + \frac{K_r[\mathrm{I}]}{K_I}\right) + [\mathrm{S}]}$$

$$v_{z,E} = \frac{V[\mathrm{S}]}{K_m + [\mathrm{S}]}, \quad K_I = \frac{k_{-3}}{k_3}, \quad v_{z,S} = \frac{V[\mathrm{S}]}{K_m + [\mathrm{S}]} \times \frac{1}{1 + xK_r[\mathrm{I}]/K_I}$$

The reaction sequence for the inhibition of serine and cysteine peptidases by peptide aldehydes is shown in Fig. 8.18b. Following the formation of an adsorptive complex between enzyme and inhibitor, the nucleophile at the active site of the peptidase reacts to form a hemiacetal (serine peptidases) or hemithioacetal (cysteine peptidases), which has a tetrahedral structure close to that of the transition-state

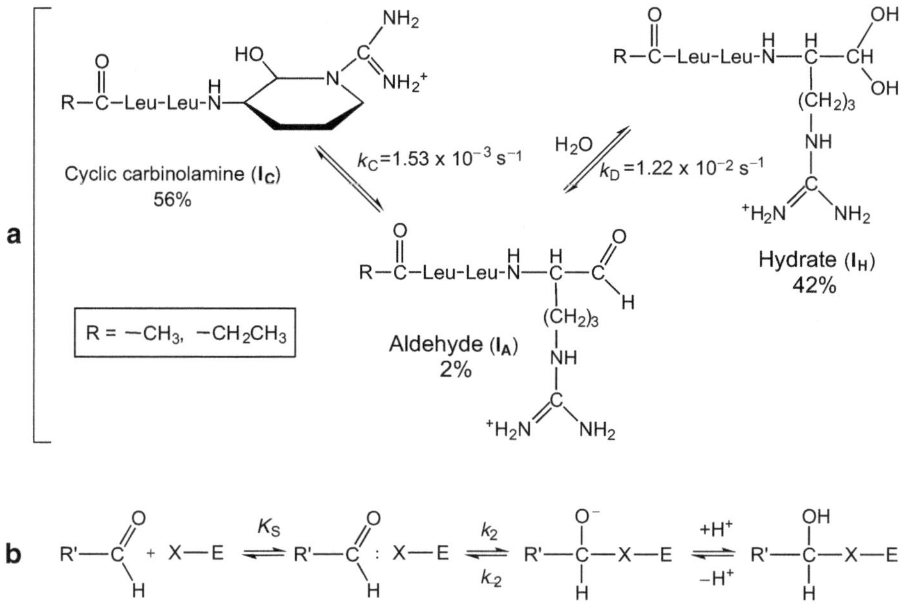

Fig. 8.18 Equilibrium forms of leupeptin in aqueous solution (**a**) and scheme for the reaction of peptide aldehydes with serine (X = -OH) and cysteine (X = S^-) peptidases (**b**). Only the free aldehyde (I_A) in panel (**a**) reacts with the enzyme to form a hemiacetal or hemithioacetal that resemble the structure of the tetrahedral transition-state [27]

during catalysis. Slow-onset inhibition of cathepsin B by leupeptin, first attributed to a rate-limiting step at the level of the tetrahedral intermediate analog [1], was later challenged on evidence that the inhibitor acts slowly because of its low concentration and a small rate-off constant [27]. Two situations thus arise as special cases of mechanism SO-LSpI-RareI depending on the total concentration of leupeptin, which includes its three forms. At nanomolar total leupeptin concentrations, considering that the active aldehyde form is only 2 % of the total, the inhibitor concentration in excess over the enzyme concentration may be very modest if $[E]_t$ is also in the low nM range, resulting in slow-onset inhibition. With total leupeptin and enzyme in the micromolar range, biphasic binding was shown to reflect the rate constants k_C and k_D for the conversion of the carbinolamine and hydrate inhibitor forms into the aldehyde form, respectively [27].

The values of k_3 and k_{-3} in Scheme 8.7 calculated for the actual concentration of leupeptin as aldehyde were $1.2 \times 10^7\,M^{-1}\,s^{-1}$ and $1.2 \times 10^{-3}\,s^{-1}$, respectively. The resulting inhibition constant $K_i = k_{-3}/k_3$ of 0.1 nM, recalculated for the total leupeptin concentration, was 5 nM, i.e., the same value measured previously by Baici and Gyger-Marazzi [1]. Indeed, this is the inhibition constant that applies in practice because of the impossibility of isolating the true inhibitor from its equilibrating isomers.

This example emphasizes once more the importance of identifying the true mechanism of inhibition for at least two reasons: (1) to gather a scientifically correct

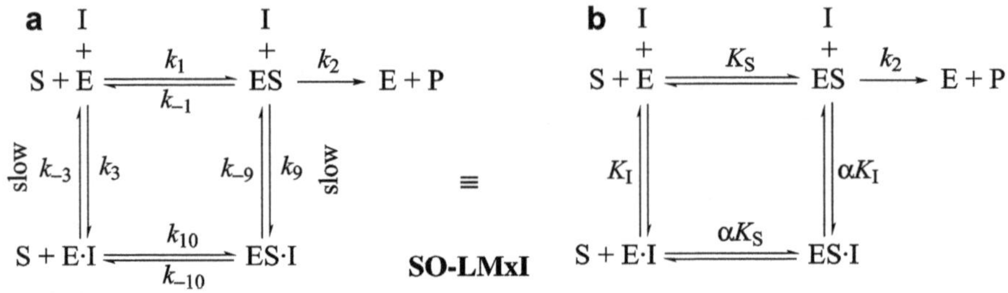

Scheme 8.8 Slow-onset mixed inhibition (SO-LMxI). In the alternative, equivalent form (**b**) K_I, αK_I, and αK_S replace the rate constants in (**a**) respecting microscopic reversibility

evaluation of what is really going on; (2) to pass medicinal chemists the necessary information for fine tuning the synthesis of new compounds or to plan appropriate structural modifications of already existing molecules of pharmacological interest.

8.10 Slow-Onset, Linear Mixed Inhibition

The kinetic properties of slow-onset mixed inhibition for non-tight-binding systems have been first analyzed by Cha [4, 5], who derived the integrated rate equation with all related expressions for the dependence of rates and frequency constant on inhibitor concentration. In a subsequent paper, the treatment was extended including an integrated rate equation for the tight-binding condition [6]. As already mentioned, despite being the fruit of intensive mathematical labor and intellectual work, it is felt that this type of equations can be substituted today with the less restrictive method of numerical integration coupled to nonlinear regression (Sect. 8.2). The mechanism is sketched in Scheme 8.8 and the relevant equations for a non-tight-binding condition are shown in the group of equations (8.25). Here, the right-hand form of the frequency constant λ and v_s are written using symbols of the general modifier mechanism as in Chaps. 2 and 3. The dependences of v_s, $v_{z,E}$ and λ on [I] are exploited for calculating the sought parameters.

As in linear mixed inhibition (Sect. 3.3.1), slow-onset, linear mixed inhibition comprises three mechanisms depending on which character, specific or catalytic, predominates or both characters are equally represented (balanced inhibition). The way for differentiating among the three mechanisms is based on the dependence of K_m^{app} and V^{app} (k_{cat}^{app}) on inhibitor concentration as illustrated in Fig. 3.3. However, in slow-onset inhibition the necessary calculations are limited to steady-state velocities (v_s), in which the changes of the apparent Michaelis constant with inhibitor concentration are minimal and possibly masked by experimental errors. Therefore, another strategy must be invoked, which performs well with data simulated with or without error.[1] Some additional experimental work is necessary for fully characterizing

[1] Unfortunately, a real example with all necessary data for performing the calculations was not available.

slow-onset mixed inhibitors with respect to other systems that manifests similar dependences of the frequency constant on inhibitor concentration. Collecting an insufficient amount of primary data is equivalent to loss of information and possibly wrong interpretation, but since assiduity in the laboratory is usually compensated by fine results, slow-onset inhibitors deserve more time than their fast-acting counterparts.

Before illustrating step-by-step the proposed procedure, let us discuss the two sets of experiments in Fig. 8.19, obtained at either fixed substrate or inhibitor concentration. The mechanism in this figure is SO-LMx(Sp>Ca)I, an acronym that can be read out without effort after having learnt the properties of linear mixed, predominantly specific inhibition [LMx(Sp>Ca)I] in Chaps. 3 and 5. Plots as in Fig. 8.19, with progress curves and all related parameters, will not be shown for the mechanisms SO-LMx(Sp=Ca)I and SO-LMx(Sp<Ca)I to give rather space to the frequency constant and its dependence on substrate and inhibitor concentration.

SO – LMxI (8.25)

SO-LMx(Sp>Ca)I ($\alpha > 1$)

SO-LMx(Sp=Ca)I ($\alpha = 1$)

SO-LMx(Sp<Ca)I ($\alpha < 1$)

$$\text{assumption}: \quad K_{\mathrm{m}} \approx K_{\mathrm{S}}$$

$$\text{required}: \quad K_{\mathrm{m}} \gg K_{\mathrm{I}}\,, \quad k_{-1} \gg k_{-3}\,, \quad k_{-1} \gg k_{-9}$$

$$\text{required}: \quad k_1\,[\mathrm{S}] \gg k_3\,[\mathrm{I}]\,, \quad k_1\,[\mathrm{S}] \gg k_9\,[\mathrm{I}]$$

$$\lambda = \frac{k_{-3} + \dfrac{k_{-9}k_{10}}{k_{-10}}[\mathrm{S}]}{1 + \dfrac{k_{10}}{k_{-10}}[\mathrm{S}]} + \frac{k_3 + k_9\dfrac{[\mathrm{S}]}{K_{\mathrm{m}}}}{1 + \dfrac{[\mathrm{S}]}{K_{\mathrm{m}}}}[\mathrm{I}] = \frac{k_{-3} + k_{-9}\dfrac{[\mathrm{S}]}{\alpha K_{\mathrm{m}}}}{1 + \dfrac{[\mathrm{S}]}{\alpha K_{\mathrm{m}}}} + \frac{k_3 + k_9\dfrac{[\mathrm{S}]}{K_{\mathrm{m}}}}{1 + \dfrac{[\mathrm{S}]}{K_{\mathrm{m}}}}[\mathrm{I}]$$

$$K_{\mathrm{I}} = \frac{k_{-3}}{k_3}\,, \quad \alpha K_{\mathrm{I}} = \frac{k_{-9}}{k_9}\,, \quad \alpha K_{\mathrm{m}} = \frac{k_{-10}}{k_{10}}\,, \quad K_{\mathrm{m}} = \frac{k_{-1} + k_2}{k_1}$$

$$v_{\mathrm{z,E}} = \frac{V\,[\mathrm{S}]}{K_{\mathrm{m}} + [\mathrm{S}]}\,, \quad v_{\mathrm{s}} = \frac{V^{\mathrm{app}}\,[\mathrm{S}]}{K_{\mathrm{m}}^{\mathrm{app}} + [\mathrm{S}]}\,, \quad k_{\mathrm{off}} = \lambda\frac{v_{\mathrm{s}}}{v_{\mathrm{z,E}}} = \frac{k_{-3} + k_{-9}\dfrac{[\mathrm{S}]}{\alpha K_{\mathrm{m}}}}{1 + \dfrac{[\mathrm{S}]}{\alpha K_{\mathrm{m}}}}$$

$$K_{\mathrm{m}}^{\mathrm{app}} = K_{\mathrm{m}}\frac{1 + \dfrac{[\mathrm{I}]}{K_{\mathrm{I}}}}{1 + \dfrac{[\mathrm{I}]}{\alpha K_{\mathrm{I}}}}\,, \quad V^{\mathrm{app}} = \frac{V}{1 + \dfrac{[\mathrm{I}]}{\alpha K_{\mathrm{I}}}}$$

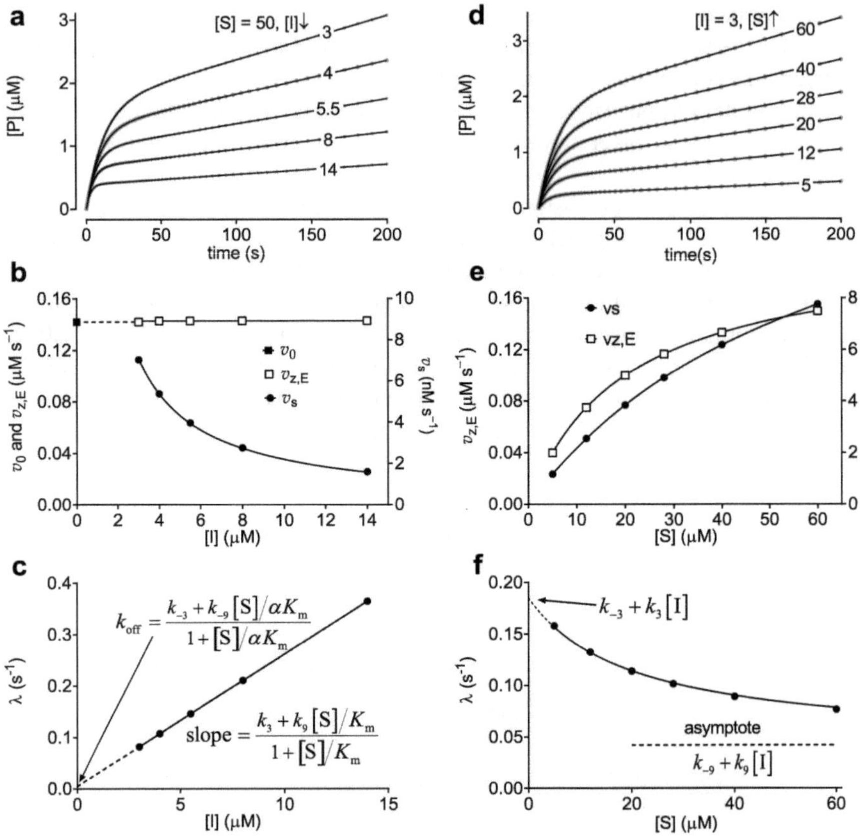

Fig. 8.19 Slow-onset, predominantly specific linear mixed inhibition, SO-LMx(Sp>Ca)I. The progress curves were simulated with the following common constants: $\alpha = 3.0$, $k_1 = 100\,\mu M^{-1}\,s^{-1}$, $k_{-1} = 1{,}990\,s^{-1}$, $k_2 = 10\,s^{-1}$, $k_3 = 0.06\,\mu M^{-1}\,s^{-1}$, $k_{-3} = 0.005\,s^{-1}$, $k_9 = 0.012\,\mu M^{-1}\,s^{-1}$, $k_{-9} = 0.003\,s^{-1}$, $k_{10} = 30\,\mu M^{-1}\,s^{-1}$, $k_{-10} = 1791\,s^{-1}$, $[E]_t = 0.02\,\mu M$. In (**a**) [S] was kept constant at 50 μM and [I] varied as indicated (μM). In (**b**) [I] was fixed at 3 μM and [S] varied as indicated (μM). All reactions were started by adding enzyme. Fitting (8.10) to simulated progress curves afforded v_s, $v_{z,E}$ and λ that were then plotted against the varied reactant: [I] in (**b, c**) and [S] in (**e, f**). In (**a**) and (**d**) simulated points on a logarithmic timebase are in *gray* and fitted curves as *black lines*. Check that microscopic reversibility applies: $k_1 k_9 k_{-10} k_{-3} = k_{-1} k_3 k_{10} k_{-9} = 1.075 \times 10^{13}\ M^{-2}\ s^{-4}$

The requirements for turning mixed inhibition into a slow-onset process are listed in (8.25) together with the assumption of quasi-equilibrium for substrate, which implies $K_m \approx K_S$ and allows deriving a practicable integrated rate equation. Should this not be the case, the same type of deviations can be expected for the calculated parameters as discussed in Sect. 2.7. The vertical reactions in Scheme 8.8 are those that determine the slowness of the inhibition process, in which the rate constants are constrained by the principle of microscopic reversibility. This is why the rate at time zero in reactions started by adding enzyme, $v_{z,E}$, is independent of inhibitor concentration (Fig. 8.19b): before the enzyme reacts with inhibitor ES is already in place. Also, in panel (b) the rate at steady state is shown to decrease with inhibitor concentration but, as stated above, the range of values for this rate is of a few percent residual activity that hampers its utilization to extract statistically sound information.

In the reactions started by adding substrate (panels d–f), both $v_{z,E}$ and v_s display hyperbolic behavior. The Michaelis–Menten equation can be fitted to both rate dependences on [S]: from the dependence of $v_{z,E}$, K_m and V can be calculated and compared with the same parameters obtained in measuring the reaction in the absence of inhibitor, K_m^0 and V^0. In this and other sections of this chapter, the Michaelis constant and the limiting rate are denoted K_m and V since there is no ambiguity in this case when their values in the presence of inhibitor are called K_m^{app} and V^{app} (8.25).

The frequency constant depends on the concentrations of substrate and inhibitor. When measured at fixed substrate and various inhibitor concentrations, the dependence is linear with a positive slope for the three mechanisms (predominantly specific, predominantly catalytic, or balanced) as illustrated in Fig. 8.19c, which shows the ordinate intercept as overall k_{off} and the slope as a function of the second-order rate constants of the vertical steps in Scheme 8.8a. In experiments performed at fixed inhibitor and variable substrate concentration, the frequency constant exhibits a hyperbolic dependence on [S] (panel f). In this case, the intercept with the ordinate does not correspond to k_{off}, being instead a function of the two rate constants for the binding of inhibitor to the free enzyme, k_3 and k_{-3}. The asymptote of the hyperbola for $[S] \to \infty$ is a function of the rate constants for the binding of inhibitor to the ES complex, i.e., k_9 and k_{-9}. An important diagnostic criterion is that λ *decreases* hyperbolically with [S] in the mechanisms LMx(Sp>Ca)I and LMx(Sp=Ca)I, while it *increases* with increasing substrate concentration for LMx(Sp<Ca)) (Fig. 8.20).

Without *additional experiments*, despite the rich information present in the dependences of λ on [I] and [S], these *are insufficient* for discriminating slow-onset mixed inhibition from other slow-onset mechanisms. For instance, the frequently occurring SO-LSpI-1 mechanism described in Sect. 8.5 manifests the same dependences of λ on [I] and [S] as does SO-LMx(Sp>Ca)I. Accurate determination of the basic mechanism, specific, catalytic, or mixed, is thus compelling.

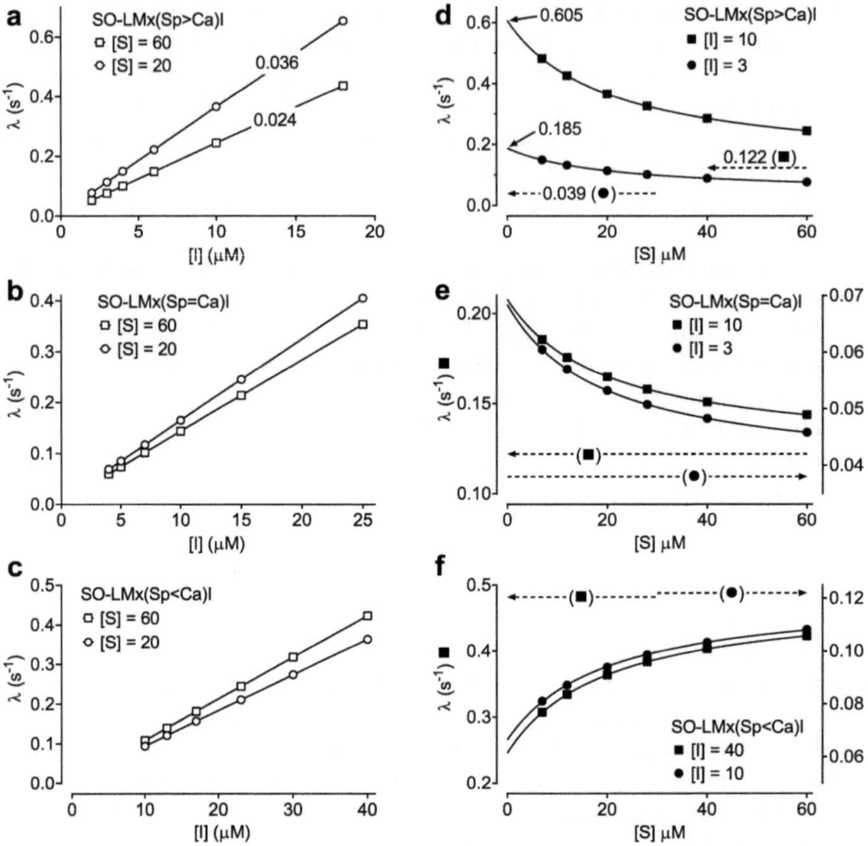

Fig. 8.20 Dependences of the frequency constant on inhibitor (**a–c**) and substrate concentration (**d–f**) for the three slow-onset linear mixed mechanisms. The numbers on the lines in (**a**) indicate the slopes. The asymptotes in (**d–f**) are shown by *dashed arrows* and in (**d**) the values of the intercepts and of the asymptotes are indicated. All concentrations are μM. Common constants used to simulate the progress curves, from which the λ-values were extracted: k_1 = 100, k_{-1} = 1,990, k_2 = 10, K_m = 20, $[E]_t$ = 0.02. The remaining constants varied with the mechanism. SO-LMx(Sp>Ca)I: k_3 = 0.06, k_{-3} = 0.005, k_9 = 0.012, k_{-9} = 0.003, k_{10} = 30, k_{-10} = 1791. SO-LMx(Sp=Ca)I: k_3 = 0.02, k_{-3} = 0.005, k_9 = 0.012, k_{-9} = 0.003, k_{10} = 30, k_{-10} = 597. SO-LMx(Sp<Ca)I: k_3 = 0.006, k_{-3} = 0.005, k_9 = 0.012, k_{-9} = 0.003, k_{10} = 30, k_{-10} = 179.1. The units of the second-order and first-order rate constants are $\mu M^{-1}\ s^{-1}$ and s^{-1}, respectively

Since slow-onset mixed inhibition cannot be unambiguously diagnosed from the rates at steady-state, the information contained in the frequency constant can be exploited. To this end, four sets of experiments are necessary, i.e., progress curves measured at fixed substrate and variable inhibitor and vice versa, each one at two values of the fixed reagent, from which the dependences of λ on [I] and [S] are evaluated as shown in Fig. 8.20 with the steps explained here below.

1. The analysis starts with the dependences on [I] shown in panels (a–c) for the three slow-onset mixed mechanisms. Linear regression of the straight lines affords intersections on the ordinate, corresponding to k_{off} (Fig. 8.19c) that are valueless because too close to the origin of the axes and can thus be forgotten. Nevertheless, the slopes can be evaluated precisely. They contain two rate constants and the known values of [S] and K_{m}. One slope is insufficient to determine the two unknowns k_3 and k_9, but with a second slope this is possible. Calling σ_1 and σ_2 the ratio $[\text{S}]/K_{\text{m}}$ for two different substrate concentrations, see λ in (8.25), the expressions of the two slopes of each plot in Fig. 8.20a–c can be rearranged as

$$\left(\frac{1}{1+\sigma_1}\right) k_3 + \left(\frac{\sigma_1}{1+\sigma_1}\right) k_9 = \text{slope}_1$$
$$\left(\frac{1}{1+\sigma_2}\right) k_3 + \left(\frac{\sigma_2}{1+\sigma_2}\right) k_9 = \text{slope}_2\ , \tag{8.26}$$

which is a system of linear equations that can be solved for the unknowns k_3 and k_9 since all other coefficients are known. The values of the slopes are given in Fig. 8.20a as an exercise for the calculation.
2. We turn now our attention to the plots in Fig. 8.20d–f that display the dependences of λ on [S]. The only fit possible is that of the expression of λ in (8.25), a hopeless task with the five parameters k_3, k_{-3}, k_9, k_{-9}, and α. However, the values of k_3 and k_9 determined from the linear plots in panels (a–c) can be declared as constants while leaving the three remaining parameters free to float in nonlinear regression. Yet, the normalized covariance matrix reveals in every case that k_{-3} and k_{-9}, k_{-3} and α, as well as k_{-9} and α are strongly intertwined. Also, the analysis of dependency reveals that each one of these parameters is intertwined with the remaining two. The outcome is that the values of k_{-3}, k_{-9} and α cannot be guessed at this stage of the analysis. Nevertheless, the resulting curve fits neatly the data and, in simulated examples with and without added error, yields the expected intercept with the ordinate and the expected value of the limit for $[\text{S}] \to \infty$. The explanation is that changing the value of one parameter causes the remaining parameters to adjust to produce exactly the same curve. The values of the ordinate intercepts and the asymptotes determined in this way are indicated in Fig. 8.20d for readers willing to practice the method numerically. Any other fitting procedure to data as those in panels (d–f) using polynomial functions or splines should be avoided because the resulting numbers are meaningless.

3. The ordinate intercepts of the plots in Fig. 8.20d–f are now given and correspond to $k_{-3} + k_3\,[\mathrm{I}]$. With two such values obtained at two inhibitor concentrations $[\mathrm{I}]_1$ and $[\mathrm{I}]_2$, k_{-3} and k_3 can be calculated by solving the following system of linear equations

$$\begin{aligned} k_{-3} + k_3[\mathrm{I}]_1 &= \mathrm{intercept}_1 \\ k_{-3} + k_3[\mathrm{I}]_2 &= \mathrm{intercept}_2\ , \end{aligned} \tag{8.27}$$

   from which k_3 can be compared for internal consistency with the value determined in point 2 of this list.
4. The asymptotes for $[\mathrm{S}] \to \infty$ in Fig. 8.20d–f correspond $k_{-9} + k_9\,[\mathrm{I}]$, which yield k_{-9} and k_9 by solving the system

$$\begin{aligned} k_{-9} + k_9[\mathrm{I}]_1 &= \mathrm{asymptote}_1 \\ k_{-9} + k_9[\mathrm{I}]_2 &= \mathrm{asymptote}_2\ . \end{aligned} \tag{8.28}$$

5. The two inhibition constants K_I and αK_I can now be calculated from the rate constants:

$$K_\mathrm{I} = \frac{k_{-3}}{k_3} \qquad \alpha K_\mathrm{I} = \frac{k_{-9}}{k_9}\ , \tag{8.29}$$

   whose precision *depends on experimental error*. The value of the reciprocal allosteric constant α obtained from the ratio of the inhibition constants indicates the nature of the mixed mechanism: predominantly specific ($\alpha > 1$), balanced ($\alpha = 1$) or predominantly catalytic ($\alpha < 1$).

An ingenious method has been exploited by Takai and coworkers for characterizing the inhibition of [Phosphatase 2A protein]-leucine-carboxy methyltransferase (EC 2.1.1.233) by okadaic acid [31]. This compound, a derivative of a C_{38} fatty acid produced by some algae, accumulates in bivalve mollusks and sponges acting as cytotoxin. As inhibitor of serine and threonine protein phosphatases, okadaic acid is used in the cell laboratory for modulating cellular processes regulated by phosphorylation. Takai et al. determined high affinity of okadaic acid for the enzyme mentioned above (30–40 pM) and identified the mechanism as slow-onset linear mixed. The value of α can be calculated from the published constants as 0.87 that corresponds to SO-LMx(Sp<Ca)I. After allowing equilibration between the enzyme and okadaic acid, the dissociation rate constants k_{-3} and k_{-9} were determined by trapping the inhibitor with a monoclonal antibody raised against this substance. This is an efficient method that can be used whenever an antibody or any chemical trap against the inhibitor is available.

8.11 Slow-Onset Catalytic Inhibition

Since slow-onset specific and mixed inhibition have been documented in the literature, slow-onset catalytic inhibition is a plausible mechanism that deserves a short comment to complete the palette of the five basic linear inhibition mechanism in the slow-onset variant. Similar to slow-onset linear inhibition, Scheme 8.9 considers a two-step mechanism, in which the adsorptive ESI complex equilibrates rapidly before isomerizing to the final ES·I complex (a), and a single-step mechanism in which the ES·I complex is formed directly in one slow step (b). The acronyms assigned the two mechanisms are SO-LCaI-2 and SO-LCaI-1. The integrated rate equation is given by (8.10) and the associated parameters are grouped in (8.30) and (8.31) for SO-LCaI-2 and SO-LCaI-1, respectively.

Slow-onset catalytic inhibition can be recognized already by comparing families of progress curves measured at fixed substrate or inhibitor concentration (Fig. 8.21). In panel (a) the substrate concentration is constant (20 μM) while [I] varies in the range 10–120 μM; in panel (b), [I] is fixed and [S] varies in a range of concentrations

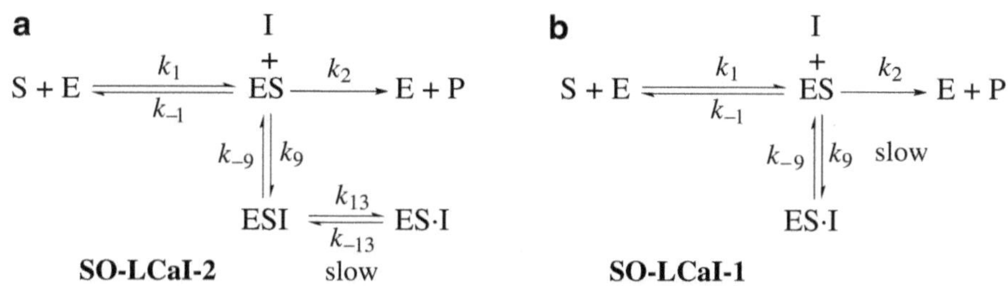

Scheme 8.9 Slow-onset, linear catalytic inhibition. In (**a**) the adsorptive ESI complex is formed rapidly before isomerizing in a slow reaction to ES·I, while in (**b**) ES·I is formed in a single slow step

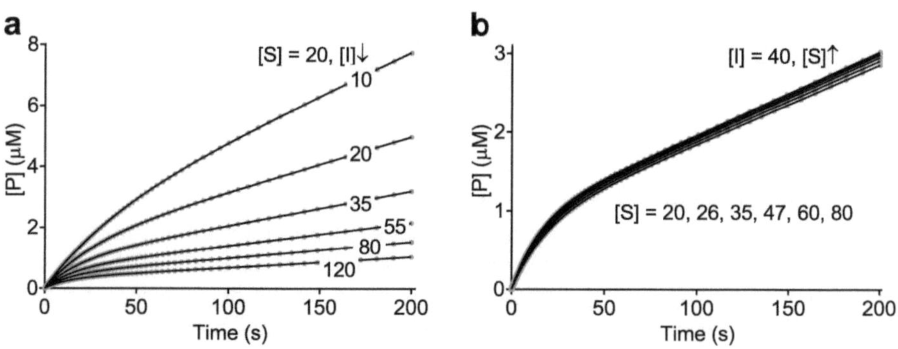

Fig. 8.21 Slow-onset linear catalytic inhibition, SO-LCaI-2. Progress curves simulated at fixed substrate (**a**) or inhibitor concentration (**b**). The constants used to simulate the progress curves were: $k_1 = 100$, $k_{-1} = 1{,}990$, $k_2 = 10$, $k_9 = 20$, $k_{-9} = 500$, $k_{13} = 0.1$, $k_{-13} = 0.01$, $K_m = 20$, $[E]_t = 0.02$. The units of the second-order and first-order rate constants are $\mu M^{-1}\,s^{-1}$ and s^{-1}, respectively

corresponding to K_m–$4K_m$. In (a) the progress curves are spaced allowing to explore a sufficiently wide range of frequency constants and steady-state slopes but in (b) both values are close together, varying in a narrow range when the substrate concentration changes. The curves in (b) almost coalesce into a single curve when [I] is fixed at 120 μM. The reason is precisely because of the fast reactions of S with E and of I with ES and because $K_I^* < K_m$: in the steady-state, the preponderant species is ES·I, [ES] is very small and [ES · I] varies extremely little when [S] is varied.[2]

Figure 8.22 illustrates the dependences of λ, $v_{z,E}$ and v_s on the concentrations of inhibitor at two fixed substrate concentrations (a–c) and at variable [S] at two fixed values of [I] (d–f). The rates at steady-state in (panel c) do not allow to extract useful information because they represent low levels of residual activity. However, the data in panel (f) yield good values of the apparent limiting rate of the overall reaction, $V^{*\mathrm{app}}$. The values of $K_m^{*\mathrm{app}}$ from the same curves in (f) cannot obviously be guessed with precision.

$$\textbf{SO} - \textbf{LCaI} - \textbf{2} \tag{8.30}$$

$$\text{required}: \quad K_m > K_I^*, \quad k_{-1} \gg k_{13}, \quad k_{-1} \gg k_{-13}$$

$$\lambda = k_{-13} + k_{13}\frac{[\mathrm{I}]}{K_I\left(1 + \dfrac{K_m}{[\mathrm{S}]}\right) + [\mathrm{I}]}, \qquad k_{\mathrm{off}} = \lambda\frac{v_s}{v_{z,E}} = k_{-13}$$

$$K_I = \frac{k_{-9}}{k_9}, \quad K_I^* = K_I\frac{k_{-13}}{k_{13} + k_{-13}}, \quad K_m = \frac{k_{-1} + k_2}{k_1}$$

$$v_{z,E} = \frac{V^{\mathrm{app}}[\mathrm{S}]}{K_m^{\mathrm{app}} + [\mathrm{S}]}, \quad V^{\mathrm{app}} = \frac{V}{1 + \dfrac{[\mathrm{I}]}{K_I}}, \quad K_m^{\mathrm{app}} = \frac{K_m}{1 + \dfrac{[\mathrm{I}]}{K_I}}$$

$$v_s = \frac{V^{*\mathrm{app}}[\mathrm{S}]}{K_m^{*\mathrm{app}} + [\mathrm{S}]}, \quad V^{*\mathrm{app}} = \frac{V}{1 + \dfrac{[\mathrm{I}]}{K_I^*}}, \quad K_m^{*\mathrm{app}} = \frac{K_m}{1 + \dfrac{[\mathrm{I}]}{K_I^*}}$$

The dependences of λ and $v_{z,E}$ on [I] and [S] allow to calculate $k_{\mathrm{off}} = k_{-13}$ and k_{13} (panels a and d). The inhibition constant K_I of the first, rapid binding step of I to ES can be calculated from the dependence of $v_{z,E}$ on [I] (panel b), while the diagnosis of catalytic inhibition is given by the dependence of $v_{z,E}$ on [S] in panel (e). Here, $V^{\mathrm{app}}/K_m^{\mathrm{app}}$ and hence k^{app} is independent of [S], the hallmark of catalytic inhibition (Fig. 3.3). After correction for any blank, when fitting the

[2]Verifying by simulation how the concentrations of ES, ESI, and ES·I vary with [S] at constant [I] is very instructive to grasp the essence of this mechanism: try it!

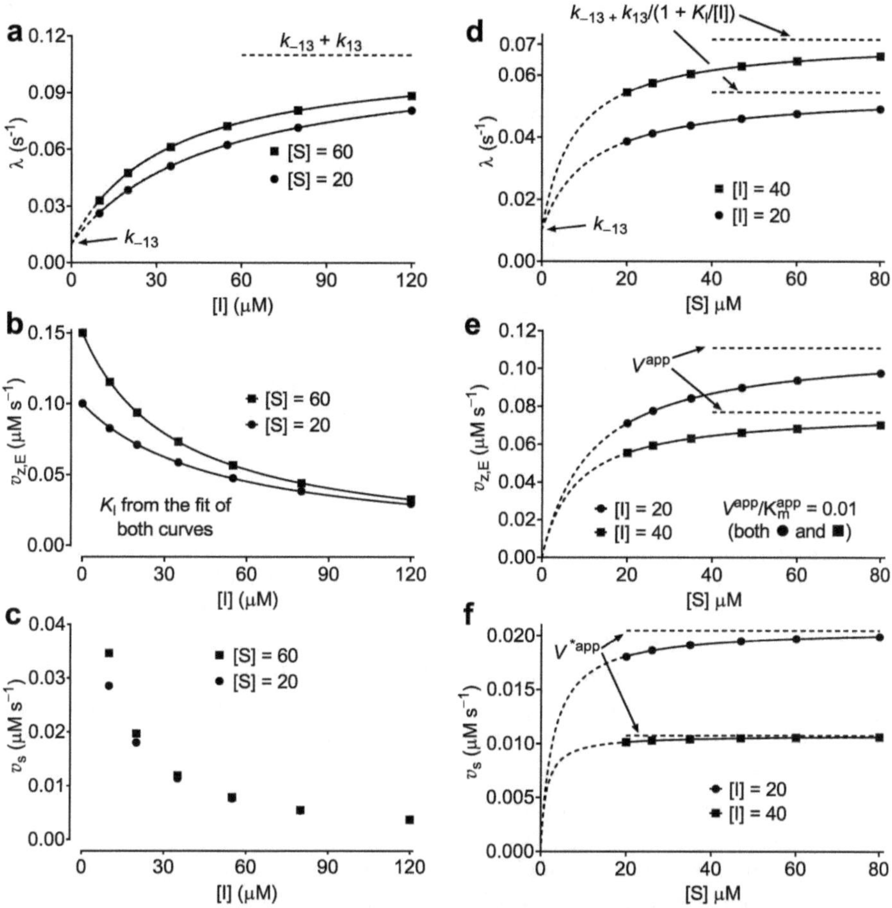

Fig. 8.22 Slow-onset linear catalytic inhibition, SO-LCaI-2. Dependences of λ, $v_{z,E}$ and v_s on inhibitor (**a–c**) and substrate concentration (**d–f**). All concentrations are μM. Common constants used to simulate the progress curves, from which the λ-values were extracted: $k_1 = 100$, $k_{-1} = 1{,}990$, $k_2 = 10$, $k_9 = 20$, $k_{-9} = 500$, $k_{13} = 0.1$, $k_{-13} = 0.01$, $[E]_t = 0.02$. The units of the second-order and first-order rate constants are $\mu M^{-1}\,s^{-1}$ and s^{-1}, respectively. The asymptotes in (**a, d, e, f**) are indicated by *dashed lines*

Michaelis–Menten equation to data points in panel (e) it is legitimate to constrain the rate at zero for [S] = 0. This allows a better fitting aimed at obtaining estimates of K_m^{app}. The overall inhibition constant K_I^*, expression in (8.30), can now be safely calculated from K_I, k_{13} and k_{-13}.

Independently of being or being not represented in real examples, mechanism SO-LCaI-2 is worth of being taken into consideration when analyzing the behavior of inhibitors that manifest slow-onset properties. This mechanism has been discovered in a case of substrate inhibition that will be discussed below in Sect. 8.12.

$$\textbf{SO} - \textbf{LCaI} - \textbf{1} \tag{8.31}$$

$$\text{required}: \quad K_m \gg K_I , \quad k_{-1} \gg k_9 , \quad k_{-1} \gg k_{-9}$$

$$\lambda = k_{-9} + \frac{k_9\,[\mathrm{I}]}{1 + \dfrac{K_m}{[\mathrm{S}]}} , \qquad k_{\text{off}} = \lambda \frac{v_s}{v_{z,E}} = k_{-9}$$

$$v_{z,E} = v_0 = \frac{V\,[\mathrm{S}]}{K_m + [\mathrm{S}]} , \qquad K_I = \frac{k_{-9}}{k_9} , \qquad K_m = \frac{k_{-1} + k_2}{k_1}$$

$$v_s = \frac{V^{\text{app}}\,[\mathrm{S}]}{K_m^{\text{app}} + [\mathrm{S}]} , \qquad V^{\text{app}} = \frac{V}{1 + \dfrac{[\mathrm{I}]}{K_I}} , \qquad K_m^{\text{app}} = \frac{K_m}{1 + \dfrac{[\mathrm{I}]}{K_I}}$$

For particular combinations of the rate constants, slow-onset catalytic inhibition SO-LCaI-2 can conceivably degenerate to a single, slow step of inhibition as shown in Scheme 8.9b. The equations for this system are collected in (8.31).

The rate at time zero for the reaction started by adding enzyme, $v_{z,E}$, does not depend on inhibitor concentration as can be expected from the reaction scheme: upon adding enzyme, the ES complex is immediately in place before ES and I react appreciably. Thus, unlikely SO-LCaI-2, $v_{z,E}$ cannot be exploited for diagnosing the catalytic character of the inhibition.

Figure 8.23 illustrates the properties of SO-LCaI-1, which is expected to be a difficult mechanism regarding experiment design as can be deduced from the two families of progress curves in panels (a) and (d). Namely, at fixed [S] and variable [I], the rate at steady-state, v_s, varies very little with [I] and after triplicating the substrate concentration the values of this rate do not practically change (panel c). At fixed [I] and variable [S] (panel d) the progress curves tend to coalesce and become virtually one curve at high inhibitor concentrations. $v_{z,E}$ follows Michaelis–Menten kinetics (not shown) while v_s varies in a tiny range, shown in panel (f) for two inhibitor concentrations just for illustrative purposes. Note the ordinates of panels (c) and (f) that have units nM s^{-1}. Lowering the inhibitor concentration in the measurement of reaction progress, very long times are necessary to reach the steady-state that result in excessive substrate turnover invalidating the use of the integrated equation (8.10).

Although not being useful for calculating the inhibition constant, the small values of v_s, preferably those from progress curves measured at variable inhibitor and fixed substrate (panel c) can be used to calculate the off-rate k_{-9} according to (8.14). In fact, the fit of (8.10) to data provides v_s, $v_{z,E}$ and λ for each inhibitor concentration. These values are used to calculate $k_{\text{off}} = k_{-9}$, which is the same at any inhibitor concentration, by averaging all data.

At variable inhibitor concentration (panel b), λ depends linearly on [I]. The ordinate intercept corresponds to k_{-9}, which is too small to be determined in this way.

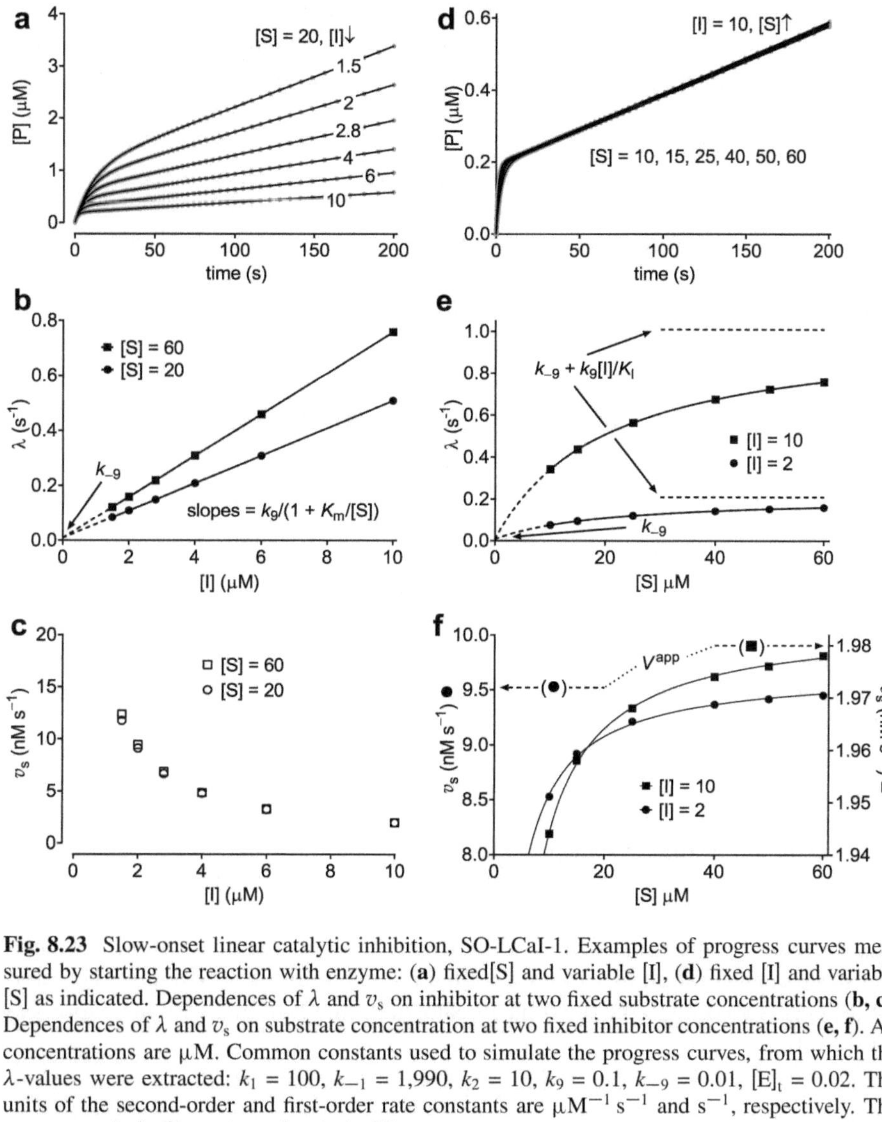

Fig. 8.23 Slow-onset linear catalytic inhibition, SO-LCaI-1. Examples of progress curves measured by starting the reaction with enzyme: (**a**) fixed[S] and variable [I], (**d**) fixed [I] and variable [S] as indicated. Dependences of λ and v_s on inhibitor at two fixed substrate concentrations (**b, c**). Dependences of λ and v_s on substrate concentration at two fixed inhibitor concentrations (**e, f**). All concentrations are μM. Common constants used to simulate the progress curves, from which the λ-values were extracted: $k_1 = 100$, $k_{-1} = 1{,}990$, $k_2 = 10$, $k_9 = 0.1$, $k_{-9} = 0.01$, $[E]_t = 0.02$. The units of the second-order and first-order rate constants are $\mu M^{-1}\,s^{-1}$ and s^{-1}, respectively. The asymptotes in (**e, f**) are shown by *dashed lines*

The slope can however be calculated precisely: it corresponds to $k_9/(1 + K_m/[S])$ and is therefore ideal to calculate k_9. The inhibition constant is finally calculated as $K_I = k_{-9}/k_9$.

For the reasons discussed above demonstrating the catalytic character of inhibition is difficult using the rates at steady-state for determining the dependence of the apparent specificity constant on substrate concentration. Therefore, we can exploit the dependences of the frequency constant on [I] and [S] and proceed by exclusion comparing the behavior of SO-LCaI-1 in Fig. 8.23b,e with that of all other mechanisms described in this chapter. There are only two mechanisms that display the dependence of λ on [I] as a straight line with positive slope *and* a hyperbolic dependence on [S] that increases with [S]. These are the mechanisms SO-LMx(Sp<Ca)I in Fig. 8.20c,f and SO-LCaI-1 in Fig. 8.23b,e. The discrimination between the mixed and the catalytic mechanism can be made by computing k_{off} from all triplets of values v_s, $v_{z,E}$ and λ obtained by nonlinear regression fitting of (8.10) to data: the *same value* of k_{off} is obtained at any substrate concentration for mechanism SO-LCaI-1, while *different values* result at different substrate concentrations for mechanism SO-LMx(Sp<Ca)I.

8.12 Slow-Onset Catalytic Substrate Inhibition

Scheme 8.10 depicts a slow-onset catalytic inhibition mechanism by substrate that is analogous to SO-LCaI-2 described in Sect. 8.11 and is assigned therefore the acronym SO-LCaI-S2, where the S before 2 stays for substrate. This mechanism has been identified by Novinec et al. in the hydrolysis of the dipeptide substrate *N*-carbobenzoxy-L-Phe-L-Arg-7-amino-4-methylcoumaryl amide catalyzed by human cathepsin K [23]. The expressions of the rates and parameters for slow-onset inhibition by substrate collected in (8.32) are analogous to those of SO-LCaI-2, with substrate replacing inhibitor.

The frequency constant increases with [S] with a curve that has a sigmoidal shape. This behavior can be better understood from the equivalent expression on the right of λ in (8.32) that shows the presence of a squared substrate concentration. The maximum in the dependences of v_z and v_s on [S] is due to the same reason. Of the two equivalent expressions for the rates in (8.32), the left forms are useful for understanding the *catalytic character* of substrate inhibition because V' and K'_m decrease by the same factor, while the right forms explicitly show a squared substrate concentration that implies a function exhibiting a maximum. The values of [S] corresponding to the maxima are shown in (8.32) in line with the expression of the rates.

Scheme 8.10 Slow-onset, linear catalytic substrate inhibition. This mechanism is analogous to slow-onset catalytic inhibition SO-LCaI-2 in Scheme 8.9a

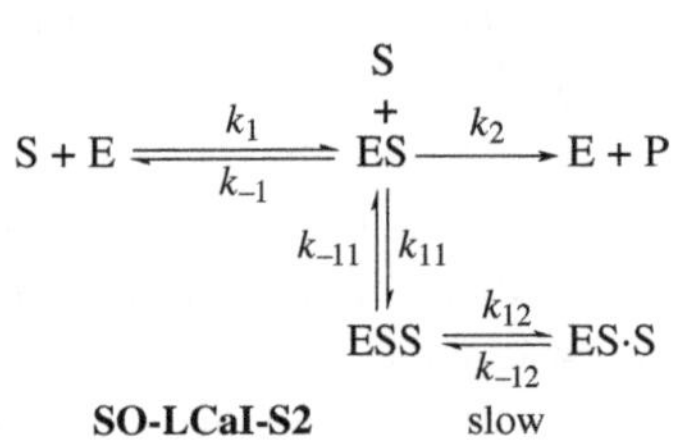

SO-LCaI-S2

$$\textbf{SO}-\textbf{LCaI}-\textbf{S2} \qquad (8.32)$$

$$\text{required}: \quad K'_{\text{m}} > K^*_{\text{Si}}, \quad k_{-1} \gg k_{12}, \quad k_{-1} \gg k_{-12}$$

$$\lambda = k_{-12} + k_{12}\frac{[\text{S}]}{K_{\text{Si}}\left(1+\dfrac{K'_{\text{m}}}{[\text{S}]}\right)+[\text{S}]} = k_{-12} + \frac{k_{12}}{1+\dfrac{K_{\text{Si}}}{[\text{S}]}+\dfrac{K_{\text{Si}}K'_{\text{m}}}{[\text{S}]^2}}$$

$$K_{\text{Si}} = \frac{k_{-11}}{k_{11}}, \quad K^*_{\text{Si}} = K_{\text{Si}}\frac{k_{-12}}{k_{12}+k_{-12}}, \quad K'_{\text{m}} = \frac{k_{-1}+k_2}{k_1}, \quad k_{\text{off}}=\lambda\frac{v_{\text{s}}}{v_{\text{z}}} = k_{-12}$$

$$v_{\text{z}} = \frac{\dfrac{V'}{1+\dfrac{[\text{S}]}{K_{\text{Si}}}}[\text{S}]}{\dfrac{K'_{\text{m}}}{1+\dfrac{[\text{S}]}{K_{\text{Si}}}}+[\text{S}]} = \frac{V'[\text{S}]}{K'_{\text{m}}+[\text{S}]+\dfrac{[\text{S}]^2}{K_{\text{Si}}}}, \quad \text{maximum at } [\text{S}] = \sqrt{K'_{\text{m}}K_{\text{Si}}}$$

$$v_{\text{s}} = \frac{\dfrac{V'}{1+\dfrac{[\text{S}]}{K^*_{\text{Si}}}}[\text{S}]}{\dfrac{K'_{\text{m}}}{1+\dfrac{[\text{S}]}{K^*_{\text{Si}}}}+[\text{S}]} = \frac{V'[\text{S}]}{K'_{\text{m}}+[\text{S}]+\dfrac{[\text{S}]^2}{K^*_{\text{Si}}}}, \quad \text{maximum at } [\text{S}] = \sqrt{K'_{\text{m}}K^*_{\text{Si}}}$$

The analysis of the characteristic progress curves for mechanism SO-LCaI-S2 poses some problems. A family of such curves is shown in Fig. 8.24a,b, where various concentrations of the substrate do not produce an uniform increase or decrease of the velocity. Since no modifiers have been added to the system, this suggests that the substrate acts at the same time as such and as inhibitor, as demonstrated by the decreasing rates v_z and v_s after a *certain threshold* of substrate concentration, which for v_s can be located by inspection of the slopes in panels (a) and (b) between 6 and 8 μM. After fitting (8.10) to data that yields v_z, v_s and λ, the plots of the rates against [S] show two distinct maxima indicated in panels (c) and (d). In the same panels there are two curves: the continuous lines represent fits of the equations of v_z and v_s given in (8.32) to the points from the preceding fitting procedure in panel (a), while the dashed lines are theoretical curves *plotted* from the same functions and the rate constants listed in the legend of Fig. 8.24.

For the rate at time zero shown in panel (c), the fitted and calculated curves overlap in the range of the points to a satisfactory extent and the theoretical dashed line shows what an experiment cannot reveal. The curve fitted to the rate at steady-state (panel d) does however not overlap the points as the theoretical (dashed)

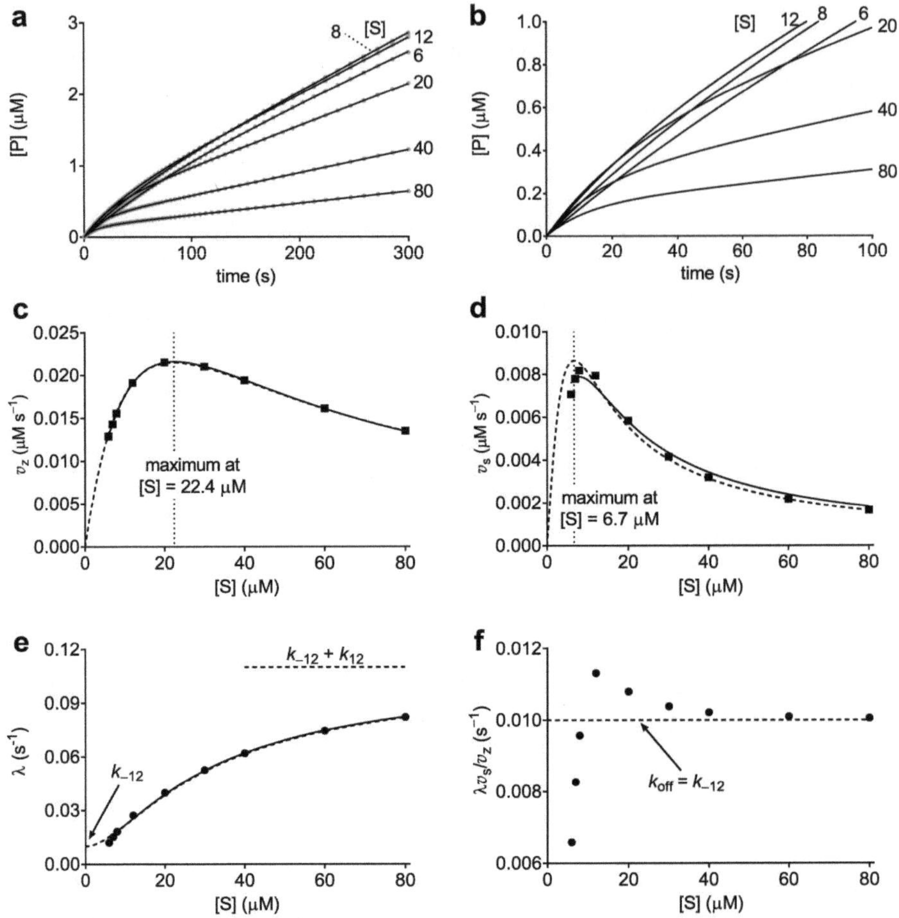

Fig. 8.24 Slow-onset linear catalytic substrate inhibition, SO-LCaI-S2, according to Scheme 8.10. The progress curves ((**a**) and (**b**)) were simulated by numerical integration on a logarithmic timebase with the following constants: k_1 = 100, k_{-1} = 1,990, k_2 = 10, k_{11} = 20, k_{-11} = 500, k_{12} = 0.1, k_{-12} = 0.01, $[E]_t$ = 0.006. The units of the second-order and first-order rate constants are $\mu M^{-1} s^{-1}$ and s^{-1}, respectively. (**b**) Shows the first 100 s of the curves in (**a**) without the simulated points. Panels ((**c**)–(**f**)) display the dependences of the calculated parameters on substrate concentration. The asymptotes in (**e, f**) are shown by *dashed lines* and all concentrations are μM. In this simulated example, nine progress curves were generated, of which those at [S] = 7, 30, and 60 μM were omitted in panels (**a**) and (**b**) for clarity but were considered in the remaining panels

line. Since we are dealing with an error-free simulated example, the origin of the discrepancy must be sought in the *curvature* of the progress curves at substrate concentrations below the critical point that demarcates the beginning of manifest inhibition. These curves do not have the clear profile of an exponent followed by a linear steady-state because *there is no steady-state*. Therefore, the integrated rate equation searches in vain for a straight line that does not exists and places

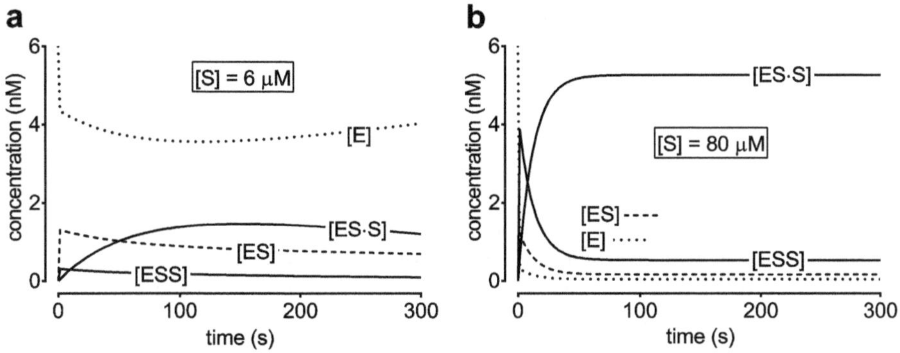

Fig. 8.25 Slow-onset linear catalytic substrate inhibition, SO-LCaI-S2. The curves show the change with time of the four enzyme-containing species in Scheme 8.10 at low (**a**) and high substrate concentration (**b**). The constants used for simulating the curves are those in Fig. 8.24

there a fitted curve just to satisfy the eye of the investigator. The outcome is a scrambled fit of the data points of v_s because, despite those measured at substrate concentrations greater than the critical [S] are correct, those below the critical point are not, resulting in deviations from the model. The discrepancy just discussed does not concern v_z because it belongs to a time range in which inhibition is not yet appreciable.

The problem at low substrate concentrations concerns also the frequency constant. As shown in panel (e), the correspondence between the fitted curve (continuous) and the plotted function (dashed) is good at [S] > 20 but not so at smaller values. The *sigmoid* beginning, which in any case remains inaccessible experimentally,[3] can be drawn only theoretically.

If any doubt remains on the explanation given above, Fig. 8.25 provides the evidence of the nonexistence of a steady-state at low substrate concentrations in the slow-onset mechanism SO-LCaI-S2. The curves show how the concentrations [E], [ES], [ESS] and [ES · S] change with time at the lowest (6 μM) and highest (80 μM) substrate concentrations used to simulate the progress curves in Fig. 8.24a. None of the concentrations of the enzyme-containing species reaches a constant value at [S] = 6 μM but all four concentrations reach a steady-state at [S] = 80 μM. This is the case at any [S] > 22 μM, the substrate concentration that corresponds to the maximum in the curve of Fig. 8.24c.

Having clarified the reasons of the jumble that arises with the progress curves in the presence of slow-onset substrate inhibition, we ask if evaluating the necessary parameters that characterize this system is still possible using the integrated rate equation. The answer is yes for two reasons: (1) experimental errors that accompany typical experiments are likely to overwhelm the deviations from the model at low

[3]Because at very low concentrations the substrate is depleted to a great extent invalidating the third assumption on p. 374.

substrate concentrations; (2) we can exploit the dependence of v_z on [S] to calculate K'_m, V' and K_{Si}, as well as the dependence of λ on [S] to calculate the isomerization constants k_{12} and k_{-12}. A suggested sequence for the calculations is the following:

1. Measure the progress curves possibly with a stopped-flow apparatus using a logarithmic timebase and averaging several progress curves for each substrate concentration.
2. Fit (8.10) to the ensemble of progress curves taking into account any displacement of the measured signal, i.e., the factor d in (8.10).
3. With the triplets v_s, v_z and λ calculate the ratio $\lambda v_s/v_z$ for every substrate concentration and plot $\lambda v_s/v_z$ against [S] as shown in Fig. 8.24f. There is no fit for this plot, which appears as a curve that first increases and then decreases toward a constant value that represents $k_{off} = k_{-12}$. In practice, the value of $\lambda v_s/v_z$ calculated at the highest possible concentration of substrate represents a good estimate of k_{-12}.
4. Plot λ against [S] and fit the equation of λ in (8.32) to the points by nonlinear regression using a *robust fit* option while constraining k_{-12} to the value determined in step 3. Without caring of the remaining parameters, ask the nonlinear regression programm to calculate the asymptote at infinite substrate concentration, which corresponds to $k_{-12} + k_{12}$. With simulated data, the robust fit has been shown to be useful because less weight is given the anomalous points at low substrate concentration. Good estimates of k_{12} and $(k_{12} + k_{-12})$ are now available for calculating the factor $k_{-12}/(k_{12} + k_{-12})$.
5. Plot v_z against [S] and fit the equation of v_z in (8.32) to the points. The parameter of interest is K_{Si}, which is used to calculate the overall substrate inhibition constant K^*_{Si} after multiplication with the factor $k_{-12}/(k_{12} + k_{-12})$ from the calculations under point 4.

Comment simulations with added errors have shown that this long-winded method yields a substrate inhibition constant with small deviations from the theoretical value. Errors of the type commonly observed in slow-onset inhibition experiments cause even larger deviations. When the substrate acts as inhibitor, performing experiments by increasing [S] means increasing the concentration of the substrate and of the inhibitor at a constant ratio. Therefore, there is no way of complementing the experiments by measuring progress curves at fixed substrate or fixed inhibitor in order to determine independently a particular rate constant of the mechanism.

Examining the covariance matrix after nonlinear regression of v_z against [S] data reveals that the parameters V', K'_m and K_{Si} are intertwined. Trying to look more in depth by numerical analysis of simulated data with added experimental errors does not help further. In this case, all individual rate constants must be passed the algorithm and there is no way of constraining one or more of them to gain meaningful results. The situation is different with inhibitors that are not the substrates themselves because the substrate reaction can be analyzed separately. The obtained parameters can be further used as known constants when analyzing the properties of the inhibitors.

8.13 Liberation from Slow-Onset Substrate Inhibition

The action of the *liberator* has been described in the chapter dedicated to multiple enzyme-modifier interactions (Sect. 7.2) recalling the scarce documentation of this regulation mechanism by practical examples. Redirected from Chap. 7, a case of liberation from substrate inhibition is described here because of its slow-onset character.

While exploring the properties of cathepsin K in extracellular proteolysis [24], Novinec and collaborators described a novel function of *clusterin*, a glycoprotein with a plethora of activities because it forms complexes with numerous partners fulfilling the function of an extracellular chaperone. With regard to cathepsin K, clusterin was found to protect and to stabilize the enzyme against thermal denaturation and inactivation by autoproteolysis. Moreover, clusterin was able to counteract the inhibition by substrate described in Sect. 8.12 without affecting itself the catalytic properties of the enzyme in any way, i.e., acting as activator or inhibitor.

The proposed liberation mechanism is shown in Scheme 8.11, where the liberator L is clusterin. I apologize for not using the same numbering of the rate constants of the original publication [24]: it is to maintain internal consistency with the numbering in this chapter. In Scheme 8.11, a reaction path in two steps is added to the reactions of slow-onset substrate inhibition to show the strategy of the liberator in subtracting enzyme molecules from the path that includes inhibition and deviating the reaction toward *regular* catalysis. The integrated rate equation for the liberation mechanism in Scheme 8.11 is again (8.10) with the following expressions of v_z, v_s and λ:

$$v_z = \frac{V'\,[\mathrm{S}] + V^{\mathrm{L}}\,[\mathrm{S}]\,\dfrac{K_{\mathrm{m}}\,[\mathrm{L}]}{K_{\mathrm{m}}^{\mathrm{L}} K_{\mathrm{L}}}}{K'_{\mathrm{m}}\left\{1 + \dfrac{[\mathrm{L}]}{K_{\mathrm{L}}}\left(1 + \dfrac{[\mathrm{S}]}{K_{\mathrm{m}}^{\mathrm{L}}}\right)\right\} + [\mathrm{S}]\left(1 + \dfrac{[\mathrm{S}]}{K_{\mathrm{Si}}}\right)} \tag{8.33}$$

$$v_s = \frac{V'\,[\mathrm{S}] + V^{\mathrm{L}}\,[\mathrm{S}]\,\dfrac{K_{\mathrm{m}}\,[\mathrm{L}]}{K_{\mathrm{m}}^{\mathrm{L}} K_{\mathrm{L}}}}{K'_{\mathrm{m}}\left\{1 + \dfrac{[\mathrm{L}]}{K_{\mathrm{L}}}\left(1 + \dfrac{[\mathrm{S}]}{K_{\mathrm{m}}^{\mathrm{L}}}\right)\right\} + [\mathrm{S}]\left(1 + \dfrac{[\mathrm{S}]}{K_{\mathrm{Si}}^{*}}\right)} \tag{8.34}$$

$$\lambda = k_{-12} + k_{12}\frac{[\mathrm{S}]}{K_{\mathrm{Si}}\left\{1 + \dfrac{K'_{\mathrm{m}}}{[\mathrm{S}]} + K'_{\mathrm{m}}\dfrac{[\mathrm{L}]}{K_{\mathrm{L}}}\left(\dfrac{1}{[\mathrm{S}]} + \dfrac{1}{K_{\mathrm{m}}^{\mathrm{L}}}\right)\right\} + [\mathrm{S}]} \tag{8.35}$$

Equations (8.33)–(8.35) apply to Scheme 8.11 as written but, according to the definition of liberator, such a molecule does not affect the activity of the enzyme per se. Therefore, in Scheme 8.11 $k_4 = k_1$, $k_{-4} = k_{-1}$ and $k'_2 = k_2$, which imply $K'_{\mathrm{m}} = K_{\mathrm{m}}^{\mathrm{L}}$ and $V' = V^{\mathrm{L}}$. Considering these equalities, (8.33)–(8.35) simplify to the forms displayed in the group of equations (8.36).

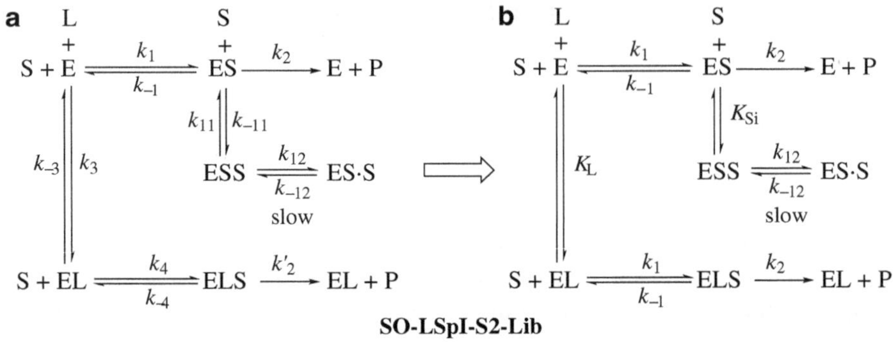

Scheme 8.11 Liberation from slow-onset substrate inhibition. (**a**) Slow-onset liberation from two-step inhibition by substrate shown as mechanism SO-LCaI-S2 in Scheme 8.10. (**b**) Representation of the same mechanism taking into account the equalities between rate constants

SO – LCaI – S2 – Lib (8.36)

$$\text{required}: \; K'_{\mathrm{m}} > K^*_{\mathrm{Si}} \,,\; k_{-1} \gg k_{12} \,,\; k_{-1} \gg k_{-12}$$

$$\text{definition (Scheme 8.11 a)}: \; k_4 = k_1 \,,\; k_{-4} = k_{-1} \,,\; K'_{\mathrm{m}} = K^{\mathrm{L}}_{\mathrm{m}} \,,\; k'_2 = k_2$$

$$\lambda = k_{-12} + k_{12}\frac{[\mathrm{S}]}{K_{\mathrm{Si}}\left(1+\dfrac{K'_{\mathrm{m}}}{[\mathrm{S}]}\right)\left(1+\dfrac{[\mathrm{L}]}{K_{\mathrm{L}}}\right)+[\mathrm{S}]} \,,\quad V' = k_2[\mathrm{E}]_{\mathrm{t}}$$

$$K_{\mathrm{Si}} = \frac{k_{-11}}{k_{11}} \,,\quad K^*_{\mathrm{Si}} = K_{\mathrm{Si}}\frac{k_{-12}}{k_{12}+k_{-12}} \,,\quad K'_{\mathrm{m}} = \frac{k_{-1}+k_2}{k_1} \,,\quad k_{\mathrm{off}} = \lambda\frac{v_{\mathrm{s}}}{v_{\mathrm{z}}} = k_{-12}$$

$$v_{\mathrm{z}} = \frac{V'[\mathrm{S}]\left(1+\dfrac{[\mathrm{L}]}{K_{\mathrm{L}}}\right)}{K'_{\mathrm{m}}\left\{1+\dfrac{[\mathrm{L}]}{K_{\mathrm{L}}}\left(1+\dfrac{[\mathrm{S}]}{K'_{\mathrm{m}}}\right)\right\}+[\mathrm{S}]\left(1+\dfrac{[\mathrm{S}]}{K_{\mathrm{Si}}}\right)}$$

$$v_{\mathrm{s}} = \frac{V'[\mathrm{S}]\left(1+\dfrac{[\mathrm{L}]}{K_{\mathrm{L}}}\right)}{K'_{\mathrm{m}}\left\{1+\dfrac{[\mathrm{L}]}{K_{\mathrm{L}}}\left(1+\dfrac{[\mathrm{S}]}{K'_{\mathrm{m}}}\right)\right\}+[\mathrm{S}]\left(1+\dfrac{[\mathrm{S}]}{K^*_{\mathrm{Si}}}\right)}$$

Liberation from substrate inhibition is illustrated in Fig. 8.26 using the rate constants of Fig. 8.24 and adding those of the liberator. The progress curve with the lowest rate is that without added liberator. Increasing the concentration of the liberator results in the increase of v_{z} and v_{s} until the exponential vanishes resulting in linear increase of [P] (the curve at [L] = 100 μM). After fitting (8.10) to simulated

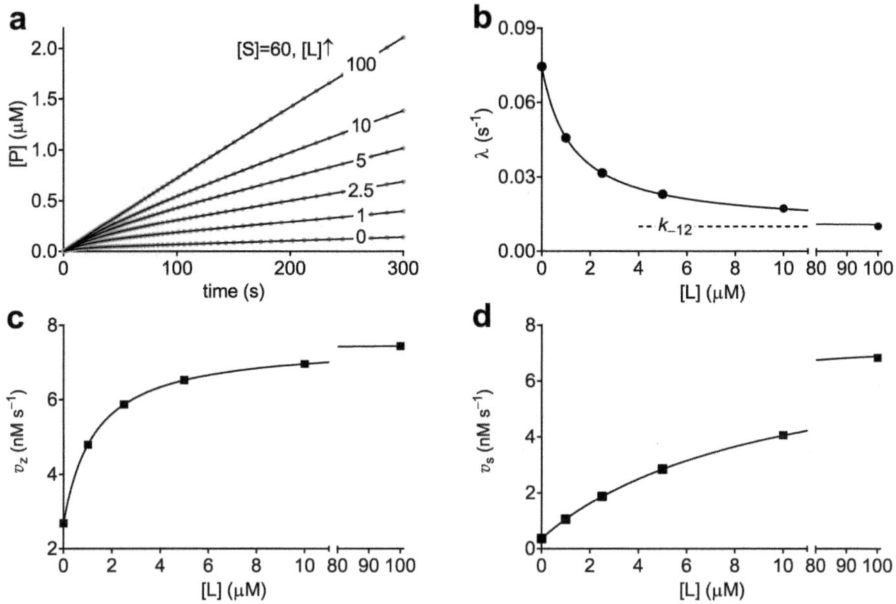

Fig. 8.26 Liberation from slow-onset linear catalytic substrate inhibition. The acronym assigned this action is SO-LCaI-S2-Lib. The progress curves (**a**) were simulated by numerical integration with: $k_1 = k_4 = 100$, $k_{-1} = k_{-4} = 1{,}990$, $k_2 = k_2' = 10$, $k_3 = 80$, $k_{-3} = 36$, $k_{11} = 20$, $k_{-11} = 500$, $k_{12} = 0.1$, $k_{-12} = 0.01$, $[E]_t = 0.001$, $[S] = 60$. The units of the second-order and first-order rate constants are $\mu M^{-1}\,s^{-1}$ and s^{-1}, respectively. The rate constants are those of SO-LCaI-S2 in Fig. 8.24 with the addition of liberator, L, whose concentrations are indicated in panel (**a**). Equation (8.10) was fitted to the progress curves in (**a**) to calculate v_z, v_s and λ that were then plotted against the concentration of the liberator as shown in panels (**b–d**). The *solid lines* through the points in (**b–d**) represent fits using the corresponding equations in (8.36): all parameters but K_L were constrained as constants known from the fits in Fig. 8.24. The best-fit value of K_L was 0.45 μM

data in (a), the rates at time zero, at steady-state and the frequency constant were plotted against the concentration of the liberator as shown in panels (b–d).

The relatively complex expressions of v_z, v_s, and λ in (8.36) do not pose problems in this case when fitted to data because all parameters but the dissociation constant of the liberator, K_L, are known from the calculations performed in Fig. 8.24. Having K_L as the sole parameter to be calculated in panels (b–d) the success is assured.

For illustrating slow-onset substrate inhibition and its liberation, simulations have been performed here and in the preceding section using fictional rate constant despite available data by Novinec et al. [23, 24]. The reason is that for simulations by numerical integration, *all rate constants* are needed, while the analysis of real data yields only the rate constants k_{12} and k_{-12} and otherwise equilibrium constants.

8.14 Slow-Onset Inhibition When the Inhibitor Binds Faster Than the Substrate

Slow-onset inhibition can result from a combination of slow association between the enzyme and the substrate and fast association between enzyme and inhibitor. The reaction shown in Scheme 8.12 is the same as mechanism SO-LSpI-1 with inverted roles of the slow and fast steps. Since *the substrate* is slow this time, the acronym SO-LSpI-SlowS has been chosen. While investigating the inhibition of hydroxymethylglutaryl-CoA reductase (NADPH) (EC 1.1.1.34) by compactin, Nakamura and Abeles observed that the inhibitor associated with the enzyme with a nearly diffusion-controlled rate while the association of enzyme and substrate was two orders of magnitude slower [22]. Many thanks to Lizbeth Hedstrom and Karen Allen for bringing to my attention the essence of this interesting paper that I had read too superficially. This work will be commented below after some considerations of the mechanism.

Hydroxymethylglutaryl-CoA reductase (HMGR) catalyzes the reduction of (*S*)-3-hydroxy-3-methylglutaryl coenzyme A (HMG-CoA) to (*R*)-mevalonate. The second substrate of the reaction is $NADP_{red}$:

$$(S)\text{-HMG-CoA} + 2\text{NADP}_{\text{red}} \xrightleftharpoons{\text{HMGR}} (R)\text{-mevalonate} + 2\text{NADP}_{\text{ox}} + \text{CoASH}$$

Mevalonate is the precursor of isoprenoids that are further used in the synthesis of cholesterol, heme, and farnesyl-pyrophosphate. Compactin, the prototype of *statins*, a group of drugs that lower cholesterol levels, is a linear specific inhibitor of HMGR with respect to HMG-CoA. If applied to the HMGR-catalyzed reaction, S in Scheme 8.12 represents HMG-CoA, while $NADP_{red}$ is kept constant at a saturating concentration and thus formally omitted from the reaction scheme (see the comments on simplified reaction schemes on p. 44).

The analysis of this mechanism poses some problems inherent in the slow substrate association with the enzyme and, for understanding the properties of the reaction, Fig. 8.27 shows the changes with time of [P] and the concomitant changes of [E], [ES] and [E · I]. The reactions were simulated with enzyme as

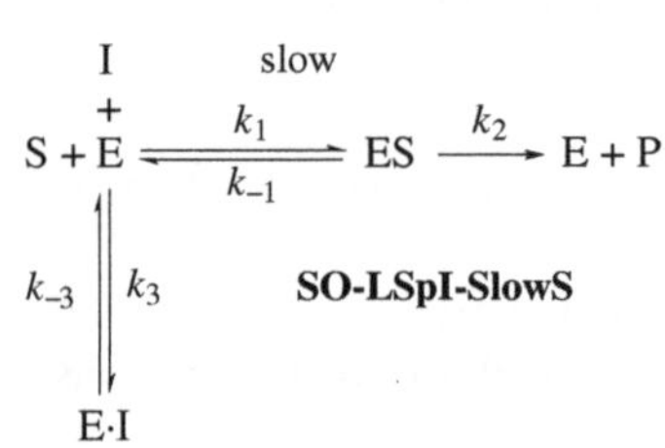

Scheme 8.12 Slow-onset inhibition: E·I is formed faster than ES (SO-LSpI-SlowS)

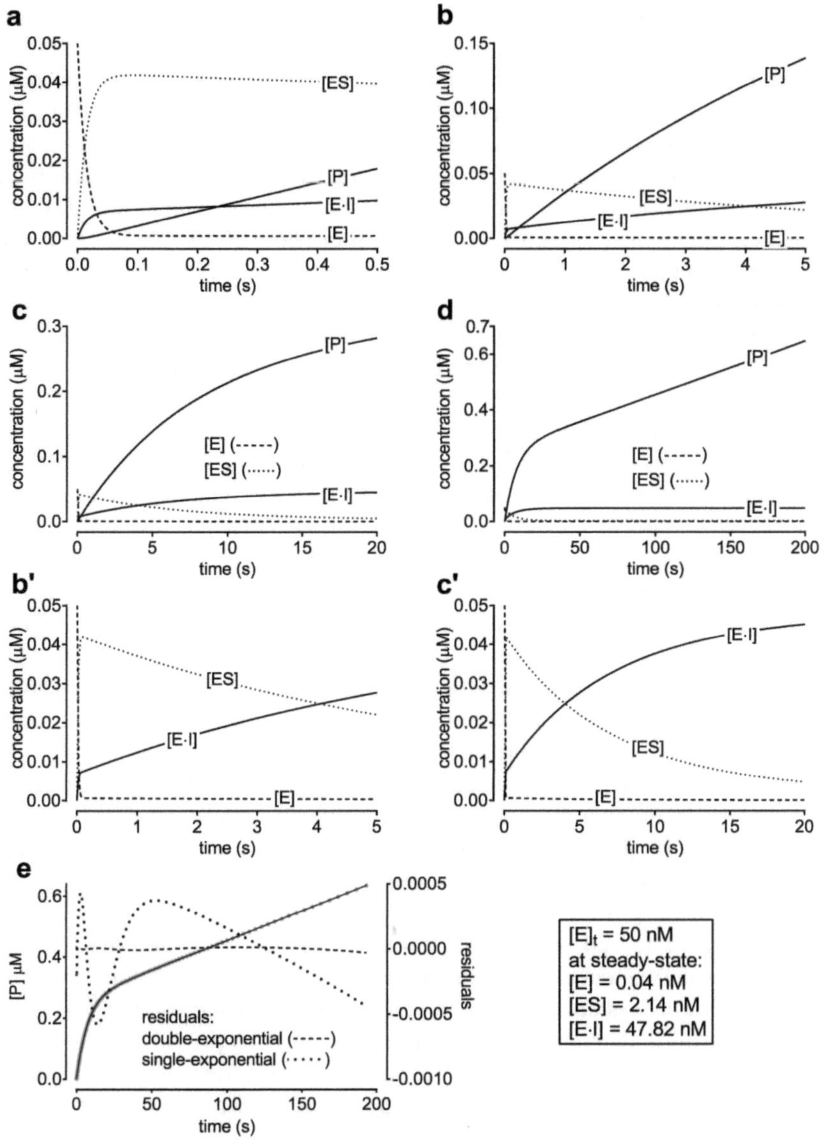

Fig. 8.27 Slow-onset inhibition SO-LSpI-SlowS based on the model in Scheme 8.12. Changes with time of [E], [ES], [E·I] and [P] in four time intervals ((**a**)–(**d**)). With the exception of k_2, the rate constants were taken from [22]: $k_1 = 0.19\,\mu\text{M}^{-1}\,\text{s}^{-1}$, $k_{-1} = 0.11\,\text{s}^{-1}$, $k_2 = 0.9\,\text{s}^{-1}$, $k_3 = 27\,\mu\text{M}^{-1}\,\text{s}^{-1}$, $k_{-3} = 0.0065\,\text{s}^{-1}$, $[\text{E}]_t = 0.05\,\mu\text{M}$, $[\text{S}]_t = 300\,\mu\text{M}$, $[\text{I}]_t = 0.35\,\mu\text{M}$. Panels (**b′**) and (**c′**) show the same results as the corresponding panels (**b**) and (**c**) omitting the traces of [P] and expanding the scale of the ordinate. Panel (**e**) shows the curve of product, simulated with 300 points on a logarithmic timebase, fitted with ether a single (8.8) or a double exponential function (8.9): the residuals indicate the double-exponential fit as appropriate. At steady-state, attained about 50 s after starting the reaction, the concentrations of the enzyme-containing species are those shown in the comment box

starter using the rate constants[4] of Nakamura and Abeles for the inhibition of hydroxymethylglutaryl-CoA reductase by compactin [22]. Panels (a–d) in Fig. 8.27 represent four time intervals of the same reaction to illustrate its relevant properties. At the beginning, the enzyme is partitioned between ES and E·I. In panel (a), the initial burst of [ES] is larger than that of [E · I] because, despite the second-order rate constant for the binding of inhibitor is greater by two orders of magnitude, the rate of ES-formation is larger because $[S]_t \gg [I]_t$. Following the large initial increase, [ES] decreases exponentially to 4.3 % of $[E]_t$ after about 50 s while [E · I] increases exponentially to 95.6 % of $[E]_t$. The entire progress of these concentrations can be followed in panels (b–c) and in panels (b′ and c′) with enlarged ordinate. Thus, in this example, the steady-state is established about 50 s after start. The free enzyme concentration drops rapidly to a very low level that corresponds 0.08 % of $[E]_t$ at steady-state.

The progress curves of [P] consist of *two exponential phases* and a linear steady-state. This property is not immediately perceptible from the trace of [P] but is revealed after comparing the fits of a single or a double exponential, (8.8) and (8.9), to data as illustrated in panel (e). It may be argued that the residuals of fitting the single-exponential function to data are small. However, this is the reason why fitting (8.10) to progress curves, as done with the mechanisms discussed in the preceding sections, yields inconsistent and erratic results between curves obtained at different inhibitor concentrations. In particular, there is no frequency constant corresponding to a *single relaxation time* and rate at time zero showing systematic trends that could be used for diagnostic and quantitative purposes. The rates at steady-state can however be measured precisely.

Established that the pre-steady-state of the progress curve in Fig. 8.27e contains two exponentials, the two frequency constants, which correspond to the reciprocal relaxation times of the system, can however not be resolved as representatives of two distinct phases of the reaction (see Sect. 8.2). Briefly and using a simple language, *both* relaxation times characterize the transient that ends up with the onset of steady-state and we cannot decide where one starts and the other ends because the two frequency constants are mutually *coupled* and *both* contain the rate constants of the system. A mathematically rigorous explanation of this concept, which is however understandable without effort, has been provided by Bernasconi [3, Chap. 3]. For the interested reader, examples of rate constants calculated from relaxation times can be found in three papers unrelated to slow-onset inhibition [13, 25, 26].

For the reasons discussed in this section and Sect. 8.2, a *practical* integrated rate equation of the type (8.10) that takes into account the *entire reaction* cannot be derived for the slow-onset mechanism SO-LSpI-SlowS. Additional problems may accompany the analysis of this mechanism, such as depletion of inhibitor (tight-

[4] With the exception of the catalytic constant k_2 that was not specified. This has however no consequences for the discussion.

binding) and possibly substrate consumption when a series of reagent concentrations is explored. Considering these facts, the analysis of this mechanism is best accomplished using a numerical integration approach while renouncing to the more *classical* application of the integrated rate equation because this underestimates the complexity of the progress curves.

The nature of the mechanism requires the use of a rapid-mixing device for data collection on a logarithmic timebase. Any stopped-flow apparatus is certainly adequate to measure with high accuracy, beyond the dead-time of the instrument, the progress of [P] to reveal the lag at the beginning (Fig. 8.27a), the inversion of tendency at higher times (panels b and c) and the steady-state (panel d).

Nakamura and Abeles analyzed the inhibition of hydroxymethylglutaryl-CoA reductase by compactin on the portions of the progress curves that follow the initial burst of ES using a graphical method based on a steady-state assumption [22]. This was a logical approach considering that the reactions were monitored by conventional photometry and manual pipetting of the reagents that did not allow the exact determination of the starting point of the reaction from one experiment to the other. The measurements were performed at saturating concentrations of substrate and coenzyme and three inhibitor concentrations that permitted to determine the association and dissociation rate constants for the inhibitor and hydroxymethylglutaryl-CoA. In the same paper [22], the authors provided a mathematical treatment of the model and derived an integrated rate equation containing one constant and two exponential terms that was used to calculate theoretical curves that were compared with experimental data. The agreement was not good but anyway within reasonable limits considering that only 7–9 rates were measured at time intervals [22, Fig. 3A]. If this work should be performed nowadays, the authors would certainly use a stopped-flow apparatus and modern software for data analysis as discussed below.

Two experiments simulated with the rate constants and concentrations of Nakamura and Abeles [22, Fig. 1] are shown in Fig. 8.28. Here only the catalytic constant is arbitrary because it was not specified by the authors and the signal refers to the increase of the product (*R*)-mevalonate instead of the decrease of $NADP_{red}$. Two progress curves, started with either enzyme or substrate, are *measured* as described in panels (a) and (b). In (a), only the enzyme is introduced in one syringe of the stopped-flow apparatus, while the other contains substrate and inhibitor. In this example a preincubation of 100 s is chosen and nothing happens during this time until the two solutions are rapidly mixed at the time indicated by the arrow. Following mixing, the reaction is recorded with 500 points on a logarithmic timebase with some error added. The enzyme partitions between ES and E·I as described above leaving very little free enzyme (panel c). In the 200 s time interval after mixing, the product concentration increases as shown in trace 1 of panel (e), where the pre-steady-state consists of a double exponential due to [ES] first increasing and then decreasing.

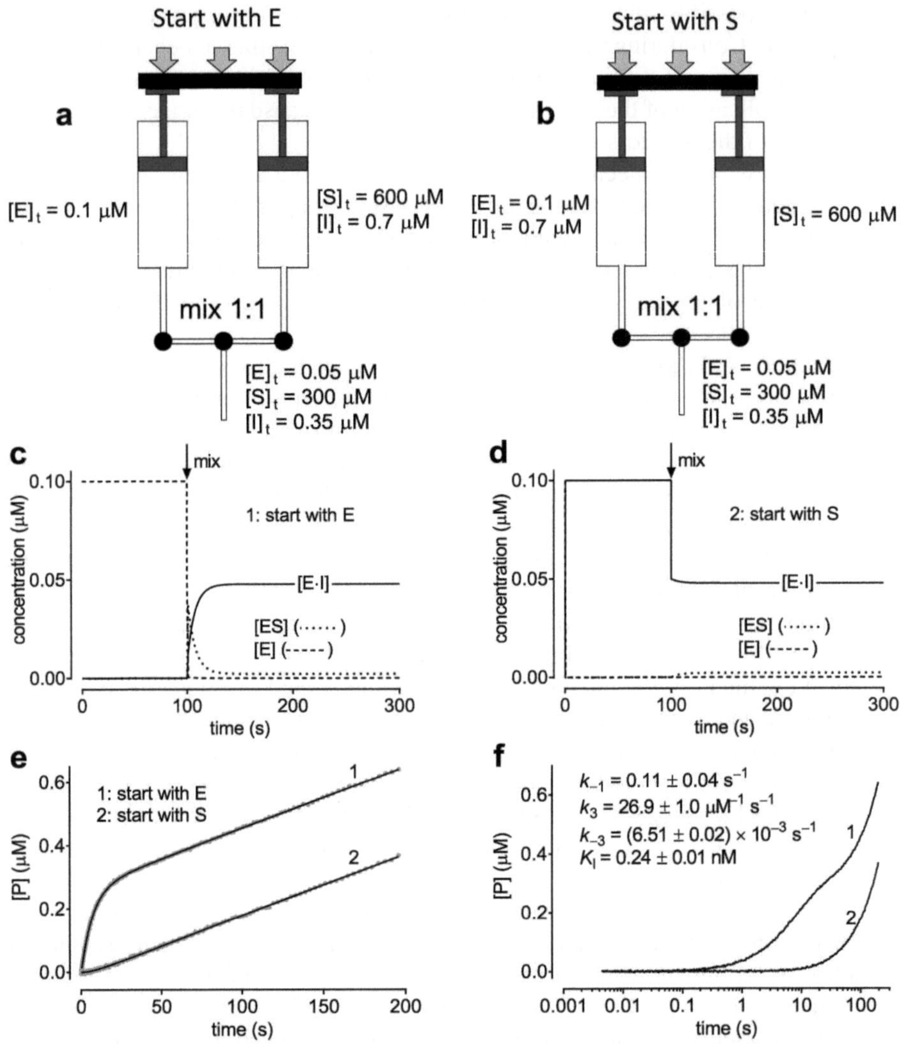

Fig. 8.28 Slow-onset inhibition mechanism SO-LSpI-SlowS. Simulated experiments by starting the reaction with either enzyme (**a**) or substrate (**b**). The rate constants but k_2 are from [22]: $k_1 = 0.19\,\mu\text{M}^{-1}\,\text{s}^{-1}$, $k_{-1} = 0.11\,\text{s}^{-1}$, $k_2 = 0.9\,\text{s}^{-1}$, $k_3 = 27\,\mu\text{M}^{-1}\,\text{s}^{-1}$, $k_{-3} = 0.0065\,\text{s}^{-1}$ and concentrations as shown in panels (**a, b**). The concentrations of the enzyme-containing species during the reaction are shown in panels (**c**) and (**d**). The progress curves, consisting of 500 points collected on a logarithmic timebase, are plotted on a linear (**e**) or on a logarithmic time scale (**f**)

Figure 8.28b shows the experimental setup to measure a reaction started by substrate. In this case, the enzyme preincubated with inhibitor forms a complex with the inhibitor immediately, leaving less than 0.04 % free enzyme. Upon mixing, the system enters in the steady-state without experiencing the large overshoot followed by decrease of [ES] observed in the reaction started with enzyme. Instead, [ES] increases by a tiny amount to its steady-state value and, as shown in panel (d), [E · I] decreases slightly. The traces in panel (d) and trace 2 in panel (e) should not suggest that the pre-steady-state is completed faster than in the reaction started with enzyme since the steady-state is also established in about 50 s. The slope of trace (2) in panel (e) is the same as that of trace 1. Finally, in panel (f) the same curves in (e) are plotted on a logarithmic time scale to better reveal the different ways the steady-state is approached when the reaction is started by adding the enzyme or the substrate.

Assuming to have previously determined the rate constants of substrate binding and catalysis by analyzing the reaction in the absence of inhibitor, the rate constants k_1 and k_2 are constrained as known values during simultaneous numerical integration of the two progress curves in panel (e) while optimizing the remaining three constants. Although being *known*, k_{-1} is included in the optimization because various simulations revealed that fixing this rate constant negatively influences the optimization of k_3 and k_{-3} suggesting that they control together the transient that precedes the steady-state. There is in fact a certain degree of dependence between the intertwined constants k_3 and k_{-3}, which are however optimized to values that give the same ratio in experiments simulated with variable added error. The results are summarized in the inset of panel (f), where the inhibition constant calculated as the ratio k_{-3}/k_3 is that expected from the values used for generating the fake data and *coincides* with the value of Nakamura and Abeles [22, Table II].

Appendix 1

Slow-Onset Inhibition: Experiment Design Using a Stopped-Flow Apparatus

The following Figs. 8.29 and 8.30 illustrate the experimental design that may be used for measuring progress curves of slow-onset inhibition using a stopped-flow apparatus in its simplest version equipped with two syringes. In Fig. 8.29 the syringes have the same volume and deliver the same amount of reagents to the

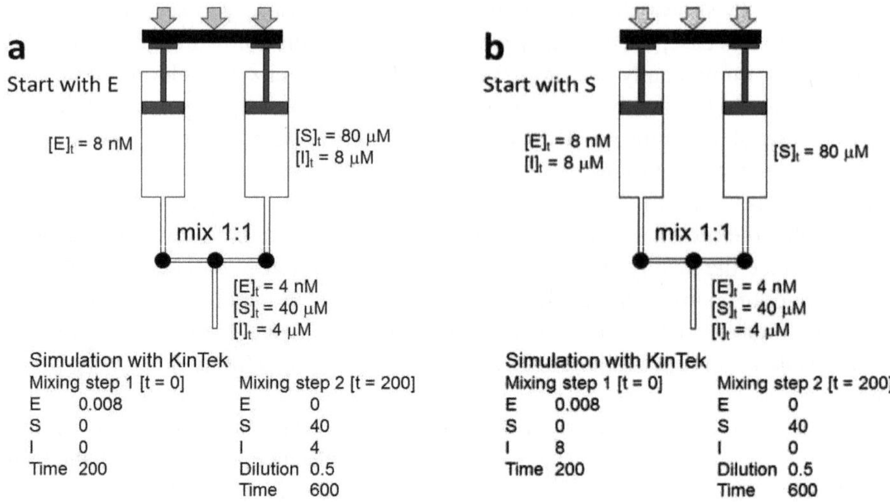

Fig. 8.29 Slow-onset inhibition: experimental design for measuring progress curves using a stopped-flow apparatus equipped with two syringes that deliver the same volume to the measuring cell. With this arrangement, the solutions of the *left* and of the *right* syringe are mixed in a ratio of 1:1, which corresponds to dilution factors of 0.5 for the reagents contained in the *left and right* syringe. The concentrations of the reagents shown in the figure refer to the example illustrated in Fig. 8.5. (**a**) Reaction started by mixing enzyme with a solution containing substrate and inhibitor. (**b**) Reaction started by mixing substrate with enzyme preincubated with the inhibitor. In the KinTek entries concentrations are μM and time seconds

measuring cell. The volume of the two syringes is not the same in Fig. 8.30, in this case a 0.2 mL and a 5 mL syringe. The concentrations of the reagents in the two syringes are shown on their left and right, while the concentrations after mixing are shown beneath the valves. The reagents and the final concentrations in both figures are those used for simulating the data in Fig. 8.5.

For readers interested in performing this and other simulations using KinTek software, information is provided beneath the pictures of the syringes. The dilution undergone by the reagent in the first mixing is specified in mixing step 2, while the concentrations of the reagents in mixing step 2 are entered with their final values.

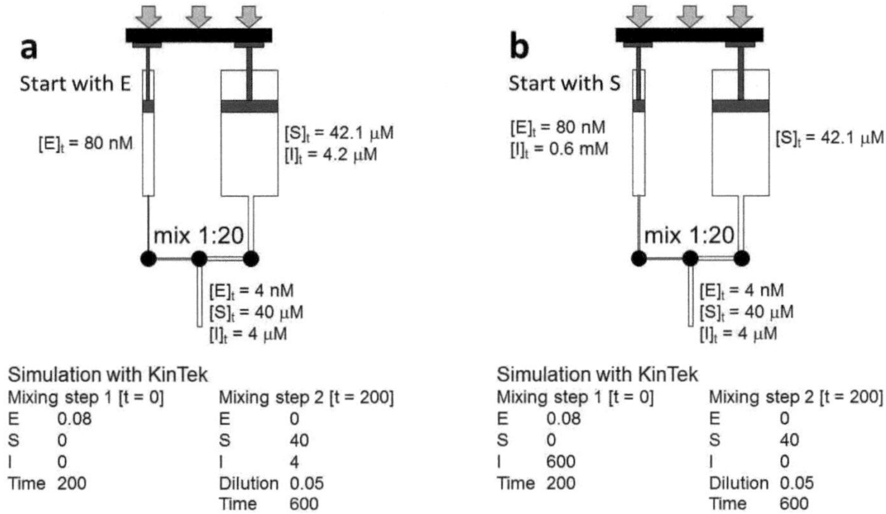

Fig. 8.30 Slow-onset inhibition: experimental design for measuring progress curves by avoiding undesired dilution of the enzyme in the preincubation phase. Scheme of the reagent mixing unit of a stopped-flow apparatus equipped with sample syringes of different size. The *left and right* syringes have volumes of 0.2 and 5.0 mL, respectively. With this arrangement, the solutions of the *left* and of the *right* syringe are mixed in a ratio of 1:20, which corresponds to dilution factors of 0.05 and 0.95 for the reagents contained in the *left and right* syringe, respectively. This is a variant of the experiment design shown in Fig. 8.29 and the final concentrations of the reagents are again those of the example illustrated in Fig. 8.5. (**a**) Reaction started by mixing enzyme with a solution containing substrate and inhibitor. (**b**) Reaction started by mixing substrate with enzyme preincubated with inhibitor. This experiment is possible if the solubility of the inhibitor does not pose problems to reach the necessary concentration in the small syringe. In the KinTek entries concentrations are μM and time seconds

Appendix 2

Derivation of the Integrated Rate Equation for Mechanisms SO-LSpI-2Temp and SO-LSpI-2

These mechanism are described in Sects. 8.4 and 8.6 with the reactions in Schemes 8.2 and 8.4. The properties of SO-LSpI-2Temp are the same as those of SO-LSpI-2 with the difference that in the temporary mechanism an alternative substrate mimics the action of an inhibitor because it is turned over very slowly. An additional assumption besides the points 1–5 on p. 374 is necessary, namely $[I^*] \ll [I]_t$, i.e., $[I]_t$ is sufficiently large to act as a sink to satisfy this condition. The derivation that follows is performed for a reaction started by adding enzyme to a mixture of substrate and inhibitor, while the reaction started by adding substrate will be considered at the end of the derivation for mechanism SO-LSpI-2.

The concentrations of free enzyme [E] and of [EI] are calculated from the steady-state expression of [ES] and from the rapid equilibrium for EI formation, respectively, while [E·I] is not in rapid equilibrium:

$$\frac{\mathrm{d}\,[\mathrm{ES}]}{\mathrm{d}t} = k_1\,[\mathrm{E}]\,[\mathrm{S}] - (k_{-1} + k_2)\,[\mathrm{ES}] = 0\,, \quad [\mathrm{E}] = \frac{k_{-1} + k_2}{k_1}\frac{[\mathrm{ES}]}{[\mathrm{S}]}\,,$$

$$[\mathrm{E}] = \frac{K_\mathrm{m}}{[\mathrm{S}]}\,[\mathrm{ES}]\,, \tag{8.37}$$

$$\frac{\mathrm{d}\,[\mathrm{EI}]}{\mathrm{d}t} = k_3\,[\mathrm{E}]\,[\mathrm{I}] - k_{-3}\,[\mathrm{EI}] = 0\,, \quad [\mathrm{EI}] = \frac{k_3\,[\mathrm{E}]\,[\mathrm{I}]}{k_{-3}} = \frac{[\mathrm{E}]\,[\mathrm{I}]}{K_\mathrm{I}}\,,$$

$$[\mathrm{EI}] = \frac{K_\mathrm{m}}{[\mathrm{S}]}\frac{[\mathrm{I}]}{K_\mathrm{I}}\,[\mathrm{ES}]\,, \tag{8.38}$$

$$\frac{\mathrm{d}\,[\mathrm{E}\cdot\mathrm{I}]}{\mathrm{d}t} = k_4\,[\mathrm{EI}] - (k_{-4} + k_5)\,[\mathrm{E}\cdot\mathrm{I}]\,. \tag{8.39}$$

The conservation of the enzyme is given by (8.40) and, since $\mathrm{d}[\mathrm{E}]_\mathrm{t}/\mathrm{d}t = 0$, it can be written as (8.41), where the species in steady-state or in rapid equilibrium (here E, ES and EI) are separated from those that are not in equilibrium (here only E·I):

$$[\mathrm{E}]_\mathrm{t} = [\mathrm{E}] + [\mathrm{ES}] + [\mathrm{EI}] + [\mathrm{E}\cdot\mathrm{I}]\,, \tag{8.40}$$

$$\frac{\mathrm{d}\,[\mathrm{E} + \mathrm{ES} + \mathrm{EI}]}{\mathrm{d}t} = -\frac{\mathrm{d}\,[\mathrm{E}\cdot\mathrm{I}]}{\mathrm{d}t}\,. \tag{8.41}$$

The expressions in (8.37)–(8.39) are introduced in (8.41) and then [E·I] is extracted after rearrangement:

$$\frac{\mathrm{d}}{\mathrm{d}t}\left(\frac{K_\mathrm{m}}{[\mathrm{S}]}\,[\mathrm{ES}] + [\mathrm{ES}] + [\mathrm{ES}]\,\frac{K_\mathrm{m}\,[\mathrm{I}]}{[\mathrm{S}]\,K_\mathrm{I}}\right) = -k_4\,[\mathrm{EI}] + (k_{-4} + k_5)\,[\mathrm{E}\cdot\mathrm{I}]\,,$$

$$\frac{\mathrm{d}\,[\mathrm{ES}]}{\mathrm{d}t}\left(1 + \frac{K_\mathrm{m}}{[\mathrm{S}]} + \frac{K_\mathrm{m}\,[\mathrm{I}]}{[\mathrm{S}]\,K_\mathrm{I}}\right) = -k_4\frac{K_\mathrm{m}\,[\mathrm{I}]}{[\mathrm{S}]\,K_\mathrm{I}}\,[\mathrm{ES}] + (k_{-4} + k_5)\,[\mathrm{E}\cdot\mathrm{I}]\,,$$

$$[\mathrm{E}\cdot\mathrm{I}] = \frac{\mathrm{d}\,[\mathrm{ES}]}{\mathrm{d}t}\frac{1}{k_{-4} + k_5}\left(1 + \frac{K_\mathrm{m}}{[\mathrm{S}]} + \frac{K_\mathrm{m}\,[\mathrm{I}]}{[\mathrm{S}]\,K_\mathrm{I}}\right) + \frac{k_4}{k_{-4} + k_5}\frac{K_\mathrm{m}\,[\mathrm{I}]}{[\mathrm{S}]\,K_\mathrm{I}}\,[\mathrm{ES}]\,. \tag{8.42}$$

Equations (8.37), (8.38), and (8.42) are now put in (8.40) to give

$$[\mathrm{E}]_\mathrm{t} = \frac{K_\mathrm{m}}{[\mathrm{S}]}[\mathrm{ES}] + [\mathrm{ES}] + \frac{K_\mathrm{m}[\mathrm{I}]}{[\mathrm{S}]K_\mathrm{I}}[\mathrm{ES}]$$

$$+ \frac{\mathrm{d}[\mathrm{ES}]}{\mathrm{d}t}\frac{1}{k_{-4}+k_5}\left(1 + \frac{K_\mathrm{m}}{[\mathrm{S}]} + \frac{K_\mathrm{m}[\mathrm{I}]}{[\mathrm{S}]K_\mathrm{I}}\right) + \frac{k_4}{k_{-4}+k_5}\frac{K_\mathrm{m}[\mathrm{I}]}{[\mathrm{S}]K_\mathrm{i}}[\mathrm{ES}]\ . \quad (8.43)$$

With rearrangement and factorization, $\mathrm{d}[\mathrm{ES}]/\mathrm{d}t$ is extracted from (8.43) giving

$$\frac{\mathrm{d}[\mathrm{ES}]}{\mathrm{d}t} = \frac{(k_{-4}+k_5)[\mathrm{S}]}{K_\mathrm{m}\left(1+\frac{[\mathrm{I}]}{K_\mathrm{I}}\right)+[\mathrm{S}]}[\mathrm{E}]_\mathrm{t} - [\mathrm{ES}]\left\{k_{-4}+k_5+\frac{k_4[\mathrm{I}]}{K_\mathrm{I}\left(1+\frac{[\mathrm{S}]}{K_\mathrm{m}}\right)+[\mathrm{I}]}\right\} \quad (8.44)$$

but to proceed with less cumbersome expressions the term in curled braces is called λ and the term multiplying $[\mathrm{E}]_\mathrm{t}$ is called α:

$$\alpha = \frac{(k_{-4}+k_5)[\mathrm{S}]}{K_\mathrm{m}\left(1+\frac{[\mathrm{I}]}{K_\mathrm{I}}\right)+[\mathrm{S}]}, \quad (8.45)$$

$$\lambda = k_{-4}+k_5+\frac{k_4[\mathrm{I}]}{K_\mathrm{I}\left(1+\frac{[\mathrm{S}]}{K_\mathrm{m}}\right)+[\mathrm{I}]}. \quad (8.46)$$

λ in (8.46) corresponds to the frequency constant that appears in the final integrated rate equation. Equation (8.44) can thus be rewritten in the simpler form (8.48) and integrated after rearrangement:

$$\frac{d[\mathrm{ES}]}{dt} = \alpha[\mathrm{E}]_\mathrm{t} - \lambda[\mathrm{ES}]\ , \quad (8.47)$$

$$-\ln\left(\frac{\alpha}{\lambda}[\mathrm{E}]_\mathrm{t} - [\mathrm{ES}]\right) = \lambda t + C\ . \quad (8.48)$$

Since the derivation is being performed for a reaction stated with enzyme, the appropriate boundary condition for calculating the integration constant C in (8.48) must consider that at $t = 0$, $[\mathrm{E\cdot I}] = 0$. The mass-conservation expression (8.40) reduces to $[\mathrm{E}]_\mathrm{t} = [\mathrm{E}] + [\mathrm{ES}] + [\mathrm{EI}]$ and $[\mathrm{ES}]_\mathrm{z}$, the concentration of the ES complex at $t = 0$, can be obtained from a modified form of (8.43):

$$[\mathrm{E}]_\mathrm{t} = \frac{K_\mathrm{m}}{[\mathrm{S}]}[\mathrm{ES}]_\mathrm{z} + [\mathrm{ES}]_\mathrm{z} + \frac{K_\mathrm{m}[\mathrm{I}]}{[\mathrm{S}]K_\mathrm{I}}[\mathrm{ES}]_\mathrm{z} = [\mathrm{ES}]_\mathrm{z}\left\{1+\frac{K_\mathrm{m}}{[\mathrm{S}]}\left(1+\frac{[\mathrm{I}]}{K_\mathrm{I}}\right)\right\}$$

$$[\mathrm{ES}]_{\mathrm{z}} = \frac{[\mathrm{E}]_{\mathrm{t}}\,[\mathrm{S}]}{K_{\mathrm{m}}\left(1+\dfrac{[\mathrm{I}]}{K_{\mathrm{I}}}\right)+[\mathrm{S}]}\,. \tag{8.49}$$

The integration constant is obtained by putting (8.49) in (8.48). Afterwards, the resulting expression (8.50) is introduced into (8.48) and rearranged to extract the complete expression of [ES] with the following passages:

$$C = -\ln\left\{\frac{\alpha}{\lambda}[\mathrm{E}]_{\mathrm{t}} - \frac{[\mathrm{E}]_{\mathrm{t}}\,[\mathrm{S}]}{K_{\mathrm{m}}\left(1+\dfrac{[\mathrm{I}]}{K_{\mathrm{I}}}\right)+[\mathrm{S}]}\right\} \tag{8.50}$$

$$-\ln\left(\frac{\alpha}{\lambda}[\mathrm{E}]_{\mathrm{t}} - [\mathrm{ES}]\right) = \lambda t - \ln\left\{\frac{\alpha}{\lambda}[\mathrm{E}]_{\mathrm{t}} - \frac{[\mathrm{E}]_{\mathrm{t}}\,[\mathrm{S}]}{K_{\mathrm{m}}\left(1+\dfrac{[\mathrm{I}]}{K_{\mathrm{I}}}\right)+[\mathrm{S}]}\right\}$$

$$e^{-\lambda t} = \frac{\dfrac{\alpha}{\lambda}[\mathrm{E}]_{\mathrm{t}} - [\mathrm{ES}]}{\dfrac{\alpha}{\lambda}[\mathrm{E}]_{\mathrm{t}} - \dfrac{[\mathrm{E}]_{\mathrm{t}}\,[\mathrm{S}]}{K_{\mathrm{m}}\left(1+\dfrac{[\mathrm{I}]}{K_{\mathrm{I}}}\right)+[\mathrm{S}]}}$$

$$[\mathrm{ES}] = \frac{\alpha}{\lambda}[\mathrm{E}]_{\mathrm{t}} - e^{-\lambda t}\left\{\frac{\alpha}{\lambda}[\mathrm{E}]_{\mathrm{t}} - \frac{[\mathrm{E}]_{\mathrm{t}}\,[\mathrm{S}]}{K_{\mathrm{m}}\left(1+\dfrac{[\mathrm{I}]}{K_{\mathrm{I}}}\right)+[\mathrm{S}]}\right\}. \tag{8.51}$$

After reintroducing (8.45) and (8.46) for α and λ, respectively, to give (8.52), (8.51) can be rewritten in its complete form

$$\frac{\alpha}{\lambda}[\mathrm{E}]_{\mathrm{t}} = \frac{[\mathrm{S}]}{K_{\mathrm{m}}\left\{1+\dfrac{[\mathrm{I}]}{K_{\mathrm{I}}}\left(1+\dfrac{k_4}{k_{-4}+k_5}\right)\right\}+[\mathrm{S}]}[\mathrm{E}]_{\mathrm{t}} \tag{8.52}$$

$$[\mathrm{ES}] = \frac{[\mathrm{S}]}{K_{\mathrm{m}}\left\{1+\dfrac{[\mathrm{I}]}{K_{\mathrm{I}}}\left(1+\dfrac{k_4}{k_{-4}+k_5}\right)\right\}+[\mathrm{S}]}[\mathrm{E}]_{\mathrm{t}} \tag{8.53}$$

$$- e^{-\lambda t}\left\{\frac{[\mathrm{S}]}{K_\mathrm{m}\left\{1+\frac{[\mathrm{I}]}{K_\mathrm{I}}\left(1+\frac{k_4}{k_{-4}+k_5}\right)\right\}+[\mathrm{S}]}[\mathrm{E}]_\mathrm{t}-\frac{[\mathrm{E}]_\mathrm{t}[\mathrm{S}]}{K_\mathrm{m}\left(1+\frac{[\mathrm{I}]}{K_\mathrm{I}}\right)+[\mathrm{S}]}\right\}.$$

The reaction rate is now calculated from (8.53) recalling that $v = \mathrm{d}\,[\mathrm{P}]/\mathrm{d}t = k_2\,[\mathrm{ES}]$ and to simplify the expressions the term multiplying $[\mathrm{E}]_\mathrm{t}$ in (8.53) is called A:

$$A = \frac{[\mathrm{S}]}{K_\mathrm{m}\left\{1+\frac{[\mathrm{I}]}{K_\mathrm{I}}\left(1+\frac{k_4}{k_{-4}+k_5}\right)\right\}+[\mathrm{S}]}, \tag{8.54}$$

$$\frac{d\,[\mathrm{P}]}{dt} = k_2A[\mathrm{E}]_\mathrm{t} - e^{-\lambda t}\left\{k_2A[\mathrm{E}]_\mathrm{t} - \frac{k_2[\mathrm{E}]_\mathrm{t}[\mathrm{S}]}{K_\mathrm{m}\left(1+\frac{[\mathrm{I}]}{K_\mathrm{I}}\right)+[\mathrm{S}]}\right\},$$

$$\frac{\mathrm{d}\,[\mathrm{P}]}{\mathrm{d}t} = VA - e^{-\lambda t}\left\{VA - \frac{V\,[\mathrm{S}]}{K_\mathrm{m}\left(1+\frac{[\mathrm{I}]}{K_\mathrm{I}}\right)+[\mathrm{S}]}\right\}, \tag{8.55}$$

with limiting rate $V = k_2[\mathrm{E}]_\mathrm{t}$. The rate at time zero when the reaction is started by adding the enzyme, $v_{\mathrm{z,E}}$, is calculated from (8.55) by setting the time $t = 0$, which means $\exp(-\lambda t) = 1$, while the rate at steady-state, v_s, is obtained at $t = \infty$, meaning $\exp(-\lambda t) = 0$. Thus, $v_{\mathrm{z,E}}$ corresponds to the last term on the right of (8.55) and $v_\mathrm{s} = VA$, with A defined in (8.54):

$$v_{\mathrm{z,E}} = \frac{V\,[\mathrm{S}]}{K_\mathrm{m}\left(1+\frac{[\mathrm{I}]}{K_\mathrm{I}}\right)+[\mathrm{S}]}, \tag{8.56}$$

$$v_\mathrm{s} = \frac{V\,[\mathrm{S}]}{K_\mathrm{m}\left\{1+\frac{[\mathrm{I}]}{K_\mathrm{I}}\left(1+\frac{k_4}{k_{-4}+k_5}\right)\right\}+[\mathrm{S}]},$$

$$= \frac{V\,[\mathrm{S}]}{K_\mathrm{m}\left\{1+\frac{[\mathrm{I}]}{K_\mathrm{I}}\frac{k_4+k_{-4}+k_5}{k_{-4}+k_5}\right\}+[\mathrm{S}]},$$

$$= \frac{V\,[\mathrm{S}]}{K_\mathrm{m}\left\{1 + \dfrac{[\mathrm{I}]}{K^*_{\mathrm{I,\,temp}}}\right\} + [\mathrm{S}]} \,. \tag{8.57}$$

The overall inhibition constant $K^*_{\mathrm{I,\,temp}}$ is defined as

$$K^*_{\mathrm{I,\,temp}} = K_\mathrm{I}\frac{k_{-4} + k_5}{k_4 + k_{-4} + k_5} \,. \tag{8.58}$$

With (8.56) and (8.57) the rate equation (8.55) can be now rewritten and integrated. The integration constant C' is calculated for $t = 0$, i.e., when $[\mathrm{P}] = 0$:

$$\frac{\mathrm{d}\,[\mathrm{P}]}{\mathrm{d}t} = v_\mathrm{s} - (v_\mathrm{s} - v_\mathrm{z})\,e^{-\lambda t} \tag{8.59}$$

$$\mathrm{d}\,[\mathrm{P}] = v_\mathrm{s}dt - (v_\mathrm{s} - v_\mathrm{z})\,e^{-\lambda t}\mathrm{d}t \tag{8.60}$$

$$\int \mathrm{d}\,[\mathrm{P}] = v_\mathrm{s}\int \mathrm{d}t - (v_\mathrm{s} - v_\mathrm{z})\int e^{-\lambda t}\mathrm{d}t \tag{8.61}$$

$$[\mathrm{P}] = v_\mathrm{s}t - \frac{1}{\lambda}(-v_\mathrm{s} + v_\mathrm{z})\,e^{-\lambda t} + C' \tag{8.62}$$

$$C' = \frac{1}{\lambda}(-v_\mathrm{s} + v_\mathrm{z}) \tag{8.63}$$

$$[\mathrm{P}] = v_\mathrm{s}t - \frac{1}{\lambda}(-v_\mathrm{s} + v_\mathrm{z})\,e^{-\lambda t} + \frac{1}{\lambda}(-v_\mathrm{s} + v_\mathrm{z}) \tag{8.64}$$

$$[\mathrm{P}] = v_\mathrm{s}t - \frac{1}{\lambda}(v_\mathrm{s} - v_\mathrm{z})\left(1 - e^{-\lambda t}\right) \,. \tag{8.65}$$

This integrated rate equation, derived for temporary slow-onset, linear specific inhibition, mechanism SO-LSpI-2Temp, but valid for other mechanisms, is reproduced in the main text as (8.10), to which a constant displacement term, d, was added for any necessary baseline correction.

The expressions of λ and v_s for mechanism SO-LSpI-2 do not need a separate derivation because they can be obtained from (8.46) and (8.57), (8.58) by deleting the rate constant k_5. A summary of equations for the mechanisms SO-LSpI-2 and SO-LSpI-2Temp is given in the collective groups (8.15) and (8.19), respectively.

The procedure outlined in this section can be applied to any other slow-onset inhibition mechanism compatible with assumptions 1–5 on p. 374. The goal is to obtain mechanism-dependent expressions of v_z, v_s and λ, while the integration steps in Eqs. (8.59)–(8.65), common to all mechanisms, are done once and for all. Besides checking all passages not to forget or loose terms during the derivation, it is essential to define correctly the *boundary conditions* necessary for calculating the integration constants.

Reaction Started by Adding the Substrate to Preincubated Enzyme and Inhibitor

This derivation applies to the mechanism SO-LSpI-2, not to its temporary counterpart SO-LSpI-2Temp, because in the last the inhibitor is in reality a substrate that is transformed during the preincubation. The extent of transformation of the alternative substrate depends on the preincubation time and thus this experiment is quantitatively meaningless. Conversely, in the mechanism SO-LSpI-2, after preincubation for a sufficient time E and I equilibrate to form the EI and E·I complexes. When starting the reaction by adding substrate, the fraction of enzyme available for catalysis depends on E and EI because EI dissociates fast giving E + I, while E·I cannot deliver free enzyme immediately because it dissociated slowly. Thus, the fraction of free enzyme at the start of the reaction is given by

$$f_{\mathrm{E}} = \frac{[\mathrm{E}] + [\mathrm{EI}]}{[\mathrm{E}]_{\mathrm{t}}} = \frac{[\mathrm{E}] + [\mathrm{EI}]}{[\mathrm{E}] + [\mathrm{EI}] + [\mathrm{E} \cdot \mathrm{I}]} . \tag{8.66}$$

[EI] is now calculated from the equilibrium constant K_{I}, $[\mathrm{E} \cdot \mathrm{I}]$ as function of the isomerization rate constants and the results are introduced in the conservation equation of the enzyme:

$$[\mathrm{EI}] = [\mathrm{E}] \frac{[\mathrm{I}]}{K_{\mathrm{I}}} , \tag{8.67}$$

$$k_4 [\mathrm{EI}] = k_{-4} [\mathrm{E} \cdot \mathrm{I}] \quad \therefore \quad [\mathrm{E} \cdot \mathrm{I}] = [\mathrm{EI}] \frac{k_4}{k_{-4}} = [\mathrm{E}] \frac{[\mathrm{I}]}{K_{\mathrm{I}}} \frac{k_4}{k_{-4}} , \tag{8.68}$$

$$[\mathrm{E}]_{\mathrm{t}} = [\mathrm{E}] + [\mathrm{E}] \left(1 + \frac{[\mathrm{I}]}{K_{\mathrm{I}}}\right) + [\mathrm{E}] \left(1 + \frac{[\mathrm{I}]}{K_{\mathrm{I}}}\right) \frac{k_{-4}}{k_4} ,$$

$$[\mathrm{E}]_{\mathrm{t}} = [\mathrm{E}] \left\{1 + \frac{[\mathrm{I}]}{K_{\mathrm{I}}} \left(\frac{k_4 + k_{-4}}{k_{-4}}\right)\right\} . \tag{8.69}$$

Defining

$$K_{\mathrm{I}}\left(\frac{k_{-4}}{k_4+k_{-4}}\right)=K_{\mathrm{I}}^{*}\,, \tag{8.70}$$

the sought concentrations of enzyme-containing species and the fraction of free enzyme are given by:

$$[\mathrm{E}]_{\mathrm{t}}=[\mathrm{E}]\left\{1+\frac{[\mathrm{I}]}{K_{\mathrm{I}}^{*}}\right\}\,, \tag{8.71}$$

$$[\mathrm{E}]+[\mathrm{EI}]=[\mathrm{E}]\left(1+\frac{[\mathrm{I}]}{K_{\mathrm{I}}}\right)\,, \tag{8.72}$$

$$f_{\mathrm{E}}=\frac{[\mathrm{E}]\left(1+\frac{[\mathrm{I}]}{K_{\mathrm{I}}}\right)}{[\mathrm{E}]\left(1+\frac{[\mathrm{I}]}{K_{\mathrm{I}}^{*}}\right)}=\frac{1+\frac{[\mathrm{I}]}{K_{\mathrm{I}}}}{1+\frac{[\mathrm{I}]}{K_{\mathrm{I}}^{*}}}\,. \tag{8.73}$$

The rate at time zero for the reaction started by adding substrate is then calculated by multiplying the fraction of available free enzyme (8.72) with the Michaelis–Menten equation:

$$v_{\mathrm{z,S}}=f_{\mathrm{E}}\frac{k_2[\mathrm{E}]_{\mathrm{t}}\,[\mathrm{S}]}{K_{\mathrm{m}}+[\mathrm{S}]}=\frac{k_2[\mathrm{E}]_{\mathrm{t}}\,[\mathrm{S}]}{K_{\mathrm{m}}+[\mathrm{S}]}\times\frac{1+\frac{[\mathrm{I}]}{K_{\mathrm{I}}}}{1+\frac{[\mathrm{I}]}{K_{\mathrm{I}}^{*}}}\,,$$

$$v_{\mathrm{z,S}}=\frac{V\,[\mathrm{S}]}{K_{\mathrm{m}}+[\mathrm{S}]}\times\frac{1+\frac{[\mathrm{I}]}{K_{\mathrm{I}}}}{1+\frac{[\mathrm{I}]}{K_{\mathrm{I}}^{*}}}\,. \tag{8.74}$$

Sometimes the preincubation of enzyme and inhibitor is performed in a given volume and an aliquot of substrate is added to start the reaction. If a very small volume of concentrated substrate solution is added, (8.74) is sufficiently precise. However, if the added substrate solution is relatively large, a factor x must be inserted into (8.74) for compensating the dilution

$$v_{\mathrm{z,S}}=\frac{V\,[\mathrm{S}]}{K_{\mathrm{m}}+[\mathrm{S}]}\times\frac{1+x\frac{[\mathrm{I}]}{K_{\mathrm{I}}}}{1+x\frac{[\mathrm{I}]}{K_{\mathrm{I}}^{*}}}\,. \tag{8.75}$$

References

1. Baici A, Gyger-Marazzi M (1982) The slow, tight-binding inhibition of cathepsin B by leupeptin. A hysteretic effect. Eur J Biochem 129:33–41
2. Baici A, Szedlacsek SE, Früh H, Michel BA (1996) pH-dependent hysteretic behaviour of human myeloblastin (leucocyte proteinase 3). Biochem J 317:901–905
3. Bernasconi CF (1976) Relaxation kinetics. Academic Press, New York
4. Cha S (1975) Tight-binding inhibitors - I. Kinetic behavior. Biochem Pharmacol 24:2177–2185 [See corrections, ibid. 25 (1976) 1561]
5. Cha S (1976) Erratum: Tight-binding inhibitors-I. Kinetic behavior. Biochem Pharmacol 24: 2177 (1975). Biochem Pharmacol 25:1561 (only)
6. Cha S (1976) Tight-binding inhibitors - III. A new approach for the determination of competition between tight-binding inhibitors and substrates. Inhibition of adenosine deaminase by coformycin. Biochem Pharmacol 25:2695–2702
7. Cha S (1980) Tight-binding inhibitors - VII. Extended interpretation of the rate equation. Experimental designs and statistical methods. Biochem Pharmacol 29:1779–1789
8. Cha S, Agarwal RP, R E Parks J (1975) Tight-binding inhibitors - II. Non-steady state nature of inhibition of milk xanthine oxidase by allopurinol and alloxanthine and of human erythrocytic adenosine deaminase by coformycin. Biochem Pharmacol 24:2187–2197
9. Cha S, Kim SYR, Kornstein SG, Kantoff PW, Kim KH, Naguib FNM (1981) Tight binding inhibitors - IX. Kinetic parameters of dihydrofolate reductase inhibited by methotrexate, an example of equilibrium study. Biochem Pharmacol 30:1507–1515
10. Cunningham RT, Mangold SE, Spruce LW, Ying QL, Simon SR, Wieczorek M, Ross S, Cheronis JC, Kirschenheuter GP (1992) Synthesis and evaluation of CE-0266: a new human neutrophil elastase inhibitor. Bioorg Chem 20:345–355
11. Duggleby RG, Attwood PV, Wallace JC, Keech DB (1982) Avidin is a slow-binding inhibitor of pyruvate carboxylase. Biochemistry 21:3364–3370
12. Frieden C (1970) Kinetic aspects of regulation of metabolic processes. The hysteretic enzyme concept. J Biol Chem 245:5788–5799
13. Koller MF, Baici A, Huber M, Christen P (2002) Detection of a very rapid first phase in complex formation of DnaK and peptide substrate. FEBS Lett 520:25–29
14. Kuzmič P (2008) A steady state mathematical model for stepwise "slow-binding" reversible enzyme inhibition. Anal Biochem 380:5–12. doi:10.1016/j.ab.2007.11.027
15. Laskowski M, Wu FC (1953) Temporary inhibition of trypsin. J Biol Chem 204:797–805
16. Machleidt W, Nägler DK, Assfalg-Machleidt I, Stubbs MT, Fritz H, Auerswald EA (1995) Temporary inhibition of papain by hairpin loop mutants of chicken cystatin. Distorted binding of the loops results in cleavage of the Gly(9)-Ala(10) bond. FEBS Lett 361:185–190
17. Meh P, Pavšič M, Turk V, Baici A, Lenarčič B (2005) Dual concentration-dependent activity of thyroglobulin type-1 domain of testican: specific inhibitor and substrate of cathepsin L. Biol Chem 386:75–83
18. Monod J, Wyman J, Changeux JP (1965) On the nature of allosteric transitions: a plausible model. J Mol Biol 12:88–118
19. Morrison JF (1982) The slow-binding and slow, tight-binding inhibition of enzyme-catalysed reactions. Trends Biochem Sci 7:102–105
20. Morrison JF, Stone SR (1985) Approaches to the study and analysis of the inhibition of enzymes by slow- and tight-binding inhibitors. Comments Mol Cell Biophys 2:347–368
21. Morrison JF, Walsh CT (1988) The behavior and significance of slow-binding inhibitors. Adv Enzymol Relat Areas Mol Biol 61:201–301
22. Nakamura CE, Abeles RH (1985) Mode of interaction of beta-hydroxy-beta-methylglutaryl coenzyme A reductase with strong binding inhibitors: compactin and related compounds. Biochemistry 24:1364–1376
23. Novinec M, Kovačič L, Lenarčič B, Baici A (2010) Conformational flexibility and allosteric regulation of cathepsin K. Biochem J 429:379–389. doi:10.1042/BJ20100337

24. Novinec M, Lenarčič B, Baici A (2012) Clusterin is a specific stabilizer and liberator of extracellular cathepsin K. FEBS Lett 586:1062–1066
25. Pierpaoli EV, Sandmeier E, Baici A, Schönfeld HJ, Gisler S, Christen P (1997) The power stroke of the DnaK/DnaJ/GrpE molecular chaperone system. J Mol Biol 269:757–768
26. Schmid D, Baici A, Gehring H, Christen P (1994) Kinetics of molecular chaperone action. Science 263:971–973
27. Schultz RM, Varma-Nelson P, Ortiz R, Kozlowski KA, Orawski AT, Pagast P, Frankfater A (1989) Active and inactive forms of the transition-state analog protease inhibitor leupeptin: explanation of the observed slow binding of leupeptin to cathepsin B and papain. J Biol Chem 264:1497–1507
28. Sculley MJ, Morrison JF (1986) The determination of kinetic constants governing the slow, tight-binding inhibition of enzyme-catalysed reactions. Biochim Biophys Acta 874:44–53
29. Sculley MJ, Morrison JF, Cleland WW (1996) Slow-binding inhibition: the general case. Biochim Biophys Acta 1298:78–86
30. Shapiro R, Riordan JF (1984) Inhibition of angiotensin converting enzyme: dependence on chloride. Biochemistry 23:5234–5240
31. Takai A, Ohno Y, Yasumoto T, Mieskes G (1992) Estimation of the rate constants associated with the inhibitory effect of okadaic acid on type 2A protein phosphatase by time-course analysis. Biochem J 287 (Pt 1):101–106
32. Tamura A, Kanaori K, Kojima S, Kumagai I, Miura K, Akasaka K (1991) Mechanisms of temporary inhibition in Streptomyces subtilisin inhibitor induced by an amino acid substitution, tryptophan 86 replaced by histidine. Biochemistry 30:5275–5286
33. Umezawa H (1973) Chemistry of enzyme inhibitors of microbial origin. Pure Appl Chem 33:129–144
34. Walmsley AR, Bagshaw CR (1989) Logarithmic timebase for stopped-flow data acquisition and analysis. Anal Biochem 176:313–318
35. Williams JW, Morrison JF (1979) The kinetics of reversible tight-binding inhibition. Meth Enzymol 63:437–467
36. Williams JW, Morrison JF, Duggleby RG (1979) Methotrexate, a high-affinity pseudosubstrate of dihydrofolate reductase. Biochemistry 18:2567–2573

Chapter 9
Enzyme Inactivation with a Note on the Significance of Slow Modification Processes

··· covalent irreversible inhibitors developed a negative flavor over time and became almost synonymous with toxicity in some organizations. The skepticism toward irreversible drugs may evaporate as more examples of irreversible drugs progress clinically that demonstrate good efficacy and safety margins.
Barf T, Kaptein A (2012) J Med Chem 55:6243–6262

Abstract Irreversible enzyme modification appears typically as a time-dependent process that can be analyzed with methods similar to those used in slow-onset inhibition. Although full inactivation can be recognized at a glance, molecules designed as irreversible modifiers behave frequently as competing substrates manifesting kinetic properties that are typical of reversible modifiers. Chemical instability of the inactivators introduces additional complexity to the interpretation of raw data. This chapter discusses the strategies that can be exploited to recognize and interpret enzyme inactivation in the presence of unstable modifiers and of modifiers with substrate-like properties.

9.1 Introduction

This book had been originally planned to deal only with reversible enzyme-modifier interactions. Renouncing to an extensive treatment of irreversible interactions is in line with the policy stated in the Preface not to include topics that are available from sound published accounts unless something new should be added. However, experimental situations exist, in which judging the nature of an alleged *irreversible* modifier basing the analysis on progress curves is not immediate. More often than in reversible slow-onset inhibition, a *temporary character* of this action is found among substances designed as irreversible modifiers. Since the progress curves of temporary irreversible inhibition cannot be distinguished from those of reversible, slow-onset inhibition without accurate investigation, users of the dichotomous keys in Chap. 10 would be at risk of misinterpretations. Therefore, the kinetics of time-dependent irreversible inhibition will be discussed here *briefly* without renouncing to the necessary diagnostic details. On the other side, mechanism-based inhibitors (suicide substrates) will not be discussed here. This topic has been treated by Waley [26, 27] and in a more rigorous form by Burke and coworkers [9, 10] and Wang [28].

A. Baici, *Kinetics of Enzyme-Modifier Interactions*,
DOI 10.1007/978-3-7091-1402-5_9

Irreversible inhibition is also called enzyme inactivation or simply inactivation. The two terms are equivalent and their use is left to personal taste. In order to distinguish at a glance reversible from irreversible inhibition using only one word, *inactivation* will be used, which is also practical for the abbreviation *Ina* in forming the acronyms of mechanisms.

Enzyme inactivation has received early attention by pioneers of this branch of enzymology who were studying the hydrolysis of acetylcholine within *the mechanism of nervous action* by the intervention of a not yet identified enzyme. In 1932, Edgar and Ellen Stedman, together with Leslie Easson, succeeded in purifying an enzyme from horse blood serum they named *choline-esterase* [18]. This enzyme that today retains its original name, just spelled *cholinesterase* (EC 3.1.1.8), and the related acetylcholinesterase (EC 3.1.1.7) stimulated a wealth of studies with inactivators and inhibitors that are experiencing new impulses for the association made between acetylcholinesterase and Alzheimer's disease (see Chap. 5 using the Index).

For the analysis of raw data in the majority of the studies in this field, the methods of Easson and Stedman [13], Kitz and Wilson[14] and Malcolm and Radda [15] have been used. Particularly the latter two methods are still popular and there are no objections to their use when applied correctly [11]. Reversible inhibitors that bind very tightly their target enzymes may be misidentified as inactivators because they behave very similarly as commented by Straus and Goldstein [19] and by Ackermann and Potter [1].

Enzyme inactivation is usually a *time-dependent* phenomenon that can be analyzed with the same methods based on entire progress curves described in Chap. 8. The epithet time-dependent has been used by many authors and therefore the mechanisms featured in this chapter will be referred to as *slow-onset inactivation* in line with the corresponding reversible counterparts. As an alternative to discontinuous methods, the possibility of continuously monitoring inactivation reactions in the presence of a suitable substrate of the investigated enzyme offers many advantages for the large number of data points that can be collected in entire progress curves. Using a rapid-mixing device also avoids the indeterminateness of the starting point of reactions, a problem that can negatively affect the calculations of the inactivation parameters [31].

The theoretical basis for exploiting progress curves and studying *substrate protection* of enzyme inactivation has been provided by Tian and Tsou [20], a contribution that has been followed by additional theoretical considerations [29, 30]. Tian and Tsou remarked that irreversible enzyme-modifier interactions can be competitive, noncompetitive, or uncompetitive (in the sense of specific, balanced and catalytic, respectively, in the nomenclature of this book). Although starting from a reaction scheme corresponding to the general modifier mechanism [20, A1], the final integrated rate equation has been derived simplifying the reactions to *linear* mechanisms, whose properties can be deduced from the expressions of the slow-onset inhibition mechanisms treated in Chap. 8 by setting the dissociation rate constants of the inhibited complex to zero.

A special mention goes to a paper by Christopher Topham [23], a recommended reading for all readers interested in the kinetics of enzyme-modifiers interactions beyond the surface. The author broaches a thorny subject with total confidence and provides expressions that can be used for nonlinear regression analysis of the general modifier mechanism of Botts and Morales in the presence of an unstable inactivator or activator. Thus, any one of the seventeen basic modifiers mechanisms described in Chap. 3 (summary in Fig. 3.2) is amenable of *extension* to irreversible enzyme-modifier interactions even when the modifier is chemically unstable.

Other authors dealt with irreversible enzyme modification, to whom I apologize for not discussing their valuable contributions in this short account. Scientific debates are often an excellent source of information for giving access to the most relevant literature and putting in focus problems by discussing them in their very essence [17, 21, 22].

The interpretation of progress curves of enzyme inactivation in the presence of substrates can be largely inherited from the treatment of slow-onset inhibition after adding the important issues of chemical instability of the inactivator and the frequently seen transformation of the inactivator to an inert species by the enzyme itself as discussed in this chapter. In a study dedicated to inactivators of acetylcholinesterase, in collaboration with Michael Wächter, Peter Rüedi, and Patricia Schenker, I had the opportunity to analyze original data of a large number of compounds. The modes of action covered a palette of eight mechanisms of inactivation, including those with temporary character and chemically unstable modifiers [5], with further specific cases described in an accompanying paper [25]. These mechanisms, as representatives of the most frequently encountered cases in the literature, are discussed below highlighting the analogies with their reversible counterparts.

9.2 Slow-Onset Linear Specific Inactivation

Two common inactivation mechanism are shown in Scheme 9.1 with assigned acronyms SO-LSpIna-2 and SO-LSpIna-1, again for practical and unequivocal referencing. They are the counterparts of the slow-onset inhibition mechanisms SO-LSpI-2 and SO-LSpI-1 described in Sects. 8.4 and 8.5, respectively, in which there is no dissociation of the E-I complexes back to EI or E. Note the representation of the covalent compound between enzyme and inactivator as E-I as distinction toward the non-covalent complex E·I in slow-onset inhibition. SO-LSpIna-2 has been named *specific* irreversible inhibition[1] or also *complexing type* irreversible inhibition [20], while SO-LSpIna-1 is known as *nonspecific* irreversible inhibition or *direct-encounter* irreversible modification [23].

[1]This is one of the reasons why, in this and other circumstances, the epithets *linear* and *hyperbolic* as part of the nomenclature and acronyms adopted in this book help avoiding confusion.

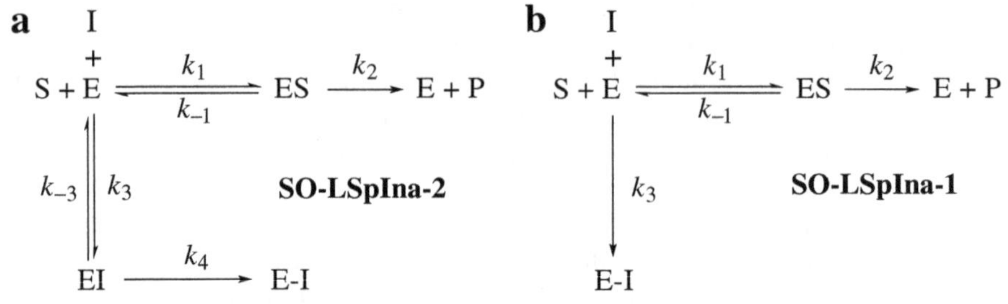

Scheme 9.1 Slow-onset linear specific inactivation. (**a**) SO-LSpIna-2 with rapidly equilibrating first step. (**b**) SO-LSpIna-1 in one step

Making the same assumptions listed on p. 374, the integrated rate equation common to SO-LSpIna-2 and SO-LSpIna-1 consists of a single exponential that decays to a plateau with increasing time. In general, however, a double exponential should be expected for SO-LSpIna-2 and verifying whether this is or is not the case is possible by fitting (9.1) and (9.2) to data, the equivalents of (8.8) and (8.9), respectively, with lacking linear term

$$[\mathrm{P}] = A_1\left(1 - e^{-\lambda_1 t}\right) + d \tag{9.1}$$

$$[\mathrm{P}] = A_1\left(1 - e^{-\lambda_1 t}\right) + A_2\left(1 - e^{-\lambda_2 t}\right) + d\,. \tag{9.2}$$

The factor d accounts for any baseline signal proportional to the measured signal proportional to [P]. If the comparative fitting of (9.2) and (9.1) reveals that the single-exponential equation is superior, the integrated rate equation can be used with satisfactory approximation to gather the frequency constant and the rate at time zero in serial measurements at fixed [S] and variable [I] and/or vice versa. Otherwise, numerical integration methods can be invoked to globally fit ensembles of progress curves on the basis of the reaction in Scheme 9.1a.

SO – LSpIna – 2 | **SO – LSpIna – 1**

$$[\mathrm{P}] = \frac{v_{\mathrm{z}}}{\lambda}\left(1 - e^{-\lambda t}\right) + d \tag{9.3}$$

$$\lambda = \frac{k_4\,[\mathrm{I}]}{K_{\mathrm{I}}\left(1 + \dfrac{[\mathrm{S}]}{K_{\mathrm{m}}}\right) + [\mathrm{I}]} \qquad \lambda = \frac{k_3\,[\mathrm{I}]}{1 + \dfrac{[\mathrm{S}]}{K_{\mathrm{m}}}} \tag{9.4}$$

(continued)

$$v_z = \frac{V\,[\mathrm{S}]}{K_m\left(1+\frac{[\mathrm{I}]}{K_\mathrm{I}}\right)+[\mathrm{S}]} \qquad v_z = v_0 = \frac{V\,[\mathrm{S}]}{K_m+[\mathrm{S}]} \tag{9.5}$$

$$K_\mathrm{I} = \frac{k_{-3}}{k_3}, \quad k_{\mathrm{ina}} = \frac{k_4}{K_\mathrm{I}} \qquad k_{\mathrm{ina}} = k_3 \tag{9.6}$$

The relevant equations for mechanisms SO-LSpIna-2 and SO-LSpIna-1 are collected in the group (9.3)–(9.6), where it can be seen that the integrated rate equation (9.3), common to both mechanisms, does not contain the linear term present in reversible slow-onset inhibition. Also, λ (9.4) does not contain the dissociation constants k_{-4} (SO-LSpIna-2) and k_{-3} (SO-LSpIna-1). The properties of the two mechanisms in Scheme 9.1 are compared in Fig. 9.1. When, experimentally, progress curves as those shown in panels (a) and (d) are recorded, their handling is *friendly* because diagnosing the underlying mechanisms is usually straightforward.

The frequency constant depends hyperbolically on [I] in the two-step mechanism and linearly in SO-LSpIna-1 as shown in panels (b) and (c), respectively. In both cases the curves intercept the ordinate at coordinates (0, 0) because there in no off-reaction with the enzyme covalently trapped in E-I. The reversible, rapid formation of an adsorptive EI complex in SO-LSpIna-2 determines the dependence of v_z on [I] because it is already in place immediately upon adding enzyme to start the reaction, while E-I is formed slowly, whereas the rate at time zero in the one-step mechanism does not depend on the concentration of the inactivator. The inactivator dissociation constant of the reversible step of mechanism SO-LSpIna-2, $K_\mathrm{I} = k_{-3}/k_3$, can be determined by fitting (9.4) (left) to data as shown in Fig. 9.1b and/or by fitting (9.5) (left) to the v_z-values determined from the progress curves (panel c).

Besides determining the inactivation mechanism, the parameter that characterizes enzyme inactivation is the second-order *inactivation constant* k_{ina} that is given as the ratio k_4/K_I for SO-LSpIna-2 and coincides with k_3 in So-LSpIna-1.

9.3 Slow-Onset Linear Specific, Temporary Inactivation

A frequently observed property of molecules designed to act as irreversible enzyme modifiers is their temporary action on the target enzyme behaving very much like competing substrates as is the case of slow-onset inhibitors discussed in Sect. 8.6. The reactions are shown in Scheme 9.2, where the sole differences from the corresponding reversible mechanisms SO-LSpI-2Temp and SO-LSpI-1Temp are the irreversible steps for E-I formation. I* denotes a transformed modifier molecule that, for the correct use of the integrated rate equation (9.7), is supposed to be inert toward the enzyme. In mechanism SO-LSpIna-2Temp the decay of E-I to free enzyme and I*, governed by the rate constant k_6, may be slower or faster than the

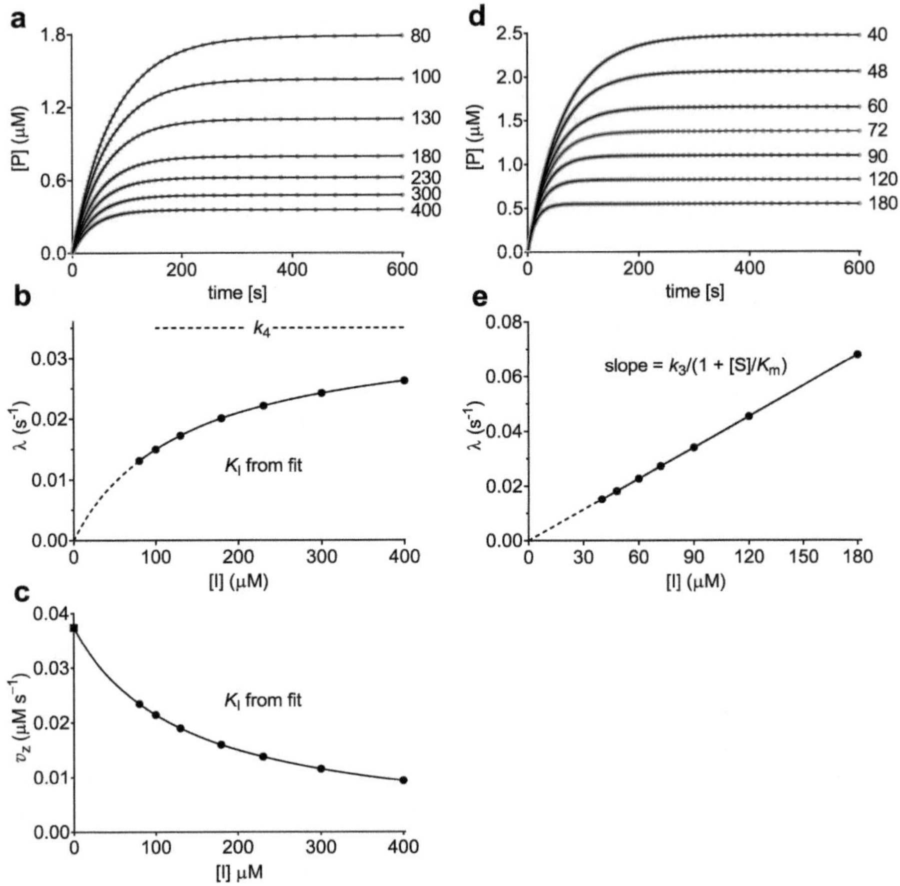

Fig. 9.1 Slow-onset linear specific inactivation. (**a–c**) Mechanism SO-LSpIna-2, (**d, e**) mechanism SO-LSpIna-1 (Scheme 9.1). Constants and concentrations used for simulating the progress curves on a logarithmic timebase in panels (**a**) and (**d**) that were common to both mechanisms: $k_1 = 10\,\mu\text{M}^{-1}\,\text{s}^{-1}$, $k_{-1} = 1670\,\text{s}^{-1}$, $k_2 = 10\,\text{s}^{-1}$, $[\text{E}]_t = 5\,\text{nM}$, $K_m = 168\,\mu\text{M}$, $[\text{S}] = 500\,\mu\text{M}$. For (**a–d**): $k_3 = 9.2\,\mu\text{M}^{-1}\,\text{s}^{-1}$, $k_{-3} = 310\,\text{s}^{-1}$, $k_4 = 0.035\,\text{s}^{-1}$. For (**d–e**): $k_3 = 1.5 \times 10^{-3}\,\mu\text{M}^{-1}\,\text{s}^{-1}$. The numbers that label the curves in panels (**a**) and (**d**) are the concentrations of $[\text{I}]_t$ (μM). (9.3) was fitted to simulated data (*gray points*) to calculate λ as a function of [I] (panels **b** and **e**) and v_z (panel **c**). The best-fit curves are shown as continuous *black lines* in (**a**) and (**d**)

unimolecular conversion of EI to E-I governed by k_4. The same relationships apply also to mechanism SO-LSpIna-1Temp.

The integrated rate equation (9.7) in the group (9.7)–(9.10) must now contain the contribution of the enzyme recycled into the system with the rate constant k_6 that leads to I*. This contribution is represented by the linear term $v_\infty t$ that is indistinguishable from the steady-state rate multiplied with time present in (8.10). The subscript ∞ indicates that this is the rate of the system observed after completion of the exponential phase, a condition that is mathematically expressed

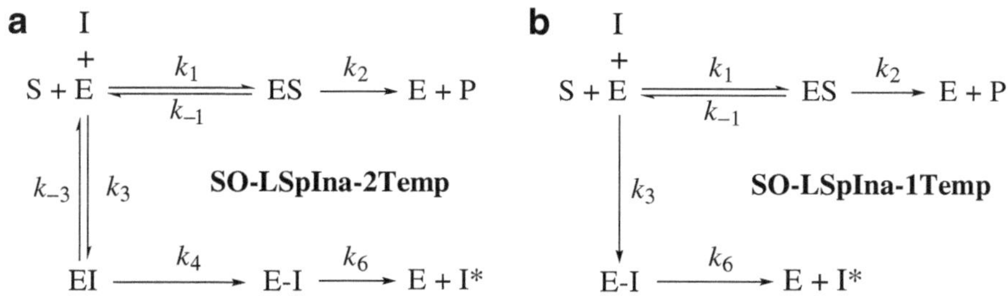

Scheme 9.2 Slow-onset linear specific, temporary inactivation in which the inactivator behaves as a competing substrate, SO-LSpIna-2Temp (**a**) and SO-LSpIna-1temp (**b**). The temporary nature of the inactivation is due to recycling of free enzyme that follows catalytic transformation of I into I* that dissociates from and does no longer interact with the enzyme

by setting $t = \infty$ in $e^{-\lambda t}$. In essence, the reaction in Scheme 9.2b represents a Van Slyke-Cullen mechanism of substrate turnover in which the inactivator plays the role of a substrate [24].

Graphical representations of progress curves and derived properties are shown in Fig. 9.2, where the mechanism SO-LSpIna-2Temp and SO-LSpIna-1Temp are compared. A noticeable feature, not expected from a putative enzyme inactivator, is the straight lines with positive slope that follow the exponential transient (panels a and d), which is a property of the reversible slow-onset inhibition mechanisms discussed in Chap. 8.

SO – LSpIna – 2Temp	**SO – LSpIna – 1Temp**	

$$[\mathrm{P}] = v_\infty t + \frac{v_z - v_\infty}{\lambda}\left(1 - e^{-\lambda t}\right) + d \tag{9.7}$$

$$\lambda = k_6 + \frac{k_4\,[\mathrm{I}]}{K_\mathrm{I}\left(1 + \frac{[\mathrm{S}]}{K_\mathrm{m}}\right) + [\mathrm{I}]} \qquad \lambda = k_6 + \frac{k_3\,[\mathrm{I}]}{1 + \frac{[\mathrm{S}]}{K_\mathrm{m}}} \tag{9.8}$$

$$v_z = \frac{V\,[\mathrm{S}]}{K_\mathrm{m}\left(1 + \frac{[\mathrm{I}]}{K_\mathrm{I}}\right) + [\mathrm{S}]} \qquad v_z = v_0 = \frac{V\,[\mathrm{S}]}{K_\mathrm{m} + [\mathrm{S}]} \tag{9.9}$$

$$v_\infty = \frac{V\,[\mathrm{S}]}{K_\mathrm{m}\left\{1 + \frac{[\mathrm{I}]}{K_\mathrm{I}}\left(1 + \frac{k_4}{k_6}\right)\right\} + [\mathrm{S}]} \qquad v_\infty = \frac{V\,[\mathrm{S}]}{K_\mathrm{m}\left(1 + \frac{k_3}{k_6}[\mathrm{I}]\right) + [\mathrm{S}]} \tag{9.10}$$

(continued)

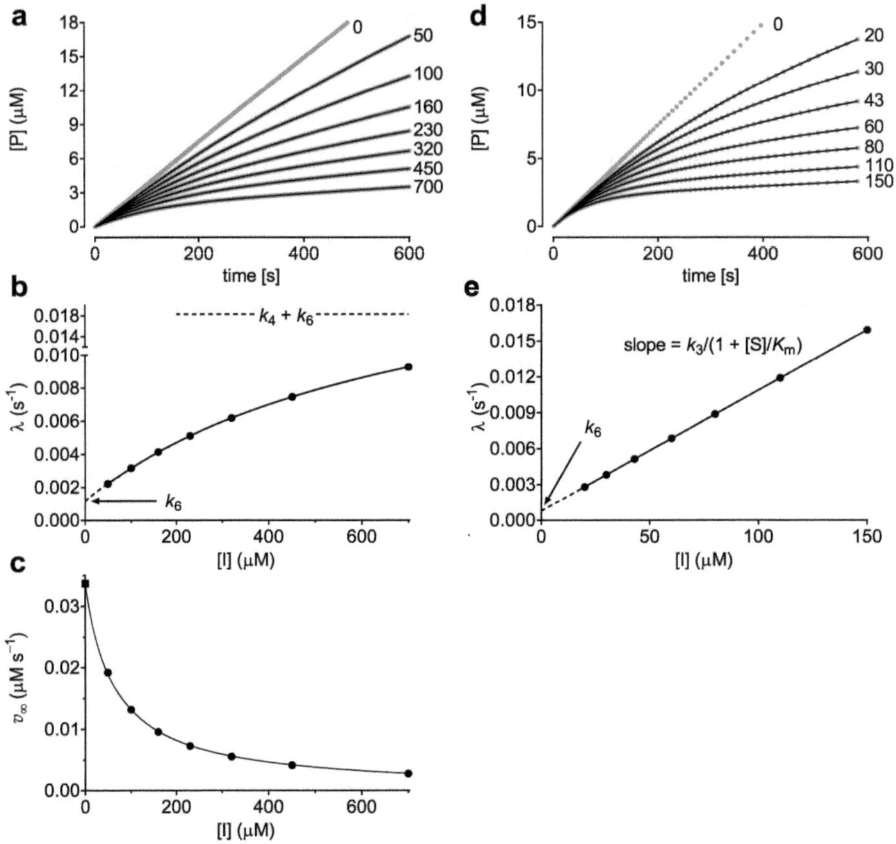

Fig. 9.2 Temporary enzyme inactivation. (**a–c**) Mechanism SO-LSpIna-2Temp, (**d**, **e**) mechanism SO-LSpIna-1Temp (Scheme 9.2). Constants and concentrations used for simulating the progress curves on a logarithmic timebase in panels (**a**) and (**d**) that were common to both mechanisms: $k_1 = 10\,\mu M^{-1}\,s^{-1}$, $k_{-1} = 1670\,s^{-1}$, $k_2 = 10\,s^{-1}$, $[E]_t = 5\,nM$, $K_m = 168\,\mu M$, $[S] = 500\,\mu M$. For (**a–d**): $k_3 = 1\,\mu M^{-1}\,s^{-1}$, $k_{-3} = 200\,s^{-1}$, $k_4 = 0.017\,s^{-1}$, $k_6 = 0.0013\,s^{-1}$. For (**d–e**): $k_3 = 4 \times 10^{-4}\,\mu M^{-1}\,s^{-1}$, $k_6 = 4 \times 10^{-8}\,s^{-1}$. The numbers that label the curves in panels (**a**) and (**d**) are the concentrations of $[I]_t$ (μM). (9.7) was fitted to simulated data (*gray points*) to calculate λ as a function of [I] (panels **b** and **e**) and v_∞ (panel c). The best-fit curves are shown as continuous *black lines* in (**a**) and (**d**)

$$k_{\text{off}} = \lambda \frac{v_\infty}{v_z} = k_6$$

$$k_{\text{off}} = \lambda \frac{v_\infty}{v_z} = k_6 \tag{9.11}$$

$$K_I = \frac{k_{-3}}{k_3} \tag{9.12}$$

Also unexpected is the intercept with the ordinate in the plots of λ against [I], that is located at coordinates (0, 0) in fully irreversible mechanisms and has instead a positive value. This intercept represents the *off-rate constant*, k_{off}, of both mechanisms in Scheme 9.2 that corresponds to k_6. Since this rate constant can be very small and therefore cannot be determined with accuracy as the ordinate intercept, the relationship $k_{\text{off}} = \lambda\,(v_\infty/v_z)$ can be exploited. Conversely, the sum $(k_4 + k_6)$ can be calculated with precision during nonlinear regression using (9.8) (left) as the asymptote of λ against [I] for mechanism SO-LSpIna-2Temp (Fig. 9.2b). Similarly, k_3 can be calculated from the slope of the λ against [I] plot for mechanism SO-LSpIna-1Temp by taking into account the factor $1 + [\text{S}]/K_\text{m}$ as shown in Fig. 9.2e.

The fit of (9.8) (left) in panel (b) also yields a good estimate of the dissociation constant of the equilibration step for mechanism SO-LSpIna-2Temp, i.e., $K_\text{I} = k_{-3}/k_3$. This value can be checked for internal consistency from the dependence of v_∞ on [I] by fitting (9.10) (left) to data. Important: in this fitting procedure it is compelling to consider k_4 and k_6 as constant values determined previously from the data in panel (b) because they are *strongly intertwined* with K_I.

Apparently, the analysis of the two temporary inactivation mechanisms in Scheme 9.2 does not pose particular problems if it were not for the dilemma that arises in front of data as those in Fig. 9.2 that shows, for SO-LSpIna-2Temp, progress curves indistinguishable from those of SO-LSpI-2 and SO-LSpI-2Temp. The same problem applies to SO-LSpIna-1, whose progress curves do not differ from those of SO-LSpI-1 and SO-LSpI-1Temp. Since kinetic methods cannot solve the puzzle, *chemical support* is required to characterize the behavior of molecules designed as inactivators. Assuming that the structure of the molecule suggests a behavior as inactivator, the transformation undergone by the latter upon incubation with the enzyme, obviously in the absence of the *official* substrate, can be unraveled by chemical analysis of the mixture, e.g., by high-performance liquid chromatography of nuclear magnetic resonance.

9.4 Unstable Inactivators and Those That Are Both Unstable and Temporary

Besides behaving as substrates, the molecules designed as enzyme inactivators may undergo nonenzymatic decomposition, e.g., by hydrolysis, to an inert species I′. The *prime* distinguishes a chemically transformed molecule from I* generated from the decay of E-I (Sect. 9.3). The chemical reaction governed by the first-order rate constant k_5 runs in parallel with the inactivation reaction as shown in Scheme 9.3, with apologies for the awkward acronyms that are nevertheless useful for cross-referencing. In the acronyms, dec signifies the *decomposition* of I to I′.

Since the chemical side reaction continuously removes the inactivator, which is not replaced in a closed system, an analytical integrated rate equation of the

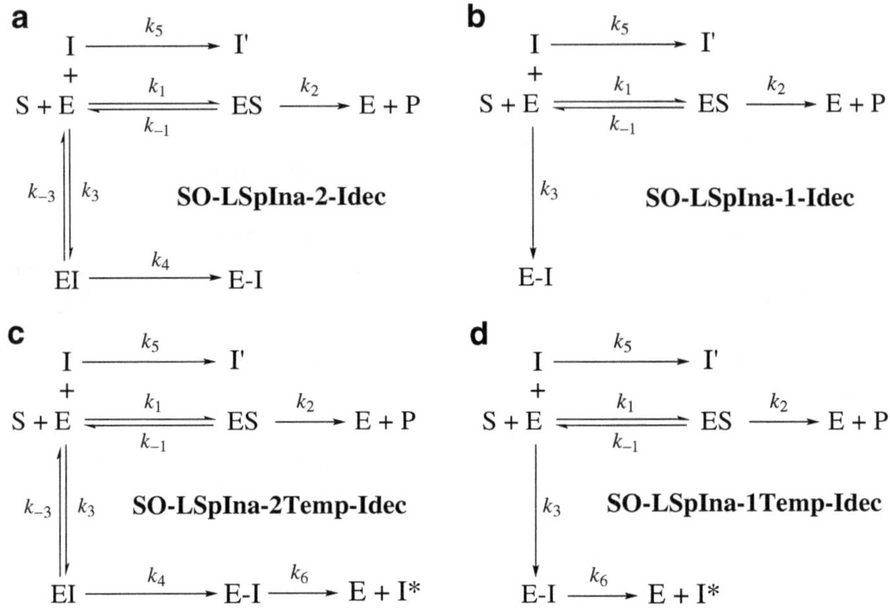

Scheme 9.3 Slow-onset linear specific inactivation in which the inactivator undergoes chemical transformation to an inert species I′ (**a, b**). Two cases in which the inactivator is both chemically unstable and undergoes transformation by catalysis are shown in (**c, d**). When the inactivator behaves as competing substrate, the enzyme is recycled into the system. The side-reactions that lead to I′ as well as catalytic transformation of I to I* deplete the pool of available inactivator

form (9.7) cannot be derived. However, help comes from the paper by Topham mentioned in the Introduction, which contains an analytical solution by integration upon Maclaurin series expansion [23]. Topham's solutions, which apply to any basic modifier mechanism in the presence of an unstable modifier such as the two cases in Scheme 9.3a,b, are not replicated here. The method, which does not apply to the more complex temporary mechanisms with additional instability of the modifies shown in Scheme 9.3c,d, is illustrated below for the mechanism in panel (b).

The rate equation integrated by Maclaurin series expansion that specifically applies to SO-LSpIna-1-Idec is given by

$$[\mathrm{P}] = \frac{v_{\mathrm{z}}}{k_5} \exp(-B) \left\{ k_5 t + \sum_{i=1}^{\infty} B^i \frac{\left[1 - (\exp(-k_5 t))^i\right]}{i \times i!} \right\}, \tag{9.13}$$

where B corresponds to

$$B = \frac{k_3 K_{\mathrm{m}} [\mathrm{I}]_{\mathrm{t}}}{k_5 (K_{\mathrm{m}} + [\mathrm{S}])}, \tag{9.14}$$

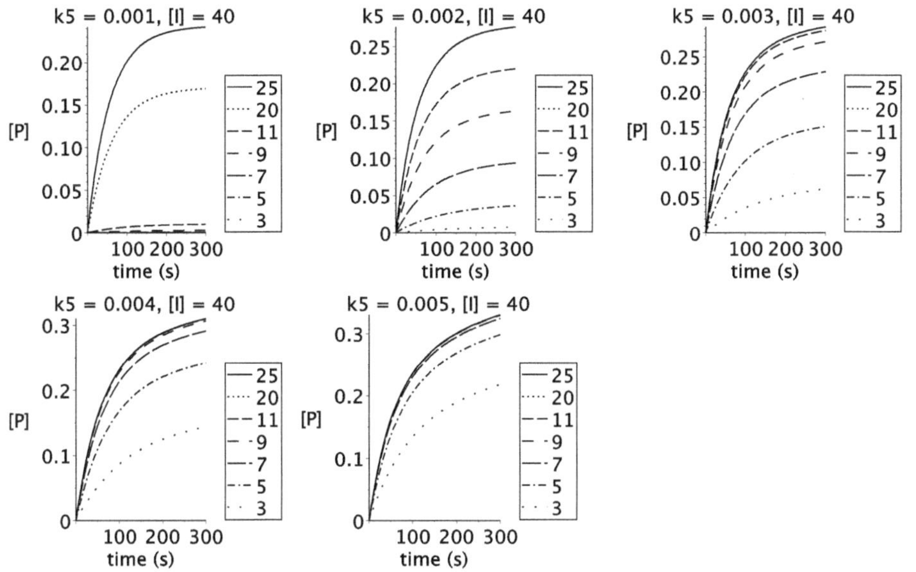

Fig. 9.3 Plots of the function (9.13) for five values of the rate constant k_5 and $[\mathrm{I}]_\mathrm{t} = 40\,\mu\mathrm{M}$. The curves were generated with Maple software using the following rate constants: $k_1 = 100\,\mu\mathrm{M}^{-1}\,\mathrm{s}^{-1}$, $k_{-1} = 1990\,\mathrm{s}^{-1}$, $k_2 = 10\,\mathrm{s}^{-1}$, $k_3 = 0.001\,\mu\mathrm{M}^{-1}\,\mathrm{s}^{-1}$ and variable k_5 as shown. $[\mathrm{E}]_\mathrm{t} = 1\,\mathrm{nM}$, $V = 0.01\,\mu\mathrm{M}\,\mathrm{s}^{-1}$, $[\mathrm{S}] = K_\mathrm{m} = 20\,\mu\mathrm{M}$. The numbers that label the seven curves in each panel indicate the number of terms in the Maclaurin expansion, i in (9.13). All concentrations are μM

$v_\mathrm{z} = v_0$, $K_\mathrm{m} = (k_{-1} + k_2)/k_1$ and $[\mathrm{I}]_\mathrm{t}$ is the total concentration of inactivator present at the beginning of the reaction. The rate constants k_3 and k_5 can be evaluated by nonlinear regression fit of (9.13) to data and the precision of the fit depends on the number of terms i used in the Maclaurin series expansion. The choice of the minimum number of terms depends on the value of the rate constant k_5 and on the concentration of the modifier as illustrated in Figs. 9.3 and 9.4 by theoretical plots of (9.13). This fact must be considered when performing global fits to series of progress curves at fixed substrate and variable modifier concentration. Given the set of constants in Fig. 9.3 at $[\mathrm{I}]_\mathrm{t} = 40\,\mu\mathrm{M}$, if k_5 is small, more terms are needed than for larger values of this constant: at least twenty-five terms are necessary if $k_5 = 0.001\,\mathrm{s}^{-1}$, while nine terms are sufficient if $k_5 = 0.005\,\mathrm{s}^{-1}$.

With the same constants used for drawing the plots in Fig. 9.3 but at $[\mathrm{I}]_\mathrm{t} = 10\,\mu\mathrm{M}$, a minimum of eleven terms is required if $k_5 = 0.001\,\mathrm{s}^{-1}$, which are reduced to three if $k_5 = 0.005\,\mathrm{s}^{-1}$ (Fig. 9.4). Since k_5 is one of the unknowns, the number of terms to be used in regression analysis can be increased until reaching a stable result. Setting this number at a high value as default will cover an ample range of needs.

To test the performance of (9.13), an error-free simulated example of mechanism SO-LSpIna-1-Idec is shown in Fig. 9.5, where eleven terms were necessary to obtain stable estimates of k_3 and k_5 in the Maclaurin expansion during nonlinear regression analysis that matched the theoretical values. When using this procedure, all variables

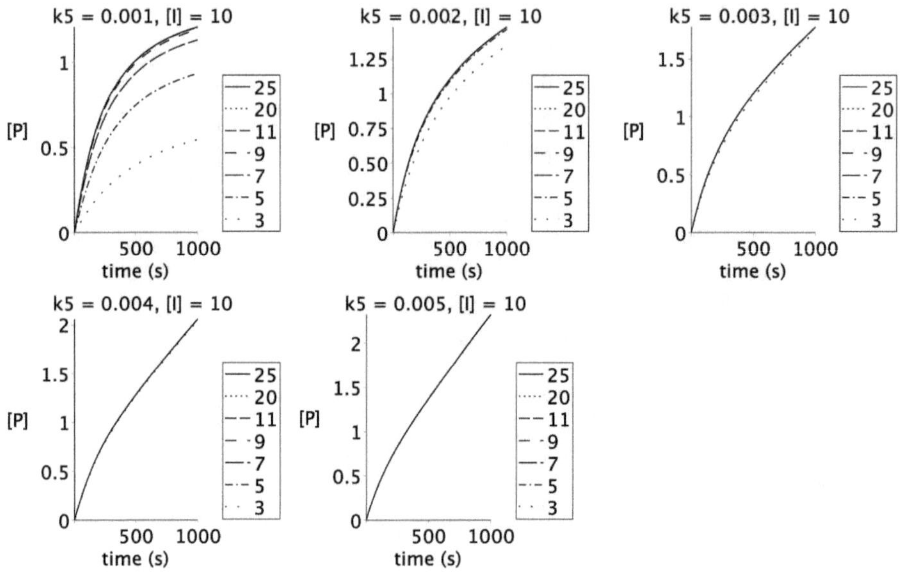

Fig. 9.4 Plots of the function (9.13) for five values of the rate constant k_5 at constant $[\mathrm{I}]_t = 10\,\mu\mathrm{M}$. All other constants are the same as in Fig. 9.3. Note that for $k_5 \geq 0.003$ the curves are all the same when the Maclaurin expansion is calculated with $i \geq 3$

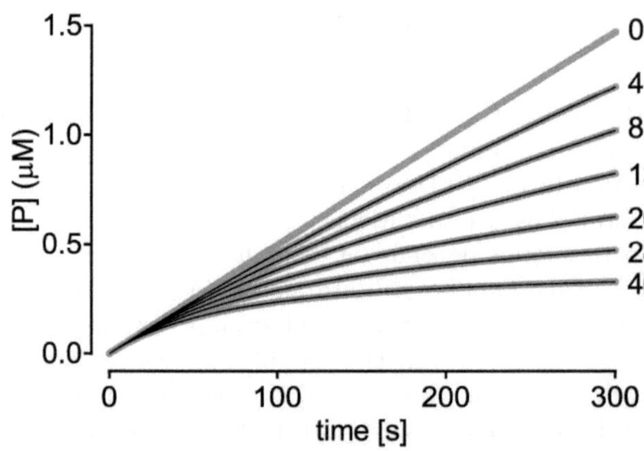

Fig. 9.5 Simulation of mechanism SO-LSpIna-1-Idec (Scheme 9.3b). Curves (*thick gray*) generated by numerical integration with $k_5 = 0.005\mathrm{s}^{-1}$ and the same constants and concentrations in Fig. 9.3 at variable $[\mathrm{I}]_t$ (numbers labeling the curves, μM). The *black curves* are best fits of (9.13) to data

but k_3 and k_5 are kept constant as known values because they can be estimated from measurements in the absence of modifier. The parameters of molecules that obey mechanism SO-LSpIna-2-Idec (Scheme 9.3a) can be analyzed in the same way using the appropriate form of (9.13) [23].

When, besides being chemically unstable (I → I′), a modifier behaves concomitantly as an alternative substrate (E-I → E + I*) by either one of the mechanisms in Scheme 9.3c,d, neither equation of the types (9.13) or (9.7) can be used to fully characterize the system. The primary analysis can be performed by fitting (9.7), formally equivalent to (8.10), to raw data. This allows to establish if v_z, the rate at time zero, does or does not depend on the concentration of the modifier and thus to discriminate between mechanisms SO-LSpIna-2Temp-Idec and SO-LSpIna-1Temp-Idec. Otherwise, the dependence of the frequency constant on [I] is in this case inconsistent as verified in practice. A method for extracting at least qualitative information on the rate constants is to fit (9.13) to data to gather an estimate of k_5 and then to proceed by numerical integration using the model in Scheme 9.3d. This procedure has been applied to the inactivation of acetylcholinesterase by a 3-fluoro-2,4-dioxa-3-phosphadecalin derivative [5]. In this study, the chemical transformation of the modifier was monitored by ^{31}P-NMR in solutions containing the modifier alone, which was partly converted to an epimer and to a hydrolysis product.

A legitimate question is if the labor for determining the mechanism and calculating the rate constants of modifiers that obey any one of the four mechanisms is Scheme 9.3 is worth the effort in view of their limited half-life that makes them less attractive for practical use. The answer is yes: this work makes sense because, even in complicated cases, at least approximate estimates of the mechanisms and rate constants help in making predictions on any practical usefulness of modifiers notwithstanding their reduced half-life.

9.5 Physiological and Pharmacological Implications of Slow-Onset Inhibition

Chapter 8 and the preceding sections in this chapter aimed at discussing the methods for accessing as much information as possible from experiments with slow-onset inhibitors and inactivators. There are in fact critical reasons to pay close attention to the very nature of the underlying mechanisms for their bearing in interpreting the physiological and pharmacological significance of slow-onset enzyme-modifier interactions. In fact, the overall inhibition constant or the second-order inactivation constant are insufficient to predict what really happens when an enzyme meets an inhibitor or an inactivator. For reversible interactions, is really a modifier with picomolar K_I superior to one having a micromolar K_I? In many physiological and pharmacological issues it is namely not only important that the target enzymes are inhibited to the desired degree, but also that the inhibition or inactivation process is *sufficiently fast*.

The concept *delay time of inhibition* has been put forward by Joseph Bieth on the basis of the association rate constant of slow-onset inhibition and can be applied as well to irreversible interactions [7, 8]. This idea has been further developed in the context of pharmacological applications of enzyme inhibitors by considering the properties of the diseases and of the enzymes in relation to those of the inhibitors [2–4]. Below, the delay time of inhibition/inactivation is discussed in relation to the kinetic mechanism of the process. To simplify the discussion, reference will be made to *inhibition* as a term that represents both reversible and irreversible interactions.

The time an enzyme needs for binding a substrate and starting to transform it into product must be compared with the time needed by an inhibitor to bind the enzyme with the aim at avoiding unwanted substrate turnover. The half-time for the onset of inhibition can be calculated from the frequency constant, the parameter that contains the information of the mechanism and the rate constants for association and, when applicable, dissociation of the enzyme-inhibitor complex. For a pseudo-first-order process such as the reaction of a slow-onset inhibitor in great excess over the enzyme concentration, the half-time is given by

$$t_{1/2} = \frac{\ln 2}{\lambda} = \frac{0.693}{\lambda}, \tag{9.15}$$

where λ is the frequency constant. After seven half-times, a first-order process is more than 99 % complete and the delay time of inhibition (DTI) can be calculated as

$$t_{1/2} = \frac{7 \times 0.693}{\lambda} \approx \frac{5}{\lambda}. \tag{9.16}$$

Equation (9.16) is useful for making predictions on the delay time of inhibition once the mechanism is known from measurements in vitro and the substrate and inhibitor concentrations at the target site have been estimated. Of course such calculations are imprecise but can help for guessing at least the order of magnitude of the delay time of inhibition as illustrated in Fig. 9.6 for the inhibition of human elastase-2 by its natural inhibitor α-1-peptidase inhibitor.

Elastase-2, a serine peptidase stored in the azurophil granules of human polymorphonuclear leukocytes, has been implicated in the pathogenesis of lung emphysema for its ability to degrade elastin during phagocytosis. To protect the tissues from being degraded by elastase-2 and other peptidases accidentally released in the extracellular space, nature provided a palette of efficient inhibitors that are proteins themselves. In plasma and in the lungs, the main target of the α-1-peptidase inhibitor is elastase-2 that is immediately captured thus avoiding tissue damage. The prompt intervention of the inhibitor is due to a favorable association second-order rate constant, $6.5 \times 10^7\,M^{-1}\,s^{-1}$, the virtually irreversible inhibition process under physiological conditions and the high inhibitor concentration estimated between 30 and 70 μM.

A strategic methionine residue of α-1-peptidase inhibitor, necessary for inhibition, is oxidized by cigarette smoke with concomitant lowering of the association rate constant to only $3.1 \times 10^4\,M^{-1}\,s^{-1}$ [6]. As seen in Fig. 9.6, the delay time of inhibition is dramatically increased for the oxidized inhibitor. Considering the

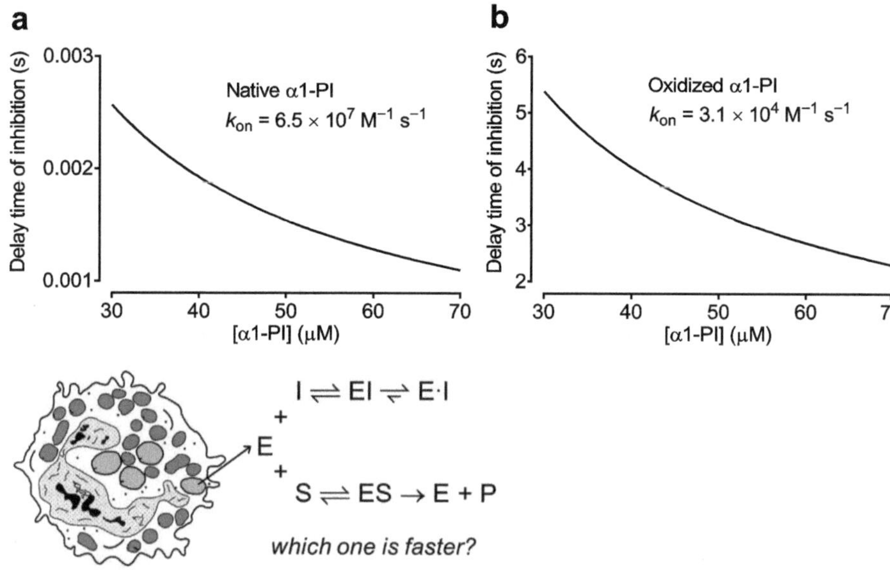

Fig. 9.6 Delay time of inhibition of human elastase-2 by native (**a**) and oxidized α-1-peptidase inhibitor (**b**). The oxidized inhibitor is about 2,000 times slower in forming the complex with its target enzyme. The *lower picture* illustrates a polymorphonuclear leukocyte leaking elastase-2 that partitions between inhibitor and substrate

lowest physiological concentration of α-1-peptidase inhibitor, $\approx 30\,\mu\text{M}$, the delay time of inhibition is ≈ 2.6 ms for the native inhibitor and ≈ 5.4 s for the oxidized counterpart, a 2,000 times difference. The failure to form a complex with elastase-2 *in due time* before significant tissue damage occurs is reputed to be responsible for elastin degradation in lung emphysema.

The same reasoning applies to synthetic inhibitors used as drugs. Figure 9.7 illustrates a hypothetical situation simulated for the slow-onset inhibition mechanisms SO-LSpI-2 (Sect. 8.4) and SO-LSpI-1 (Sect. 8.5) using the same rate constants of Figs. 8.8 and 8.10, respectively. In both cases $[S] = K_m$ and $[E]_t = 4\,\text{nM}$. The simulation was performed for an *open system* in which the substrate is maintained at a constant level. Panels (a) and (b) in Fig. 9.7 illustrate substrate turnover as product concentration against time at two concentrations of the slow-onset inhibitors. In panels (c) and (d) the delay times of inhibition are calculated from the frequency constants of the two mechanisms that are functions of the substrate (fixed) and inhibitor concentrations (variable).

Determining the mechanism of inhibition is necessary for making predictions on the efficiency in vivo because the frequency constant of mechanism SO-LSpI-2 depends in a hyperbolic fashion on [I] while it depends linearly on [I] in mechanism SO-LSpI-1 as shown in the equation groups (8.15) and (8.16), respectively. This behavior is reflected in the delay times of inhibition, which, as seen in Fig. 9.7, is saturable at a finite value when the inhibitor concentration is increased to infinity in

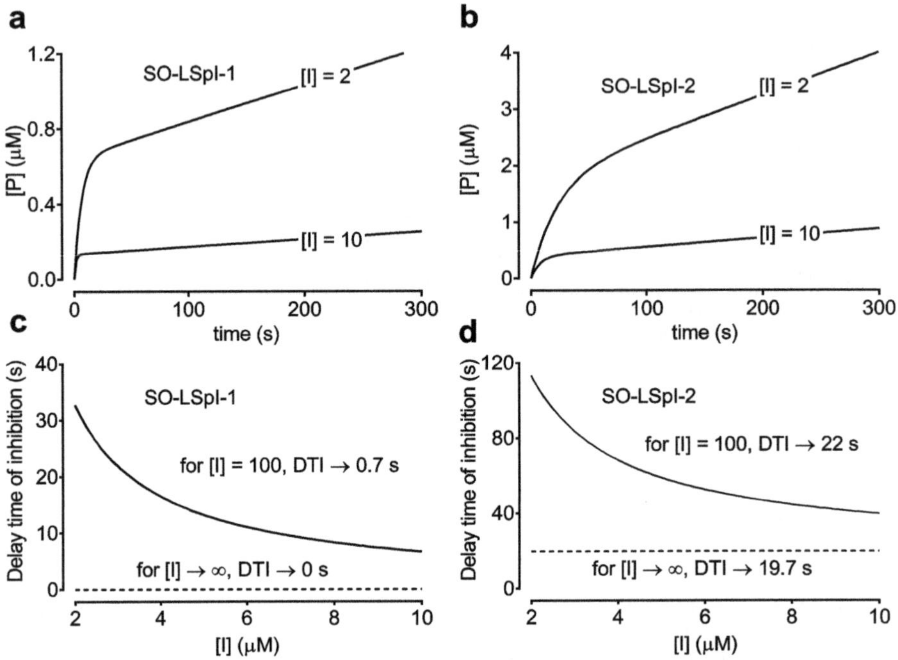

Fig. 9.7 Substrate turnover and delay time of inhibition for the slow-onset inhibition mechanisms SO-LSpI-1 (**a, c**) and SO-LSpI-2 (**b, d**). The progress curves were simulated with the same kinetic constants of Figs. 8.8 and 8.10 reproduced here for convenience. Common to all panels: $k_1 = 100\,\mu M^{-1}\,s^{-1}$, $k_{-1} = 1950\,s^{-1}$, $k_2 = 50\,s^{-1}$, $[E]_t = 4\,nM$, $[S]_t = 20\,\mu M$, $[I]_t = 2$ or $10\,\mu M$. For mechanism SO-LSpI-1 (panels **a, c**): $k_3 = 0.15\,\mu M^{-1}\,s^{-1}$, $k_{-3} = 0.003\,s^{-1}$. For mechanism SO-LSpI-2 (panels **b, d**): $k_3 = 2.3\,\mu M^{-1}\,s^{-1}$, $k_{-3} = 12\,s^{-1}$, $k_4 = 0.25\,s^{-1}$, $k_{-4} = 0.004\,s^{-1}$

mechanism SO-LSpI-2 (panel d), while this is not the case of SO-LSpI-1 because, increasing the concentration of the inhibitor, the delay time continues to decrease toward a theoretical limit of 0 s (panel c). At any concentration of inhibitor that obeys mechanism SO-LSpI-1, the delay time of inhibition is smaller than that of mechanism SO-LSpI-2. With the rate constants used to simulate these two examples and imagining a concentration of inhibitor at the target site of 100 μM, the delay time of inhibition of mechanism SO-LSpI-2 is 31 times larger than that of mechanism SO-LSpI-1.

The same considerations can be extended to all inhibition and inactivation mechanisms described in this chapter and in Chap. 8. In summary, it can be said that the efficiency of slow-onset inhibitors in vivo does not only depend on the overall inhibition or inactivation constant but also on the ability to avoid unacceptable substrate turnover during the time necessary to form the inhibited enzyme complex(es) at the inhibitor concentration achievable at the target site.

Examples of slow-onset inhibitors used as drugs are the molecules specifically designed to inhibit peptidyl-dipeptidase A (the angiotensin converting enzyme,

EC 3.4.15.1) used in the treatment of hypertension. Captopril, the prototype of these inhibitors, was the first successful result of *drug design* based on the structure of the active site of the target enzyme [12, 16]. Similar molecules synthesized with the same goal continue to be used as efficient drugs against hypertension behaving as slow-onset inhibitors with the mechanism SO-LSpI-2. Favorable rate constants for the interaction with the target enzyme and good bioavailability after oral intake were the key to success.

References

1. Ackermann WW, Potter VR (1949) Enzyme inhibition in relation to chemotherapy. Proc Soc Exp Biol Med 72:1–9
2. Baici A (1988) Criteria for the choice of inhibitors of extracellular matrix-degrading endopeptidases. In: Glauert AM (ed) The control of tissue damage. Elsevier, Amsterdam, pp 243–258
3. Baici A (1990) Interaction of human leukocyte elastase with soluble and insoluble protein substrates. A practical kinetic approach. Biochim Biophys Acta 1040:355–364
4. Baici A (1998) Inhibition of extracellular matrix-degrading endopeptidases: problems, comments, and hypotheses. Biol Chem 379:1007–1018
5. Baici A, Schenker P, Wächter M, Rüedi P (2009) 3-Fluoro-2,4-dioxa-3-phosphadecalins as inhibitors of acetylcholinesterase. A reappraisal of kinetic mechanisms and diagnostic methods. Chem Biodivers 6:261–282
6. Beatty K, Bieth J, Travis J (1980) Kinetics of association of serine proteinases with native and oxidized a1-proteinase inhibitor and a1-antichymotrypsin. J Biol Chem 255:3931–3934
7. Bieth JG (1980) Pathophysiological interpretation of kinetic constants of protease inhibitors. Bull Eur Physiopathol Respir 16(Suppl):183–195
8. Bieth JG (1984) In vivo significance of kinetic constants of protein proteinase inhibitors. Biochem Med 32:387–397
9. Burke MA, Maini PK, Murray JD (1990) On the kinetics of suicide substrates. Biophys Chem 37:81–90
10. Burke MA, Maini PK, Murray JD (1993) Suicide substrate reaction-diffusion equations: varying the source. IMA J Math Appl Med Biol 10:97–114
11. Cornish-Bowden A (1979) Validity of a 'steady-state' treatment of inactivation kinetics. Eur J Biochem 93:383–385
12. Cushman DW, Ondetti MA (1991) History of the design of captopril and related inhibitors of angiotensin converting enzyme. Hypertension 17:589–592
13. Easson LH, Stedman E (1936) The absolute activity of choline-esterase. Proc Roy Soc London B 121:142–164
14. Kitz R, Wilson IB (1962) Esters of methanesulfonic acid as irreversible inhibitors of acetylcholinesterase. J Biol Chem 237:3245–3249
15. Malcolm ADB, Radda GK (1970) The reaction of glutamate dehydrogenase with 4-iodoacetamido salicylic acid. Eur J Biochem 15:555–561
16. Ondetti MA, Rubin B, Cushman DW (1977) Design of specific inhibitors of angiotensin-converting enzyme: new class of orally active antihypertensive agents. Science 196:441–444
17. Rakitzis ET (1987) Kinetics of protein modification, and/or enzyme inactivation, reactions by an unstable modifying agent. Biochem J 246:803–804
18. Stedman E, Stedman E, Easson LH (1932) Choline-esterase. An enzyme present in the blood-serum of the horse. Biochem J 26:2056–2066
19. Straus OH, Goldstein A (1943) Zone behavior of enzymes. Illustrated by the effect of dissociation constant and dilution on the system cholinesterase-physostigmine. J Gen Physiol 26:559–585

20. Tian WX, Tsou CL (1982) Determination of the rate constant of enzyme modification by measuring the substrate reaction in the presence of the modifier. Biochemistry 21:1028–1032
21. Topham CM (1987) Alternative methods for the determination of rate constants describing enzyme inactivation by an unstable inhibitor. Biochem J 246:804–806
22. Topham CM (1988) Ill-conditioning associated with the "end-point" method for the determination of kinetic parameters describing irreversible enzyme inactivation by an unstable inhibitor. J Theor Biol 135:169–173
23. Topham CM (1990) A generalized theoretical treatment of the kinetics of an enzyme-catalysed reaction in the presence of an unstable irreversible modifier. J Theor Biol 145:547–572
24. Van Slyke DD, Cullen GE (1914) The mode of action of urease and of enzymes in general. J Biol Chem 19:141–180
25. Wächter M, Rüedi P (2009) Synthesis and characterization of the enantiomerically pure cis- and trans-2,4-dioxa-3-fluoro-3-phosphadecalins as inhibitors of acetylcholinesterase. Chem Biodivers 6:283–294. doi:10.1002/cbdv.200800335
26. Waley SG (1980) Kinetics of suicide substrates. [See correction: Biochem. J. 203:807 (1985)]. Biochem J 185:771–773
27. Waley SG (1985) Kinetics of suicide substrates. Practical procedures for determining parameters. Biochem J 227:843–849
28. Wang ZX (1990) Kinetics of suicide substrates. J Theor Biol 147:497–508
29. Wang ZX (1990) Two theoretical problems concerning the irreversible modification kinetics of enzyme activity. J Theor Biol 142:551–563
30. Wang ZX, Tsou CL (1990) An alternative method for determining inhibition rate constants by following the substrate reaction. J Theor Biol 142:531–549
31. Wimalasena K, Haines DC (1996) A general progress curve method for the kinetic analysis of suicide enzyme inhibitors. Anal Biochem 234:175–182

Chapter 10
Dichotomous Keys to Enzyme-Modification Mechanisms

The identification of objects is a fundamental human activity. ⋯ For a biologist, identification usually means finding the name for a specimen of animal or plant, and the specimen to be identified is usually assigned to a species. ⋯ and when a new object is examined, and it is decided that it belongs to one of the existing groups, then it has been identified.
Richard J. Pankhurst (1991) Practical taxonomic computing. Cambridge University Press, p. 1

Abstract The analysis of experiments with enzyme modifiers is aimed at extracting information on the mechanism of action and at calculating kinetic constants. Some mechanisms are characterized by unique combinations of characters and their identification does not pose particular problems. However, mechanisms exist which share common features and are not distinguishable from one another at a glance. The attribution of the kinetic behavior of an enzyme modifier to a known type and eventually to a particular mechanism of action is similar to the identification performed by biologists when assigning a specimen to a family, genus, and finally to a species. This task, which consists in evaluating all elements available and in analyzing them until finding a unique combination of traits, is facilitated by the use of diagnostic keys. The dichotomous keys in this chapter aim at supporting the identification of kinetic mechanisms of enzyme modifiers.

10.1 Introduction

The use of diagnostic keys in animal and plant taxonomy dates back to the seventeenth century [1, Chap. 6]. These keys, typically of the dichotomous type, constitute a precious support for both specialists and occasional users. Of the two main categories of keys, the parallel (or bracketed) type and the yoked (or indented) type the second will be used here. Before proceeding to the differential diagnosis of kinetic mechanisms of enzyme-modifier interaction, experiments of the best possible quality must have been performed on the basis of accurate experimental design. In some cases just one set of data is sufficient, while in other cases two or more sets of data obtained under different conditions are required for the identification of a mechanism.

A. Baici, *Kinetics of Enzyme-Modifier Interactions*,
DOI 10.1007/978-3-7091-1402-5_10

The keys below will ask specific questions to ascertain if the essential elements for the determination of a mechanism are at hand. If this is not the case, collecting the requested data is imperative. Therefore, reading at least once the keys even not having a concrete experiment to analyze, may be useful to familiarize with this method. While enzyme kinetics, when interpreted correctly, offers valuable mechanistic information, there are situations in which even the best experiments are insufficient for assigning data to a mechanism unambiguously, in which case help from other methods, e.g., chemical or spectroscopic analysis, must be invoked.

10.2 Memorandum for the Acquisition of Kinetic Data

Every investigator has her/his own approach and preferences with work in the laboratory. However, there is no room for different opinions on fundamental concepts such as temperature, pH, ionic strength, and other factors. The success of enzyme kinetic experiments very much depends on carefully designed working protocols, on accurate data acquisition, and on the choice of appropriate detection methods as suggested in the list below.

- TEMPERATURE control is fundamental in kinetic measurements because reaction velocities depend on it. After deciding the temperature for the envisaged measurements, this must be kept constant within narrow limits, e.g., $\pm 1\,^{\circ}$C. Buffer solutions and stable reagents can be kept throughout in a water bath at the same temperature of the measurements. If enzyme solutions and temperature-sensitive reagents are kept on ice, it is important to assure that the temperature does not undergo important changes when adding the reagent that starts the reaction. Needless to say that performing measurements at *room temperature* is not admissible in any protocol involving kinetic measurements.
- BUFFER composition should be declared without ambiguity by specifying the total concentration of the buffering species and of any added non-buffering component, ionic strength, and pH to the second decimal, e.g., pH = 7.40. pH-meters should be appropriately calibrated taking care of the temperature at which the buffer solution is prepared and the temperature at which the buffer will be used. While some buffers have a negligible temperature coefficient. e.g., phosphate, others do not, e.g., Tris and other "Good" buffers.
- ORGANIC SOLVENTS are often added to dissolve reagents that are poorly soluble in water. If this is the case, it is important that the final concentration of the added solvent is kept constant throughout in the final assay solution.
- WHENEVER POSSIBLE, the use of a rapid mixing device to quickly mix reagents and monitor the reaction from the first milliseconds or seconds is preferable to a conventional apparatus. This avoids loss of information during the dead time necessary to mix the solution and start the measurement. See Sect. 1.9 on the measurement of initial rates. Modern conventional instruments not specifically indicated for rapid mixing are often equipped with a stirring accessory that can

be optimized to a speed that ensures reasonably rapid mixing while avoiding turbulence in the observing cell. Dead times down to 1 s can be realized if the starting reagent is added by direct injection into the observing cell using a precision microsyringe fitted to the lid of the instrument, whose needle reaches the reacting solution avoiding opening and re-closing the lid.

- TIME CONSTANT and its proper use. In spectroscopic measurements, the electronic signal, which is proportional to the concentration of the measured species, is always accompanied by electronic noise. The time constant, also called response time, filter time, damping, or attenuation of an electronic circuitry can be set on recording instruments to limit the noise. If the software of an instrument can set the time constant automatically, using this option is often the best choice. Otherwise, manual setting of the time constant should be done carefully. This is particularly important for fast reactions measured with flow-apparatus, where a hundredfold increase of the time constant causes a tenfold reduction of noise but can also cut away any fast-changing signal and produce artifacts.
- LOGARITHMIC TIMEBASE for data acquisition. This method has been introduced to intensify data sampling in the first part of a reaction, with sampling time decreasing on a logarithmic scale as the reaction proceeds [2]. The logarithmic timebase is very helpful when measuring progress curves containing multiple exponentials at the beginning of progress curves. The large number of data points in the exponential phase(s) is advantageous for subsequent nonlinear regression or numerical integration analysis. The logarithmic timebase, which has been introduced for supporting data acquisition of fast reactions by stopped-flow measurements, is very helpful also for much slower reactions, notably slow-onset inhibition. In this case, progress curves consist of one or two exponentials followed by a steady-state release of product. Nonlinear regression analysis benefits from a large number of data points in the first, exponential region and less points in the steady-state region in place of equispaced data points acquired with a linear timebase.
- INNER FILTER in fluorescence measurements. The attenuation of the fluorescence signal due either to direct absorbtion of incident light at the excitation wavelength or to reabsorption of emitted light at the emission wavelength can be and should *always* be corrected. The inner filter can simulate inhibitions that do not exist and in any case produce serious distortions to data making them useless. See the comments on p. 202.
- ENZYME STABILITY during the time of measurements. This is a fundamental issue that should never be forgotten, in any enzyme kinetic experiment. Long-lasting measurements such as those typically necessary in the acquisition of slow-onset inhibition data may be compromised if loss of enzyme activity, independent of the ongoing reaction with modifiers, is important. Enzyme instability can be conveniently ascertained using the Selwyn test (Sect. 1.8).

Support to Key 1 (Essential Activation) and to the Primary Key Beginning on the Facing Page

Key 1 is necessary for distinguishing catalysis that is possible only in the presence of an essential activator from all other modification mechanisms. The primary key redirects the search to specific keys or conveys the appraisal straight to mechanisms that, for logical reasons, cannot be included in other keys. The symbol X is used for all types of modifiers, either inhibitors or activators with the exception of the liberator, called L.

The following shortcuts and abbreviations are used to optimize the available space:

- (E+X) ⏱ ⇓S: enzyme and modifier are preincubated for a known time and the reaction is started by adding substrate(s) and any cofactors or coenzymes
- (S+X) ⇓ E: reaction started by adding enzyme to a solution containing substrate and modifier
- S ⇓ E: enzyme added to substrate to start a reaction in the absence of modifiers
- *hyperbolic*↑, *hyperbolic*↓: increasing or decreasing hyperbolically with the concentration, or reciprocal concentration, of the indicated reagent, respectively
- *independent*: the considered parameter remains unchanged at any concentration, or reciprocal concentration, of the indicated reagent
- *linear*↑: the considered parameter increases linearly with the concentration, or reciprocal concentration, of the indicated reagent

Suggestion for This and the Subsequent Keys in going through the keys a piece of paper may be used. It helps in following the lines and jumping from a numbered option to another, covering the options beneath that do not apply.

In All Keys the numbers in brackets indicate the page that discusses the relative mechanism.

10.3 Dichotomous Keys

Key 1: Essential Activation Against the Remaining Modifier Mechanisms

Support on the facing page

1. Adding enzyme to a solution containing its substrate(s) *no reaction* is detectable unless a third substance is also present
   2. k_{cat}^{app} *independent* of (1/[X]) AND K_m^{app} *linear*↑ with (1/[X]) ↑ . **LSpA [147]**
   2. k_{cat}^{app} *hyperbolic*↓ with (1/[X]) ↑
      3. K_m^{app} depends hyperbolically on (1/[X])
         4. K_m^{app} *hyperbolic*↓ with (1/[X]) ↑ . . . . **LMx(Sp>Ca)A [143–147]**
         4. K_m^{app} *hyperbolic*↑ with (1/[X]) ↑ . . . . **LMx(Sp<Ca)I [143–147]**
      3. K_m^{app} *independent* of (1/[X]) . . . . **LMx(Sp=Ca)A [143–147]**
1. Enzyme and substrate form a complex and catalysis occurs without the intervention of a third substance . . . . **go to point 1 below**

Primary Key

Support on the facing page

1. (S + X) ⇃⇂ E. The Selwyn test (Sect. 1.8) does not pass
   2. Using any detection method the results are erratic, not reproducible and are sensitive to the concentration of the enzyme AND/OR to experimental conditions. The same problems are observed when the modifier is assayed in the presence of nontarget enzymes AND/OR aggregates are detected by dynamic light scattering, typically in the micromolar concentration range of the alleged modifier . . . . **promiscuous inhibitor [200]**
   2. Results are reproducible, promiscuous inhibition can be excluded **unstable enzyme, design alternative assay conditions**
1. (S + X) ⇃⇂ E. The Selwyn test passes
   3. Initial rates can be obtained from progress curves that are linear from the beginning till end of the measurement OR any deviation from linearity, appreciable by inspection, is uniform and *does not* show a burst OR a lag followed by a linear segment. If the dependence of the initial rates on [S] shows a maximum consider substrate inhibition (p. 221 and 252), otherwise proceed with this key. Enzyme activity can be substantially affected at $[X]_t \gg [E]_t$ OR at $[X]_t \approx [E]_t$
      4. The modifier affects enzyme activity when $[X]_t \approx [E]_t$: tight-binding modifier. Analysis of the mechanism can be performed with Keys 2 AND/OR 3 followed by refinement . . . . **Refinement using** (4.35)
      4. The modifier does not appreciably affect activity when $[X]_t \approx [E]_t$

5. Analysis of primary data using the specific velocity plot (Sect. 3.6) **Key 2 [471]**
5. Analysis of primary data using the dependence of the apparent parameters on modifier concentration . . . . **Key 3, [473]**

3. (S + X) $\Downarrow$ E OR S $\Downarrow$ E: progress curves consist of one or two exponential phases plus a straight line that can have a positive slope or decline to slope zero (trace parallel to the time axis)
   6. Fitting (8.8) and (8.9) to progress curves clearly suggests that the slow transient consist of *two exponentials*. Further analysis using (8.10) is not recommended. Analysis by numerical integration is the method of choice.
      7. S$\Downarrow$E (flow methods!): There is evidence that the substrate *does not* bind the enzyme in a diffusion-controlled process or nearly so (small second-order association rate constant), while binding of the modifier may either be diffusion-controlled or anyway its second-order association rate constant is larger than that of the substrate **SO-LSpI-SlowS [428]**
      7. S$\Downarrow$E (flow methods!): the substrate appears to bind the enzyme in a fast process WHILE (S + X)$\Downarrow$E reveals a sluggish process at the level of the modifier. Some of the mechanisms in Chaps. 8 and 9 based on LSpI, LCaI, or LMxI can in principle be extended to include two relaxation times in the formation of the inhibited enzyme and will in this case key out here. Slow-onset inhibition, based on LSpI and exhibiting two-exponentials in the pre-steady-state, is the starting point for analyzing progress curves by numerical integration methods. . . **SO-LSpI [369]**
   6. The fit of (8.8) and (8.9) to progress curves confirms *only one exponential* in the slow transient
      8. S$\Downarrow$E: a *scrambled* family of progress curves consisting of an exponential phase plus steady-state is observed in the absence of modifiers. The expected sequence of the progress curves is not respected at low substrate concentrations (Fig. 8.24 a,b). The substrate itself acts as a slow-onset modifier binding at a secondary site of the enzyme
         9. v_z AND v_s exhibit a maximum, λ *hyperbolic*$\uparrow$ with [S] $\uparrow$ shows irregularities at low [S] . . . . **SO-LCaI-S2 [420]**
         9. (S + L)$\Downarrow$E, where L is a modifier that has no effects of its own in the absence of substrate: v_z AND v_s do no longer show a maximum AND λ *hyperbolic*$\downarrow$ with [L] $\uparrow$ . . . . **SO-LCaI-S2-Lib [425]**
      8. S$\Downarrow$E is characterized by progress curves linear with time AND (S + X) $\Downarrow$E exhibits progress curves that *vary regularly* as the modifier concentration is increased
         10. (E+X) $\Downarrow$S: adding substrate to enzyme that had been incubated for a *sufficiently long* time with the modifier, a reaction takes place OR nothing happens
            11. no reaction takes place, enzyme inactivation . . . . . **Key 5 [476]**

11. a reaction takes place, progress curves show a lag plus a straight line with positive slope
    12. (E+X) ⏱ $\Downarrow$S: preincubation for various time intervals, start the reaction by adding substrate. Finely tuning the experiments allows to discriminate between temporary inactivation from regular and temporary slow-onset inhibition: the second option is valid . . . . **Key 4 [475]**
    12. (E+X) ⏱ $\Downarrow$S: preincubation for various time intervals excludes temporary slow-onset inhibition in favor of temporary enzyme inactivation . . . . **Key 5 [476]**

10. (S + X)$\Downarrow$E: progress curves consist of a concave-down exponential phase (burst) followed by a straight line parallel to the time axis, enzyme inactivation . . . . **Key 5 [476]**

Support to Key 2 on the Facing Page

Refer to Sect. 3.6 for the description of the specific velocity plot. The combinations for a given mechanism in Key 2 consist of a number, corresponding to the *primary* specific velocity plot and an uppercase letter corresponding to the *secondary* plot. For instance, **10-K** identifies hyperbolic specific activation (HSpA). In the primary plots, the arrows indicate the direction of increasing modifier concentration. In the secondary plots, **a** corresponds to $a/(a-1)$ and **b** corresponds to $b/(b-1)$. *Primary* intersection means the abscissa coordinate of the intersection point of all curves, while that of the ordinate is always 1. The *secondary* intercepts are those with the ordinate. The numbers in brackets next to the acronyms of the mechanisms refer to the pages where they are described.

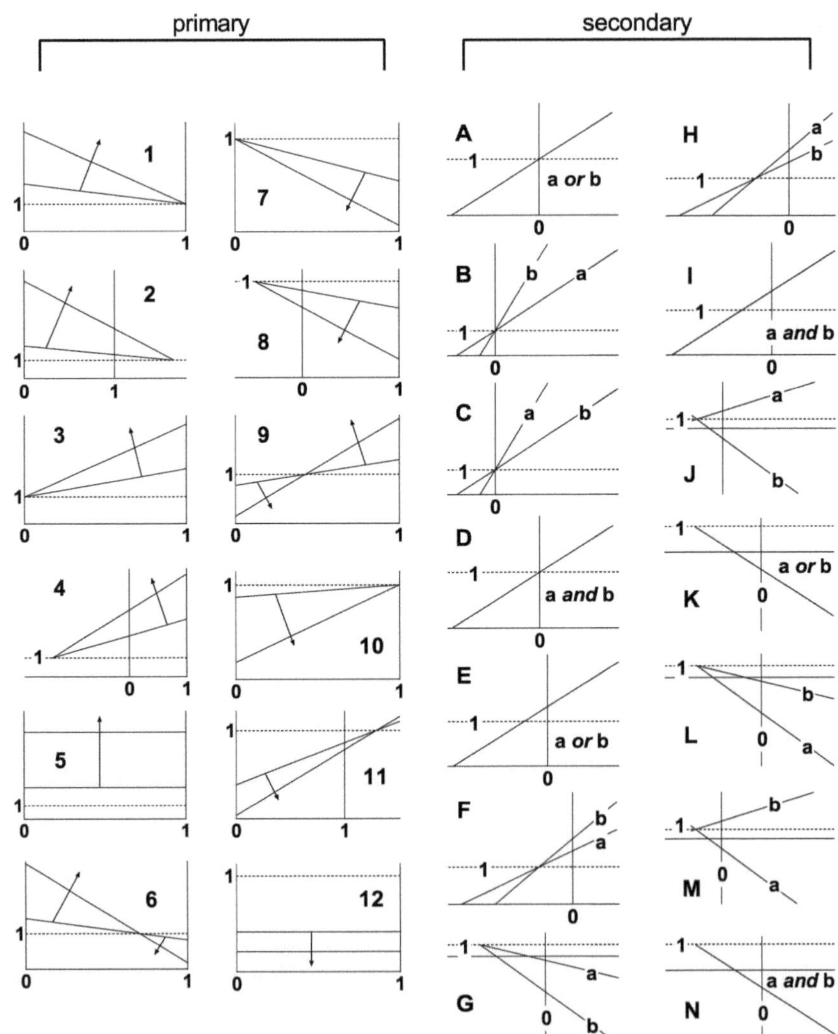

Key 2: Basic Modification Mechanisms, Key Based on the Specific Velocity Plot

Support on the facing page

1. *Primary* slopes $\neq 0$: lines converge to a common intersection point
    2. *Primary* intersection in the interval (0,1)
        3. *Secondary* a-slope > 0 AND b-slope < 0: **6-J** . . . . **HMxD(I/A) [257]**
        3. *Secondary* a-slope < 0 AND b-slope > 0: **9M** . . . . **HMxD(A/I) [268]**
    2. *Primary* intersection in the intervals $(-\infty,0]$ OR $[1,\infty)$
        4. *Primary* intersection in the interval $(-\infty,0]$
            5. *Primary* slopes > 0 AND *secondary* intercepts > 0
                6. *Primary* intersection = 0
                    7. *Secondary* b-intercept = 1: **3-A** . . . . **LCaI [217]**
                    7. *Secondary* b-intercept > 1: **3-E** . . . . **HCaI [244]**
                6. *Primary* intersection < 0
                    8. *Secondary* a- AND b intercepts = 1, AND a-slope > b-slope: **4-C** **LMx(Sp<Ca)I [227]**
                    8. *Secondary* a- AND b intercepts > 1, AND a-slope > b-slope: **4-H** **HMx(Sp<Ca)I [249]**
            5. *Primary* slopes < 0 AND *secondary* intercepts < 0
                9. *Primary* intersection = 0: **7-K** . . . . **HCaA [263]**
                9. *Primary* intersection < 0 AND *secondary* a-intercept < b-intercept: **8-L** . . . . **HMx(Sp>Ca)A [265]**
        4. *Primary* intersection in the interval $[1,\infty)$
            10. *Primary* slopes < 0 AND *secondary* intercept(s) > 0
                11. *Primary* intersection = 1
                    12. *Secondary* a-intercept = 1: **1-A** . . . . **LSpI [212]**
                    12. *Secondary* a-intercept > 1: **1-E** . . . . **HSpI [235]**
                11. *Primary* intersection > 1
                    13. *Secondary* a- AND b-intercepts = 1, AND b-slope > a-slope: **2-B** **LMx(Sp>Ca)I [223]**
                    13. *Secondary* a- AND b intercepts > 1, AND b-slope > a-slope: **2-F** **HMx(Sp>Ca)I [238]**
            10. *Primary* slopes > 0 AND *secondary* intercept(s) < 0
                14. *Primary* intersection = 1 : **10-K** . . . . **HSpA [276]**
                14. *Primary* intersection > 1: **11-G** . . . . **HMx(Sp<Ca)A [279]**
1. *Primary* slopes = 0 AND *secondary* a-slope = b-slope
    15. *Secondary* intercept = 1: **5-D** . . . . **LMx(Sp=Ca)I [230]**
    15. *Secondary* intercept $\neq 1$
        16. *Secondary* intercept > 1: **5-I** . . . . **HMx(Sp=Ca)I [254]**
        16. *Secondary* intercept < 0: **12-N** . . . . **HMx(Sp=Ca)A [281]**

Support to Key 3 on the Facing Page

The necessary primary data consist of initial rates measured at several fixed values of [S] and [I]. A minimum of each five values of [S] and [I] is recommended and any additional data will improve the analysis. See for instance the design in Fig. 2.3 (c). The Michaelis–Menten equation in the form

$$v = \frac{k_{\mathrm{cat}}^{\mathrm{app}}[\mathrm{E}]_{\mathrm{t}}\,[\mathrm{S}]}{K_{\mathrm{m}}^{\mathrm{app}} + [\mathrm{S}]} = \frac{V^{\mathrm{app}}\,[\mathrm{S}]}{K_{\mathrm{m}}^{\mathrm{app}} + [\mathrm{S}]} \tag{10.1}$$

is then fitted to data to calculate $k_{\mathrm{cat}}^{\mathrm{app}}$, $K_{\mathrm{m}}^{\mathrm{app}}$ and $k^{\mathrm{app}} = k_{\mathrm{cat}}^{\mathrm{app}}/K_{\mathrm{m}}^{\mathrm{app}}$. Instead of $k_{\mathrm{cat}}^{\mathrm{app}}$, V^{app} can of course be used in the following analysis. The dependence on modifier concentration, [X], of $k_{\mathrm{cat}}^{\mathrm{app}}$, $1/k_{\mathrm{cat}}^{\mathrm{app}}$, $K_{\mathrm{m}}^{\mathrm{app}}$, k^{app} and $1/k^{\mathrm{app}}$, can assume four forms that are denoted in Key 3 key by the following shortcuts in *emphasized* typeface and depicted in the plots below:

- straight line parallel to the [X]-axis (*independent*)
- linearly increasing with [X] (*linear* ↑)
- hyperbola decreasing for increasing [X] (*hyperbolic* ↓)
- hyperbola increasing for increasing [X] (*hyperbolic* ↑).

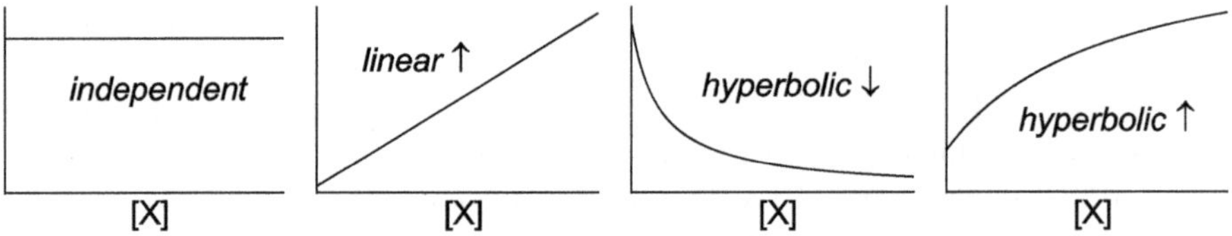

These figures refer all to dependences on [X]. Therefore, to avoid long definitions in the key that would require sometimes two lines and render the key less legible, the concentration of X is omitted.

Key 3: Basic Modification Mechanisms, Key Based on the Dependence of Kinetic Parameters on Modifier Concentration

Support on the facing page

1. k_{cat}^{app} *hyperbolic*
    2. k_{cat}^{app} *hyperbolic* ↑
        3. K_m^{app} *hyperbolic*
            4. K_m^{app} *hyperbolic* ↑
                5. k^{app} *hyperbolic*
                    6. k^{app} *hyperbolic* ↑ . . . . **HMx(Sp>Ca)A [265]**
                    6. k^{app} *hyperbolic* ↓ . . . . **HMxD(I/A) [257]**
                5. k^{app} *independent* . . . . **HCaA [263]**
            4. K_m^{app} *hyperbolic* ↓ AND k^{app} *hyperbolic* ↑ . . . . **HMx(Sp<Ca)A [279]**
        3. K_m^{app} *independent* . . . . **HMx(Sp=Ca)A [281]**
    2. k_{cat}^{app} *hyperbolic* ↓
        7. K_m^{app} *hyperbolic*
            8. K_m^{app} *hyperbolic* ↑
                9. k^{app} *hyperbolic* ↓ AND $1/k^{app}$ *hyperbolic* ↑ . **HMx(Sp>Ca)I [238]**
                9. k^{app} *hyperbolic* ↓ AND $1/k^{app}$ *linear* ↑ . . . . . **LMx(Sp>Ca)I [223]**
            8. K_m^{app} *hyperbolic* ↓
                10. k^{app} *independent*
                    11. $1/k_{cat}^{app}$ *independent* . . . . **LCaI [217]**
                    11. $1/k_{cat}^{app}$ *hyperbolic* ↑ . . . . **HCaI [244]**
                10. k^{app} *hyperbolic*
                    12. k^{app} *hyperbolic* ↓
                        13. $1/k^{app}$ *hyperbolic* ↑ . . . . **HMx(Sp<Ca)I [249]**
                        13. $1/k^{app}$ *linear* ↑ . . . . **LMx(Sp<Ca)I [227]**
                    12. k^{app} *hyperbolic* ↑ . . . . **HMxD(A/I) [268]**
        7. K_m^{app} *independent*
            14. k^{app} *hyperbolic* ↓ AND $1/k^{app}$ *hyperbolic* ↑ . . . . **HMx(Sp=Ca)I [254]**
            14. k^{app} *hyperbolic* ↓ AND $1/k^{app}$ *linear* ↑ . . . . **LMx(Sp=Ca)I [230]**
1. k_{cat}^{app} *independent*
    15. K_m^{app} *linear* ↑ . . . . **LSpI [212]**
    15. K_m^{app} *hyperbolic*
        16. K_m^{app} *hyperbolic* ↑ . . . . **HSpI [235]**
        16. K_m^{app} *hyperbolic* ↓ . . . . **HSpA [276]**

Support to Key 4 on the Facing Page and Key 5

- (S+X)↓E: the substrate(s) and the modifier are preincubated and the reaction is started by adding the enzyme. Possible ways to perform experiments with a stopped-flow apparatus are suggested in Appendix 8.1
- (E+X) ⏱ ⇊S: enzyme and modifier are preincubated for a know time and the reaction is started by adding substrate(s) any cofactors or coenzymes. Suggestions when using a stopped-flow apparatus are given in Appendix 8.1
- S↓E: reaction measured in absence of modifiers
- X⏱ ↓buffer: dilution of the modifier in the same buffer used for the enzymatic assays. Analysis of any chemical modification by HPLC, NMR, or other methods at time intervals, identification of the molecule(s) produced
- (E+X)⏱ ↓buffer: incubation of the enzyme with the modifier in the absence of substrate. Analysis of any reaction products that may differ from those of the incubation of modifier alone in buffer (X⏱ ↓buffer)
- ►: kinetic information alone is insufficient to assign a mechanism, additional methods are necessary to complement kinetic measurements

Key 4: Slow-Onset Enzyme Inhibition

Support on the facing page

1. S $\Downarrow$ E: there is no lag OR burst preceding the steady-state
  2. (E+X) $\Downarrow$S: progress curves *do not change* with preincubation time
    3. λ *hyperbolic*$\uparrow$ with [I] $\uparrow$
      4. λ *hyperbolic*$\downarrow$ with [S] $\uparrow$ . . . . **SO-LSpI-2 [382]**
      4. λ *hyperbolic*$\uparrow$ with [S] $\uparrow$ . . . . **SO-LCaI-2 [415]**
    3. λ *linear*$\uparrow$ with [I] $\uparrow$
      5. λ *hyperbolic*$\uparrow$ with [S] $\uparrow$
        6. k^{app} *hyperbolic*$\downarrow$ with[I] $\uparrow$ . . . . **SO-LMx(Sp<Ca)I [408]**
        6. k^{app} *independent* of [I] . . . . **SO-LCaI-1 [415]**
      5. λ *hyperbolic*$\downarrow$ with [S] $\uparrow$, k_{cat}^{app} *independent* AND K_m^{app} *linear*$\uparrow$ with [I] $\uparrow$
        7. ► Additional information besides kinetics is necessary for fine-tuning the discrimination between the three following mechanisms
          8. ► The inhibitor exists as single chemical species AND the enzyme does not isomerize between different forms . . **SO-LSpI-1 [388]**
          8. ► The inhibitor exists in equilibrium between different forms and only a minor, *rare form* acts as inhibitor OR the enzyme exists in equilibrium between different forms
            9. A *rare form* of the inhibitor (minor population in a mixture of different forms) is the true inhibitor . . **SO-LSpI-RareI [405]**
            9. Only a *rare form* of the enzyme reacts with the inhibitor **SO-LSpI-RareE [396]**
        7. k_{cat}^{app} *hyperbolic*$\downarrow$ with [I] $\uparrow$
          10. K_m^{app} *hyperbolic*$\uparrow$ with[I] $\uparrow$ . . . . **SO-LMx(Sp>Ca)I [408]**
          10. K_m^{app} *independent* with[I] $\uparrow$ . . . . **SO-LMx(Sp=Ca)I [408]**
  2. (E+X) $\Downarrow$S: progress curves *change* with preincubation time, λ *hyperbolic*$\downarrow$ with [S] $\uparrow$
    11. λ *hyperbolic*$\uparrow$ with [I] $\uparrow$ . . . . **SO-LSpI-2Temp [391]**
    11. λ *linear*$\uparrow$ with [I] $\uparrow$ . . . . **SO-LSpI-1Temp [391]**
1. S $\Downarrow$ E: *small* lag or *small* burst precede the steady-state AND S + I $\Downarrow$ E at fixed [S] and variable [I], progress curves may show a lag, a burst OR be a straight line
  12. S $\Downarrow$ E, λ *hyperbolic*$\downarrow$ for [S] $\uparrow$
    13. [S] $\uparrow$: a small burst changes into a small lag with intermediate progress curves indistinguishable from a straight line . **SO-IsoE, [E] $\geq$ [*E] [400]**
    13. [S] $\uparrow$: a manifest lag is always present . . . . . **SO-IsoE, [E] < [*E] [400]**
  12. S + I $\Downarrow$ E, λ *hyperbolic*$\downarrow$ for [I] $\uparrow$ at fixed [S] AND λ *hyperbolic*$\downarrow$ for [S] $\uparrow$ at fixed [I]
    14. [I] $\uparrow$: progress curves change from an almost straight line to a burst followed by steady-state . . . . **SO-LSpI-IsoE, [E] $\geq$ [*E] [403]**
    14. [I] $\uparrow$: progress curves show a lag at low [I], an almost straight line at intermediate [I] and a burst followed by steady-state at high [I] **SO-LSpI-IsoE, [E] < [*E] [403]**

Key 5: Slow-Onset Enzyme Inactivation

Support on p. 474

1. ►, X⏱ ↓buffer: no side reaction(s) detectable, the modifier is stable in the buffer solution of the enzymatic assay
   2. progress curves level off at *slope zero* after the initial transient
      3. the rate at time zero is *independent* of [X] AND λ increases *linearly* with [X] ↑ .... **SO-LSpIna-1 [447–449]**
      3. the rate at time zero *depends* on [X] AND λ increases *hyperbolically* with [X] ↑ .... **SO-LSpIna-2 [447–449]**
   2. progress curves display a *positive slope* after the initial transient AND the plot of λ against [X] has a positive intercept on the ordinate
      4. the rate at time zero is *independent* of [X] AND λ increases *linearly* with [X] ↑ .... **SO-LSpIna-1Temp [449–453]**
      4. the rate at time zero *depends* on [X] AND λ increases *hyperbolically* with [X] ↑ .... **SO-LSpIna-2Temp [449–453]**
1. ►, X⏱ ↓buffer: analysis at time intervals of a solution containing X in the buffer of the enzymatic assay reveals time-dependent transformation of X to one or more different molecules
   5. ►, E+X⏱ ↓buffer: analysis of the mixture at time intervals does not reveal additional products with respect to X⏱ ↓buffer BUT consider the possibility that I′ is the same as I* in Scheme 9.3, in which case go to items 7 and *good luck!*
      6. progress curves exhibit a more or less pronounced positive slope after the initial transient AND the rate at time zero is *independent* of [X] **SO-LSpIna-1-Idec [453–457]**
      6. progress curves exhibit a more or less pronounced positive slope after the initial transient AND the rate at time zero *depends* on [X] **SO-LSpIna-2-Idec [453–457]**
   5. ►, E+X⏱ ↓buffer: analysis of the mixture at time intervals *reveals* additional products with respect to X⏱ ↓buffer
      7. progress curves exhibit a positive slope AND the rate at time zero is *independent* on [X] .... **SO-LSpIna-1Temp-Idec [453–457]**
      7. progress curves exhibit a positive slope AND the rate at time zero is *depends* on [X] .... **SO-LSpIna-2Temp-Idec [453–457]**

References

1. Pankhurst RJ (1991) Practical taxonomic computing. Cambridge University Press, Cambridge
2. Walmsley AR, Bagshaw CR (1989) Logarithmic timebase for stopped-flow data acquisition and analysis. Anal Biochem 176:313–318

Index

To avoid overly long descriptions, the basic and slow-onset modification mechanisms are indicated by their acronyms.

A. Baici, *Kinetics of Enzyme-Modifier Interactions*,
DOI 10.1007/978-3-7091-1402-5

MIX
Papier aus verantwortungsvollen Quellen
Paper from responsible sources
FSC® C105338

If you have any concerns about our products, you can contact us on
ProductSafety@springernature.com

In case Publisher is established outside the EU, the EU authorized representative is:
Springer Nature Customer Service Center GmbH
Europaplatz 3, 69115 Heidelberg, Germany

Printed by Libri Plureos GmbH
in Hamburg, Germany